1st Nordic Conference on Zero Emission and Plus Energy Buildings 2019

IOP Conference Series: Earth and Environmental Science
Volume 352

Trondheim, Norway
6 - 7 November 2019

ISBN: 978-1-7138-0148-1
ISSN: 1755-1307

Printed from e-media with permission by:

Curran Associates, Inc.
57 Morehouse Lane
Red Hook, NY 12571

proceedings.com

Some format issues inherent in the e-media version may also appear in this print version.

Institute of Physics
Dirac House, Temple Back
Bristol BS1 6BE UK

Phone: 44 1 17 929 7481
Fax: 44 1 17 920 0979

techtracking@iop.org

Additional copies of this publication are available from:

Curran Associates, Inc.
57 Morehouse Lane
Red Hook, NY 12571 USA
Phone: 845-758-0400
Fax: 845-758-2633
Email: curran@proceedings.com
Web: www.proceedings.com

TABLE OF CONTENTS

PAPERS FOR ORAL PRESENTATION

PAPERS FOR POSTER PRESENTATION

IOP Conf. Series: Earth and Environmental Science **352** (2019) 011001 doi:10.1088/1755-1315/352/1/011001

The first Nordic Conference on Zero Emission and Plus Energy Buildings (ZEB+) was arranged in Trondheim, Norway, on November 6-7 2019. The conference presented state-of-the art within research and practical solutions that pave the way towards a low carbon built environment, with a focus on Nordic countries. The proceedings include 69 papers addressing the following main topics:

- Lessons learnt from design, construction and use of ZEB+
- Planning and procurement strategies
- Energy supply systems
- Energy flexibility
- Policy and business models
- Life cycle analysis and carbon footprint
- Scenarios and potential studies
- Energy performance simulation
- Method and tools for design
- Existing buildings: renovation and management
- Performance criteria and certification
- Building physics
- Indoor environment
- Heating, cooling and ventilation
- Carbon footprint

Peer review statement

All papers published in this volume of *IOP Conference Series: Earth and Environmental Science* have been peer reviewed through processes administered by the proceedings Editors. Reviews were conducted by expert referees to the professional and scientific standards expected of a proceedings journal published by IOP Publishing.

IOP Conf. Series: Earth and Environmental Science **352** (2019) 012001 doi:10.1088/1755-1315/352/1/012001

Users' satisfaction of indoor environmental quality conditions in ZEB+ at high latitudes

Veronika Zaikina[1], Claudia Moscoso[2*], Barbara Matusiak[2], Ali Chelli[3], Ilangko Balasingham[3]

1 Department of Optometry, Radiography and Lighting Design, USN. Hasbergsvei 36, 3616 Kongsberg, Norway.

2 Department of Architecture and Technology, NTNU. Alfred Getz vei 3, 7491 Trondheim, Norway.

3 Department of Electronic Systems, NTNU. O.S. Bragstads plass 2b, 7491 Trondheim, Norway.

* Corresponding author: claudia.moscoso@ntnu.no

Abstract. Energy efficiency policy forces architects to design buildings with increasingly well-sealed building skin. With the minimized outdoor connection, the indoor environment factors depend strongly on technical systems and control. In these scenarios, occupant dissatisfaction indicates a need for improvements of indoor environment and the way it is controlled. The aim of the article is to contribute to the discussion about the user perspective of indoor environmental quality in ZEB in the Nordic region. The focus on daylight as a factor for visual comfort, and on low outside temperature as an aspect of thermal comfort was dictated by this choice. An experimental study was conducted with 75 participants, in which the thermal, acoustic and visual conditions (controlled factors) together with a view out, humidity and CO2 level (monitored factors) were assessed by them and quantified via sensors. In most studied settings, the thermal comfort was the most determinant factor, followed by the acoustic and visual comfort. Other significant factors were mean illuminance in the room, mean temperature at the participants' desks and a mode value of the noise level. The daylight levels much lower than recommended in regulations were accepted by participants as comfortable if they were sitting by the window. Also, participants preferred a higher indoor temperature than the recommended in Norway.

1. Introduction

Humans are extremely adaptive to the environment, which allows us to survive as a species. But the repetitive adaptation to uncomfortable conditions is difficult to combine with health maintenance and a high level of performance expected nowadays e.g. from office workers. Also, the opportunities to individually adjust the indoor conditions are limited in most new office buildings today. Windows cannot be open for natural ventilation. Automatically operated sun-shading devices cannot be individually adjusted for optimal daylighting. Similar restrictions apply to ventilation, heating and electric light systems. Fixed working days and fixed workstations reduce temporal and spatial adaptive opportunities, meaning that employees are not able to change their working place and hours. In addition, office occupants rarely change their clothes during the working day to adapt to internal environmental conditions [1]. As the automated control of the indoor environment quality (IEQ) parameters becomes a standard, more knowledge is necessary to optimize it, especially deeper knowledge of occupants' comfort/discomfort experience is very much needed. In this regard, comfort can be defined as the subjective sensation of an indoor environment, including physical comfort (e.g. temperature, illumination etc.), functional comfort (e.g. disturbances, interruptions from work) and psychological comfort (e.g. privacy and territoriality) [2]. The scope of the present paper focuses on the physical comfort of the users. More specifically, on the thermal, acoustic and visual conditions at high latitudes.

Regarding thermal comfort, in a study comparing office layouts with high level of thermal control (traditional cellular office) and low level of thermal control (open plan office) found that individual thermal control resulted in 35% higher user satisfaction and 20% higher user comfort compared to open plan offices. At the same time, individual thermal comfort significantly increased the energy consumption of these offices. Researchers suggested, that individual user control over both temperature and ventilation, instead of using a centrally controlled air conditioning for fresh air and user control of radiator for temperature control, can help to reduce the energy use [3].

Regarding acoustic comfort, and considering that an office work requires high concentration, good acoustic conditions are needed. Speech, noise from ventilation systems, office machines and from the outside may cause annoyance and reduce the productivity of the occupants. Interestingly, high noise levels contribute to thermal discomfort and a temperature change of 1° C may have the same effect on productivity as a change in the noise of 2.6 – 2.9 dB [4]. Frequently experienced noise at work (e.g. loud air conditioning, internal machines such as fax machines and telephones) can produce annoyance that may result in significantly higher stress levels, and thus overall environmental discomfort.

Regarding visual conditions, it is well-known that vision cannot operate without light. Daylight is the natural and the most appropriate light for humans. It enables perfect colour rendering, natural directionality and supports our circadian rhythms [5]. Therefore, the daylight temporal availability and the daylight level are the most important aspects of providing a healthy and comfortable visual environment [6;7]. Previous studies showed also that office occupants strongly prefer daylight over artificial lighting [8]. Other factors as glare, light uniformity and colour rendering are frequently associated with visual and overall comfort [9]. Nevertheless, the light level remains the most relevant factor affecting visual comfort. In the Nordic region, with short daytime during the winter, it may become even more critical to provide the required amount of natural illumination supporting employees' physical and mental health and productivity. A too low light level that is a rather usual case in the Nordic region, may cause visual discomfort, as it won't provide an adequate amount of light required for visual task accomplishment.

Researchers of the European project OFFICAIR on indoor environmental quality assessments of modern office buildings in eight European countries found that the highest association with occupants' overall comfort was for "noise" factor, followed by "air quality", "light" and "thermal" satisfaction [10]. In a previous study carried out in 26 offices across five countries over North-Central and South Europe, including 4655 respondents, thermal comfort was identified as the most important factor [11]. Another study performed in China examined the IEQ satisfaction in cellular offices with various thermal conditions (18°-32° C), acoustic conditions (45-65 dB) and lighting conditions (100-1300 Lux). Authors found that thermal and acoustic conditions were the most important factors for the overall comfort of the occupants [12].

Although the literature regarding indoor environmental quality can be considered vast, to the authors' knowledge, there is little research considering daylight, especially daylight at high latitudes, as a factor in IEQ studies. In most of the studies, electrical lighting was used as a factor of visual comfort and glare caused by electrical lighting as a factor of visual discomfort.

The main objective of this study was to investigate IEQ satisfaction through the examination of thermal, acoustic, and visual conditions. The specific aim was to examine IEQ satisfaction during wintertime (November) at the location representing the Nordic region (Trondheim, Norway, latitude 63°43'). Consequently, the visual comfort has been examined for low daylight levels and the thermal comfort in the presence of low outside temperatures. The experiment was carried out in overcast sky conditions (no direct sunlight) which is the dominating weather type most of the year, and especially

IOP Conf. Series: Earth and Environmental Science **352** (2019) 012001 doi:10.1088/1755-1315/352/1/012001

during the winter, in the Nordic region. The results will contribute to knowledge about the impact of different factors on IEQ satisfaction in the location and development of control systems that better meet occupants' comfort needs.

2. Method

The experiment was carried out in the Test Cell Laboratory (SINTEF/NTNU) built according to Zero Emission Buildings (ZEB) standard (see Figure 1a). The cells are 4.35 m × 2.5 m in size; are equipped with a single South-West oriented window (1.10 m x 1.10 m) and a single standard door. Since the cells are identical, both cells were used simultaneously in the same way, which enabled researchers to streamline the experiment time. The cells allowed efficient control of heating and ventilation. The experiment was performed during November.

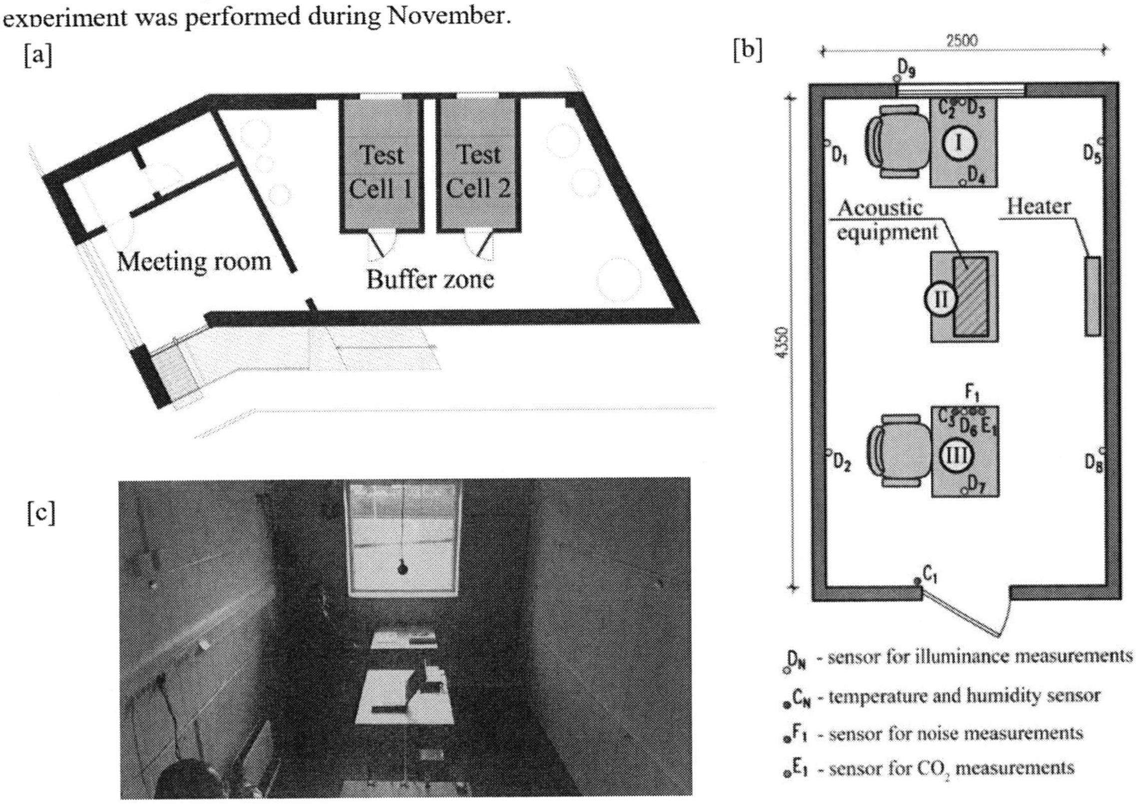

Figure 1. Plan of the Test Cell Laboratory [a], detail plan of Test Cell 1 [b], and picture of one experimental session on Test Cell 1 [c].

2.1. Experiment set-up

Two workstations were placed in each cell, they consisted of simple working desks and office chairs. The desks were positioned at the room (and window) axis, the first one (desk I) was placed next to the window, while the second one (desk III) was placed 2.75 m far from the window. One more desk (desk II) was positioned between these two workplaces and was used for a laptop and the acoustic equipment (see Figure 1b). Windows of both rooms are equipped with shading systems controlled from a central server; however, due to the overcast weather conditions, these were not used. Sensors collecting the data were deployed at the walls, at the desks (I and III) and at the wall outside the building. These were sensors for temperature and humidity registration (C1 – C3), inside illuminance registration (D1 – D8), outside illuminance registration (D9), CO2 level registration (E1) and noise registration (F1). The data from these sensors were collected during the whole experiment 24 hours a day, whereas only relevant time periods were analysed.

2.2. *Participants*

This experiment involved 75 participants. 80% of the participants were 19-29 years old; 13.3% were 30-39 years old; rest of the participants represented other age groups. 50.7% of the participants were males and 49.3 % were females. Norwegians accounted for 24.66% of the participants, the rest of the group represented other nationalities. Clothing insulation of the participants varied from 0.31 to 1.34 clo, with an average value equal to 0.78 clo and mode value equal to 0.65 clo. This was registered via questionnaires filled out by the participants (see section 2.5). 53.3% of the participants had normal vision, the rest of them used either glasses (37.3%) or lenses (9.3%). 45.3% of the participant consumed coffee prior to the experiment. 95% of the participants reported that they have good hearing.

2.3. *Experimental procedure*

Due to the short-day length at high latitudes in November, only one experimental session a day could be carried out i.e. between 09:30 and 13:00. Each session lasted approximately 2.5 hours, including introduction (where procedure, terminology and other details were explained to the participants), and a short lunch pause. Four people participated in the experiment at a time (one pair in each experimental cell) testing the same stimuli combination (SC) and being exposed to eight consecutive SCs arranged in randomized order (see Table 1).

2.3.1. Thermal conditions. The city of Trondheim, Norway is located at 63°43' N, and 10°40' E. The Köppen Climate Classification subtype for this climate is "Dfc" - Continental Subarctic Climate [13]. This type of climate characterizes by having long, usually very cold winters. Although November is not considered a common winter month for other parts of the world, it is not uncommon to experience winter features during this month in Trondheim. Accordingly, the registered outside temperature conditions during November in Trondheim presented an average of 1° C and 80% humidity. Indeed, the ground was covered by snow during the experiments. The tested temperature levels (21°C as comfortable and 25°C as uncomfortable) were selected following Norwegian recommendations on thermal conditions at working places specifying permissible temperature range from 19 to 26 degrees but recommended to keep it below 22 degrees most of the operational time [14]. The required temperature in the cells was regulated and adjusted via computer; it was done based on temperature levels measured by the globe-thermometer installed in the centre of each cell (see Figure 1c). This computer system regulated the temperature of the air coming into the cell through ventilation, therefore some minor deviations from the set temperature could occur. The temperature was homogeneous across the cells i.e. the temperature at both seating places was equal. To shorten the experiment duration and accelerate room heating (from 21°C to 25°C). a simple electrical heater was used in addition to ventilation. They were positioned at the floor and approximately in the middle of each cell. Heaters were activated during the short break when the participants were sitting in the Meeting room. The heaters did not have any fan, so no additional and unrequired noise was produced.

2.3.2. Acoustic conditions. The noise level equal to 38 dB presented comfortable conditions, and 50 dB presented uncomfortable acoustic conditions. These quantities were chosen based on Norwegian recommendations on the acoustics conditions at the work places with low noise levels, where recommended highest noise levels should not exceed 45 dB during the working hours. These levels were achieved through a play of recorded fan sound via the advanced acoustic system, which enabled precise volume control.

2.3.3. Visual conditions. In the scope of the present research, daylight was the only light source in the study, not including artificial lighting in the experiments. This study was focused on light level, excluding glare as additional factor. Considering that sun angles in the winter months tend to be low and can produce glare, sunlight was thus not present in the experiments, and overcast sky conditions were used. Weather conditions in the Nordic region during the autumn-winter period provide very restricted access to daylight (short daytime length, low outside illuminance levels) with a low frequency

IOP Conf. Series: Earth and Environmental Science **352** (2019) 012001 doi:10.1088/1755-1315/352/1/012001

of sunlight. As direct sunlight was not present during the experiment there was no significant risk of glare, and no glare was reported by any participant. The comfortable and uncomfortable visual environment were enabled through the choice of the seating position (short/long distance from the window). The participant sitting at the desk close to the window (desk I) had a higher light level and a rather large view through the window. At the same time, the participant sitting at the desk III had the significantly lower light level and much more limited view out (see Figures 1b and 1c).

Table 1. Eight combinations of indoor environmental conditions tested during the experiment.

Stimuli combination	Corresponding indoor environmental condition
SC 1	T_c+A_c+V_c
SC 2	T_c+A_u+V_c
SC 3	T_c+A_c+V_u
SC 4	T_c+A_u+V_u
SC 5	T_u+A_c+V_c
SC 6	T_u+A_u+V_c
SC 7	T_u+A_c+V_u
SC 8	T_u+A_u+V_u

* *Table notes: Subscript "c" means comfortable conditions, subscript "u" means uncomfortable conditions. Letter T means temperature, Letter A means acoustic conditions, letter V means visual or lighting conditions.*

2.4. Data collection
To monitor and register physical indoor environmental conditions, several sets of the sensors were deployed. These sensors included: one external light sensor measuring outdoor illuminance on the window wall every 10 seconds (range 10 to 100 000 Lux), eight indoor light sensors measuring illuminance level at horizontal and vertical surfaces in the room every 5 seconds (range 1 to 40 000 Lux), three sensors measuring temperature (range -40°C to 125°C) and humidity in the room (range 0% to 100%) every 5 seconds, and one sensor measuring noise level in the room every 5 seconds (range 8 to 80 dB).

2.5. Written questionnaires
The main questionnaire form included sets of questions to be answered on a 5-point Likert-typed scale. These questions included the evaluations on *i.* overall comfort (evaluation of all three main factors of IEQ), *ii.* thermal comfort, acoustic comfort and visual comfort separately, *iii.* satisfaction with the air quality in the room, and *iv.* quality of the view from their seating positions. Additionally, there were questions of choice for overall comfort among temperature, noise, lighting and other factors including *i.* choice of the most important factor, and *ii.* choice of most disturbing factor. Moreover, two preference questions were added: *i.* preferred changes to the perceived thermal-, acoustic-, and visual conditions (measured in 5-point scale), and *ii.* preferred change of seat location in the room (to be marked in a drawing). Finally, the participants were also asked to provide demographic (age, gender, country of origin), personal (height and weight, profession/occupation, vision- and hearing ability), and supplementary information (coffee consumption, evaluation of the psychological atmosphere at home and work/school, and clothing). This information could be important for their subjective evaluation of environmental comfort [15], e.g. caffeine can potentially affect the cognitive performance of the participants at the experiment, and thus was used for statistical analysis.

3. Results
3.1. Statistical analysis
A two-level ordinal regression analysis using the software STATA was applied. This statistical approach allowed to analyse the multiple studied factors together with the sensors' data readings. The analysis was performed for each SC separately and for all the SCs together, resulting in nine analysis blocks. The studied factors of this study are in Tables 2 and 3.

3.2. Results

3.2.1. Overall comfort satisfaction across all stimuli combinations. The results indicate that the participants' satisfaction with thermal conditions is the strongest factor affecting IEQ satisfaction in this experiment. The more satisfied with the temperature inside the cell, the higher the participant rated the overall comfort in the cell. Other two important factors that almost equally strongly affected the assessment of IEQ are the acoustic comfort and visual comfort. They also affected the perception of indoor environmental comfort in a positive manner, so higher rates of acoustic and visual comfort resulted in higher rates of overall comfort. Other significant factors are found in Table 2.

Table 2. Results of the two-level ordinal regression performed based on combined data of all eight SCs.

	Odds Ratio	Standard Error	z	P>|z|
View satisfaction	1.055	0.118	0.48	0.634
Thermal comfort satisfaction	**2.428**	**0.193**	**11.17**	**0.000**
Acoustic comfort satisfaction	**1.669**	**0.147**	**5.81**	**0.000**
Visual comfort satisfaction	**1.521**	**0.138**	**4.61**	**0.000**
Indoor air quality (IAQ) satisfaction	1.146	0.105	1.48	0.139
Psychological atmosphere at work	1.236	0.147	1.79	0.073
Illuminance at the desk (E_{desk})	0.997	0.003	-1.30	0.192
Illuminance in the room (E_{room})	**1.009**	**0.004**	**2.48**	**0.013**
Temperature at the desk (T_{desk})	**0.605**	**0.152**	**-2.00**	**0.046**
Temperature in the room (T_{room})	1.544	0.412	1.63	0.103
Noise$_{(mode)}$	**0.952**	**0.018**	**-2.65**	**0.008**
CO_2	1.000	0.001	0.23	0.820
Humidity in the room (H_{room})	0.972	0.023	-1.19	0.234
Level 2 variance	0.901	0.288		

* *Table note: Alpha level 0.05, z = z-value, P>|z| = p-value for z statistic.*

3.2.2. Comfort satisfaction for the individual SCs. The separate results for each stimuli combination were analysed based on the descriptive statistics and the two-level ordinal regression analysis. The simplified results are on Table 3. More complete numerical results are available from the authors upon request.

Table 3. Table showing simplified statistical results as significance of each tested factor for each particular SC, where *** means highly significant, ** - moderate significance, * - just significant.

Factors tested:	Significance of each factor for particular stimuli combination							
	SC1 T_c+A_c+V_c	**SC2** T_c+A_u+V_c	**SC3** T_c+A_c+V_u	**SC4** T_c+A_u+V_u	**SC5** T_u+A_c+V_c	**SC6** T_u+A_u+V_c	**SC7** T_u+A_c+V_u	**SC8** T_u+A_u+V_u
Thermal comf. satisf.	***	***	***	***	***	***	***	***
Visual comf. satisf.	***		**	**			***	***
Acoustic comf. satisf.	***	***		**		***	**	***
Noise, mode value	**							
Air quality satisfaction								
CO_2 level (registered)		*						
View out satisfaction								
M illuminance room			*					
Prior coffee consumption					*			
M temp. room					**	***		***
M temp. desk								***
M_{lux} desk					**	**		
M_{lux} wall front partic.					*			
Ratio[M_{lux} desk/M_{lux} room]					*			
Psych. comf. work/univ						***	*	
Clo value								*

* *Table notes: M - Mean Mlux - Mean illuminance, Clo – clothing.*

4. Discussion

Regression analysis of the results of all stimuli combination showed objective results. It is important to note that in the tested situations with moderate discomfort conditions that frequently exist in real offices the thermal comfort was the determinant factor for overall comfort satisfaction. The visual and acoustic comfort satisfaction were other two statistically significant factors. Mean illuminance of the room (rather than illuminance on the desk), mean temperature at the participants' desks (rather than in the room) and noise mode were factors that were of importance according to analysis. However, it is interesting that thermal conditions that were expected to be uncomfortable (25° C), were, in fact, more preferred by participants than conditions with 21 degrees that were considered to be comfortable. The factor that could have affected temperature preferences was the outside weather conditions (around 1° C, snow). The participants exposed to the cold air on the way to the Test Cell Lab appreciated warmer conditions indoors more than could be expected.

These results seem to be in accordance with research stating that cold climatic conditions, such as the ones existing in Nordic regions, have been found to be an important factor that influences the need to count with higher temperature levels indoors [16;17]. Thus, within SC1 (air temperature 21°) 56% responded that air temperature in the room was either comfortable or somewhat comfortable. But, within SC5 (air temperature was 25°) 65% of respondents positively evaluated the air temperature in the room. As it was noted earlier, the temperature range in Norwegian offices may vary from 19 to 26 degrees, but it is recommended to not exceed 22 degrees. Thereafter, even though this is dictated by energy saving, the recommended level is not always comfortable, as our results show.

Interestingly, even though the average illuminance at the working place situated by the window was low (oscillated around 100 lux), it has been perceived as comfortable by most participants This testifies about the easiness of visual adaptation of young people, but also that they enjoy low light levels in the vicinity of the window where they may perceive low light level outdoors.

From the numerical results and comparing SC6 (Tu+Au+Vc), SC7 (Tu+Ac+Vu) and SC8 (Tu+Au+Vu) we may observe that significant dissatisfaction with sound in SC6 or light in SC7 alone does not result with a large number of subjects dissatisfied with IEQ. If the uncomfortable conditions occur for both sound and light simultaneously, as in SC8, the number of subjects dissatisfied with IEQ (14%) is higher than the sum of dissatisfied with IEQ because of sound (3%) and dissatisfied with IEQ because of light (4%).

5. Conclusions

In this study, the overall comfort of occupants in small offices built according to Zero-emission building principles was examined. Three major components of indoor environmental quality (thermal, acoustic and visual conditions) together with the view out, humidity and carbon dioxide level were included into the experiment. These parameters were assessed by experiment participants and monitored via sensors. Participants had no individual adaptive opportunities, i.e. the ability to open the windows, turn on or off the air ventilation, the source of the noise or radiators, change of the clothes. This policy of deprivation of individual control over aspects helping to adapt to IEQ becomes more and more popular in energy efficient buildings widely designed and promoted nowadays, also in Norway. Daylight, a natural and healthy light, was chosen as the only light source for experimental rooms; it is a common knowledge that daylight if used at most during the working hours, contributes to energy saving.

The results showed that in most studied stimuli combinations when the participants experienced moderate discomfort conditions similar to real office conditions, the thermal comfort was the most determinant factor of indoor environmental comfort. Acoustic and visual comfort were the next two factors that influenced participants' assessments equally strong. Other factors that were statistically significant were: *i.* mean illuminance in the room, *ii.* mean temperature at the participants' desks, and *iii.* mode value of the measured noise level in the room. Nevertheless, the importance of light should not be overseen. When asked which environmental condition was the most important for the experience of overall comfort, lighting scored considerably higher than noise in all stimuli combinations; temperature scored highest. The results suggest that at office workplaces occupied by young people in the close vicinity of windows, the lower limit of the average illuminance at the desk may be reduced from 500 lux to 100 lux. This implies a potential for energy saving if applied in the control system for electric lighting.

The study points also that the recommendation for maximum 22 degrees for a thermal comfort in buildings situated in cold climate during winter is too strong, as the temperature of 25 degrees (used as uncomfortable in the study) was appreciated more than 21 degrees (used as comfortable) by the majority of participants.

IEQ is a multilateral concept affected by physical and psychosocial aspects. Being in a small office in a modern energy-efficient building with minimized adaptive opportunity, thermal, visual and acoustic conditions are highly important. Workstations placed by the window may be sufficiently illuminated exclusively by daylight over a longer time than it happens in practice creating good conditions for cognitive work, view out and energy saving. The results of this study shed new light to office design, contributing to discussions about daylighting and temperature effects on users' satisfaction.

Acknowledgements

This study was funded by The Research Council of Norway, under the project "HOME – Holistic Monitoring of Indoor Environments". We greatly thank Arild Gustavsen from the Research Centre on Zero Emission Buildings, NTNU, and Christian Schlemminger and Einar Bergheim from SINTEF for all their help with the usage of the Test Cell Laboratory in Trondheim, Norway. We would also like to show our gratitude to Professor Peter Svensson and Tim Cato Netland, both from Department of Electronic Systems at NTNU, who lent us acoustics instruments and guided us regarding usage.

References

[1] Barlow, S., & Fiala, D. (2007). Occupant comfort in UK offices—How adaptive comfort theories might influence future low energy office refurbishment strategies. Energy and Buildings, 39(7), 837-846. doi:10.1016/j.enbuild.2007.02.002

[2] Al Horr, Y., Arif, M., Kaushik, A., Mazroei, A., Katafygiotou, M., & Elsarrag, E. (2016). Occupant productivity and office indoor environment quality: A review of the literature. Building and Environment, 105, 369-389. doi:10.1016/j.buildenv.2016.06.001

[3] Shahzad, S., Brennan, J., Theodossopoulos, D., Hughes, B., & Calautit, J. K. (2017). Energy and comfort in contemporary open plan and traditional personal offices. Applied Energy, 185, 1542-1555. doi:10.1016/j.apenergy.2016.02.100

[4] Pellerin, N., & Candas, V. (2004). Effects of steady-state noise and temperature conditions on environmental perception and acceptability. Indoor Air, 14(2), 129-136. doi:10.1046/j.1600-0668.2003.00221.x

[5] Wirz-Justice, A. (2017). Seasonality in affective disorders. General and Comparative Endocrinology. doi:http://dx.doi.org/10.1016/j.ygcen.2017.07.010

[6] Edwards, L., & Torcellini, P. (2002). A Literature Review of the Effects of Natural Light on Building Occupants. (NREL/TP-550-30769). Golden, Colorado: National Renewable Energy Laboratory.

[7] van Bommel, W., & van den Beld, G. (2004). Lighting for work: a review of visual and biological effects. Lighting Research and Technology, 36(4), 255-269.

[8] Galasiu, A. D., & Veitch, J. A. (2006). Occupant preferences and satisfaction with the luminous environment and control systems in daylit offices: a literature review. Energy and Buildings, 38(7), 728-742. doi:10.1016/j.enbuild.2006.03.001

[9] Carlucci, S., Causone, F., De Rosa, F., & Pagliano, L. (2015). A review of indices for assessing visual comfort with a view to their use in optimization processes to support building integrated design. Renewable and Sustainable Energy Reviews, 47, 1016-1033.

[10] Sakellaris, I. A., Saraga, D. E., Mandin, C., Roda, C., Fossati, S., de Kluizenaar, Y., . . . Bluyssen, P. M. (2016). Perceived Indoor Environment and Occupants' Comfort in European "Modern" Office Buildings: The OFFICAIR Study. Int J Environ Res Public Health, 13(5). doi:10.3390/ijerph13050444

[11] Humphreys, M. A. (2005). Quantifying occupant comfort: are combined indices of the indoor environment practicable? Building Research & Information, 33(4), 317-325. doi:10.1080/09613210500161950

[12] Huang, L., Zhu, Y., Ouyang, Q., & Cao, B. (2012). A study on the effects of thermal, luminous, and acoustic environments on indoor environmental comfort in offices. Building and Environment, 49, 304-309. doi:10.1016/j.buildenv.2011.07.022

[13] Peel, M. C., Finlayson, B. L., & McMahon, T. A. (2007). Updated world map of the Köppen-Geiger climate classification. Hydrology and Earth System Sciences, 11(5), 1633-1644. doi:doi.org/10.5194/hess-11-1633-2007

[14] Arbeidstilsynet. (2016). Veiledning om klima og luftkvalitet på arbeidsplassen. (444).

[15] Tuomaala, P., Holopainen, R., Piira, K., & Airaksinen, M. (2013, August 26-28). Impact of individual characteristics - Such as age, gender, BMI, and fitness - on human thermal sensation. Paper presented at the 13th Conference of International Building Performance Simulation Association, Chambéry, France.

[16] Becker, R., & Paciuk, M. (2009). Thermal comfort in residential buildings – Failure to predict by Standard model. Building and Environment, 44(5), 948-960. doi:10.1016/j.buildenv.2008.06.011

[17] Zhang, Y., & Altan, H. (2011). A comparison of the occupant comfort in a conventional high-rise office block and a contemporary environmentally-concerned building. Building and Environment, 46(2), 535-545. doi:10.1016/j.buildenv.2010.09.001

IOP Conf. Series: Earth and Environmental Science **352** (2019) 012002 doi:10.1088/1755-1315/352/1/012002

User engagement with smart home technology for enabling building energy flexibility in a district heating system

Simon Peter Larsen[1] **and Hicham Johra**[2]

[1]The Danish Building Institute, Aalborg University, A.C. Meyers Vænge 15, DK-2450, København SV, Denmark
[2]Department of Civil Engineering, Aalborg University, Thomas Manns Vej 23, DK-9220, Aalborg Øst, Denmark

Abstract. The future low-carbon emission societies rely on energy systems bearing an increasing share of renewable energy sources (RES). Consequently, demand-side management and energy flexibility become a key solution to compensate for the intermittent nature of RES. District heating systems hold a large potential for energy flexibility if households are actively integrated. While previous research and local policies have applied demand-side management such as smart meters, new smart home technology envisions full employment of the flexibility potential of the building stock. Morning energy demand peak is a major concern for district heating systems in Nordic countries. Demand-side management for district heating has thus mainly focused on morning hours peak-shaving. While integrating smart home technology as a demand-side management solution, the household becomes a flexible energy hub for thermal energy storage. While the technical potential of achieving such flexibility has been investigated, less research has been carried out concerning how users engage with smart home technology and how this influences the possibilities for load-shifting of the indoor space-heating demand. By conducting qualitative studies (interviews and 'show and tell' home tours) in 16 Danish households, this paper explores how users engage with smart home technology and how this influences the possibilities for load-shifting in a district heating system. The study provides insight into how the occupants interact with different smart technologies providing space-heating control. Results show that engagement with smart home technology must be understood as part of people's everyday practices. The flexibility in energy demand must be generated by understanding and changing practices to make them more flexible during peak hours. While smart home technology holds the potential for adding flexibility within the district heating system, the technology is rarely used as intended by occupants. Smart home technology is disrupting, and users rapidly create workarounds in order to perform everyday practices. Load-shifting during morning hours is thus a technical possibility, but the dominant techno-economic paradigm embedded in smart home technologies remains a barrier, as the latter must adapt to the everyday practices.

1. Introduction

Traditionally, the building stock is perceived as a passive end-user within the energy grids. Because the building stock accounts for one-third of the final energy usage and CO_2 emissions globally, large efforts during these last decades were dedicated to drastically improve the energy efficiency of buildings [1]. Although this is a major target to achieve sustainability of future societies, there is a paradigm shift concerning the role of buildings in the energy grids. In the context of Smart Energy Grid systems [2], the building stock should be considered as an important active element which can help to reduce the

mismatch between instantaneous energy production and usage by means of energy load-shifting, peak-shaving and valley filling. These demand-side management methods are commonly called "building energy flexibility" strategies [3]. Activated by a penalty signal (energy spot price, CO2 intensity of current energy production, local state of the grid, marginal production cost, etc.), this building energy flexibility can significantly ease the establishment of a reliable and sustainable energy supply network with massive penetration of intermittent renewable energy sources (RES). Moreover, it can solve local or global congestion issues occurring at critical periods in the bottleneck weak points of a grid.
The recent studies on the topic have investigated the potential of buildings providing flexibility services to the grid by smart control of thermal and electrical energy loads. Electricity which is produced locally (for example with photovoltaic panels) or imported from the grid, can be stored in the building's batteries [4] or in the electrical vehicles connected to it [5]. Furthermore, the building stock has a large potential for thermal storage in hot water tanks (heating system and/or domestic hot water tank) [6] and in the indoor environment thermal mass [7]. The latter can be used as a cost-effective thermal storage solution when employing indoor temperature setpoint modulation [8]. Excess RES production can thus be stored in the built environment and allows significant space-heating energy shifting over few hours for poorly insulated buildings, and up to more than 24 hours for well-insulated buildings with large effective thermal inertia without jeopardizing the thermal comfort [9].
These demand-side management and building energy flexibility strategies are based on advanced control systems, enabling a certain communication between the Smart Grid, the building systems and the occupants. In practice, the activation of the building energy flexibility is performed by so-called "smart home technologies". These technologies also insure occupants' comfort and awareness about the status of their building. However, there is a clear lack of understanding whether these smart home technologies will be largely accepted by households, how they will interact with the it and whether occupants will allow control signals from the outside (penalty signals from the grid, energy providers and distributors) to interfere with their own private indoor environment. The households' perspective is a key topic of investigations, as it can be a strong barrier to the successful implementation of energy flexibility measures. The study presented in this article is part of the transdisciplinary InterHUB project. The InterHUB project aims at understanding the interactions between the different actors and stakeholders of the building energy flexibility in a smart energy grid system [10]. The current paper focusses on assessing the engagement of occupants in dwellings with smart home technology for enabling building energy flexibility in Denmark. Following the details of the methodology and the different study cases, the preliminary results of this ongoing investigation are presented. The authors hope that this study can give a clearer insight into user's acceptability to other researchers working on the topic. The article closes with the main conclusions and suggestions for future work.

2. Methodology

The overall methodological approach is qualitative in design. Semi-structured interviews and home tours have been conducted in 16 households of the Copenhagen urban area (Denmark). Applying an open character, the interviews aims at understanding the 'doings and sayings' of the occupants, and thus assess how occupants engage with smart home technologies. As space-heating is largely considered 'invisible' by occupants [11], special attention has been put on everyday activities and routines performed by the occupants. This has given valuable insights into how space-heating practices are re-configured when the smart home technologies are integrated within the domestic sphere. Prior to the interview, a semi-structured interview guide has been drafted. The interview guide consists of themes concerning space-heating, comfort and engagement with smart home technology. The interview guide is structured so that the interviewee can 'pursue' reflections occurring during the interview. Each interview had a duration of 1 ½ to 2 hours. During the interview, the occupants were asked to give a home tour, explaining their use of each room in the house. This approach is valuable for generating insight into mundane everyday activities such as adjusting space-heating or airing, which might not have been clear if the interview was staged otherwise. Furthermore, occupants were asked to 'show and tell' how they operated their smart home technologies. This often led to a conversation about their general experiences with technology. The methodological approach is largely in line with the research conducted on energy consumption in the domestic sphere [12,13,14,15,16]. Such methodology highlights the complexity of everyday life and underlines the fact that research on user engagement with technological equipment and energy consumption must be a user-centric approach. The approach relies on an epistemological assumption that, in order to generate knowledge on user engagement, research must examine how occupants themselves perceive their own practices. While other studies have looked into user engagement with smart home technologies and their potential benefits (including the possibility for energy savings [17]), only a few of them have assumed that everyday life practices are the smallest unit of analysis and that any research enquiry must, therefore, uncover it [18]. Qualitative methods are thus preferred as they are able to capture the complexity and unpredictability of everyday life. Following Bent Flyvbjerg [19], the use of case studies has also been valued as a fruitful way of both generating insight and hypothesis for further investigations, but also testing such hypothesis and contributing to theory building. As this short paper only presents preliminary results, the focus is placed on the trends of user engagement with smart home technology for enabling space-heating flexibility. Consequently, this article acts as a starting point for further analysis of user engagement with such technologies.

3. Study cases

The current analysis incorporates four different study cases [20]. Each study case represents a residential area (varying in size) which applies the same smart home technology for enabling energy flexibility. Four interviews were conducted in each study case and all present adult family members (18+) were involved. While some interviews were conducted with only one person, the others involved multiple people and, thereby, captured the negotiation of everyday life between household members. Furthermore, small kids were also present at some of the interviews, but they were not interviewed. However, their presence unfolded the engagement and practices of kids, and how the adult members of the household reacted to them (e.g. keeping a higher floor heating temperature so that the kids could play on the floor). The selection of the households was made to gain a broad representation of different households, varying in terms of size, age of occupants, educational background and gender. As shown in Table 1, the interviews include a 50/50 gender variation, a wide age distribution (21-58 year old) and variation in terms of household size (1-5 occupants). However, the data is slightly biased in terms of the occupational background of the informants. Indeed, the vast majority of the interviews has senior-level jobs. This is with the exception of students who were represented in 1/3 of the interviews. In all study cases, the energy for indoor space-heating and domestic hot water production is supplied by a district heating network. The study cases are either apartments or terrace houses and are all recently built with a good level of building envelope thermal performance. A short description of each study case can be found in the following sections.

3.1. Måneparken, Trekroner, Roskilde: Måneparken was constructed in 2004 and located in Trekroner, in close proximity to higher education institutions and 30 km from Copenhagen. The estate consists of 31 apartments: 24 one-bedroom apartments (35 m^2) and 7 two-bedroom apartments (39 m^2). The apartment block consists of two floors, with apartments on each level. The rents range from 500 to 800 euros per month. The estate is a student residence administered by the local social housing organization. The apartments are thus only occupied by students. Most of them are in their early twenties. The allocation of the apartments is based on a waiting list, meaning that the occupants represent different social spheres of the society. Most occupants live in the apartments for short time periods (½ - 5 years). Most apartments are only occupied by one person. The seven two-bedroom apartments accommodate small families or flatmates living together. Within the apartments, radiators are installed in the kitchen/entrance and the bedroom(s). In the bathroom, floor-heating is installed. Control of space-heating is enabled by two different types of devices. Digital thermostats (Danfoss Living Connect©) are placed on every radiator, allowing the occupant to adjust setpoints directly on each of them. The thermostats are connected to a central Danfoss CC link© In-Home-Display (IHD) located in the kitchen. The occupants can control all thermostats and set rules or schedules for the entire apartment space-heating. Different sensors in the house can be used for the control rules, e.g. opening windows automatically shuts the space-heating off until the window is closed again. The thermostats and the IHD are also accessible to the occupants via an application installed on their smartphone, allowing the remote management of the space-heating. Besides the Danfoss technology, a model predictive control (MPC) system called LeanHeat© is in use. By measuring both indoor (occupancy and consumer profiles) and outdoor conditions (temperature and humidity), the system optimizes the heat flow going into the apartment.

Table 1. Overview of the study cases.

Name and location	Building	People	Technology	Interview details
Måneparken Trekroner Roskilde	31 apartments 1 and 2 bedrooms Built in 2004	Social housing for students. Mostly single living tenants, the rent is 500 - 800 euros per month. Few apartments with couples.	District heating. Radiators in rooms, floor heating in the bathroom. Control by Danfoss CC Link© and Leanheat© technology.	4 interviews: 1 male and 3 females. Occupants: 1-2 Age: 21-25 All students
Frikvarteret, Nordhavn, Copenhagen	29 terrace houses 100 - 200 m^2 Built in 2016	Ownership – prices start at 800,000 euros. Mostly families from upper-middle-class. Also couples.	District heating. Floor heating in all rooms. Control by digital on-wall thermostats, and fuel-shift controlled by the utility.	4 interviews: 4 males and 3 females. Occupants: 2-5 Age: 35-58 Senior-level jobs
Sundmolehusene Nordhavn Copenhagen	72 apartments and 11 terrace houses 45 - 210 m^2 Low-energy building Built in 2017	Ownership – prices from 270,000 euros to over 1 million euros. Differentiated household composition, but mainly upper-middle-class and upper-class.	District heating. Floor heating in all rooms. Control by digital on-wall thermostats and external control (centralized) during a limited time period.	4 interviews: 3 males and 3 females. Occupants: 2-4 Age: 21- 58 Senior-level jobs and 2 students.
Havnekanten Nordhavn Copenhagen	86 apartments 55 - 146 m^2 Low-energy building Built in 2016	Ownership – prices from 300.000 to 1.4 million euros. Differentiated household composition, but mostly families from upper-middle-upper class.	District heating. Floor heating in every room. Control by Danfoss CC link© system with the possible scheduling. Sensors in the whole apartment.	4 interviews: 3 males and 2 females. Occupants: 1-4 Age: 35-56 Senior-level jobs.

3.2. Frikvarteret, Nordhavn, Copenhagen: Frikvarteret is a newly built (2016) residential area, in the recently redeveloped area of Nordhavn. Nordhavn is located nearby Copenhagen city center. Since the beginning of the 2010s, the Nordhavn district has undergone an extensive transformation, from being an old harbour area to an attractive residential and business area. The prices in Frikvarteret starts at approximately 800,000 euros. Frikvarteret was one of the first residential complexes of Nordhavn and consists of 29 single-family terrace houses, ranging from 100 to 200 m^2. The terrace houses are in 2 - 3 floors and rather spacious for Danish standards. The occupants are, to a large extent, middle-class to upper-class families, typically with small children. An under-floor heating system is installed in all rooms of all houses. A hot water tank with an electric heating unit has been installed. It allows "fuel-shifting" by selecting either district heating network and electricity as heat source. Control of when to conduct fuel-shift is made by the district heating company. Occupants can control and monitor space-heating by using digital on-wall thermostats. They also have access to a web-based platform to visualize their energy consumption related to space-heating on a day-to-day basis.

3.3. Sundmolehusene, Nordhavn, Copenhagen: The Sundmolehusene was built in 2017, and consists of 72 apartments and 11 terrace houses. The size of the apartments ranges from 45 to 210 m^2. The building block is considered as a low-energy building with energy for heating (indoor space and hot water) and ventilation below 20 kWh/m^2 per year. The prices for apartments in Sundmolehusene range from 270,000 to over 1 million euros. In general, Sundmolehusene is intended for upper-middle-class. Occupant composition varies significantly, consisting of both families, couples and single-person households. An under-floor heating system is installed in all rooms of all dwellings. Control of space-heating is performed by digital on-wall thermostats. Information about the humidity level and CO_2 concentration are also displayed. On-wall thermostats are placed in every room. Opening of the doors and windows are monitored to automatically shut heating off. In 10 of the apartments, enabling flexible heating demand has been experimented. The space-heating of those dwellings was, for a short period, controlled from the outside by researchers from the Technical University of Denmark [21].

3.4. Havnekanten, Nordhavn, Copenhagen: Havnekanten, constructed in 2016, is also located in Nordhavn. The estate consists of 86 apartments, ranging from 55 to 146 m^2. They are considered low-energy buildings with minimum space-heating needs and good indoor comfort. Like most of the housings in Nordhavn, the price for Havnekanten dwellings range from 300,000 to 1.4 million euros. This makes Havnekanten one of the most expensive areas in Nordhavn. Consequently, its occupants are mostly people from the upper-middle and upper class. The different apartment sizes in Havnekanten induce a differentiated household composition, and both families, couples and people living alone reside in the apartments. Floor heating is installed in every room of all the apartments. Additionally, temperature sensors are installed throughout the apartment, in order to evaluate the possibilities for enabling heat storage in the building. Control of space-heating is performed by a decentralized system: Danfoss CC link©. Within the apartment, digital on-wall thermostats are placed in every room, allowing the occupants to adjust indoor temperature setpoints. In addition, users can centrally monitor, control and schedule the space-heating from the IHD. Sensors detect if doors or windows are opened, and automatically turn off space-heating. Space-heating can also be controlled and monitored remotely, using a smartphone application.

4. Results and discussion

As previously mentioned, this paper only presents the preliminary results and trends observed during the qualitative studies in four different cases. Overall, the interviews showed that the young informants were more satisfied with the technology compared to the older informants. Despite educational background or familiarity with technology or the energy system, older informants were generally more concerned about the functionality and accuracy of the technology. In terms of household composition, larger households had a more diverse engagement with the technology (e.g. different settings in different rooms or more frequently shifting setpoints). This was often the results of continuous negotiations between the household members. That being said, there was a clear trend towards males being the main user of the technology. They were responsible for the setup and the continuous operation (also in terms of technological failures).
In general, three main trends for user engagement with smart home technology for enabling a flexible space-heating demand have been observed:

- The importance of embodied know-how
- The notions of comfort evolve
- The smart home technologies as scripted material arrangements

4.1. The importance of embodied know-how:
Throughout the 16 qualitative interviews and home tours, one trend which stood out was the importance of the so-called embodied know-how. It appeared that informants who were very familiar with the use of smartphones and other network-connected technologies had greater ease for controlling smart home technologies, and hence they engaged more actively and continuously with the systems. This was especially evident among the young occupants who approached the technologies with confidence and quickly made use of the advanced features of the system. It was also this group of users who expressed the largest satisfaction and interest in the system. As one occupant expressed: *"Yes, but it is also because it is easier. (...) On an ordinary radiator, what the hell does "5" mean? Does it mean 50 °C, does it mean 5 °C, what does 5 mean, what does the number 5 mean? I have no idea what that means. So I do not know if I should put it on 3 to get the temperature I want or if I should put it on 5, so I have never*

figured out what the number means in terms of degrees. Wherewith Danfoss, I know exactly that it is 20 °C if I set it (The system) *to 20 °C. So it is much easier to control now"* (Occupant A3).
On the contrary, occupants who had less experience with 'smart' technologies, expressed less interest and sometimes had dissatisfaction with the system. These occupants were generally more uncertain of whether or not the system operated correctly and performed different experiments to test it. Such experiments often included material arrangements and tools which the occupants trusted and were familiar with, such as thermometers: *"Yes, I bought it* (a thermometer) *because I just had to see if it said the same thing. I think I bought it because the problem with the heat control in the bathroom didn't really work, so I brought it there to investigate what the temperature really was"* (Occupants E3).
Other occupants felt a disempowerment by the smart home technology, resulting in rare or no engagement with the systems. The former thus adapted other practices for achieving their desired level of comfort, such as putting on blankets or wearing more clothes. These results emphasize the importance of occupants' prior experience with similar technologies. This is often an embodied experience in which previous hand-on experience is more important than e.g. the user having read information about the technology. The results highlight the fact that competency comes in different forms and that tacit knowledge is important for enabling user engagement with smart home technologies and thus the possibilities for enabling a flexible heating demand. While occupants with embodied competencies related to these technologies might be more competent practitioners in order to provide flexibility within the current system, these results also highlight that future systems have to incorporate how users without such embodied knowledge can contribute to flexibility. One solution might be an increased automation and less user control.

4.2. The notions of comfort evolve
A second trend was the tendency to increased notions of comfort, as smart home technologies were integrated within the household. This result has also been found in other studies e.g. [22]. It appears that, as smart home technologies offer the occupants a convenient way of controlling space-heating, this is often translated into an increased notion of comfort. Several occupants thus expressed that their indoor comfort requirements had increased, since moving into a house with smart home technologies. This often resulted in higher indoor temperature setpoints, equivalent to a comfort level for which being able to walk barefoot within the household was pleasant. This was clearly evident in households which had floor heating installed. Occupants liked the notion of warm feet, and quickly adapted to this as a notion of comfort throughout the house by considering that being able to walk barefoot in the entire house was now a normal level of comfort for them. One expressed this sensory notion as follows: *"It (the temperature) is somewhere between 22 - 23 °C. I think it is the most comfortable. I like that you can walk around without having to wear a big sweater or that you can basically walk around in socks or barefoot"* (Occupant C2).
Similar investigations have previously highlighted the sensory feelings that floor heating generates among occupants, and how this tends to be translated into increased notions of comfort [14]. However, such findings are not directly connected to the engagement with smart home technology, but are evident in low-energy buildings in general. The smart home technologies and the remote control were generally well perceived by the occupants, who were largely positive towards the increased automation of the domestic sphere. They were mostly valued because of their ability to increase the convenience of controlling space-heating. In relation to generating flexibility for the district heating system (e.g. peak-shaving), the occupants were generally more sceptical about enabling such feature themselves. They preferred that this was done either automatically (MPC, Machine Learning) or centrally by a third party such as the utility or the building manager. However, occupants expressed that such control must not decrease comfort within the dwelling. This was furthermore challenged by the other space-heating practices which occupants seemed to have. Most of them were e.g. used to opening windows for a variety of reasons. In general, a cold bedroom was preferred during the night, and opening windows in the morning and throughout the day was considered a healthy practice. This can be problematic for enabling space-heating flexibility and using the indoor environment for heat storage. For instance, one informant expressed his practice of airing the bedroom while keeping the floor-heating turned on as follows: *"I like the feeling that there is a bit of heat in the room, while there is fresh airing"* (Occupants C2).

4.3. The smart home technologies as scripted materials

As well emphasized in the scientific literature, technologies are scripted towards a certain use. Through a careful design process, developers hope that they will be used as intended. However, literature shows that material arrangements (equipment) are not always used as intended. This was also evident throughout the 16 interviews and home tours. Most noticeably, the differentiated use and engagement with smart home technologies and understandings of how it worked. This stresses the importance of careful design and implementation of smart home technologies for enabling a flexible space-heating demand, considering the differences in engagement. One clear trend was that while the technology scripted a certain use, occupants often created workarounds, as the technology interfered with other practices in their everyday life. Such workarounds included moving the IHD to other locations within the household (away from the front door), resulting in a decreased engagement with the system. Others did not like the LED lighting that the smart home technologies produced, and thus turned the whole system off. The technologies were also used very differently than their intended script, and some occupants used IHD as a wall-clock or a simple thermometer.

The use of smart home technology was also often scripted by the occupants themselves, and engagement with it varied from being used as a simple on/off device to a precise scheduling tool. Occupants often negotiated between the script of the smart home technology and that of other material arrangements in the household. Occupants did not perceive the technology as flexible as it was initially scripted, and felt that it did not operate fully with the other home materiality. For instance, the occupants of low-energy dwellings did not perceive floor heating as flexible, as it would take too long to heat up if turned off. Consequently, they felt that the smart home technologies were not compatible with such purpose.

5. Preliminary conclusions

In future low-carbon societies, domestic energy consumption must be more flexible, as energy will increasingly have to rely on intermittent energy sources such as wind and solar power. While energy consumption within the domestic sphere is rather mundane, as occupants perform everyday practices (e.g. getting ready for work and school in the morning), peak-demand problems are increasing. The possibilities for time-shifting such peaks have recently received important interest. In that regard, smart home technologies have often been highlighted as the possible mean for enabling energy flexibility by use of heat storage in the indoor environment and its thermal mass. However, there is a clear need for deeper knowledge concerning the smart home technologies acceptability by the households, how the latter interact with the smart home technologies and if they will allow external control signal to alter their private indoor environment. This paper presents preliminary conclusions from an ongoing study on occupant engagement with smart home technologies in 16 Danish households. By applying qualitative methods, i.e. in-depth interviews and home tours with occupants, the results highlight three trends of occupant engagement. First, the importance of embodied knowledge, showing that occupants who have previous and hands-on experience with same or similar technologies increasingly engage with smart home technologies and take advantage of the more complex features. These occupants are also generally more positive about the system and more competent at enabling flexibility. Second, a trend of increasing notions of comfort is evident. As smart home technology provides the occupants with a more convenient way of controlling space-heating, such convenience is often translated into higher comfort norms such as higher indoor temperature setpoint, challenging the potential for increasing flexibility in the system. Third, although technology is often scripted towards a certain use, i.e. enabling flexibility, they are often not used by the occupants as intended. Engagement with smart home technologies proved to be very diverse and ranged from taking advantage of the complex features to merely being used as a wall clock or simply hidden away. This research project will continue until the spring of 2021, and during that period the sample size will be increased. The focus will be on broadening the variety of occupants and buildings included in this investigation.

Acknowledgments

This work was financed by the InterHUB project which is funded by Aalborg University's strategic funding for interdisciplinary research. InterHUB: Intermittent energy – Integrating Households, Utilities, and Buildings; for more info on the project see www.interhub.aau.dk.

IOP Conf. Series: Earth and Environmental Science **352** (2019) 012002 doi:10.1088/1755-1315/352/1/012002

References

[1] International Energy Agency 2013 *Transition to Sustainable Buildings: Strategies and Opportunities to 2050*. IEA (Paris)

[2] Bamberger Y, Baptista J, Belmans R, Buchholz BM, Chebbo M, Doblado JLDV et al. *Vision and Strategy for Europe's Electricity Networks of the Future: European Technology PlatformSmartGrids*. Luxembourg: Office for Official Publications of the European Communities, 2006. 22 p. (Community research. Energy; No. EUR 22040)

[3] Jensen S Ø, Marszal-Pomianowska A J, Lollini R, Pasut W, Knotzer A, Engelmann P, Stafford A and Reynders G 2017 IEA EBC Annex 67 Energy Flexible Buildings *Energy and Buildings*. **155** 25–34

[4] Cao S, Hasan A and Sirén K 2013 Analysis and solution for renewable energy load matching for a single-family house *Energy and Buildings*. **65** 398-411

[5] Clement-Nyns K, Haesen E and Driesen, J 2010 The impact of charging plug-in hybrid electric vehicles on a residential distribution grid *IEEE Trans Power Syst*. **25** 371-380

[6] Arteconi A, Hewitt N J and Polonara F 2013 Domestic demand-side management (DSM): Role of heat pumps and thermal energy storage (TES) systems *Applied Thermal Engineering*. **51** 155-165

[7] Johra H and Heiselberg P 2017 Influence of internal thermal mass on the indoor thermal dynamics and integration of phase change materials in furniture for building energy storage: A review *Renewable and Sustainable Energy Reviews*. **69** 19-32

[8] Hedegaard K, Mathiesen B V, Lund H and Heiselberg P 2012 Wind power integration using individual heat pumps – Analysis of different heat storage options *Energy*. **47** 284-293

[9] Johra H, Heiselberg P and Le Dréau J 2019 Influence of envelope, structural thermal mass and indoor content on the building heating energy flexibility *Energy and Buildings*. **183** 325-339

[10] InterHUB: Intermittent energy – Integrating Households, Utilities and Buildings. https://www.strategi.aau.dk/Forskning/Tv%C3%A6rvidenskabelige+forskningsprojekter/interHUB/

[11] Hargreaves T, Nye M and Burgess J 2010 Making energy visible: A qualitative field study of how householders interact with feedback from smart energy monitors *Energy Policy*. **38**(10) 6111-6119

[12] Royston S(2014 Dragon-breath and snow-melt: Know-how, experience and heat flows in the home *Energy Research & Social Science*. **2** 148-158

[13] Madsen L V and Gram-Hanssen K 2017 Understanding comfort and senses in social practice theory: Insights from a Danish field study *Energy Research & Social Science*. **29** 86–94

[14] Gram-Hanssen, K 2010 Residential heat comfort practices: understanding users *Building Research & Information*. **38**(2) 175-186

[15] Hargreaves T and Wilson C 2017 *Smart Homes and Their Users*. Cham: Springer.

[16] Nicholls L and Strengers Y 2015 Peak demand and the 'family peak' period in Australia: Understanding practice (in)flexibility in households with children *Energy Research & Social Science*. **9** 116-12

[17] Wilson C, Hargreaves T and Hauxwell-Baldwin R 2015 Smart homes and their users: a systematic analysis and key challenges *Personal and Ubiquitous Computing*. **19**(2) 463-476

[18] Reckwitz A 2002 Toward a Theory of Social Practices *European Journal of Social Theory*. **5**(2) 243-263

[19] Flyvbjerg B 2006 Five misunderstandings about case-study research *Qualitative Inquiry*. **12**(2) 219–245

[20] Larsen S, Gram-Hanssen K and Marszal-Pomianowska A 2019 *Smart Home Technology Enabling Flexible Heating Demand: Implications of Everyday Life and Social Practices* Proc. of the 2019 ECEEE Summer Study, Energy efficiency first, but what next?, Belambra Presqu'île de Giens, France, 03–08 June 2019

[21] EnergyLAB Nordhavn – Work Package 3. http://www.energylabnordhavn.com/wp3.html

[22] Strengers Y 2008 Comfort expectations: The impact of demand-management strategies in Australia *Building Research & Information*. **36**(4) 381-391

New town hall in Freiburg (D): concept, performance and energy balance after the first year of monitoring of a large net plus-energy building

Nicolas Réhault, Peter Engelmann, Manuel Lämmle, Leo Munzinger
Fraunhofer Institute for Solar Energy Systems, Heidenhofstr. 2, 79110 Freiburg, Germany
E-mail: nicolas.rehault@ise.fraunhofer.de

Abstract. The town-hall building (Rathaus im Stühlinger: RIS) of the city of Freiburg (D) has been designed and built with the objective of a plus-energy balance and has been handed over to the city in November 2017. With a net floor area (NFA) of 22.650 m^2, it is to date one of the largest designed and built plus-energy building in Europe. The boundaries considered in the annual energy balance are limited to the loads of the heating, ventilation and cooling systems (HVAC), of the domestic hot water (DHW) and of the lighting; user-dependent energy demands are not considered. To achieve a positive primary energy balance, the building envelope is highly insulated, the energy supply is based on a low-exergy concept and onsite energy is generated by a large Building Integrated Photovoltaic (BIPV) plant combined with photovoltaic-thermal combined collectors (PVT). The central component of the heat generation is a ground-water coupled heat pump system supplying thermal activated concrete slabs, heating ceilings and air handling units (AHUs). Cooling is ensured over the geothermal well. To the date of this publication, the building has been intensively monitored for one-and-a-half year. Here, we present the energy concept as well as the results of the first monitoring period and provide an assessment of the energy balance in comparison to the plus-energy objective as well as an initial analysis of the load curves and of the heating and cooling systems.

1. Introduction

The German government has recently recognized that the greenhouse gas emissions have hardly been reduced in Germany between 2009 and 2017 and that the objective of 40% CO_2 emissions reduction for 2020 based on the reference year 1990 will not be met. Whereas the share of renewable energies in the fuel mix has continuously increased during the same period, up to 36% in 2017, the success of the German energy transition is almost concentrated on the power system. In general, the carbon footprint of the heat consumed in the household, services and industry sectors and corresponding to about half of the delivered energy consumption of Germany, has not been reduced [5]. Nevertheless, studies have shown that the sector coupling between power and heat by the use of heat pumps, besides the increase of the energy efficiency and of the local production of energy by means of renewable energy systems, can play an important role to reach the goal of the German government to achieve an "almost climate neutral" building sector by 2050 [4, 6, 7]. Furthermore, about 300.000 of the approx. 1.7 million non-residential buildings in Germany are municipal property and are thus subject to special attention by the German government.

In this context, the city of Freiburg (D), already known for the pioneering and sustainable character of the Vauban and Rieselfeld neighborhoods, voted climate protection targets in 2014 consisting of CO_2 emissions reduction of at least 50% compared to 1992. Moreover, the municipality is striving to reach climate neutrality by 2050. Especially municipal buildings are in the focus of energy saving measures, aiming among others at building up renewable energy systems and increasing the energy efficiency of buildings by implementing ultra-low-energy concepts. With the design, construction and monitoring of its new administrative center, the city of Freiburg wanted to implement a new concept for large net plus-energy buildings based on the combination of a highly isolated building envelope, of Building Integrated Photovoltaics (BIPV) and on a heat pump system. Furthermore, the objective of the City was to collect experience on the design, the commissioning and the exploitation phase of this type of buildings before replication to further buildings on a whole new administrative campus.

The design phase of the building started in 2012 and the research project, whose initial results are presented here, focusses on the operation phase with a scientific monitoring running until the end of 2019. In this paper, we present the building concept and the targeted energy performance indicators as well as the methodology applied to assess the plus-energy objective. We then provide the results of the first monitoring period and provide an assessment of the energy balance in comparison to the plus-energy objective as well as an initial analysis of the load curves and of the heating and cooling systems.

2. Building concept

2.1. Architectural concept

The initial motivation of the City of Freiburg was to centralize several departments that were scattered over 16 sites in the city to a new location organized as a campus and thus to streamline its organization and services offer for the citizens. In 2013, a European architecture competition led to the selection of a designer team led by ingenhoven architects. The RIS building, which represents the first element of the new administrative campus, has been handed over to the City of Freiburg in summer 2017. A total of 840 employees are working in the building, which contains mainly offices but also a canteen and server rooms. The architectural design is characterized by an oval shape and a wooden facade with vertical overhanging BIPV elements that serve as fixed exterior shading devices (see figure 1). The total NFA is distributed over six floors above ground and one basement floor. The thermal performance of the building envelope is high with a specific heat loss by transmission H'_T of 0.45 $W/m^2\Delta K$, thermal transmittance values of 0.1 $W/m^2\Delta K$ for the opaque surface and of 0.8 $W/m^2\Delta K$ for the triple-glazed windows. Air permeability limit values have been defined for each component of the budiling envelope. The specific primary energy demand within the boundary defined in the German Energy Saving Ordinance (EnEV: Energieeinsparverordnung) amounts to 61.1 $kWh/m^2_{NFA_{EnEV}} \cdot y$ (see also Table 1).

2.2. Technical boundaries for the energy supply concept and PV plant

Large buildings in dense urban zones require compact building shapes that enable optimizing the land area use. However, large compact buildings have limited possibility to achieve a positive energy balance with the use of on-site photovoltaic power generation as the roof area to envelope area and the envelope area to netto ground floor area ratios are low and lead to a restricted area for PV installations and thus to low specific photovoltaic yields [4]. The figures 2, 3 and 4 show the different characteristics for the RIS building compared to benchmark data from the tabula building typology [8]. These technical boundaries involve very high requirements with regard to the performance of the HVAC and lighting systems. Onsite power is generated here by a photovoltaic plant composed of monocristalline solar panels with a total peak power of 682

kW_p installed on the roof (461 kW_p) and in the facade (221 kW_p). The specific installed PV power amounts thus to approx. 30 W_p/m^2_{NFA}. Each solar module is equipped with a Maximum Power Point tracker, which continuously controls the point of the current-voltage characteristic curve at which the solar module provides the highest performance. A compound of 22 inverters converts the direct current of the solar modules into alternating current and feeds it into the building or in the public electricity grid accordingly to the current load of the building.

Figure 1: External view of the new City Hall building. Copyright: Stadt Freiburg, Y. Zerdoun

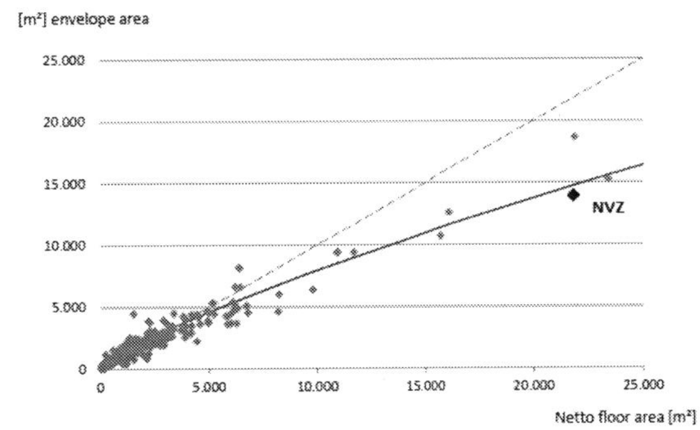

Figure 2: Building envelope area/NFA characteristic

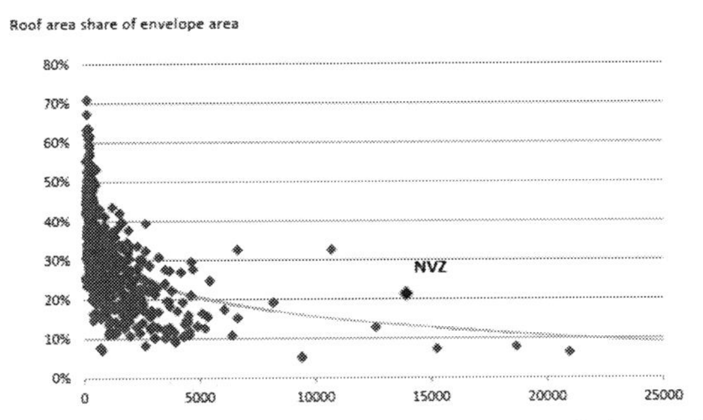

Figure 3: Roof area to building envelope area ratio/NFA characteristic

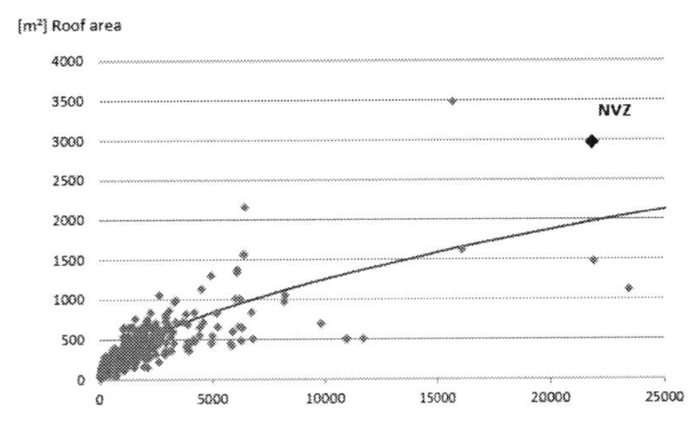

Figure 4: Specific PV gain/NFA characteristic

2.3. Hybrid photovoltaic thermal solar collectors

The heat demand for domestic hot water is mainly required for the canteen. Glazed hybrid photovoltaic-thermal solar collectors (PVT) provide low-temperature heat to cover 15% of the domestic hot water demand. PVT collectors potentially resolve the conflict between the application of photovoltaic or solar thermal systems on limited roof areas. While PVT collectors have a slightly lower electrical yield per square meter than PV modules, they simultaneously generate heat on the same surface area and thus achieve higher specific primary energy yields. Originally, conventional solar thermal collectors with an area of 29 m^2 were planned with a designed solar fraction of 15% or an annual thermal output of 22 MWh/a. The PVT collector array was designed to achieve the same solar fraction. System simulations were conducted with TRNSYS to assess the annual electrical and thermal yields of the PVT collectors. To compensate for their lower thermal efficiency, the area of the PVT collector array amounts to 43 m^2. Thus, the substitution of the solar thermal collectors with glazed PVT collectors is planned to increase

IOP Conf. Series: Earth and Environmental Science **352** (2019) 012003 doi:10.1088/1755-1315/352/1/012003

the electrical output of 180% on the given area, while maintaining the same thermal output [11]. The PVT system is subject to a detailed monitoring to assess the thermal and electrical energy output and to identify optimization potentials. The detailed analysis is however beyond the scope of this paper and will be featured in a future publication.

2.4. Building services concept

The heat generation is realized by a combination of two ground-water coupled heat pumps with a total thermal power of 400 kW that supply thermal activated concrete slabs (TACS) and radiant ceilings with $28°C$ hot water, radiators in the offices with $60°C$ hot water,as well as five air-handling units (AHU) with $35°C$ hot water. Each AHU has a heat recovery system as well as heating and cooling coils. A gas boiler with a thermal power of 100 kW operates as back-up to cover peak loads in periods of high heat demand. The specific installed thermal power amounts to 22 W/m^2_{NFA}. The cooling of the building is realized by a free-cooling system using the ground-water brine and providing the TACS with $18°C$ chilled water and AHUs with $15°C$ chilled water. The TACS are controlled on the basis of weather forecast data to guarantee indoor air temperatures of $20°C$ in Winter and $26°C$ in summer. The artificial lighting is realized with LED lights. An automated sunshade system completes the vertical BIPV overhangs and controls the daylight and solar radiation gains in the different building zones accordingly to the azimuth angle of the sun.

3. Quality insurance and energy balance methodology

3.1. Energy balance methodology

At the publication date of this paper, the German government has still not approved a normative methodology for balancing the energy of net zero-energy and net energy-plus buildings. Here, the applied methodology follows a net-zero energy balance over one whole year [2, 3]. The reasons for this approach is first that local power generation through the PV-plant and power demand of the building do not match most of the time and secondly because seasonal energy storage was not relevant on an economical and technical point of view for a building of that type. The boundary for the assessment of the plus-energy balance used in our calculations accounts for the final and primary energy consumption for HVACs, domestic hot water and lighting accordingly to the EnEv rules. Usage-specific consumptions, such as those of plug-loads, employees personal computers, the restaurant and the IT-servers, as well as their corresponding floor areas are not taken into account. The reference NFA used in our calculations amounts to $\text{NFA}_{EnEV} = 21.819$ m^2. For the calculation of the primary energy consumption, we have considered the primary energy factors used in the design phase of 2.5 for the electricity from the public and into the grid, of 1.1 for fossil gas and of 0.1 for biogas. The results of the monitoring are provided as total and specific final and primary energy consumptions in Table 1 and compared against the target values of the design phase for information. We deliberately did not carry out a climate adjustment that takes into account the climate data used in the model of the design phase and the measured climate data. Reason for this choice is that the adjustment would have only considered temperature variations and neglected other sensitive variables like sun radiation and user behaviour. Focus here is on the verification of the plus-energy target. We plan to conduct a climate adjustment between the first and the second monitoring period on the basis of measured data after the second year of operation.

3.2. Energy performance specifications and monitoring concept

The design and commissioning engineers defined a list of Key Performance Indicators (KPI) in the owner specifications that have been checked systematically through simulations during the design phase and measurements after the commissioning [9]. In the design phase, a detailed monitoring concept aiming at measuring, reporting and analysing the performance of the building and of its building services has been developed. On this basis, heat and power meters and submeters have been installed to measure and quantify the energy flows into and out of the building and for each main system like the heat pumps, the AHUs and the hydronic loops. In total, over 9.000 data points are measured with a 5 min resolution and stored in a monitoring database. A major drawback remains that, for technical reasons, no submeters could be installed to measure the electrical consumption of the lighting system separately from the plug-loads of the office storeys. Also the power consumption of the building automation systems (BAS) has been accounted as a fixed value as it has not been measured separately. Thus, the reference values from the design phase have been used to take account for the energy consumption of these systems in the real energy balance. For the considered balance boundary and under consideration of a test reference year for the region of Freiburg, the objective was to obtain a specific primary energy surplus of 0.2 $kWh/m^2_{NFA_{EnEV}} \cdot y$.

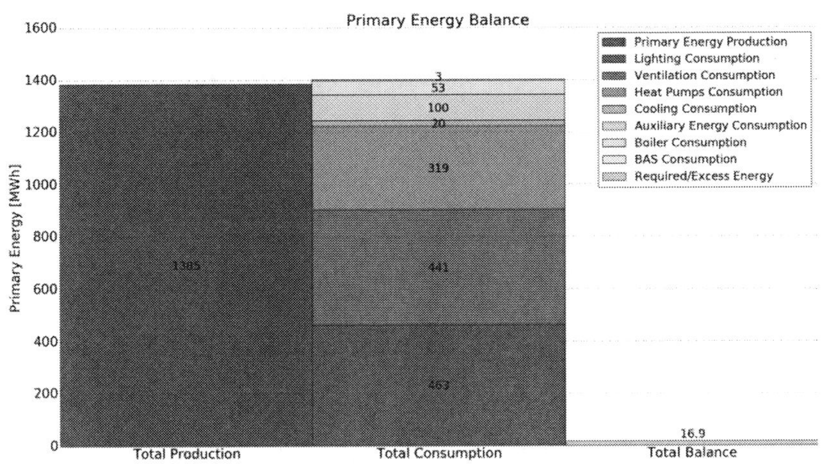

Figure 5: Yearly primary energy balance within the EnEV-boundary.

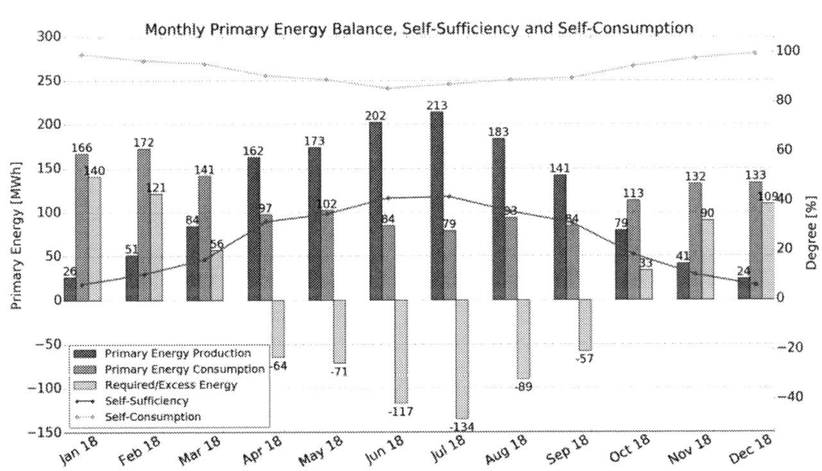

Figure 6: Monthly primary energy balance, self-sufficiency and self-consumption

4. Results

4.1. Primary energy balance

For 2018, the results of the energy performance for the main services heating, cooling, lighting, auxiliary energy for pumps, AHUs are presented and compared with planned values in Table 1 and Figure 5, which shows the primary energy balance on a yearly basis. The photovoltaic plant produced 554.1 MWh of electricity in 2018, which corresponds to a specific gain of 25.4 $kWh/m^2_{NFA_{EnEV}}$. These results exceed the targeted values for the PV production whereas it is important to mention that the solar radiation was particularly high in Freiburg in 2018 (reference: 1050 $kWh/m^2.y$, measured: 1243 $kWh/m^2.y$). The roof PV plant produced the highest share of the electricity production with 81%, whereas the facade PV plant and the PV-T plant contributed respectively to 18% and 1%. On the other side, the main consumers are the respectively the lighting (33% of the total consumption), the AHUs (32%) and the heating system (23%). The part of the primary energy comsumption for cooling is very low with 2% of the total consumption. The results reveal a slightly negative balance with a gap of about + 16.7 MWh, representing 1.2 % of the total primary energy consumption, which amounted to 1402.0 MWh for 2018. In the same time, the photovoltaic plant produced 1385.2 MWh. After one year of operation, this result is encouraging as the net plus-energy objective could have be reached without a failure in a BIPV-inverter and if bio-gas instead of fossil gas would have been purchased by the building owner, leading to a plus-energy balance of about +30 MWh or +1.3 $kWh/m^2_{NFA_{EnEV}}.y$. Nevertheless, as 2018 was especially sunny and warm in Freiburg, additional measurements are necessary to assess the performance and the robustness of the concept in the long run.

As shown in figure 6, during six months of the year, from April to September, the building produces more energy than it consumes within the balance boundary and feeds surplus energy into the grid. From January to March and from October to December, the building energy provision is strongly dependent of the public grid. As expected under this latitude, July is the month with the highest energy excess (- 135 MWh) and January with the largest gap (+ 140 MWh).

Table 1: Planned and measured KPIs in $kWh/m^2_{NFA_{EnEV}}.y$

Service	Delivered energy target	Primary energy target	Delivered energy result	Primary energy result	deviation
Heating Total	4.3	10.7	8.1*	17.1	+60%
Cooling	0.9	2.2	0.4	0.9	-57%
AHUs	7.0	23.3	8.1.6	20.2	-13%
Aux. Energy	1.4	3.5	1.8	4.6	+32%
BAS **	0.1	0.2	0.1	0.2	
Lighting **	8.50	21.3	8.5	21.3	
Total Demand		61.1		64.3	+5%
Gain PV-Roof	20.2	50.6	20.6	51.4	+2%
Gain PV-Facade	4.0	10.0	4.5	11.1	+11%
Gain PVT (electrical)	0.3	0.7	0.3	0.7	2%
Gain PV-Total	24.5	61.3	25.4	63.5	+4%
Energy balance		+0.2		-0.8	

*:power consumption heat pumps = 5.9, gas consumption: 2.2, **: not measured/design value

4.2. Performance assessment of the photovoltaic plant

The specific PV gain of the roof installation amounted to 981.6 kWh/kW_p, the facade to 447.9 kWh/kW_p and the PVT to 869.0 kWh/kW_p. The maximum power generated by the whole PV plant amounted to 430.8 kW on June, 22nd at 2 pm. It is noteworthy to mention here that an inverter controlling one facade BIPV group was defect during summer 2018. This breakdown led to an eletrical gain loss estimated in a range of 5.0 to 10.0 MWh, which could not be accounted for in the energy balance [10]. The self-consumption reached values above 90% from January to April and from October to December as depicted in figure 6. These results are above typical values for residential buildings equipped with PV-plants. It can be explained by the fact that the offices, the canteen and the servers, whose energy demand is high compared to the PV gain during the winter months, are mostly used during the day and can thus directly benefit from

IOP Conf. Series: Earth and Environmental Science **352** (2019) 012003 doi:10.1088/1755-1315/352/1/012003

the solar power production. The mean yearly value of the degree of electrical self-sufficiency for the whole building amounted to 23.7 % and was above 30% from April to September 2019.

4.3. Performance assessment of the heat and cold generation

In 2018, the total heat consumption was higher by +78% than the targeted value. This overshooting is due to a higher consumption of the main heat consumers - the TACS and the AHUs - whose control settings (time-schedules, temperature set points,...) needed being adjusted during the first months of operation. The performance of the heat pumps are below the expectations. The Seasonal Performance Factors (SPF) of the heat pumps 1 and 2, calculated for the system boundary of the heat pump units, reached 4.6 and 4.0 respectively and lie below the targeted SPF of 4.8. The heat pumps contributed to 87% of the total heat production amounting to 579 MWh and heat from the gas boiler has only been consumed to a minor part, corresponding to approx. 44 MWh. A detailed analysis of the heat pump operation revealed that heat pump 1 has run about one and a half times longer than heat pump 2 due to an issue in the alternating operation control, which could lead to earlier wear and efficiency loss of heat pump 1. Furthermore, the mean temperature level of the heat supply for the AHUs lied with $35°C$ above the design value of $32°C$ and reached a values above $40°C$ at some operation points in winter. An optimization of the temperature levels could lead to higher SPF of the heat pump system and thus enhance the total performance. The operation of the cooling system was very high in 2018 with a seasonal energy efficiency ratio (SEER) of 45.0 and a measured specific primary energy that undercut the targeted values by -51%.

Even if optimization potentials are remaining, the monitoring results confirm that this kind of heat and cold generation system based on a low-exergy principle is a cornerstone for the success of the plus-energy concept.

4.4. Load curves analysis

Figure 7 shows a typical profile for a working day in summer. First, the base load of the whole building at night is about 120 kW and the maximum load is about three times larger at 1 pm with a peak at 335 kW. The load profile of the facade shows a local maxima of about 40 kW at 10 am and at 4 pm corresponding respectively to the maximum power of the BIPV groups of the east and of the west oriented facades. This specific construction of the BIPV installation with BIPV groups almost all around the facade (with the exception of the North orientation) contributes to a maximum yield. During that typical working summer day, power from the grid is drawn at night and with a consumption peak at around 6 am with the start of the AHUs and lighting. Power is fed into the grid from 10 am to 5 pm with a peak at 3 pm. The figure also shows clearly that the load curve of the services within the EnEV-boundary is almost constant due to the fact that lighting is not measured but estimated with a constant consumption and that the power consumption of the AHUs is almost steady. Furthermore, one notes that the load curve within the EnEV-boundary is far beneath the total load curve of the building due to the power consumption of the canteen, the plug loads and the servers.

For the winter, Figure 8 shows that the power is mainly consumed from the grid as expected. The PV plant alleviates the energy balance to only to a small part. The curve load of the EnEV-boundary is dominated in winter mainly by lighting and the heat pumps.

IOP Conf. Series: Earth and Environmental Science 352 (2019) 012003 doi:10.1088/1755-1315/352/1/012003

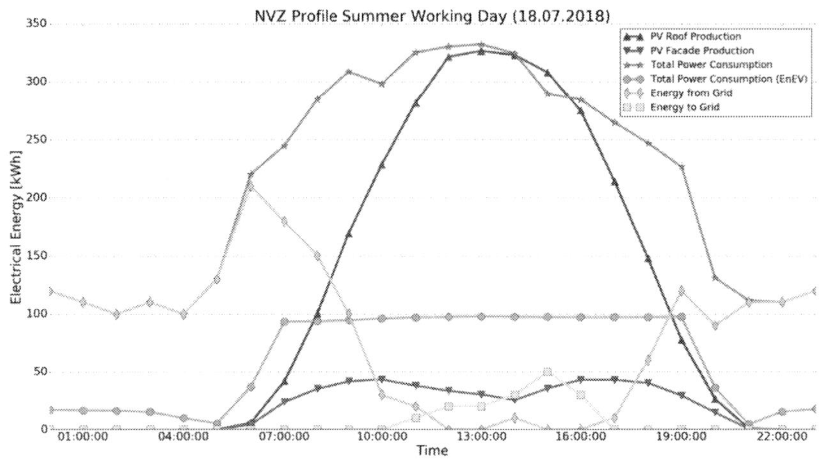

Figure 7: Load curves during a working day in summer

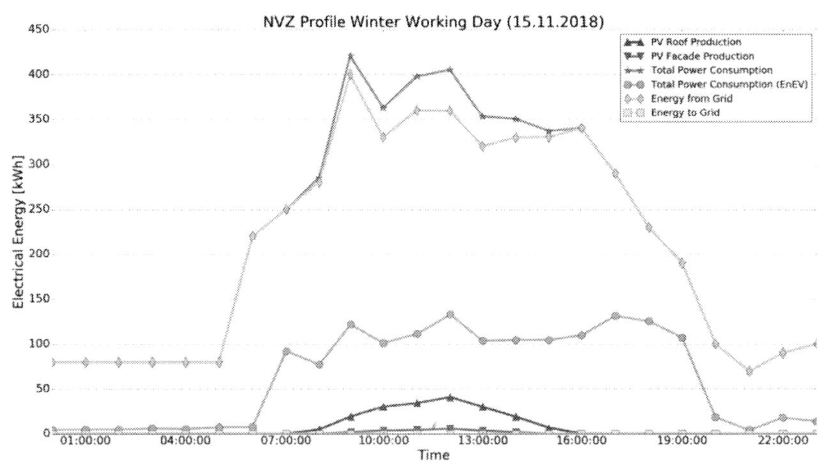

Figure 8: Load curves during a working day in winter

5. Conclusion and outlook

The results of the first full year of monitoring of the new town-hall of the city of Freiburg, designed as a net plus-energy building, have shown that a plus energy balance has almost been reached within the defined boundary, under the limitation that the electricity consumption of the lighting has been estimated as it could not be measured. The analysis of the energy balance showed that the BIPV system on the facade, even though it only contributed to approx. 18% of the total solar gain in addition to the PV installation of the roof, is indispensable to reach this result for this type of building. Also, the performance of the building services and especially the very low specific energy consumption for cooling contributed significantly to a very high energy efficiency compared to mechanical cooling with classical chillers. The performance of the heat pump system revealed optimization potentials with lower COP as expected. After one year of operation, the presented results are encouraging as the net plus-energy objective could have been reached without a flaw in one BIPV group, if bio-gas instead of fossil gas would have been used and through fine-tuning of the heat and cold generation system. The city of Freiburg is currently implementing corrective measures. Nevertheless, the solar radiation was particularly high and the heating degree days were 5% below the reference in Freiburg in 2018. Additional measurements are thus necessary to assess the robustness of the concept over the years. The design phase of the second net plus-energy building of the future campus with an approx. similar area will start in 2019 and the results presented here will serve as reference in the design process. Additional analyses on hygrothermal comfort, on the performance of the PVT-collectors and of the building services are currently being conducted and will be published at a later date. A special emphasis will also be set on the interaction of the building with the local power grid, and especially on the role and control strategies of the heat pump system that realizes the desired sector coupling between heat and electricity sectors.

Acknowledgements

The authors gratefully acknowledge the City of Freiburg for its cooperation in the project and the Federal Ministry for Economic Affairs and Energy (BMWi) for financial support due to an enactment of the German Bundestag under Grant No 03ET1316A.

References

[1] R. Grob, "Überprüfung von Automatisierungsfunktionen heiz- und raumlufttechnischer Anlagen", PhD Thesis (2003), Universität Stuttgart IKE Lehrstuhl für Heiz- und Raumlufttechnik
[2] I. Sartori et al., "Criteria for Definition of Net Zero Energy Buildings", EuroSun Conference, Graz, Austria, 2010.
[3] A.J. Marszal et al.,"Zero Energy Building – A review of definitions and calculation methodologies", Energy and Buildings 43 (2011), pp. 971–979, doi:10.1016/j.enbuild.2010.12.022
[4] T. Kuhn et al., "Active solar facade (PV and solar thermal)", 3rd International Conference on Passive & Low Energy Cooling for the Built Environment, PALENC 2010. Conference Proceedings., 29 September - 1 October 2010, Rhodes Island, Greece, ISBN: 978-960-6746-08-6
[5] B. Oschatz et al., "Building Energy Law (GEG) and Nearly Zero Energy Buildings", GI Gebäudetechnik in Wissenschaft & Praxis (2019), volume 01-19
[6] T. Pregger et al., "Long-term scenarios and strategies for the deployment of renewable energies in Germany," Energy Policy (2013), pp. 350-360, volume 59, ISSN 03014215, DOI 10.1016/j.enpol.2013.03.049
[7] A. Palzer et al., "A comprehensive model for the German electricity and heat sector in a future energy system with a dominant contribution from renewable energy technologies – Part II: Results", Renewable and Sustainable Energy Reviews (2014), pp. 1019-1034, volume 30, ISSN 13640321, DOI 10.1016/j.rser.2013.11.032
[8] T. Loga et al., "Use of Building Typologies for Energy Performance Assessment of National Building Stocks. Existent Experiences in European Countries and Common Approach", First TABULA Synthesis Report (2010), IWU—Institut Wohnen und Umwelt, Darmstadt (D)
[9] R. Grob, "Prozessleitfaden Energetische Inbetriebnahme - Neubau Verwaltungszentrum Stadt Freiburg", internal report (2018), DS-Plan Ingenieurgesellschaft für ganzheitliche Bauberatung und Generalfachplanung mbH, Stuttgart (D)
[10] K. Baisch et al., "Abschlussbericht Monitoring und Betriebsoptimmierung über 12 Monate (2019), DS-Plan Ingenieurgesellschaft für ganzheitliche Bauberatung und Generalfachplanung mbH, Stuttgart (D)
[11] Lämmle, M., "PVT-Kollektoren - die solare Kraft-Wärme-Kopplung", Ingenieurspiegel 1/2016 (2016)
[12] M. Liu,"Improving Building Energy System Performance by Continuous Commissioning", 1999, Energy Engineering, 96:5, 46-57, DOI: 10.1080/01998595.1999.10530472

IOP Conf. Series: Earth and Environmental Science **352** (2019) 012004 doi:10.1088/1755-1315/352/1/012004

Living labs in a zero emission neighbourhood context

R Woods[1], **T Berker**[2]

[1, 2] Department of Interdisciplinary Studies of Culture, Norwegian University of Science and Technology, Trondheim, Norway

ruth.woods@ntnu.no

Abstract. The Research Centre for Zero Emission Neighbourhoods in Smart cities has established a concept for living labs. Central to the concept is an experimental format that offers some control over the social and physical environment, as well as opportunities to observe and engage during a limited period. According to Bulkeley et al. 2018, an experiment offers a means to make sense of the present whilst also providing a vision of the future [1]. The paper asks, what are the implications for users when engaging with experiments and presents results from a ZEN living lab experiment that took place at Campus Evenstad in Hedmark, Norway.

Technical management at Evenstad proposed the experiment, they wanted to test if it was possible to reduce campus energy consumption and the starting point was the old administration building, which has the highest energy consumption on campus. The energy use reduction was to be achieved by turning off the building's heating and ventilation systems during a limited four-week period. This took place in July 2018, when the building users were expected to be on holiday or doing fieldwork. A workshop to anchor the experiment among building users took place a month before the experiment started. During the workshop challenges associated with the experiment and with the building, for users, became apparent. However, building users agreed to participation in the experiment because they saw it as an opportunity to highlight what they understood as necessary changes to the building. The experiment achieved the energy saving potential that the building managers envisioned, but the results for the building users are less tangible.

From a pragmatist approach, living labs and their experiments are about providing solutions, but the Evenstad example highlights the challenge of providing tangible solutions and how we engage users with more intangible future solutions. We discuss therefore the limitations and potentials associated with the experimental format. Moving beyond demonstrating what a sustainable future should look like and include [1p.1], and instead noting opportunities for the translation of societal learning into concrete actions that serve the user groups engaged as well as demonstrating, the potential to influence wider sustainable transitions.

1. Introduction

Living labs are a means to "gain experience, demonstrate and test ideas", that have the potential to be scaled up across systems [2]. They have therefore become an essential part of any sustainable transition toolbox. The starting point for this paper is the living lab concept developed by the Research Centre for Zero Emission Neighbourhoods in Smart Cities (ZEN). Using experiences with a ZEN living lab on Campus Evenstad, we ask what are the implications for end users when engaging with experiments in living labs? The ZEN Centre has chosen living labs as the main format to secure user engagement and to organise user involvement in pilot projects. The specific demands and conditions encountered in ZEN pilot neighbourhoods, require a tailored ZEN concept that provides structure for actions taking place, but is broad enough to deal with the various challenges arising in each neighbourhood. A ZEN living lab includes, representatives from different user groups affected by the sustainable neighbourhood transition proposed by ZEN. Secondly, it has a clearly defined geographical place. In ZEN, this is supplied by pilot projects. Thirdly, a ZEN living lab should develop an in-depth qualitative understanding of the social and physical context based on information collected by for example, interviews, workshops, and participant observation. Finally, an experiment arising from the qualitative understanding takes place, testing ideas, based on the challenges, technology, and needs, supplied by the neighbourhood.

Campus Evenstad is part of the Inland Norwegian University of Applied Sciences, one of three campuses that are part of the Faculty of Applied Ecology, Agricultural Sciences and Biotechnology. The campus has 60 employees and approximately 200 students and is one of nine ZEN pilot projects. The main aim behind the actions taking place in the pilot project is "*optimisation of energy management*". The living lab at Campus Evenstad took this aim as its starting point and the living lab has therefore a technology and innovation focus. Statsbygg who is responsible for the technical management of the campus buildings established this aim[1]. The campus has buildings dating from the late 1800's to 2017. Some buildings are energy efficient such as a ZEB administration building (ZEB-com) and the student dorms (Norwegian passive house standard). It also has buildings from the 1970's and 1980's that consume large amounts of energy to heat and ventilate. One of the least efficient buildings is the old administration building built in 1987, which every day is responsible for one third of the campus's energy requirements. From 01.07- 01.08.2018 an experiment took place based on one main action, the heating and ventilation systems in the old administration building were closed down.

The paper starts with a brief presentation of current living lab theory and practice. It then describes the different activities associated with the living lab at Evenstad. The next section takes a detailed look at the building users' response to the experiment, comparing their tangible expectations with their response to intangible aims for energy optimisation supplied by the ZEN Centre. The paper concludes with a brief discussion and conclusions.

2. What are living labs?

Living labs vary according to the problems they focus on, and because of the variety of problems they deal with there are large number of different kinds of living lab. It is however possible to identify two main types, the first is the technology and innovation based living lab, which have a limited social and physical concept and primarily focus on product development within a co-creative environment that is as close to real-life as possible. This kind of living lab stems from the original living lab developed in the 1990's by Mitchell at MiT. The second kind is the citizen centred or civic urban living lab that aims at sustainable urban transitions. Bulkeley et al. 2018 has identified three main kinds of design and practice within urban living labs, strategic, civic and organic, that are distinguished by the differences in control and contingency associated with how they are designed and practiced [1 p.6]. A civic urban

1 Statsbygg is a public sector administration company and advisor to the Norwegian government in construction and property affairs. Statsbygg was a ZEB partner and is a ZEN Centre partner. The campus was also at the centre of pilot activity between 2009 and 2017 because the Centre for Zero Emission Buildings (ZEB Centre) developed one of its eight pilot projects, a new administration building, on campus https://www.zeb.no/index.php/en/pilot-projects.

lab includes a larger and more complex physical, social and political context and a wider range og user groups than a technical and innovation based living lab [3].

Certain qualities characterise both the two main kinds of living labs, these qualities have inspired the ZEN concept. A living lab should, according to Pierson and Lievens 2005, ideally be "an experimental field" that deals with a social and technical challenge, have clear goals and structure and take place within the framework of everyday life [4]. This is relevant for the ZEN Centre because it requires the means of bridging the gap between the technical and social context within the pilot projects. In addition to being experimental living labs are interventions that are designed to improve, are characterised by the intention to observe and they can if successful offer the means to manipulate existing socio-material conditions, creating new forms of conduct [5]. These qualities are presented here in what we describe as two stages, and they can be applied in both civic and technical living labs.

2.1. The first stage: qualitative understanding of the social and physical context

Living labs live when they include people. A user or user group may be a company, employees, researchers, residents, planners or activists, include different age groups, cultures or social groups. Discovering the interests and values existing within a context are part of the first phase of any living lab, because aim-based interaction between different individuals and/or user groups is required, also the success of an experiment is often based on whether the aims of participants have been achieved [6].

Veeckman and van der Graaf (2015) propose that successful living labs organise a process that encourages people "to make sensible decisions through reasoned deliberation" [7]. However, the involvement of users is not a neutral process. Individuals working in government and universities often dominate knowledge production in living labs [8] and they influence the choice of problems being dealt with. This is a top down process and not the citizen empowerment or co-creation that the term living lab implies. Including a broad group of different users and emphasising their interests and values is one way to avoid this.

A living lab should have a clearly defined geographical place. A place that offers "a material, sensory, social and experimental environment" [9]. The place provides the boundaries, defining who is included and therefore sets the scene for the problems engaged with. Understanding place clarifies challenges and values of the people associated with it and supports an in-depth qualitative understanding of the context. This understanding can be said to support the introduction and use of technologies, because they are best "assessed in their relations to the sites of their production and use" [10]. However, although living labs are context dependent, a tension exists because of an underlying expectation that the interventions and experiments could have taken place anywhere [1 p.6]. Making their results relevant outside the context where they were gathered.

2.2. The second stage: experiment

"*experimentation provides the means through which diverse actors seek to navigate and make sense of the present whilst giving concrete form to particular visions of the future*" [11].

Living labs aim to impose control over a social and physical context. It is not enough that people simply meet and engage; a framework is required that enables the meetings to produce new knowledge. Experiments provide a framework by supplying a planned set of parameters and the opportunity for meticulous recording of conditions and outcomes. Sengers (2016) suggests that living labs are a form of transition experiment where the aim is to stimulate complex processes of social and technical co-evolution, but where the focus is on who participates, what is learned and who appropriates what is learned [5 p.21].

The two stages include a broad set of qualities, and these are useful within the two main types of living lab, civic and urban and technology and innovation driven. There can exist an overlap between the two different types, but although both are user centred, they are not interchangeable. They have

IOP Conf. Series: Earth and Environmental Science **352** (2019) 012004 doi:10.1088/1755-1315/352/1/012004

different relationships with the context and aims, and the kinds of problems they deal with are different. This has implications for the format and the kind of experiments that take place. Keeping sight of the difference is important because living labs are most useful when they are aim based and have a clear research focus.

3. Evenstad living lab method and actions

This section presents the methodology applied in the technology and innovation living lab at Campus Evenstad. In addition to the flexible ZEN living lab concept presented above, a ZEN living lab has an anthropological approach to data collection and analysis. The ethnographic method applied by anthropologists produces insights into the informal aspects of social and cultural life, emphasising "the contrasts between what people say and what they do", and the more formal structures provided by the surrounding society [12]. Anthropology offers a perspective that does not take preconceived assumptions about human societies for granted and is sensitive to both similarities and differences. This makes its use pertinent within the context of a ZEN pilot project, where the difference between what ZEN plans and expects and what the local context requires is not always in focus. The anthropological approach applied in Evenstad, is not based on classical observational methods implied by participant observation, which due to limited access to places or practices can be "impractical and inappropriate" in a contemporary context [9]. New approaches to gaining access to people's lives are required. The Zen living lab concept provides an innovative approach for an anthropologist.

3.1. *The experiment*

Some of the actions associated with the living lab at Campus Evenstad were associated with general pilot project activities taking place on campus, such as the partner workshop in November 2017, which presented the plans for Evenstad to stakeholders involved with ZEN. The workshop provided the official start the living lab activities and was followed up by a ZEN living lab week in March 2018, that was tailormade for the living lab, and aimed to introduce ZEN and it's living lab to broad group of users on campus. Interviews, the questionnaire, the workshop in May followed up on ideas that came to light during the living lab week (see table 1). The workshop in May included representatives from different employee groups. Plans for the experiment were discussed, the intention was to uncover new challenges or issues associated with the experiment.

The idea for experiment in the old administration building arose during the living lab week and was proposed by the technical management who regarded it as concrete action that would not impose too many challenges on the social context. The building has three floors, four different ventilation systems, houses the canteen and library, as well as offices, a laboratory and storage space. From 01.07-01.08.2018 the heating and ventilation systems in the old administration building were shut down (this did not include the ventilation system used in the canteen on the ground floor). The pragmatic aims associated with the experiment were to highlight the energy costs associated with different buildings on campus and to test a low-tech energy saving action. As an experimental action it brought to the forefront the difference between everyday activities and living lab activities, enabling engagement with building users.

The paper describes the response to the experiment and is primarily based on interviews and participant observation after the experiment took place in July 2018. A group of seven engineers who had offices on the second floor of the building and two librarians located on the first floor took part. Interviewees are presented in the text below as Office 1 to 7, or Librarian 1 to 2.

IOP Conf. Series: Earth and Environmental Science **352** (2019) 012004 doi:10.1088/1755-1315/352/1/012004

Table. 1. Overview of ethnographic activities, time-frame and roles associated with the activity

Action	Role	Number of participants
ZEN partner workshop	ZEN partners, technical management Evenstad, Campus administration, ZEN researchers	Ca. 50 people
Living lab week 05.03-08.03.2018 Lectures, workshops and four interviews	Campus administration, scientific personnel, students, technical management, ZEN researchers	Ca. 27 people
Questionnaire	Campus administration, scientific personnel, students, technical management, ZEN researchers	17 people
Meeting 22.05.2018	Campus administration, technical management, Statsbygg, ZEN researchers	12 people
Workshop with site visit in the old administration building 23.05.2018	Campus administration, scientific personnel, technical management, ZEN researchers	10 people
Experiment in the old administration building 01.07 - 01.08.2018	Technical management, scientific personnel and others with offices in the building	Ca. 20 people
Interviews: skype and face to face November 2018	Scientific personnel with offices in the building	9 people (Offices 1-7 and Librarian 1-2)
Participant observation 12.- 16.11. 2018	Office in the old administration building	Informal meetings, coffee breaks, lunch, 2-5 participants

4. Experimental impact, building problems and the role of ZEN

The pragmatic aim behind the experiment was achieved, 12 000 kWt (or 12 MWt) were saved, during the four-week period. There was approximately 12,000 kWt greater energy consumption in the building in August compared with July. This is due in part to the shutdown of the ventilation system, but it is also due to higher activity in the building in August. This is regarded as a success and Statsbygg plans to apply the same low-tech method in other buildings that it manages[2]. However, the focus of the paper is the building user's response to the experiment and not on the amount of energy saved. This section is divided into three parts, starting with the impact of the experiment on building users, then moving on to problems with the old administration building and concluding with ZEN's connection to Campus Evenstad.

The impact: The experiment at Evenstad took place during the summer when the old administration building is largely empty, the main summer holiday in Norway is traditionally in July. We proposed that building users did not use their offices during the four-week period the heating and ventilation was shut down, but we did not ban them from using their offices. Temporary offices in the new zero-emission administration building were made available if they needed an office. Only Office 6 and Librarian 2 were at work during the four-week period. This is a period when campus buildings require cooling. Office 6 who chose to use his own office and not the temporary office spaces, struggled with the building temperature. The summer in 2018 was the hottest ever recorded and it is therefore unclear if the temperature was due to the experiment or the weather conditions. When possible, he worked from home.

[2] The application of the strategy has not been discussed with Statsbygg, but the implication is that Statsbygg is responsible for the management of a number of public buildings that are not in full use during holiday periods and weekends. Looking into the possibilities for shutting down HVAC systems has therefore been proposed. Other issues into addition to energy savings would have to be considered if this low-tech action is to be applied in other buildings. These include the impact on air quality due to the potential increase in air particles that may have implications for the health of building users, and wellbeing and functionality for building users. Not everyone is able to have holidays at the same time of year.

Librarian 2 claimed to be happy with the experiment, expressing satisfaction with the freedom to open the windows, something she is not usually encouraged to do by technical management. In addition, she stated that it was "wonderfully quiet" without the noise from the ventilation systems. The other seven building users were either on holiday or doing fieldwork.

The building: Prior to the experiment the building users were sceptical. They maintained that the energy saving benefits would not weigh up for the inconvenience that they would experience. Any energy savings made during the shutdown benefit the university centrally, not Campus Evenstad and not the building users. They would be helping technical management and the university, but what was in it for them? After the experiment, the seven building users who were away during the four-week period stated that they had not been inconvenienced, but when told about the energy savings the building users still could not see any benefits, Office 2 stated,

> "The university likes to appear as sustainable… but the wallet comes before sustainability. The whole Evenstad concept, which equals minimal emissions, the solar panels and zero emission buildings, is about a signal effect."

The experiment although less inconvenient than expected only supplied intangible results and the building users who took part in the preliminary workshop stated that they wanted tangible results, in the form of building improvements when they agreed to the experiment. They hoped that the experiment would highlight functional issues associated with the building. Particularly the problems with the second floor south facing offices and the window systems. Office windows have external sunscreens installed but when the sunscreens are lowered, it is not possible to open the windows more than two centimetres. When the sun is shining the offices become very warm and the ventilation system does not have enough cooling capacity. Opening the window would allow natural ventilation. Building users must choose between keeping the sun out of the office and ventilating using the window. There is agreement among all the office users that the sun creates a difficult working environment,

> Office 6 "There is a great difference between summer and winter. I can regulate the temperature myself in the winter. In the summer I can regulate early in the morning until the sun starts burning through the window. I had a thermometer in the office (August) and around lunch time it's about 30 degrees. The solution during the summer is to go home, but it is often not possible to go home. We have tried to create an air-flow by opening doors. I often leave the window open when I go home, in the morning I close it and lower the sun screening. Birds have come into the office twice."

The example shows that the building users apply various strategies to dealing with the problem of overheated offices and experience a variety of inconveniences. Despite the problems they are not interested in moving to offices in other buildings, not even during the experiment. This is because of two main aspects; the social environment is considered good and they have individual offices. Office 3 stated, "it's not too bad here. The climate is one thing, but we have become a small enclave." The group of engineers with offices on the floor work within the same field and none of the building users was interested in sharing an office or working in an open landscape. They moved into the old administration building two years ago, previously they were in an off-campus laboratory building. Their current offices are an improvement and despite periodically having overheated offices, not having a modern coffee machine or comfortable sofas like the ones in the zero-emission administration building, they like where they are.

The problems with the windows are well-known on campus, the building's previous users experience the same problems. During the workshop in May two of the building users proposed installing roller blinds on the inside of the window as an immediate solution to the problem, but nothing changed in the four-month period after the experiment. The technical management team stated that improving the window systems were on "the "to do" list of measures on campus, but low down on the list.

ZEN's relevance: The ZEN Centre's main activity on Campus Evenstad is the testing and introduction of zero emission technology, examples are a CHP biogas burner, battery systems for energy storage and vehicle to grid storage systems. These technical innovations have a wider potential in other neighbourhoods. The experiment offered the opportunity to ask building users what they knew about ZEN, its activities on campus and the technologies. Half the interviewees said that they knew very little about ZEN, two stated that involvement in the ZEN Centre takes place on a leadership level, and one that this was also the case with the ZEB Centre. The ZEN Centre is not the first research centre to be active on Campus Evenstad. The Research Centre for Zero Emission Buildings (ZEB) was involved with the design, development and building of the new campus administration building, and installing of the biogas burner. Few users or groups on campus are involved in the planning or development of buildings and technology, the process is top down. Librarian 1 "The leadership talks about it when they talk about Evenstad. We (the employees) don't care about it so much."

However, building users are in general positive to being part of ZEN. Although there is the lack of detailed knowledge, and some confusion about what the technology is useful for. Building users are proud that being a zero emission neighbourhood is part of Evenstad's profile.

> Office 5 – "I hadn't noticed ZEN or the technology. Now that we have the CHP system I thought that power cuts would not be a problem (the biogas burner will only supply heat, not electricity). But the chip burner and solar production are a plus. Its renewable, a plus plus really."

However, the lack of connectivity between users and ZEN on campus, when combined with a faith in the sustainable potential associated with zero emission technology, causes a deeper criticism of the activity on campus, because ZEN cannot simply be written off as irrelevant for them or uninteresting. Disappointment is expressed because ZEN and Statsbygg have not placed more emphasis on involving students or staff. The building physics and energy management associated with ZEN on campus is not part of the subject matter that students and staff from the Faculty of Applied Ecology, Agricultural Sciences and Biotechnology at Evenstad usually working on, but as Office 1 pointed out

> "The two things complement each other. Here on campus we could produce food and deal with waste products for a zero-emission neighbourhood……. I have never been involved in any brainstorming. It is very technical and outside our field, but ecosystem services are within it. There ought to be some kind of collaboration."

5. Discussion: The third stage – intangible results

The paper presents the different phases of a ZEN living lab, starting with the in-depth qualitative phase, where information necessary to develop and activate the second experimental stage was gathered. The second stage focused on the experiment and the building user's response to it. This final section represents the third phase, a discussion of the results from the experiment.

The ZEN living lab on Campus Evenstad is a technology and innovation based living lab. The practical aim was to enable the campus's technical management to reduce campus energy use and to highlight the energy costs associated with different buildings on campus. This was achieved, but when faced with this goal and the resulting energy savings the building users asked, "What's in it for us?" They required more tangible results and in a living lab whose aims are associated with products and solutions, the building user's expectation that rational choice be involved should not come as a surprise [14]. Building users asked for building improvements, tangible results, these are associated with building efficiency, but although the experiment helped to highlight the well-known problems with the indoor climate, changes have not been made. A reason for the lack of improvements may be because the building users like their offices and would not willing move from them. The problems on sunny days are out-weighed for by a good social environment and the "luxury" of individual offices. This limits their bargaining power.

6. Conclusions

The paper asks, what are the implications for users when engaging with experiments. Living labs are a useful format for demonstration. In a ZEN living lab, the focus is on securing engagement with the development of zero emission technologies or neighbourhoods, the challenge is to connect ZEN aims with the aims of the user groups associated with pilot neighbourhood. Outside the ZEN partner group and stakeholders involved in the development of zero emission technology and buildings, interest in the zero emissions is limited. Therefore, within the social and technical context supporting the living labs there is a danger of the process being primarily top-down. The experiment did raise awareness about energy use in buildings and the role of the ZEN Centre on campus. However, faith in both the technology and centre activities may be challenged if the engagement with living labs is not combined with tangible results for building users. A living lab can raise awareness among user groups affected by the sustainable transitions planned by ZEN, and it can offer insight in how user groups would like to be involved. At Evenstad some interest is already in place and has the potential to be built upon. Not involving a broad group of users is a missed opportunity in the work towards sustainable transitions.

The social and physical place provided by Campus Evenstad is associated with its own set of challenges and needs. Following on from the experiment at Evenstad, the question is how the actions in one context be upscaled or translated so that they can be applied in other pilot locations? In addition, can we make sure that the results are tangible for users outside the main ZEN stakeholder groups? Finally, an issue arose about the role of the ventilation system in the experiment, after tentative results were presented to representatives from the construction industry in a blog text [14]. It was proposed that greater focus should have been on the air quality. The existence of particles within the air circulated could have been measured before and after the experiment, and the impact on users studied. This suggests that a broader interdisciplinary focus within the engineering team is useful when planning the technology and innovation inspired experiments in living labs.

References

[1] Bulkeley, H., Marvin, S., Palgan, Y. Voytenko, McCormick, K., Breitfoss-Loidl, M., Lindsay, M., von Wirth, T., Frantzeskaki, N. 2018 Urban living laboratories: conducting the experimental city? European and Regional Studies. Sage.
[2] Marvin, S., Bulkeley, H., Mai, L., McCormick, K., Voytenko Palgan, Y. 2018 Urban living labs: Experimenting with city futures. London. Routledge.
[3] Woods, R., Berker, T., Baer, D., Bø, L. A. 2019 ZEN Living Labs Definition, Ideas and Examples. ZEN report forthcoming.
[4] Pierson, J., Lievens, B. 2005 Configuring Living Labs for a "Thick" Understanding of Innovation. EPIC 2005, pp. 114-127. American Anthropological Association.
[5] Bulkeley, H., Castan-Broto, V., Edwards, G. 2015 An Urban Politics of Climate Change: Experimentation and the Governing of Socio-Technical Transitions, Routledge, London.
[6] Sengers, F., Berkhout, F., Wieczorek, A.J., Raven, R. 2016 Experimenting in the City: unpacking notions of experimentation for sustainability. In Evans, J., Karvonen, A., Raven, R. (Edt.) The Experimental City. London. Routledge.
[7] Veeckman, C., & van der Graaf, S. 2015. The City as Living Laboratory: Empowering Citizens with the Citadel Toolkit. Technology Innovation Management Review, 5(3): 6–17
[8] Evans, J. Karvonen, A., 2014. Give Me a Laboratory and I Will Lower Your Carbon Footprint! Urban Laboratories and the Governance of Low- Carbon Futures. International Journal of Urban and Regional Research, 38(2), 413-430.
[9] Pink, S., Mackley, L. K. 2012 Video and a Sense of the Invisible: Approaching Domestic Energy Consumption through the Sensory Home. Sociological Research Online 17 (1) 3.
[10] Suchman, L.A. 1987 Plans and situated actions: the problem of human-machine communication. Cambridge: Cambridge university press.
[11] Bulkeley, H., Marvin, S., Voytenko Palgan, Y., McCormick, K., Breitfuss-Loidl, M., Mai, L., von Wirth, T., Frantzeskaki, N. 2018 Urban living laboratories: Conducting the experimental city? European Urban and Regional Studies 1 –19
[12] Hylland Eriksen, T. 2010 The Challenges of Anthropology. Int. J. Pluralism and Economics Education, Vol. 1, No. 3, 2010
[13] Parker, G., Murray, C. 2012 Beyond tokenism? Community-led planning and rational choices. TPR, 83(1) 2012 doi:10.3828/tpr.2012.1

[14] https://forskning.no/energi-forskeren-forteller-innovasjon/halverte-stromforbruket-med-et-enkelt-lavteknologisk-grep/1266291

IOP Conf. Series: Earth and Environmental Science **352** (2019) 012005 doi:10.1088/1755-1315/352/1/012005

The FlexNett Simulator

Kristoffer Tangrand, Bernt Bremdal

PhD Research Fellow: IVT: UiT Narvik: Norway, Professor: UiT Narvik, Smart Innovation Norway: Norway

E-mail: kristoffer.tangrand@uit.no, bernt.bremdal@uit.no

Abstract. This paper documents research conducted in the Norwegian FLEXNETT project. It describes a new tool that was developed to study the future impact of prosumers with PV panels on the grid in Norway and the potential energy flexibility that lies with residential prosumers. Systematic use of energy flexibility can be an important instrument for managing peak loads and voltage problems in weak power grids. The influx of distributed energy resources can amplify this problem, but also help to resolve it. Self-balancing neighborhoods can be very attractive. This implies that loads related to energy demands can be curtailed and leveled out by different controllable devices or managed by using local energy production in the area to reduce the impact on the general distribution grid. The simulation tool is GIS based and can be applied to study the situation related to a single household, a neighborhood or in a specific transformer area. Unlike similar tools that address production yields over a period, the FLEXNETT Simulator addresses production and energy dynamics down to every 10 minutes. Due to the relatively low solar angle in Norway and rapidly changing weather these dynamics can be very prominent and induce local impact that is specific to a house or a neighborhood. The paper further describes how a recurrent neural network has been used as an engine to produce realistic values for the simulator.

1. Introduction

This work was conducted as part of the FLEXNETT project [1] lead by SINTEF Energi and with contribution of more than 20 industrial partners. The overall focus of the project was to investigate the present and future role of prosumers in the distribution grid. In particular, the role of private prosumers as instruments of energy flexibility for the local Distribution System Operator(DSO) was investigated. Energy flexibility relates to the possibility of reducing and leveling loads in order to avoid congestions in the electricity infrastructure and to maintain a high quality electricity supply. One of the central research questions specified was how prosumers with roof top based photovoltaic (PV) panels, with and without batteries, can help reduce peak loads in the grid. The grid was never designed to handle a two-way flow of energy with production facilities at its terminal points. The advent of distributed, renewable energy sources has called for research to understand how production at the grid's terminal points best should be catered for. As the use of PVs are still in its infancy in Norway it was considered important to determine new policies and methods early and be proactive to avoid situations that can cause problems for the operation of the grid. Experiences from southern Germany, where certain parts of the distribution grid have suffered uncontrolled and rapid peak accumulations of feeds, alternated by significant aggregated consumption loads triggered part of the research in FLEXNETT. High loads and rapid peak shifts can be a challenge. Congestions and voltage problems related to the

IOP Conf. Series: Earth and Environmental Science 352 (2019) 012005 doi:10.1088/1755-1315/352/1/012005

influx of distributed resources are not uncommon. Besides it has been shown that rapid peak shifts can cause life degradation on transformers. One important cause for these problems are related to the intermittency of solar and wind production. Another is that PV based production typically tops when consumption is low. This is quite common for residential areas where people spend time away from their house when the sun is at its highest and PV based production reaches its maximum.

A proactive investigation of PV based production and the role of prosumers in a Norwegian context was therefore called for. Some of the challenges experienced in southern Europe could be amplified further north due to geographical latitude and the low azimuth levels of the sun during larger parts of the year. The impact of local topography, vegetation and house architecture could cause greater local variations and together with cloud conditions cause even a higher frequency of alternating, aggregated loads in the distribution grid. At the same time opportunities were spotted. If local production could be organized so that local load balance could be maintained this could benefit the grid operations. Demand-response programs where loads are moved to periods of sunshine have been introduced and tested [2]. As the quality and price levels of batteries are improving they could also make a difference. Finally, pro-active planning could help to eliminate or alleviate the load issues, especially for new residential areas to be built where solar power is considered as an integral part of the construction project. But the same procedure could also be used to for existing residential areas in combination with a distributed set of batteries [3]. All of these issues were embraced by the FLEXNETT project. To answer part of this it was decided to create a novel simulation tool. Members of UiT tied up with Smart Innovation Norway to produce a solution that will be presented here. In the next paragraphs we will first explain some prerequisites for the work undertaken.

Then we will describe the tool created and how it was applied to produce relevant results. Finally, we will discuss the benefits and how we see the way forward.

2. Empirical Studies

Prior to the development of the FLEXNETT simulator empirical studies were conducted. The data used were harvested from the municipality of Hvaler in the county of Østfold in Norway. Hvaler was the first municipality in Norway where every residence was equipped with a smart meter connected to an AMS infrastructure deployed by Norgesnett, the local DSO. This happened in 2011. A substantial data history with a fair resolution was therefore available. This data allowed us to study the hourly load profiles of many different residences across the municipality. The double hump consumption profiles were quite uniform around the year with a distinct morning peak and a similar one in the late afternoon or early evening. However, there were certain differences with respect to when these peaks appeared. On the level of the principal substation these differences tended to even out, suggesting non-uniform behavior across areas and individual residences. However, within a single neighborhood the opposite seemed to be the case. Here, acute peaks were more common since the origin of the data received was not disclosed for the study we could only compare with consumption profiles of people we were able to get closer too. These were residential owners who took a special interest in the project. A statistical analysis showed, not surprisingly, that there was a relationship between demographic parameters and the type of consumption profile obtained. But we also found that many of the neighborhoods represented in the area were populated by people in the same age group and with matching daily routines. This could be related to the fact that many of these neighborhoods, mostly single residences and chained houses, were established at distinct points in time as the needs for housing emerged. Thus, older neighborhoods were typically inhabited by older and retired people, while recent property developments were dominated by families with smaller children. Hvaler was also the first municipality in Norway with a distinct population of residences with roof top panels. An active and facilitating policy introduced in

IOP Conf. Series: Earth and Environmental Science **352** (2019) 012005 doi:10.1088/1755-1315/352/1/012005

2015 created a set of incentives for residents to invest in panels. A small subset of these shared their production data with the project from the start, thus making it possible to create a history on local production, albeit not more than records for 20 units. To determine the conditions for production we studied the local topography and weather conditions for each one. For half of these we were allowed to inspect the property physically to specify features like height and angle of the roof and its orientation with respect to vegetation and other structures. The records harvested, and the observations made were used for a wider set of project purposes, some of which have been published elsewhere [4, 5]. One important conclusion from these studies was that production varied significantly for different neighbors in the same area and with the same type and size of roof top PV panels. It was apparent that local conditions causing shadows, both permanent, arbitrary and periodic made an impact. On an aggregated level this increased the intermittency, although much of this could be considered periodic.

One of the objectives of FLEXNETT was to carry out analyses to create the basis for the FLEXNETT simulation tool. The basic task was to examine empirical data and to use it to develop a tool to study the non-observed and the non-existent. In other words, we would use the tool to study what would happen if whole new areas were equipped with rooftop panels. What impact would a high-density deployment of PVs cause and how can it potentially be exploited for the benefit of the grid? A simulation tool was an obvious answer, if not the sole option. There was enough time series to power the generation of records for the simulations. But the set with production data was significantly leaner. These records represented far less households, despite significant histories for each of them. Standard statistical methods were considered but were rejected in favor of machine learning techniques based on a form of recurrent neural networks called the Long Short Term Method (LSTM).

3. Tool Making

One reason that standard statistical methods were rejected for the simulation was the inclusion of circumstantial data such as local weather and contextual elements representing causes for non-permanent shadows. All of this influence the rate of production and thus the degree of intermittency. Hence, the current state of production may, in the general case, not honor the Markov assumption. In other words, the current input does not hold all the information needed to compute the next event. LSTMs represent a type of deep learning techniques [6] that have proven useful for different applications, including time series and text analysis. Like most regression techniques these type of neural networks are typically used for prediction. And as such they can also be used for generating series of data based on the history [7]. The FLEXNETT problem was treated as a Partially Observable Markov Decision Process (POMDP). Each house in the municipality and their future production history could be considered the full state space were only part of it could be observed. Like other vanilla flavor neural networks, a recurrent network can be seen as a function approximator for such a universe. But a recurrent network like LSTM produces a dense state specification as a function of the entire history. This is due to their ability to carry out operations over sequences of vectors. In addition, it combines the vector containing the input with a state vector that in addition to the output vector creates a new state vector. Hence neural networks like the LSTM can be effectively used as a world model generator. The LSTM in FLEXNETT was constructed accordingly. The output $y(t, R)$, representing the production at any given time of the year, **t**, of any residence **R**, equipped with a solar panel would then be approximated with the LSTM representing the function $\mathbf{f}(P_1, P_2, P_3, P_4, ..., P_n, R(PV, x_1, x_2, ..., x_n))$. Here, P_i, PV, x_i represent historic production at specific times, peak panel production and a vector of geographical and topographical parameters x_i, respectively. Different versions of the part vector $R(PV, x_1, x_2, ..., x_n))$ were attempted, but problems with convergence was experienced. The final version of R was limited to PV panel size and physical orientation in 3 dimensions. The LSTM was trained on the time series collected and tested on a part of the collected data history separated from the training set.

Since PV production follows the sunlight, it has an inherent Sin Wave-like property. As such, variations in production would come as a consequence of physical objects existing between the sun rays and the PV panel. Such objects could come in the form of weather, such as clouds, and stationary or mobile ground objects. The LSTM should be able to account for ground objects simply from the recurring change in production. For the weather, including data from nearby PV panels should include information about changing weather conditions. A multi-dimensional training data consisting of production times series from multiple PVs was tested to see if it could provide better predictions than using only the time series from one single PV. For any given installation P_i, its training data consisted of time series $P_{i_1}, P_{i_2}, P_{i_3}, ...$ as well as time series from other PVs P_j, P_k, P_l in the area. The training data was organized in sliding windows(batches) of 36 time steps (6 hours X 6 values per hour), where the first (35 x number of PVs) values were input. The sliding windows was shifted by 1. The output then, was a single floating point value that represents the predicted next production value.

Real consumption data was applied. The rich data set made available to the project permitted this. Time series from the complete data set was chosen at random. Bias was further reduced by introducing multiple replications for the simulations. One issue was the difference in time resolution. As the training data was sampled on a 10-minute interval this became the standard for the simulations too. Since the meters provide on hourly data for the consumption we used that. For the purpose at hand it sufficed to use average data across the hour. In the future higher resolution consumption data is desired. A graphical interface was developed to control the simulation. It included a Geographical Information System (GIS) section that also provided information to the simulation engine (see Figure 1). By selecting a house or a group of houses linked to the same part of the grid the tool would generate the dynamics of loads for a single household as well as a neighborhood or a larger area. Similarly, it would be possible to choose simulations for a particular period i.e. day, month, season or a full year and the results could be shown in real time or produced as a batch.

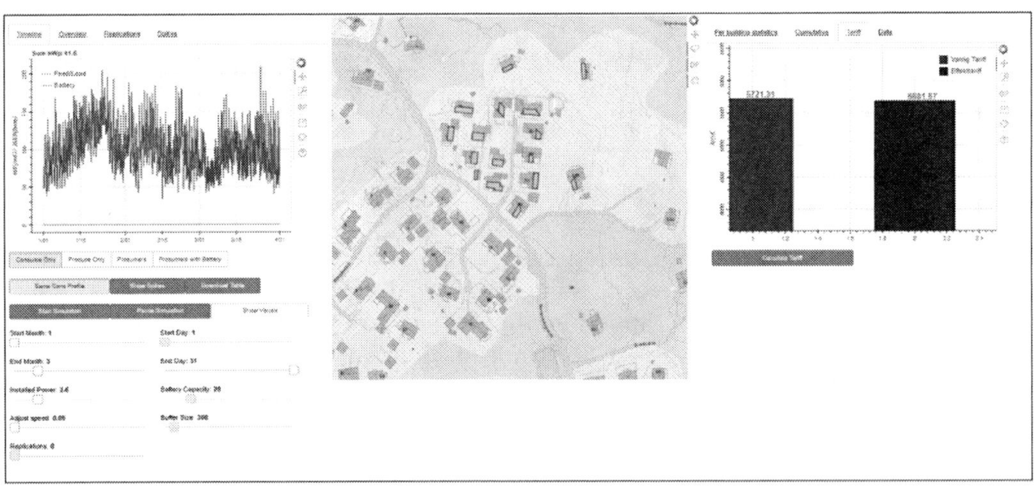

Figure 1. The FLEXNETT user panel showing time series for consumption and production on the left, the GIS panel in the middle where a selection of houses belonging to the same grid connection is investigated. Right a set of tabs can display aggregated power and cost data for the group and per building.

A selection of houses could also be equipped with batteries. For the economic calculations, battery degradation was taken into account, but the battery management was held very simple.

IOP Conf. Series: Earth and Environmental Science **352** (2019) 012005 doi:10.1088/1755-1315/352/1/012005

Battery would always have priority to surplus when fully or partly discharged. Degradation was based on the rainfall method [8]. Beyond this we applied a Coloumb Counting [9] method to monitor the state of charge. Economic calculations used the principal grid tariff at Hvaler introduced by Norgesnett as default. This includes a power part [kW]. The tariff for regular households is shown in Table 1 below. But the tool can be also used to analyze economic impacts due to tariffs with fixed and energy related fees only.

Table 1. Tariffs introduced by Norgesnett at Hvaler. Note the power related part

Fixed fee(NOK/year)	Period	Energy part NOK/100*kWh	Power part NOK pr. max hour per month [NOK/100*kW]
625	May-Oct	26.36	61.25
0	Nov-April	28.23	61.25

4. Results

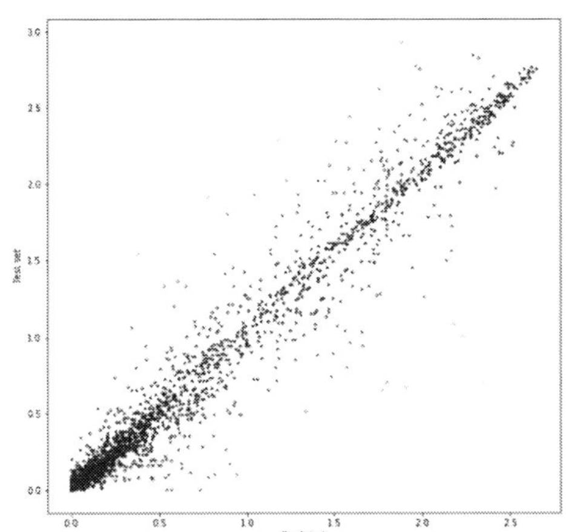

Figure 2. Scatter plot shows predicted values (x-axis) vs. actual values in test set (y-axis).

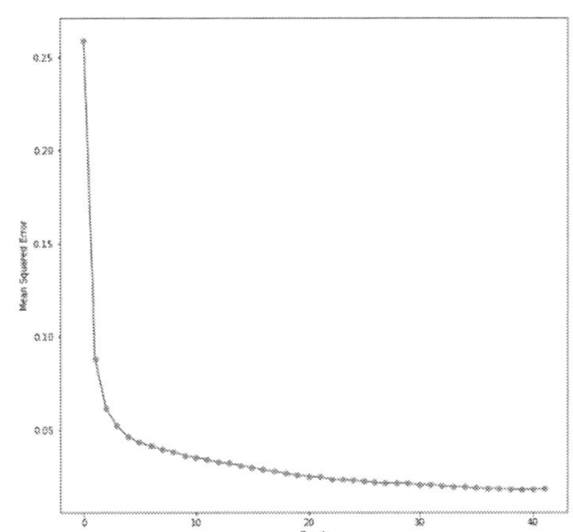

Figure 3. : Learning rate- showing the reduction of the Mean Square Error (y-axis) per training epoch (x-axis).

Figure 2 and Figure 3 shows the result of the training of the LSTM. The method showed good fit and acceptable error rates (MSE). It should be noted that the potential for overfit was present due to the limited sample size available. The network was trained until standard overfitting cutoff measures kicked in. Figure 4 shows the graphs produced for aggregated loads across time for a group of houses connected to the same point of coupling.

The tool was used to determine the impact of PV size on the grid and the potential role of the battery. The orientation of PV-panels on self-consumption was also investigated with the tool

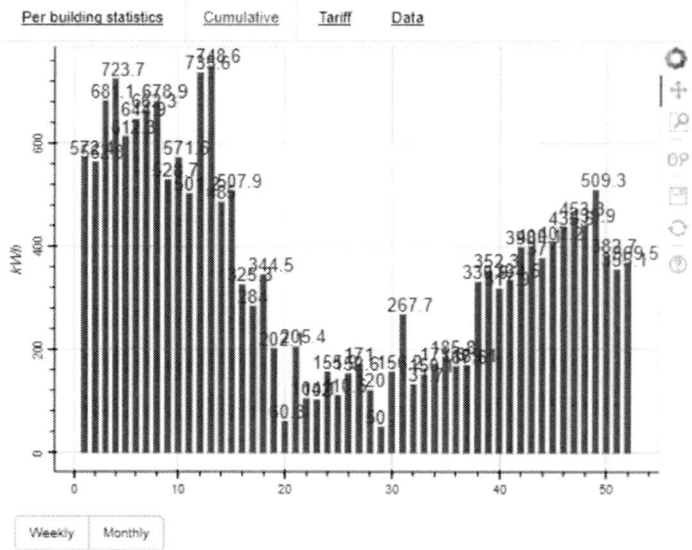

Figure 4. The cumulative net loads (kWh/h) for different hours for a period of app. 50 hours.

[3]. For the selection of households shown in Figure 2 the maximum loads on the neighboring transformer was calculated. Examples are given for three cases: No production facility, all houses equipped with 3,1kWp and 9,3kWp panels.

Table 2. Aggregated net loads for a selection of houses (kWh/h) (see Figure 1) equipped with roof top panels. Min specifies the minimum net load. Max the hour during the year with highest accumulated consumption.

Panel Size	0 kWp	3.1 kWp	9.3 kWp
Min	-9.94	31.58	86.61
Max	-118.93	119.43	-129.45

As can be observed from Table 2, the panel sizes need to be large in order to reach the magnitude of negative loads typically during the month of January with no production. During the warm and sunny part of the year the simulations also showed huge differences in max and min from minute to minute. The tool thus provided evidence that the local infrastructure for the grid areas studied at Hvaler were robust enough to handle a high density of large rooftop panels. Another issue studied was the impact of house orientation. Figure 5 shows the production profiles based on actual metering in the month of June for three different houses in the same neighborhood at Hvaler with PV panels oriented at 106, 182 and 200 degrees. In addition, and arbitrary consumption profile is shown for comparison in order to highlight self-consumption issues and the possibility of self-balancing by design. As can be observed, the easterly oriented panel meet the morning peak of the consumption better than the other two. But the westerly and southerly oriented profile meets the afternoon/evening peak better. The westerly oriented panel could have better absorbed the consumption if it was not cutoff as the curve in Figure 5 shows. This illustrates the susceptibility to local topography and vegetation that obstruct the sun at early and late hours during the summer. Using the tool to investigate the degree of self-consumption related to PV orientation the easterly or westerly evening peaks did not match the consumption as well as expected. One explanation is that the neighborhoods investigated was inhabited by a several retired couples. (See Table 3). The dominant morning and evening peaks in the pilot areas at Hvaler came later than expected compared to the average records for the country and that generally show an earlier morning peak and a a more distinct afternoon peak. This fact favored the panels facing south. With the power tariffs introduced the economic benefit of the PV panels increased significantly for the prosumers. The tool further highlighted the need for diversity in orientation. An average neighborhood in Norway with a high density of panels facing south are more likely to accumulate high peaks at noon than areas with roof tops pointing slightly different ways. Moreover, with power tariffs it makes economic sense to adjust panels according to the households consumption profile. But it is important that production does curb the consumption at the hour with the highest energy use. For more on the economic part of FLEXNETT see [10].

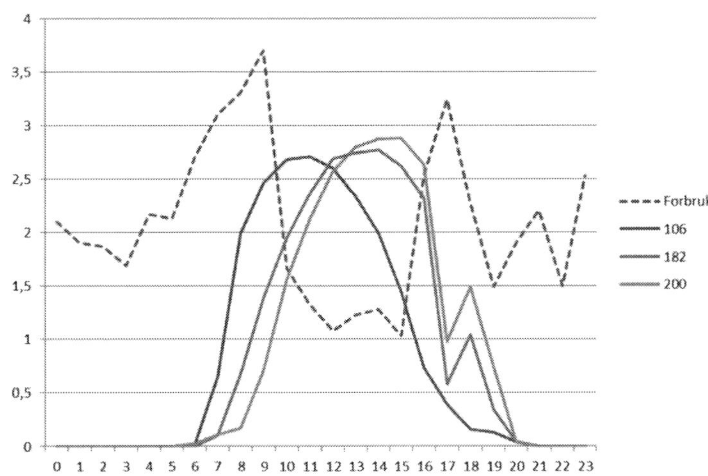

Figure 5. Shows production profiles for an easterly oriented rooftop PV (106), a panel pointing south (182) and a westerly oriented one (200). A consumption profile is shown for comparison.

Table 3. Max and min power for the average household equipped with a 3.1kWp panel. Max net consumption is obviously experienced during winter time. Max net consumption is positive during the summer.

Compass direction of PV	Max power consumption[kW]	Month	Min power (feed-in) [kW]	Month
South	-12.82	january	2.41	june
East	-12.82	january	2.06	june
Southeast	-12.82	january	1.99	june
West	-12.82	january	1.91	june

5. Conclusive Remarks

The FLEXNETT tool for analysis of load dynamics imposed by consumption and rooftop solar panels have been explained. UiT is currently maintaining the software. The simulation tool created applies a machine learning approach for the purposes. A LSTM neural network was trained and applied to simulate the consequences of high-density deployment of solar panels in different areas at Hvaler. The primary idea was to use the tool to investigate the impact of prosumers on the local distribution grid. The tool proved to be useful to determine the magnitude of peaks caused by local feeds and consumption and how rapidly changes in loads can occur. Local conditions play an important role on the production profile and yield at all hours of the day. The tool accommodates this by means of the GIS. The tool developed made it possible to analyse and conclude that the grid sections investigated would be able to absorb a high number of prosumers. However, the tool also showed its strengths in exposing potential vulnerabilities where grid weaknesses can be a reality. The tool can help analyses where collective self-consumption is sought. This would be a kind of self-balancing "by design" where residential areas can be planned to minimize the impact on the grid and to avoid potential congestions. The results from the simulations combined with the empirical studies, suggest that local demography can impact the aggregated, net profiles for sections of the grid. When time of production does not match consumption well demographic homogenity may be an important reason for this. The tool was also useful to determine the cost for single households exposed to Norgesnetts power tariff. The benefit of rooftop PVs that meet hours with maximum consumption can be economically significant if power tariffs are introduced. The tool includes options where batteries can be deployed, and their effects studied. This part has not yet been fully developed and validated. But UiT is currently addressing the issue. Furthermore, the tool is considered as an engine for future studies of local energy markets. The plan is also to extend it to be able to address other areas in Norway. It is believed that the tool will be especially important for studies in the northern parts of the country where the average azimuth is even lower.

6. Acknowledgements

The FLEXNETT project was supported by the Norwegian Research Councils ENERGIX program. The research reported here was further supported by Smart Innovation Norway and GeoData (GIS system).

References

[1] https://www.sintef.no/prosjekter/flexnett/ (accessed April 2019)
[2] Andrzej Ozadowicz, A New Concept of Active Demand Side Management for Energy Efficient Prosumer Microgrids with Smart Building Technologies Energies (accessible at https://www.mdpi.com/)
[3] Stig Ødegaard Ottesen, Pol Olivella-Rosell, Pau Lloret, Ari Hentunen, Pedro Crespo del Granado, Sigurd Bjarghov, Venkatachalam Lakshmanan, Jamshid Aghaei, Magnus Korpås and Hossein Farahmand, Simplified Battery operation and control algorithm, Deliverable D5.3 v.1, INVADE Project, 20/12/2017 (Available at https://h2020invade.eu/)
[4] Bernt A Bremdal, Hanne Sæle , Geir Mathisen , Merkebu Zenebe Degefa; Flexibility offered to the distribution grid from households with a photovoltaic panel on their roof. IEEE: Energycon Conf, Limassol, June 2018.
[5] Bernt A. Bremdal and Kristoffer Tangrand: Investigations of prosumers aggregated loads and energy flexibility at Hvaler(In Norwegian): Undersøkelse av aggregerte last- og fleksibilitetsbidrag fra plusskunder på Hvaler, FLEXNETT rapport, Smart Innovation Norway. Published by SINTEF Energi, March 2018
[6] Jürgen Schmidhuber: Deep learning in neural networks: An overview in Neural Networks 61(2015) p. 85-117
[7] Xiaocheng Feng, Ming Liu, Jiahao Liu, Bing Qin , Yibo Sun, Ting Liu, Topic-to-Essay Generation with Neural Networks Proceed. of 21st International Joint Conference on Artificial Intelligence (IJCAI-18)
[8] Thomas Martinsen, A Business Model for an EV Charging Station with Battery Energy Storage, Paper 0374 CIRED Workshop, Helsinki, June 2016
[9] Nugroho, Asep. Battery State of Charge Estimation by Using a Combination of Coulomb Counting and Dynamic Model With Adjusted Gain. 10.1109/ICSEEA.2015.7380745.
[10] Hanne Sæle, Bernt A. Bremdal: Economic Evaluation of the Grid Tariff for Households with Solar Power Installed. Paper 0556, CIRED Conference, Glasgow, 12-15 June 2017

IOP Conf. Series: Earth and Environmental Science **352** (2019) 012006 doi:10.1088/1755-1315/352/1/012006

Zero energy at the neighbourhood scale: Regulatory challenges regarding billing practices in Norway

Magnus Askeland[1,2], Stian Backe[3] and Karen Byskov Lindberg[2,4]

[1] Energy Systems, SINTEF Energy Research, Trondheim, Norway
[2] Department of Electric Power Engineering, Norwegian University of Science and Technology, Trondheim, Norway
[3] Department of Industrial Economics and Technology Management, Norwegian University of Science and Technology, Trondheim, Norway
[4] Energy and Environment, SINTEF Building and Infrastructure, Oslo, Norway

E-mail: Magnus.Askeland@sintef.no

Abstract. Buildings are becoming an increasingly active part of the power system due to the ongoing deployment of decentralized energy resources. To reap the added value that may be realized by zero emission neighbourhoods, it is important that the regulatory framework promotes an efficient development with buildings as an integrated and active part of the power system. When considering energy resources at the neighbourhood level and energy flows within neighbourhoods in Norway, the regulatory framework is challenged by innovative technical solutions. Therefore, it is necessary to explore how deployment of energy resources in neighbourhoods fit together with existing regulations and market mechanisms. Challenges concerning decentralized energy resources are identified based on discussions with stakeholders in Norway and a review of relevant literature in the scientific and regulatory domain. Key challenges for the deployment of energy resources in ZEN are identified, explained through examples, and related to ongoing projects in Norway. It is found that incentives regarding decentralized energy resources are highly dependent on the ownership structure, and therefore a distinction between two major classes of ZEN is made.

1. Introduction

The primary goal of the Norwegian energy law [1] is to ensure economic efficiency in the power system by creating a level playing field for competition in the power market. One major factor is the principle of individual metering to allow every consumer to decide on their electricity supplier, which increase the competition in the power market. The energy law was introduced in the 1990s, at a time where demand was viewed as fixed and the potential for cost reductions were on the supply side. However, there is an ongoing increase in energy-related measures including energy generation resources deployed at the building level. As we move from a goal of zero-energy on a building level towards the neighbourhood scale, it is essential that the regulatory framework continues to promote economic efficiency when buildings become an integrated and active part of the power system.

1.1. Zero-energy buildings

Buildings constitute about 40% of total primary energy consumption in Europe and the US [2]. To reduce the amount of energy required by buildings, the European Union (EU) has

IOP Conf. Series: Earth and Environmental Science **352** (2019) 012006 doi:10.1088/1755-1315/352/1/012006

set ambitious targets through the energy performance of buildings directives (EPBD), most recently in a 2018 revision [3]. Among other things, the EPBD promote development towards cost-effective nearly zero energy buildings (NZEBs) by 2020.

Although the exact definition of NZEB is not clearly defined, the development towards NZEBs in the building sector means that we move from focusing on reducing energy needs through passive buildings towards also generating energy at the distributed level to accommodate increasingly ambitious targets [4]. Due to the flexible definition of what a zero-energy building is, different interpretations exist in the scientific literature, and despite different interpretation for the 'zero' balance in the EU [5], national policy are driving the development of ZEB.

1.2. Zero-energy at the neighbourhood scale

The concept of zero energy can also be considered at other scales than individual buildings, for example, neighbourhood, district or city level. Marique and Reiter [6] articulate three main energy uses at the neighbourhood level: building energy consumption, the production of on-site renewable energy and transportation energy. When the scope of zero energy is extended beyond individual buildings, the system boundary change from individual buildings to groups of buildings. In the following, such groups of buildings of any size will be referred to as 'zero energy neighbourhoods' (ZEN).

In the following, we categorize ZEN into two different classes based on owner structure: S-ZEN, which is owned by a single entity, and M-ZEN, where multiple owners are present. To illustrate the difference, we provide two examples:

- Campus Evenstad (S-ZEN): A university campus, including student housing, offices and teaching facilities. It is developed, owned and operated by a single institution (Statsbygg).
- Verksbyen Fredrikstad (M-ZEN): One of the largest residential neighborhood development projects in Fredrikstad, Norway [7]. The project is developed by Arca Nova, and the area include several owners of the buildings.

By extending the system boundary to several buildings, it is possible to obtain additional benefits as compared to considering individual buildings separately. Benefits of considering the ZEN scale compared to ZEB include possibilities to build energy resources at the most favorable locations [8], decreased unit costs compared to smaller systems [9], and coordinated balancing of the energy needs of buildings to achieve a more flexible cumulative load profile [10]. In addition to improved conditions for deployment of energy resources, ZEN also facilitates investments that are not available for individual buildings, such as large-scale solar plants in [9].

Along with the benefits, ZEN also introduce challenges when compared to ZEB. Technically, an increased amount of buildings means more complex systems need to be designed and operated compared to the case of one individual building. Furthermore, when buildings become a more integrated and active part of the power system, regulations, taxes, and tariffs need to be designed in a way that facilitates decentralized decisions that are also system optimal.

1.3. Remuneration models for local energy generation

Investments in renewable energy generation assets is increasingly being made by non-utility type stakeholders, e.g. home owners and public institutions. Traditional consumers thus become *prosumers*: consumers that also generate energy. The growth of prosumers is partly a result of policies, and has been especially prominent for solar photovoltaic (PV) panels. For a prosumer, there are two common categories of remuneration models based on generation: (1) a net-metering policy based on (volumetric) energy [11] and (2) a Feed-in Tariff (FiT) [12].

The net-metering policy allows a prosumer to offset electricity consumption from the grid with local production defined over a measuring period (e.g. hourly, daily, monthly, yearly). The

IOP Conf. Series: Earth and Environmental Science **352** (2019) 012006 doi:10.1088/1755-1315/352/1/012006

longer the measuring period, the more likely the net-metering policy is to reduce the electricity bill of the prosumer. The FiT policy offers a fixed payment for prosumers and requires metering of the local generator. Prosumers subject to a FiT face no price risk except the future removal of the policy [12].

Net-metering policies are found to be financially bad for utilities and end-users without local generation as the lost revenue for utilities cannot be balanced by the saved costs from reduced power grid usage (referred to as the 'revenue erosion effect' [13]). Some fear this could lead to a utility 'death spiral' [14] where using the power grid becomes increasingly expensive as more end-users partly produce their own energy, which would lead more end-users to produce their own energy, and so on. Adjusting remuneration models to be partly based on peak demand will likely be a better measure to allocate costs among grid connected consumers and prevent the utility death spiral [11].

In Norway, the policy framework is favourable for S-ZEN. The prosumer policy ('plusskundeordningen') is a net-metering policy that applies to one meter per prosumer. The prosumer policy grants an exception from the conventional need for a regulatory concession to be an electricity producer in Norway, and it is subject to a requirement that the delivered power to the grid does not exceed 100 kW. The payment to prosumers for surplus energy is usually higher than wholesale electricity spot price, but can vary depending on the retailer contract. Prosumers can get additional remuneration from the joint certificate market [15] for Norway and Sweden if a production plant is commissioned before 2022. Since 2019, it is also more economically viable for small-scale producers (< 1 MW installed capacity) to apply for producer concession after a change in the producer tariff from being dependent on installed capacity to being dependent on net energy delivered to the grid [16].

1.4. Content of the paper
The feasibility of achieving a zero energy balance is dependent on rules and regulations for consumption, production, and pricing of energy. The present paper aims to provide an overview of current regulatory challenges concerning energy resources in ZEN with a particular focus on Norway. Topics covered include why ZEN poses challenges compared to ZEB, and how the concept of ZEN fits current regulations, tariffs, and incentives.

2. Regulatory challenges for ZEN
The following section describes two regulatory challenges that need to be addressed for consistent integration of ZEN in the Norwegian power system: (1) Balancing energy between buildings and (2) Sharing energy resources. These challenges have been identified based on discussions with stakeholders in Norway and a review of relevant literature both in the scientific and regulatory domain.

2.1. Balancing energy between buildings
Different types of buildings (residential buildings, schools, offices) may not have their peak energy usage at the same moment in time due to 'the coincidence factor' [17]. The coincidence factor means that the capacity connecting a neighbourhood to the rest of the power system generally is less than the sum of individual peaks in the neighbourhood. This cumulative load profile can be further improved by coordinating energy usage within a ZEN. Coordination of loads can potentially lead to cost reductions since the total amount of capacity connecting the ZEN to the rest of the power system can be reduced. The view that (a) the zero energy concept should be applied to a higher level and (b) energy should be shared between buildings is supported by e.g. [4, 18, 19, 20, 21]. In this context, [21] argue for an energy hub concept to extend the system boundary and realize a system with higher flexibility due to the availability of multiple energy sources and sharing of energy among various consumers and producers.

IOP Conf. Series: Earth and Environmental Science **352** (2019) 012006 doi:10.1088/1755-1315/352/1/012006

We now turn to the Norwegian context and the incentives towards consumers for sharing energy between buildings.

In the case of S-ZEN, all the energy resources and the demand are owned by the same stakeholder although they could be spread across several buildings. An example of this is Campus Evenstad which was described in Section 1.2. In S-ZEN projects, the energy usage within the ZEN can be efficiently balanced before the surplus or deficit is traded with the rest of the power system. Since grid charges and taxes are applied at the interface between the ZEN and the rest of the power system, the owner of S-ZENs has incentives to balance out their energy as much as possible before interacting with the grid since this is most economical.

The other class of ZEN, namely the M-ZEN, is more complicated than the S-ZEN when considering energy flows between buildings. The difference occurs because the energy needs to be transferred from one owner to another. Such transfer of energy requires some trading between the two owners, either directly or through a third party.

We use an illustrative example with two buildings as depicted in figure 1 to compare the case of the two classes of ZEN. For simplicity, we only consider electric energy and one hour of operation. During this hour, Building 1 has a surplus of 10 kWh, while Building 2 has a deficit of 10 kWh. Furthermore, we assume that the grid tariff and energy prices are both 0.50 NOK/kWh resulting in a total cost of 1 NOK/kWh for imports and an income of 0.50 NOK/kWh for exports, assuming the grid cost does not apply for exported energy.

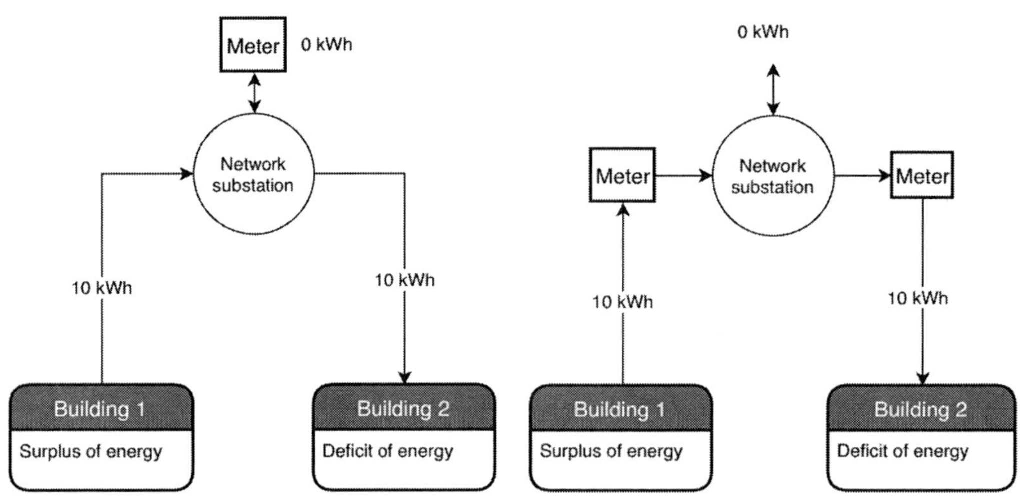

(a) Single-owned ZEN (S-ZEN) (b) ZEN with multiple owners (M-ZEN)

Figure 1: Illustrative example of energy sharing

In the case of S-ZEN, both buildings are owned by the same stakeholder, so these buildings are metered together at the interface to the rest of the power system (see figure 1a). Costs charged are zero as calculated in (1):

$$Cost^a = 1\frac{\text{NOK}}{\text{kWh}} * (10 - 10)\text{kWh} = 0 \text{ NOK} \tag{1}$$

Next, we consider the M-ZEN case where two different stakeholders own these two buildings. We now calculate the costs for each of the buildings separately since they are not metered together (see figure 1b). The result is a total cost of 5 NOK as calculated in (2):

$$Cost^b = 0.5\frac{\text{NOK}}{\text{kWh}} * (-10)\text{kWh} + 1\frac{\text{NOK}}{\text{kWh}} * 10\text{kWh} = 5 \text{ NOK} \tag{2}$$

IOP Conf. Series: Earth and Environmental Science **352** (2019) 012006 doi:10.1088/1755-1315/352/1/012006

Although our example is simplified, it can be observed that energy balancing between buildings in ZEN is not supported by the regulatory framework in the case of M-ZEN. The problem occurs as the individual owners do not have any incentives to cooperate in shifting their demand to balance the energy within the ZEN. The regulatory framework provide incentives to consume locally produced energy behind the meter, which in M-ZEN means avoiding surplus energy for single buildings. One might ask if it is possible for several owners located geographically close to each other to establish one common interface towards the rest of the power system, but this is prohibited by Norwegian regulations [22] requiring individual costumers to be metered separately. The current regulatory framework in Norway only allows trading of electricity between two consumers producing energy (prosumers) through a third party (an energy retailer and the local distribution system operator).

This is indeed a challenge for projects in Norway such as Verksbyen in Fredrikstad which will consist of multiple owners that could potentially be able to balance their energy usage locally, but lack incentives for doing so with the current regulatory framework. As an extension of this argument, [23] note that products and markets for demand response should be developed further in several EU countries.

2.2. Shared energy resources

The concept of ZEN facilitates several stakeholders pooling their financial resources together to be able to build larger power generating facilities. One typical example would be an apartment block in which the different apartment owners build a shared PV-plant on the roof or building facade.

So far, with the concept of ZEB, the distinction of on-site vs. off-site resources has been clear due to the well-defined system boundary. In the context of ZEB, [24] argue that if a generation system is behind the meter, it is on-site. Otherwise, it is off-site. This distinction is challenged by the concept of shared energy resources in a ZEN since the system boundary is not as well defined as for a single building. In some cases, it might be optimal to build larger shared plants within the ZEN, but located outside the meters of the individual households.

To illustrate the economics of shared energy resources, we will consider an illustrative example as depicted in figure 2. The example has been based on the business model of energy resources in Verksbyen Fredrikstad excluding the effect of a Feed-in Tariff (FiT). The reason for excluding FiT is that such policies represent an artificial market price and therefore inherent uncertainty regarding the development of future policies. In our example, we consider an operational hour in which a household and the shared consumption (e.g. EV-charging) requires 10 kWh while the shared generation facility generates 20 kWh.

We now look at this system in more detail. First, we consider the shared facilities in which both the consumption and the generation is behind the same metering point. Such a billing practice means that for any energy consumed directly, grid charges are avoided. A result of this is that it is possible to define a local energy price which is higher than the spot price and lower than the retail electricity price which include spot price added grid charges and taxes:

$$\text{spot price} \leq \text{LCOE} \leq \text{local energy price} \leq \text{retail electricity price} \tag{3}$$

The levelized cost of energy (LCOE) is the lifetime costs divided by energy generation and therefore represent the minimum remuneration per unit of energy to pay back an investment in energy resources. The electricity spot price, e.g. at Nord Pool Spot [25] for nordic countries, is usually lower than the LCOE for distributed energy resources as stated in (3) so it is not profitable to invest in such resources solely to feed it into the electricity grid. Furthermore, (3) states that the LCOE can be lower than the retail electricity price. Therefore, if (3) holds, it is possible to define a local energy price which is larger than the LCOE for energy that is consumed directly. This means that even if distributed energy resources in ZEN are not competitive to

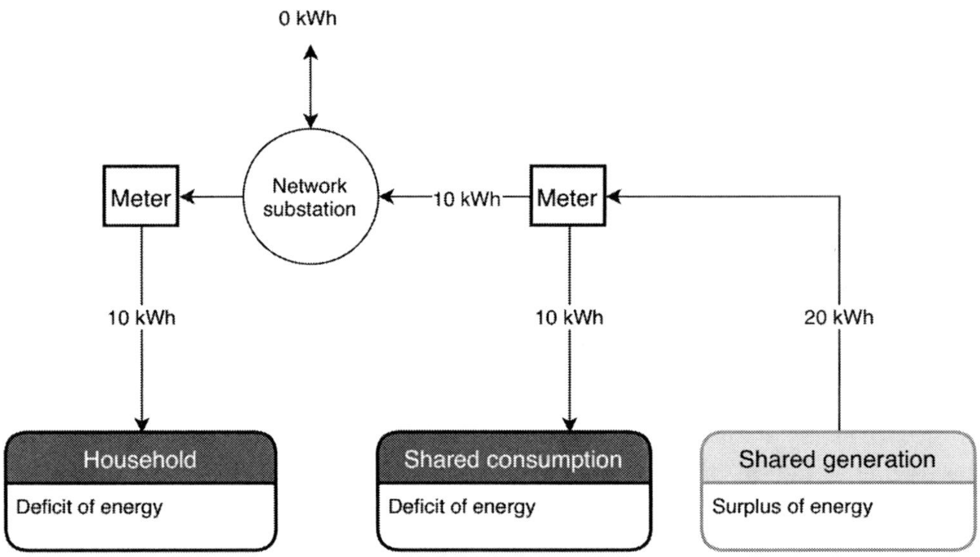

Figure 2: Illustrative example of shared energy resources

large-scale resources in the power market, they may have a place in the power system since they can provide a locational value due to the effect of reducing load as proposed in [26]. This is supported by e.g. [27] which argue that the payment received for exporting excess energy to the grid is insufficient. The underlying reason for this is that distributed generation in ZEN lack the economies of scale which reduce the LCOE for large-scale generation facilities.

We now look at the incentives of the household in this ZEN, which is not behind the same meter as the shared generation. It might be possible for these households to shift their load to some extent in order to match their consumption to the generation. However, due to the pricing policies, the households do not have any incentives for engaging in such behavior. This finding pose a problem since consumers are not incentivized to alleviate grid stress and defer grid investments since their local pricing information do not include the state of the local energy resources.

The issue of shared energy resources is highly relevant for Verksbyen which currently has a plus costumer scheme in place with with an agreement for 0.8 NOK/kWh for any surplus energy fed into the grid. However, the developer of the project consider this price to be highly uncertain since it can change on short notice while investments in energy resources have a long lifetime. Since the project needs to make their decisions subject to regulatory uncertainty, the net present value calculations are based on the assumption that no such agreement is in place since it is an artificial market price. This is in principle a robust optimization approach, and the result is that investments decisions in Verksbyen are based on current market prices and regulatory conditions as depicted in our example.

3. Discussion and conclusions

We have through examples and economic principles shown how current pricing policies in Norway do not offer proper incentives to align behavior by individual stakeholders in ZEN with multiple owners with efficient operation of the overall ZEN. We have demonstrated that when demand response and efficient sharing of energy resources is possible, the current Norwegian billing practices do not offer incentives for the activation of such potential. Furthermore, it should also be noted that grid companies will need to recover their sunk costs, but efficient pricing policies to properly incentivize consumer behavior can activate the potential of local coordination and reduce the need of grid upgrades. The findings of this paper indicate that the current technical possibilities are ahead of the regulatory framework. Regulatory innovation to fill the gap of missing local incentives can take a multitude of forms, ranging from adapting the network tariffs to implementing other kinds of market mechanisms.

Incentives for energy balancing between buildings and sharing energy resources depends on the ownership structure of the ZEN. The critical distinction is between a single-owned ZEN (S-ZEN) and ZEN with multiple owners (M-ZEN). For S-ZEN, the owner has incentives for balancing the energy needs of the various buildings and generation facilities, while this is not the case for M-ZEN. The difference occurs because of billing practices since the interface to the rest of the power system is different for the two classes. In Norway, there is a lot of interest in local energy generation, but to deploy such assets efficiently for M-ZEN, it is necessary to design market mechanisms that enable utilization of such assets at the time of generation. In the case of M-ZEN, this raises the need for local energy trading between stakeholders within ZEN while also maintaining individual metering to facilitate economic efficiency in the overall power system. Therefore, it is worth considering how the locational value of energy can be exploited while maintaining economic efficiency in the overall power system.

The issues addressed in this paper is largely based on the premise that self-consumption is the most important factor to make decentralized energy resources profitable in Norway. However, the incentives vary across Europe as found in [28, 29] which compared support policies for decentralized photovoltaic systems. For instance, the Flemish policy differs from Norwegian billing practices and do not promote self-consumption since the electricity usage is netted out over a period of time. Germany employs on a Feed-in Tariff to guarantee a minimum selling price, which currently is lower than the electricity price resulting in incentives promoting self-consumption similarly as the situation in Norway. However, if the Feed-in tariff rate approaches the electricity retail price, such as in France, the incentives for self-consumption will disappear. Despite the varying incentives facing individual consumers in different countries, the issues concerning M-ZEN are relevant in a more general sense since there is currently no best practice regarding sharing of energy when multiple owners are present.

This article provides insight into how current pricing policies may need to evolve due to the on-going deployment of ZEN in the power system. Regarding policy development, a recent report from the Norwegian Water Resources and Energy Directorate (NVE) emphasize the need for a regulatory sandbox regime to allow testing of policies by providing temporary regulatory exceptions [30]. The findings in this article provide a starting point for further research on how we can design efficient market mechanisms and pricing policies to incentivize decentralized decisions that are also beneficial for the larger power system.

4. Acknowledgements

This article has been written within the Research Centre on Zero Emission Neighbourhoods in Smart Cities (FME ZEN). The authors gratefully acknowledge the support from the ZEN partners and the Research Council of Norway. Furthermore, the authors gratefully acknowledge the discussions with Arca Nova regarding the Verksbyen Fredrikstad project.

References

[1] The Norwegian Ministry of Petroleum and Energy. The Energy Law, 1990.
[2] Xiaodong Cao, Xilei Dai, and Junjie Liu. Building energy-consumption status worldwide and the state-of-the-art technologies for zero-energy buildings during the past decade. *Energy and Buildings*, 128:198–213, 2016.
[3] European Commission. Directive (EU) 2018/844. *Official Journal of the European Union*, 2018.
[4] Ilaria Vigna, Roberta Pernetti, Wilmer Pasut, and Roberto Lollini. New domain for promoting energy efficiency: Energy Flexible Building Cluster. *Sustainable Cities and Society*, 38(February):526–533, 2018.

[5] Eleonora Annunziata, Marco Frey, and Francesco Rizzi. Towards nearly zero-energy buildings: The state-of-art of national regulations in Europe. *Energy*, 57:125–133, 2013.

[6] Anne Françoise Marique and Sigrid Reiter. A simplified framework to assess the feasibility of zero-energy at the neighbourhood/community scale. *Energy and Buildings*, 82:114–122, 2014.

[7] Multiconsult. Verksbyen – en komplett bydel i Fredrikstad, 2019.

[8] R. Compagnon. Solar and daylight availability in the urban fabric. *Energy and Buildings*, 36(4):321–328, 2004.

[9] Jay Burch, Jason Woods, Eric Kozubal, and Aaron Boranian. Zero energy communities with central solar plants using liquid desiccants and local storage. *Energy Procedia*, 30:55–64, 2012.

[10] Luigi Martirano, Emanuele Habib, Giuseppe Parise, Giacomo Greco, Matteo Manganelli, Ferdinando Massarella, and Luigi Parise. Demand Side Management in Microgrids for Load Control in Nearly Zero Energy Buildings. *IEEE Transactions on Industry Applications*, 53(3):1769–1779, 2017.

[11] Sharon JW Klein and Caroline L Noblet. Exploring Sustainable Energy Economics: Net Metering, Rate Designs and Consumer Behavior. *Current Sustainable/Renewable Energy Reports*, 4(2):23–32, 2017.

[12] Yuliya Karneyeva and Rolf Wüstenhagen. Solar feed-in tariffs in a post-grid parity world: The role of risk, investor diversity and business models. *Energy Policy*, 106:445–456, 2017.

[13] Andrew Satchwell, Andrew Mills, and Galen Barbose. Quantifying the financial impacts of net-metered PV on utilities and ratepayers. *Energy Policy*, 80:133 – 144, 2015.

[14] Kenneth W Costello and Ross C Hemphill. Electric utilities' 'death spiral': hyperbole or reality? *The Electricity Journal*, 27(10):7–26, 2014.

[15] The electricity certificate market. `https://www.nve.no/electricity-certificates/the-electricity-certificate-market/`. [Online; accessed 20-March-2019].

[16] Høringsdokument nr 6-2018: Forslag til endring i forskrift om kontroll av nettvirksomhet. `https://www.nve.no/om-nve/regelverk/forskriftsendringer-pa-horing/horing-forslag-til-endringer-i-kontroll-av-nettvirksomhet-avsluttet/`. [Online; accessed 16-April-2019].

[17] Constantine Bary. Coincidence-factor relationships of electric-service-load characteristics. *Transactions of the American Institute of Electrical Engineers*, 64(9):623–629, 1945.

[18] Majid Ahmadi, Jay M. Rosenberger, Wei Jen Lee, and Asama Kulvanitchaiyanunt. Optimizing Load Control in a Collaborative Residential Microgrid Environment. *IEEE Transactions on Smart Grid*, 6(3):1196–1207, 2015.

[19] Andreas Koch, Jason Woods, Eric Kozubal, and Aaron Boranian. Towards a neighbourhood scale for low- or zero-carbon building projects. *Building Research & Information*, 40(4):527–537, 2012.

[20] Rui Dai, Mengqi Hu, Dong Yang, and Yang Chen. A collaborative operation decision model for distributed building clusters. *Energy*, 84:759–773, 2015.

[21] Kristina Orehounig, Ralph Evins, and Viktor Dorer. Integration of decentralized energy systems in neighbourhoods using the energy hub approach. *Applied Energy*, 154:277–289, 2015.

[22] The Norwegian Ministry of Petroleum and Energy. Regulation 302 on economical and technical reporting, income limits for the electricity distribution sector and tariffs, 1999.

[23] Tracey Crosbie, Michael Short, Muneeb Dawood, and Richard Charlesworth. Demand response in blocks of buildings: opportunities and requirements. *Entrepreneurship and Sustainability Issues*, 4(3):271–281, 2017.

[24] Karsten Voss, Igor Sartori, and Roberto Lollini. Nearly-zero , Net zero and Plus Energy Buildings. *REHVA Journal*, 49(6):23–28, 2012.

[25] Nord Pool Spot. Day-ahead prices. `https://www.nordpoolgroup.com/Market-data1/Dayahead/Area-Prices`. [Online; accessed 08-August-2019].

[26] Scott P Burger, Jesse D Jenkins, Sam C Huntington, and Ignacio J Pérez-arriaga. Why distributed? A critical review of the tradeoffs between centralized and decentralized resources. *IEEE Power and Energy Magazine*, In Press(april):16–24, 2019.

[27] Grant Allan, Igor Eromenko, Michelle Gilmartin, Ivana Kockar, and Peter McGregor. The economics of distributed energy generation: A literature review. *Renewable and Sustainable Energy Reviews*, 42:543–556, 2015.

[28] L. Dusonchet and E. Telaretti. Comparative economic analysis of support policies for solar PV in the most representative EU countries. *Renewable and Sustainable Energy Reviews*, 42:986–998, 2015.

[29] L. De Boeck, S. Van Asch, P. De Bruecker, and A. Audenaert. Comparison of support policies for residential photovoltaic systems in the major EU markets through investment profitability. *Renewable Energy*, 87:42–53, 2016.

[30] The Norwegian Water Resources and Energy Directorate. Descriptive study of Local Energy Communities. Technical report, 2019.

IOP Conf. Series: Earth and Environmental Science **352** (2019) 012008 doi:10.1088/1755-1315/352/1/012008

Analysing electricity demand in neighbourhoods with electricity generation from solar power systems: A case study of a large housing cooperative in Norway

Åse Lekang Sørensen[1]*, Igor Sartori[1], Karen Byskov Lindberg[1], Inger Andresen[2]

[1]SINTEF Building and Infrastructure, Oslo, Norway; [2]Norwegian University of Science and Technology (NTNU), Trondheim, Norway.

ase.sorensen@sintef.no

Abstract. An energy management system can be introduced on a neighbourhood level, to achieve energy goals such as increased self-consumption of locally produced energy. In this case-study, electricity generation from photovoltaic (PV) systems is simulated at Risvollan housing cooperative, a large housing cooperative in Norway. The electricity generation from PV systems of different orientations and capacities are analysed with the electricity load. Key performance indicators (KPIs) such as self-generation, self-consumption and generation multiple are described, based on hourly values. The electricity generation from the south-oriented building façade PV systems are about 5-6% higher than for the east-west oriented rooftop PV systems on an annual basis, since the façade PV systems generate more electricity in the spring and autumn. The self-consumption factor is the most important KPI in Norway, due to the national tariff structure. For the total housing cooperative, a PV capacity of about 1,000 kW_p seem suitable, giving a self-consumption factor of 97% for a rooftop system, based on 2018 electricity and climate data. From the perspective of the housing cooperative, it is financial beneficial to aggregate electricity loads for common areas and apartments, since a higher share of the electricity can be used by the cooperative. For this to be possible, also housing cooperatives with PV must be facilitated for in the prosumer agreement. Comparing a single 1,100 kW_p PV system providing electricity to the total cooperative with 22 PV systems of 50 kW_p behind 22 garage meters, the self-consumption factor decreases from 95% to average 14%, resulting in a 41% lower financial value for the PV electricity.

1. Introduction

In zero emission neighbourhoods, thermal and electric energy should be managed in a flexible way, to achieve reduced power peaks, reduced energy use, reduced CO_2-emissions and increased self-consumption of locally produced energy [1]. Further, smart energy management systems (EMS) with building loads can provide energy flexibility services to distribution system operators (DSOs) and district heating companies, both on a building and a neighbourhood level.

A prosumer agreement exists in Norway, for locally produced electricity [2]. AMS meters (Advanced Metering System) at each customer measure net electricity export and import on an hourly basis. Financially, consumers normally receive less payment for electricity sold to the energy company than what they pay for buying electricity. This makes it beneficial to maximise self-consumption, i.e. minimising export of electricity to the grid.

IOP Conf. Series: Earth and Environmental Science **352** (2019) 012008 doi:10.1088/1755-1315/352/1/012008

Risvollan housing cooperative is a large housing cooperative in Trondheim, built in the 1970s. Risvollan cooperates with energy companies and researchers to develop a neighbourhood EMS. In Risvollan, there are 1,058 apartments with in total 93,713 m^2 heated floor area, distributed on 121 similar apartment blocks, as shown in Figure 1. In total 2,321 residents live in the apartments: 53% female and 47% male [3], as shown in Figure 2. Previously, measurements of electricity and heat loads of Risvollan in 2018 have been analysed in respectively [4] and [5]. The electricity loads also include around 55 electric vehicles (EVs) in the parking houses, which is expected to increase within the next years. Space heating and domestic hot water (DHW) is provided by district heating.

To be partly self-sufficient with electricity, the housing cooperative considers installing photovoltaic (PV) systems on some of their buildings. This article analyses the electricity demand at Risvollan together with possible electricity generation from the PV systems. In this article the electricity delivery is also referred to as electricity use, demand or load.

Figure 1. Example of apartment blocks at Risvollan housing cooperative, with 121 similar building blocks.

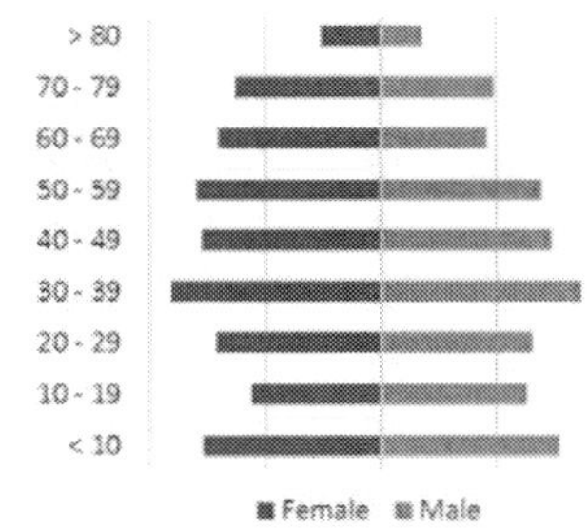

Figure 2. Age and gender distribution of the 2,321 residents.

2. Methods

Future scenarios for PV systems are developed, with varying installed PV capacity. PV generation is simulated with hourly resolution, using the software PVsyst [6]. Input data and system information is shown in Table 1. The PV systems are placed in two directions: 1) rooftop PV systems on flat roofs, with a 15° tilt orientated east and west, and 2) building façade PV systems, south oriented with a 90° tilt. No shadings are defined for the systems. Snow cover is considered by increasing the albedo values during the winter months, as shown in the table. Two system sizes are simulated in each orientation, to develop a scalable 1 kW$_p$ PV system based on the average, since variations can occur between simulated systems in PVsyst. The area suitable for PV on the 121 building blocks varies. Analysing internet maps, it seems that around 600 m^2 roof area may be available on the most suitable buildings, while other building might be evaluated as not suitable at all, due to shadings or roof conditions. In this article, it is roughly estimated that a 50 kW$_p$ system can be placed on one third of the buildings, giving a total of 2,000 kW$_p$ on the roofs. For the systems placed on the façades, less suitable area is available. Assuming that a 12 kW$_p$ system can be placed on one third of the buildings, the total potential is about 500 kW$_p$ on the façades.

Climate data for 2018 collected from eKlima [7] is imported to PVsyst. The outdoor temperatures are mainly from a weather station at Risvollan, where a few missing values are replaced with data from weather station Voll, 2.5 km away. The wind data are also from Voll, while the global radiation is from the weather station Gløshaugen, 3 km away. Based on the 2018-climate data, PVsyst creates hourly meteo-data for the simulations, where the annual horizontal global irradiation is 868.3 kWh/m^2, the horizontal diffuse irradiation 432.96 kWh/m^2 and ambient temperature 5.49°C.

Based on analysis of electricity loads in 2018 [4], this analysis uses load data from 1,009 apartments (95%), 22 electricity meters in garages (88%) and 82 electricity meters in other common areas (92%), excluding metering points with less than 7000 hours of data. Still some missing measurement periods remain, mainly in January, where only 67% of garages, 76% of other common areas and 72% of the apartments are measured. From February, most AMS meters are installed.

To evaluate the results, the self-generation, self-consumption and generation multiple factors are calculated based on hourly values. The 'self-generation factor', is the percentage of the electrical demand that is covered by on-site electricity generation [8]. The 'self-consumption factor' is defined as the percentage of the on-site generation that is used by the buildings [8]. 'Generation multiple factor' is the ratio between exported and imported peak powers [8].

A range of PV capacities are chosen, up to the maximum of 2,000 kW_p rooftop and 500 kW_p façade PV systems, with capacity steps of 0, 50, 100, 500, 1,000 and 2,000 kW_p. The aim of the chosen steps is to illustrate changes in the key performance indicators (KPIs) with changing PV capacities. In the analysis, the main focus in on the KPI self-consumption, since this is a financial important KPI with the Norwegian tariff structure. When comparing KPIs for several smaller PV systems to a large PV-system, the smallest capacity step of 50 kW_p is chosen for the 22 individual systems, with an aggregated capacity of 1,100 kW_p, which is the capacity used for the single large system. Both the electricity loads and the simulated PV electricity generation are analysed using the statistical computing environment R [9].

Table 1. Input data and system information for the simulated PV systems, with climate data for 2018.

Location	Latitude 63.39° N, Longitude 10.44° E, Altitude 116 m					
Horizon	From GVGIS website API					
Monthly albedo values	Dec, Jan, Feb, Mar, Apr: 0.4, May, Jun, Jul, Aug, Sep, Oct, Nov: 0.2					
PV module	Si-poly, 285 W_p, 72 cells (generic), 14.78% efficiency at STC					
Inverter	12 kWac inverter (generic)					
Orientations, tilts/azimuths	**Rooftop**: 15°/-90° and 15°/90°			**Façade**: 90°/0°		
PV capacity (kW_p)	42.8	68.4	**1**	12.82	42.8	**1**
Nb. modules	150	240	-	45	150	-
Module area (m^2)	291	466	**6.8**	87.3	291	**6.8**
Nb. inverters	3	5	-	1	3	-
Inverter power (kW_{ac})	36	60	-	12	36	-
Produced electricity (MWh/year)	32.25	51.58	**0.75**	10.26	34.18	**0.80**
Specific prod. (kWh/kWp/year)	754	754	**754**	800	800	**800**

3. Results

The total average specific electricity use in 2018 was approx. 56.7 kWh/m^2, used in apartments and common areas [4]. Average specific use of district heating was approximately 139 kWh per heated floor area [5]. In this article, only the electricity use is analysed with simulated PV electricity.

3.1. Analysing electricity generation from PV with electricity use in the housing cooperative

Table 2 summarizes KPIs for analysing electricity demand at Risvollan with simulated PV electricity. Both electricity and climate data are from 2018. The results are for the housing cooperative in total, where hourly electricity load from several AMS meters are aggregated.

In the following, the results are analysed for the common areas only, followed by the total Risvollan. For the common areas it is simulated that a 50 kW_p, 100 kW_p or a 2,000 kW_p PV system on the roof could cover about 6.5%, 12.3% or 35.3% respectively, of the electricity use on an hourly basis (self-generation factor). For the façade systems, the self-generation factor for a 50 kW_p PV system is 6% higher than for the rooftop system, and 30% lower for the larger 500 kW_p system. For the 50 kW_p rooftop PV system, nearly all (99.9%) of the generated electricity can be used by the common areas (self-consumption factor). For larger rooftop systems of 500 kW_p or 2,000 kW_p, the self-generation factor is declined to 41.5% or 13.5% respectively. For the façade systems, the range of the self-consumption factor for the common areas is from 100% for the smallest PV system at 50 kW_p, down to 27.6% for the largest PV system at 500 kW_p. For both the roof and wall systems, the ratio between exported and imported peak powers is 0.1 for the 50 kW_p PV system and 0.3 for the 100 kW_p PV system (generation multiple factor). For a 500 kW_p system the generation multiple factor increases to 2.4 for a rooftop system and to 3.1 for a building façade system. For the 2,000 kW_p rooftop system the generation multiple factor is increased to 10.4.

For the total Risvollan, the three KPI factors change. It is simulated that a 500 kW_p PV system would cover about 8% of the loads and a 2,000 kW_p PV system would cover about 23.3%, on an hourly basis. For the façade systems, with maximum capacity of 500 kW_p PV, the self-generation factor is slightly higher than for the rooftop system. The self-consumption factor for a 500 kW_p PV system on both roof and the façades is about 100%. For a 2,000 kW_p PV system, the self-consumption factor is around 77% for a rooftop system, with a ratio between exported and imported peak powers of 0.8.

Table 2. KPIs for analysing electricity use with simulated PV electricity (2018- electricity/climate data).

Type of user (# el meters)	Electricity demand (MWh/y)	Max. load (kWh/h)	PV capacity (kW_p)		Simulated gen. (MWh/y)		Self-gen. (%)		Self-cons. (%)		Gen. multiple	
			Roof	Façade	Roof	Façade	Roof	Façade	Roof	Façade	Roof	Façade
Electricity common areas (hourly sum of 104 el meters)	576	129	50	50	38	40	6.5	6.9	99.9	100	0.1	0.1
			100	100	75	80	12.3	12.7	94.0	91.3	0.3	0.3
			500	500	377	400	27.2	19.1	41.5	27.6	2.4	3.1
			1,000		754		31.9		24.4		5.0	
			2,000		1,507		35.3		13.5		10.4	
Total electricity, apartments and common areas (hourly sum of 1,113 el meters)	4,977	1,126	50	50	38	40	0.8	0.8	100	100	-	-
			100	100	75	80	1.5	1.6	100	100	-	-
			500	500	377	400	7.6	8.0	100	100	-	-
			1,000		754		14.7		96.8		0.2	
			2,000		1,507		23.3		77.1		0.8	

Figure 3 shows analysis of electricity demand with simulated PV generation, using electricity and climate data from 2018. The electricity demand shown is for the common areas (figures in column 1), and in total, also including apartments (figures in column 2). The sizes of the PV systems shown are 100 or 500 kW_p for the common areas and 500, 1,000 or 2,000 kW_p for the total housing cooperative. The electricity load and PV generation on a monthly basis is shown in Figure 3 a) and b), for PV systems on the roof or façade. The figures show that the façade-placed south oriented systems generate more electricity during the swing seasons, compared to the rooftop east-west oriented systems, but have a lower electricity generation during the summer months. In Figure c) and d), hourly duration curves are shown, for net electricity load (positive values: import from grid, and negative values: export to grid). The figures show how the export increases, if the PV system is large compared to the electricity demand, giving a high generation multiple factor. Figure e) and f) shows example of hourly load and generation during a week in April, showing daily variations in electricity use and PV generation. Average daily electricity profiles is shown in Figure 3 g) and h), for load and PV generation on roofs or façades during spring (Mar, Apr, May) and summer (Jun, Jul, Aug).

3.2. Comparing KPIs for a large PV-system to several smaller PV systems

Table 3 summarizes KPIs for several smaller PV system, compared to one large PV system with the same aggregated PV capacity. AMS measurements from the garages are analysed individually, where hourly electricity generation from 22 rooftop PV systems, each with a capacity of 50 kW_p, are located behind the meter of each of the 22 garages. As a comparison, a single large rooftop PV system with a capacity of 1,100 kW_p is delivering electricity to the garages aggregated, to the common areas (including garages), or to the total Risvollan housing cooperative (including common areas and apartments). Figure 4 shows hourly duration curves for net electricity from or to the grid, comparing the 22 rooftop 50 kW_p PV systems with a single rooftop 1,200 kW_p PV system.

Due to the tariff structure in Norway, it is normally financially beneficial to maximise self-consumption, i.e. minimising export of electricity to the grid. Table 4 compares the financial values of four system solutions; 1) 22 PV systems of 50 kW_p, providing electricity to 22 garages only, 2) 1 PV system of 1,100 kW_p providing electricity to 22 garages only, 3) 1 PV system of 1,100 kW_p providing electricity to all common areas (incl. garages) and 4) 1 PV system of 1,100 kW_p providing electricity to all of Risvollan (incl. apartments and common areas). The tariff estimations are based on [10] and [11], and is 1 NOK/kWh for self-consumed PV-electricity, which is the estimated end-user cost for electricity

IOP Conf. Series: Earth and Environmental Science **352** (2019) 012008 doi:10.1088/1755-1315/352/1/012008

from the grid, and 0.5 NOK/kWh for exported PV-electricity. Only electricity costs are considered, assuming that the choice of PV system solution would not change the investment costs.

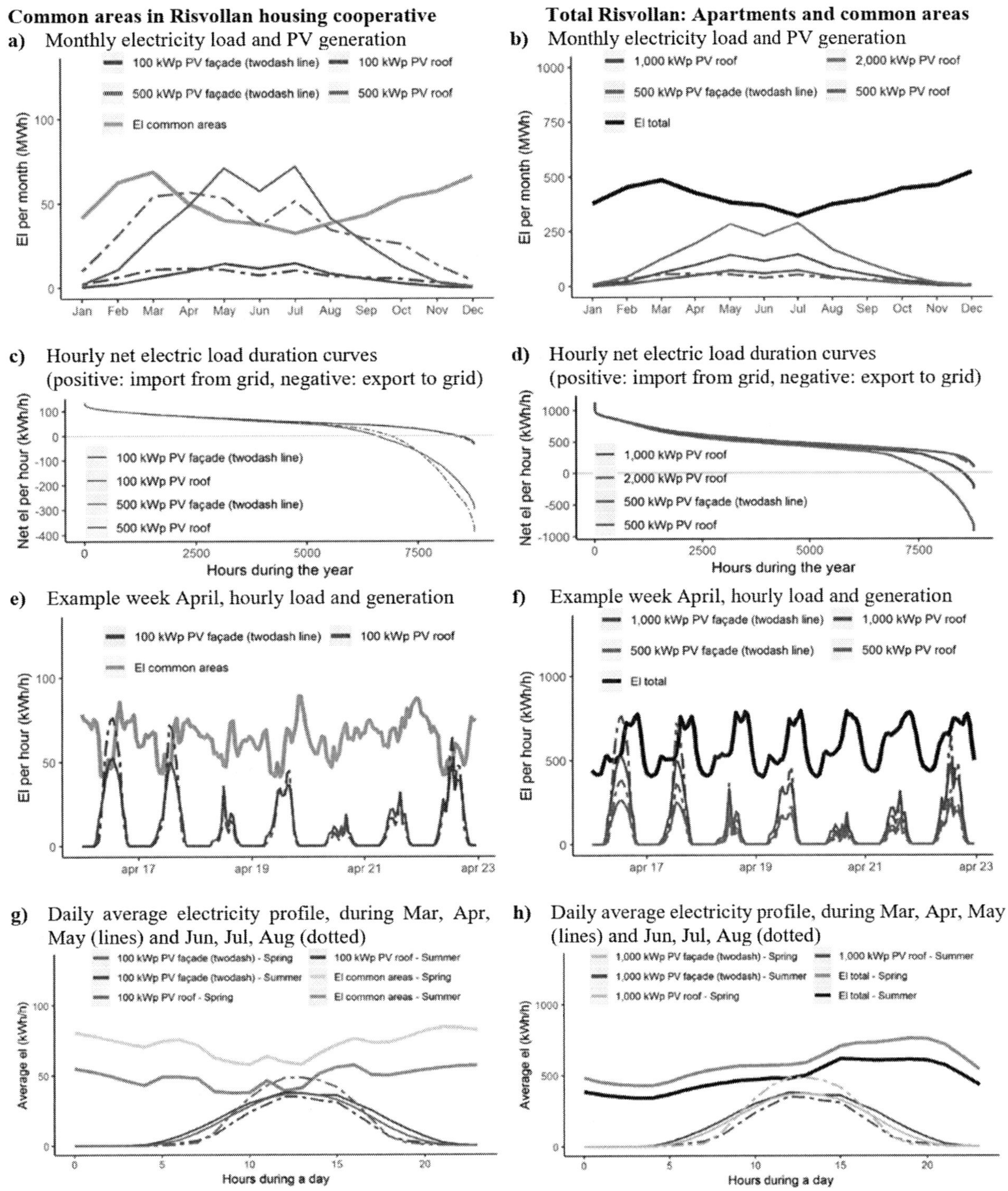

Figure 3. Analysing electricity use in Risvollan housing cooperative with simulated PV generation.

Table 3. KPIs for analysing electricity use with simulated PV electricity (2018- electricity/climate data). All the PV systems are east-west oriented rooftop systems with a total capacity of 1,100 kW_p.

KPIs for 22 individual PV systems, located behind 22 garage meters:

	Electricity (MWh/y)	Max. load (kWh/h)	PV capacity (kW_p)	Simulated gen. (MWh/y)	Self-gen. (%)	Self-cons. (%)	Gen. multiple
El garages (22 el meters)	Per garage: Max: 56 Mean: 17 Min: 5	Per garage: Max: 34 Mean: 7 Min: 3	22·50 Tot: 1,100	22·38 Tot: 829	Per garage: Max: 46.4 Mean: 34.9 Min: 17.1	Per garage: Max: 31.8 Mean: 14.3 Min: 3.2	Per garage: Max: 11.0 Mean: 6.0 Min: 1.0

KPIs for one large PV system, with aggregated load:

	Electricity (MWh/y)	Max. load (kWh/h)	PV capacity (kW_p)	Simulated gen. (MWh/y)	Self-gen. (%)	Self-cons. (%)	Gen. multiple
El garages	363	91	1,100	829	35.0	15.3	8.1
El common areas	576	129	1,100	829	32.5	22.6	5.6
El total Risvollan	4,977	1,126	1,100	829	15.8	95.0	0.3

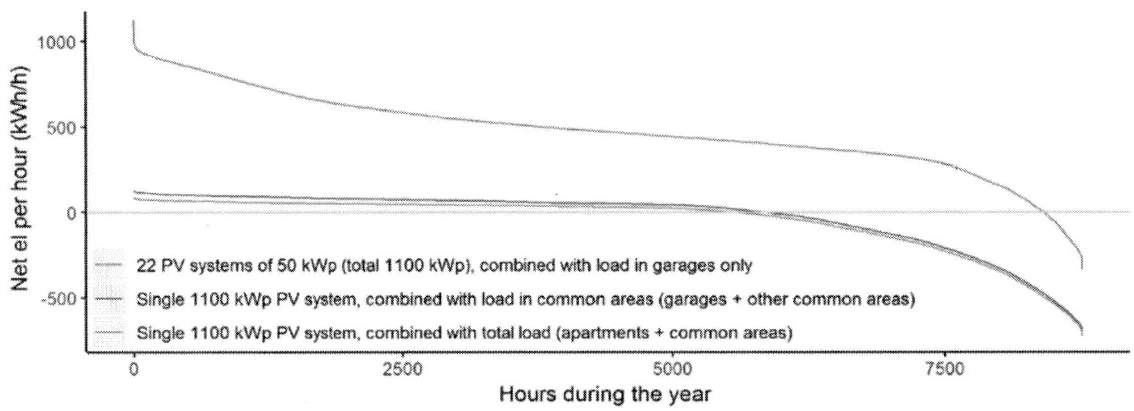

Figure 4. Hourly duration curves for net electricity from/to grid, with 22 rooftop 50 kW_p PV systems or a single rooftop 1,100 kW_p PV system. PV generation from the 22 PV systems cover electricity load in 22 garages. PV generation from the single large PV system cover aggregated electricity loads in all common areas or in the total housing cooperative, including apartments and common areas.

Table 4. Estimation of financial value of PV electricity, comparing four cases with different electricity use, each with a total capacity of 1,100 kW_p rooftop PV system.

Case	Simulated gen.	Self-cons. (%)	Value self-use	Value export	Total annual value
22 PV systems a 50 kW_p, providing electricity to 22 garages only	22·38, Tot: 829 MWh/y	Per garage: Mean: 14.3	121 MWh: 121,000 NOK	708 MWh: 354,000 NOK	475,000 NOK
1 PV system a 1,100 kW_p, providing electricity to 22 garages aggregated	829 MWh/y	15.3	127 MWh: 127,000 NOK	702 MWh: 351,000 NOK	478,000 NOK
1 PV system a 1,100 kW_p, providing electricity to all common areas	829 MWh/y	22.6	187 MWh: 187,000 NOK	642 MWh: 321,000 NOK	508,000 NOK
1 PV system a 1,100 kW_p, providing electricity to all of Risvollan (incl. apartments and common areas)	829 MWh/y	95.0	788 MWh: 788,000 NOK	41 MWh: 20,000 NOK	808,000 NOK

4. Discussion

As a basis for energy management in apartment blocks, this article analyses the electricity demand at Risvollan together with simulated electricity generation from several different PV systems. The total electricity use included in the analysis is 4,977 MWh, where 12% is used in common areas and 88% in 1.009 apartments. It is estimated that the total electricity delivery to Risvollan in 2018 was 5,318 MWh [4], meaning that 6% of the annual electricity use is excluded or missing from this analysis. If all measurements had been available, the KPI factors would have changed slightly. Electricity generation from PV is simulated based on climate data from 2018. The climate data is therefore not general, and

2018 was a year with higher temperatures and less rain than normally [12]. In addition, no shadings were assumed in the simulations, which is rather optimistic. The simulated PV generation may therefore be higher than can be expected in real life.

Comparing the simulated electricity generation on roofs and walls, the façade systems generate about 5-6% more electricity than the rooftop systems on an annual basis. The systems on the façades generate more electricity in the spring, autumn and winter, because of the steeper PV array angle [13]. A south-oriented 100 kW_p system on the façade generates 11% of its electricity during winter, 41% in the spring, 31% in the summer and 17% in the autumn, while the seasonal division for the east-west oriented rooftop system is 4%, 40%, 45% and 11% accordingly. This is positive for the self-generation and self-consumption factors, since PV electricity can cover electricity demand in the swing seasons. However, the wall-placed system is orientated towards south, which gives a higher midday peak than the rooftop east-west oriented systems. More PV power is available during morning or afternoon hours for east west oriented systems [13]. A higher electricity generation early and late in the day is positive for the matching with electricity use. The systems with highest self-generation factor therefore varies: For the smaller PV capacities it is the rooftop systems and for the larger PV capacities it is the façade systems. For the self-consumption factors, the rooftop systems have the highest values. More electricity is exported with the façade systems, giving a higher generation multiple factor. In general, the performance of the façade systems is somehow better than the rooftop systems. However, the area available for façade systems is limited, and it may be advisable with a combination of the two orientations.

The KPIs are calculated based on hourly values. If calculating the factors based on 15 minutes, daily or monthly intervals, the results would differ. For example, for the 500 kW_p PV system providing electricity to all common areas, the self-consumption factor of 27.2% based on hourly measurements, increases to 45% or 48% if calculated based on daily or monthly values. The self-generation factor is 41.5% based on hourly values and increases to 69% or 74% based on daily or monthly values. The generation multiple factor of 2.4 based on hourly values, decreases to 1.0 or 0.6 based on daily or monthly values. It is expected that the self-consumption and self-generation factors would be somewhat lower, using 15 minutes instead of hourly measurements. For load and generation power flows, shorter time steps give more realistic values for how a real system works. However, when estimating financial values, it is usually more relevant to use time steps from the tariff structure.

The self-consumption factor is the most important KPI in Norway, due to the Norwegian tariff structure. When the self-consumption factor is close to 100%, the self-generation factor is around 10% and the generation multiple factor is close to zero, since very little electricity is exported. From the perspective of the housing cooperative, it is therefore beneficial to locate the PV system and the load behind the same AMS meter, or to aggregate electricity loads in the common areas and the apartments. To aggregate electricity load from several AMS meters is currently not possible in Norway, but the authorities plan to facilitate also for housing cooperatives with PV in the prosumer agreement [14]. In principle, all electricity loads in common areas can be behind one AMS meter, but at Risvollan there are 104 such meters, 22 in garages and 82 in other common areas. The annual financial value of the 22 single PV systems of 50 kW_p, providing electricity to 22 garages only was at 475,000 NOK, whereas a single 1,100 kW_p PV system, providing electricity to all common areas (incl. garages), the total value increases with 6%. For the housing cooperative, the best financial option would be to provide PV electricity both to apartments and common areas, increasing the financial value with 70%, to approximately 808,000 NOK per year.

Energy flexibility of Risvollan housing cooperative will be a topic in the further work. According to Annex 67 [15], Energy Flexible Building Clusters should *demonstrate the capacity to react to forcing factors in order to minimize CO_2 emissions and maximize the use of Renewable Energy Sources (RES)*. Electricity loads can be adapted to PV generation, by increasing use of electricity during sunny periods or by storing electricity in heat storages or EVs [13]. EV charging is a main source of flexible electricity use in Norwegian apartment buildings. Besides often being flexible in starting time, duration and charging power [16], EV charging infrastructure is the responsibility of the Risvollan cooperative. A neighbourhood battery could also increase the self-consumption of PV generated electricity.

5. Conclusion

This article analyses the electricity use at Risvollan housing cooperative together with possible electricity generation from PV systems. Risvollan is a large housing cooperative in Norway, built in the 1970s, with in total 1,058 apartments. The study shows that the electricity generation from south-oriented systems on the building façades are about 5-6% higher than for east-west oriented rooftop systems on an annual basis, since the façade systems generate more electricity in the spring and autumn. However, more PV power is available during morning and afternoon hours for the rooftop east-west oriented systems. A combination of PV systems on the roofs and façades seem advisable. The self-consumption factor is the most important KPI in Norway, due to the national tariff structure. For the total housing cooperative, a PV capacity of about 1,000 kW_p seem suitable, giving a self-consumption factor of 97% for a rooftop system, based on 2018 electricity and climate data. From the perspective of the housing cooperative, it is financial beneficial to aggregate electricity loads for common areas and apartments, since a higher share of the electricity can be used by the cooperative. For this to be possible, also housing cooperatives with PV must be facilitated for in the prosumer agreement. Comparing a single 1,100 kW_p PV system providing electricity to the total cooperative with 22 PV systems of 50 kW_p behind 22 garage meters, the self-consumption factor decreases from 95% to average 14%, resulting in a 41% lower financial value for the PV electricity. The analysis will be used in further work, together with analysis of electricity and heat load patterns at Risvollan, aiming to play a role answering how effective management of power and energy at neighbourhood level can be realized.

Acknowledgements

The work presented in this paper was developed within the Institute PhD grant (701703) at SINTEF, financed by the Research Council of Norway. Contributions from Risvollan housing cooperative, NTE Marked, Enoco, Fosen Innovasjon and TrønderEnergi Nett are highly appreciated. The study is part of the Research Centre on Zero Emission Neighbourhoods in Smart Cities (FME ZEN). The authors gratefully acknowledge the support from the ZEN partners and the Research Council of Norway.

References

[1] Jensen S Ø O, Marszal-Pomianowska A, Lollini R, Pasut W, Knotzer A, Engelmann P, Stafford A and Reynders G 2017 IEA EBC annex 67 energy flexible buildings *Energy Build.* **155** 25–34
[2] NVE 2016 Plusskunder
[3] National registry in Norway 2019 Residents in Risvollan (dataset received on request)
[4] Sørensen Å L, Lindberg K B, Sartori I and Andresen I 2019 Electricity analysis for energy management in neighbourhoods: Case study of a large housing cooperative in Norway *CISBAT 2019 (unpublished)*
[5] Sørensen Å L, Lindberg K B, Walnum H T, Sartori I, Aakenes U R and Andresen I 2019 Heat analysis for energy management in neighbourhoods: Case study of a large housing cooperative in Norway *IAQVEC 2019 (unpublished)*
[6] PVsyst SA 2019 PVsyst 6.7.9
[7] Norwegian Meteorological Institute 2019 eKlima: Weather- and climate data in Norway
[8] Salom J, Marszal A J, Widen J, Candanedo J and Lindberg K B 2014 Analysis of load match and grid interaction indicators in net zero energy buildings with simulated and monitored data *Appl. Energy* **136** 119–31
[9] The R Foundation for Statistical Computing Platform 2018 R version 3.5.1
[10] Multiconsult og Asplan Viak 2018 Solcellesystemer og sol i systemet
[11] Anon 09387: Kraftpris, nettleie og avgifter for husholdninger, etter statistikkvariabel og kvartal
[12] Skaland R G, Colleuille H, Solveig A, Andersen H, Mamen J, Grinde L, Therese H, Tajet T, Lundstad E, Sidselrud L F and Tunheim K 2019 *Tørkesommeren 2018*
[13] Krauter S 2018 Simple and effective methods to match photovoltaic power generation to the grid load profile for a PV based energy system *Sol. Energy* **159** 768–76
[14] Anon Mulig å bli plusskunde i boligselskap - NVE
[15] Vigna I, Pernetti R, Pasut W and Lollini R 2018 New domain for promoting energy efficiency: Energy Flexible Building Cluster *Sustain. Cities Soc.* **38** 526–33
[16] Knezovic K 2016 Active integration of electric vehicles in the distribution network - theory, modelling and practice *Ph.D. Thesis, DTU*

IOP Conf. Series: Earth and Environmental Science **352** (2019) 012009 doi:10.1088/1755-1315/352/1/012009

LCA of the Zero Emission Neighbourhood Ydalir

K M Lund[1], C Lausselet[1]*, H Brattebø[1]

[1]Norwegian University of Science and Technology, NTNU, Trondheim, Norway

*carine.lausselet@ntnu.no

Abstract. Buildings represent a critical piece of a low-carbon future and their long lifetime necessitates urgent adoption of state-of-the-art performance standards. So far, LCA studies have assessed buildings, mobility and energy systems mainly individually. Zero Emission Neighbourhoods (ZEN) give a unique chance to combine these elements. In Norway, the Research Centre on ZEN has as a goal to enable the transition to a low carbon society by developing sustainable ZENs.

In this study, a LCA model for neighbourhood based on a modular structure with five physical elements; buildings, mobility, infrastructure, networks and on-site energy was applied on Ydalir, a pilot project of the ZEN Centre. Revealing that regardless of which scenario considered, the ZEN Ydalir does not achieve their ambitious goal of zero emissions. Further, the results show that the operation of mobility is a major source of the total greenhouse gas (GHG) emissions, accounting for 21-46%. Considering the life cycle stage materials, buildings are the largest contributor representing 24% of all GHG emissions. Thus, these two areas have been highlighted as the best options for improvement. Parameters related to uncertainties or are large contributors to the environmental impact are included in a sensitivity analysis.

1. Introduction

In Paris in December 2015, the United Nations Climate Change conference was held. Here the main goal of limiting the global warming to a maximum of 2 degrees compared to pre-industrial time, was defined. This has led to a growth of climate awareness and as the building sector is responsible for 40% of the total energy consumption and 30% of all energy-related greenhouse gas (GHG) emissions in the European Union (EU) [1], decreasing the emissions from this sector is critical. To improve the environmental aspect of the building sector, several leading international organizations have taken measures. In Norway, the Research Centre on Zero Emission Neighbourhoods (ZEN) has been founded with a goal of reducing the emissions on neighbourhoods to a minimum level through combining local production of local renewable energy, storage and interacting systems. The ZEN Centre will run over the period 2017-24 with the vision *"Sustainable neighbourhoods with zero greenhouse emissions"* [2]. By using life cycle analysis (LCA) as a helping tool, researchers have consequently had to acknowledge the new challenges that arise regarding functional unit, system boundaries and sensitivity parameters.

To map and assess the source of the emissions from buildings the well-established tool LCA is commonly used, as it looks at the entire life span of the building [3]. LCA systematically goes through each life cycle from raw materials acquisition, production of energy and materials, usage and end-of-life processing [4]. LCA is commonly used in assessing both buildings and neighbourhoods. However, when performing LCAs on an urban scale, the focus becomes quite different from when assessing individual buildings. Other aspects as density, transportation, infrastructure and consumption have to be included and different system boundaries and framework are needed to be able to assess critical environmental impacts of neighbourhoods [5].

Defining the system boundaries is crucial for the LCA's results and their reliability. The system boundaries define which life cycle stages and physical elements (i.e. buildings, open spaces and mobility) that are to be included in the analysis. Some LCA studies only include the buildings [6], while others also consider the mobility of the inhabitants [7, 8]. The most comprehensive and complex LCA studies are the ones that include several elements such as building, mobility and other elements as networks and infrastructure [9, 10]. There are also variations in what life cycle stages are considered, from studies that only look at the usage to the opposite side of the scale where construction and deconstruction are included [11, 12]. These differences create challenges in comparing results from LCA studies. However, some key points are worth noting.

Studies have shown that transportation of the inhabitants has significant impact on the total emissions. Nichols and Kockelman [9] found that 44-47% of the total emissions from the use stage came from transportation. Similarly, when including materials in construction, usage stage and transportation, Bastos, Batterman [8] found that transportation contributed with 51-57% of the total emissions. These findings indicate that further research on the field of emissions related to mobility in neighbourhoods is necessary.

The predictions and assumptions of the future scenarios are crucial when performing LCA. The service lifetime of each element in a neighbourhood makes the forecasting challenging and a source of uncertainties. Several studies highlight this challenge, and emphasize evolving technology, time distribution of environmental impact and emission intensity of electricity as key factors [8, 10, 12]. These factors can have great impact on the predictions of future scenarios and long-term decisions.

Further research on the field on ZENs is obviously required, on both critical factors for the results and what life cycle stages and physical elements contribute considerably to the environmental impact categories. This insight should build the foundation in future ZEN projects.

2. Method

This study is based on the LCA model developed by Lausselet, Borgnes [13] with some minor modifications, and further adapted and applied to the case study at hand, the ZEN Ydalir. Ydalir is a pilot project for the ZEN Centre, located in Elverum.

2.1. Model

The LCA study is designed as a modular structure defined by three dimensions; the physical elements (buildings, mobility, infrastructure, networks and on-site energy), the life cycle stages and time. To match the neighbourhood at interest, the elements and life cycle stages can be adjusted for each assessment.

The total emissions from the neighbourhood is calculated by Equation 1.

$$E_{tot} = E_{b,mat} + E_{b,oper} + E_{m,mat} + E_{m,oper} + E_{o,mat} + E_{o,oper} \quad (1)$$

Where $E_{b,mat}$, $E_{m,mat}$ and $E_{o,mat}$ are the emissions from the materials from buildings, mobility and infrastructure, respectively, and $E_{b,oper}$, $E_{m,oper}$ and $E_{o,oper}$ are the emissions from the energy use in operation respectively from buildings mobility and infrastructure.

The Norwegian standard on method for greenhouse gas calculations in buildings suggests that two different energy intensity scenarios are to be assessed, namely scenario 1 (NO) and scenario 2

(EU28+NO). These are based on the assumed evolution of the Norwegian and European electricity mix, respectively.

2.2. Case Study, Ydalir

The LCA model has been applied on the ZEN Ydalir, including the elements building, mobility, infrastructure, networks and on-site energy. For each element the life cycle stages production stage (A1-A3), replacement (B4) and energy use in operation (B6) are included. In addition, construction (A5) is included for infrastructure, while for networks energy use in operation (B6) is excluded. The analysis period is 60 years, and this study focuses on the GHG emissions associated to each element over this period. At Ydalir three different sources of energy have been chosen; district heating, combined heat and power (CHP) machines and photovoltaic (PV) panels. Both electric and heat excess energy not used by the neighbourhood is exported to the external grid and credited with negative emissions.

2.2.1. Buildings

The building stock at Ydalir consists of 1000 residential buildings and two non-residential buildings; a school and a kindergarten. Resulting in a total area of 108 614 m^2. The residential buildings are still under planning and the ZEB 1 from a concept analysis by Kristjansdottir, Houlihan-Wiberg [14] has been chosen for its resemblance with the present design of the residentials. It is also assumed that the residential buildings will have the same amount of emissions per area as for the ZEB 1, while the materials for the school and the kindergarten are collected from EDPs. Further, the annual energy use in operation are 78 kWh/m^2 for the residentials and 109 kWh/m^2 for the non-residential buildings. These numbers are based on the passive house standard NS 3700 [15] and NS 3701 [16] for the residential and non-residential buildings respectively. The total thermal load for the buildings at Ydalir is 4.81 GWh/year and the electrical load is 3.93 GWh/year. The loads are assumed to stay constant over the building's lifetime.

2.2.2. Mobility

At Ydalir three means of transportation have been assessed; personal vehicle, bus and light rail. The travel habits of the inhabitants is based on the National travel survey [17] and further adapted to specific measures taken at Ydalir regarding parking space availability. Car-sharing has also been included as a scenario where the total travel distance has been halved.

The evolution of the vehicle stocks is adapted to two different scenarios; a trend path and a ultra-low emissions path [18]. Both scenarios assess the evolutions of the personal vehicle and bus stock, looking at several fuel/energy carriers. The forecast lasts over the period from 2010 to 2050 and is then expected to stay constant until 2080.

Considering the embodied emissions from each mean of transportation and the emissions intensity of the fuel/energy carrier, the initial emissions have been collected from Vestlandsforskning [19] after suggestions from the Norwegian standard NS 3720 [20]. Future evolution has then been adopted from a scenario analysis by Lausselet, Ellingsen [21].

2.2.3. Infrastructure

The infrastructure includes embodied emissions from roads and sidewalks, energy use in operation from public lighting and the diesel consumption from construction. The embodied emissions are collected from Arda, while the operation of the public lighting has been calculated from an average of dark hours per day.

2.2.4. On-site Energy Production District Heating

The district heating system will cover the total heating demand for the buildings at Ydalir. The heat production is off-site, but still within the system boundaries. The embodied emissions related to the pipes are included, but not the emissions of the heat production plant. There will be some excess heat that will be exported from the neighbourhood to the external grid and contribute as negative emissions.

2.2.5. *On-site Energy Production Photovoltaic*

The PV-panels will be mounted to roofs, facades and in the surrounding area. There will be 18m^2 of PV- per residential building. The embodied emissions have been collected from Ecoinvent 3.2. The panels produce an average of 130 kWh/m^2, which result in an annual total of 2.34 GWh. The emissions associated to the operation are assumed to have a symmetric weighting to the emission intensity of electricity respectively for each scenario. However, as the electricity produced from the PV-panels are assumed to replace the electricity from the scenarios, emissions are seen as negative contributors (i.e avoided emissions). The electricity is either consumed by the neighbourhood itself or exported to the external grid.

2.3. *Scenarios*

Four scenarios on mobility have been created in order to analyse the impact mobility have on the total emissions from the neighbourhood. Scenario A represents the base case where the travel distance is for Ydalir and the technology development path for mobility follows the trend path. Scenario B is also for travel distance Ydalir, but the technology follows the ultra-low emission path. Scenario C and D illustrate the results for Ydalir and car-sharing for the trend path and ultra-low emission path respectively.

2.4. *Sensitivity analysis*

A sensitivity analysis was carried out with the goal of revealing the critical parameters in the LCA model. The factors expected to have significant impact to the results or associated with large uncertainties were chosen and increased with 25%. The parameters analysed were mobility energy use in operation, area of PV panels, energy load (thermal and electric), emissions embodied in building materials, emissions associated with vehicle production, travel distance/inhabitant/year, emission intensity district heat and emission intensity electricity. Further, the sensitivity ratio was calculated using Equation 2.

$$SR = \frac{\Delta R/R_0}{\Delta P/P_0} \qquad (2)$$

$\Delta R/R_0$ represent the relative change in the results while $\Delta P/P_0$ is the relative change in the input parameters.

3. Results

3.1. *General Results*

The results from using the described methodology with the emissions associated to the included physical elements (buildings, mobility, infrastructure, networks and on-site energy) and the life cycle stages (A1-3, B4 and B6) were 162 ktonne CO_2-eq over the lifetime of 60 years (scenario A). Equivalent to 1.1 tonne CO_2-eq/capita/year or 24.8 kg CO_2-eq/m^2/year. Figure 1 illustrates the different results depending on the different mobility scenarios. As shown in Figure 3.1, scenario D is the most optimistic scenario regarding mobility.

IOP Conf. Series: Earth and Environmental Science **352** (2019) 012009 doi:10.1088/1755-1315/352/1/012009

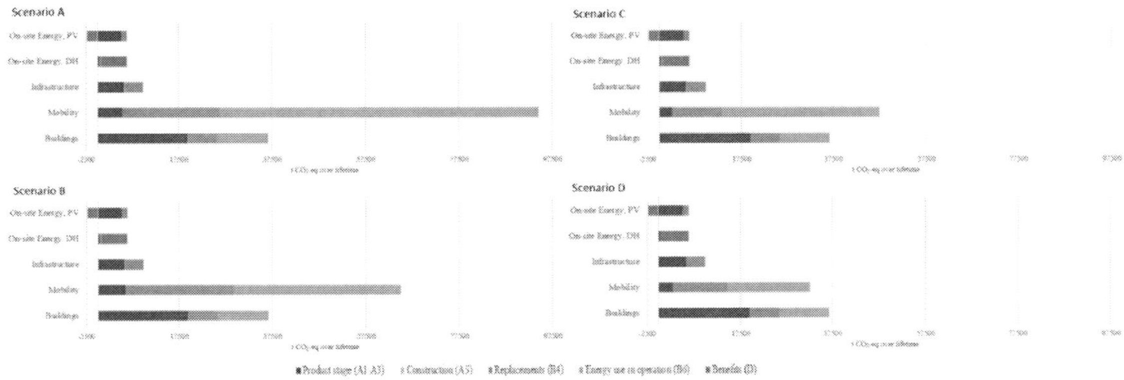

Figure 3.1 Results of total emissions over lifetime, A-Ydalir & trend path, B-Ydalir & ultra-low emission path, C-Ydalir + car-sharing & trend path D-Ydalir + car-sharing & ultra-low emission path.

The results for the scenario A (base case), show that when all emissions considered, the energy use in operation stage contribute with 54% of the total emissions, and 46% of these are from the element mobility, making energy use in operation of mobility the greatest source of emissions. However, when looking at scenario D, it is the product stage of buildings that becomes the greatest source of emissions, representing 22% of the total emissions. By comparing scenario B and C, it can be concluded that it is the daily travels of the residents that have the greatest impact on the total emissions.

3.2. Sensitivity Analysis.

Figure 3.2 shows the results from the sensitivity analysis and the impact from four fundamental assumptions. It is revealed that the two most critical parameters are connected to mobility, namely the mobility energy use in operation and travel distance/habitant/year. The assumptions related to mobility, the technology development path and travel distances, are also shown to have major impact on the results, by both increasing and decreasing the total emissions from the neighbourhood. The assumption regarding the emission intensity for scenario 2 is also worth noticing as this increase the total emissions from the neighbourhood by 19.4%.

IOP Conf. Series: Earth and Environmental Science **352** (2019) 012009 doi:10.1088/1755-1315/352/1/012009

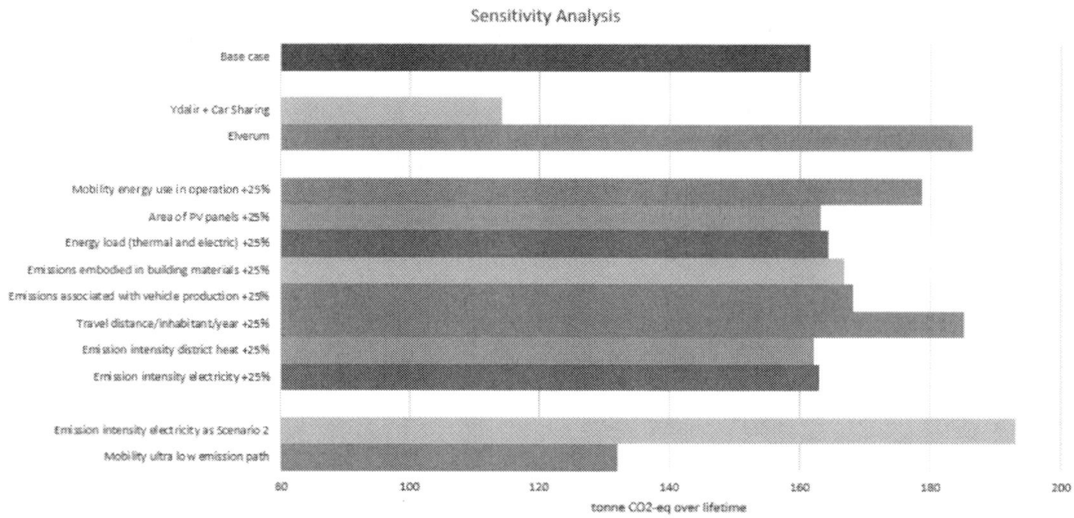

Figure 3.2 Results of the sensitivity analysis and critical parameters relatively to the base case. (Note that the axis does not start at zero)

4. Discussion

4.1. LCA Modelling on Neighbourhood Scale

The placement of the system boundaries to decide which life cycle stages and physical elements to include in an LCA, appears to have significant impacts on the results. The model revealed that mobility contributes with 38-64% of the emissions depending on the scenario and its preconditions. This states the major impact mobility have on the resulting emissions., similarly to results of Nichols and Kockelman [9] and Bastos, Batterman [8]. The variations are the results of optimistic or pessimistic assumptions of the future evolutions of mobility. In the most optimistic assumptions (scenario D), the buildings become the largest contributor with 43% of the total emissions.

4.2. Uncertainties and Limitations

In association with the emissions from Ydalir, several assumptions on the emission intensities have been made in order to calculate the total emissions over the lifetime. When expanding LCA models from individual building to complex neighbourhoods, the impacts of the assumptions, and which life cycle stages and physical elements to include, become more uncertain.

Regarding buildings, it is assumed that the buildings in the neighbourhood will be built over a period of 15-20 years. However, in this model it is assumed that all the buildings will be built immediately with no considerations towards evolution of technology, change in the Norwegian building regulations (TEK) or an establishment of a ZEN definition. The emissions associated to the materials used in the residential buildings are also related to some uncertainties as these are not based on the actual buildings that will be built at Ydalir, but on ZEB 1.

The results clearly state that mobility has the most significant impact on the total emissions, and it is the energy use in operation and the daily travels of the inhabitants that affects the results. There are high uncertainties connected to predicting the evolution of the vehicle stocks. The model only assumes changes until 2050 and then constant values. This is an unrealistic prediction, but as further assumptions are associated to high uncertainties, it is chosen not to be included. Another important notion is that mobility connected to the users of the school and the kindergarten that do not live at Ydalir, is not considered.

The use of PV to locally produce electricity can be questioned for the particular case of Norway. Yet, while the efficiency is lower than in Southern European countries, the use of PV at those latitude is a proven concept [22]. Also, as long as the carbon intensity of the European electricity mix is higher than electricity produced by PV placed in Norway, to produce electricity from PV in Norway will help decarbonize the European grid. It will as well liberate and redirect low carbon hydropower to other purposes such as for example as to help electrifying the mobility sector.

There are also some uncertainties related to the included infrastructure. The embodied emissions from the materials have been highly simplified and higher emissions from this element is therefore expected. Regarding the operation of the infrastructure, only public lighting has been included. Road maintenance, snow clearance and other operational elements have been excluded, which is a limitation to the model.

4.3. Further Work

The modular structure of this model has the advantages to highlight the source of emissions related to both life cycle stages and physical elements. However, there are still limitations that weaken the model and some parameters that need further attention.

Regarding the materials replacement in buildings, infrastructure and networks, the model does not consider any improvements in the technology development. As the emissions from materials replacement represent 22-27% of the total emissions, when all elements are included, this should be looked closer at. Another area that needs further research is the behaviour of the inhabitants. The travel habits and energy use of the inhabitants are based on national numbers and is a source of uncertainties.

An LCA often include several impact categories to show a comprehensive picture of the product or process at hand. In this study, only greenhouse gas equivalents related to climate change has been analysed and discussed. In order to avoid problem shifting phenomena, other impact categories as land use change, acidification and human toxicity should also be included into the model. However, the lack of data on other impact categories bring challenges and further work on this field is therefore suggested.

5. Conclusion

In order to research the dominant drivers both related to the physical elements and the life cycle stages in the ZEN Ydalir, a modular structure was chosen for the LCA. The modular structure consists of five physical elements; buildings, mobility, infrastructure, networks and on-site energy, as well as three life cycle stages; product stage, replacement stage and the energy use in operation stage.

The results revealed that regardless of which scenario considered, the ZEN Ydalir does not manage to achieve their ambitious goal of zero emissions with the present plan. Nevertheless, the results highlight some important take-away messages on areas for further improvement on the field of ZEN. The results show the major impact the energy use in operation of mobility has on the total emissions from the neighbourhood, contributing with 21-46% of the total emissions, depending on the scenario. Looking at scenario A, the scenario assumed to be the most realistic regarding the evolution of the vehicle stock, 24% of the emissions comes from the product stage, and 22% from the replacements. On the other hand, for scenario D, the most optimistic scenario regarding the vehicle stock, the product stage and replacement stage contribute with 39% and 27% of the emissions. As the operation of mobility and materials from buildings are the two major contributors, these two areas are highlighted as options for improvement on the way to the goal of zero emission societies.

References

1. UNEP-SBCI, *United Nations Environment Programme Sustainable Buildings and Climate Initiative*. 2017.
2. The Research Centre on Zero Emission Neighborhoods, *The Research Centre on Zero Emission Neighborhoods*. p. https://fmezen.no.
3. Rossi, B., et al., *Life-cycle assessment of residential buildings in three different European locations, basic tool.* Building and Environment, 2012. **51**: p. 395-401.
4. ISO, *ISO 14040:2006 Environmental management - Life cycle assessment - Priciples and framework*. 2006.
5. Anderson, J.E., G. Wulfhorst, and W. Lang, *Energy analysis of the built environment - A review and outlook.* Renewable and Sustainable Energy Reviews, 2015. **44**: p. 149-158.
6. Davila, C.C. and C. Reinhart. *Urban energy lifecycle: An analytical framework to evaluate the embodied energy use of urban developments*. in *Proceedings of BS 2013: 13th Conference of the International Building Performance Simulation Association*. 2013.
7. Anderson, J.E., G. Wulfhorst, and W. Lang, *Expanding the use of life-cycle assessment to capture induced impacts in the built environment.* Building and Environment, 2015. **94**: p. 403-416.
8. Bastos, J., S.A. Batterman, and F. Freire, *Significance of mobility in the life-cycle assessment of buildings.* Building Research and Information, 2016. **44**(4): p. 376-393.
9. Nichols, B.G. and K.M. Kockelman, *Life-cycle energy implications of different residential settings: Recognizing buildings, travel, and public infrastructure.* Energy Policy, 2014. **68**: p. 232-242.
10. Stephan, A., R.H. Crawford, and K. de Myttenaere, *Multi-scale life cycle energy analysis of a low-density suburban neighbourhood in Melbourne, Australia.* Building and Environment, 2013. **68**: p. 35-49.
11. Mastrucci, A., et al., *Life Cycle Assessment of building stocks from urban to transnational scales: A review.* Renewable and Sustainable Energy Reviews, 2017. **74**: p. 316-332.
12. Lotteau, M., et al., *Critical review of life cycle assessment (LCA) for the built environment at the neighborhood scale.* Building and Environment, 2015. **93**: p. 165-178.
13. Lausselet, C., V. Borgnes, and H. Brattebø, *LCA modelling for Zero Emission Neighbourhoods in early stage planning.* Building and Environment, 2019. **149**: p. 379-389.
14. Kristjansdottir, T.F., et al., *Is a net life cycle balance for energy and materials achievable for a zero emission single-family building in Norway?* Energy and Buildings, 2018. **168**: p. 457-469.
15. NS, *NS 3700:2013, Criteria for passive houses and low energy buildings, Residential buildings*. 2013.
16. NS, *NS 3701:2012, Criteria for passive houses and low energy buildings, Non-residential buildings*. 2012.
17. Hjorthol, R., Ø. Engebretsen, and T.P. Uteng, *The Norwegian National Travel Survey 2013/14*. 2014, Institute of Transport Economics: Oslo.
18. Fridstrøm, L. and V. Østli, *Vehicle fleet forecast based on stock-flow modelling*. 2016, Institute of Transport Economics: Oslo.
19. Vestlandsforskning, *Transport, energi og miljø*. 2017. p. http://transport.vestforsk.no/.
20. NS, *NS 3720:2018 Method for greenhouse gas calculations for buildings*. 2018.
21. Lausselet, C., et al., *A novel LCA model for the zero emission neighbourhood concept.* 2019.
22. Kristjansdottir, T.F., et al., *Embodied greenhouse gas emissions from PV systems in Norwegian residential Zero Emission Pilot Buildings.* Solar Energy, 2016. **133**: p. 155-171.

IOP Conf. Series: Earth and Environmental Science **352** (2019) 012010 doi:10.1088/1755-1315/352/1/012010

Energy and indoor climate measurements in Denmark's first energy neutral block of flats

Kim B. Wittchen

Danish Building Research Institute,
Aalborg University
A.C. Meyers Vænge 15
DK-2450 Copenhagen SV
Denmark

Søren Østergaard Jensen

Danish Technological Institute
Gregersensvej 1
DK-2630 Taastrup
Denmark

kbw@sbi.aau.dk

Abstract. Denmark's first energy neutral multi-family house, BOLIG+, is constructed at Søborg near Copenhagen and consists of 10 flats divided on 4 floors. BOLIG+ was the first attempt to construct a multi-family block of flats that complies with a set of criteria and dogmas that defines the BOLIG+ concept. Energy production on the building needed e.g. to meet not only the buildings energy demand, but also the residents' electricity energy use for light and appliances. Additionally when exchanging energy with the grid, energy "sold" should have at least the same exergy level as energy "bought". Building multi-family houses that are energy neutral on annual basis, is, therefore, much more difficult than for a single-family house. One of the reasons for this is the small amount of roof area available (per. dwelling) for local electricity production compared to the roof area of a single-family house. Last, but not least, the buildings was constructed under normal economical market conditions, i.e. without any subsidies for the construction.

When designing the building, the normal standard conditions for calculating the buildings energy performance was offset in order to ensure better coherence between the design energy demand and the following energy measurements. The indoor temperature was set higher, and so was the domestic hot water demand. Additionally, free loads (gains) from persons, light and appliances was set lower than the standard.

To meet the goal of being energy neutral, several low-energy solutions are used e.g. compact thermal envelope; highly insulated constructions; highly insulating windows; hybrid decentral ventilation; heat recovery on grey wastewater; PV on facades and roof; buffer zones. Additionally, an electric battery was installed to improve the economy of the PV installation by optimizing the amount of electricity used inside the meter (tackling the feed-in tariff issue).

This paper presents results from 1½ year of measurements of energy performance and indoor climate carried out in the building during the period 2017-2018.

Measurements of energy consumption showed that if the PV installation has produced as expected and the residents had used electricity as expected, the building would have been only 11% from being energy neutral.

Indoor climate measurements show high temperatures in a large share of the period.

IOP Conf. Series: Earth and Environmental Science **352** (2019) 012010 doi:10.1088/1755-1315/352/1/012010

1. Introduction

The BOLIG+ [1] concept was initiated at the Danish Energy Camp 2005 and developed through 2006-2007 and concentrates around six dogmas:

- Zero Energy Building (ZEB) on annual basis, including auxiliary electricity for private household and lightning,
- intelligent and user-friendly,
- flexible – in use and over time,
- good and healthy indoor climate,
- high architectural quality, adapted to the local context,
- constructed on ordinary economic market conditions.

2. BOLIG+

In the design process, there has been a close collaboration between architects and engineers, plus other relevant parties, e.g. utility companies, urban planners, sociologists, etc. This resulted in a thoroughly processed building concept that focused on all six dogmas. Additionally, all aspects of the building had been scrutinized for possible cost and energy optimization, sometimes exploiting alternative measures such as a lift as replacement of an elevator. This resulted in a design for a block of flats that should produce the same amount of primary energy as is used for operating the building plus the occupants' use of electricity for light and appliances.

2.1. Design and construction

From the developer, it was a demand to have balconies at each flat, however due to heavy traffic the municipality required closed balconies. This challenged the overall economy and the available space for PV on the facades. The architectural design of the facades was then a game in which windows, PV and balconies must be part of a harmonious whole. The PV installation on the roof will only be able to produce just over half of the building's energy needs, so PV panels on facades towards the southeast and southwest are a necessity. This challenge helped to foster the idea of designing the building's facade as a game between light and dark fields – similar to a chessboard.

Figure 1. BOLIG+ seen from Southeast.

2.2. Obtaining energy neutrality

The secret behind a very low energy consumption is integrated design. When looking at the building it looks like many other modern buildings, but all solutions have been selected carefully and can be summarised as:

IOP Conf. Series: Earth and Environmental Science **352** (2019) 012010 doi:10.1088/1755-1315/352/1/012010

- compact and well insulated thermal envelope,
- airtight building,
- effective decentralized mechanical ventilation,
- hybrid ventilation in the summer,
- heat recovery from grey wastewater,
- electricity production via multiple PV panels,
- connection to local district heating grid.

Up front, it was decided to keep the roof surface free of installations, allowing maximum space for PV panels placed on the roof with minimum stands. This affected several other installations, e.g. exhaust and inlet of the ventilation were located in the facades and the elevator was replaced by a lift, i.e. no elevator shaft raise over the roof. A lift also had some energy and economic effects, as it was both cheaper and more energy-efficient due to less expected use.

During the contest and the early design process many different solutions were investigated, such as ground connected channels, electricity production from rainwater running in barrel drains, dynamic facades, high-efficient insulating materials, solar walls etc. However, they were all found to be economically heavy to integrate. The project team experienced that, in order for BOLIG+ to meet the economic constraint, it was necessary to utilize passive solutions instead of new and often costly technology. The limitations on economy thus had the effect that integrated design was fully exploited.

Two heat supply solutions were investigated, namely district heating and heat pump. With an estimated heat pump COP of 3.5, it would be the optimum primary energy solution. A heat pump solution would need less primary energy and thus less PV. Taking into account that a district heating solution required more PV area, the two solutions were estimated to have approximately the same economy (plant plus operating costs). In the Table 1, varying primary energy factors over time are shown. However, the local district heating company had a provision which stated that in connection with the establishment of new main lines, no connection fee had to be paid. This made the district heating solution the most economically advantageous solution. And since the space needed for PV was highly critical, district heating was then the possible, reliable and, in a Danish context, politically opportune choice.

Table 1 Varying primary energy factors in Denmark

	2008-2010	2010-2015	2018-
Electricity	2.5	1.8	1.9
District heating	1.0	0.6	0.85
Other	1.0	1.0	1.0

Due to the new Danish feed-in tariff structure, there is a large price difference between selling and buying electricity (approx. 0.27 €/kWh) from the grid. Therefore, an experimental vanadium redox flow-battery [2] was installed to improve the overall private economy for operating the building, by storing the electricity from time of production to time of use.

3. Energy and indoor climate measurements

Measuring energy and indoor climate in a building is a comprehensive and costly task and without support from the ELFORSK programme [3], it would not have been possible to establish a measuring campaign. Never the less, measurements also relied on the needed measurement systems in the building to be able to bill the occupants and to comply with requirements for metering stated in the European directive on energy efficiency [4].

In the flats, indoor climate was logged using WiFi loggers in 2 rooms, logging indoor temperature, humidity and CO_2. In addition to this, the total electricity consumption, electricity from every outlet (incl. decentral ventilation unit) were measured. Space heating was measured, incl. forward and return temperature of the heat circulation. The consumption cold water and the position of the door to the balcony were further measured.

Centrally, in the boiler room, the following measurements were carried out: total district heating delivered to the building (incl. forward and return temperatures) and district heating to the domestic hot water tank; heat recovery from grey wastewater; electricity to hot water tank to fight Legionella; electricity to and from the PV flow battery; total electricity in and out of the building; billing meters for the 10 apartments; electricity used on common spaces; electricity production from 4 PV systems; water temperature in top of tank and flow of space heating in flats; outdoor temperature.

Weather was taken from an existing nearby weather station located at Danish Technological Institute in Taastrup.

3.1. Comparing calculated and measured energy

It is often seen that new buildings have a different and often higher energy consumption than calculated in the design phase. In addition, the indoor climate in the buildings is often not as good as expected. It is a commonly known fact, that a building in use has a higher energy consumption than calculated during the design phase. There are many reasons for this:

1. input error in the calculation program used or that the calculation program cannot handle specific components of the building correctly,
2. changes in the design building and/or constructions and installations during the construction phase,
3. other use of the building, including other occupational behaviour than assumed in the calculations, e.g. increased room temperature,
4. real weather differs from the design year used,
5. lack of focus on indoor climate both in the design and construction phase,
6. faults in and poor commissioning of structures and installations.

Because of the above, it is typically not possible to assess whether a building functions as calculated simply by comparing the energy consumption calculated in the design phase with the measured annual energy consumption during operation. It is necessary to adjust the calculations relative to the actual conditions. How, a calculation program is adjusted to the current conditions, is described in detail in [5] and [6]. In [5] is also described, how to troubleshoot to determine which components and/or installations which results in a higher measured energy consumption for the building also after adjusting the calculation program.

4. Results

To meet the BOLIG+ dogmas, a building must comply with the requirements for being a Building class 2020 building according to the Danish Building regulations 2015 [7], but without contribution from locally produced electricity, i.e. a highly insulated passive solution. Additionally – to accommodate points 3 and 5 above - more realistic indoor temperature, internal loads, and domestic hot water consumption should also be part of the calculated energy demand.

Besides the demand for energy performance, there were also demands for focus on the indoor climate, including punishment by overheating and requirements for daylight conditions.

In the following, mainly the heating demand of the buildings is evaluated.

The design tool Buildings energy demand 2010 (Be10) [8] estimated an annual heating demand of 43.5 MWh while the real annual heating usage was measured to 47.3 MWh. This is only a 9 % higher heating demand than anticipated during the design. So, within the uncertainty of the calculation and the measurements the building performs as it should – or does it? This is investigated in the following.

Table 1 shows some of the input parameters to Be10 during the design phase (left column) compared to the same parameters measured in the actual building (right).

Table 2. Input parameters to Be10 during the design phase compared with the actual measured values. 1) Volume of DHW is decreased because of heat recovery on the grey wastewater from bathrooms.

	Design	Measured
Room temperature	20 °C	22.8 °C
Heat gains from persons	1.5 W/m²	1.03 W/m²
Heat gains from appliances	3.5 W/m²	2.4 W/m²
Mechanical ventilation	0.32 l/s/m²	0.347 l/s/m²
Infiltration	0.07 l/s/m²	0.064 l/s/m²
DHW	175 l/m² 1)	162 l/m² 1)
Temperature of DHW	55°C	54°C

Table 1 shows that the indoor temperature was higher than foreseen, while the heat gains were lower. The mechanical ventilation airflow rate was a bit higher, while the infiltration rate was a bit lower. Both DHW consumption and the temperature of the DHW was a bit lower in the actual building. Furthermore, the total solar radiation was almost the same as in the weather file in Be10, while the ambient temperature most of the time was higher. The number of heating degree-days in the actual year was 3033, while in the weather file in Be10 it is 3200.

When introducing the measured values from table 1 and a milder climate in Be10 the calculated heating demand changes from 43.5 MWh to 36.8 MWh, which is much lower than the measured: 47.3 MWh.

When investigating the measurements it is observed, that the actual buildings has a heat consumption for floor heating in the bathrooms during the summer, which normally isn't considered in a Be10 calculation for determination if a building comply with the Danish Building Regulation. It was further observed, that the length of the pipes were large and that the heat exchangers of the ventilation systems was slightly less efficient in the real building compared to the original Be10 model. When including the summer floor heating and the heat losses from the extra piping/a bit less efficient heat exchangers the calculated heating demand increased to 43.8 MWh. As the model now only calculates a 7 % too low heating demand, it was judged that it was not necessary to adjust the model further.

Some of the input values are now set to the standard values: room temperature to 20°C, heat gains to 1.5 and 3.5 W/m² (see table 1) and a higher DHW demand of 250 l/m² (minus the effect of heat recovery on the grey wastewater from the bath rooms) and without floor heating in the bathrooms during the summer. This leads to an annual heat demand of 29 kWh/m² while the heat demand in the design case was 28 kWh/m². So, overall the building performs as expected with regard to heating demand although the demand was distributed differently when compared to the original calculation. The calibration of the model hereby led to an insight, which otherwise would not have been possible. The electricity use for operating the building was a bit higher in the real building compared to the design calculation: 2.1 vs. 1.8 kWh/m². The combination of the heating demand of 29 kWh/m² and an electricity demand of 2.1 kWh/m² leads to a primary energy demand of $29*0.6+2.1*1.8 = 21.2$ kWh/m², which is only 6 % higher that the aim of 20 kWh/m². So, within the uncertainties, the building did perform as expected.

Was the building energy neutral with all the PV panels on the facades and roof? Well, the PV panels produce 26 % less than expected and the occupants used a bit more electricity in the apartments, so the building was not energy neutral. However, if the PV panels had produces as designed, and the electricity use in the apartments was as defined in the design phase, the building would only be 11 % from being energy neutral, which is actually a very good result.

More information on BOLIG+ and the performed energy calculations can be found in [3].

5. Discussion

BOLIG+ is the first example of how far, in terms of energy neutrality, it is possible to go regarding energy neutrality for a block of flats, under Danish climate and market conditions. However, not all energy conservation measures on the building may in itself be cost optimal and better societal energy economics may arise if some of the investments were moved to central energy production instead of local sub optimization.

Introduction of an in house electrical battery storage is far from optimal from an overall energy efficiency or CO_2 elimination point of view. But, due to the current feed-in tariff structure in Denmark, it may be economical feasible to optimise the electricity consumption inside the meter.

6. Conclusions

Did the building meet the BOLIG+ dogmas?

The building would almost have been energy neutral, given the photovoltaic system had produced the planned amount of electricity and the residents had used a little less electricity, in accordance with the BOLIG+ requirement of 1725 kWh/dwelling per year. The building meets the Building class 2020 energy performance requirement as defined in the Danish Building regulations 2010/2015 without contribution from local renewable electricity production. This highlights that the building itself has a very good energy performance, created by utilizing simple and well-known measures. However, the evaluation also highlights that without the resident's involvement in correct use of the building, it is not possible to meet the target.

During winter, there is a good indoor climate in the building in terms of in-door temperature and CO_2 level. However, some residents complain about dry air during the heating season, which unfortunately is a natural consequence of having an adequate air-change. During summer, the moisture and CO_2 level are satisfactory, but overheating occurs in several of the rooms. The rooms most exposed to overheating are the ones that faces northeast and thus get a large amount of morning sun. It is expected that the overheating issue could have been reduced by introduction of external solar shading – especially as the natural ventilation proved insufficient to reduce the problem.

It is possible for the residents to control several systems in their flats; however, an insufficient information level has been an issue for some systems. Especially, information on the ventilation system was insufficient resulting in often-wrong use of the systems. As a contrast, information about the intelligent power outlets and control of these has been sufficient and several residents have customised the programming of the system.

The building fits the local energy infrastructure as it is heated by district heating. The building can in the future provide service to the surrounding electricity grid, as it is equipped with a battery (connected to the PV system) that can be charged and discharged according to the amount of electricity from renewables in the grid. Weather the building, with its black and white sections on the facade, fits into the local architecture dominated by red masonry houses is of course a matter of taste, however, it is undiscussable that the building is a landmark at the top of the hill where the building is situated.

By utilizing simple and well-known technologies, it has been possible to construct BOLIG+ at Søborg on ordinary economic market conditions.

Has BOLIG+ reached its goal regarding energy neutrality?

For BOLIG+ the direct comparison of the designed energy demand and the actual energy consumption led to the same overall conclusion as the calibration exercise, i.e. that the building performs as expected. However, the calibration exercise gave an insight to the actual energy flows in the building, which otherwise would have remained hidden. The actual use of the building and a milder climate lead to a lower energy demand, which was counterbalanced by not originally considered floor heating in the bathrooms during the summer and more heat losses from piping and slightly less efficient heat exchangers in the ventilation systems than in the original model of the building.

Bearing in mind that it is the first attempt to construct a multi-family block of flats in accordance with the BOLIG+ dogmas, the results is quite remarkable. Not all dogmas are fully met, but sufficient to demonstrate that the BOLIG+ dogmas can be good references for residential buildings in the future.

7. Acknowledgements

BOLIG+ has over the lifetime of the collaboration (2006-2018) been supported financially by: Danish Energy Saving Trust, Ecological Council Denmark, Realdania, Danish Energy Technology Development and Demonstration Program (EUDP), and Realdania Byg who was developer.

The program for measurements, evaluation and documentation of BOLIG+ was supported by the Danish Energy Research and Development Program (ELFORSK).

The following institutions have all been actively involved in the BOLIG+ process: Danish Society of Engineers; Architects` Association of Denmark; Danish Energy Saving Trust; Ecological Council Denmark; Danish Building Research Institute; Danish Technological Institute.

Arkitema Architects designed the building in close collaboration with MOE consulting engineers.

References

[1] BOLIG+ 2019 official web site: www.boligplus.org
[2] visblue 2018. "Case Studies." Retrieved September 27, 2018 (https://www.visblue.com/case-studies.html).
[3] Jensen S Ø, Wittchen K B & Knudsen H N 2018. BOLIG+ energy neutral block of flats – Measurements and evaluation (in Danish: BOLIG+ energineutrale etageboliger - Måling og evaluering) Danish Building Research Institute, Aalborg University, Copenhagen. SBi report 2018:12.
[4] European Parliament 2012. Directive 2012/27/EU of the European Parliament and of the Council of 25 October 2012. Brussels, Belgium.
[5] Jensen S Ø 2013. Guideline on Documenting the Performance of Built Low Energy Buildings. Danish Technological Institute, Taastrup, Denmark.
[6] Jensen S Ø 2015. Documentation of buildings' energy consumption (In Danish: Dokumentation af bygningers energiforbrug). Danish Technological Institute, Taastrup, Denmark.
[7] Danish Transport and Construction Agency 2015. Danish Building Regulations 2015 (BR15).. Copenhagen, Denmark.
[8] Agerholm S and Grau K 2014. Buildings energy demand, 3rd ed. (In Danish: Bygningers energibehov). SBi Direction 213. Danish Building Research Institute, Aalborg University, Copenhagen, Denmark.

IOP Conf. Series: Earth and Environmental Science **352** (2019) 012011 doi:10.1088/1755-1315/352/1/012011

Parametric study for the long term energetic performance of geothermal energy piles

A Ferrantelli[1], J Fadejev[1,2] and J Kurnitski[1,2]

E-mail: andrea.ferrantelli@taltech.ee
[1] Tallinn University of Technology, Department of Civil Engineering and Architecture, 19086 Tallinn, Estonia
[2] Aalto University, Department of Civil Engineering, P.O.Box 12100, 00076 Aalto, Finland

Abstract.
Geothermal energy constitutes an important renewable resource that will become increasingly prominent in future constructions. A common method of extraction and usage consists of installing, inside the foundation piles of buildings, U-shaped heat exchangers called "energy piles".

In this paper such installations are addressed by means of a full parametric study, performed for a hall-type commercial building in a cold climate. By computing the transient heat transfer between energy piles and ground for a period of 20 years, guidelines for a preliminary sizing of the geothermal system as a whole are provided. These are valid for this specific building and climate, for a clay-type soil and without assuming thermal storage.

A highly nonlinear behaviour of the expected yield in relation to pile separation and evaporator extraction power is observed. Furthermore, 15m-long piles are found to be more efficient than those with double length, a smaller extraction power seems to be more favourable and differences in the pile diameter have little impact for heat transfer.

A geothermal system sizing guide, which is useful for a preliminary quantitative test prior to any installation, is introduced. Even though our specific results are valid only for a commercial hall-type building in Finland, our procedure is qualitatively general and can be utilized for any given building type and climate zone.

1. Introduction

Energy efficient buildings have become one of the primary concerns of the actual construction industry and research. Since future structures will be required to satisfy increasingly stringent constraints in terms of energy efficiency [1], the usage of renewable sources is now established as a prominent field of investigation.

In this respect, geothermal energy [2, 3] is one of the most sought-after solutions, due to the feasibility of geothermal systems and the possibility of underground (namely, invisible) installations. A particularly intriguing solution is given by ground-source heat pumps, which utilize ground heat exchangers (GHE) [4] to extract the heat stored inside the ground under a building. Some widespread practice consists indeed of installing heat exchange piping into the pile foundations of buildings, thus creating a GHE system known as geothermal energy piles [5, 6]. This is similar to a method that employs boreholes for geothermal energy extraction [7, 8], with the economical advantage that drilling a new borehole can be avoided, since the heat exchange piping can be installed into an already existing foundation pile.

IOP Conf. Series: Earth and Environmental Science **352** (2019) 012011 doi:10.1088/1755-1315/352/1/012011

Heat transfer processes between the energy piles and surrounding soil have been the object of various investigations for decades [9]; these include conduction through the ground within large volumes and convection inside the pipes, which can be often longer than 30m. The according calculations are therefore rather involved, due to the system size and to the inhomogeneous thermal properties of materials and soil layers, which can exhibit large variations according to location and depth [10, 11].

Due to these technical difficulties, numerical simulations have proven to be very effective in the energetic assessment of concrete geostructures for energy extraction and storage in the ground [12]. For instance, the authors of [10] formulate a mixed 1D-3D approach, to simulate both the 3D heat transfer occurring in pile and surrounding ground as well as the 1D heat and fluid flow inside the probe. Additionally, Cecinato and Loveridge develop in [13] an FEM model to investigate the impact of different design choices for shorter operational times (for about one week), and find the number of pipes to be the most influential design parameter, when compared to increasing the pile dimensions or acting on the fluid flow.

The study of heat transfer under the building is unfortunately only one part of the story: for a realistic full energy performance assessment of the geothermal system, one needs also to consider how the heat transfer in the foundation combines with the heat exchangers and pumping system above [10, 11, 14], a challenging task that is going to be investigated in the following sections.

The paper is organized as follows: the building model is discussed in Section 2.1 and our modelling of the heat pump load profile is described in Section 2.2. Section 3 features our results, together with a geothermal plant sizing guide in Section 3.1. A discussion of our findings is given in Section 4, and conclusions are finally drawn in Section 5.

2. Method

In the present study the coupling between GHE installation design and pumping system was investigated through parametric studies, by simulating the one year performance of different energy piles configurations with an IDA-ICE 4.61 model [15] that was validated in [16]. The thermal energy that can be extracted either monthly or yearly for a commercial hall-type building (Figure 1) located in the cold climate of Finland was computed. The thermal performance of the full system was calculated by considering both the heat pump profile and the thermal mass of the soil medium in which the piles were buried.

Our results are listed in both graphic and table form, addressing two different pile lengths and reporting the evaporator sizing power and expected yield per ground surface area in function of the pile separation. The energy demand covered by the heat pump was also calculated, and a geothermal system sizing guide that can be applied to the construction design at preliminary stages is provided.

2.1. Building model and case study parameters

Our model was entirely developed in IDA-ICE, Figures 1 and 2. Our case study corresponded to the building geometry and envelope parameters reported in Table 1. For the indoor air, the fresh air flow was 1.1 $l/(sm^2)$ and the setpoint 18 °C. The heating demand of the building at the design temperature of -26°C amounted to about 465 kW, and Helsinki-Vantaa weather data for the year 2012 were used. The simulated clay had the following thermal properties: thermal conductivity 1.1 W/mK, porosity 56%, saturation 100%, bulk density 1250 kg/m^3, wet density 1812 kg/m^3, volumetric heat capacity 3343 $kJ/(m^3K)$ and heat capacity 1845 J/(kgK). Heat gains and occupancy instead held as follows: each occupant generated 2 W/m^2, the lighting 8 W/m^2, the equipment 1 W/m^2. The occupancy periodwas 8:00-21:00 for 6 days out of 7, and the occupancy rate was 1.0. Finally, the AHU heat recovery was 80%.

Figure 1. IDA ICE model for the commercial hall.

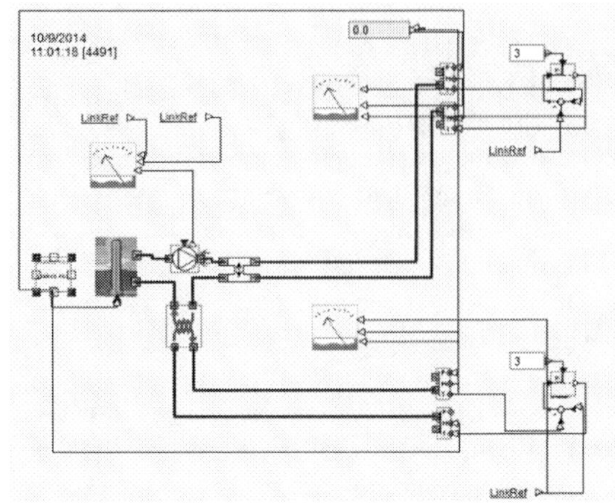

Figure 2. IDA-ICE simulation model based on the source file.

The operation performance of energy piles and boreholes is typically assessed on a simulation period of 20 years [5]. In order to make all the considered cases comparable, in the parametric study the ground source heat pump was sized to 50% of the overall heating demand for each and every simulation. One should also account for the extraction capability of the piles system. The building model size will thus vary according to heat pump power, energy piles total length and specific heat extraction rate (given in W/m). This generated a large number of different building models to be implemented.

In IDA-ICE, a detailed computation for a single case of this type may take up to 3 days, therefore in order to save time and eliminate the risk of erroneous model settings, a simplified approach was developed. The piles were not directly connected to the detailed plant with heat pump and building, which were replaced by an hourly time-step based on the heat pump evaporator load data.

Running an energy simulation for the initial model defined in Table 1 then generated an hourly time-step load data for the heat pump evaporator for one year, which could be used for each single case. The upper limit of the building design heat load corresponded to T=-26°C. In other words, each case could be generated by using only one input parameter – the design heat load. The profile used for generating the heat pump evaporator load data assumed a constant annual T=15.8°C for the ground.

Table 1. Building model parameters for the reference simulation.

Building size	66 x 137.4	m
Roof (310mm) U	0.12	$W/(m^2K)$
Floor (EPS100) U	0.09	$W/(m^2K)$
Walls (Sandwich 230mm) U	0.16	$W/(m^2K)$
Windows (SHGC 0.51) U	1.0	$W/(m^2K)$
Air tightness q_{50}	2	$m^3/(m^2h)$

IOP Conf. Series: Earth and Environmental Science **352** (2019) 012011 doi:10.1088/1755-1315/352/1/012011

2.2. Heat pump load profile and operation

In general, there exist two possible options for the heat pump operation, one at constant (on/off) load and the other with partial (inverter) load. In the first case, the heat pump is turned ON and will operate at full load in the presence of heat demand in the building. In contrast, the inverter heat pump allows for a partial load operation, typically within the range 30%-100%.

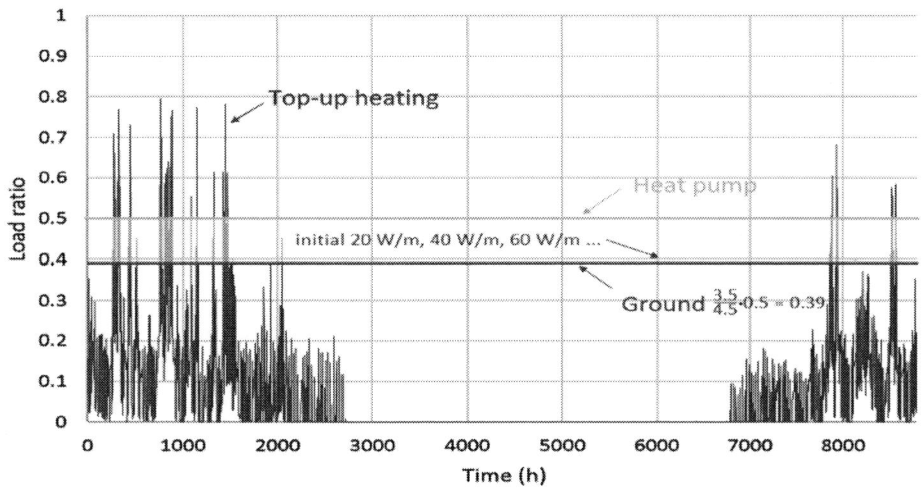

Figure 3. Annual load profile for the heat pump evaporator.

In our assessment, the hourly average (dynamic) data previously generated from the building load profile in Section 2.1 were used as input. Accordingly, the heat pump was allowed to operate at partial load, meaning that our results are applicable only to plants provided with inverter heat pumps.

The heat pump evaporator load data were generated from the building load data (which were based on the annual average COP of the heat pump). It was assumed that the mean annual ground temperature and resulting average brine temperature would remain in the 5°C to 10°C range during the heat pump operation, thus an annual average heat pump COP of 4.5 was applied. The COP was determined by means of a heat pump performance map data with a condenser side outlet temperature of 45°C. In Figure 3, the evaporator load profile is presented in green colour and the heat pump condenser load in yellow, while the remaining amount in red should be met with top-up heating.

The simulation started by feeding liquid mass flow data into the input link of the energy piles model, then a feedback controller measured the brine outlet T and fed a new inlet T value. The latter temperature was computed according to a pre-set ΔT value (in this particular study, $\Delta T = 3K$). As the brine inlet temperature hit the limited pre-set value (-1°C here), it stayed constant until the brine outlet exceeded 1°C. Such control logic guaranteed non-freezing ground temperatures, and simulated the heat pump turning off due to a low T of the evaporator inlet fluid.

Finally, the additional so-called "free thermal storage" effect that is provided by the floor heat loss to the ground was also implemented, by feeding the floor surface temperature to the energy piles model in IDA-ICE as a ground surface T (with thermal properties of the floor structure defined in Section 2.1). The free thermal storage effect can substantially increase the piles' performance, as illustrated in Table 2 and discussed in Section 4.

The geothermal plant sizing guide relies on the simulation results. The heat pump evaporator sizing is given by $Q_{\text{evap}} = Q_{\text{cond}}(\text{COP} - 1)/\text{COP}$, with COP=4.5 as discussed above, which

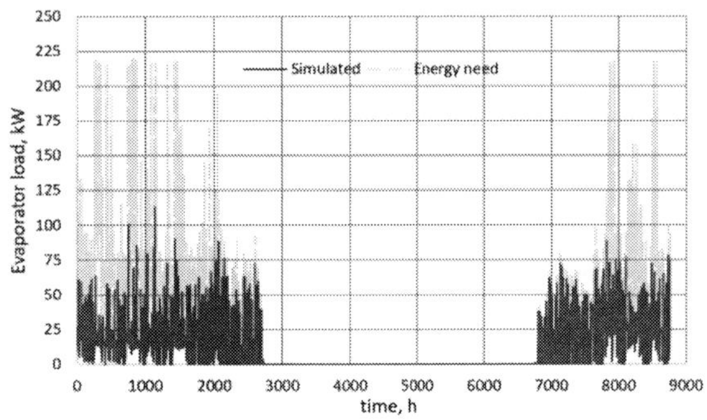

Figure 4. Energy need vs simulated yield for 200 W/m, 6m spacing and 30m length, first year of simulation.

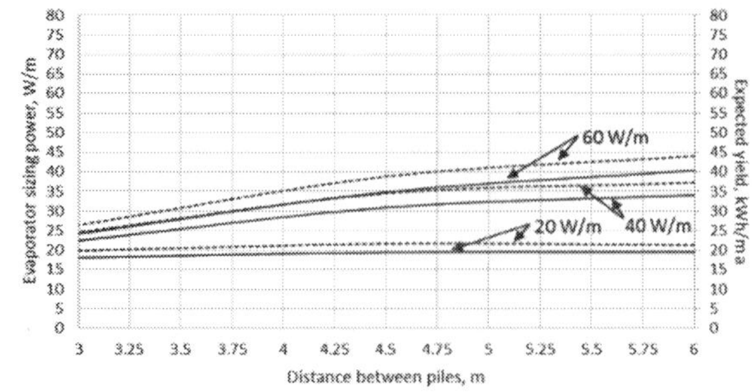

Figure 5. Results for evaporator power (solid) and condenser yield (dashed) of 30m-long energy piles.

determines the energy piles length $L = Q_{\text{evap}}/(W/m)$. The maximum power found during a 20-year simulation in this study is represented as "evaporator sizing power W/m" in Figure 5.

3. Results

The piles modelled in our simulations were either 15m or 30m long and buried in clay, with thermal properties given in Section 2.1. The design heat load was 360 kW and the annual heating energy demand was 168 MWh, consistently with the benchmark simulation discussed in Section 2.1. A spacing of 6m, 4.5m and 3m corresponded to 36, 48 and 121 piles respectively.

To illustrate, Figure 4 shows the results of a simulated case where the heat pump evaporator was sized at ca 215 kW. The pink line describes the *expected* evaporator yield (energy need), with the *simulated* performance of the heat pump in blue (case 200 W/m for 6m spacing and 30m length, only the first year was computed). The simulation results are summarized in Figure 5, where the evaporator power per unit length (in [W/m]) and expected area yield (in [kWh/m^{2}a]) are plotted in function of the pile spacing.

The effect of different pile lengths (either 15m or 30m) is compared in Figures 6 and 7, which report the expected yield resulting from the initial heat pump evaporator sizing power (20, 40 or 60 W/m), again in function of the pile spacing. Our results for every configuration are finally summarized synthetically in Table 3 (the contribution of thermal storage was not addressed, it will be included in a future work).

3.1. Geothermal plant sizing guide and example

In this section we are going to estimate the approximate energy piles length (or number) and their performance depending on the building design heat load Q and annual energy need E. The procedure can be extended to include different soil types, and also thermal storage when this is required. The steps to be performed are the following:

(i) **Determination of building design heat load and annual heating energy need.** Consider a commercial hall-type building with the following initial parameters: design heat load (design temperature -26°C) Q = 360 kW, annual energy need E ~183 MWh.

(ii) **Sizing the heat pump evaporator.** First of all, one should estimate the heat pump condenser sizing Q_{cond}. In this particular study, according to the building load profile (Figure 3), a condenser power sized to ca 50% of the building design heat load is able to cover up to 98.9% of the annual demand, with only 1.1% covered by top-up heating. This means that a heat pump with ca 50% less output

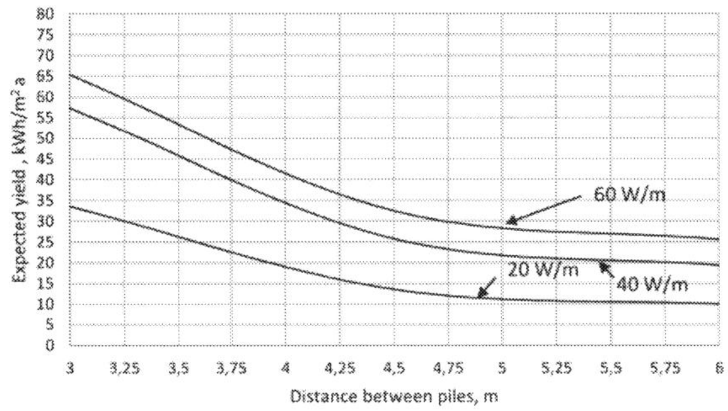

Figure 6. Results of yield per ground surface area of 15m-long energy piles.

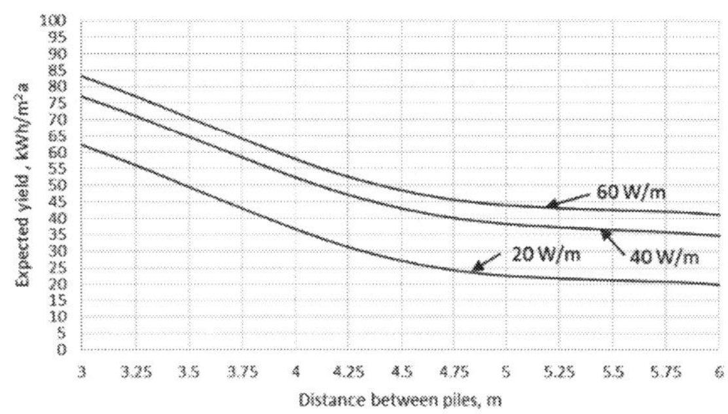

Figure 7. Results of yield per ground surface area of 30m-long energy piles.

power can be installed. The heat pump condenser is then sized accordingly as 180 kW and the evaporator as $Q_{\text{evap}} = 140$ kW (see Section 2.2), with an annual average for the COP.

(iii) **Estimation of the total pile field length and condenser yield.**
Let us assume that each pile is 30m long. The simulated energy piles performance under the conditions at points (i) and (ii) above is plotted in Figure 5, thus we can collect simulated performance data for three different initial evaporator sizing values: 20 W/m, 40 W/m and 60 W/m. The total energy piles length L is computed as in Section 2.2, thus it is only a function of the system sizing and geometry.
The specific yield per unit length E/L [kWh/m] is instead given directly by the simulations, finally returning the total expected condenser yield as $E = E/L \times L$ [kWh]. The results for an evaporator sizing of 20 W/m, 40 W/m and 60 W/m and 3m, 4.5m and 6m pile spacing are given in Table 2.
In other words, for e.g. 60 W/m the maximal energy yield without thermal storage is 103 MWh for 6m pile spacing, if the building annual energy need is 168 MWh. We can accordingly conclude that since the demand (168 MWh) is larger than what is produced by the energy piles (103 MWh), one should either install more piles, or consider the contribution of thermal storage.
To calculate how many piles are needed, we should simply compute the total length by dividing the annual need by the condenser yield. This gives $168000\,[kWh] : 44\,[kWh/m] \sim 3818\,m$, corresponding to n = 3818m/30m = 127.3∼127 energy piles with spacing 6m. The heat pump condenser power should be 180kW and the evaporator power 140W. With this amount of piles (which should be certainly rounded to e.g. 128 for a feasible pile field application), thermal storage is not needed.

Table 2. Simulated condenser and total yield for 60 W/m evaporator sizing, L=30m.

Pile spacing	Condenser yield	Total yield
6m	44 kWh/m	103 MWh
4.5m	39 kWh/m	91 MWh
3m	27 kWh/m	63 MWh

3.2. Impact of pile outer diameter on performance

For all the configurations reported in Table 3, an energy pile with double U-pipe and 170mm outer diameter was considered. To assess the impact of the pile's outer diameter, 20 simulations with smaller piles (outer diameter 125mm) were conducted. These resulted in 125mm piles being less performing by 2% compared to 170mm piles, which is not very significant. We accordingly conclude that the pile diameter is not important for the heat transfer efficiency, in agreement with earlier studies [13, 17].

Table 3. Summary of the study results in table format, extraction power 20 W/m, 40 W/m and 60 W/m (the empty column is due to an oversized system).

		step 3m		step 4.5m		step 6m	
		15m	**30m**	**15m**	**30m**	**15m**	**30m**
20 W/m	evaporator sizing power, W/m	20	18	20	19		20
	yield, kWh/m	21	20	22	22		21
	ground area yield, kWh/m2a	34	62	14	27	-	20
	demand covered by the heat pump	97%	90%	97%	96%		97%
40 W/m	evaporator sizing power, W/m	33	22	37	31	38	34
	yield, kWh/m	37	25	41	35	41	37
	ground area yield, kWh/m2a	57	77	26	43	19	35
	demand covered by the heat pump	83%	56%	92%	76%	94%	84%
60 W/m	evaporator sizing power, W/m	38	24	47	35	50	40
	yield, kWh/m	42	27	52	39	55	44
	ground area yield, kWh/m2a	65	83	32	48	26	41
	demand covered by the heat pump	63%	40%	77%	57%	83%	66%

4. Discussion

The set of our results can be analysed by investigating Figures 4 to 7 and Table 3. We recall that these are valid for a period of 20 years and for a pile field buried in clay, without assuming thermal storage.

First of all, the energy performance is not proportional to the initial evaporator extraction power. From Figure 4 we see that for most of simulation hours the energy need for a largely sized heat pump evaporator (200 W/m) is much higher than the geothermal system yield. This is evident also when comparing the 60 W/m to the 20 W/m curves in Fig.5, i.e. a three times larger evaporator heat extraction rate does not return three times the power and yield.

Secondly, the piles' separation seems not to be a crucial factor for energy performance if the initial evaporator heat extraction rates are low. Fig.5 shows a hardly linear performance/spacing ratio for piles with same length, since for 20 W/m both evaporator sizing power and expected yield even stay constant although the spacing is doubled.

On the other hand, Figures 6 and 7 show instead very clearly that a high yield per ground surface area [kWh/m2a] is strongly correlated with a smaller spacing. A low 20 W/m evaporator power can indeed return more than three times the yield with 3m spacing compared to a 6m spacing. Note here the nonlinearity of yield versus heat extraction rate and pile length (30m-long piles return a roughly double yield only for 20 W/m).

Table 3 reports a sample of these results, summarizing them in a more immediate form. The added information is the demand covered by the heat pump, calculated as the % difference between required and produced heat, which is clearly larger for the lowest 20 W/m initial evaporator power. We conclude that 15m long piles performed better than 30m long piles, due to proximity to the building floor boundary, where floor heat losses provide a free thermal storage effect. Besides, a larger spacing is generally preferable. The more distant the piles, the less they interacted with each other, thus the more heat could be extracted without assuming thermal storage. According to the simulated results, the evaporator yield varied from 20 to 55 kWh/ma, depending on the expected initial evaporator sizing power.

5. Conclusions

The study at hand considered a parametric study of performance for a geothermal energy pile field providing heating to a commercial type building. Our computer simulations addressed the heat transfer processes occurring inside the soil surrounding the energy piles, and quantified their effect on the performance of the entire heat extraction system connected to the building.

Running simulations pertaining a period of 20 years, including stress tests with very low external temperatures, various system parameters were tracked. For instance, the effect of different piles length, their spacing and performance with respect to the extraction system efficiency were assessed.

A high nonlinearity of the expected yield in relation to spacing and evaporator extraction power was observed, leading to favour a 15m length and a smaller extraction yield. The parametric simulations were conducted for different extraction powers and energy pile field configurations, thus it was possible to present the results in the form of specific power [W/m] and yield [kWh/m] which allow a preliminary sizing for any building. To illustrate how the developed method works, a geothermal system sizing guide, including a preliminary assessment of both plant and energy pile field, was also presented.

One should keep in mind though that any quantitative results included in our investigation are only valid for a commercial hall-type building in a cold climate. The impact of thermal storage in the ground surrounding the energy pile field was also not addressed here, which constitutes another limitation. Moreover, comparing the role of different soil types could be very interesting. Finally, experimental validation and performing a theoretical cross check of our results on physical grounds would be very valuable. All these considerations provide motivation for future work.

Acknowledgments

The authors acknowledge support by the European Regional Development Fund via the Estonian Centre of Excellence in Zero Energy and Resource Efficient Smart Buildings and Districts ZEBE, grant 2014-2020.4.01.15-0016. They are also grateful to the Estonian Research Council through grant IUT1-15.

References

[1] European Parliament 2010 *Official Journal of the European Union 53 (2010) 13-35*
[2] Agemar T, Weber J and Moeck I S 2018 *Energies* **11** ISSN 1996-1073
[3] Park K S and Kim S 2018 *Energies* **11** ISSN 1996-1073
[4] Spitler J 2005 *HVAC&R Research* **11** 165–167
[5] Fadejev J, Simson R, Kurnitski J and Haghighat F 2017 *Energy* **122** 390 – 407 ISSN 0360-5442
[6] Sani A K, Singh R M, Amis T and Cavarretta I 2019 *Renewable and Sustainable Energy Reviews* **106** 54 – 78 ISSN 1364-0321
[7] Lamarche L, Kajl S and Beauchamp B 2010 *Geothermics* **39** 187 – 200 ISSN 0375-6505

[8] Zhao J, Li Y and Wang J 2016 *Energy Procedia* **104** 413 – 418 ISSN 1876-6102 clean Energy for Clean City: CUE 2016–Applied Energy Symposium and Forum: Low-Carbon Cities and Urban Energy Systems

[9] Faizal M, Bouazza A and Singh R M 2016 *Renewable and Sustainable Energy Reviews* **57** 16 – 33 ISSN 1364-0321

[10] Carotenuto A, Marotta P, Massarotti N, Mauro A and Normino G 2017 *Applied Thermal Engineering* **124** 1492 – 1504 ISSN 1359-4311

[11] Ferrantelli A, Fadejev J and Kurnitski J 2019 *Energies* **12** 770 ISSN 1996-1073

[12] Dupray F, Laloui L and Kazangba A 2014 *Computers and Geotechnics* **55** 67 – 77 ISSN 0266-352X

[13] Cecinato F and Loveridge F A 2015 *Energy* **82** 1021 – 1033 ISSN 0360-5442

[14] Fadejev J, Simson R, Kurnitski J, Kesti J, Mononen T and Lautso P 2016 *Energy Procedia* **96** 489 – 502 ISSN 1876-6102 sustainable Built Environment Tallinn and Helsinki Conference SBE16

[15] EQUA 2013 IDA ICE - Indoor Climate and Energy Tech. rep. EQUA Stockholm, Sweden

[16] Fadejev J and Kurnitski J 2015 *Energy and Buildings* **106** 23 – 34 ISSN 0378-7788 sI: IEA-ECES Annex 31 Special Issue on Thermal Energy Storage

[17] Loveridge F and Powrie W 2014 *Geothermics* **50** 122 – 135 ISSN 0375-6505

IOP Conf. Series: Earth and Environmental Science **352** (2019) 012012 doi:10.1088/1755-1315/352/1/012012

A tool for the parametric assessment of operational energy use and embodied GHG emissions in a single-family house concept study

Marianne Kjendseth Wiik[a], Nicola Lolli[b], Selamawit Mamo Fufa[a]

[a]SINTEF Building and Infrastructure, Forskningsveien 3b, 0373, Oslo, Norway.
[b]SINTEF Building and Infrastructure, Høgskoleringen 7b, 7465 Trondheim, Norway.

marianne.wiik@sintef.no

Abstract. Present day European legislation focuses upon the reduction of greenhouse gas (GHG) emissions in buildings through better design and technological solutions, with a view to reduce operational energy use and embodied GHG emissions. Currently, environmental performance assessment tools for buildings lack in assessing these parameters over the entire lifecycle of a building. Life cycle assessment (LCA) is a well-established methodology that gives a clear insight into the potential environmental impacts during the service life of a building. However, architect, engineer and constructor (AEC) professionals consider LCA as complex and time consuming. This paper builds upon a methodology for the development of a parametric analysis tool (PAT), which comprehensively assesses operational energy use and embodied GHG emissions during the lifecycle of a building. In this phase of the PAT's development, the complexity of the tool has been increased by expanding the number of parameters from four (i.e. insulation thickness, window types, North façade glazing area and South façade glazing area) to seven (i.e. climatic zones, solar shading and electricity emission factors). This has increased the amount of parametric permutations from 1,372 to 12,348. The PAT has been applied to a conceptual two-storey single-family house, developed by the Norwegian ZEB Research Centre, which is assumed to be in either Oslo (Norway) or Lecce (Italy). The results show that the choice of insulation thickness influences total energy use less than the selection of shading types or glazing areas. The results also show the parametric selection with the least amount of operational energy use (34kWh/m^2/yr) in the Lecce climate consists of triple-glazed windows (0.5W/m^{2}k), 10m^2 of glazing on the north façade, 20m^2 of glazing on the south facade. The results show the parametric selection with the lowest total GHG emissions in the Oslo climate (7.8kgCO_{2eq}/m^2/yr) consists of triple glazing (0.5W/m^{2}k), 10m^2 of glazing on the north facade and 20m^2 glazing on the south facade. This is because a lower electricity emission factor (132gCO_{2eq}/kWh for Norway and 290gCO_{2eq}/kWh for Italy) was used for converting operational energy use to GHG emissions, even though the Oslo climate has a higher heating demand compared to Lecce. In conclusion, this paper shows how complex design options can be evaluated in a holistic way through PAT to ascertain the best selection of design criteria for low GHG emissions and low operational energy use.

1. Introduction

Present day European legislation focuses upon the reduction of greenhouse gas (GHG) emissions in buildings through better design and technological solutions [1]. Existing approaches for the reduction of operational energy and emissions in buildings include passive house, zero emission, energy-plus buildings and zero emission neighbourhoods [2-13]. These approaches are generally achieved by implementing energy efficiency measures such as, using effective insulation and better performing windows, reducing infiltration losses, recovering heat loss from ventilation systems and using renewable energy resources [14, 15]. Regardless, as buildings achieve lower operational energy demand, the relative proportion of embodied material emissions increases [16-19].

Life cycle assessment (LCA) is a well-established methodology used for the systematic assessment of environmental burdens during the lifecycle of a building and can be used to assess the environmental performance of buildings [20-24]. LCA appraises all resource inputs, in order to calculate the environmental impacts of a building at either the material, product or whole building level [25].

Currently, there are numerous environmental building assessment tools (e.g. BREEAM, LEED), energy modelling tools (e.g. SIMIEN, IDA-ICE, EnergyPlus [26]) and parametric analysis tools (PATs) (e.g. Grasshopper, Daysim) available for a range of building parameters such as optimal structure, daylighting or comfort. Previous studies have shown environmental building assessment tools [27-30] cover a wide range of building typologies and emphasise various phases of a building's lifecycle. Previous studies that assess both embodied GHG emissions and operational energy use in Norwegian buildings include a study on lessons learnt from embodied GHG emission calculations in ZEBs [31], a study on strategies for reducing embodied impacts from buildings [32], a study on analysing methodological choices in embodied energy and GHG emission calculations from buildings [33], and a study that investigates if a net life cycle balance for energy and materials is achievable in a single family ZEB in Norway [34]. However, none of this research includes PATs. One literature review of parametric LCA models and algorithms used in the building and construction industry highlighted that few previous studies have focused on parametric design [35]. Previous international PAT studies have appraised complex structural systems in tall buildings, [36] optimised daylighting for occupants in a Grasshopper based PAT, [37] and assessed energy efficiency measures [38]. Another study documented an MS Excel-based PAT, which assessed the performance of building envelopes in residential buildings across climatic regions in India. However, this tool did not assess embodied emissions of building envelopes against operational energy [39]. To the author's knowledge there is only one previous study that has used parametric design to minimise the embodied GHG emissions in a ZEB [40], and only one other study that assesses ventilated timber wall constructions using parametric LCA [41]. One advantage of PATs is that they facilitate for the dynamic comparison of various scenarios simultaneously, in order to ascertain the best design alternative within a modest timeframe and at a relatively low cost. A disadvantage of market-available PATs is that they are difficult for inexperienced users to operate [36].

The objective of this paper is to continue the previous body of work explored by the authors in [42]. This work parametrically assessed insulation thickness, and window area for wall and window components in buildings and facilitated for the first steps of development and testing of a PAT that optimises operational energy use, embodied GHG emissions and embodied energy use. This paper takes the parametric tool a step further by expanding the scope of the PAT to include climate, electricity emission factors and solar shading. The tool uses the Norwegian ZEB Research Centre's single-family house (SFH) concept building as a case study, and assesses wall and window components in a building, to reduce operational energy use and embodied GHG emissions [43, 44].

This paper starts by outlining the methodology, parameters and case study used. This is followed by a discussion of the results, concerning the impacts parameters have on operational energy use and embodied GHG emissions. Finally, the conclusion surmises the work and results so far, and outlines scope for further research. It should be acknowledged that the development of this PAT is part of an ongoing process, and the results from this paper contribute to its further development.

2. Methodology

An MS Excel-based PAT was previously developed and tested for the ZEB SFH concept building [42]. In the previous version of the PAT, seven insulation thicknesses (50-350mm), four window types (3-ply 0.71 W/m^2K g-value 0.55, 2-ply 2.6W/m^2K g-value 0.78, 3-ply 0.50 W/m^2K g-value 0.5, 2-ply 1.30 W/m^2K g-value 0.62), seven North façade glazing areas [10-35m^2] and seven south façade glazing areas

IOP Conf. Series: Earth and Environmental Science **352** (2019) 012012 doi:10.1088/1755-1315/352/1/012012

(10-35m^2) were tested. This generated 1,372 parametric permutations. Further details about the original methodology and parameters can be found in [42].

This study builds upon that research and methodology in MS Excel, and facilitates for the testing of additional parameters, namely two different climates (Oslo, Norway and Lecce, Italy), two electricity emission factors (132gCO_{2eq}/kWh and 290gCO_{2eq}/kWh) and three solar shading options (none, curtains and venetian blinds). In this version of the PAT, some of the original parameters have been adjusted according to lessons learnt. Firstly, there are now six insulation thickness parameters ranging from 100 to 350mm. Secondly, the North and South glazing area parameters have been reduced to 10, 15 and 20m^2. Lastly, the four window types have been adjusted to three window types (2 ply 2.8W/m^{2}K g-value 0.78, 2-ply 1.0W/m^{2}K g-value 0.63, 3-ply 0.50 W/m^{2}K g-value 0.50). These changes have been made to portray more realistic building scenarios and reflect available market products. In this study, the updated array of parameters increases the PAT's complexity and generates up to 12,348 permutations. An overview of the parameters considered in this study are shown in Table 1. In this study, the parametric assessment has been carried out in two stages. The first stage generates operational energy use results for all parametric variations listed in Table 1, whilst the second stage takes a selection of parametric variations and calculates embodied GHG emissions from operational energy use and material use (Table 2). Next is a description of the operational energy use calculations, embodied GHG emission calculations and case study.

Table 1. First selection of parametric variations used in the operational energy use analysis.

Parameter	Option A	Option B	Option C	Option D	Option E	Option F
Insulation	100mm	150mm	200mm	250mm	300mm	350mm
Window type	2-ply 2.8 W/m^{2}k g-value 0.78	2-ply 1.0 W/m^{2}k g-value 0.63	3-ply 0.50 W/m^{2}k g-value 0.50	-	-	-
North window area	10m^2	15m^2	20m^2	-	-	-
South window area	10m^2	15m^2	20m^2	-	-	-
Solar shading type	No shading	Curtain 0.5 global radiation	Venetian blind 0.08 global radiation	-	-	-
Climate	Oslo	Lecce	-	-	-	-
Electricity mix	Norwegian	European	-	-	-	-

Table 2. Second selection of parametric variations used in the emissions analysis.

	Window type	Insulation thickness (mm)	Shading strategy	Electricity to emission factors (gCO_{2eq}/kWh)	Climate
Option 1a	Double glazed [a]	100	No shading	132	Oslo
Option 1b	Double glazed [a]	100	No shading	290	Oslo
Option 1c	Double glazed [a]	100	No shading	290	Lecce
Option 2a	Double glazed [a]	350	Venetian blinds	132	Oslo
Option 2b	Double glazed [a]	350	Venetian blinds	290	Oslo
Option 2c	Double glazed [a]	350	Venetian blinds	290	Lecce
Option 3a	Triple glazed [b]	100	No shading	132	Oslo

Option 3b	Triple glazed [b]	100	No shading	290	Oslo
Option 3c	Triple glazed [b]	100	No shading	290	Lecce
Option 4a	Triple glazed [b]	350	Venetian blinds	132	Oslo
Option 4b	Triple glazed [b]	350	Venetian blinds	290	Oslo
Option 4c	Triple glazed [b]	350	Venetian blinds	290	Lecce

[a] U-value 2.8 W/m^2k, g-value 0.78, $10m^2$ glazed area in N façade, $20m^2$ glazed area in S façade.
[b] U-value 0.50 W/m^2k, g-value 0.50, $10m^2$ glazed area in N façade, $20m^2$ glazed area in S façade.

2.1. Operational energy use

Operational energy use has been calculated for the whole building, on a monthly basis, and includes heating and cooling lighting appliances, ventilation internal heat gains from people and solar heat gain. More details on the calculation of operational energy use can be found in [42].

2.2. Embodied GHG emissions

Embodied GHG emissions have been calculated for the whole building according to EN 15978 and NS 3720 [22, 23] and includes emissions from the production phase and operational energy use. Calculations are structured according to NS 3451 [45] and includes groundwork and foundations, outer and inner walls, floors, roof, stairs and technical equipment. A functional unit of 'per m^2 of heated floor area per year of building lifetime' has been used whereby the heated floor area is $160m^2$ and the building lifetime is 60 years. Emissions are calculated according to [46] and are measured in terms of GWP ($kgCO_{2eq}/m^2/yr$) and use material emission factors from [24, 47, 48]. Operational energy use emissions are calculated by multiplying operational energy use ($kWh/m^2/yr$) with either a Norwegian electricity-to-emission factor of $132gCO_{2eq}/kWh$ or an average European factor of $290gCO_{2eq}/kWh$ [49, 50], depending on the geographical location of the case study. More information on the calculation of embodied GHG emissions can be found in [42].

2.3. Case study

The ZEB SFH concept building has been used as a case study. The building is a two-storey, zero emission single-family house. It is characterised by a well-insulated building envelope, solar thermal collectors, grid-connected photovoltaics, air-to-water heat pump, balanced mechanical ventilation system with heat recovery. All data inputs have been transferred into the PAT for assessment. More information on the case study can be found in [42, 43, 44].

3. Results and Discussion

Figures 1 and 2 show the results for operational energy use in Oslo and Lecce respectively. Both figures simulate 486 iterations each for the parameters listed in Table 1. In Figures 1 and 2, area 1 refers to the parametric results for the three shading options and the three glazing areas to the North and South facades; area 2 refers to the same combination of parameters as in Area 1 and includes the parametric results for the six insulation thicknesses; whilst area 3 includes the three window types, which are identified by areas 3a, 3b and 3c respectively.

By comparing Figures 1 and 2, we observe that the choice of insulation thickness influences total energy use less than the selection of shading type or glazing area. For insulation thickness, the difference in total energy use is approximately $3kWh/m^2/yr$. Variations in glazing area leads to a difference in total energy use of approximately $15kWh/m^2/yr$. By using a highly insulated window (Option C in Table 1, which corresponds to Area 3c), it is observed that operational energy use decreases. Figure 1 shows in Area 1a (100mm insulation) and Area 1f (350mm insulation) that moving the building into a warmer climate results in a smaller impact from insulation thickness on operational energy use. The glazing areas and type of solar shading is predominant in determining the level of operational energy use.

Table 3 shows the operational energy use, embodied material emission and total emission results for the parametric variations shown in Table 2. For example, Option 4c has the lowest operational energy use of 34kWh/m^2/yr) in the Lecce climate and consists of triple-glazed windows (0.5W/m^{2}k), 10m^2 of glazing on the north façade, 20m^2 of glazing on the south façade; whereby there is no heating demand and little cooling demand because of effective solar shading, and a well-insulated building envelope. In contrast, Option 4a has the lowest total emissions of 7.8kgCO_{2eq}/m^2/yr in the Oslo climate and consists of triple glazing (0.5W/m2k), 10m^2 of glazing on the north facade and 20 m^2 glazing on the south façade. This is because a lower electricity emission factor (132gCO_{2eq}/kWh) has been used for converting operational energy use to emissions, even though the Oslo climate has a higher heating demand than Lecce. This shows that total emissions are largely sensitive to the electricity emission factor used.

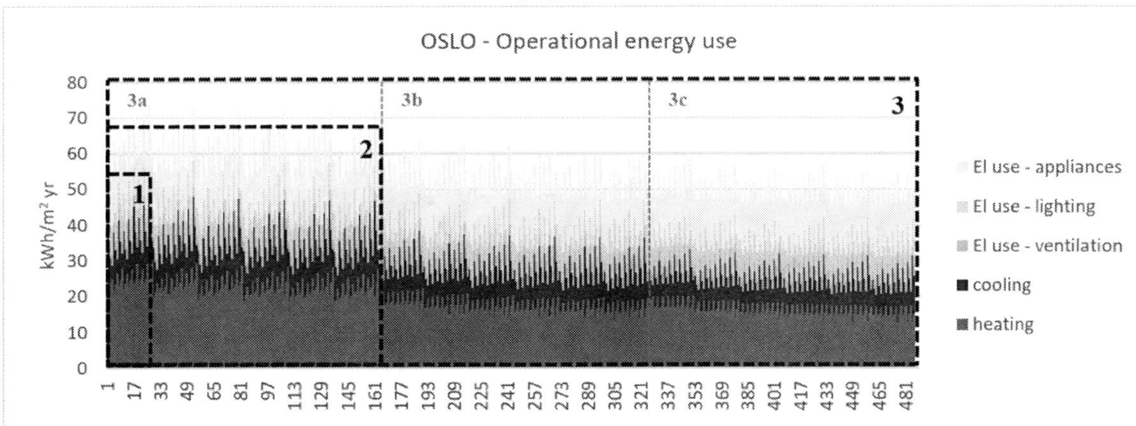

Figure 1. Operational energy use for all parametric iterations, calculated for the Oslo climate.

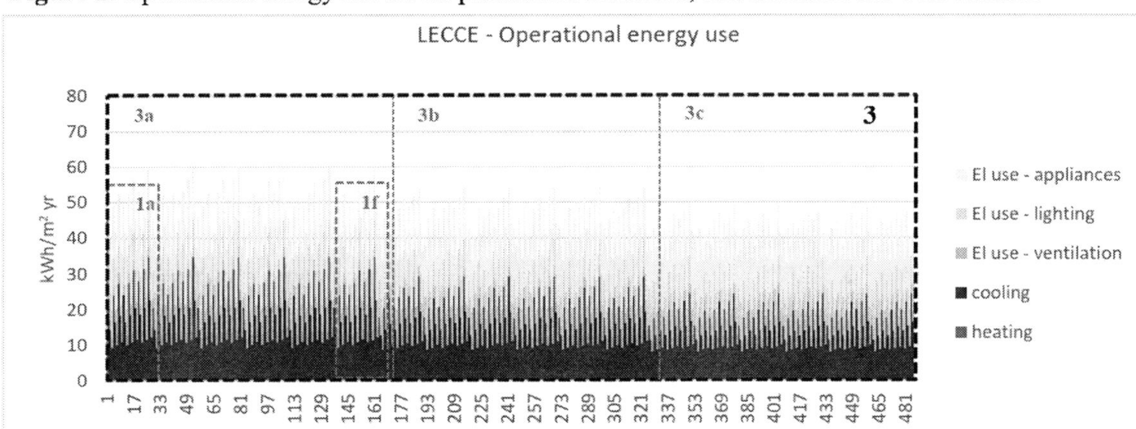

Figure 2. Operational energy use for all parametric iterations, calculated for the Lecce climate.

Table 3. Operational energy use and the embodied material emission results from the parametric analysis.

	Heating demand (kWh/m^2/yr)	Cooling demand (kWh/m^2/yr)	Total operational energy use (kWh/m^2/yr)	Operational energy use emissions (kgCO_{2eq}/m^2/yr)	Embodied material emissions (kgCO_{2eq}/m^2/yr)	Total emissions (kgCO_{2eq}/m^2/yr)
Option 1a	25	19	69	9.1	1.6	10.6
Option 1b	25	19	69	19.9	1.6	21.5
Option 1c	0	27	53	15.2	1.6	16.8
Option 2a	25	4	55	7.2	1.7	8.9

Option 2b	25	4	55	15.8	1.7	17.5
Option 2c	0	11	36	10.4	1.7	12.1
Option 3a	18	13	57	7.5	1.6	9.1
Option 3b	18	13	57	16.5	1.6	18.1
Option 3c	0	20	45	13.0	1.6	14.6
Option 4a	17	4	46	6.1	1.7	7.8
Option 4b	17	4	46	13.4	1.7	15.1
Option 4c	0	9	34	9.9	1.7	11.6

This study has generated 12,348 parametric permutations, a 10-fold increase on the previous study. It demonstrates that PAT is a powerful tool which can handle an increasing level of complexity. This study has emphasised the enormity of analysing and evaluating all possible permutations and has therefore selected 12 variants (Table 3) for discussion. Future work may investigate the introduction of algorithms to provide either analytical or decision support, for example through multicriteria decision-making. Further development may also include expanding the tool to the neighbourhood level, whereby the leap from building to neighbourhood increases the number of parameters and variables exponentially. In the future, this PAT could be used for setting energy and emission performance goals and benchmarks, optimising renewable energy and passive systems, evaluating building envelope schemes, integrating architectural features and integrating building systems.

4. Conclusion

This study has demonstrated that parametric analysis can be a useful tool for AEC professionals when assessing operational energy use and embodied GHG emissions in buildings. The results show that the choice of insulation thickness influences total energy use less than the selection of shading types or glazing areas. The results show the parametric combination with the least amount of operational energy use (34kWh/m^2/yr) and lowest total emissions (7.8kgCO_{2eq}/m^2/yr) do not have the same parameters. This is because a lower electricity emission factor (132gCO_{2eq}/kWh) was used for converting operational energy use to emissions, even though the Oslo climate has a higher heating demand compared to Lecce. The results show that the parametric results are sensitive to climate and electricity emission factor. In conclusion, this paper shows how complex design options can be evaluated in a holistic way through PAT to ascertain the best selection of design criteria for low operational energy use and GHG emissions.

References

[1] Directive 2010/31/EU of the European Parliament and of the Council of 19 May 2010 on the energy performance of buildings 2010
[2] NS 3700: 2013. Criteria for passive houses and low energy buildings - residential buildings (in Norwegian). Standards Norway, Oslo, Norway
[3] NS 3701: 2012. Criteria for passive houses and low energy buildings - non-residential buildings (in Norwegian). Standards Norway, Oslo, Norway
[4] Kristjansdottir T et al. 2014 *A Norwegian ZEB definition: embodied emissions* Oslo: ZEB Project Report (17) SINTEF Academic Press ISBN: 978-82-536-1398-7
[5] Fufa SM, Dahl Schlanbusch R, Kari Sørnes, Inman M, Andresen I. 2016 *A Norwegian ZEB definition guideline* Oslo: ZEB Project Report (29) SINTEF Academic Press ISBN: 978-82-536-1513-4
[6] Marszal AJ et al. 2011 Zero Energy Building – A review of definitions and calculation methodologies. *Energy Build* **43** (4) 971-9
[7] P. Torcellini, Pless S, Deru M, Crawley D 2006 Zero energy buildings: a critical look at the definition Pacific Grove, California; Contract No.: NREL/CP-550-39833

[8] Fjeldheim H, Kristjansdottir T, Sørnes K 2015 Establishing the life cycle primary energy balance for powerhouse kjørbo Passivhus Norden, Sustainable Cities and Buildings; Copenhagen, 20-21 August 2015
[9] PHI 2019 passivhaus institut [Available from: http://passiv.de/en/].
[10] ZEB 2019 The research centre on zero emission buildings [Available from: http://www.zeb.no/].
[11] Powerhouse 2019 Powerhouse [Available from: http://www.powerhouse.no/en/].
[12] Wiik MK et al. 2018 *Zero emission neighbourhoods in smart cities: definition, key performance indicators and assessment criteria: version 1.0 billingual version* Oslo: ZEN Project Report (7) SINTEF Academic Press ISBN: 978-82-536-1593-6
[13] Wiik MK, Fufa SM, Bær D, Sartori I, Andresen I 2019 *A ZEN guideline for the ZEN pilot areas. version 1.0.* Oslo: ZEN Project Report (11) SINTEF Academic Press ISBN: 978-82-536-1608-7
[14] Sartori I, Hestnes AG 2007 Energy use in the life cycle of conventional and low-energy buildings: a review article. *Energy Build* **39** (3) 249-57
[15] Lechner N 2009 *Heating, cooling, lighting : sustainable design methods for architects* 3rd ed. Hoboken, N.J John Wiley & Sons
[16] Ibn-Mohammed T, Greenough R, Taylor S, Ozawa-Meida L, Acquaye A 2013 Operational vs. embodied emissions in buildings - a review of current trends. *Energy Build* **66** 232-45
[17] Dixit MK, Fernández-Solís JL, Lavy S, Culp CH 2012 Need for an embodied energy measurement protocol for buildings: A review paper. Renewable and Sustainable Energy Reviews 16(6):3730-43
[18] Lützkendorf T, Foliente G, Balouktsi M, Wiberg AH. 2014 Net-zero buildings: incorporating embodied impacts. *Building Research & Information* **43** (1) 62-81
[19] Thiel C, Campion N, Landis A, Jones A, Schaefer L, Bilec M. 2013 A materials life cycle assessment of a net-zero energy building. *Energies* **6** (2) 1125
[20] ISO 14040: 2006 Environmental management -- life cycle assessment -- principles and framework. Geneva, Switzerland: International Organization for Standardization
[21] ISO 14044: 2006 Environmental management - life cycle assessment - requirements and guidelines. Geneva, Switzerland: International Organization for Standardization
[22] NS 3720: 2018 Method for greenhouse gas calculations for buildings. Standard Norge
[23] EN 15978: 2011 Sustainability of construction works - assessment of environmental performance of buildings - calculation method European Committee for Standardization, Brussels, Belgium
[24] EN 15804: 2012 + A1: 2013. Sustainability of construction works - environmental product declarations - core rules for the product category of construction products. Oslo, Norway: Standard Norway, European Standard
[25] Atmaca A, Atmaca N 2015 Life cycle energy and carbon dioxide emissions assessment of two residential buildings in Gaziantep, Turkey. *Energy Build* **102** 417-31
[26] AIA 2012 *An architect's guide to integrating energy modelling in the design process*
[27] Cole RJ, Howard N, Ikaga T, Nibel S. Building environmental assessment tools: current and future roles.
[28] Reed R, Wilkinson S, Bilos A, Schulte K-W 2011 A comparison of international sustainable building tools – an update. The 17th Annual Pacific Rim Real Estate Society Conference, Gold Coast 16-19 January 2011
[29] Haapio A, Viitaniemi P. 2008 A critical review of building environmental assessment tools. *Environ Impact Assess Rev* **28** 469-82
[30] McKay J, editor 2007 Green assessment tools: the integration of building envelope durability. 11th Canadian Conference on Building Science and Technology Banff, Alberta
[31] Wiik MK, Fufa SM, Kristjansdottir T, Andresen I 2018 Lessons learnt from embodied GHG emission calculations in zero emission buildings (ZEBs) from the Norwegian ZEB research centre *Energy Build* **165** 25-34 https://doi.org/10.1016/j.enbuild.2018.01.025
[32] Malmqvist T et al. 2018 Design and construction strategies for reducing embodied impacts from buildings - case study analysis. *Energy Build* **166** 35-47

[33] Rasmussen FN, Malmqvist T, Moncaster A, Wiberg AH, Birgisdottir H 2018 Analysing methodological choices in calculations of embodied energy and GHG emissions from buildings. *Energy Build* **158** 1487-98
[34] Kristjansdottir TF et al. 2018 Is a net life cycle balance for energy and materials achievable for a zero emission single-family building in Norway? *Energy Build* **168** 457-69
[35] Skaar C, Labonnote N, Gradeci K 2018 From zero emission buildings (ZEB) to zero emission neighbourhoods (ZEN): a mapping review of algorithm-based LCA *Sustainable Built Environment* **10** (7)
[36] Park SM, Elnimeiri M, Sharpe DC, Krawczyk RJ 2004 Tall building form generation by parametric design process CTBUH 2004; Seoul
[37] Ramsden J, Keeling T, Shepherd P, Shea A, Sharma S 2015 Smart building analyser: a parametric early-stage analysis tool for multi-objective building design CIBSE Technical Symposium; 16-17 April 2015; London, UK
[38] Parker A, Kyle Benne LB, Hale E, Macumber D, Schott M, Weaver E 2014 A parametric analysis tool for building energy design workflows: application to a utility design assistance incentive program. ACEEE Summer Study on Energy Efficiency in Buildings
[39] Jaboyedoff P, Cusack K, Bhanware P, Ganesan K, Chetia S, Maithel S, editors. 2015 Parametric analysis using dynamic energy simulation tools to evaluate the performance of building envelope in the residential buildings of composite climatic region of India. 14th Conference of International Building Performance Simulation Association; 7 - 9th December 2015; Hyderabad, India
[40] Lobaccaro G, Wiberg AH, Ceci G, Manni M, Lolli N, Beradi U 2018 Parametric design to minimize the embodied GHG emissions in a ZEB. *Energy Build* **167** 106-23
[41] Fufa SM, Skaar C, Gradeci K, Labonnote N 2018 Assessment of greenhouse gas emissions of ventilated timber wall construction based on parametric LCA. *J Clean Prod* **197** (1) 34-46
[42] Lolli N, Fufa S, Inman M 2017 A parametric tool for the assessment of operational energy use, embodied energy and embodied material emissions in buildings. *Energy Procedia* **111** 21-30
[43] Houlihan Wiberg A, et al. 2014 A net zero emission concept analysis of a single-family house. *Energy Build* **74** 101-10
[44] Houlihan Wiberg A, Georges L, Fufa Mamo S, Risholt B 2015 A zero emission concept analysis of a single family house: part 2 sensitivity analysis. ZEB Project report no 21, The Research Centre on Zero Emission Buildings (ZEB)
[45] NS 3451: 2009 Table of building elements. Oslo, Norway: Standard Norge
[46] PRé Consultants. 2007 IPCC 2007 method GWP 100a V.1.02. Netherlands: PRé
[47] EPD Norge. 2016 Environmental product declarations [Available from: http://www.epd-norge.no/].
[48] Ecoinvent. 2014 Ecoinvent database v3.1. Swiss Centre for Life Cycle Inventories, Dübendorf, Switzerland
[49] Dokka TH, Sartori I, Thyholt M, Lien K, Lindberg KB, editors 2013 A Norwegian zero emission building definition. Passivhus Norden 15-17 October 2013; Göteborg, Sweden.
[50] Capros PP, Mantzos DL, Tasios N, Vita AD, Kouvaritakis N 2010 EU energy trends to 2030 — UPDATE 2009, in: EU energy trends to 2030, EUROPEAN COMMISSION. Directorate-General for Energy, Luxembourg

Acknowledgements
This work was financed thanks to the contribution of the EU Horizon 2020 NERO - Nearly Zero Energy Wooden Buildings in Nordic Countries grant agreement 754177.

IOP Conf. Series: Earth and Environmental Science **352** (2019) 012013 doi:10.1088/1755-1315/352/1/012013

Innovative public procurement (IPP) – Implications and potential for zero-emission neighborhood (ZEN) projects?

Hasan A. M. Hamdan and Luitzen de Boer

Department of Industrial Economics and Technology Management,
NTNU, Norway

Hasan.a.hamdan@ntnu.no

Abstract. As a component of the urban fabric system, zero-emission neighborhood (ZEN) projects represent an opportunity to boost the sustainable performance of cities. However, overlooking the potential of different actors or underestimating the complexity of interactions among them may threaten the projects themselves. Public procurement is a powerful tool that potentially enables national and local authorities to achieve sustainable development goals while procuring necessary products and services. This paper aims to understand the potential of innovative public procurement (IPP) to reduce some of the project complexity in ZEN. Besides literature on sustainable neighborhoods (SN), empirical insights are drawn from an ongoing ZEN project to map the primary sources of complexity in such projects. Afterward, the potential for dialogue with suppliers is mainly discussed in light of these sources. Our findings suggest that using IPP may assist in reducing the complexity imposed on ZEN projects.

1. Introduction

Cities and urban areas occupy just 3% of the planet's surface, but consume 60 to 80 % of the world's energy and produce around 75% of carbon emissions [1]. Although the integrated nature of neighborhood-scale projects entails an opportunity to achieve sustainable development, see, i.e. [2, 3], by nature such projects comprise a diverse group of actors. This imposes high complexity and uncertainty on the projects. Many studies have addressed the positive influence of end-user engagement (or citizen participation) in sustainable neighborhood (SN) projects [4-7], concerning both new developments and retrofitting. However, to the authors' knowledge existing literature largely ignores dialogue with suppliers in SN projects. Public procurement offers various interaction opportunities with suppliers before, during and after the formal tender stage. In this paper, we will focus on interaction during the pre-tender stage (market dialogue). The increasing trend towards using dialogue with suppliers in the public procurement process [8-12] deserves attention in regards to the impact on SN projects and comparison to end-user engagement activities.

The potential of dialogue with suppliers in the context of SN projects is addressed primarily due to two reasons. First, previous work has called for research aimed at reducing the influence of market forces in neighborhood development, since they contributed to the limited success of planning movements [13]. However, under the right circumstances, market forces could be seen as an asset rather than a liability. For example, extensive research shows that interaction with suppliers can improve efficiency, learning, and innovation [14-16]. Second, SN projects constitute a large and diverse group of stakeholders. This brings the idea of complexity into the picture. Therefore, dialogue at the outset of the project and public procurement could be harnessed to reduce, or at least understand, the increased complexity in such projects. The purpose of this paper is to study the potential of dialogue with suppliers before the formal tender stage in the public procurement process, and its potential for complexity mitigation. Thus, this article intends to fill the void in the literature on SN through dealing with the following central question: "How can we conceptualize the potential of innovative public procurement (IPP) and in particular pre-tender market dialogue, for mitigating the added project complexity in SN (and ZEN) projects?"

This study treats ZEN as a specific type of SN (or SN 2.0) due to (1) its inherent complexity from SN; where both contain a broad group of actors, and (2) its quest to advance the sustainable performance of cities through more ambitious goals. Our contribution is twofold. First, we aim to describe the sources of complexity in SN and ZEN projects. Second, we assess the potential of (IPP) to reduce this complexity. In the following section selected theories and concepts are reviewed for this study. Section 3 describes the methodology and empirical material. In section 4, we discuss the impact of pre-tender market dialogue on the complexity and attempt to answer the research question. Section 5 concludes the study and provides suggestions for further research.

2. Relevant literature

2.1. SN projects and complexity

To be considered sustainable, cities need to meet "the needs of the present without compromising the ability of future generations to meet their own needs" [17]. However, no city can become sustainable without the contributions of its smaller urban parts, such as neighborhoods. Due to the social and environmental risks associated with cities and urban areas, SN development is receiving more attention from both governments and industry [18]. Several studies, see for example [13, 19, 20], describe the key points that planners and local authorities should take into consideration when planning and developing SN projects. In those studies, the environmental, social and economic aspects are prevalent. Those aspects have direct benefits on citizens, i.e., public health, clean air, inclusive community, self-sufficiency, and social relationships. It is natural to see that SN criteria are centered on citizen needs as they represent the ultimate end-users who will live, work and enjoy those neighborhoods. Some authors treat stakeholder engagement as part of the institutional aspect in SN projects [13]. Besides governments and citizens, the private sector, represented by contractors, developers, R&D firms, utility and service providers and material suppliers, could participate in developing new standards on SN inspired by their expertise [18]. For simplicity, we use the word supplier(s) in this study to refer to all private sector's actors.

Although there is no single, generally accepted definition of project complexity, it has received attention due to its influence on projects; it can manipulate objectives, organizing forms, management processes, and procurement arrangements [21-24]. A complex system can be defined as "a large number of parts that interact in a nonsimple way" [25]. Project complexity can be characterized by the number of different parts and the degree of interrelatedness among those parts [21-23]. Williams [23] named this type of complexity as structural complexity, where it applies to any project dimension, such as physical and organizational dimensions of projects. In project management studies, concepts of complexity and uncertainty are found connected [23, 26]. We will follow Williams [23] who included uncertainty as a further element to project complexity.

Concerning complexity, project management scholars distinguish between two types of uncertainties; uncertainty in 1) goals and 2) methods [22, 23]. The first concerns how well-defined the project's outcome up front against an extensive collection of requirements and needs (i.e., user requirements and customer needs). The second deals with new methods and technology to achieve those goals. In the context of SN projects, barriers such as unsustainable planning, poor project management capabilities, difficulties in applying sustainable technologies, and difficulties in coordination among stakeholders and their interests [18], are all examples of how ill-defined methods can impede the development of SN. In line with Davies and Mackenzie [26], we acknowledge the previously described uncertainty as internal sources of uncertainty, while external sources originate from the environment surrounding and affecting the project. SN projects intend to improve the social, environmental and economic characteristics of the urban system. Therefore, such projects are expected to be impacted by the external forces located in the surrounding environment, such as political and business players. For example, lack of supporting policies and national standards as barriers to SN development [18], can increase the uncertainty in SN projects.

In this paper, we conceptualize complexity in SN projects in terms of 1) structural complexity and 2) uncertainty, see figure 1. First, we operationalize structural complexity in SN projects from physical and organizational standpoints. Physical complexity is the number of neighborhood physical parts (sub-projects) and the interdependencies between them, while organizational complexity represents the number and diversity of project actors and the interactions between them. Second, as described above, we also distinguish between internal and external sources of uncertainty in SN projects.

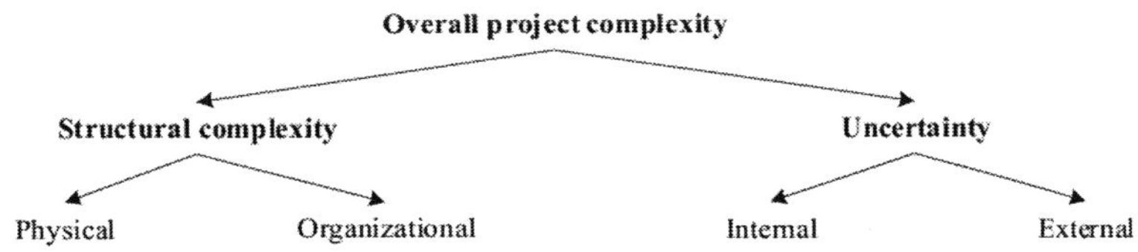

Figure 1. Complexity in SN projects.

2.2. *Innovative public procurement (IPP)*

Public procurement is a tool available at the disposal of public institutions (i.e., governments and state-owned enterprises) to procure goods, services, and works. As a demand-side innovation policy instrument, public procurement for innovation (PPI) can contribute to "satisfying unsatisfied human needs and solving societal problems" [10]. IPP, on the other hand, happens when public institutions conduct procurement activities in an unconventional way [27]. This includes changes to traditional strategies, guidelines, processes, legal frameworks, and practices. For example, a partnership approach is an example of an untraditional public procurement [28].

We increasingly witness more dialogue-based procurement activities [8, 10, 12], in which dialogue happens before a formal contract exists between the public buyer and supplier(s); either before (i.e., market dialogue) or during the formal tender stage (i.e., competitive dialogue, innovative partnership). This paper regards dialogue-based procurement as a branch of the IPP process since it empowers public institutions to procure unconventionally in order to meet unmet needs [10, 27]. The pre-tender stage in the public procurement process is the least regulated stage compared to the remaining formal tender and post-tender stages. In the EU, directives on public procurement allow public institutions to interact with the market during the pre-tender stage for purposes such as preparation of specifications, sharing of plans and visions, and seeking advice from the market. Hence, the pre-tender stage can be exploited to overcome some aspects of the agency problem, such as information asymmetry or 'market failure' described by Edler and Georghiou [29]. Moreover, as a part of an IPP process, dialogue with suppliers during this stage may facilitate the public institutions' mission to procure innovatively. For this paper, we summarize the rationales to conduct a dialogue with suppliers during the pre-tender stage in public procurement as the following:

- *Needs mapping*. Public institutions do not always have precise, well-defined needs. Current or potential stakeholders can validate already, identified needs or identify new ones [8].
- *Improving requirements and specs*. Poorly-formed specs can undermine procurement outcomes. Consulting the market helps buyers to formulate more functional and innovation-driven (yet achievable) specs in the call for tender [8, 10, 30].
- *Supply market access*. Interaction allows purchasers to get information about potential suppliers and available solutions [8, 10, 11, 29], including capacities and capabilities. In the case of new solutions, a market dialogue may be utilized as a means to plan and coordinate the path forward.
- *Market visibility*. Early signals regarding particular public demands enable suppliers to prioritize which technology (or markets) they should tackle first [29]. For example, market dialogue may

reduce some of the market risk perceived by suppliers once they become aware of the buyers' needs and willingness to buy [11].

- *Market link creation.* Due to the diverse group of actors, market dialogues may trigger different levels of interaction through exchanging knowledge (not necessarily relevant to the procurement in hand), or even new, spin-off collaborations. For example, third actors can provide buyers with the support [31] necessary to link with the market and conduct the dialogue (see, i.e. www.innovativeanskaffelser.no). Furthermore, such dialogues may provide better conditions (i.e., trust) for future collaboration.

3. Methodology

In the theory section, we investigated several studies regarding SN, where two of which are reviews, to understand its characteristics and barriers. Next, we looked at the complexity, particularly project complexity, to define the primary sources of complexity in SN projects. Afterward, we outlined the potential value of dialogue with suppliers in the pre-tender stage from IPP literature. While in the discussion, we follow the framework illustrated in figure 2. This approach allows us to elaborate on the sources that make ZEN projects such complex endeavors, based on theoretical insights from literature on SN projects and empirical insights from the Research Centre on Zero Emission Neighborhoods in Smart Cities (ZEN Centre). In which we studied an ongoing ZEN project in Norway, Ydalir project. The ZEN center was established in collaboration with the Research Council of Norway in 2017 to accelerate the transition to a low carbon society by developing sustainable neighborhoods with zero greenhouse gas emissions, see (https://fmezen.no/). Lastly, we discussed how pre-tender market dialogue (as part of the IPP process) can potentially contribute to reducing the project complexity.

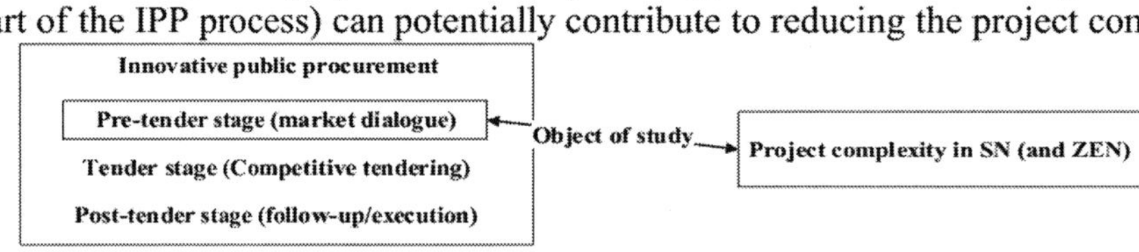

Figure 2. Analytical framework.

Following Dubois and Gadde [32] recommendation, looking into real world observation was necessary to understand how IPP's potential can be conceptualized to mitigate project complexity and develop better theoretical perspective. For example, the continuous interplay between complexity theory and project data helped us to define the sources of complexity in SN and ZEN projects. Data concerning Ydalir project was collected from project documents and a group interview conducted with two project managers. Moreover, both authors are part of the research team in the ZEN center.

3.1. *Illustrative case study - Ydalir project*

The project aims to develop a new neighborhood with high ambitions regarding energy and emission. The estimated timeframe for project completion is 2030, in which around 1,000 residential units (a combination of detached houses and apartment) are planned to be built around a school and kindergarten. Currently, the project is in the design and construction phases; the building of the first residential unit is planned to start within the end of 2019. The project owner (hereafter ETS) is a local development agency that aims to enable population growth by developing lands for housing and businesses at a reasonable price. The local municipality owns this agency; however, ETS operates as a private entity.

ETS owns 80% of the project land, where it sells land plots for private developers who wish to invest in building residential units. The remaining 20% belong to two private landowners, who also have plans to develop residential units. Only one of those landowners agreed to follow ZEN ambition plans. The school and kindergarten are being built by the municipality and planned to open autumn 2019. Among the stakeholders, several public players represented by the local municipality, energy utility company, waste management company, and transportation agency. Private actors include housing developers, consultancies and engineering firms, law firm, and contractors. Also, the project receives research support from the ZEN center.

During the strategic planning phase between autumn 2016 and spring 2017, ETS held five dialogue rounds that took place in the form of workshops between interested stakeholders, in order to frame and formulate project goals. The output of these workshops put into a master plan that functions now as a roadmap to realize the ZEN ambitions of the project. It translates the project's vision into functional requirements addressing issues related to material selection, design, energy, and mobility. The masterplan's requirements are anchored in the purchase contracts among ETS and developers. Citizens were not involved during the workshops, primarily because the project is a new development and takes place on a non-populated area; however, it intends to run participation rounds with future citizens.

4. Discussion

In section 2, we conceptualized two primary sources of complexity in SN projects and described five rationales to conduct pre-tender market dialogues. In what follows, we discuss the potential of market dialogues in regards to project complexity in ZEN projects, as illustrated in figure 3.

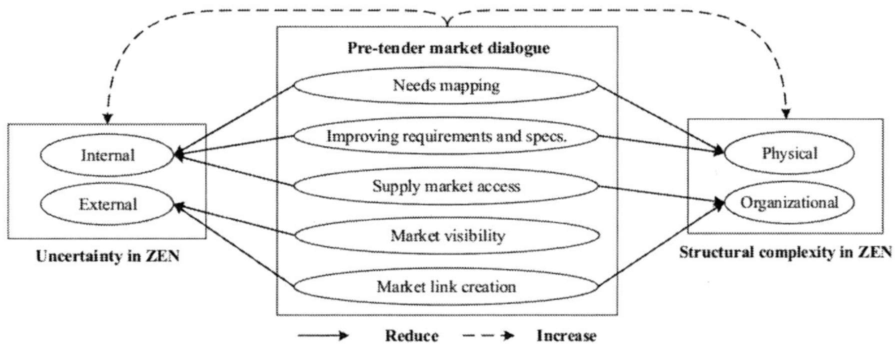

Figure 3. The impact of pre-tender market dialogue on ZEN project complexity

4.1. Physical complexity

Neighborhood-scale projects consist of many interrelated buildings with associated infrastructure, located within a confined geographical area. Traditionally, changes in the design of one component may release sequential impacts to other components, and affect the design (or construction) of the whole neighborhood [23]. Nevertheless, in the wake of SN projects, like ZEN, the interrelatedness of parts exceeds the traditional engineering view to include smart grids, shared renewables and batteries, shared transport and shared utilities. Those shared parts contribute in reducing the carbon footprint of the neighborhood as a whole, but they add more complexity to the project. Thus, in a complex project like ZEN, the project itself functions as a system of systems (or array) [22], where the ZEN's smaller parts function as standalone (sub-projects), yet interdependent.

Early market dialogue with suppliers may provide insights on what is sufficient to build an integrated and sustainable neighborhood. This may result in reducing the number of variant parts and suggest more collective-based solutions, including, for example, mobility and parking solutions. Using their market expertise, suppliers can also inspire public actors to formulate more standardized specifications, i.e., energy and emission targets, which contributes in reducing the number of variant parts or subparts.

4.2. Organizational complexity

Different actors are expected to exist in a neighborhood project since it contains several different parts (or sub-projects) that require varying degrees of specialization (i.e., resource, knowledge, and so on.). Some actors may even hold more than one role. For instance, the local municipality in Ydalir project is responsible for (1) developing of the school and kindergarten and (2) supporting private developers in their zoning plans. In order to understand the level of interactions in a complex system like ZEN, we use the notion of near decomposability by Simon [25], where intense interaction exists between actors operating in the same subproject (i.e., one building) and less interaction between actors operating in different subprojects. Weak interaction between the actors responsible for different parts or belongs to different subprojects may have adverse effects as joint planning and coordination are required to realize the neighborhood's vision (and shared solutions). In Ydalir case, for instance, the competitive nature of private developers is seen as a challenge that can affect coordination and knowledge exchange practices.

Market dialogues give public actors – who lead or have an influential role in neighborhood projects –access to the supply market. They learn not only about which solutions the market can deliver today but also the organizational capacities of suppliers. This empowers public actors to choose their partners based on relevant capabilities (e.g., communication), which may serve the interactive and exploratory nature of ZEN projects. Furthermore, market links are essential, especially when multiple tiers of suppliers are expected to exist. Companies that participate in market dialogues get the chance to meet their peers who have a similar interest in ZEN projects. It also may encourage some players to discuss potential collaboration plans (i.e., co-operative procurement) or even formalize their collaboration into an alliance or joint-venture. So consequently, market dialogue can contribute to building better conditions for collaboration, such as mutual trust and knowledge sharing.

4.3. Internal uncertainty

ZEN projects have a variety of goals and require a variety of methods to meet the emission targets related to material, energy, and mobility. Uncertainty in project goals results in ambiguous needs and specs, while uncertainty in methods (and technologies) may result in changes to project deliverables and challenges to project actors. Ydalir tried to mitigate this issue through conducting a series of dialogue workshops early in the planning stage, where interested parties participated in formulating a high-level master plan describing the project's long-term (zero-emission) vision. The master plan provides guidelines, boundaries, and a set of functional requirements targeting, i.e., mobility and energy, which leaves some room for developers to decide how to proceed based on their expertise.

Although IPP was not followed formally in Ydalir project; since ETS uses private purchasing contracts and the local municipality uses traditional public procurement process, we see that some elements of IPP were present in the strategic planning phase, such as market dialogues. The master plan translated the project's goals into functional requirements addressing issues related to material selection, design, energy, and mobility. Functional specs allow developers to apply their expertise and propose new solutions and technologies that buyers might be unaware of [33]. Thus, dialogue with the market, including developers and engineering companies in Ydalir project provided better specs and requirements to cope with the uncertainty in the project goals. Another benefit of the master plan is creating one standard for all interested private actors who wish to work in the project. This may help in reducing the complexity of project specs. Consulting the supply market seems rational because it also allows buyers to learn what the market has to offer in terms of new methods and latest technologies. Moreover, if a specific solution is not available in the market, a market dialogue may be utilized as a means to plan a future R&D with specialized suppliers.

4.4. External uncertainty

Some contextual factors are found to be related to project complexity, such as policies and regulations, and perceived risk from the market. They are crucial to maintaining internal project stability and continuity. Policies, for instance, facilitate sustainable development [18]. In the case, ETS planned to include ZEN requirement (master plan) in the purchase contracts between itself and developers, and in the zoning plans between the municipality and developers. However, while the purchase contracts covered all requirements, the municipality could not include all of them (i.e., passive house standards) in the zoning plans since the national building laws do not cover them. This political factor impacted the project and created several challenges, such as less control over the stakeholders who are reluctant to follow high environmental standards.

Linking with multiple actors from different public institutions can help address such issues up front, and hence avoiding being trapped in current regulations or leading to develop supporting policies. Moreover, market dialogues give suppliers visibility over the market and allow them to rethink their business strategies and consider new opportunities – based on which market or technology is emerging.

This may attract new, experienced suppliers and encourage less experienced suppliers to catch up. For example, companies lacking knowledge about neighborhood sustainability assessment tools (i.e., BREEAM Communities) may want to follow their peers to avoid losing a potential seat in ZEN markets.

4.5. Upfront increase for overall decrease?

Project complexity might spike at the outset of the project due to the pre-tender market dialogues (by inviting the market players). However, this temporary increase can be tolerated if it led to a reduction in overall project complexity and improved project outcomes. Furthermore, unlike the first three rationales, *market visibility* and *market links creation* serve buyers and suppliers to prepare for the procurement of future projects. This implicates that a ZEN project can use market dialogues as a means to diffuse ZEN-based knowledge and develop its ecosystem of actors.

5. Conclusions

We began our paper with the observation that SN literature has focused on end-user engagement more than market engagement activities, and stressed the need to explore in greater depth IPP to understand its potential in reducing complexity in ZEN projects. Our findings are twofold. First, we described four different sources of project complexity in ZEN. Second, IPP is shown to be an attractive governance tool that could potentially manage some of the project complexity in ZEN. Future research should explore in-depth case studies of IPP with dialogue practices, in order to measure the impact on complexity. Furthermore, we encourage practitioners and scholars to address the issue of complexity in the context of SN as it may assist in uncovering new challenges.

6. Acknowledgment

This article has been written within the Research Centre on Zero Emission Neighborhoods in Smart Cities (FME ZEN). The authors gratefully acknowledge the support from the Research Council of Norway and the ZEN partners, in particular Anna-Thekla Tonjer from Elverumvekst.

References

[1] UNDP 2018 Sustainable development goals. Available from: http://www.undp.org/content/undp/en/home/sustainable-development-goals/goal-11-sustainable-cities-and-communities.html.
[2] Wiik MK, Fufa SM, Kristjansdottir T and Andresen I 2018 Lessons learnt from embodied GHG emission calculations in zero emission buildings (ZEBs) from the Norwegian ZEB research centre. *Energy and Buildings* **165**:25-34.
[3] Koch A, Girard S and McKoen K 2012 Towards a neighbourhood scale for low- or zero-carbon building projects. *Building Research and Information* **40**(4):527-37.
[4] Valkering P, Beumer C, De Kraker J and Ruelle C 2013 An analysis of learning interactions in a cross-border network for sustainable urban neighbourhood development. *Journal of Cleaner Production* **49**(August 2012):85-94.
[5] Purtik H, Zimmerling E and Welpe IM 2016 Cooperatives as catalysts for sustainable neighborhoods – a qualitative analysis of the participatory development process toward a 2000-Watt Society. *Journal of Cleaner Production* **134**(Part A):112-23.
[6] Strasser H, Kimman J, Koch A, Mair am Tinkhof O, Müller D, Schiefelbein J and Slotterback C 2018 IEA EBC annex 63–implementation of energy strategies in communities. *Energy and Buildings* **158**:123-34.
[7] Oliver A and Pearl DS 2018 Rethinking sustainability frameworks in neighbourhood projects: a process-based approach. *Building Research and Information* **46**(5):513-27.
[8] Alhola K and Nissinen A 2018 Integrating cleantech into innovative public procurement process–evidence and success factors. *Journal of Public Procurement* **18**(4):336-54.
[9] Alhola K, Ryding S-O, Salmenperä H and Busch NJ 2018 Exploiting the Potential of Public Procurement: Opportunities for Circular Economy. *Journal of Industrial Ecology* **0**(0).

[10] Edquist C and Zabala-Iturriagagoitia JM 2012 Public Procurement for Innovation as mission-oriented innovation policy. *Research Policy* **41**(10):1757-69.
[11] Pelkonen A and Valovirta V 2015 Can service innovations be procured? An analysis of impacts and challenges in the procurement of innovation in social services. *Innovation: The European Journal of Social Science Research* **28**(3):384-402.
[12] Uttam K and Le Lann Roos C 2015 Competitive dialogue procedure for sustainable public procurement. *Journal of Cleaner Production* **86**:403-16.
[13] Sharifi A 2016 From Garden City to Eco-urbanism: The quest for sustainable neighborhood development. *Sustainable Cities and Society* **20**:1-16.
[14] Håkansson H 1982 An interaction approach. In: Håkansson H, editor. International marketing and purchasing of industrial goods: An interaction approach. Chichester: Wiley. p. 10-26.
[15] Araujo L, Dubois A and Gadde L-E 1999 Managing interfaces with suppliers. *Industrial Marketing Management* **28**(5):497-506.
[16] Donahue JD and Zeckhauser R 2011 *Collaborative Governance : Private Roles for Public Goals in Turbulent Times*. Princeton, N.J.: Princeton University Press.
[17] World Commission on Environment and Development 1987 *Our common future*. Oxford: Oxford press. 8-9 p.
[18] Shi Q, Yu T, Zuo J and Lai X 2016 Challenges of developing sustainable neighborhoods in China. *Journal of Cleaner Production* **135**:972-83.
[19] Luederitz C, Lang DJ and Von Wehrden H 2013 A systematic review of guiding principles for sustainable urban neighborhood development. *Landscape and Urban Planning* **118**:40-52.
[20] Choguill CL 2008 Developing sustainable neighbourhoods. *Habitat International* **32**(1):41-8.
[21] Baccarini D 1996 The concept of project complexity—a review. *International Journal of Project Management* **14**(4):201-4.
[22] Shenhar AJ and Dvir D 2007 *Reinventing project management: the diamond approach to successful growth and innovation*. Harvard Business Review Press.
[23] Williams TM 1999 The need for new paradigms for complex projects. *International Journal of Project Management* **17**(5):269-73.
[24] Yugue RT and Maximiano AC Year editor^editors. Project complexity and management processes2012: Project Management Institute; Published.
[25] Simon HA 1962 The Architecture of Complexity. *Proceedings of the American Philosophical Society* **106**(6):467-82.
[26] Davies A and Mackenzie I 2014 Project complexity and systems integration: Constructing the London 2012 Olympics and Paralympics Games. *International Journal of Project Management* **32**(5):773-90.
[27] Obwegeser N and Müller SD 2018 Innovation and public procurement: Terminology, concepts, and applications. *Technovation*.
[28] Lawther WC and Martin LL 2005 Innovative practices in public procurement partnerships: The case of the United States. *Journal of Purchasing and Supply Management* **11**(5):212-20.
[29] Edler J and Georghiou L 2007 Public procurement and innovation—Resurrecting the demand side. *Research Policy* **36**(7):949-63.
[30] Torvatn T and De Boer L 2017 Public procurement reform in the EU: start of a new era? *IMP Journal* **11**(3):431-51.
[31] Edler J and Yeow J 2016 Connecting demand and supply: The role of intermediation in public procurement of innovation. *Research Policy* **45**(2):414-26.
[32] Dubois A and Gadde L-E 2002 Systematic combining: an abductive approach to case research. *Journal of Business Research* **55**(7):553-60.
[33] Van Weele AJ 2018 *Purchasing & supply chain management*. 7th edition ed: Cengage Learning EMEA.

IOP Conf. Series: Earth and Environmental Science **352** (2019) 012014 doi:10.1088/1755-1315/352/1/012014

Validation of a Flexibility Assessment Methodology for Demand Response in Buildings

S O'Connell[1], G Reynders[2,3], F Seri[1] and M Keane[1]

[1]National University of Ireland, Galway, Ireland
[2]EnergyVille, Genk, Belgium
[3]Vito, Mol, Belgium

E-mail: s.oconnell19@nuigalway.ie

Abstract. Flexibility in Buildings is an integral part of the solution to address the electrical grid's challenges of grid balancing and renewable generation hosting capacity. One of the main barriers to greater participation of commercial and residential buildings in demand response schemes is the complexity and cost associated with assessing the flexibility of buildings. To overcome these barriers, an early stage flexibility assessment methodology was developed to provide stakeholders with actionable information in a concise and relevant way, so they can effectively evaluate the flexibility of their building and negotiate with aggregators for demand response participation. This paper validates the early stage flexibility assessment methodology in multiple buildings, demonstrating its ease of implementation and scalability. The validation was conducted at five pilot sites, in different geographical regions, activating a range of flexible sources through experiments on site.

1. Introduction

Buildings typically have power flexibility in the kW range, therefore, in most jurisdictions, they need an intermediary, known as an aggregator, to participate in demand side programmes. To do this, building operators must first assess the flexibility of the building and define its available range so they can negotiate with aggregators for contracts to deliver demand side services. At present, aggregators primarily target large users, e.g. industrial sites [1] offering few sources of flexibility such as backup generators [2]. However, to achieve the EU goals for renewable integration of more than 50% by 2030 and over 90% by 2050 [3], participation from large numbers of smaller buildings providing flexibility from multiple sources is required.

An early stage standardised four step flexibility assessment methodology was created and implemented by the author [4] as an enabler to increase participation of the building sector in demand response programmes. Increased participation is required for balancing the electrical grid and increasing renewable generation hosting capacity [5]. The methodology addresses the barriers of i) lack of clarity around energy flexibility potential, specifically the quantification of flexibility and the financial cost or technical effort required to activate it [1] and ii) difficulty in identifying, implementing and actuating numerous small sources of energy flexibility rather than a few large systems [6]. The methodology provides an easily implementable protocol for assessing the power flexibility of buildings, overcoming the requirement for hiring building energy experts or conducting detailed on-line data acquisition. Explicit and systematic source selection ensures flexible systems are not missed during cursory assessments and avoids time wasted on non-flexible systems in detailed assessments. The output of the methodology is a KPI label which may for example have applicability in the development of the Smart Readiness Indicator (SRI) for buildings [7] included in the 2018 re-cast of the EU Energy Performance of Buildings Directive (EPBD). The flexibility assessment methodology has the capability to operationalise the concept of building flexibility to a wider spectrum of society, enabling smart grid demand response roll-out to residential and small commercial customers.

To date, the majority of the available research in flexibility focuses on simulation [8] (Reynders 2018) [9] and published demonstration studies documenting implementation of demand response in real buildings are rare. This paper addresses the need for more published demonstration studies to enable the creation of benchmarks. Experimental validation was conducted in a single building previously [4] demonstrating the viability of the methodology. That work is extended in this paper by demonstrating the scalability and ease of implementation of the methodology through implementation and validation in multiple buildings and districts. Use cases from price-based and market-based demand response programs both current (peak shaving) and future (CO2 minimization) are used in the validation experiments to activate the available flexibility in the buildings and districts.

This paper is structured as follows: Section 2 outlines the flexibility assessment method; Section 3 details the experimental setup including the use cases; Section 4 presents and analyses the results of the validation and Section 5 documents the conclusions.

2. Flexibility Assessment Method

A four-step flexibility assessment methodology was developed by the author [4]. This section provides an overview of the methodology and demonstrates i) its scalability and ii) ease of implementation. The four steps in the methodology are Step 1: Systems, Loads, Storage & Generation Identification; Step 2: Flexibility Characterisation; Step 3: Scenario Modelling and Step 4: KPI Label, shown in Figure 1. To create a structure underpinning the methodology, relevant elements of the energy auditing standard ISO 50002 [10] were selected and modified by the author to address power and energy flexibility assessment rather than energy auditing.

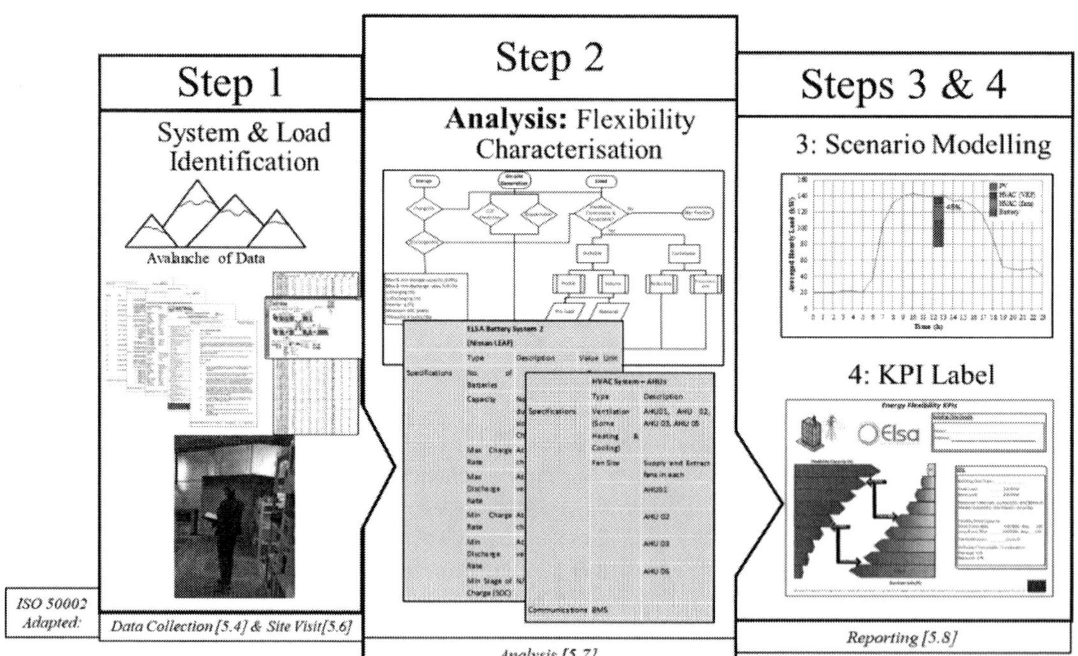

Figure 1. Four-step Flexibility Assessment Method

Step 1 identifies the power and energy systems in the building or site which have flexibility potential. To evaluate these, a site visit in conjunction with collecting data from as-built engineering drawings, documents and specifications is performed. Systems are assessed at a high level only. A detailed analysis of the flexible systems is then conducted in Step 2 following the flexibility characterization process. A triage process allows to select flexible sources in a systematic way and reduce the volume of building data, drawings and specifications to be analysed is significantly compared with standard energy auditing procedures. The power, in kW and energy, in kWh, characteristics for each flexible system are identified. In step 3, scenario modelling uses this data to visualise flexibility ranges These illustrate what impact flexibility will have on the power profile of the building being assessed which facilitates building operators and aggregators understanding of what a demand response event will look like for their building. Finally, step 4 captures the flexibility ranges of the building on a KPI label, providing a comprehensive summary of the key technical criteria which indicate the demand response potential of the site. This enables contract negotiation between building operators and aggregators and allows aggregators to quickly and easily assess buildings when creating portfolios of sites to meet the participation thresholds of grid operators.

2.1. Scalability

The scalability of the assessment method is demonstrated by its implementation at five pilot sites. The types of sites range from commercial buildings to a cluster of buildings to residential districts and so cover a wide range of current and future participants in demand response markets. To operationalise the concept of building flexibility to residential and small commercial customers, flexibility assessment needs to be quick, cost effective, repeatable and technically accurate. The validation of the methodology at five pilot sites demonstrates this, enabling roll out of demand response services to a wider spectrum of society.

2.2. Ease of Implementation

Assessments at each of the five pilot sites were performed by technical evaluators who were not expert in flexibility, using off-line data already available at each site. Training on the implementation of the methodology was provided by the author which consisted of a one-hour presentation, followed by two half-hour online technical support sessions. Evaluators were provided with documentation on guidelines for implementation, templates for capturing flexible system variables and a calculation tool for generating the KPI label. A survey of the evaluators determined that time for assessment was 1-2 weeks and level of difficulty was scored 3 on a scale of 1 -5 with 1 being very easy and 5 being very difficult. The evaluators found the level of training and support sufficient and would recommend the methodology to others.

The ease of implementation demonstrated through these multiple implementations by technical evaluators enables building operators to easily and cost effectively evaluate the flexibility of their building. The time frame for implementation is compared with that of a traditional Type 2 energy audit in which all energy systems are analysed in detail. The duration of a Type 2 audit varies depending on the building but based on the author's experience, conducting energy audits for the pilot sites in this work would take approximately 3 to 4 weeks [11]. By comparison, the flexibility assessment methodology reduces the time for assessment to 1-2 weeks, a reduction of approx. 60%.

Cost may be reduced by up to 80% as the methodology provides a systematic means of capturing the key technical information for flexibility without the individual implementing it requiring a detailed technical knowledge of the domain. This enables implementation by a technical person, e.g. a junior engineer, instead of a flexibility expert, achieving a cost reduction in excess of the time decrease.

While the implementation of the methodology did not require installation of additional equipment, validation of the flexibility ranges predicted by the methodology did. Details of the pilot sites and the experimental setup required for the validation are described in Section 3.

3. Experimental Setup

The flexibility method in Section 2 was implemented at five pilot sites. The types of sites include commercial buildings, a residential district and a cluster of DSO operation's buildings. A brief description of each site is given below. The development of an ICT Platform was required to enable simulation of demand response events, control of the flexible systems from a remote location and data acquisition from the building. Use cases were selected to demonstrate that based on this simple assessment procedure, the potential of buildings in a wide range of demand response schemes could be quantified.

3.1. *Pilot Sites*

Commercial Building, Sunderland, UK: The Skills Academy for Sustainable Manufacturing and Innovation (SASMI) building in Sunderland, UK is a 5,500 m^2 mixed-use commercial building. Its peak power load is of the order of 140 kW and its base load is between 20 to 40 kW. Flexible loads consist of HVAC loads coming from the Variable Refrigerant Flow (VRF) heat pump system and Air Handling Unit (AHU) fans; storage consists of 2nd life EV battery system, with an installed capacity of 48 kWh and on-site renewable generation consisting of a 50 kWp PV array.

Building Cluster, Terni, IT: The ASM Terni pilot site is a cluster of buildings comprised of ASM Terni's electricity Distribution System Operator (DSO) three operations buildings. Flexibility is provided by 180 kWp and 60 kWp PV arrays and 2nd life battery storage with an operating capacity of 36 kWh. Base load varied between 50 kW and 90 kW and typical peak load was 150 kW.

Commercial Building, Paris, FR: The Ampère building is a 14,000 m^2 commercial office building located in the La Defence area of Paris. It recently underwent a deep retrofit and is now certified as a sustainable building with HQE and BREEAM certification. Flexibility was provided by a 22 kWh capacity 2nd life battery system. Peak load is of the order of 250 kW.

Commercial Building, Aachen, DE: The E.ON ERC building is located on the RWTH university campus. Load flexibility is provided by HVAC AHU fans. The building has a peak load of 250 kW.

Residential District, Kempten, DE: The test site of Kempten consists of six apartment buildings hosting a 2nd life battery storage system with 66 kWh installed capacity and a 37 kWp PV array. There are no flexible loads. Space heating and hot water are not provided from electrical sources hence the electrical load is low compared with the commercial buildings. Combined peak load is 8.5 kW with an average daily consumption of 45 kWh.

3.2. *ICT Platform & Simulation of Grid Signals*

An ICT platform was installed at each of the pilot sites and used to validate the flexibility as predicted by the assessment methodology. In an ideal future scenario, the methodology would be implemented and validated on-line in an automated way. However, at present, a technology gap exists which prevents this being implemented in large numbers of buildings in a cost effective and scalable way. ICT platform integration with existing building systems is bespoke, complex, time consuming and expensive. Until ICT platforms for building flexibility reach plug-and-play capability at TRL 9, implementation as described in this paper is required.

The ICT platforms control the loads and storage on site to provide flexibility services in response to simulated demand response signals. The pilot site ICT system architecture for two of the buildings, SASMI in the UK and Ampère in France, is shown in Figure 2. The development of this system architecture was led by the author for previous work in optimising energy management in buildings [12] and districts [13]. For demand response applications, the ICT platform was adapted to incorporate simulation of grid or aggregator signals using the OpenADR protocol and the integration of the 2nd life battery management system using a web service API. Meter and equipment data are extracted from the site Building Management System (BMS) using an OPC (OLE for Process Control) server and may be read in real time. In addition, this data is continuously stored in a database in the middleware layer. Set points are sent from the ICT platform to equipment in the building via the BMS. The ICT platform is physically installed at the building and controlled remotely using a remote desktop application. All experiments were conducted remotely.

3.3. 2nd Life EV Batteries

Each of the pilot sites has a stationary battery storage system consisting of 2nd life EV batteries installed. These are early prototype systems at TRL 5/6. For example, for the system installed at the Sunderland pilot site, first life capacity was 72 kWh and 2nd life capacity was 48 kWh. For further information on the 2nd life EV battery system, see www.elsa-h2020.eu.

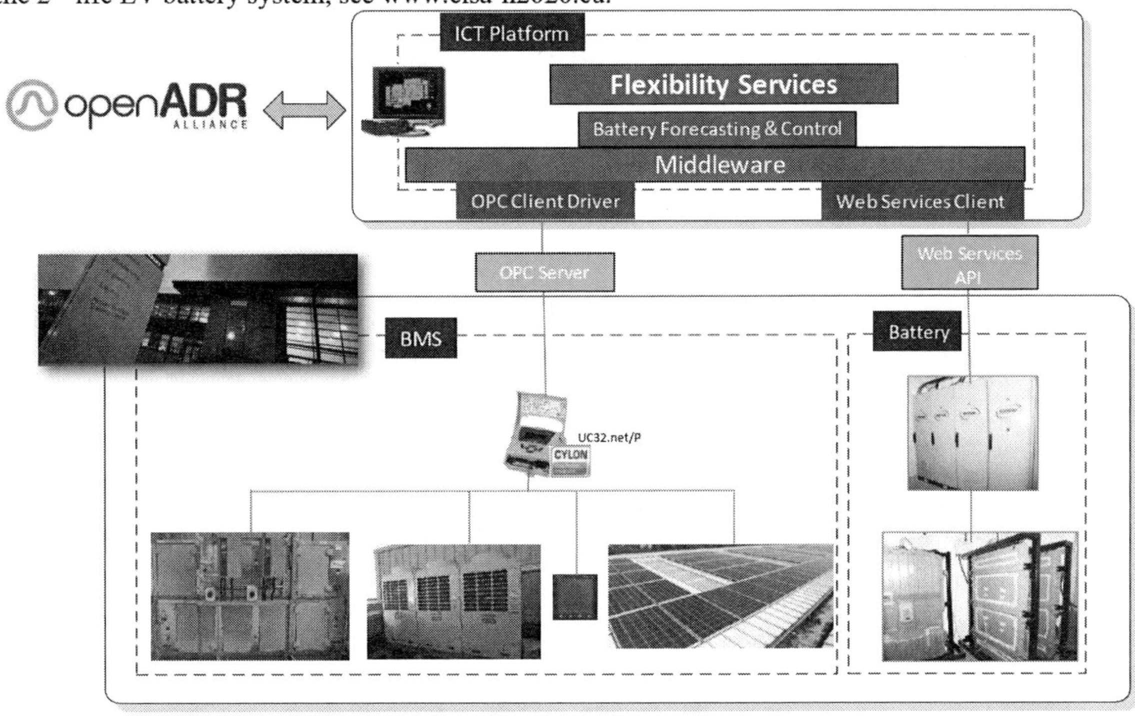

Figure 2. ICT Platform System Architecture

3.4. Use Cases

Demand response use cases selected for the validation experiments spanned a range of services and included peak shaving, PV power smoothing, CO2 minimisation and a market-based programme that requires the building to respond to a specific grid request.

Peak shaving is a price-based programme. It is the most widely used demand response service globally with Ireland [14], the US [15], France [16] and China [17] including it in their demand side services. It involves reducing grid import of electricity during periods of peak consumption, e.g. between 11am and 3pm. This use case was implemented at the Commercial building in Paris and the cluster of buildings in Terni.

A market-based programme that requires the building to respond to a grid request intra-day within a short timeframe is a more challenging use case than peak shaving. With peak shaving the building operator knows the price and time schedule sometimes more than a year in advance. Market based programmes are more dynamic, as buildings may be called to respond within an hour, for example in Ireland's Short Term Active Response (STAR) [14] programme. This use case was implemented at the Commercial building in Sunderland.

CO_2 based demand response signals have been proposed as an alternative to price based market signals [18] to incentivise electricity use or reduction in times of high or low renewable generation on the grid, respectively. For the CO_2 minimisation use case proposed, average CO_2 emissions per kWh are estimate at 416.58 g/kWh based on 2017 ENTSO-E generation mix for Germany. In a future grid scenario where hourly or real time generation emissions are available, this may be used by businesses who wish to minimise their carbon footprint or by grid operators to maximise renewable generation consumption. This use case was implemented at the Aachen building.

PV power smoothing is used to mitigate PV generation variability [19]. It requires storage coupled with PV and is activated at the request of the grid operator. The objective of this use case is to smooth PV peak production by storing excess renewable electricity generated to reduce grid export. This use case was implemented at the Kempten Residential District.

In validating the flexibility assessment methodology with the above use cases, the available capacity of the 2nd life EV batteries was used in the calculation of the predicted value. As the 2nd life battery management system is an early prototype at TRL 5/6, still under development, during many of the experimental use case validations the available capacity varied.

4. Results

Validation of the flexibility assessment method was conducted at each of the pilot sites. The output of the flexibility methodology is a range of flexibility the building has the capability of offering. To validate this, predicted flexibility for the particular use cases were calculated based on a selection of the building systems appropriate to that use case, and a calculation of the maximum flexibility achievable for the use case time period. The actual flexibility is the flexibility measured for a specific use case event and one realisation of that event [20]. The ICT platform was used to trigger demand response events for each use case and measure the corresponding load reduction or increase at the building, cluster or district. The actual results were compared with the predicted flexibility for and both are shown for each pilot site in Table 1. The sources of flexibility are categorised by load, storage and generation as evaluated in Step 1 and Step 2 of the assessment methodology. Load flexibility is denoted as F^L, storage flexibility as F^S and renewable energy flexibility as F^{RES}. For the pilot sites, F^{RES} is provided by PV, F^S by the 2nd life EV battery system and F^L by HVAC systems. Flexibility is expressed as a percentage of peak load.

Table 1. Results for Pilot Sites

Pilot Site Location	**Type**	**Sources**	**Use Case**	**Use Case Flexibility (%)[1]**		
				Predicted	**Actual**	Error
Sunderland, UK	Building	F^{RES}, F^S, F^L	Intra-day Grid Request	36%	33%	9%
Terni, IT	Cluster of Buildings	F^{RES}, F^S	Peak Shaving	90%	81%	10%
Paris, FR	Building	F^S	Peak Shaving	9%	7%	22%
Aachen, DE	Building	F^L	CO_2 Minimisation	3%	3%	0%
Kempten, DE	Residential District	F^{RES}, F^S,	PV Power Smoothing	103%	106%	3%

1. Use case flexibility is expressed as a percentage of typical peak load for each site. The predicted and actual flexibility for each use case event was calculated based on the load increase or decrease divided by the typical peak load.

4.1. Analysis of the Results

In Table 1, the predicted and actual flexibility was largest for the Terni site and Kempten residential district. For Kempten, flexibility was over 100% of the site's peak load at 103% predicted and 106% actual. The PV output in Kempten during the event was greater than predicted, with the result that the actual flexibility was 3% higher than predicted. For all the other pilot sites, actual flexibility was lower than predicted. Both Terni and Kempten had large PV installations and, in Kempten's case, a large battery system relative to the electrical load of the site. As heating in Kempten was not provided by electrical sources, the electrical load was low, thus, even though the battery capacity was similar to that of Sunderland, it was able to provide a much larger percentage of flexibility. The Sunderland building was the only building to combine all three sources - load, storage and generation, and had predicted and actual flexibility of approximately one-third of its peak load. The sites with only one source, Paris and Aachen buildings, had the lowest flexibility at 7% and 3% of peak load respectively. It is worth noting that load flexibility, particularly HVAC, is generally linked to building physics and floor area whereas there may be more scope to add additional flexibility through generation and storage at the building scale. From the results in can be seen that i) number of available flexible systems, ii) size of installed renewable generation and iii) installed storage capacity are the three factors which most influence the flexibility of buildings or sites.

4.2. Validation

The error between prediction and accuracy for each pilot site is given on the right-hand side of Table 1. This show flexibility prediction within a 10% error range for four out of the five pilot sites. The site with the highest error, the commercial building in Paris, had a very low predicted flexibility, 9% of peak load, therefore the error was large relative to the quantity of power available to flex. For this building, only one source of flexibility was used, the battery system. It was found by [21] that prediction accuracy improved when the number of sources, in their case dwellings, was increased, which correlates with the validation results.

For smart grid applications, prediction accuracy of +/- 10% is considered optimal [22] [23]. However, errors of up to 36% in flexibility prediction for heat pumps [24] were considered acceptable. All of the sites are within the 36% range while the majority of the pilot sites are within the 10% error range.

4.3. Benchmark Comparison

Benchmarking is required to understand how the pilot site's flexibility compares to that of a typical site. As previously mentioned, standardised benchmarks are not yet available, therefore demonstration studies with i) a large number of buildings and ii) systems providing flexibility similar to the pilot sites were selected as benchmarks for the pilot sites. Benchmark 1 involved 28 buildings providing HVAC flexibility who participated in a utility led demand response programme in California [17]. Benchmark 2 consisted of a demonstration project with similar sources of flexibility to the pilot sites, namely PV, battery storage and loads [25]. These studies demonstrated a maximum range for flexibility between 18% to 56% with average flexibilities between 7% to 9% for Benchmark 1.

Comparing the flexibility ranges with the benchmarks in Table 2, the sites with multiple sources have flexibility equivalent to, or greater than, the benchmarks. The Sunderland building was within the range of Benchmark 1, which was expected as it is a commercial building similar to those in the comparison study, and exceeded the range for Benchmark 2 which was surprising as the systems installed were similar. The Terni site and the Kempten residential district far exceeded the benchmarks. However, this is not surprising as the installed capacity of the PV and storage systems is large compared with the peak load of the sites. The Paris and Aachen sites with are not within the maximum range for either benchmark but the Paris building is within the average range for Benchmark 1, this is due to only one source of flexibility being activated in each, and additionally in Paris's case, the battery system being small relative to the peak load of the building.

Table 2. Benchmarking Comparison

	Max flexibility [%]	Average Flexibility [%]	Sunderland UK	Terni, IT	Paris, FR	Aachen, DE	Kempten, DE
Benchmark 1	28 - 56	7 – 9	Within max	> max	Within average	< average	> max
Benchmark 2	~18	-	> max	> max	< max	< max	> max

5. Conclusions

Validation of the early stage, four step flexibility assessment methodology was conducted at five pilot sites using an ICT platform whereby grid signals were simulated using OpenADR protocols. Specific demand response use cases were implemented to trigger flexibility events at each of the sites and the resulting reduction in grid import electricity was measured. Sites with multiple sources of flexibility and large storage or renewable generation systems delivered higher levels of flexibility than those with single sources. Comparing actual flexibilities achieved to the predicted values from the methodology, four out of the five pilot sites were within 10% of the predicted flexibility. Benchmarking the results against other demand response demonstration studies indicated that three of the four sites were within or above the maximum range of flexibility, one was within the average range and one below average.

Acknowledgements
This research was conducted under the scope of the International Energy Agency's (IEA) Energy in Buildings and Communities (EBC) Annex 67 'Energy Flexible Buildings' and was funded by the Sustainable Energy Authority of Ireland and the EU project ELSA (Energy Local Storage Advanced system) which received funding from the European Union's Horizon 2020 research and innovation programme under grant agreement number 646125.

References

[1] Ofgem 2016 *Aggregators – Barriers and External Impacts* (London: UK Office of Gas and Electricity Markets / PA Consulting Group)
[2] Yi Y, Zheng A, Shao X, Cui G, Wu G, Tong G and Gao C 2018 *IEEE Int. Conf. on Energy Internet (ICEI), Beijing, China,* 72-76
[3] European Climate Foundation 2010 EU Roadmap 2050: practical guide to a prosperous, low-carbon Europe (The Hague: European Climate Foundation)
[4] O'Connell S, Reynders G, Seri F, Sterling R and Keane M 2019 *Intl. J. of Building Pathology and Adaptation* (in press)
[5] European Commission 2016 *European Commission Decision C (2016) 4614 of 25 July 2016*
[6] Annala S, Lukkarinen J, Primmer E, Honkapuro S, Ollikka K, Sunila K and Ahonen T 2018 *J. of Cleaner Production* **195** 1139-1148
[7] Verbeke S et al. 2018 *Final Report. Study accomplished under the authority of the European Commission DG Energy* 2017/SEB/R/1610684 (Brussels: DG Energy)
[8] Reynders, G, Amaral Lopes R, Marszal-Pomianowska A, Aelenei D, Martins J and Saelens D 2018 *Energy & Buildings* **166** 372–390
[9] Grønborg Junker R, Azar A, Amaral Lopes R, Byskov Lindberg K, Reynders G, Relan R and Madsen H 2018 *Applied Energy* **225** 175–182
[10] ISO 50002:2014 International Organization for Standardization (Geneva: ISO)
[11] O'Connell S 2010 Development of new Commercial Buildings Energy Audit Methodology based on the EED principle *Report for the Commercial Buildings Special Working Group* (Dublin: SEAI)
[12] Valdivia V et al 2014 *IEEE 5th Int. Symposium on Power Electronics for Distributed Generation Systems (PEDG), (Galway)* (New York: IEEE Xplore)
[13] Monti A, Pesch D, Ellis K and Mancarella P 2017 *Energy Positive Neighborhoods and Smart Energy Districts: Methods, Tools and Experiences from the Field* ed A Monti et al (Amsterdam: Academic Press/Elsevier) Chapter 5
[14] Eirgrid 2018 *The DS3 Programme: Delivering a Secure, Sustainable Electricity System* (Dublin: Eirgrid)
[15] Piette M A, Watson D, Motegi N, Kiliccote S and Linkugel E 2006 *14th National Conference on Building Commissioning (San Francisco)*
[16] RTE 2014 *Mecanisme de capacite guide pratique* (Paris: Le reseau de transport d'electricte)
[17] Li H, An Q, Yu B, Zhau J, Cheng L and Wang Y 2017 *Energy Procedia* **105** 2207 – 2213
[18] Stoll P, Brandt N, Nordström L 2014 *Energy Policy* **65** 490-500

[19] Stroe D, Zaharof A and Iov F 2018 *Energy Procedia* **155** 464-477
[20] Lapedra A, Croce V, Ziu D, LeCam M, Bode G, Gross S and Eberl T 2018 *D6.3 Results of service evaluation* ELSA project report (Brussels: European Commission)
[21] Wang A, Li R and You S 2018 *Applied Energy* **232** 89-100
[22] Borelli D, Repetto S and Schenone C *Thermal Science and Engineering Progress* **6** 436-446
[23] MacDougall P, Ran B, Huitema G and Deconinck G 2017 *Energies* **10** 1673
[24] Neupane B 2017 *PhD Thesis* (Aalborg: Aalborg University)
[25] Siebert N et al 2015 *IEEE PowerTech (Eindhoven)* (New York: IEEE Xplore)

Microclimate analysis of a university campus in Norway

Johannes Brozovsky[1], Sara Corio[2], Niki Gaitani[1], Arild Gustavsen[1]

[1] Department of Architecture and Technology, Norwegian University of Science and Technology – NTNU, 7034 Trondheim, Norway
[2] Department of Civil, Chemical and Environmental Engineering (DICCA), University of Genoa, 16145 Genoa, Italy

Corresponding email: johannes.brozovsky@ntnu.no

Abstract. Climate responsive urban design that makes use of passive heating and cooling strategies and enhances pedestrian comfort is a key element to a sustainable society. Knowing the impact of different materials, building typologies and arrangements and a building's sensitivity to microclimatic conditions, the energy balance can be enhanced. This study aims to evaluate the microclimatic conditions at the campus of the Norwegian University of Science and Technology, located in Trondheim, Norway. A numerical model was used and validated with punctual measurements on site. The results for air temperature showed a considerable seasonal variation with highest spatial temperature differences in summer and lowest in winter. Regarding the analysis of the wind field, east-west passages were identified as problematic. The influence of the seasonal changing leaf area density of the vegetation had a visible but minor influence on the wind field. The calibrated model will be used for further research on the influence of new building bodies on microclimate and thus building energy demand and pedestrian comfort.

1. Introduction

Ongoing urbanisation, population growth and a steadily warming climate put enormous pressure on cities worldwide. Environmental conditions and anthropogenic pollution of the air, soil and water pose an increasing threat not only to the ecological equilibrium but also to people's health [1]. Through their fabric, consumption of resources, emissions, shape and pattern, urban areas are a massive interference with nature. They are able to significantly change meteorological conditions at the microscale (< 2 km; microclimate) and mesoscale (< 200 km; mesoclimate) [2]. One of the most well-known phenomena of human-induced modification of climate is the urban heat island effect (UHI). It describes the distinct warmth of an urban area compared to its rural surroundings [3]. Many scientific publications have reported the UHI across virtually all climate zones and settlement sizes [4]. Urban-rural temperature differences (UHI magnitude) typically range from 1–3 K. Occasionally, UHI magnitudes of 10–14 K were reported during anticyclonic conditions [5, 6].

The summer heat wave of 2003 in Europe is an alarming example of how devastating climatic conditions can affect our lives. In France alone, 15 000 excess deaths were registered during only three weeks in August [7]. Seven years later, a heat wave struck Moscow with close to 11 000 excess deaths during the 44-day heat wave period [8]. Therefore, and because of increasing city populations, the evaluation of the urban environment has drawn ever-increasing attention in the past decades to make cities more resilient to elevated temperatures. Especially with the advancement of computational power, numerical tools are increasingly used for microclimate analyses [9]. They allow for instance investigating the impact of new building projects within the development of urban districts on local microclimatic conditions [10], the effectiveness of mitigating measures for the urban heat island UHI [11], or the evaluation of outdoor comfort [12].

However, not only temperate or warm-climate regions are affected by climate change and rising temperatures. In 2018, for the first time since the beginning of the recordings in 1920, the arctic city of Tromsø in northern Norway experienced a "tropical night" where temperatures did not drop below 20 °C. Additionally, only two out of the last 30 years had a lower mean annual temperature than the mean of the norm period from 1961 to 1990 in Norway [13]. Studying urban microclimate thus also gains interest in cold climate countries where summers are usually cooler. While numerical studies on urban microclimate are quite common especially since the beginning of the 2000s, in high latitude locations close to the Arctic Circle they are scarce [9].

The 0.26 km² Gløshaugen university campus in Trondheim (Norwegian University of Science and Technology, NTNU), Norway (63.4° N, 10.4° E) is currently in the early stages of redevelopment. About 90 000 m² of gross floor area are planned to be added to the current building stock until 2027. This study aims to investigate the current microclimatic conditions of a part of the campus (see Figure 1 and 2).

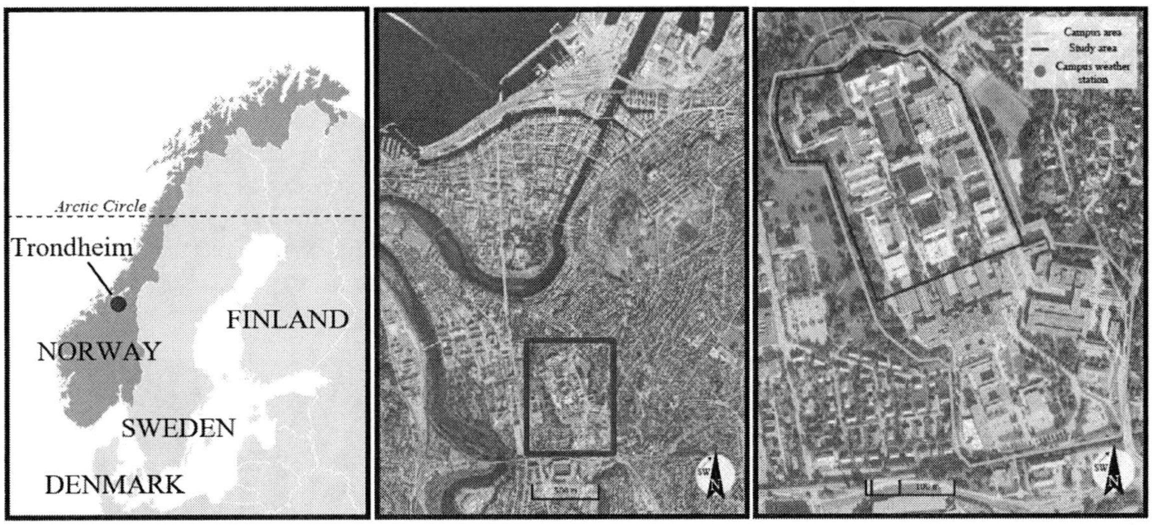

Figure 1. Location of the study area of NTNU campus within Norway and Trondheim (aerial photographs from the Norwegian Mapping Authority www.kartverket.no).

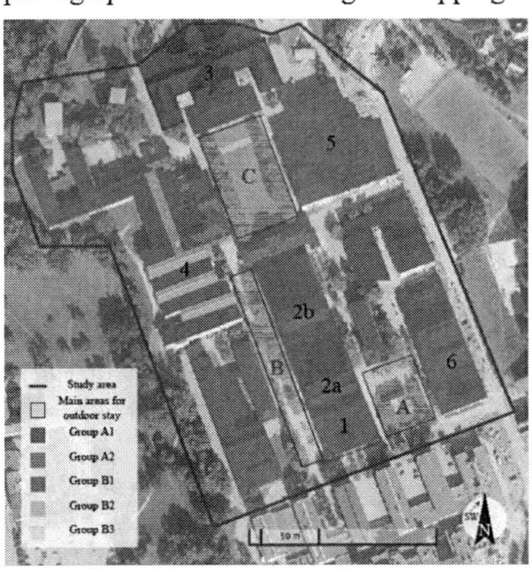

Figure 2. The study area with the building categories and the location of the main areas of outdoor stay: a small grass area (*A*) between the cafeteria (*1*) and the *Berg*-building (*6*) with seating possibilities; a pedestrian zone (*B*) west of the central buildings 1 and 2 (*2a* and *2b*) with several seating possibilities close to the cafeteria; a grass area (*C*) behind the main building (3), east of the *Elektro* building (*4*) and west of the VATL (*5*) with deckchairs during the summer.

IOP Conf. Series: Earth and Environmental Science **352** (2019) 012015 doi:10.1088/1755-1315/352/1/012015

2. Methodology

In this paper, two methods are being applied: field measurements and numerical simulations. The measurements are used to record the punctual microclimate conditions and to validate the simulation model of the current campus layout. Validation is defined as the process of determining if a computational simulation represents the real world and how accurately the computational results compare with experimental data [14, 15]. For the numerical simulation, the widely applied three-dimensional non-hydrostatic model for surface-plant-air interaction ENVI-met v. 4.4 is used [16].

2.1. Field measurements

The on-site field survey consisted of two parts: continuous meteorological measurements at a weather station and mobile infrared (IR) thermography. The meteorological measurements are obtained from a weather station which is located on the roof of the ZEB Test Cell Laboratory (TCL), south of the study area, see Figure 1. The weather station records air temperature, relative humidity, global horizontal radiation, wind speed and wind direction in one-minute intervals. The data is then compiled into one-hour averages to serve as validation input in the simulation model.

Surface temperature measurements were derived using an infrared camera (FLIR E60). The IR measurements were taken during calm, clear-sky conditions on the morning of 11 April 2019. The ambient temperature that day ranged from a low -1 °C to a high +3 °C. The IR camera is able to operate in a temperature range between -20 °C to +120 °C and has an accuracy of ±2 °C or ±2 % of reading for an ambient temperature between 10 °C and 35 °C [17]. The surface temperature of the most representative buildings for each category (see Table 1) and ground surface temperatures have been measured (see Figure 3). For the validation process, the five groups of buildings (*A1, A2, B1, B2, B3*) were sub-categorised based on the main characteristics such as the main orientation, the surroundings, the presence of trees, and the closeness and size of the plants. For each sub-category, a reference building was chosen to be analysed specifically (see Figure 4).

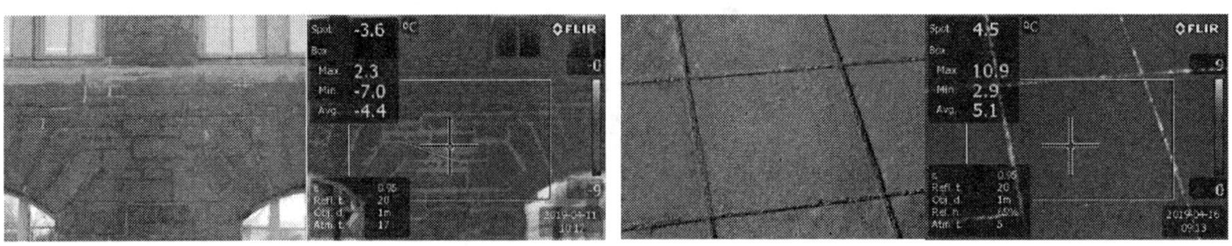

Figure 3. IR measurements of a building surface with granite stone (left) and concrete pavement (right).

2.2. Study area

Trondheim is Norway's third largest city by population with close to 200 000 inhabitants and is located at a 130 km long and 10–20 km wide fjord in Central Norway. The location's climate is subarctic (Dfc) according to the Köppen-Geiger classification system [18]. The 0.23 km² Gløshaugen campus of NTNU is located about 2 km south of Trondheim's city centre on a hill plateau (see Figure 1). In order to reduce the simulation time, a smaller portion of the campus (0.10 km²) was selected. This decision is supported by the fact that the main areas for people are also located in this smaller subarea (see Figure 2). This part has a relatively uniform elevation of about 47 m above mean sea level.

The mean annual temperature of Trondheim's official weather station in Voll is 4.8 °C. The average annual precipitation is 855 mm and the prevailing wind direction is south west (225°). As the official weather station is not in the vicinity of the campus, meteorological recordings from the ZEB TCL, located 400 m south of the study area, are used (see Figure 1 and Figure 2).

2.3. Model and computational domain

Aerial photographs from the Norwegian Mapping Authority (www.kartverket.no) were used to determine the positions of the different surface and vegetation cover types in the 3D model. Figure 4 shows the ENVI-met simulation model with view from the south. On-site inspections on the measurement date provided specific characteristics of the buildings, materials, and vegetation (including species) and confirmed that the information from the aerial photographs was up to date.

The grid of the calculation domain has dimensions of 416(x)_400(y)_94.5(z) metres and consists of 104_100_32 cells at each axis respectively. The lowest grid cell is split into five sub-cells with 0.6 m height each. Concerning the interference coming from boundaries to the flow and for the flow to develop, the distance between the objects and the domain boundaries was arranged based on ENVI-met specifications. For that, the height of the domain was modelled more than two times the highest building (Central Buildings' height: 45 m) and the horizontal distance to the model boundaries was at least equal to the closest building's height. In addition, four nesting grids were added around the core of the 3D model. For turbulence closure, ENVI-met uses the Mellor-Yamada turbulence model from 1982 [19].

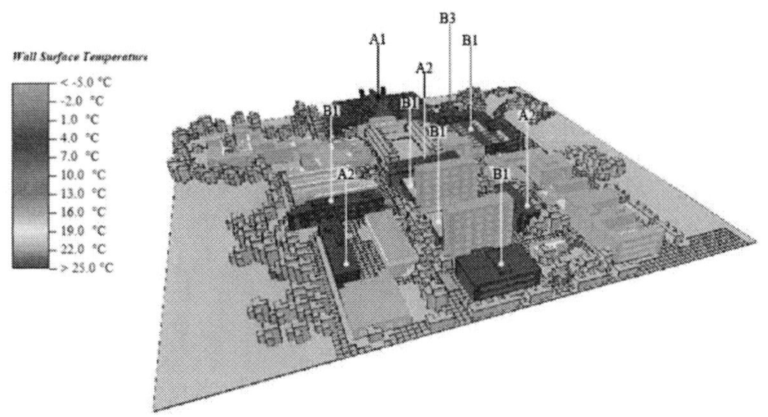

Figure 4. View of the ENVI-met simulation model from the south. The reference buildings for each category are indicated in brighter colours and are representative of sub-categories based on different conditions of orientation and surroundings.

2.3.1. Soils and surfaces. Three different surface types predominate in the area of interest. This includes the basic soil type and sealed surfaces like streets and pavements. Soil maps of the region showed that the model area's soil is mainly characterised by its clay content [20]. The remaining areas of the domain which are not covered by vegetation or buildings are impermeable surfaces. They were modelled based on the ENVI-met standard library values (surface albedo, aerodynamic roughness length, emissivity) for *Asphalt Road* and *Pavement (Concrete), used/dirty* [21].

2.3.2. Vegetation. Gløshaugen campus is embedded within a park-like environment and features green spaces and a great number of trees, shrubs and hedges throughout the area. For simplification purposes, vegetation types were limited to six different types of trees (of only three species), two hedge types, and one grass type. The existing vegetation from the program database has been scaled to fit the plant sizes on campus. Following tree species were modelled: *Norway Maple* (10, 15 and 20 m), *Larch* (20 m) and *Scots Pine* (5 and 10 m). The two hedge-types were 1 m high (one modelled as a conifer, one as deciduous) and the grass height was set to 0.1 m. The deciduous trees were assigned an annual schedule to regulate the leaf area density (LAD) according to the seasonality of foliation.

2.3.3. Buildings. The buildings at the campus were grouped into five categories (see Figure 2; Table 1) according to the wall's outermost thermally relevant material layer: stone, brick, concrete, glass and wood. Most of the buildings were built with concrete as an external wall material, followed by brick and wood. It is worth noting that the wooden buildings are rather small and situated on the north-eastern and north-western side of the campus. Moreover, for some of the buildings, glass is the main material for both walls and roofs (*Elektro* buildings). The roof materials were identified for each category. The properties of the materials used in the ENVI-met model are the default ones for concrete and glass, while for granite stone, solid brick and wood, new material database entries were created. All the materials have an emissivity of 0.90 except for the wood (0.86).

IOP Conf. Series: Earth and Environmental Science **352** (2019) 012015 doi:10.1088/1755-1315/352/1/012015

Table 1. Categories of the buildings in the ENVI-met model.

Category	Year of construction	Wall material	Roof Material
A1	1900-1950	Stone: granite	Wood: spruce
A2	1900-1950	Solid brick	Wood: spruce
B1	1951-2000	Concrete: filled blocks	Concrete: default
B2	1951-2000	Glass: clear float	Glass: clear float
B3	1951-2000	Wood: spruce	Wood: spruce

2.4. Boundary conditions

For the evaluation of the microclimatic conditions, simulations were performed for four days: 21 March (vernal equinox), 21 June (summer solstice), 23 September (autumnal equinox) and 21 December (winter solstice). The boundary conditions of the model are based on the records of the year 2018 from the weather station on campus. The hourly values of air temperature and relative humidity, the mean daily wind speed and the prevailing SW (225°) wind direction for Trondheim, were applied (Table 2). As the program does not allow for incorporating snow covered surfaces, freezing or thawing of water, the same surface properties and physical conditions for all simulated dates were used [22].

Table 2. Boundary conditions and statistical meteorological values of the simulated days during simulation time (T_{min}, T_{max} = temperature min./max. and time of occurrence in local standard time (LST); $\bar{v}_w$ = mean wind speed; $\bar{\phi}$ = mean relative humidity; t_{sr} = LST of sunrise; t_{ss} = LST of sunset; $t_{sim,s}$ = LST of simulation start; $t_{sim,e}$ = LST of simulation end; $t_{sim,tot}$ = total time of simulation).

	T_{min} (LST) [°C]	T_{max} (LST) [°C]	$\bar{v}_w$ [m/s]	$\bar{\phi}$ [%]	t_{sr} (LST)	t_{ss} (LST)	$t_{sim,s}$ (LST)	$t_{sim,e}$ (LST)	$t_{sim,tot}$ [h]
21.03.	2.6 (05:00)	4.2 (16:00)	1.7	85.1	6:17	18:35	6:00	19:00	13.0
21.06.	8.3 (06:00)	12.2 (18:00)	1.8	73.5	03:02	23:37	03:00	24:00	21.0
23.09.	5.5 (19:00)	8.0 (15:00)	2.5	77.7	07:03	19:16	07:00	20:00	13.0
21.12.	0.2 (06:00)	0.9 (12:00)	2.6	55.0	10:01	14:31	06:00	15:00	9.0

3. Results and discussion

3.1. Validation

The obtained data from the field measurements of the reference buildings were used to validate the results derived from the ENVI-met simulation from 10 April at 20:00 to 11 April at 13:00. The average difference between measured and simulated surface temperatures of the thirty measurement points was 4.9 K. However, ENVI-met presents a tendency to overestimate the surface temperature values, where IR-measured values were below or close to 0 °C (see Figure 5, eight values in total). For those values, the simulations surpassed the IR measurements by 7.9 K on average. The average deviation from the measurements of all other validation points was 2.8 K.

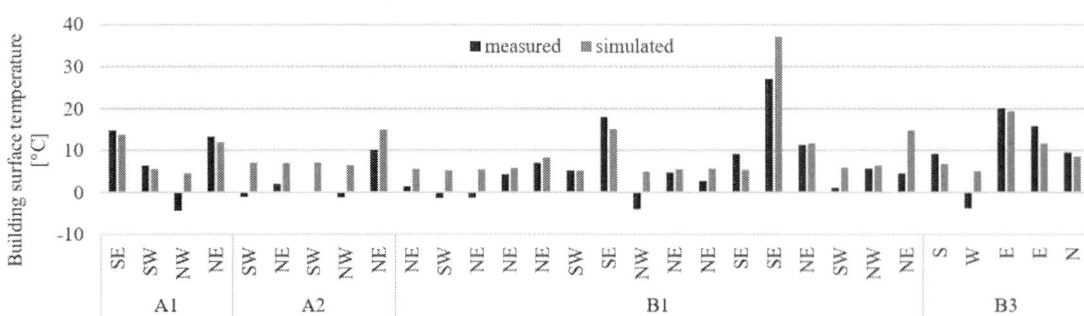

Figure 5. Measured (IR) and simulated building surface temperatures of the building categories on 11 April for different orientations.

Apart from the larger deviations at low temperatures, the magnitude of error is scattered arbitrary across the measurement points. Therefore, no systematic error for example regarding orientation, surface material, closeness to vegetation or point in time could be identified.

3.2. Microclimate analysis

Figure 6 illustrates the simulated air temperature field at 1.5 m height (pedestrian level). The unshaded south and west-facing surfaces get heated up by the exposure to solar radiation the most. Therefore, areas close to those surfaces show the highest air temperatures at pedestrian level. Especially the inner corner in the north west of the campus, where wind speeds are particularly low, represents a hot spot throughout the year, except for winter. The simulated values of air temperature correlate to the different surface types (albedo, emissivity) together with the wind field pattern and solar access. The materials with high heat storage capacity (i.e. asphalt, concrete pavement) present higher air temperatures than the vegetated areas.

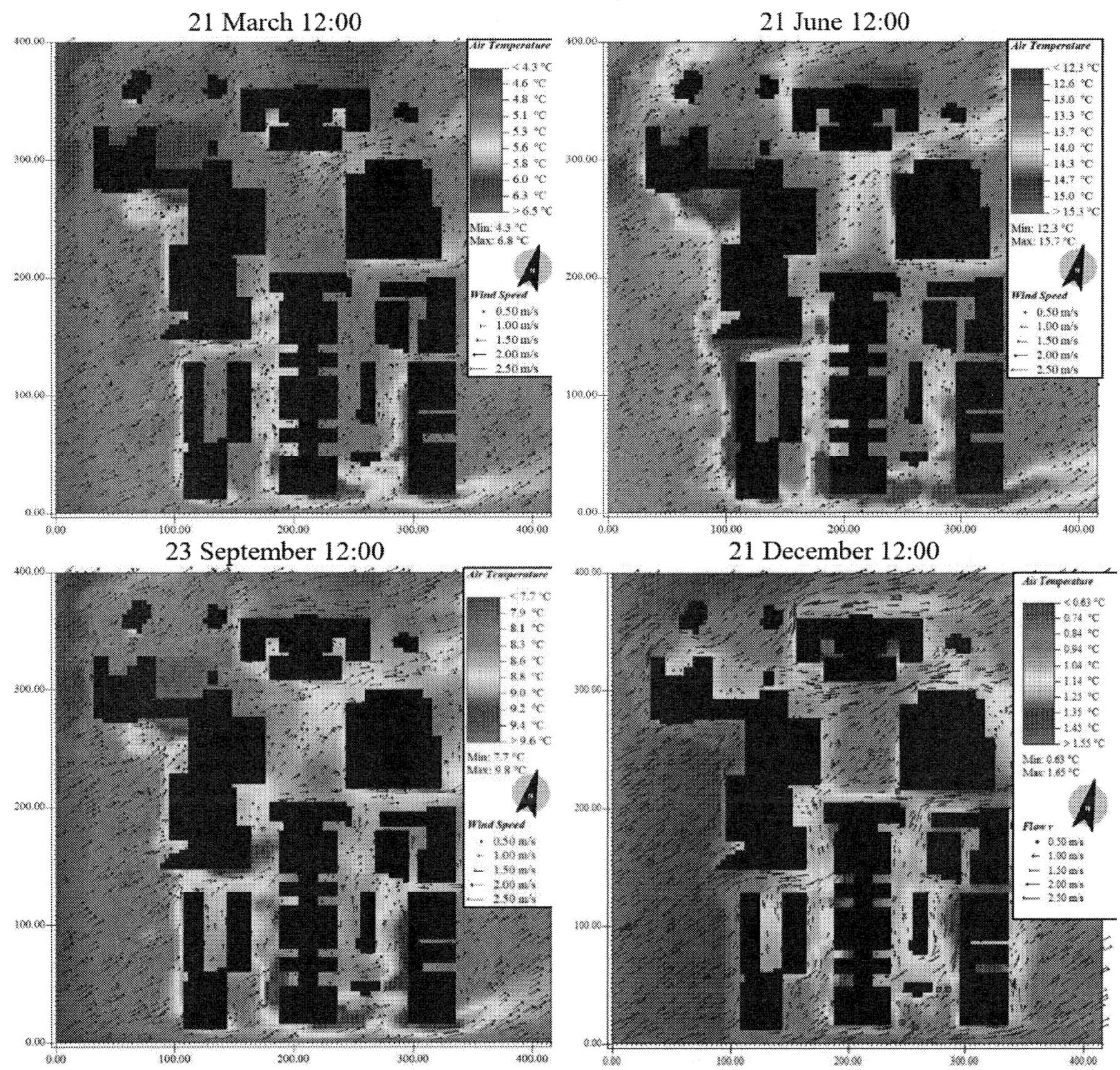

Figure 6. Air temperature and wind field at 12:00 at pedestrian level (1.5 m above ground) for the four simulated days.

IOP Conf. Series: Earth and Environmental Science **352** (2019) 012015 doi:10.1088/1755-1315/352/1/012015

In the colour map of 21 December, the grassed areas behind the main building (*C*), east of the cafeteria (*A*) and the smaller patch west of the cafeteria are recognisable in terms of contrasting colour. On the contrary, the paved and asphalted surfaces around the central buildings (*2a, 2b*), and south of the VATL (*5*) show the highest temperatures. The cooler patches east of the central buildings are the areas where hedges and trees lead to lower air temperatures. The grassed areas surrounding the campus are generally cooler all year round and with exception of 21 December, the asphalted areas north of the main building (*3*) and north of the *Elektro* building represent the cold spots.

The highest absolute temperature difference (ATD) within the study area was simulated on 21 June with 3.4 K between the coolest and warmest spot. On 21 December on the other hand, it is lowest with only 1.2 K. The vernal and autumnal equinoxes have an ATD of 2.5 K and 2.1 K respectively. It is supposed that the ATD is mainly influenced by the amount of solar radiation reaching the surfaces.

Regarding the wind field, the input data for the wind speed varied between the simulated dates. However, the prevailing wind direction (225°) was kept the same. Thus, in this study, the proportional relation of local wind speeds to the undisturbed reference inflow at the same height level is of higher interest than absolute values. The resulting so-called wind speed change (WSC) showed only minor differences between the simulated dates. The maximum value for the WSC was lowest for the vernal equinox (138.0 %) and highest for the winter solstice (157.1 %). Summer solstice and autumnal equinox had a maximum WSC of 140.5 % and 152.0 % respectively.

On all the simulated dates, the south east corner of the campus (*Berg*-building) showed the highest WSC values. The southern part of the study area is close to the boundaries of the model, so even if it is an equally densely built-up area, the simulated results for this location present higher values due to the fact that the model considers there an undisturbed approaching flow. Furthermore, the simulation results indicate windy conditions at east-west passages. Those areas are for example the passage east of central building 2 (*2b*), south of the VATL (*5*), and between the main building (*3*) and the VATL. Additionally, the passage between the main building (*3*) and the small wooden building in the north west exhibits a high WSC with values close to the maximum on all simulated dates. The influence of the seasonal changing LAD of the vegetation had a visible but minor influence on WSC, especially in the tree-lined patch of grass behind the main building (*C*). The lowest wind speeds occurred in the wake of the larger buildings and especially in the small backyards south of the main building (*3*) and east of the central buildings (*2a, 2b*).

4. Conclusion

In this study, the simulation tool ENVI-met was used for analysing the microclimatic conditions at a university campus in Trondheim, Norway. The analysis focused particularly on local air temperatures and the wind field at pedestrian level. The validation revealed an average difference of 4.9 K between the measured and the simulated building surface temperatures. However, ENVI-met tended to overestimate the low surface temperatures, especially where IR-measured values were below or very close to 0 °C.

The simulated values of air temperature correlate to the different surface types (albedo, emissivity) together with the wind field pattern and solar access. The materials with high heat storage capacity (i.e. asphalt, concrete pavement) present higher air temperatures than the vegetated areas.

The areas in front of south-facing and sunlit surfaces presented elevated local air temperature values. On 21 December, when almost no solar radiation is reaching the area due to very short days and low solar angles, the influence of the surface materials becomes more evident. At this time of the year, vegetated surfaces cause lower air temperatures at pedestrian level compared to heavier materials like concrete and asphalt. The highest ATD occurred on 21 June with 3.4 K between the warmest and coolest spot, the lowest on 21 December with 1.2 K.

The analysis of wind speed change identified east-west passages on campus as problematic. There, local wind speeds exceeded the reference wind speed at the inlet by up to 57.1 %. On the other hand, the lowest wind speed values occurred in the wake of the larger buildings and in particular in the small backyards south of the main building and on the east side of the central buildings.

The model has limitations as to the coarse grid resolution and the uniform discretisation, as the software does not allow refinements. Notably, the regular grid structure that does not admit of transverse, sloped or curved geometries, resulting in inaccuracies of simulated air flows. Additionally, the program does not allow for considering snow covered surfaces [22], which especially in December and March can have significant influence on physical properties and thus the microclimatic conditions [23]. However, the model can be of advantage and informative at the early stage of microclimate assessments.

In the future, a more comprehensive study will be conducted by using sophisticated computational fluid dynamics programs that allow a finer and more demand-actuated discretisation. The results of this study, in combination with a specific validation process for the air flows, will further deepen the knowledge of the microclimatic situation on the NTNU campus. This knowledge can then be applied in the planning of the campus' redevelopment, for instance to decide on where to plant new vegetation, which surface materials to use or where new buildings decrease the potential for natural ventilation.

5. Acknowledgements

This paper has been written within the Research Centre on Zero Emission Neighbourhoods in Smart Cities (FME ZEN). The authors gratefully acknowledge the support from the ZEN partners and the Research Council of Norway.

References

[1] Rhind S M 2009 *Philosophical transactions of the Royal Society of London. Series B, Biological sciences* **364** 3391–401
[2] Orlanski I 1975 *Bull. Amer. Meteor. Soc.* **56** 527–30
[3] Oke T R, Mills G, Christen A and Voogt J A 2017 *Urban climates* (Cambridge: Cambridge University Press)
[4] Stewart I D and Oke T R 2012 *Bull. Amer. Meteor. Soc.* **93** 1879–900
[5] Kłysik K and Fortuniak K 1999 *Atmos. Environ.* **33** 3885–95
[6] Bowling S A and Benson C S 1978 *Study of the Subarctic Heat Island at Fairbanks, Alaska (Environmental Monitoring Report* EPA-600/4-78-027*)* (United States Environmental Protection Agency)
[7] Fouillet A, Rey G, Laurent F, Pavillon G, Bellec S, Guihenneuc-Jouyaux C, Clavel J, Jougla E and Hémon D 2006 *Int. Arch. Occup. Environ. Health* **80** 16–24
[8] Shaposhnikov D *et al* 2014 *Epidemiology (Cambridge, Mass.)* **25** 359–64
[9] Toparlar Y, Blocken B, Maiheu B and van Heijst GJF 2017 *Renewable and Sustainable Energy Reviews* **80** 1613–40
[10] Berardi U and Wang Y 2016 *Sustainability* **8** 822
[11] Wang Y and Akbari H 2016 *Sustainable Cities and Society* **26** 438–46
[12] Chokhachian A, Santucci D and Auer T 2017 *Buildings* **7** 113
[13] Meteorologisk Institutt 2017 *Klima siste 150 år [Climate in the last 150 years]* https://www.met.no/vaer-og-klima/klima-siste-150-ar (accessed 18 Apr 2019)
[14] Oberkampf W L and Trucano T G 2002 *Progress in Aerospace Sciences* **38** 209–72
[15] American Institute of Aeronautics and Astronautics 1998 *AIAA guide for the verification and validation of computational fluid dynamics simulations* (Reston, VA: American Institute of Aeronautics and Astronautics)
[16] Bruse M and Fleer H 1998 *Environ. Model. Softw.* **13** 373–84
[17] FLIR Systems I 2016 *User's manual Flir Exx series*
[18] Peel M C, Finlayson B L and McMahon T A 2007 *Hydrol. Earth Syst. Sci. Discuss.* **4** 439–73
[19] Mellor G L and Yamada T 1982 *Rev. Geophys. Space. Phys.* **20** 851–75
[20] Sigmond E M O 1985 *Nasjonalatlas for Norge: National atlas of Norway* (Oslo: Statens kartverk)
[21] ENVI_MET GmbH 2019 *ENVI_MET V4.4.2: Winter1819* (Essen, Germany)
[22] ENVI_MET GmbH 2019 *Under Zero Temperature* personal email communication
[23] Malevich S B and Klink K 2011 *J. Appl. Meteor. Climatol.* **50** 1884–94

Perimeter Blocks in Nordic Towns - How latitude affect daylighting

Bengt Sundborg[1,2], Barbara Szybinska Matusiak[1] and Shabnam Arbab[1]

[1]Faculty of Architecture and Design, Department of Architecture and Technology, Norwegian University of Science and Technology (NTNU), Trondheim, Norway. [2]Research Institutes of Sweden, RISE, Sweden

Abstract: One of the design principles for future sustainable towns is compactness. The densification of cities is very much needed, but it usually compromises the access of daylight. Densification is especially challenging in the Nordic region characterized by low angled sunlight, something that also limits daylight distribution and restricts its intensity. The higher the latitude, the greater is the difficulty in the distribution. Perimeter blocks give shelter from wind and often create semi-public courtyards which have been seen to be attractive in many Nordic settlements during history. In the present study, alternative design to the conventional perimeter blocks are explored and geometric options such as chamfered corners, strategically varied building heights and differently positioned openings in a broken perimeter block are analyzed. The yearly simulations as well as simulations for May 1st have been carried out for the same perimeter blocks located at four different latitudes (decimal coordinates):

1. **65.0** Oulu (similar to Mo i Rana 66.3, Jokkmokk 66.6 and Rovaniemi 66.5)
2. **63.4** Trondheim (similar to Reykjavik 64.1, Östersund 63.2 and Vaasa 63.1)
3. **59.3** Stockholm (similar to Oslo 59.9, Helsinki 60,2, Tallinn 59.4, Saint Petersburg 59.9 and Anchorage 61.2)
4. **55.7** Copenhagen (similar to Malmö 55.6, Glasgow 55.9 and Moscow 55.8)

The choice of evaluation criteria is based on scientific discourse in the field of daylighting. According to the new European standard, solar radiation is included. Computer-based daylighting simulations are performed for different designs of the perimeter blocks with equal density, FAR = 1.33. The further north a city is located, the lower the houses in a perimeter block must be to maintain a certain level of daylight. The study confirms that latitude affects daylighting and that geometrical change can improve the conditions for daylight in the perimeter blocks.

1. Introduction

Around 56 million people are living with the daylight conditions which in art and literature is denoted as the Nordic Light - from southern Denmark up to the geographic North Pole (about 54.0°N to 90.0°N). In the southern hemisphere, the corresponding latitudes with the same type of daylight, are not that much populated, only around 1000 people live there, mainly researchers at Antarctica. The most characteristic feature of the Nordic Light region is low angled sunlight that contributes also to long twilights and white nights in summer. This study aims to show how high latitudes hamper the daylight and how the shapes of urban blocks can be developed to improve daylighting in urban settlements.

This study about the impact of the latitude on daylighting is part of a series of studies; one article about visible access to the sky and the views in different perimeter blocks has been presented in an earlier paper[2]. Another article is planned about different compass directions and one about different variations of the heights in the blocks. Theoretical knowledge and practical experiences have been combined in the present study aiming to develop a method to compare town planning from different latitudes considering daylighting. For this purpose, alternative shapes of perimeter blocks have been developed. The research question was if improved quality of daylight can be reached with the same density Floor Area Ratio (FAR) as for the conventional perimeter block.

1.1 The impact of heights
The height is a well-studied parameter. The higher the building, the longer are the shadows. Assumed height in the latitude simulations have been five floors (= 15 meters) except alternative 4 and 5 which has 4, 5, 6 and 7 floors, see figure 1.

1.2 The impact of configurations
An important study by Mark DeKay describes how the different geometry of a standard atrium building impacts daylighting conditions [1]. A configuration study was presented in Graz by the actual research team [2]. The tested urban configuration alternatives have considerable advantages regarding daylight compared to more conventional blocks.

2. The Alternative Types of Perimeter Blocks
International comparisons over rectangular street grids show large variation from town to town. Some have quadratic shapes but most of the grids have rectangular shapes with considerably different dimensions (length versus width of the block). The variation of the street grids within a town is smaller. In specific city district the measures of length and width often are standardised.

Five alternatives to the conventional perimeter blocks in street grids are studied. Following sustainable design recommendation, the blocks are oriented east-west, see general guidelines as https://greenpassivesolar.com. The size of the blocks is 100 meter in east-west direction and 60 meter in north-south direction. The scale is common in the Nordic countries as well as the rest of Europe though the average size of blocks is larger.

The design principle for all alternatives is keeping the same floor area ratio as in the standard block. The first alternative to the standard perimeter block is the same block with chamfered corners, a modification which is sometimes used in practise. Other options are blocks consisting of openings between the streets and the courtyard. The second alternative has openings positioned in the corners of the block and the third in the middle of the blocks. In frequent practical use these two types vary both in proportions and in scale. The two last alternatives with variated heights, 4 and 5, are developed within this project to find new strategies. All alternatives have been developed in Sketch Up-drawing, see figure 1 and later transformed to models for simulations in DIVA.

The total floor area is the same in all alternatives, which also means that it is possible to regard every block quarter as a module which can be combined with others to form new alternative hybrids. The Floor Area Ratio (FAR) is 1.33, which is a relative low value in an international context but relevant for most Nordic towns. The six alternatives are:

0. The conventional perimeter block (NULL), 0-alternative.
1. Chamfered corners in the outer corners of the building as well as in the courtyard.
2. Openings in a broken perimeter block positioned in the four midpoints of the building.
3. Openings in a broken perimeter block positioned in the four corners of the building.
4. Varied building heights around the courtyard with fewer floors in the corners of the block.
5. Varied building heights around the courtyard with more floors in the corners.

IOP Conf. Series: Earth and Environmental Science **352** (2019) 012016 doi:10.1088/1755-1315/352/1/012016

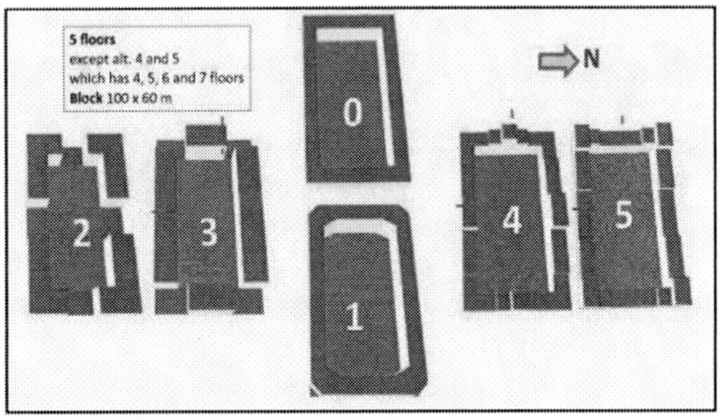

Figure 1. The six different types of urban blocks.

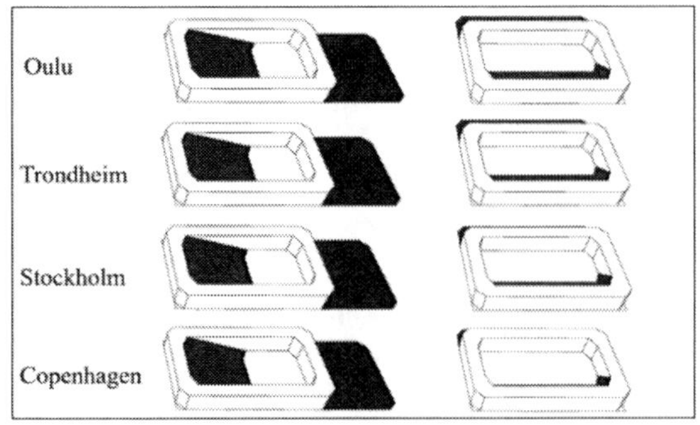

Figure 2. Shadows the 1st of May Alt. 1. Evening, sun in the west. Noon, sun in the south.

3. Daylighting Strategies in Urban Settlements

It is impossible to create satisfactory daylighting conditions during the whole day in dense settlements. It is always necessary to resolve the priorities between outdoor and indoor spaces and to consider which parts of the day and the year are most important. Especially difficult to handle is the sunlight at high latitudes due to the large variation of the azimuth angle (compass direction) during the day and across the seasons, as well as low average solar elevation angles through the year.

To develop good strategies requires a focus on the different times of the day. In people's every-day life the needs and desires vary depending on the type of buildings such as buildings for seniors, offices or kindergartens as well as the activity, underlined in a handbook for town planners [3]. A midwinter strategy leads to openings between the buildings in a North/South direction which is opposite to a strategy for the extended summer when the sunlight from east in the morning and west in the evening is highly appreciated. We have many possibilities to formulate goals and evaluation criteria connected to them, and it is easy to calculate many relevant metrics by computers. However, joining strategies together (e.g. making calculations on the yearly basis) often fails because the iterations will converge instead of diverge.

During many years' equinox studies have been very popular among practitioners as well as researchers. But the equinoxes are not suitable to analyze sunlight from east or west simply because in those two days the sun is positioned precisely at the east or west during sunrise or sunset, also it is on the horizon. That gives strong arguments for choosing other days between spring equinox and summer solstice. "The Sitting Outdoor – Period" in Nordic towns is roughly the warmest part of the year between late spring and early autumn. However, the temperature in the air is not high enough to bring comfort in most cases. The direct sunlight with its heating by solar radiation is also needed.

We look to a representative average place in the Nordic countries - the Oslo/Stockholm/Helsinki latitude. For that latitude the period around the 1st of May is the time when people start to sit outdoors during good weather. The 11th of August happens to be the point on the other side of the summer solstice. The period in between has good possibilities for outdoor activities due to good daylighting. However, the proposed period should not be regarded as an exact specification of the actual conditions. Since the shadows for the 1st of May also describe the conditions at the 11th of August, during the 101 days in between the sunlight distribution is better with a maximum at the summer solstice.

3.1 The Shadow Maps

Shadow Maps are often used tools in town planning. The shadows from surrounding obstructions as mountains, trees and buildings affect the conditions at the outdoor spaces as well as in apartments. Most of the obstructions are permanent. The shape, size, orientation and position of the building must therefore be developed during careful considerations. The shadow map also specifies surfaces on ground and facades which have sunlight qualities outdoors and indoors. That information can be used by planners to create daylight qualities for the residents.

The time of comfortable sitting outdoors in the sunshine on the balconies and in the courtyard can also be expanded by sunscreens in hot days and, in cold days by windscreens. The fundamental needs of daylight are different, both in quantity and quality, between different groups of people and even between individuals. Anyway, we may roughly agree that in residential areas interesting periods to study are in the morning and in the evening because almost all residents are at home, contrary to during noon when people are at work, schools and on weekend trips, etc. This is also the most difficult time for daylighting due to low sun angles. Besides the shadow maps for the western sun we also show the less critical shorter shadows in the middle of the day. The shadows from the western sun illustrates also the shadows in the morning because the shadows are equal although mirrored in the opposite direction (assuming symmetrical block shape). Depending on the clockwise movement of the sun the shadows also move. The minutes just before and just after the exact cardinal directions for the sun are especially interesting. All the facades orientated to the south can be in sunshine at 18:07 PM but on the contrary at 18:08 PM the facades orientated to the north are in sunshine (example from Stockholm, 1st of May, DST = Daylight Saving Time).

4. Methodology

The three main steps in the research were: Constructions of the Urban Models, Calculations and Simulations of the Models and Comparisons, Analyses and Conclusions. The second step has two parts, sunlight radiation (shown in table 1 and 2) and shadow maps (shown in figure 2). According to the new European standard solar radiation should be included [4].

Part one: simulation of solar radiation at a specific day and for the whole year. Facades and courtyards. Comparison between the latitudes.

- Sunlight radiation on the first floor of the facades, average values for the 1st of May (kWh/m^2) and for the whole year (kWh/m^2).
- Sunlight radiation on courtyards, average values for the 1st of May (kWh/m^2) and for the whole year (kWh/m^2).

The simulations of the solar radiation are performed with DIVA for Rhino, an often-used tool for climate based and static daylighting calculations. The DIVA (Design, Iterate, Validate and Adapt). A plug-in for Rhinoceros software, enables effective calculations of daylight metrics, e.g. daylight factor, using the Radiance/Daysim engine. The facades on the upper floors have good distribution of sunlight in all alternatives. The challenge is to create settlements with good daylight for the worst cases, the facades along the first floors. To look only at the first floor means critical studies focusing on the worst case. We focus therefore on the differences and possible improvements. The studied model has been placed in a street grid with surrounding blocks performed in the same way. The street canyons have a width of twenty meters façade to façade.

Part two: Sunlight distribution on two different hours at a specific day. Shadow Maps of the 3D-models. Comparison between the latitudes.

- Simulations by SketchUp.

4.1 Assumption of the facades as representative for the windows

The generated data is for the facades except for the ground values. Solar radiation on facades and daylight level on facades are values in our studies to describe conditions at the windows. We assumed that the windows are evenly distributed along the facades. This assumption is common in general studies as ours because no special details are known.

4.2 The impact of the latitudes on the daylight conditions
To quantify the impact of the latitudes on the daylight conditions simulations of daylight have been done over a year and a day for towns located at four different latitudes. The rays from the sun at more northern latitudes have lower altitude angles but also larger range of azimuth angles. The investigation has been done with the specific angles from the actual latitude and all other parameters the same including the same clear sky type.

4.3 Clear Sky simulations
The parameter, Cloudiness, is dependent of the local climate in the calculated towns. The climates vary a lot depending on differences in topography, water surfaces, the surrounding physical geography, etc. In analyzing sunlight at different latitudes, we reduce distortions from different climate and amount of clouds in different cities by assuming one similar type of climate. Then we can study the changes in one parameter, the latitude. The alternative to handle the parameters by multi-regression analysis or similar method was rejected due to more complicated processes. The same climate means the same weather and cloudiness for all simulations. That can be created in many ways; as an average weather file representing the four cities (1) or a "clear sky" weather file without clouds (2). In this study the latter, "clear sky", is chosen for three reasons;

a. The choice of "clear sky"-simulations is prepared for a second step: impact of different cloud conditions.
b. The Shadow Patterns are consistent with "clear sky"-simulations.
c. "Clear sky" facilitates the analysis depending on it is precise.

The "Clear sky" weather files are epw-files made by Meteonorm 7 based on a clear sky for the whole year for respective city. The content of ozone and particles in the atmosphere (long distance transport from volcano eruption and other particles) gives small variation in radiations from year to year. However, the relative differences between the cities are about the same regardless which year. The Meteonorm software contains worldwide climate data. This can be an Excel file for specific analysis and be directly imported to photovoltaic, solar thermal or building simulation software. The data base contains observations from 8 325 weather stations and five geostationary satellites. On this basis, state of the art interpolation models provide data for every site with high accuracy. The clear sky radiations are calculated as maximum global radiation and corresponding diffuse radiation for clear days (cloudless sky) at hourly intervals. This affects the automatic selection of the clear day temperature model (warmest possible temperature). Eventual improvements of the method for clear sky can be done if a future evaluation will show a specific need for more precision.

4.4 The impact of clouds
A comparison between the simulations in Stockholm with or without climate factors shows:
- 15-19 % reduction of radiation in adding the clouds by a weather-file based on climate data for the actual place, see table 1.
- The same ranking order for the alternatives with the only a very small difference, that No. 1 and No.3 with local climate at the first-floor simulation came in split first place. The comparison shows that it is possible to use the "clear sky" result instead of the complicated climate simulation for the ranking of the best daylight strategies in Stockholm.
- It is probable that the conclusion about the "clear sky" is valid even for other towns - a reasonable conclusion even if a very odd distribution of clouds during day could affect the ranking in some way. Further test in other towns must be done to confirm the hypothesis (tests are planned in the autumn 2019).

Radiation (kWh/m2) Model:	Cloudy Sky. first floor.	Clear Sky. first floor.	Cloudy as % of Clear first floor	Cloudy Sky. Courtyard.	Clear Sky. Courtyard.	Cloudy as % of Clear Courtyard
Alt. 0	2.41	2.96	81	3.42	4.02	85
Alt. 1	2.47	3.03	82	3.41	3.97	86
Alt. 2	2.45	2.98	82	3.65	4.29	85
Alt. 3	2.47	3.04	81	3.07	3.81	81
Alt. 4	2.32	2.85	81	3.32	3.88	86
Alt. 5	2.33	2.86	81	3.55	4.17	85

Table 1. Daylight Radiation. 1st of May in Stockholm. Cloudy and Clear Sky.

A single "Clear Sky" simulation is valid as such for every place along the actual latitude. If the hypothesis above about a strong covariance with "Cloudy Sky" simulation can be proved in detail or at least have few exceptions it will lead to a general conclusion;
- The ranking of the best daylight strategies at a specific place can be done from a "Clear Sky" simulation from every place along the same latitude.

5. Results - The Daylight Values for different latitudes

The solar radiations are lower at higher latitudes which also are confirmed in the table below. The most important explanation is the lower sun angles. The two highest radiations for each town are presented on a blue background. The model with highest radiation is alternative 3 with openings in the corners. The differences between the models are relatively small which means that windy conditions and other aspects can be crucial. In windy cases the enclosed perimeter block is recommended, and it should be the chamfered version alternative 1 which has clearly higher radiation than alternative 0. Two deviations from the typical patterns are marked with stars and explained below.

Radiation (kWh/m2)	Model	1st of May	Year
Oulu 65.0	Alt 0	2.63	519.69
	Alt 1	2.71	536.62
	Alt 2	2.67	538.09*
	Alt 3	2.72	551.63
	Alt 4	2.49	502.47
	Alt 5	2.52	499.38
Trondheim 63.4	Alt 0	2.88	570.37
	Alt 1	2.97	591.52
	Alt 2	2.92	596.92*
	Alt 3	2.98	613.73
	Alt 4	2.74	562.58
	Alt 5	2.77	561.55
Stockholm 59.3	Alt 0	2.96	613.1
	Alt 1	3.03	634.68
	Alt 2	2.98	629.27
	Alt 3	3.04	633.09
	Alt 4	2.85	611.17
	Alt 5	2.86	605.99
Copenhagen 55.7	Alt 0	2.85	650.75
	Alt 1	2.91	672.33
	Alt 2	2.86	669.85
	Alt 3	2.94	684.71
	Alt 4	2.82	632.59
	Alt 5	2.83	627.12

Table 2. Sunlight radiation – average values along the facades on the first floor

The ranking order considering high solar radiation of the different alternatives are the same for all towns with two unexpected exceptions. In the northern cities, Oulu and Trondheim, the yearly radiation at the first floor is higher for alternative 2 (•) with central openings than for alternative 1 with chamfered corners. An explanation is the lower average altitude angle for the sun compared to more southern cities which will be especially difficult for alternative 1 with no openings. Although for the first of May and related period the alternative 1 is better than alternative 2 for all latitudes.

If the yearly values or the first of May and related period of 101 days should be in focus is a tricky question which depends on which conditions are most relevant for the people. But the figures above point at the same model except the second choice for Oulu and Trondheim. Looking to the whole year for northern cities, including the darkest part of the year - in contrast to the first of May with its higher sun angles - the alternative 2 is natural to prioritise before alternative 1. The local wind conditions must always be considered. At windy northern places the conclusion will be the opposite - the more closed alternative 1 is natural to prioritise before the more open alternative 2.

The higher the latitude, the greater are the shadows is a basic fact, but despite of that, often forgotten in town planning. In the development of northern city districts, the practice from southern projects is often used. The height of the perimeter blocks in the figure 2 are five floors (= 15 meters). To keep the same size of shadow in Oulu as in Copenhagen requires a significantly lower building 3.6 floors instead of 5.0 floors. For Malta it is 11.2 floors. For comparison across latitudes, see figures 3 and 4.

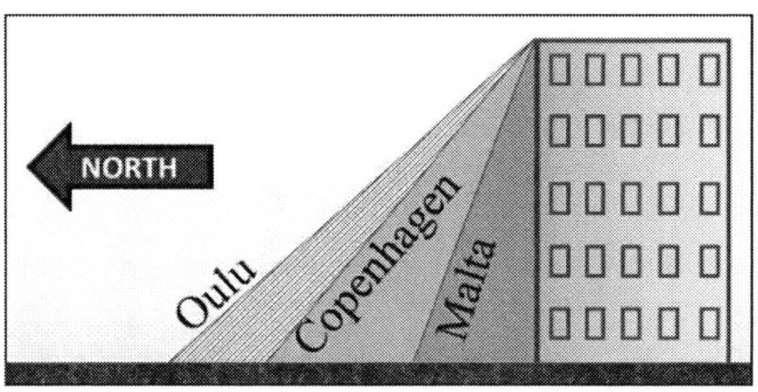

Figure 3. The length of the shadows depends on the latitude (17,6 m, 12,8 m and 5,7 m).

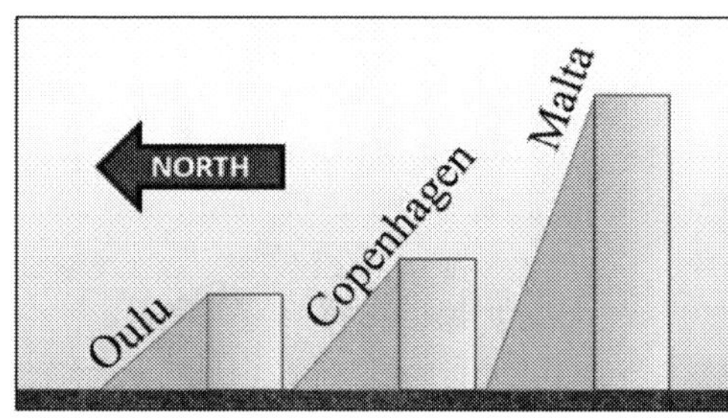

Figure 4. The heights on the buildings at different latitudes for equal length of shadows (3.6, 5.0 and 11.2 floors with three meters each).

Some efforts can be done to support daylight at the "dark" northern latitudes:

1. A mathematical guideline - with the latitude as input - limiting the heights of the buildings gradually the further north has relevance. However, more important than a specific formula, is a restrictive approach to higher buildings at the northern latitudes.
2. To preserve openings in the settlements against the sky and, if possible, against the horizon.
3. To choose types of settlements with configurations which give much daylight outdoor (e.g. alt. 2 and 5) and indoor (e.g. alt. 3 and 1).
4. To develop town districts of such settlements with configurations in each block which are synchronized to the next and give long openings for low angled sunlight.
5. To prioritize light colours and well-reflective material for all surfaces in the towns especially ground and facades.

6. Design Strategies

The construction of the different alternatives in the project was based on the interests to analyse the impacts of different shapes. Using the experiences from this study and other tests it is possible to develop more design strategies than the five efforts mentioned above. The design process, often including iterations, has been investigated by Donald Schön and others. In a famous study he mapped the important steps in design [5]. To improve the design of the urban blocks and adapt it to a specific location the iterative process of Create > Test > Analyse > Create >…is often used. The creations of the blocks can be done in strategic sketches or by programmes as Grasshopper using parametric design testing hundreds of possibilities. A third way is to optimise shapes and configurations by algebraic operations.

7. Energy Aspects

Many aspects of well-being are created by daylight. However, the most fundamental result of the alternative urban blocks is probably the increase of the direct sunlight in the urban settlements. The total potential of solar energy along the facades are showed in the yearly figures of radiations, see table 2. There is up to 7% difference between the most sunlit alternative - Alt 3 - and Alt. 0. In the sustainable compact city with high and dense urban blocks much of the sunlight is obstructed. Strategical geometry for the buildings has a special relevance here. In an earlier study the energy savings for the electric lighting was 13 % comparing an ordinary cranked street with a street with strategically positioned openings in the buildings [6].

The heat loss in buildings is to the large degree dependent on the shape of the building envelope (facades and the roof together). In the Nordic region, having negative heating energy balance, increasing of compactness causes a decrease of energy consumption for heating. In this respect, the ideal shape of a building is a sphere. Modifying the shape to a cube helps to lower construction costs with still high efficiency [7], stretching the shape to a linear form has a negative effect. On the other side, too large compactness means larger distance to windows (reduced view and well-being) and can have negative effect on energy consumption for lighting as the daylight level decreases dramatically with the distance from the window. Comparison of the studied alternatives in terms of compactness points at the alt. 3 as having highest compactness (lowest energy use for heating).

8. Conclusions

The same type of settlements as in southern Europe erected in the Nordic countries gives more darkness due to low sun at high latitudes. With lower heights of the buildings it is possible to adapt to the low sun angles at a price of decreased utilization of the plot. Other possibilities are to plan openings in the settlements and choose light colours and well-reflective materials for all surfaces in the towns. Settlements with configurations as chamfered corners and with openings in the corners of the perimeter block are supporting the daylight distribution and are therefore of special interest for the Nordic countries. Accurate general guidelines improve the implementation of knowledge a lot and save money and time for calculations. A guideline – as one example - limiting the heights of the buildings gradually the further north is relevant. However, general guidelines are never enough. It is always important with adaptations to local conditions as wind, noise, topography, etc.

Acknowledgments

This research is part of a larger project at NTNU, Norway which is administrated by RISE, Research Institutes of Sweden and financed by the Swedish Energy Agency. Special thanks for advice around the climate simulations to Consultant Majid Miri, Sweco Architects.

References

[1] DeKay M 2010 Daylighting and Urban Form: An Urban Fabric of Light, *Journal of Architectural and Planning Research 27:1* (Locke Science Publishing Company, Inc. Chicago, IL, USA)

[2] Sundborg B, Matusiak B S and Arbab S 2019 Perimeter blocks in different forms - aspects of daylight and view, *Smart and Sustainable Built Environments (SBE19)*. Graz.

[3] Sundborg B 2010 *Ljus i bebyggelsen* (Svensk Byggtjänsts förlag) p 70

[4] CEN 2019. 2018 *Daylight in buildings* NS-EN 17037:2018 European committee for standardization.

IOP Conf. Series: Earth and Environmental Science **352** (2019) 012016 doi:10.1088/1755-1315/352/1/012016

[5] Schön D 1983 *The Reflective Practitioner: How professionals think in action*. (London: Temple Smith).
[6] Sundborg B 2016 *Energy Savings by Using Daylight for Basic Urban Shapes – With a Case Study of Three Different Street Types* (KTH Royal Institute of Technology, Stockholm)
[7] Matusiak B 2012 Low-energy house, back to the 'årestue': a thought experiment about low-energy houses, *Architectural Science Review, Vol. 55, No. 2, May 2012,* p. 86–91

IOP Conf. Series: Earth and Environmental Science **352** (2019) 012017 doi:10.1088/1755-1315/352/1/012017

Energy performance evaluation of a nearly Zero Energy Building and the reasons for the performance gap between expected and actual building operation

Evangelia Loukou[1], Per K. Heiselberg[1]*, Rasmus L. Jensen[1] and Hicham Johra[1]

[1]Department of Civil Engineering, Thomas Manns Vej 23, 9220 Aalborg Ø, Denmark

*ph@civil.aau.dk

Abstract. As energy standards focus on reducing energy use in new buildings, attention is drawn to the gap between the expected and actual building operation and energy performance. This performance gap can be associated with the building construction, its systems, its unbalanced operation, the assumptions on occupancy profiles during the design phase, or the users' interaction with the systems' operation and control. This work focuses on a nearly zero-energy (nZEB) single-family house located in central Denmark. The analysis of indoor environment and energy use is based on year-long data monitoring. The reasons for the deviation between the expected and actual energy use are suggested. The indoor environmental quality is analyzed to verify the compliance with the standards. The European recommendation for the yearly primary energy use of new single-family houses is 50 – 65 kWh/m^2. For the current case, the simulation tool Be18 gives a result of 30.8 kWh/m^2 for the design phase. However, the actual energy use is measured to be 58.2 kWh/m^2. The sensibility of nZEBs to such imbalances can lead to houses that do not function as intended. It is thus crucial to investigate further the causes of these disparities in order to bridge the gap between expected and final energy use in dwellings.

1. Introduction

The European Union and building regulations increase their focus on the reduction of energy use and greenhouse emissions, and on the incorporation of renewable energy sources in the building sector, and especially in newly constructed buildings [1]. Hence, the need for a new generation of high-performance buildings with nearly zero energy use is strengthened. However, a performance gap between the predicted and real energy performance of the dwellings has been documented in multiple studies [2-7]. It shows that new buildings constructed in accordance with these standards can experience poor indoor environment, overheating and much higher energy use than expected [8].

This discrepancy can be connected to the assumptions made during the initial design phase, changes implemented during the construction phase, or to the operation of the dwelling. Many researches focused on the influence of human behavior and the challenges to achieve appropriate energy performance forecasting [9-13]. It is pointed out that more energy-efficient buildings are more sensitive to the influence of occupant behavior [9,11,14]. Different suggestions have been put forward in order to assess more accurately the building's actual performance. A reference occupancy and user behavior profile could be assigned, alongside the reference building model [10]. Additionally, reconsidering the energy demand calculation by the building regulations can have a significant impact on achieving nearly-zero energy buildings that perform as intended [9,11]. Finally, in order to achieve better prediction and obtain higher understanding of the occupancy patterns and behavior, it is necessary to perform extensive

IOP Conf. Series: Earth and Environmental Science **352** (2019) 012017 doi:10.1088/1755-1315/352/1/012017

occupant-related data collection, monitoring of real occupied buildings to get good insight about the performance gap [2,7,15]. Nevertheless, energy efficiency is not the only parameter that should be addressed when evaluating a building's performance. The indoor environmental conditions are an equally important aspect, with high complexity, where a similar mismatch can also be found [2].
The objective of this project is to tackle the challenges arising with the 2020 regulations and the new generation of dwellings trying to comply with them. Detailed monitoring and assessment of both the energy efficiency and indoor environment are intended to address these issues.

2. Case study

2.1. Building description

The studied building is a detached one-story, single-family house, located in Ry, in central Denmark. The building was constructed at the end of 2017 and is characterized as a nearly zero energy building, based on the Danish energy labeling standards [16].
The dwelling has a total floor area of 160 m^2. It is constructed on two levels, with a height difference of 0.5 m. It includes 4 bedrooms, 2 bathrooms, 1 kitchen, 1 dining area, 1 living room, and a utility room. The kitchen, dining area and living room are connected to form an open space. There is also an unheated storage space attached to the building. There are windows towards all directions, while the openings in the South, East and West façades are equipped with sunscreens. Figure 1 shows the plan of the dwelling, while Figures 2 and 3 give an external view of the house.
The characteristics of the building study case are summarized in Table 1. The building is occupied by a young couple. The windows are triple-pane low-energy windows, with high light transmittance and low solar heat gain factor.

Figure 1. Floor plan of the study case.

IOP Conf. Series: Earth and Environmental Science **352** (2019) 012017 doi:10.1088/1755-1315/352/1/012017

Figure 2. External northwest view.

Figure 3. External south view.

Table 1. Characteristics of the construction elements.

Building description	
External walls U-value [W/m^2 K]	0.15
Roof U-value [W/m^2 K]	0.09
Ground slab U-value [W/m^2 K]	0.08
Windows U-value [W/m^2 K]	0.9
Windows g-value [-]	0.49
Air tightness n_{50} [l/s m^2]	0.52

2.2. System description

The building is equipped with a floor heating system connected to district heating and a compact ventilation unit, which is also providing sanitary hot water. The compact ventilation system has an integrated air-to-water heat pump and a counter-flow heat exchanger with 85% efficiency. The domestic hot water is primarily produced by the ventilation unit through the recovered energy from the exhaust air, and supplemented by an auxiliary electrical heater. The ventilation control is based on indoor CO_2 and relative humidity levels. Ventilation operation can be controlled individually for each room and based on occupancy schedules. When the building is not occupied and the indoor conditions are adequate, the system runs with a minimum airflow rate of approximately 0.15 l/s m^2, compared to 0.3 l/s m^2. The ventilation set-points are summarized in Table 2.

Table 2. Ventilation control set-points.

	Damper opening	Damper closing
CO_2 [ppm]	900	700
Relative humidity [%]	70	50
Relative humidity – bathroom [%]	60	50

Natural ventilation and sun protection can be activated in order to create a pleasant indoor environment and minimize energy usage. Natural ventilation can be achieved through the skylights and a novel ventilation system with louvers incorporated in several windows in rooms with overheating risk. The external solar shadings are installed at the windows of the master bedroom and living room, with east, south and west orientations. In order to encourage the use of these systems, the solar shades are designed so that they allow good view to the outside, even when activated, while the ventilation louvers are providing theft and rain protection so that they can be used even when the dwelling is not occupied. In Figures 4 and 5 are shown the ventilation louvers and solar shades. At the end of November 2018, the natural ventilation and sun protection systems were connected to the control system, so that they can be

IOP Conf. Series: Earth and Environmental Science **352** (2019) 012017 doi:10.1088/1755-1315/352/1/012017

automatically controlled. The examined period for this report only covers the period with manual activation of these systems.

Figure 4. Ventilation louvers.

Figure 5. Solar shades and view to the outside.

2.3. Data monitoring

The data logging started in November 2017 and is still ongoing. This paper analyzes the data collected during the entire year of 2018. The measurements are recorded every 5 minutes in all rooms of the dwelling and include cold and hot water consumption, energy use for heating, ventilation, production of sanitary hot water, the status of the storage tank, the heat pump, as well as electricity use for all appliances. The operative temperature, CO_2 and relative humidity levels are monitored in all rooms. Additionally, the airflow, temperature of supply and return hot water, fan speed and damper opening in the ventilation unit are recorded. All measurements are performed by state-of-the-art sensors commonly used for indoor climate monitoring, electricity and heating metering in buildings.

3. Methodology

Over the entire period of the building's design, construction and occupancy, it has been undergoing a commissioning process to reassure its proper operation, as initially intended. During this process, a continuous readjustment of the systems' operation has taken place and correction of mistakes made during design or construction. Not all measurements have been available from the beginning, therefore the presented results include only the information that has been accessible during the period of study and is within the scope of this analysis.

The commissioning of the building has revealed several issues that had to be adjusted, revised or improved. For example, in June, it was noticed that a valve in the sanitary hot water loop was constantly open, even though it was not necessary, resulting in increased energy used for heating. Changes in the monitoring system have resulted in some missing values. The external sun shades were not installed until March 2018. The user control system, with which the occupants can automatically control or overwrite the different systems of the house, was activated at the end of October 2018.

Finally, during the data analysis process, some problems were observed. Therefore, additional sensors (IC-meters) were installed right beside the building sensors, with the purpose to verify the data accuracy. It was observed that monitored CO_2 levels and relative humidity values from the building sensors were not reliable. The sensors were installed at the end of June in the kitchen and living room. In this paper, values for CO_2 and relative humidity analysis are only those from the IC-meters for July to December.

The aim is the long-term assessment of indoor environmental quality and energy use of the dwelling. Furthermore, the monitored data analysis aims to reveal mismatches between the intended and actual operation of the systems and equipment.

The evaluation of the indoor environment is done on room level, considering the most used rooms: master bedroom, living room, and kitchen / dining room. The thermal comfort categories are taken for winter and summer conditions, with January, February, March, April, October, November and December considered winter months, and May to September as summer. Also, from the 10th to the 24th of July, the tenants were on holiday, therefore the building was not occupied.

4. Results and discussion

4.1. Energy use

Through analysis of the monitored values, the energy needs of the building are assessed and the actual yearly energy use is compared to the expected one. Figure 6 presents the measured monthly delivered energy for heating, ventilation and production of sanitary hot water, as well as the total electricity use. The electricity use of the compact ventilation unit is shown both as part of the total electricity use and separately, as it accounts for 47% of the total amount. The rest of the electricity use is given by the white goods (cooking plate, refrigerator, dishwasher, cooker – for boiling water, washing machine and dryer), storage tank, floor heating pump, control system, and all the rest connected to the occupants' activities (household appliances and lighting). The white goods contribute to the total amount with 29%, the control system with 11%, and the storage tank and floor heating pump with 10.5%. A more detailed analysis of the electricity use by the different appliances can be found in the study of Carpino et al. [17], conducted for the same building.

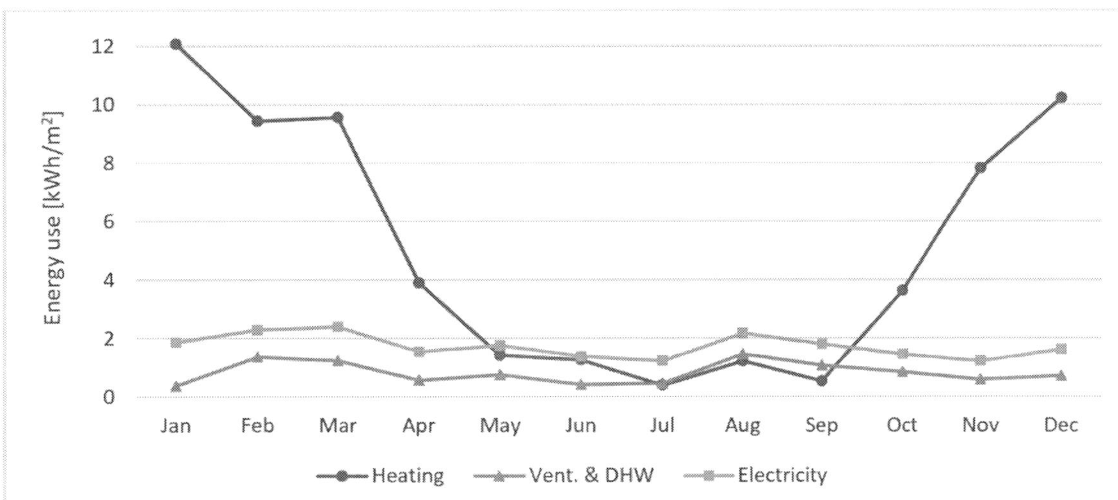

Figure 6. Energy use in 2018.

In order to compare these values with the numbers given by the European standards and simulation tool Be18 (which was used to calculate the total energy requirement during the design phase), the conversion from delivered to primary energy use is made with a primary energy factors in Denmark of 1.8 for electricity and 0.6 for district heating. The results are summarized in Table 3. As one can see, the total energy use of the building for heating, cooling, ventilation and hot water is higher than expected by 89%. This mismatch may be due to several reasons. Firstly, simulation program Be18 is not meant for detailed energy analysis, but rather as an estimation of the energy frame, using a standard occupancy profile [17]. Additionally, during the design phase, assumptions had to be made, which in several cases might have been quite far from the reality. Finally, during the first year, the building's operation does not comply entirely with the intended, due to necessary adjustments or corrections to the systems, and also due to the fact that some of the equipment or control has not been implemented yet. On the other hand, electricity use is much lower than expected, resulting in smaller internal gains and thus higher heating needs. In [17], it was estimated that the reduced effective heat gains from people and equipment load could explain an increase in the heating energy use of about 10 kWh/m^2 per year.

Table 3. Primary energy consumption [kWh/m² year].

		Measured energy use	Primary energy use		Be18 [a]	Commission recommendation [b]
	Heating	61.4	36.8			
Electricity	Ventilation & DHW	9.7	17.5	**58.2**	**30.8**	**50 - 65**
Electricity	Tank & heat pump	2.2	3.9			
Electricity	White goods	6.1	10.9			
Electricity	Control	2.3	4.1	**15.8**	**54.8**	
Electricity	Other consumption	0.4	0.8			
	Total	**82.1**	**74**		**85.6**	

[a] Energy frame BR 2015/ 2018 [b] (EU) 2016/ 1318

4.2. Indoor environment

When evaluating energy use, indoor environmental quality (IEQ) should also be considered. IEQ is assessed based on thermal and atmospheric indoor climate only. In Tables 5, 6 and 7, the time distribution in different comfort categories is presented. The categories for the different parameters are given in Table 4, as they have been formulated based on the Danish standards [18]. The percentages in each category are calculated for the period that each room is expected to be occupied. Therefore, the time between 7:00 to 17:00 is taken as unoccupied, while for the living room and kitchen the considered period is between 17:00 – 23:00, and for the bedroom between 23:00 – 7:00.

Table 4. Thermal comfort and indoor air quality categories.

	Temperature in Summer [°C]	Temperature in Winter [°C]	CO_2 above outdoor level [ppm]	Relative humidity [%]
IV+	$27 < t$	$25 < t$	> 800	$70 < t$
III+	$26 < t \leq 27$	$24 < t \leq 25$	800	$60 < t \leq 70$
II+	$25.5 < t \leq 26$	$23 < t \leq 24$	500	$50 < t \leq 60$
I	$23.5 \leq t \leq 25.5$	$21 \leq t \leq 23$	350	$30 \leq t \leq 50$
II-	$23 \leq t < 23.5$	$20 \leq t < 21$	-	$25 \leq t < 30$
III-	$22 \leq t < 23$	$19 \leq t < 20$	-	$20 \leq t < 25$
IV-	$t < 22$	$t < 19$	-	$t < 20$

Table 5. Time distribution in thermal comfort categories [%].

	IV-	III-	II-	I	II+	III+	IV+
Master bedroom	4.9	14.0	9.3	66.0	1.5	1.3	3.1
Living room	0.6	6.1	9.9	58.7	15.9	4	4.8
Kitchen / dining room	0.7	5.2	6.9	52.6	10.7	14.5	9.4

As one can observe in Table 5, thermal comfort remains in Category I and II during most of the time, with 77% and 85% for the bedroom and living room, respectively, while in the kitchen / dining room a percentage of 70% is achieved. In Figure 7, the aforementioned categories are divided into winter and summer period, in order to create a better visualization of the situation. It is worth mentioning that during the heating period, the achieved thermal comfort level is very high. That is not exactly the case for the kitchen / dining area, where the large glazed areas towards the east and west can create a bit of overheating, even during the winter. Nevertheless, a decrease in the temperature set-point could be suggested, in order for the energy consumption for heating to be reduced. On the other hand, during summer, the temperature in the bedroom and living room seem to be quite low. It can be concluded that during the cooling season, the occupants prefer lower temperatures in the bedroom, which is probably achieved through the opening of windows at night time. Consequently, the temperatures in the living room and kitchen are also affected by these practices.

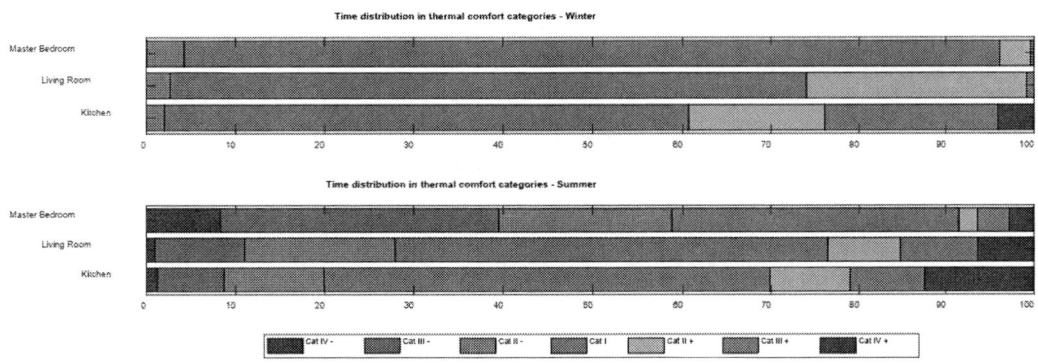

Figure 7. Time distribution in thermal comfort categories for winter and summer periods [%].

Regarding indoor air quality, CO_2 and relative humidity levels do not deviate much from Categories I and II. It can thus be concluded that the ventilation could be decreased significantly in the living room and kitchen without compromising the indoor air quality.

Table 6. Time distribution in comfort categories for CO_2 concentration [%].

Sensor	Room	I	II	III	IV
IC-meters	Living room	94.4	3.9	1.6	0.1
	Kitchen / dining room	95.7	3.0	1.2	0.1

Table 7. Time distribution in comfort categories for relative humidity level [%].

Sensor	Room	IV-	III-	II-	I	II+	III+	IV+
IC-meters	Living room	0.0	0.5	11.3	78.5	9.8	0.0	0.0
	Kitchen / dining room	0.0	3.5	12.5	77.0	7.1	0.0	0.0

5. Conclusion

Within the building industry, it has been broadly documented that there is a mismatch between the theoretical and real building energy performance, resulting in higher energy use than expected. This performance gap gives a distorted image of achievement concerning the energy reduction targets set by the European Union and building regulations. Additionally, this problem can create distrust between engineers and clients, and reluctance towards future technologies and practices [2]. In order to mitigate the problem, it is necessary to look into the reasons causing it and find ways to address these issues.

Through this analysis, it has been concluded that the imbalance is the result of the combination of three factors. The first one is the use of theoretical profiles and assumptions made during the initial stages of design, in order to calculate the expected energy use. In the present study this account for about 1/3 of the deviation (10kWh/m^2 per year of primary energy). The second factor concerns the errors commonly made during the construction and installation phase and the necessary adjustments and tuning of the systems so that they will perform as intended. These include set-point regulation, control methods, commissioning, verification, calibration or replacement of the different sensors. Finally, one of the most critical parameters is the occupants' behavior and practices. Even though the systems and automations might be operating perfectly, if the users' daily actions do not comply with energy efficiency practices, the result might be the opposite than intended. Especially when it comes to high energy performance buildings, which are more sensitive to the user's behavior. Therefore, set-point interference, window opening, or overwriting of the automatic controls can have a large impact on energy use, as well as IEQ. To bridge the performance gap, further analysis of these practices is needed to identify their impact on the building energy.

Acknowledgements

Bolig 2020 - "Bolig 2020 med godt indeklima og høj brugekomfort" is a project funded by EUDP (Energy Technology Development and Demonstration Program), J.nr. 64015-0640 [8].

References

[1] Directive 2010/31/EU of the European Parliament and of the Council of 19 May 2010 on the energy performance of buildings (recast). Official Journal of the European Union 2010. L153/13 – 35
[2] De Wilde P 2014 *Autom. in Constr.* **41** (2014) 40 – 49
[3] Yohanis Y G, Mondol J D, Wright A, Norton B 2008 *Energy and Buildings* **40** (2008) 1053 – 1059
[4] Majcen D, Itard L C M, Visscher H 2013 *Energy Policy* **54** (2013) 125 – 136
[5] Brohus H, Heiselberg P K, Simonsen A, Sørensen K 2010 *Clima 2010: 10th Rehva World Congress*
[6] Zavrl M S, Stegnar G 2017 *Procedia Environ. Sci.* **38** (2017) 52 – 59
[7] Menezes A C, Cripps A, Bouchlaghem D, Buswell R 2012 *Appl. Energy* **97** (2012) 355 – 364
[8] Odermatt B, Reutter H 2015 *Project proposal – EUDP Bolig 2020 med godt indeklima og høj brugekomfort* (64015-0640)
[9] Gram-Hanssen K, Georg S, Christiansen E, Heiselberg P K 2018 *Build. Research & Information* **46** (7) 790 – 803
[10] Carpino C, Mora D, Arcuri N, De Simone M 2017 *Build. Simul.* **10** (2017) 875 – 888
[11] De Meester T, Marique A-F, De Herde A, Reiter S 2013 *Energy and Buildings* **57** (2013) 313 – 323
[12] Andesen R V, Toftum J, Andersen K K, Olesen B W 2009 *Energy and Buildings* **41** (2009) 11 – 16
[13] Ouf M, Gilani S, O'Brien W 2019 *Ashrae journal* March 2019 28 – 36
[14] Wei S, Jones R, De Wilde P 2014 *Energy and Buildings* **70** (2014) 36 – 44
[15] Yan D, O'Brien W, Hong T, Feng X, Gunay H B, Tahmasebi F, Mahdavi A 2015 *Energy and Buildings* **107** (2015) 264 – 278
[16] Commission Recommendation (EU) 2016/1318 of 29 July 2016 on guidelines for the promotion of nearly zero-energy buildings and best practices to ensure that, by 2020, all new buildings are nearly zero-energy buildings. Official Journal of the European Union 2016. L208/46 – 57
[17] Carpino C, Loukou E, Heiselberg P K, Arcuri N 2019 Energy performance gap of a nearly zero energy building: influence of occupancy modelling and effectiveness of simulation *Preprint* buil-s-19-00105
[18] DS/EN 15251 Indoor environmental input parameters for design and assessment of energy performance of buildings addressing indoor air quality, thermal environment, lighting and acoustics 2007

IOP Conf. Series: Earth and Environmental Science **352** (2019) 012018 doi:10.1088/1755-1315/352/1/012018

Business models for nZEB renovation of small wooden buildings

Ann Kristin Kvellheim[1], Sigrid Damman[2] and Anne Gunnarshaug Lien[1]

[1]SINTEF Community, Postboks 4760 Torgarden, 7465 TRONDHEIM, Norway
[2]SINTEF Digital, Postboks 4760 Torgarden, 7465 TRONDHEIM, Norway

Ann.Kristin.Kvellheim@Sintef.no; Sigrid.Damman@Sintef.no; Anne.G.Lien@Sintef.no

Abstract. To achieve a sustainable building stock, it is necessary to focus on existing as well as new buildings. New houses built according to the Norwegian building code are close to nZEB (near Zero-Energy Building) level, but existing buildings contribute a significant share of the energy consumption, and the difference between new and older houses is increasing. As the renovation rate is low, it is important to include ambitious energy upgrading once a renovation project is undertaken. So far, however, there is almost no market for nZEB renovation of small wooden buildings. These buildings are challenging to renovate, and even more challenging to do so with ambitious climate targets. In effect, only half of the renovations in Norway include energy renovation. The demand for ambitious renovation depends on a value proposition that is understood by the customer. Novel and well-designed business models can be central tools for achieving this. In this paper, existing and potential business models for renovation of wooden housing are identified and analysed.

This paper is based on the "Energy upgrading of wooden dwellings to nearly zero energy level" (OPPTRE) project. One of the activities for researchers, industry and public partners in OPPTRE is to identify, analyse and assess current business models, and develop novel approaches through collaborative workshops and interviews. The results are documented and elaborated upon by the research team before the next step, which is to test the most promising models.

Previous research indicates that only a small share of Norwegian enterprises have changed their business model over time. Two of the reasons for this are lack of critical reflection on existing business models and fear of changing the status quo. A third reason is lack of knowledge of how to manage the transformation process. In the process of identifying existing and developing novel business models, the partners in OPPTRE explore new opportunities to promote ambitious energy upgrading. Their commercial performance is likely to increase through the focus on business model design. New business models may also help increase the uptake of innovative energy solutions in the renovation market for small wooden buildings.

1. Introduction

In order to mitigate climate change, it is necessary to create business opportunities in new and more sustainable ways. Business model (BM) innovation is seen as a prerequisite to achieving a sustainable transition to a low-carbon society. In addition to the need for new and more sustainable BMs to combat climate change, research has shown that BM innovation is economically rewarding [1]. Despite this, BM innovation is rare in practice. This is related to limited awareness in enterprises, both about their present BMs and about how challenging it can be to change the status quo [1].

In this paper, we discuss current BMs and prospects for sustainable BM innovation in the market for renovation of small wooden dwellings in Norway to nearly Zero-Energy Building (nZEB) level. The discussion is based on ongoing research which started in spring 2018 and will continue over the next two to three years. The findings must therefore be considered preliminary.

1.1 Renovation of small wooden buildings

Buildings are accountable for approximately 40% of energy consumption and 36% of CO_2 emissions in the EU. Furthermore, 35% of buildings in the EU are over 50 years old and nearly 75% of the building stock is energy inefficient [2]. Energy efficiency renovations present a tremendous opportunity to decrease energy use [3]. Energy use in Norwegian dwellings was 48 TWh in 2016. This is more than half of the total energy used in the whole building stock. Klinski et al. estimated the potential energy savings for residential buildings to be 4.5 TWh by 2020 [4]. Norwegian residential buildings are largely wooden houses. A quarter (25%) of these are privately owned detached houses built between 1960 and 1990 and will soon need repair and renovation. A study from 2015 shows that only half of the renovations in Norwegian dwellings include upgrading of the energy standard [5]. A large amount of money is, however, spent on kitchens, bathrooms and other interior upgrading.

In the EU, the delivery of energy renovations is hindered by financial, technical, process, regulatory and awareness barriers [6] [7]. Much of the previous research on nZEB renovation is on how such barriers work and how they may be reduced or eliminated [7]. The need to develop skills and capacity on the supply side has received less research attention to date [8].

1.2 The OPPTRE research project

This paper is based on the "Energy upgrading of wooden dwellings to nearly zero energy level" (OPPTRE) research project, which aims to create knowledge for cost-efficient energy renovation of wooden dwellings for homeowners and builders, and knowledge for public and private decision-makers on BMs, regulations and incentives that can lead to a significant market uptake of nZEB renovation towards 2030. The project runs from 2018 to 2021 and includes nine partners. The level of energy efficiency for nZEB renovations is not necessarily as ambitious as for new buildings for all parts of the building. In this paper, nZEB renovation is used as a term for renovation towards nZEB level and may comprise several steps of upgrading.

2. Earlier findings and relevant research projects

2.1 Previous research on business model innovation

The capability to develop new BMs is increasingly seen as key to improving sustainability performance and building sustainable competitive advantage [9]. Private business is a pivotal stakeholder in sustainability transitions, commanding most resources and capabilities. Technological advances tend to be incremental, and innovation at BM level is required to align incentives and revenue mechanisms and facilitate uptake of sustainable solutions. While there has been a lack of consensus concerning what a BM actually is [10], Geissdoerfer et al. propose a definition that is much in line with Osterwalder and Pigneur [11], seeing BMs as *"simplified representations of the value proposition, value creation and delivery, and value capture elements and the interaction between these elements within an organizational unit"* [9]. The BM canvas, a template for developing new or recording existing BMs, designed by Osterwalder and Pigneur [11] can be seen as a representation of the economic side of a BM. By adding two more layers – the environmental and the social layers – the depth of the model corresponds to the triple bottom line of the sustainability concept [12]. This model is referred to as the triple layered BM canvas [13].

BM research in the building and construction disciplines is still at an emerging stage [14]. The home renovation market is highly fragmented, and it is argued that better coordination of the supply-side actors is necessary to increase the scope for sustainable innovation. One such way of coordinating the market is through a One Stop Shop (OSS). The idea of an OSS in the context of renovating private homes is to form teams of architects, energy consultants, builders and other craftspeople that together can supply the homeowner with all the services needed for an energy renovation. Only a few researchers have adopted BM research for analysing energy renovations in detached houses, e.g. [6] [15] [16] [17].

2.2 Previous relevant research projects

Most of the researchers referred to in the previous paragraph collaborated in the EU project COHERENO, which explored collaborative BMs (e.g. OSS) as a market device for realizing single-family home energy renovations [8] [18]. The COHERENO team basically worked as change agents to facilitate collaboration between enterprises in 24 consortia across Europe. A key finding was that innovative business modelling was useful for vision formation, networking and increasing customer orientation, but did not readily lead to implementation of collaborative business approaches. Effort to support small and medium-sized enterprise (SME) collaboration and continued action-based research was therefore recommended [8]. A general observation was that SMEs in construction are not used to target-group-oriented thinking, and although participation is more mixed in other countries, architects and consultants are less involved in energy renovation of private homes in Norway [8].

In the IEE EU NeZeR project, among other projects also drawing upon COHERENO, many countries identified the OSS model as relevant for nZEB renovation of private buildings [19]. While advantages of this model are holistic solutions and provide easy access to the service, it also requires high levels of trust, and customers may be more inclined towards stepwise actions for overall energy performance and costs [19].

The Norwegian research project SEOPP (Systematic energy upgrading of small wooden dwellings built between 1960 and 1990) was carried out from 2013 to 2016. The scope was to show examples of ambitious energy renovation combined with up-to-date architectural expression and living qualities. Two detached houses were fully upgraded [20].

Picture 1: Example from SEOPP, before (1) and after (2 and 3) upgrading. Photo: Jyri Havran.

For the participating partners, these best practice projects were used to promote their experiences of energy renovation. This has up to now only resulted in a few new projects, demonstrating that a best practice project alone is not enough to induce major market change. Most of the OPPTRE partners are therefore continuing to collaborate in OPPTRE, where the focus is on affordable upgrading for a larger market share.

3. Method

The paper is based on findings developed from two workshops in OPPTRE, as well as a limited study of literature and websites. One objective has been to investigate how actors in general and the user partners in OPPTRE in particular relate to BM innovation. What do they think of it, and how does this thinking materialize? Another objective has been to develop new ideas for BM innovation for nZEB renovation. This was discussed using the BM canvas as developed by Osterwalder and Pigneur [11].

3.1 Workshops

BM innovation was discussed as part of a general workshop (partner seminar) at the project start-up, where all partners were present. Furthermore, all partners were invited to a half-day workshop on BMs in March 2019. Seven of nine partners accepted the invitation and six participated.

In the first workshop, new business ideas were discussed with a focus on the following questions: 1) How can the building industry earn money from nZEB upgrading of wooden detached houses? and 2) How can the building owners gain from upgrading of their home? The second workshop investigated the following topics: 1) How the partners relate to their present BMs, and 2) New ideas to develop innovative BMs for nZEB upgrading.

3.2 Business modelling

The partners have individual motivations for participating in OPPTRE. However, all agreed on the aims to contribute to sustainability transitions, by developing BMs, methods, products and solutions for sustainable renovation of detached, wooden houses. By focusing on innovative BMs, we would like to illuminate the possibility of changing one or more steps in a current BM. An innovative BM may only have changed, for example, the customer segment. If this change makes the BM new to the industry or market, it is simultaneously an innovation. The main questions related to the BM canvas are as follows: 1) **What** are we selling/which needs do we fulfil? 2) **Who** are we offering this to? 3) **How** do we go about bringing our offer from the "production line" to the customer? 4) **Why** do we do this?

In order to deliver nZEB renovation, it is necessary to be able to supply products and services that not only deliver economic value, but also social value as well as environmental qualities such as in the triple layered BM. Supplementary questions raised by applying this model are related to environmental and social impacts and benefits.

4. Results

In this section, preliminary findings from the workshops and the literature will be presented while answering the questions of **What, Who, How** and **Why** explained in sub-section 3.2. Furthermore, aspects of the respective BMs of partners in OPPTRE are investigated along the three layers of the sustainable BM canvas.

4.1 Current business models and market situation

In the following, the industry partners in OPPTRE are referred to by category, followed by number of partners in brackets (). Furthermore, the producers of building materials will be referred to as the producer of wood fibre insulation, the producer of roof windows, the supplier of tightening products (to aid air and water tightness) and membranes, and the producer of ventilation systems.

The partners in OPPTRE relate to business development towards nZEB renovation in different ways. The architect (1) primarily has large public projects in its portfolio, but is also working to establish a stable income from smaller private projects. They have found that a way to reach the private renovation market is to register as energy advisors in the Enova system[1]. For energy renovation, they see the largest potential among young, economically secure people who buy old houses. Architects are generally renowned for having less focus on profit, and more focus on high-quality solutions. The experience so

[1] Enova SF is a state owned enterprise which aims to reduce greenhouse gas (GHG) emissions, develop energy and climate technology, and strengthen security of supply.

far is that marketing and sale of energy renovation has not led to significant growth in demand, even with the "best practice projects" from SEOPP to refer to.

The producers of building materials (4) primarily sell to wholesalers in the Norwegian market, some of which are expanding in the value chain by vertical integration (organic growth and acquisitions). Partially because of discounts and bonus agreements, it is difficult to get an overview of prices of products and services. Craftspeople have deals with the wholesalers which give them privileges in terms of better prices compared to what private homeowners can achieve. Lack of overview hinders the end customer (the homeowner) from comparing prices and to some degree makes this "the wholesalers' market". Value creation in the existing BMs must take this into consideration.

Many producers of building materials create invisible materials inside the construction, like insulation and tightening products. This means that they compete on other aspects aside from the visible ones such as quality/durability, price and environmental performance. The producer of wood fibre insulation has sustainability as part of its vision and also has elements of the circular economy in its BM. This is mainly linked to the environmental quality of wood, as well as use of residuals (sawdust) in its production. The supplier of tightening products and membranes competes on price, quality (durability) and risk reduction. They work to bring the end users closer, for example by inviting building contractors to attend courses. Both see a small but growing focus on environmental aspects among their customers.

The producer of roof windows is selling "fresh air and daylight" and appeals to emotions when they market their products to the end user. They primarily sell to house builders and also work in close cooperation with architects. About 50% of the firm's turnover is from renovation projects. In the Netherlands, where the headquarter is situated, their products are all based on traceable, reused wood, and they also promote holistic thinking of sustainability on their international website. The producer of ventilation systems is the market leader for domestic ventilation in Norway. They also offer indoor climate control with integrated ventilation and domestic hot water, which reduces the overall energy bill by around 50%. It has been expanding by servicification: after selling a ventilation system, they can also offer regular filter changes. This market is expected to grow.

The house builders associations (2) support their member craftspeople with design and building documentation of catalogue houses, advice and competence-building activities. They also have a role as wholesalers and sell discounted building materials to their members. Both housing associations informs readers about retrofitting and energy renovation on their website. The market for new houses has been booming over recent years and the profit from new buildings is probably at the expense of prioritizing renovation projects. Renovation projects are often small and seen as infill jobs between larger projects concerning new buildings.

4.2 Business model innovation

Some aspects of the current BMs were highlighted in sub-section 4.1. In Table 1, some key characteristics of the BMs in OPPTRE are mapped according to the three aspects of the triple layered BM canvas. The table describes aspects of the current situation and focuses less on the aspirations. Unless stated otherwise, the location is set in Norway.

Table 1. Key aspects in a triple layered business model canvas related to the OPPTRE partners.

	Economic aspects	Environmental aspects	Social aspects
The architect	Customer is the end user. Provides plan and design, as well as holistic cost control. Often in cooperation with builders. Sees renovation projects as a small but stable source of income.	Provides functionality and aesthetic solutions. Promotes environmental and energy-friendly solutions on website. Eco-Lighthouse Certified.	Develops solutions in line with the client's budget.
The producer of wood fibre insulation	Sales through wholesalers and house builders. Provides wood fibre insulation and related building materials. Competition based on durability, sustainability and price.	Wood implies low GHG emissions. Continuously searching for ways to reduce energy use in the production. EPD (Environmental Product Declaration) owner.	Builds allianze with primary distributor, promoting products with low emissions. Uses waste/residual materials (sawdust) in their production.
The supplier of tightening products and membranes	Customer is primarily wholesalers and house builders. Provides tightening products and membranes for the building envelope. Competition based on durability and price.	Some information about energy efficiency on website. EPD owner.	Aims to be ahead of stricter building regulations. Reduces moisture problems which leads to a more healthy indoor climate.
The producer of roof windows	Customer is primarily house builders. Half of the market is renovation. Marketing addresses end users and appeals to emotions.	International website has extensive information on impact and benefits of roof windows.	More daylight improves well-being.
The ventilation system producer	Customers are primarily house builders and HVAC contractors. About 20% of market is in renovation of dwellings.	Reducing energy consumption and CO_2 emissions through energy efficient products for ventilation system with heat recovery, hot water production and heating solutions.	Is selling indoor comfort through fresh air. Technology to improve indoor climate using a lower power load.
Association of house builders, 1	Customer is house builders (members). Provides professional advice and competence-building activities.	Has information about renovation /refurbishments on website, with advice about energy efficiency.	
Association of house builders, 2	Customer is house builders (members). Provides professional advice and competence-building activities.	Promotes healthy indoor climate and energy efficiency by referring to its partners. Informs about renovation/refurbishments on website.	

IOP Conf. Series: Earth and Environmental Science **352** (2019) 012018 doi:10.1088/1755-1315/352/1/012018

Economic impacts and benefits are mainly dependent upon customers, the character of the product (such as visibility) and the sales channels. For some, the renovation market is a large part of their activity, although the share of *energy* renovations is low or not possible to gather information on. Environmental impacts and benefits are mainly linked to energy efficiency, and some of the partners own environmental product declarations. Social impacts and benefits are linked to what relations are built with stakeholders as well as attention to qualities such as aesthetic, fresh air and daylight. It can also be argued that all participants in OPPTRE pay attention to social responsibility by attending research projects such as OPPTRE. It shows determination to contribute to a sustainable transition.

5. Discussion and conclusion

It seems challenging to create profit from energy renovation projects. This is due to factors such as high activity (and profit) in new buildings, as well as a high level of uncertainty for the craftspeople: what will they find when they open the construction? This perceived risk is reflected in the pricing of renovation projects. The building materials market is dominated by very few and powerful actors which also affects the pricing of products. At the same time, building owners struggle to find coherent and credible advice that takes their budget into account. Where do we go from here?

The triple layered BM canvas builds on and expands the template created by Osterwalder and Pigneur [11, 13]. Economic aspects include value proposition, value creation and capture. When environmental values are added, environmental impact and benefits must be included, which could be explored by, for example, life cycle assessment including materials and energy used in the process of renovation and afterwards. In OPPTRE, the focus is on cost-effective solutions, which consequently also has social implications. In addition, the BMs and actions of each organization have consequences, not only for the customer, but also for other actors in society. These are part of the social layer in the triple layered BM canvas. Other research has shown that sustainable behaviour has ripple effects on other areas in life as well. In OPPTRE, more knowledge will contribute to risk reduction when pricing renovation projects. An architect competition (part of the OPPTRE project) will provide several examples that can be used for promotion and to increase engagement.

These preliminary findings are in line with earlier research; partners in OPPTRE are likely to have unreleased potential if they become more conscious of how to use their BM in order to trigger more activity in the renovation market. In this respect, the focus on BMs in OPPTRE is likely to give ideas for BM innovation, and already has.

5.1 Ideas for business model innovation and further research

There are several ideas for BM innovation that need to be further explored. The application of the appraised OSS model has so far not led to any boom in the renovation market. Preliminary findings indicate that it has failed, so far, in terms of affecting market growth. However, the model addresses several barriers towards growth in energy renovation and will be further explored in OPPTRE.

Could there be potential if a neighbourhood, not a separate building, was the core of attention and analysis? This has been discussed in OPPTRE and needs to be further elaborated upon. It is potentially cost-effective to expand from one to several buildings if these are from the same time period and somewhat similar, but the potential also depends on the owners of these buildings.

Expanding by new ways of approaching the market should also be further explored. Could potential customers be presented with a net-based calculation and visualization tool to explore renovation of their homes? Could the regular line of production be expanded by adding a service, like the producer of ventilation systems has, as well as the architect? New developments provide new possibilities, such as smart technology, big data, artificial intelligence and more. In connection with the aforementioned architect competition, we will investigate motivations for renovation on the demand side.

The lack of transparency in pricing of building materials/products is likely to slow the transition down. Is the "wholesalers' market" a barrier to cost-efficient nZEB renovation, and could new ways of collaboration be found to give energy renovation the necessary push? The builders have key roles in the sustainable transition of domestic buildings, but the wholesalers could also be more directly engaged in holistic renovation projects.

The primary objective of OPPTRE is to enable the development of cost-effective solutions for nZEB renovation of detached wooden houses. The role of sustainable BMs in order to reach this objective has been explored in this paper. We will continue our research by exploring and developing current as well as new ideas for BM development together with the partners in OPPTRE.

References

[1] Saebi T 2016 Fremtiden for forretningsmodellinnovasjon i Norge *Magma* vol 7 pp 33–41
[2] European Commission 2019 Energy Performance of Buildings website accessed 28th April 2019
[3] Güneralp B, Zhou Y, Ürge-Vorsatz D, Gupta M, Yu S, Patel PL, Fragkias M, Li X and Seto KC 2017 Global scenarios of urban density and its impacts on building energy use through 2050 *Proc. Of the National Academy of Sciences* 114(34) pp 8945–8950
[4] Klinski M, Lappegard Hauge Å, Godbolt Å and KS Stenerud Skeie 2017 Energioppgradering av norske boliger – Evaluering av scenariorapporter og forslag til virkemidler ZEB Project report 32 SINTEF Academic Press
[5] Bjørnstad E 2015 Rehabilitering og energioppgradering av boliger Enovarapport 2015:10
[6] Haavik T, Mlecnik E and Rødsjø A 2012 From demonstration projects to volume market of sustainable construction Energy Procedia 30 pp 1411–1421
[7] Artola I, Rademaekers K, Williams J and Yearwood J 2016 Boosting building renovation: what potential and value for Europe? Study for the European Parliament (IP/A/ITRE/2013-046)
[8] Mlecnik E, Straub A and Haavik T 2018 Collaborative business model development for home energy renovations Energy Efficiency https://doi.org/10.1007/s12053-018-9663-3
[9] Geissdoerfer M, Vladimirova D and Evanset S 2018 Sustainable business model innovation: A review *Journal of Cleaner Production* 198 pp 401–416
[10] Zott C, Amit R and Massa L 2011 The business model: recent developments and future research *J. Manag.* 37 pp 1019–1042 https://doi.org/10.1177/
[11] Osterwalder A and Pigneur Y 2010 Business Model Generation (Hoboken New Jersey Wiley & Sons)
[12] Henriques A and Richardson J 2013 The Triple Bottom Line, Does It All Add Up? (New York Earthscan) ed A Henriques and J Richardson
[13] Joyce A and Paquin R L 2016 The triple layered business model canvas: a tool to design more sustainable business models *Journal of Cleaner Production* 135 pp 1474–1486
[14] Abuzeinab A, Arif M, Quadri MA and Kulonda D 2018 Green business models in the construction sector: an analysis of outcomes and benefits Construction Innovation 18(1) pp 20–42
[15] Mahapatra K, Gustavsson L, Haavik T, Aabrekk S, Svendsen S, Vanhoutteghem L, Paiho S and Ala-Juusela M 2013 Business models for full service energy renovation of single-family houses in Nordic countries *Applied Energy* 112 pp 1558–1565
[16] Mlecnik E and Straub A 2014 Barriers and opportunities for business collaboration in the nZEB single-family housing renovation market. http://www.cohereno.eu/about/projectoutcomes
[17] Grøn Bjørnboe M, Svendsen S and Heller A 2017 Using a one-stop-shop concept to guide decisions when single-family houses are renovated ASCE *Journal of Architectural Engineering* https://doi.org/10.1061/(ASCE)AE.1943-5568.0000238
[18] COHERENO 2016 COHERENO – collaboration for housing nZEB renovation, Intelligent Energy Europe project (2012–2016) http://www.cohereno.eu/about/project-outcomes
[19] ISPE 2016 Report on the successful business models for NZEBR D5.4 Promotion of smart and integrated nZEB renovation measures in the European renovation market (NeZeR)
[20] Lien A G, Skeie K S, Bjaanes E, Hagen K and Kvalø Y 2017 Oppgradering av et 60-tallshus og et 70-tallshus. Oslo: SINTEF Fag 42

IOP Conf. Series: Earth and Environmental Science **352** (2019) 012019 doi:10.1088/1755-1315/352/1/012019

Process of Energy Master Planning of Resilient Communities for comfort and energy solutions in districts

Matthias Haase, SINTEF Building and Infrastructure, Norway*,
Ruediger Lohse, Climate Protection and Energy Agency, Germany

*Matthias.Haase@sintef.no

Abstract. The paper explains the major steps in Energy Master Planning process. It proposes a definition of target goals. Then, a number of constraints have to be analyzed in order to be able to define site specific framing goals and associated limitations. This process will narrow the numerous design options down to those that offer an optimized fit to the local conditions and the objectives for the building or community. Based on the target definition a Baseline can then be developed. This consists of a snapshot of the current energy use situation. The baseline is one reference point used to evaluate alternative futures. Then Base Cases will be developed that extends the baseline into the future and includes already-funded renovation as well as planned construction and demolition activities. The base case is a future reference point for "business as usual." Different alternatives – A selected set of scenarios that include different energy measures related to buildings, distribution systems, and generation systems will then developed. These scenarios are compared to the baseline for energy use change and to the Base Case for investment and operational costs.

1. Introduction

Until recently, most planners of public communities (military garrisons, universities, etc.) addressed energy systems for new facilities on an individual facility basis without consideration of community-wide goals relevant to energy sources, renewables, storage, or future energy generation needs. Because building retrofits of public buildings typically do not address energy needs beyond the minimum code requirements, it can be difficult if not impossible to achieve community-level targets on a building-by-building basis.

As more and more countries push to improve the efficiency, environmental impact, and the resilience of their buildings and communities, the need for early and comprehensive energy master planning is critically important. The best energy master planning is highly dependent on a thorough understanding of framing goals and constraints, both local and regional, and their associated limitations that will dictate the optimum master planning design.

This paper will analyze and contrast the framing goals and limitations that must be considered when energy master planning is conducted for communities in six different countries. The analyses will be based on findings from countries participating in the International Energy Agency's "Energy in Buildings and Communities Program Annex 73" The analysis will cover design constraints such as emissions, sustainability and resilience goals, and regulations and directives, and regional and local limitations such as available energy types, local conditions, and community objectives and illustrate how a comprehensive consideration of these can be used to guide the planner toward design options that will lead to an optimum solution for a master plan.

IOP Conf. Series: Earth and Environmental Science **352** (2019) 012019 doi:10.1088/1755-1315/352/1/012019

Lastly, the paper will propose a comprehensive table of framing goals and associated limitations and a suggested process that the master planner can use to narrow the numerous design options down to those that offer an optimized fit to the local conditions and the objectives for the building or community.

2. Background

The status quo in planning and execution of energy-related projects will not support attainment of current energy goals (Energy Performance of Buildings Directive [1] in Europe and 10CFR-433 in the United States [2]) or the minimization of costs for providing energy security.

Most national and international research and policy energy-related efforts in the built environment focus on renewable energy sources and energy efficiency in single buildings. Significant additional energy savings and increased energy security can be realized by considering holistic solutions for the heating, cooling and power needs of communities – comprising collections of buildings.

Building-centric planning falls short of delivering community-level resilience. For example, the frequency and duration of regional power outages from weather, manmade events, and aging infrastructure have increased. Major disruptions of electric and thermal energy have degraded critical mission capabilities and caused significant economic impacts at military installations.

Organizations that have made first efforts to evaluate and analyze international experiences with planning and implementation of low-energy communities include the International Energy Agency (IEA) Energy Conservation in Buildings and Community Systems (ECBCS) Annex 51, 67, 70 and 75 [3-7], the German-funded EnEff Stadt project [8] (a comprehensive approach to urban areas with local and district heating networks), the World Bank Energy Sector Management Assistance Program (ESMAP) [9] and the Energy Efficient Cities Initiative (EISA) [10]. The U.S.Army pioneered a Net Zero Installations program for selected installations [11], which goes beyond zero energy and includes zero waste and zero water initiatives. Other initiatives e.g.in the center for Zero Emission Neighborhoods (ZEN) [12] recently analyzed tools in use for stakeholder engagement in four ZEN pilot projects in Norway. The results show that the tools have different goals and involve different stakeholders, some are focusing on citizens, while others aim for engagement of professional stakeholders such as construction and energy companies [13]. Other studies aim at developing guidelines for Zero Emission Neighborhoods (ZEN) by focusing on how the definition of ZEN and its KPIs could be assessed and implemented into the planning, design, construction, and operational phases of planned and existing neighborhoods [14].

It was therefore important to collect data on EUI from existing Commercial Buildings Energy Consumption Surveys and from existing standards (ASHRAE Std 100 [15], German VDI 3807 [16], Switzerland SIA 380.1 [17] etc.). EUIs are a necessary requirement for efficient energy management and for establishing national or agency-specific energy targets.

The development of districts requires a distinct understanding of the situation now as well as a vision of the future district in order to be able develop suitable pathways for this transition. In order to be able to do that a district needs to be modelled that consists of several buildings (new, retro-fitted or a combination of both), sufficiently described so that the future district can actively manage their energy consumption and the energy flow between them and the wider energy system. The energy master planning process requires an analysis of different scenarios, which include new construction to different levels of energy efficiency, major renovation of all or some buildings comprising building stock under consideration with Deep Energy Retrofit of these buildings, minor renovations with energy-related scope of work, or demolition of some old buildings. Such analysis requires building energy modeling. In this research work we collected models of representative buildings from several countries and compared them.

However, in community-wide energy planning, it is important to understand the various constraints, which frame the planning goal and the way towards an optimized solution. The planners need to know the design constraints such as emissions, sustainability and resilience goals, and regulations and directives, and regional and local limitations such as available energy types, local conditions, and community objectives. In order to be able to apply principles of a holistic approach to community energy planning and to provide the necessary methods and instruments to master planners, decision makers, and stakeholders it is essential to identify and frame the constraints that bound the options towards an optimized energy master planning solution.

3. Energy Master Planning

The energy master planning process (EMP) for building communities is carried out in several, at least three stages starting with the concept phase, the first planning and iterations. In the big picture of a refurbishment or newly set up compound EMP is one part of the total design process which usually claims between 5 and 20% of the total cost budget. Interactions between EMP and the other constructive planning has to be set up from day one to avoid costly iterations. In a few EU countries a small number of best practice projects provides first experience in the interaction of BIM (building implementation management) systems and EMP processes. It is important to create a better understanding of the practical application of different interfering constraints on different stages of the energy master planning EMP process which addresses energy master planners, architects, spatial planners, public and private real estate management and financiers.

EMP can be carried out in a hierarchic "top down" approach or in a "bottom up" process, in [21] both methods are compared. The decision making on hard and soft constraints is often required when existing buildings and infrastructure is refurbished under cost limit constraints. The EMP process is usually carried out in two stages. In the first stage of EMP (concept phase) more holistic and even generic constraints resulting from limitations from the spatial planning and mission related indicators have to be considered. The second stage adds the assessment of constraints on the level of components. To minimize risks of failure and abundant cost budgets in the EMP process it is of advantage preparing a set of pre- selected scenarios with technical solutions for a number of selected types of compounds and neighborhoods and their specific needs.

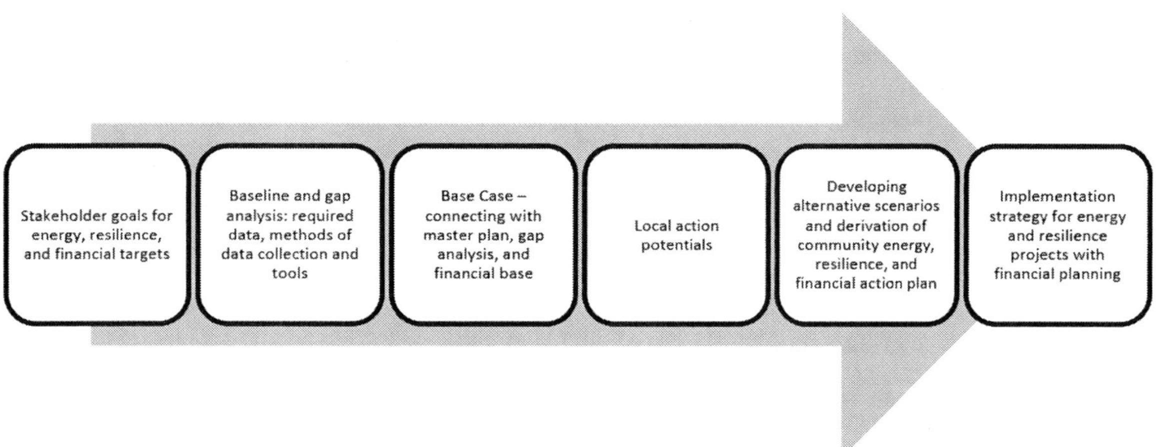

Figure: Different stages in Energy Master Planning (EMP)

3.1. Stakeholders goals and Constraints

One of the first steps in energy master planning is to determine the framing constraints. Having this as an early step will typically significantly reduce the possible solution sets for both individual buildings/structures/facilities and the entire district as a whole. Framing constraints in this context are defined as any constraints that will influence/restrict the possibilities for using or installing particular technologies or utilizing specific solutions in the district or in individual buildings/structures. The framing constraints can be divided into two subgroups; the natural constraints and the imposed constraints.

The natural constraints cover e.g. locational threats and resources. Locational threats deals with all natural threats that influence the possible choices of technologies or solutions and could be e.g. regional or local air quality, extreme temperatures or high winds. Locational resources deal with the availability of energy on-site or nearby. It covers both renewable energy sources for the location, e.g. wind, solar etc. and existing available energy infrastructure, e.g. power lines, gas pipes, district heating etc. Harnessing adequate amounts of energy from renewable energy sources usually requires quite a lot of space, e.g. it may be difficult to harness solar energy in big cities where roof or land area is not available and it may be difficult to utilize wind turbines since they require open spaces to be efficient. Therefore, the spatial possibilities are also part of the natural constraints.

The imposed constraints are constraints that for the most part is relevant for individual buildings or facilities (e.g. requirements on maximum energy consumption, emissions or requirements on specific indoor climate parameters) but the imposed constraints can also apply to the entire district (e.g. local plans or national energy targets). The energy planner, owner or operator of the district could also choose to impose special voluntary operational constraints that are more restrictive than e.g. legislative constraints, e.g. 100% renewables, possibility for islanding for a certain length of time etc.

3.1.1. List of constraints.

- Natural Locational Constraints – Resources and threats
- Distribution System & Storage Constraints
- Building and Facility Constraints:
- Indoor Environment Constraints
- Building Equipment and District System Constraints

Natural locational constraints can typically be categorized into resources and threats. In EMP processes, energy resource limits such as the unavailability of natural gas, or low biomass or wind resources will limit your technology options. As with all constraints, these should be identified and applied early. Fuel and water resource limits can of course be identified via local utility providers. Chilled water, hot water, and steam resource limits can be identified via the capacity of the local central plants that supply them. These resource limits must be considered with regard to the resource demand from any users on the district system outside the building or campus under consideration.

3.1.2. European Building Performance Directive (EPBD)

In Europe, the European Building Performance Directive (EPBD) creates the framework for the national building performance legislation of the EU member countries. EPBD set targets for the overall energy efficiency improvements in the building stock and for the partition of renewable energy to be integrated in the building sector. The pathway towards a 80- 90% carbon neutral building sector includes cross sectoral strategies such as:

- Energy demand limits for newly built and extensively renovated buildings
- Energy certification formats to provide transparency on the energy demand of t the building
- Integration of information and communication systems in order to adapt the energy supply to the energy demand of the users, create transparency on the actual performance and consumption of the building users
- E- mobility as a cross cutting approach for power use
- Building refurbishment roadmaps in order to implement strategic building refurbishment targets for the next 30 years

IOP Conf. Series: Earth and Environmental Science **352** (2019) 012019 doi:10.1088/1755-1315/352/1/012019

The EBPD has to be implemented under consideration of the subsidiarity principle on the national legislation of each EU member countries. After the recent update of the EPBD in 2018, July, member countries are requested to provide national implementation programs within the next 24 months.

In the case of Germany, the EPBD is transferred into different legislative constraints of relevance for the EMP process:

- Energy efficiency regulation (EnEG) [18]: relevant for the design of heating supply units and methodologies for cost calculation and billing of heating supply costs in the building sector; since 2005 efficiency of lighting systems has been adopted in order to provide energy performance certificates.
- Energy efficiency ordinance (EnEV) [19]: since 2002 the energy saving ordinance has been revised three times in order to adapt EPBD regulations on the national level. However, the EnEV has to be adopted into the building ordinances of each German Federal State. Newly constructed non-residential buildings are calculated by a method using a reference building in accordance with the German Standard DIN V 18599. Non-residential buildings have to achieve certain energy requirements such as the annual primary energy demand for heating, hot water, ventilation, air conditioning and lighting cannot exceed the annual primary energy consumption (QP) of a reference building, regarding the same geometry, useable floor area, orientation and utilization with the predefined technical reference execution in the EnEV. Additionally, the upper limiting values of the average heat transfer coefficient ($\bar{U}$) are not to be exceeded.
- Since the EnEV 2009, this calculation method has also to be applied for residential buildings. In this case the planned residential buildings cannot exceed the requirements of the annual primary energy consumption (QP) of the reference building and in the EnEV predefined maximum transmission heat loss (H'T) of the entire building envelope.

Table 1. End energy demand values for non- residential buildings according to EnEV 2012 (average values) in Germany and total net energy requrirements according to TEK17 [20] in Norway

Building usage	ENEV 2012		TEK17
	Heating / DHH	Electricity	Total net energy requirement
	(kWh/m²yr)	(kWh/m²yr)	(kWh/m²yr)
Middle class hotel	85	55	170
Restaurant	205	95	180
Cinema	55	80	180
Gyms	120	35	145
Multipurpose Convention Centers	240	40	180
Swimming pool (indoor)	385	105	145
Non food commerce small	135	45	180
Shopping malls	70	75	180
Hospitals	175	80	225 (265)*
Office building (heating only)	105	35	115
Office building (heating/cooling)	110	85	115
Cultural building			130
Light industry/workshop			140 (160)*
School building			110
University/university college			125
Nursing home			195 (230)*
Kindergarten			135

* Numbers in parenthes are buildings with reduced possibility for heat recovery from ventilation
For the renovation of residential and non-residential buildings, a verification according to the EnEV is not necessary, if the change of the building envelope is < 10% of the total individual component area. A complete verification, respectively a verification of the individual measures on the building envelope should be carried out for all other renovations. The thresholds for primary energy (QP) and the transmission heat loss over the envelope (H´T, resp. Ū for non-residential buildings) can exceed the requirements of the reference buildings by a maximum of 40% in the complete verification.

3.2. Local action potentials and development of alternatives

At the end of the gathering of existing framework, the number of potential technical solutions will be limited, the needs for refurbishment will be known, also energy saving potentials in the buildings, heating, ventilation and air-conditioning (HVAC), energy supply. From recent years man made and natural threats will be available for the neighbourhood which will be cross-checked with data sets from insurance companies to reframe risk scenarios from current and predictable future threats.

Usually the work and cost effort of EMP is mainly defined by the number of scenarios considered in the modelling process. Especially in Europe, costs often hamper the broader use of hourly modelling processes. An analysis of 10 refurbished neighbourhoods ("Quartierssanierung") in south west Germany shows, that the EMP is only carried out by using calculation tool which refer to monthly or even annual data sets [21]. However, the challenges created by the broader use and production of renewables require the broader application of hourly based modelling tools. One strategy of increasing the use of accurate modelling tools which is pursued in the Annex 73 is to lower the impediment "costs" by reducing the number of scenarios modelled. To minimize the modelling work effort for the EMP processes a number of pre- selected scenarios will be available which fit to the specific requirements of typical compounds and neighbourhoods. The scenarios will provide a set of simplified supply schemes and suggestions for energy efficiency and energy resilience approaches.

3.3. Implementation

In the following, the approach for the decision making is differentiated in the concept and the design phase of the EMP of a top- down or a bottom- up approach. So far, an overarching rule when to use an EMP "top down" or "bottom up" approach is not in place in literature of EMP. In management and design theory "top down", "waterfall-" or "cascading management" is mostly known in the context of hierarchic structures where a leading person defines the overall target and often a set of sub- targets. The next lower hierarchies take decreasing pieces of the activities which can be derived from the over-all target and well overseen from the respective leadership of the hierarchy.

The "bottom up" approach is known in the context of management methods which sees the top leadership in the position to define overall targets and delegate actively the design, execution and controlling of the sub- targets and work processes. This method is used in flat and matrix hierarchies and requires a high level of autonomy and also participative processes with end users, decision makers and multi- competence teams.

Transferred to the EMP it can be observed, that "top down" EMP are often used in military or comparable compounds with clearly hierarchic management structures. As no general rule exists, the assumption of "top down" in the context of EMP shall be, that a compound is refurbished or newly built with the first priority to the functionalities which can be derived from the mission of the compound after the refurbishment. Energy or environment related aspects are one out of many decision-making criteria in the overall design process. Often the design approach is to follow the minimum requirements for the efficiency of buildings, equipment and processes. In comparison the management of natural and man-made risks are considered to have more impact on the mission and have a higher priority in the design process. In consequence, the EMP process allows larger tolerance bands for energy efficiency or energy resilience in a top down process. As one proof of this assumption: In some European countries, i.e. German military buildings are even exempted from the energy targets relevant for other buildings of the Federal Government and only need to fullfill the minimum energy requirements.

Discussion

The paper explains the major steps in Energy Master Planning process. It proposes first a definition of target goals. Then, a number of constraints have to be analyzed in order to be able to define site specific framing goals and associated limitations. This process will narrow the number of design options down to those that offer an optimized fit to the local conditions and the objectives for the building or community. Based on this target definition a Baseline can be developed. This consists of a simulation of the current energy use situation. The baseline is one reference point used to evaluate alternative futures. Then Base Cases will be developed that extend the baseline into the future and include already planned/funded renovation as well as planned construction and demolition activities. In that sense, the Base Case is defined as a future reference point for "business as usual" and starting point for further economic analysis. Different alternatives will then be developed – a selected set of scenarios that include different energy measures related to buildings, distribution systems, and generation systems. These scenarios are compared to the baseline for energy use change and to the Base Case for investment and operational costs. The pre- selected scenarios contain characteristics for the constraints which have to be considered for different types of compounds, which could be for example:

- High- sensitive service areas require full power, heating, cooling back up often combined with short term storages to cover black outs caused by natural and manmade threats.
- Building installation: hygienic constraints provide limitations for the distribution of heating/cooling in the service/surgery/care areas, in most of the cases ventilative systems are preferred to static space heating systems such as radiators. Ventilative systems in surgery areas require high level air quality to minimize the distribution of pathogenic bacteria. In low intensity care and housing areas the heating/cooling can be distributed also by radiators. Sterilization requires high temperature systems such as low-pressure steam. Hot water systems have to be designed and operated on temperature levels which allow to limit the growth and distribution of pathogenic bacteria.
- Building construction: minimum legal requirements for the energy design which affects demand of building and communities by setting target or maximum values for the overall energy demand of one or a group of buildings. In the majority of the European member countries the European Building Performance Directive sets targets for source energy demand and respectively for
- Indoor environment: indoor air quality in surgery and intensive care areas is requiring systems which are able to provide conditions close to clean rooms, including filter systems, disinfection, steam based moisturization with mould-critical pathways in combined heating/cooling systems; hot water systems for care purposes also require minimal pathogenic risks in the storage and distribution.

Conclusions

The proposed approach provides a coherent method for Energy Master Planning. However, when it comes to the implementation and the two alternative approaches of bottom-up vs. top-down implementation the bottom up EMPs often appear in design challenges with ambitious targets such as energy efficiency, resilience, short implementation time or other imposed constraints. One approach for a "bottom up" approach is to initiate a "net zero" compound or other specific targets which are pursued with small tolerance bands or other limitations on the EMP process. Then the design challenge is "how can we match these ambitious energy targets under those fuel and carbon footprint constraints", the design group is interdisciplinary and integrates different hierarchies and even end users of the compound and the respective buildings.

Recent discussion on the adoption of the EPBD 2018 into German national targets show, that the focus of the discussion is moving from the single building towards targets of a community of buildings. The adoption of EPBD 2018 will have to be translated on the level of a community. This provides certain flexibilities for the constraints on the building level, e.g. if a historic building cannot fulfill the target values on the building level, buildings in the neighborhood or an energy supply based mainly on renewables can compensate the "failure" of the individual.

Further work

Individual building computer-based energy models are currently available for general use that could be exploited for Base Case development.

In the future, building models should be developed and further customized to function as archetypes to predict energy use in districts and adapted to different climate conditions and energy use requirements. To be used for community planning, all prototype models have to be fully parametrized for detailed modelling inputs in order to be able to build-in site-specific constraints.

A list of site-specific constraints is needed that can help the master planner in determining the solution room. Further, a list of building models would be useful that could be implemented for Base Case modelling.

Different baselines could then be internationally compared (for different countries as Australia, Austria, Canada, Denmark, Finland, Germany, Norway, UK and USA). This work is planned as a next step in IEA ECB Annex 73. These need to be put into context (cultural and economic) and constraints pattern should be developed. Such a database of collected models represent their national/agency building stock, that include energy systems specific to their representative climate conditions, and that have representative operation schedules. Another important step will be to develop a common approach to calibration of building models to existing energy use data available from metering and sub metering.

References

[1] EPBD. 2018. Directive (EU) 2018/844 of the European Parliament and of the Council of 30 May 2018 amending Directive 2010/31/EU on the energy performance of buildings.
[2] 10 CFR 433, U.S. Code of Federal Regulations, Part 433.
[3] IEA. 2011, 2012, 2013. International Energy Agency, Annex 51, Energy Efficient Communities. Jank, R., A Zhivov, R Liesen, et.al. 2012. Net zero building cluster energy systems analysis for US army installations, *ASHRAE Transactions*. Atlanta: American Society of Heating, Refrigerating and Air-Conditioning Engineers, Inc.
[4] Strasser, H., et.al.. 2018. IEA EBC annex 63–implementation of energy strategies in communities, *Energy and Buildings*.
[5] Annex 67 - Energy Flexibility, http://www.annex67.org/, Access date: Feb 2019
[6] Annex 70 - Building Energy Epidemiology, https://energyepidemiology.org/, Access date: Feb 2019
[7] Annex 75 - Cost-effective Building Renovation at District Level Combining Energy Efficiency & Renewables, http://annex75.iea-ebc.org/, (ongoing) Access date: Feb 2019
[8] EnEff Stadt. 2014. Germany Energy Efficient Cities Program.
[9] ESMAP. 2014. Energy Sector Management Assistance Program, https://www.esmap.org/node/55386
[10] EISA. 2007. U.S. Congress Energy Independence and Security Act of 2007
[11] Zhivov et al. 2014b. 2014. Energy Master Planning Towards Net-Zero Energy Communities/Campuses, *ASHRAE Transactions*. Atlanta: American Society of Heating, Refrigerating and Air-Conditioning Engineers, Inc.
[12] Other initiatives e.g.in the center for Zero Emission Neighborhoods (ZEN centre)
[13] Baer, D. 2018. "Tools for Stakeholder engagement in Zero Emission Neighbourhood Developments Mapping of tools in use in Trondheim, Steinkjer, Elverum and Bodø" ZEN

Report 13. Research Centre on Zero Emission Neighbourhoods in Smart Cities (FME ZEN). ISBN: 978-82-536-1617-9

[14] Wiik, M. K., Fufa, S. M., Baer, D., Sartori, I. & Andresen, I. 2011. "The ZEN Definition – A Guideline for the ZEN Pilot Areas. Version 1.0". ZEN Report 11. Research Centre on Zero Emission Neighbourhoods in Smart Cities (FME ZEN), ISBN 978-82-536-1608-7

[15] American Society of Heating, Refrigerating and Air-Conditioning Engineers, Inc. (ASHRAE) Standard 100-2015 -- Energy Efficiency in Existing Buildings. Atlanta.

[16] German VDI 3807, VDI-Richtlinie: VDI 3807 Blatt 1 Verbrauchskennwerte für Gebäude - Grundlagen. In German.

[17] Switzerland SIA 380.1 «Thermische Energie im Hochbau», Ausgabe 2009. In German.

[18] EEG, Erneuerbare-Energien-Gesetz, https://www.erneuerbare-energien.de/EE/Redaktion/DE/Dossier/eeg.html?cms_docId=73930, Access date: Feb 2019

[19] ENEV 2012, Energieeinsparverordnung, http://www.enev-online.com/index.htm, Access date: Feb 2019

[20] TEK17, Regulations on technical requirements for construction works, DiBK, https://dibk.no/globalassets/byggeregler/regulation-on-technical-requirements-for-construction-works--technical-regulations.pdf, Access date: Feb 2019

[21] Lohse et al. 2012, "Strategiepapier zur Bewertung der wirtschaftlichen Verbesserungspotentiale bei der energetischen Quartiersplanung", (in German) Working paper in the context of the German Contracting Initiative 2012- 2013

Comparison of different measures for reducing the carbon footprint of the building sector – a rental house case study

R Holopainen[1], **R Ahola**[2], **K Bergbom**[1], **H Vaarala**[1]

[1]Ecolabelling Finland, Urho Kekkosen katu 4-6 E, Helsinki, 00100, Finland
[2]Optiplan, Mannerheimintie 105, Helsinki, 00280, Finland

riikka.holopainen@ecolabel.fi

Abstract. This paper compares and estimates cost-optimal means to increase the potential of reducing the carbon footprint of future rental houses. As a case study we present a comparison between four rental apartment houses, where one house has achieved the Nordic Swan ecolabel. The carbon footprints of these buildings have been calculated for the building lifetime. According to the results, the effects of single measures are quite small in reducing the carbon footprint, but larger effects can be obtained by combining several measures together. It is essential to minimize the amount of used materials by paying attention to material efficiency under both design and construction phase. The paper also discusses the role of the voluntary ecolabelling of buildings as a tool to reduce the carbon footprint of the building sector.

1. Introduction

The built environment of a typical European country accounts for approximately 40 % of the total energy consumption, 30 % of the total emissions and 50 % of the resource use. Lowering the energy and resource use of the built environment has a key role when aiming to reduce the country-specific carbon emissions. It is therefore important both to estimate and to decrease the carbon footprint of buildings.

The life-cycle carbon footprint of a single building describes how much greenhouse gas emissions the building emits to the atmosphere during its lifetime. Main factors affecting the carbon footprint are the energy consumption of the building, the type of energy consumed and the used building materials. The most important factor is still the energy consumption, but with the increased utilization of renewable energy and better energy-efficiency of the building systems, the proportional importance of the building materials on the total carbon footprint is steadily growing.

The Finnish Ministry of the Environment is promoting low-carbon construction with several measures. The ministry has compiled recommendations for purchase criteria of public construction projects, which enable cities and communities to construct buildings with a lower carbon footprint. In line with European standards, the Ministry of the Environment is introducing a method to calculate embodied energy and CO_2 emissions of building materials during the whole life cycle. The ministry is also developing low-carbon evaluation tools and updating building regulations to support low-carbon construction. At the end of 2018 the ministry launched a common request for the opinion about a draft for the assessment method to estimate the carbon footprint of a building [1]. There may be a voluntary piloting phase before setting the boundary values by 2025.

IOP Conf. Series: Earth and Environmental Science **352** (2019) 012020 doi:10.1088/1755-1315/352/1/012020

One example of voluntary climate-conscious actions is environmental certification of buildings. The official ecolabel of the Nordic countries, the Nordic Swan includes criteria for both new and renovated buildings. The criteria have requirements for the entire building process, from cutting of trees to recycling of waste. The criteria have both direct and indirect impacts on the climate and the carbon footprint due to e.g. requirements for a tighter energy efficiency demand than the building regulations, a demand-controlled lighting diminishing the electricity consumption and over 70 % construction site waste recycling target. There are also circular economy requirements, e.g. of the window and door materials aluminium must contain at least 40 %, PVC-material at least 30 % and steel at least 20 % of recycled material. The criteria have also requirements concerning chemical load, indoor environment quality, sustainable materials and subcontractor chains.

The Nordic Swan criteria for newly constructed building types, revised in 2015, is valid for small houses, apartment buildings, buildings for pre-schools and schools/educational buildings, extensions to existing buildings and homes for the elderly [2]. Also office buildings are included in the separate criteria for renovated buildings. The Nordic Swan Ecolabel was established in 1989 by the Nordic Council of Ministers as a voluntary ecolabelling scheme for Denmark, Finland, Iceland, Norway and Sweden. Currently the Nordic Swan Ecolabel has over 60 different product groups with more than 200 different product types and more than 25.000 different products.

The criteria development process of the Nordic Swan Ecolabel is based on preliminary studies highlighting areas where the need for requirements is greatest. Development of new criteria and revision of existing criteria takes place in close contact with businesses and other stakeholders with relevant knowledge, using available literature from academic institutions and experts. The final proposal is passed by the Nordic Ecolabelling Boards and uploaded to both Nordic and national websites, together with background documents and comments from the public consultation.

The Housing Finance and Development Centre of Finland, in Finnish ARA (Asumisen rahoitus- ja kehittämiskeskus), is an agency to implement social housing policy and operates under the supervision of the Ministry of the Environment. ARA is supporting the goals of the Ministry of the Environment with a development project, where the carbon footprints of the new built houses with ARA funding are being calculated.

This paper presents the main conclusions of a ARA report [3], where the life-cycle carbon footprint was calculated for four new rental apartment houses by the Finnish consulting company Optiplan. One of these houses has achieved the Nordic Swan ecolabel. The main goal of the ARA report was to study the carbon footprint level of an ecolabelled building. The additional goal was to find out the most cost-optimal measures to reduce the carbon footprint. The report is written in Finnish.

2. Carbon footprint calculation and results

Life-cycle carbon footprint was calculated for four rental houses owned by VAV, Vantaan Asunnot Oy: Kaskelantie 1, Lipputie 14, Marsinkuja 1 and Loiskekuja 1. The houses are all located in the city of Vantaa in Southern Finland. Kaskelantie 1 has the Nordic Swan ecolabel. In Lipputie 14, also called as the Opaali house, special attention has been paid to utilize recycled material and material-effective choices. The other two houses are built with ordinary solutions. All houses had the same constructor, NCC Finland Ltd.

2.1. Calculation method and common input values

The carbon footprint was calculated using the One Click LCA calculation program. One Click LCA is a web-based cloud service for environmental and life cycle assessment calculation verified by a third actor according to EN15978.

The program includes environmental impact data of building products and materials from Finland, the Nordic countries and Europe. The tool includes emission data also from outside of Europe. The emission data of the material database is based on environmental declarations according to the standard EN-15804. The database also includes emission data for the national Finnish electricity network and the local district heating networks. The emission factors used in calculations were 0.23 kg CO_2-eq./kWh for electricity and 0.20 kg CO_2-eq./kWh for district heating (Vantaa Energy district heating network 2016).

The instructions of the Level(s) procedure by the European Commission [European Commission 2018] have mainly been used in the calculation, e.g. the calculation period was 60 years. All lifecycle phases have been included in the calculation.

All houses have a centralized mechanical inlet and exhaust air ventilation system with heat recovery. The houses are heated with district heating and they have no cooling system. Kaskelantie 1 has a 12 kWp solar electricity system. Following materials have been included in the calculation: foundations, outer and inner walls, facades, pillars, columns, horizontal structures, stairs and banisters, windows and doors, surface materials, building service systems including elevators and outdoor area materials including landmass, asphalt and concrete paving.

2.2. Case buildings

Kaskelantie 1. This case building from 2018 consists of two houses with eight floors each. The total number of the housing units is 127. The net floor area is 5 832 m^2, the gross area is 8 830 m^2 and the building volume is 28 250 m^3. The total site area is 2 726 m^2. The building foundation has reinforced concrete foundation piles and footing. The slab-on-grade base floor is embedded on site. The upper floors and roof structures are made of precast hollow-core concrete slabs and the staircase landings of massive-slab elements. The bearing walls and pillars are made of reinforced concrete elements. The partition walls are mainly concrete elements or gypsum walls with metal studs. The suspended ceilings are mainly constructed of gypsum or acoustic boards. The floor coatings are laminate inside of the dwellings, floor tiles in common areas and plastic mat in technical spaces. The building has achieved the Nordic Swan ecolabel. According to the Finnish Energy Performance Certificate the electricity consumption is 137 MWh/a (15.5 kWh/m^2 GFA,a) and the calculated district heating energy consumption is 448 MWh/a (50.7 kWh/ m^2 GFA,a).

Figure 1. Kaskelantie 1 (Photo Henrik Kettunen).

Lipputie 14. The second case building from 2015 consists of two houses with five floors each. One house has one staircase and the other house has three staircases. The total number of the housing units is 87. The net floor area is 4 274 m^2, the gross area is 6 224 m^2 and the building volume is 19 920 m^3. The total site area is 2 163 m^2. Specific targets of the construction project were to save natural resources, reduce the carbon footprint by using recycled materials and to pay special attention on material efficiency. The necessary measures taken to fulfill these targets were to lighten partition walls, reduce the amount of concrete in structures and to utilize unprocessed wood in the facades. The building foundation has reinforced concrete palisade and footing. The base floor is mainly made of hollow-slab elements and the staircase landings of massive-slab elements. The bearing walls and pillars are made of reinforced concrete elements. The partition walls are concrete, calcareous sandstone or gypsum walls with metal wire frames. The suspended ceilings are mainly constructed of gypsum or acoustic boards. The floor coatings are mainly laminate inside of the dwellings and plastic mat in common areas and technical spaces. The facades are mostly steel framed with coating of fibre cement boards or wood. The Load bearing facades are insulated concrete elements topped with plaster and the first floor of the façade facing the street has brick coating. The calculated electricity consumption is 96 MWh/a (15.4 kWh/ m^2 GFA,a) and the calculated district heating energy consumption is 377 MWh/a (60.5 kWh/ m^2 GFA,a).

Figure 2. Lipputie 14 /3/

Marsinkuja 1. The third case building from 2017 consists of four houses. The total number of the housing units is 147. The net floor area is 6 632 m^2 and the gross area is 11 675 m^2. The building foundation is slab-on-grade and there is no palisade. The footing is reinforced concrete embedded on site and the plinth is made of concrete elements. The base floor is mainly made of ventilated hollow-slab elements. The shelter has a slab-on-grade. The bearing walls are embedded on site or reinforced concrete elements. The intermediate and attic floors are mainly hollow-slab elements. The stairs are made of reinforced concrete elements. Most of the outer walls are concrete-sandwich elements and factory-plastered, smooth, painted or fully tinted concrete. The outer wall of the ventilation machinery space is coated with sheet metal profile. The light partition walls are mostly made of metal wire framed gypsum boards or expanded clay concrete blocks. The calculated electricity consumption is 224 MWh/a (19.1 kWh/ m^2 GFA,a) and the calculated district heating energy consumption is 673 MWh/a (57.6 kWh/ m^2 GFA,a).

Figure 3. Marsinkuja 1 /3/

Loiskekuja 1. The fourth case building from 2016 consists of three houses with six floors each. The total number of the housing units is 115. The net floor area is 5 197 m^2, the gross area is 8 842 m^2 and the building volume is 18 830 m^3. The building foundation has reinforced concrete palisade and footing. The base floor is mainly made of hollow-slab elements. The intermediate and attic floors are mainly hollow-slab elements and the staircase landings massive-slab elements. The bearing walls and pillars are embedded on site or reinforced concrete elements. The light partition walls are mostly made of concrete elements, metal wire framed gypsum boards or calcareous sandstone bricks. The suspended ceilings are mainly constructed of gypsum or acoustic boards. The floor coatings are mainly laminate, floor tile and plastic mat. The facades are mostly reinforced concrete with brickwork partly coated with fibre concrete board. The calculated electricity consumption is 150 MWh/a (16.9 kWh/ m^2 GFA,a) and the calculated district heating energy consumption is 517 MWh/a (58.4 kWh/ m^2 GFA,a).

Figure 4. Loiskekuja 1 /3/

2.3. Carbon footprint calculation results

The total calculated life-cycle carbon footprint is 11 700 t CO_2e for Kaskelantie 1, 9 010 t CO_2e for Lipputie 14, 16 900 t CO_2e for Marsinkuja 1 and 14 100 t CO_2e for Loiskekuja 1. The specific life-cycle carbon footprint for each case building is presented in Table 1.

Table 1. Specific life-cycle carbon footprint, kg CO_2e/ m^2 GFA

	Kaskelantie 1	Lipputie 14	Marsinkuja 1	Loiskekuja 1
Production phase	304	291	285	326
Transport to construction site	10	8	9	11
Construction phase	42	57	50	40
Maintenance and repair	83	83	81	75
Energy consumption	823	938	959	1082
Water consumption	57	57	57	53
Deconstruction	12	12	8	9
Total	**1 325**	**1 448**	**1 448**	**1 595**

The construction season, weather conditions and waste sorting have impact on the carbon footprint of the construction phase. Special attention was paid to the waste sorting at the construction site of Kaskelantie 1 because of the requirements of the Nordic Swan label. This stricter way of waste sorting lowered the carbon footprint of the construction waste, where especially the mixed waste and energy waste have significantly larger emissions than other waste types.

The renewable energy production of the solar panels at Kaskelantie 1 has been calculated to save 15.5 kg CO_2/ m^2 GFA during the life-cycle calculation period of 60 years. The carbon footprint of the solar panel system materials has been estimated as 2.0 kg CO_2/ m^2 GFA and thus the net saving is 13.5 kg CO_2/ m^2 GFA. The solar panels have been estimated to be replaced once during the calculation period of 60 years.

The life-cycle carbon footprint calculation excludes the domestic electricity use inside of the flats as a non-building related energy consumption, though it produces a considerable amount of emissions of the energy used in the whole building

2.4. Additional carbon footprint calculation results

An additional carbon footprint calculation was done for Kaskelantie 1 to estimate the effect of alternative design solutions increasing the energy-efficiency (Table 2). Here both the effect of a single measure and the total effect of several measures was compared to the base case presented in the earlier chapter. According to the calculation results, the most cost-effective solutions with the 60 years calculation period were option 8) decentralized ventilation system, option 3) decentralized demand-controlled ventilation in flats and option 1) increasing the annual ventilation heat recovery efficiency rate. Carbon footprint decrease of a single measure is relatively small, but several simultaneous measures give noticeable changes.

Table 2. Effect of alternative design solutions on the base case of Kaskelantie 1

	Effect, kg CO_2e/ m^2 GFA	Investment, €/tnCO^2e	Lifecycle cost €/tnCO^2e
1. Increasing annual ventilation heat recovery efficiency rate from 66.7 % to 78.0 %	-56	110	-30
2. Decreasing the window U-value from 0.8 to 0.6	-24	300	120

3. Increasing the amount of decentralized demand-controlled ventilation in flats	-57	90	-50
4. Decreasing the attic floor U-value from 0.13 to 0.07	-9	1 030	840
5. Decreasing the outer wall U-value from 0.17 to 0.14	-13	330	130
6. Decreasing the base floor U-value from 0.16 to 0.10	-2	1 240	1 050
7. Decreasing the air-tightness value from 2 to 1 $m^3/h,m^2$	-17	190	10
Total effect of measures 1.-7.	**-161**	**240**	**40**
8. Decentralized ventilation system instead of centralized	-97	70	-40

The effect of the energy source on the carbon footprint was studied by comparing district heating from the local energy company and ground source heating. The change from district heating to ground source heating decreased the carbon footprint by around 35 %. However, a possible future decrease of district heating emissions was not considered in the calculation.

3. Discussion

The specific lifecycle carbon footprints for the four case buildings were between 1325–1595 kg CO_2e/ m^2 GFA for the calculation period of 60 years. The share of the energy consumption in the lifecycle carbon footprints of the case buildings was between 823–1082 kg CO_2e/ m^2 GFA and the share of the building materials between 384–421 kg CO_2e/ m^2 GFA. This means that around 65 % of the lifecycle carbon footprint was caused by the energy consumption and around 20 % by the building materials.

The largest building material carbon footprints were caused by building service systems, horizontal structures, outer walls and foundations. The share of the building service system materials was around 25-31 % of the total carbon footprint of the building materials. The amount of building service materials was estimated as a rough figure based on system specific data and building area including also elevators and solar panels. The estimated amount is probably too high considering other research results, e.g. CIBSE (2013) where the estimated share was only around 10-12 % [4]. Increasing the complexity and amount of building service systems, including renewable energy production systems, increases also the carbon footprint of the building service system materials, also due to replacements during the building lifecycle. However, this increase might be compensated due increased energy-efficiency and reduced carbon footprint of the energy consumption.

Loiskekuja 1 had the largest carbon footprint due to its larger energy consumption and chosen building materials with larger emissions. Marsinkuja 1 had the lowest material carbon footprint, because no palisade was needed for the foundation of the building. The specific life-cycle carbon footprint was smallest for Kaskelantie 1, the case building with the Nordic Swan ecolabel. The main reason for this smaller carbon footprint was its better energy efficiency.

75 % of the carbon footprint of the energy consumption was caused by district heating energy consumption and 25 % of electricity consumption. According to the additional carbon footprint calculation results for Kaskelantie 1 the largest potential for decreasing the carbon footprint, compared to the base case, is the utilization of ground source heat instead of district heating. Here the probable development of technology and change of energy consumption emissions in the future have not been taken into account.

The most cost-effective alternative design solutions of the additional carbon footprint calculation were all related to decreasing the energy consumption of the ventilation system: the most cost-efficient single measure for reducing the carbon footprint was the implementation of decentralized ventilation instead of centralized ventilation. Other cost-efficient measures were decentralized demand-controlled ventilation in flats and increasing the annual ventilation heat recovery efficiency rate.

It is worth noticing that the energy emission data utilized in the lifecycle carbon footprint calculations was based on the emission values from the year 2016. Thus, the calculation did not consider the possible, foreseen reduction of the energy emissions in the future. This reduction may decrease the significance of energy-efficiency and increase the significance of the building materials in reducing the carbon footprint.

4. Conclusions

The four most important measures to minimize the carbon footprint of buildings are 1) utilization of renewable energy sources, 2) increasing the energy-efficiency, 3) increasing the material-efficiency and 4) the favour of low-emission materials.

The Finnish building regulations concerning nearly zero-energy buildings, effective since 2018, have tightened the energy-efficiency demand of buildings by around 10-15%. Even though the energy-efficiency of new buildings has been steadily increasing due to tighter building regulations and actions by building owners, the largest potential in reducing the carbon footprint of a building is still in increasing the energy efficiency of the building.

The passive, architectural solutions are the first step for minimizing the energy consumption. Ventilation system design has a large effect on the energy-efficiency of the building. Demand-controlled ventilation reduces consumption of the fan electricity, heating energy and cooling energy. The heating energy consumption can be reduced by an effective heat recovery system from the outlet air. According to Liljeström & Törnblom (2009) the annual heat recovery efficiency of decentralized ventilation units inside flats is often higher than with centralized, larger ventilation units and the annual energy costs are lower [5].

The emission factors of different energy sources have a large effect on the life-cycle carbon footprint calculation results. Therefore, real emission factors should always be used when they are available. It should also be possible to take into account if the electricity used on the construction site has been produced with renewable energy sources. One Click LCA includes also a greener electricity profile, where the production is made with 10% of solar, 30 % of water and 60 % of wind.

Even though the energy consumption currently is the most dominant factor in the lifecycle carbon footprint, the potentially decreasing carbon emissions of energy production will increase the importance of the building materials in the total lifecycle footprint, and therefore solutions for e.g. low carbon concrete and renewable and/or recyclable building materials should be developed and taken into use.

The role of circular economy solutions and minimization of building materials is increasing in new construction projects. The amount of building materials can be minimized e.g. by increasing the lifetime of the buildings and considering both the flexibility of the building use and material recycling possibilities already in the design phase.

The life-cycle carbon footprint should be considered as early as possible during the construction project for a larger impact and better cost-efficiency. The carbon footprint calculation method is a necessary step to enable the evaluation and future restriction of the life-cycle carbon footprint of a building. The calculation should guide the design of the building and enable design choices ending into a lower carbon footprint. The calculation method should also consider the existing EU standards and be in line with the future product environmental footprint (PEF) methods.

The material data in the carbon footprint calculation database should be transparent and justified and there should be a possibility to utilize own, verified emission data. This is also possible with One Click

LCA. Utilization of generic material data does not enable taking into account the utilization of recycled material, which may also reduce the carbon footprint of a building. For example, the Nordic Swan ecolabel requires that 40 % of the aluminium, 30 % of PVC and 20 % of steel used in windows is recycled. Considering this requirement in carbon footprint calculation would require a verified environmental product declaration (EPD).

Ecolabelling is one measure for guiding the design of low-carbon buildings. Kaskelantie 1, the building with the Nordic Swan ecolabel, had the lowest lifecycle carbon footprint of all the four case buildings. Even here it would still be possible to decrease the carbon footprint caused by the energy consumption by 10 % with energy-efficiency measures. The carbon footprint decrease potential for the building materials is estimated to be the same 10 %. The total effect of these both 10 % + 10 % saving potentials would be around 15 % during the whole building life cycle of 60 years. In this case the specific lifecycle carbon footprint for 60 years would be around 1 100 kg CO_2e/m^2 GFA.

According to the calculation results presented in this paper, a potential set value by the building regulations for the carbon footprint of the building materials could be around 350 kg CO_2e/m^2 GFA without delimiting the use of concrete in building. For construction with concrete, this limit would demand specific attention to increase the material efficiency and replacing concrete structures with wood when possible. The carbon footprints of alternative building service systems should also be thoroughly inspected, e.g. plastic sewers vs. cast iron sewers, copper pipes vs. plastic pipes, centralized vs. decentralized ventilation and floor heating vs. radiator heating. It would also be important to study possible low-carbon concrete alternatives and if the carbon footprint of footing could be decreased e.g. by using steel bore piles.

The role of the building regulations is to state the minimum level for construction, whereas the role of the voluntary ecolabelling of buildings is to enhance the construction industry into continuous improvement and to speed up the technological development.

Ecolabelling can therefore act as a test bench for upcoming building regulations. Here the forerunners can first test fulfilling the requirements of the ecolabel and after this piloting period specific issues from the ecolabelling criteria can be introduced by the authorities in the obligatory building regulations. The Nordic Swan ecolabel as the third party can also follow and verify that the material choices made in the design phase have been considered in the construction phase and not replaced by other alternatives.

5. References

[1] Finnish Ministry of the Environment 2018 Assessment method for the carbon footprint calculation of a building. Draft for a common request for the opinion. https://www.lausuntopalvelu.fi/FI/Proposal/DownloadProposalAttachment?attachmentId=9517

[2] Nordic Ecolabelling 2016 Small houses, apartment buildings and buildings for schools and pre-schools. Criteria document. http://www.svanen.se/Templates/Criteria/CriteriaGetFile.aspx?fileID=3927

[3] Ahola R and Liljeström K. Cost-optimal reduction of the life-cycle carbon footprint of a rental house (Rakennuksen elinkaaren hiilijalanjäljen pienentäminen kustannustehokkaasti vuokratalokohteessa). In Finnish. ARA Reports 08/2018. https://joutsenmerkki.fi/wp-content/uploads/2018/12/Hiilijalanj%C3%A4ljen-pienent%C3%A4minen-kustannustehokkaasti_2018.pdf

[4] CIBSE 2013 Embodied carbon and building services, Roger Hitchin. CIBSE Research Report 9.

[5] Liljeström K and Törnblom T 2009. A comparison on the energy-efficiency of ventilation units of residential houses (Vertailu asuinkohteiden ilmanvaihtokoneiden energiatehokkuudesta). In Finnish.

IOP Conf. Series: Earth and Environmental Science **352** (2019) 012021 doi:10.1088/1755-1315/352/1/012021

Lessons learnt from the design and construction strategies of two Norwegian low emission construction sites

Selamawit Mamo Fufa[a], Marianne Kjendseth Wiik[a], Sofie Mellegård[a] and Inger Andresen[b]

[a]SINTEF Building and Infrastructure, Forskningsveien 3B, NO-0373 Oslo, Norway
[b]Norway Norwegian University of Science and Technology (NTNU), Department of Architecture, 7491 Trondheim, Norway

selamawit.fufa@sintef.no

Abstract. Over the last couple of years, research related to fossil free and emission free construction sites has developed rapidly in Norway, with an ambition to contribute towards global, national and regional emission reduction targets. Major public players are already demanding fossil free construction sites through public procurement, whilst requirements for emission free construction sites are on the way. Even though the Norwegian construction industry is a forerunner, there is a lack of knowledge or common understanding among different stakeholders on the definition, scope and strategies needed for fossil free and emission free construction sites. The aim of this paper is to present the main challenges and opportunities from the construction phase of two Norwegian zero emission construction sites, namely Campus Evenstad in Hedmark and Lia nursery school in Oslo. Construction activities considered include transportation and installation of building materials, construction machinery, temporary works, energy use, waste management and person transport. This paper presents and discusses the lessons learnt from the design, ambition levels, inputs from stakeholders, emission reduction solutions of these two construction sites, and evaluates methods considered to address conceptual and practical issues. In conclusion, this paper suggests lessons learnt for reducing GHG emissions from Norwegian zero emission construction sites.

1. Introduction

The Norwegian construction industry is responsible for approximately 1.2% of national GHG emissions, which corresponds to around 660,000 tCO_{2eq} [1]. The significance of construction phase emissions becomes clear when one considers that these emissions occur over a short period of time during the early stages of a building's life cycle [2]. In comparison, use phase emissions occur over the lifetime of the building, typically over a 60-year period. Emissions from the construction site may be high enough to question whether new construction hinders ambitions in reaching GHG mitigation goals, no matter how energy efficient buildings are during operation [3].

Over the last couple of years, research relating to fossil free and emission free construction sites has developed rapidly in Norway [4-9], with an ambition to contribute towards international (e.g. the Paris agreement), national (for example, 40% emission reduction by 2030 and becoming a low-emission society in 2050) and regional (e.g. 95% direct emission reduction before 2030 in Oslo) emission reduction targets.

There has been a stronger focus on emission reduction, for example through the revised national transport plan, technical regulations [10] and environmental criteria from DIFI [11]. There has also been an increase in demand from public and private actors to prioritise emission reduction measures in their projects. The market has developed rapidly since Omsorgsbygg (Oslo municipality's enterprise for social service buildings) started the market dialogue with Bellona (environmental NGO) on emission free construction sites. Omsorgsbygg has since developed requirements for emission reductions on their construction sites. The authorities contribute with

IOP Conf. Series: Earth and Environmental Science **352** (2019) 012021 doi:10.1088/1755-1315/352/1/012021

national support in the form of innovation projects with targets for emission reductions, including that from construction sites.

The authors have identified a need for increased knowledge on fossil free and emission free construction sites among different actors in the construction industry, including the identification and implementation of design and construction strategies for fossil free and emission free construction sites, and to set industry standards for fossil free and emission free construction sites. There is also a need for more knowledge and expertise among contractors and suppliers on emission free alternative solutions and technologies.

The aim of this study is to present the valuable lessons learnt from two Norwegian construction sites which have implemented solutions aiming to achieve emission free or fossil free construction. The paper will discuss emission free and fossil free construction site definitions, how the case studies implemented measures to achieve these ambitions and what measures may be further considered.

2. Definition

The terms 'fossil free' and 'emission free' are often used interchangeably when discussing emissions from a building site. However, these two terms are dependent on first defining which construction activities take place on a building site. In both cases, construction activities include transport of materials, transport and operation of construction machinery, transport of construction workers, energy use, internal transport, storage, temporary works, additional materials for installation of building materials and components, transport of waste, waste treatment and disposal, as depicted in the system boundary of construction activities in Figure 1. These are good examples of construction site activities identified by harmonising EN 15804 [12], EN 15978 [13] and NS 3720 [14]. To follow, is a definition of fossil free and emission free construction sites by harmonising existing Norwegian definitions [4, 15, 16].

Firstly, a fossil free construction site is a construction site that does not use any fossil fuels in any of its on-site construction activities. Fossil fuels (i.e. diesel or propane) are often replaced with bioenergy and biofuels (i.e. HVO or wood pellets) or alternative renewable energy resources such as electricity or hydrogen.

On the other hand, an emission free construction site is a construction site that does not have any direct or indirect greenhouse gas (GHG) emissions from its construction activities. However, achieving no GHG emissions from the construction process is difficult to obtain especially when indirect, upstream emissions are included in the system boundary. Therefore, the Norwegian construction industry has adopted a stepwise approach that works towards the ambition of an emission free construction site. This stepwise approach starts with the fossil free construction site. Next, ambitions can be raised to an 'on-site emission free' construction site which covers no direct GHG emissions from construction activities taking place on-site (e.g. from internal transport, operation of construction machinery and on-site energy use). The next step involves adding emission free transport to and from the construction site, whilst the final step covers the whole system boundary depicted in Figure 1. The authors acknowledge a parallel initiative which investigates the 'waste free' construction site, however this initiative is outside the scope of this article. The authors have also noted that the Norwegian construction industry has adopted the term 'clean construction' to cover both fossil and emission free construction sites.

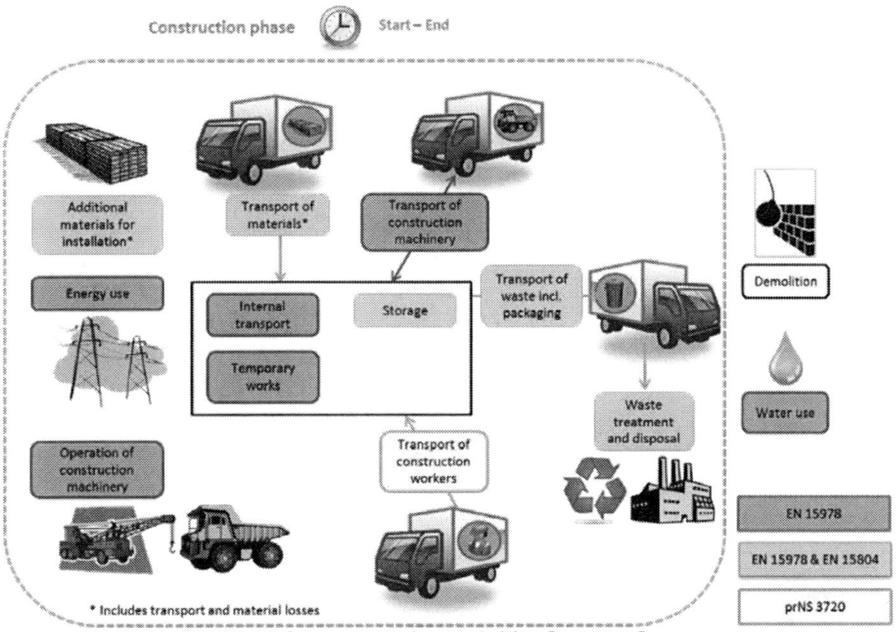

Figure 1. System boundary for construction activities [2, 12-14].

When the terms fossil free or emission free are used, it is important to clearly define the system boundary for both what is included and what is not included in the assessment. This should include, among other things, construction site activities, construction method (e.g. on-site or off-site), type of emissions considered (e.g. direct and / or indirect GHG emissions), emission factors used, as well as a clear description of construction solutions, implementation model and choice of technologies used.

An advantage to implementing emission free construction solutions includes not only zero GHG emissions, but also reduces other types of harmful environmental emissions such as nitrogen oxides (NOx), sulphur oxides (Sox), particulate matter (PM_5, PM_{10}) and noise (dB), which affect both local air quality and human health. Some examples of emission free alternatives include electric, battery-powered, or hydrogen-powered construction machinery, electricity or district heating for temporary heating and drying, use of zero emission vehicle transport to, from, and at building sites (for transport of machinery, materials, waste and personnel).

3. Case studies

The construction sites chosen for this study consist of two Norwegian building projects, namely Campus Evenstad in Hedmark and Lia nursery school in Oslo. Campus Evenstad consists of an administration and educational school building, and has the highest ZEB ambition level out of the ZEB pilot projects (ZEB-COM, which compensates for all emissions relating to operational energy use "O", embodied emissions from materials "M" and emissions relating to the construction phase "C" with local renewable energy generation) [17-19]. Lia nursery school is described as the first fossil free construction site and Norway's most environmentally friendly nursery, with a very ambitious level of BREEAM Very Good [2].

Both case studies document the same construction activities, namely: transport of building materials, transport and operation of construction machinery, energy use, temporary works, transport of waste, waste treatment and disposal and person transport. Any demolition works belonging to the previous life cycle of the existing building, and any cleaning services or water use during the construction period are not accounted for in both case studies. A summary of key information on the two case studies is given in Table 1. An overview of the success factors and challenges of the two case studies is given in sections 4 and 5.

Table 1: Key information on the two case studies

	 **Campus Evenstad, Ola Roald Arkitekter**	 **Lia nursery school**
Type of building	Administration and educational building	Nursery school
Location	Høgskolen i Innlandet, Postboks 400, 2418 Elverum, Hedmark, Norway	Harald Sohlbergs vei 19, Oslo, Norway
Heated floor area (m^2)	1141	1600
Main construction materials	Solid wood construction, wood fibre insulation, and an untreated timber cladding	Prefabricated timber elements with timber interior and exterior cladding, hollow concrete slab flooring for the first-floor construction, and light weight concrete roof elements
Energy system	Combined heat and power (CHP) unit, powered by the gasification of wood chips	Water-based ground source heat pump and photovoltaic panels on the roof
Project owner	Statsbygg	Omsorgsbygg Oslo
Construction period	15th December 2015 - 22nd December 2016 (374 days)	10th April 2016 - 27th November 2017 (166 days)
Ambition	ZEB-COM	Plus energy building (produces more energy than it uses), BREEAM-NOR Very Good; fossil free construction site

GHG emission results from construction phase (for life cycle modules A4 and A5, in accordance with EN 15978)

Construction site activities	Campus Evenstad Design phase ($kgCO_{2eq}/m^2/yr$)	Campus Evenstad As-built phase ($kgCO_{2eq}/m^2/yr$)	Lia nursery school ($kgCO_{2eq}/m^2/yr$)
Transport of building materials	0,44	0,30	0,55
Installation of building materials	0,00	0,20	N/A
Construction machinery	0,21	0,96	0,4
Energy use	0,13	0,34	0,04
Person transport	0,26	0,21	0,11
Temporary work	0,00	0,02	0,01
Waste	0,00	0,01	0,08
Uncertainity	0,09	N/A	N/A
Sum ($kgCO_{2eq}/ m^2/yr$)	**1,14**	**2,03**	**1,19**

References	[2, 17, 19]	[4, 7]

4. Lessons learnt from the two case studies

4.1. Procurement process

In Lia nursery school, Omsorgsbygg's ambition was to implement one of the first fossil-free construction sites. In order to realise this ambition, a set of requirements were set during the early procurement process. This included three focus areas, with corresponding requirements, targeting the following areas: construction machinery, heating and drying, and transport to, from and within the construction site. The goal was to use electrified machinery, where possible, as well as to use renewable solutions for heating and drying, if available. The procurement process also included quality and environmental criteria (weighted by 60%), whilst the weighting for price criteria was set to 40% [4]. First-hand experiences from Lia nursery school has enabled Omsorgsbygg to raise the quality and environmental criteria in future procurement processes to up to 75% (e.g. Tåsen nursing home project [4]). Furthermore, Omsorgsbygg used lessons learnt from Lia barnehage to set the reduction of GHG emissions and

IOP Conf. Series: Earth and Environmental Science **352** (2019) 012021 doi:10.1088/1755-1315/352/1/012021

local emissions from the construction site as award criterion in future procurement competitions (e.g. Tåsen nursing home project).

4.2. *Early planning*

It was also observed that more time and resources were used in the early project phases for the planning of both Campus Evenstad and Lia nursery school compared to other typical projects. In Lia nursery school, there was a close cooperation between actors, where the ambition definition, concepts, challenges and opportunities were discussed at an early stage in the project. Several clarification meetings were carried out to collect input, learn from previous experiences, examine accessibility of machineries and equipment in the market and discuss measures that could enable to achieve the goal of a fossil free construction site. A draft plan was created in the early phase where the various actors were given the opportunity to give input to get a more efficient and productive workflow. Lia nursery school is situated adjacent to a school area which also sets construction limitations related to the safety of school children and heavy transport restrictions to and from the building site. All these factors contributed to the project being completed one month before schedule. In summary, the project achieved a shorter construction phase as well as better transport logistics and safer construction site.

In Campus Evenstad, early planning of the construction site was performed to consider emission reduction measures. GHG emission calculations were performed to evaluate the potential source of emissions based on experience-based estimates provided by the contractor through a series of partner workshops. The early design phase emission results enabled stakeholders to plan for the reduction of GHG emissions during the construction phase. The as built phase GHG emission results showed a 33% decrease in emissions from person transport and a 25% decrease in emissions from the transport of building materials [17]. Campus Evenstad is situated in a rural area. Therefore, the contractor enabled onsite living for construction workers to help reduce emissions from person transport. Similarly, the contractor selected locally produced building materials to help reduce embodied emissions from distances travelled. The results from some construction activities showed an increase in emissions. This is due to limited previous GHG emission calculation experience and a lack of emission data. One such example of this is the emission calculations for construction machineries, whereby on-site diesel consumption was underestimated due to unknown weather conditions, and a delayed construction start which meant casting concrete foundations during a Norwegian winter. On the other hand, measures for reducing GHG emissions from transport include increasing the technological level of vehicle transport (i.e. EURO class 6).

4.3. *Choice of construction system*

In Lia nursery school, the use of prefabricated construction solutions and locally produced elements is considered as one solution which reduced the number of transports of both materials and personnel to site. Choosing off-site construction has also reduced the use of construction machinery and the handling of waste as well as a total reduction of the construction period. This is demonstrated by the fact that it only took 13 days from when the foundation was finished until the roof was mounted. Nevertheless, the scope of the project did not include the task of finding out and calculating how much material choices contribute to total GHG emissions. Quantification of the emission saving potential due to the choice of construction method is an important factor in ascertaining this consideration.

On the other hand, the design of the administration and educational building at Campus Evenstad focuses on material choices and energy systems to reduce GHG emissions relating to operational energy use and embodied material emissions. As a result, a lot of attention was given to choosing a solid wood construction system with wood fibre insulation and timber cladding to harness the biogenic carbon properties of wood.

4.4. *Energy sources*

In Lia nursery school electricity has been sufficiently provided and supplied directly from the electricity grid throughout the construction process. Electric machineries such as small electric excavators and wheel loaders have been used for specific activities. It was difficult to get fossil-free alternatives in place in the early phase of the project. There was a desire to use electric construction machines, but electric machineries were not available during the construction period of this project. As an alternative biodiesel was used instead fuelling the large construction machineries. The other challenge was that the construction machines were delivered to the

IOP Conf. Series: Earth and Environmental Science **352** (2019) 012021 doi:10.1088/1755-1315/352/1/012021

construction site with a tank full of diesel even if they were planned to use biodiesel. This would have been solved if all parties involved in the project being familiar with the goals and potential solutions.

In Campus Evenstad, electricity imported from the grid was replaced with onsite electricity generated from the CHP unit during the last four months of construction, which enabled to reduce emissions by 0.13 $kgCO_{2eq}/m^2/yr$. If the CHP system had been implemented before the construction phase started, then the grid-based electricity could have been replaced by electricity and heat generated by the CHP system, leading to even lower embodied construction emissions. Further reduction of emissions from energy use could be achieved by reducing the need for heating and drying, by keeping the building dry. For example, a temporary roof cover or tent may have been in-stalled to hinder rain from soaking down the construction. A cover would also enable for building materials being properly stored in dry places. Similarly, the seasons can be exploited to reduce onsite energy demands further. For example, installing concrete foundations during the summer months reduces the need for thawing the ground and can improve curing times. Energy consumption from lighting may be reduced by using energy-saving lightbulbs and motion sensors for security against break ins or theft outside of working hours, which means lighting is only required for 8 instead of 24 hours a day, leading to a 66% saving [2]. Another energy saving measure may be improved construction of the onsite construction cabins to include; thicker wall and roof insulation, heat recovery, thermostats, and air-to-air or water-to-air heat pumps.

4.5. *Transport and logistics*
The construction logistics was well planned in the early design phase of Lia nursery school. External factors such as an urban setting with a lack of space and storage area and the fact that the construction site was located next to a school contributed to the need for extra precise planning of deliveries and logistics to, from and within the construction site. Prefabricated elements were chosen for floors, walls and ceilings, in combination with choice of local suppliers of concrete to reduce waste and to enable consequent lower transport need leading to a shorter construction period. However, the use of prefabricated elements also may increase the construction logistics. The GHG emission calculation results show that the transport of materials (including the prefabricated elements) to the construction site is the main GHG emission contributor. It would have been interesting to perform a sensitivity analysis in order to evaluate the impact from the choice of construction method. In addition, other solutions such as use of public transport and car-pooling of workers were considered to reduce the impact from transport.

4.6. *Waste*
In both Lia nursery school and campus Evenstad, ambitious waste minimization targets were set by establishing goals on the reduction of the amount of waste generate (in kg) per square meter of the building. The waste plan was used to collect the total amount and type of onsite construction waste generated during the construction period and consider waste reduction solutions.

4.7. *GHG emission*
GHG calculation was not performed in the early planning phase of Lia nursery school project. Emission reduction measures considered and implemented were rather based on assumptions that they lead to lower emissions. Performing LCA in the early design phase would have helped to further evaluate, plan and compare GHG emission reduction measures. The LCA study performed during the construction phase however has provided experience and knowledge about how to analyze and document GHG emissions from the construction phase, defining the system boundaries, data sources, and selecting the correct calculation methodology. The only possible sensitivity analysis performed in the LCA study enables the possibility to evaluate emission reduction measures, namely the use of different fuel sources in the construction machinery. However, it was difficult to perform a complete sensitivity analysis to evaluate the emission reduction results from other measures due to lack of data.

In the Campus Evenstad project, the estimated results from early design phase were used to evaluate, plan and reduce emissions during the construction phase. The calculations in Campus Evenstad also showed discrepancies between estimated emissions and actual emissions calculation results. E.g. there was an 80% increase in emissions from construction machinery, 67% increase from energy use, 33% decrease from person transport and a 25% decrease from transport of building materials.

5. Discussions

The findings from the two case studies show that the following aspects are of high importance in succeeding with achieving fossil free or emission free construction sites. Good planning in the early phase, consideration of external factors, cooperation between all involved stakeholders, clear definitions of requirements and set goals and finally openly sharing knowledge, experience, implementation- and documentation methods.

Both case studies show the great importance and necessity of early planning and cooperation between different actors. Decisions through collaboration made in the early phase impose on several aspects in the implementation phase. External factors such as availability of machineries, sustainable fuel, electricity, available infrastructure during the construction period and choices of construction methods such as prefabricated elements also influence the results and level of ambitions. Close cooperation can enable to use previous knowledge, resources and expertise wisely. This will eventually increase the competence amongst involved actors and result in more specific requirements and award criteria. Setting environmental requirements or targets in the procurement process is a very important factor which enables to consider new and innovative solutions from the builders. The main barriers are considered to be the lack of knowledge and experience amongst stakeholders, high costs, lack of access to fossil-free or emission-free solutions and use of new technologies. Both cases clearly show examples of traditional solutions based on linear thinking, without consideration of new technologies aiming for optimization and circularity.

The need for clear definition and description of system boundary for both fossil and emission-free construction sites to create a common understanding of what is included in the term fossil-free and emission-free is identified. Requirements should also be transparent, measurable and comparable. E.g. even if both case projects are considered as low-emission projects, they have set different ambitions and emission reduction solutions. There is also a need for increased expertise among owners and builders about their expectations when they are setting the requirements and possibilities for fossil-free or emission-free alternatives.

Quantitative evaluation methods should be used in the early design phase of the project to highlight real emission savings. Utilization of LCA methodology will enable to evaluate the environmental performance of emission reduction measures. Further study is required to collect case studies of different building typologies to gain experience in evaluating and minimizing the potential environmental impacts from different types of construction sites nationally and internationally. In the LCA calculation, it would be useful to evaluate how or in what degree the emission reduction measures from the construction phase enable to meet local or regional goals and its contribution to Norway's pledged reductions. Performing cradle-to-grave LCA, including various environmental indicators, will enable to avoid the problem shifting from one life cycle phase to another or from one environmental indicator to another. Furthermore, the economic and social pillars of sustainability should be integrated using life cycle cost (LCC) and social life cycle assessment (SLCA) to evaluate the sustainability of the building in general and the construction site specifically.

Experience from emission reduction measures, evaluation and documentation methods in Campus Evenstad and Lia Nursery school can be used as a reference for other projects. There are also new guidance documents [4, 5, 15] and good examples [6] which can be used as reference. Further work on collecting of case studies and testing the guidance in actual pilot project would be useful.

6. Conclusions

This study has presented the lesson learnt from the two Norwegian construction sites. The results from the two case studies clearly show that the goal of achieving a fossil and/or emission free construction site is possible. It requires giving close attention to the early planning phase, a thorough and open collaboration between involved stakeholders where knowledge and previous experience is shared. There is also a great need for clearly defined ambitions, requirements, system boundaries and quantitative evaluations methods in order to document the actual emission reductions. This will help to compare building sites not only against reference projects but also between projects and to further develop methodologies to achieve both national and global emission reduction goals.

References

[1] NHO. Energibruk på byggeplassen (i byggefasen). Oslo: NHO; 2009.

[2] Fufa SM, Wiik MK, Andresen I, editors. Estimated and actual construction inventory data in embodied GHG emission calculations for a Norwegian zero emission building (ZEB) construction site. International Conference on Sustainability in Energy and Buildings (SEB-18), 24-26 June 2018, Gold Coast, Australia; 2018
[3] Saynajoki A, Heinonen J, Junnila S. A scenario analysis of the life cycle greenhouse gas emissions of a new residential area. Environmental Research Letters. 2012;7(3):11.
[4] Fufa SM, Mellegård S, Wiik MK, Flyen C, Hasle G, Bach L, et al. Utslippsfrie byggeplasser - State of the art. Veileder for innovative anskaffelsesprosesser. SINTEF Fag rapport nr. 49. ISBN:978-82-536-1589-9. 2018
[5] DNV.GL. Emission-reduction potential of fossil- and emission-free building and construction sites. Report no. 2018-0367, Rev. 1-ENG. 2018a.
[6] Multiconsult. Review of implementation of fossil free building sites. Document no. 10206471-TVF-RAP-001. 2018.
[7] Fufa SM. GHG emission calculation from construction phase of Lia barnehage. SINTEF notes no. 29. ISBN:978-82-536-1586-8 2018
[8] Aragones MP, Serafimova T. Zero emission construction sites: The possibilities and barriers of electric construction machinery. Bellona Europa. 2018.
[9] Wiik MK, Jon Are Suul SM, Kamal Azrague KS, Ødegård A, Haukaas N-O, Ibsen JI, et al. 30 tonns utslippsfri gravemaskin. SINTEF Fag rapport nr. 52. ISBN:978-82-536-1603-2. 2018
[10] TEK 17. The Norwegian building regulations (Byggteknisk forskrift, TEK 17). https://dibk.no/byggereglene/byggteknisk-forskrift-tek17/9/9-8/. 2017.
[11] Difi. Et verktøy for bærekraftige anskaffelser. Cited 13.03.2018 https://kriterieveiviseren.difi.no. 2018.
[12] NS-EN 15804:2012+A1:2013. Sustainability of construction works Environmental product declarations Core for the product category of construction products. Standard Norge, Oslo, Norway; 2014.
[13] NS-EN 15978. Bærekraftige byggverk - Vurdering av bygningers miljøprestasjon - Beregningsmetode / Sustainability of construction works - Assessment of environmental performance of buildings - Calculation method European Committee for Standardization, Brussels, Belgium; 2011.
[14] NS 3720. Metode for klimagassberegninger for bygninger / Method for greenhouse gas calculations for buildings. . Standard Norge; 2018.
[15] DNV GL. Fossil and emission free building sites: Guide to arranging fossil- and emission-free solutions on building sites. Report no. 2018-0418, Rev. 2-ENG. 2018b.
[16] Difi. Utslippsfrie bygg- og anleggsplasser/Emission-free construction sites [Online]. DIFI-Norway's Agency for Public Management and e-Government.. Cited 13.04.2019 https://www.anskaffelser.no/hva-skal-du-kjope/bygg-anlegg-og-eiendom-bae/utslippsfrie-bygg-og-anleggsplasser. 2019.
[17] Wiik MK, Sørensen ÅL, Selvig E, Cervenka Z, Fufa SM, Andresen I. ZEB Pilot Campus Evenstad. Administration and educational building. As-built report. The Research Centre on Zero Emission Buildings. ZEB Project report no 36.; 2017
[18] Fufa SM, Schlanbusch RD, Sørnes K, Inman M, Andresen I. A Norwegian ZEB Definition Guideline. The Research Centre on Zero Emission Buildings. ZEB Project report no 29.; 2016.
[19] Eivind Selvig, Marianne Kjendseth Wiik, Åse Lekang Sørensen. Campus Evenstad - Jakten på nullutslippsbygget ZEB-COM. 2017.

Acknowledgements
This research has received funding from the research centre for zero emission neighbourhoods in smart cities (FME ZEN). The authors are thankful for the support from ZEN partners and the Norwegian Research Council.

IOP Conf. Series: Earth and Environmental Science **352** (2019) 012022 doi:10.1088/1755-1315/352/1/012022

Wood burning habits and its effect on the electrical energy demand of a retrofitted Norwegian detached house

L. C. Felius[1], M. Thalfeldt[3], L. Georges[3], B. D. Hrynyszyn[1], F. Dessen[2] and M. Hamdy[1]

[1] Department of Civil and Environmental Engineering, Faculty of Engineering, NTNU Norwegian University of Science and Technology, Trondheim, Norway.
[2] Department of Engineering Cybernetics, Faculty of Information Technology and Electrical Engineering, NTNU Norwegian University of Science and Technology, Trondheim, Norway.
[3] Department of Energy and Process Engineering, Faculty of Engineering, NTNU Norwegian University of Science and Technology, Trondheim, Norway.

E-mail: bozena.d.hrynyszyn@ntnu.no

Abstract. Using firewood as a space heating source is a popular solution in Norwegian housing and can significantly reduce the electrical energy demand of houses. This study analysed habits and reasons for using a wood stove from survey data. From this, typical behaviour patterns were defined. These patterns were imported into a building performance simulation model of a typical Norwegian single-family detached house to evaluate the impact of the stove user behaviour on the electrical energy demand and on the overheating risk. Results showed that up to 32% of the electrical energy demand for space heating can be saved using a wood stove. The number of overheating hours increased when the wood stove was used more frequently. However, it decreased after full renovation because the stove is used less often, as the total space heating demand decreases and the indoor temperature drops less often below the temperature set-point when the stove is started. Active use of the wood stove is effective as retrofitting measure when the aim is to save electricity or fossil fuels. Nevertheless, if the stove power is not adapted to the building, it can be challenging to maintain a comfortable temperature in the room.

1. Introduction

Retrofitting the existing building stock has the potential to improve the overall energy efficiency and reduce CO_2-emissions [1]. There is a significant energy saving potential in upgrading housing as around 75% of the building stock is residential [2]. About 40% of the houses in Norway are detached single-family houses built before 1990 [3]. These houses typically have a wood burning stove in addition to electric heaters to cover the space heating demand. This makes it possible to use a renewable energy source for space heating and biofuel currently covers around 16% of the national energy consumption for households [4]. In addition, a wood stove improves the flexibility of the heating system (i.e. the possibility to choose between two heat generation systems) and decreases the dependence on non-renewable resources. Norway wants to increase its use of bioenergy by 2020, and firewood is the most promising bioenergy resource for both small houses and existing buildings without a water-borne heating system [5]. In the building code of 1969, small houses, if not connected to a central heating system, were required a chimney

IOP Conf. Series: Earth and Environmental Science **352** (2019) 012022 doi:10.1088/1755-1315/352/1/012022

so that a wood stove could be installed [6]. TEK17, the building code valid today, states that small houses must be built with a chimney unless they have a water-borne heating system or reach passive house requirements [7].

Regarding the indoor environment, a potential concern is that wood stoves produce pollutants and CO_2 emissions that could affect the indoor air quality. Hamon et al. [8] found that the emitted particles and CO_2 concentration in both old and new houses stayed below the maximum permitted levels. However, a significant drop in humidity was found when the wood stove was used. Another concern is the increase in room temperature when using a wood stove. Studies showed that the nominal power of typical wood stoves can be oversized for buildings with high-performance envelopes, which can pose a risk for overheating [9, 10, 11]. The installation of a wood stove without the proper nominal power can lead to similar overheating behaviour in less insulated buildings, as studied by Skare [11] and Thalfeldt et al. [12].

Past studies focused mainly on overheating and the modeling of the indoor thermal environment of buildings heated by wood stoves and less on user behaviour [13, 9, 10]. Finstad et al. [14] conducted a survey about fireplace behaviour in Norway in 2003. In Trondheim 27% of those who use a fireplace or wood stove used it as a primary heating source during the winter of 2002/2003. In Trondheim and Oslo, 10% and 44%, respectively, used it only for coziness. Most people used their fireplace during the evening (79-85%), followed by the late afternoon (35-73%), though the results vary a lot depending on the location. The main reasons for using the fireplace more often than the previous year were colder outdoor temperatures and a higher electricity price. However, this study is not recent and it is not possible to identify behaviour patterns from the results. Kipping et al. [15] analyzed the electric consumption of Norwegian dwellings using smart meters measurement and user surveys. They showed that using firewood for space heating can result in a significant reduction of the hourly electricity consumption. Kipping and Trømborg [15] found that 60% of the investigated households with electric heaters say they used firewood as their primary heating source during the winter and 30% used it as a supplementary heating source. Less than 10% used it for coziness only. Except Kipping et al. [15], none of these studies discussed how user behaviour may affect the electricity consumption and indoor comfort. Skare [11] investigated the effect of several energy related user behaviors related to stoves on the space-heating needs, such as the set-point temperature, the stove power modulation, batch load and internal door opening in the building. However, this study did not focus on different user customs (i.e. why and when the stove is used) which is the aim of the present work. Georges et al. [13], analyzed the space-heating needs in passive houses using the wood stove as primary heating source. In reality, it can be challenging to provide heating to the whole house using a wood stove only. In housing with a low space heating demand, such as passive houses, it was possible to cover most of the demand with the wood stove though it depends on the effectiveness of the heat distribution in the house and the desired indoor temperature in the bedrooms. If relatively warm bedrooms were required, extreme weather conditions resulted in unacceptable temperatures in parts of the house. Therefore, the wood stove should be complemented by another heating source, especially in poorly insulated buildings.

The first objective of this paper was to analyse habits and reasons for using a fireplace or wood stove as a heating source in Norway and to identify typical behaviour patterns. The second objective was to evaluate the effect of these patterns on the electrical energy demand for space heating and on the indoor comfort (i.e. overheating hours) of a typical Norwegian single-family house. The paper is organized as follows. First, the methodology is explained. Then, survey results and identified behaviour patterns are presented, followed by energy performance and indoor climate results. Finally, limitations and suggestions for future work are discussed.

IOP Conf. Series: Earth and Environmental Science **352** (2019) 012022 doi:10.1088/1755-1315/352/1/012022

2. Methodology

2.1. Survey design

An online survey about fireplace behaviour was designed to provide insight in how and why fireplaces are used in Norway and how different groups use them. In the survey and in sections 2.1 and 3.1, 'fireplace' refers to any type of wood burning place, such as open fireplaces and wood stoves. The survey was sent out in February 2019 and was available for 6 weeks. The first part of the survey contained questions about the characteristics of the participant and their house. After this, the survey focused on characteristics of the fireplace and fuel. The final part focused on the importance of a fireplace, how and why it is used (or not used) and indoor climate satisfaction. It should be mentioned that personal and professional networks were used to distribute the survey. Therefore, the survey sample cannot be considered representative for all of Norway.

2.2. Data analysis

A total of 645 valid responses were obtained and analyzed with SPSS Statistics [16]. For this paper, only the results from those with one or more fireplaces were analyzed (N=509). The association between monthly use, hours of use, and time period of use and reasons of using the fireplace was investigated. Because all dependent variables for defining behaviour patterns were categorical, Chi-square tests of independence and Cramer's V were used to find correlations. The former is used to conclude if there appears to be a relation between two variables ($p<.05$ for a significant association). The latter is used to identify the strength of the relation (weak=.15 to .20; moderate=.20 to .30; strong=.30 to .50).

2.3. Simulation-based impact assessment

The case study is a typical Norwegian detached single-family house [17], placed in the climate of Trondheim, Værnes. The house was modeled as a simplified multi-zone model in IDA-ICE [18] with five zones: the living room and kitchen, all rooms south of the living room, all rooms north of the living room, the stairs and the whole ground floor (see Figure 1). Standardized input values from NS 3031 [19] were used for occupancy and internal gains. The simulation model was validated by comparing results from the model to reference values for the same building category [20]. Blind control and window opening were added to simulate real user behaviour and to provide free cooling. Windows are opened for 30 minutes when the temperature exceeds 24°C. The blinds go down for 4 hours when the temperature exceeds 23.5°C. The doors inside the building are kept closed.

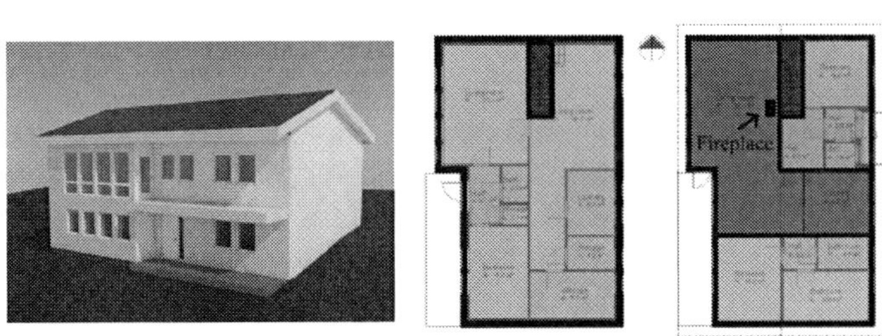

Figure 1. Appearance of a typical single-family house (built 1970-85) and layout with zoning.

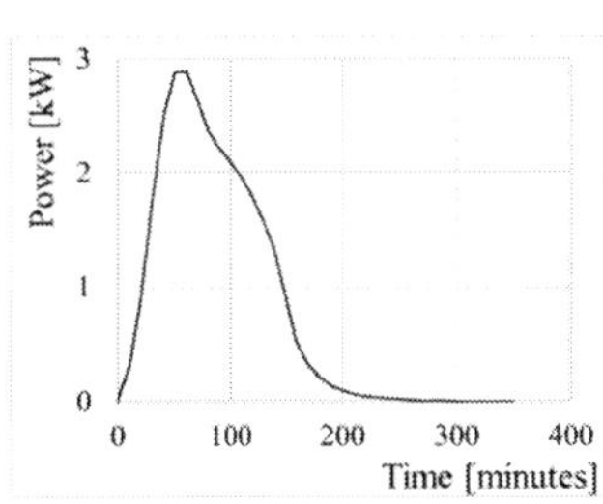

Figure 2. Combustion profile of a 6 kW stove [11]

The house is heated by electric heaters and a wood stove, which is located in the living room on the first floor (i.e. the blue zone in Figure 1). For the sake of the simplicity, this wood stove is implemented as an electric radiator where the time-varying emitted power corresponds to a wood stove. A realistic wood stove control strategy developed and implemented by Skare and Thalfeldt [11] was used. This control imposes the power emitted by the stove inside the room (i.e. the net space-heating power) and not the combustion power. The nominal power of the wood stove was selected to be 6 kW [11] and without additional thermal storage (see Figure 2). A combustion cycle starts when the temperature drops below 22.1°C, just above the NS 3031 set-point of 22.0°C for the electric heaters, and when a schedule (i.e. behaviour pattern) is active. Because most fireplaces are only used during the winter [14] and to avoid the effect of overheating during the summer, the simulation period is taken as the heating season. This is from the day the average daily temperature drops below 11°C until springtime when the outdoor temperature exceeds 9°C. In Trondheim, this is from the 6th of September to the 24th of May.

The fireplace behaviour patterns were combined with four retrofitting packages. These are not optimized for the house, but based on standards. Package R0 is the house before retrofitting. In package R1 only the windows are replaced. Package R2 represents retrofitting to TEK 17 [7]. In package R3, the envelope is retrofitted to meet Norwegian passive house criteria [21], though other energy performance requirements, such as maximum space heating demand and share of renewable energy, are not considered.

Table 1. Energy performance characteristics in the four renovation packages.

	R0 - no renovation [19]	R1 - minimum	R2 - moderate [7]	R3 - major [21]
U-value basement wall [W/m^{2}K]	1.0	1.0	0.18	0.10
U-value basement floor [W/m^{2}K]	0.5	0.5	0.10	0.08
U-value timber frame wall [W/m^{2}K]	0.6	0.6	0.18	0.10
U-value roof (loft insulation) [W/m^{2}K]	0.6	0.6	0.13	0.08
U-value windows [W/m^{2}K]	2.8	1.2	0.8	0.8
Air leakage at 50 Pa [h^{-1}]	6.0	1.4	0.6	0.6
Norm. thermal bridge value [W/m^{2}K]	0.07	0.07	0.05	0.03
SFP [kW/(m^3/s)]	2.0	1.5	1.5	1.5
Heat recovery [%]	0	80	80	80
Ventilation system	Mech. exhaust	Balanced	Balanced	Balanced

3. Results and discussion

3.1. Fireplace behaviour patterns

Associations between use characteristics (e.g. how often it is used and when) and five reasons for using the fireplace were found (for coziness, because it is cheap, because it is fast, because it is environmentally friendly, and because it heats up the whole house). These five reasons are further referred to as pattern variables. Table 2 presents the patterns that were identified based on associations between the pattern variables and how often the fireplace is used. Figure 3 shows the time periods (i.e. when during the day) associated with the pattern variables. These were converted to input schedules for the wood stove control in IDA-ICE. If no specific correlation was found, the mean value was assumed.

The data shows that when a fireplace is used for coziness or because it is a fast heating source, it is used for a couple of hours per day (i.e. one combustion cycle). When a fireplace is actively used as a main heating source or because it is environmentally friendly, it is used from the morning or early afternoon to the evening. It was not possible to find how often a pattern occurs in the population from the survey results. Figure 4 gives an indication of which defined patterns are more popular, and therefore may occur more often in this population. The grey bars show how many participants agreed the pattern variable was their reason for using the fireplace. The green bars show their level of agreement to statements about using the fireplace for each pattern variable. Interestingly, over 50% replied that one of the reasons why they used their fireplace was because it is cozy, though around 20% used it only for coziness. Most people (>75%) agreed that the fireplace is a fast way to heat up their house.

Table 2. Fireplace behaviour on weekdays (wd) and weekends (we) and weekly use patterns associated with the pattern variables, derived from the survey data.

	Cozy	Environment	Cheap	Fast	Main heat
Use per week	3-4 days	Daily	5-7 days	5-6 days	5-7 days
	$\chi^2(6)$=53.931, p=.000, φ_c=.326	$\chi^2(6)$=27.595, p=.000, φ_c=.233	$\chi^2(6)$=58.562, p=.000, φ_c=.339	$\chi^2(6)$=64.356, p=.000, φ_c=.356	$\chi^2(6)$=63.078, p=.000, φ_c=.352
Use hours (wd)	2-4 hours	10-20 hours	2-4 hours	2-4 hours	4-6 hours
	$\chi^2(6)$=26.669, p=.000, φ_c=.229	$\chi^2(6)$=29.850, p=.000, φ_c=.242	$\chi^2(6)$=21.831, p=.001, φ_c=.207	$\chi^2(6)$=50.676, p=.000, φ_c=.316	$\chi^2(6)$=28.956, p=.000, φ_c=.239
Use hours (we)	2-4 hours	10-15 hours	4-6 hours	0-2 hours	6-10 hours
	$\chi^2(6)$=42.315, p=.000, φ_c=.288	$\chi^2(6)$=16.233, p=.013, φ_c=.179	$\chi^2(6)$=25.812, p=.000, φ_c=.225	$\chi^2(6)$=34.418, p=.000, φ_c=.260	$\chi^2(6)$=40.342, p=.000, φ_c=.282

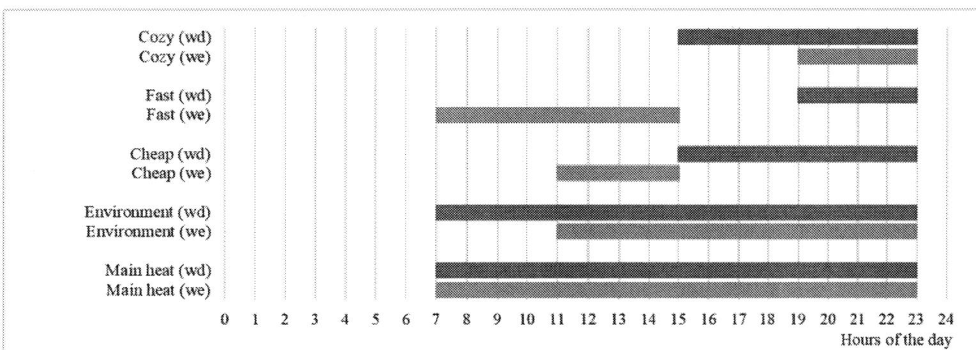

Figure 3. Time periods of use associated with the pattern variable.

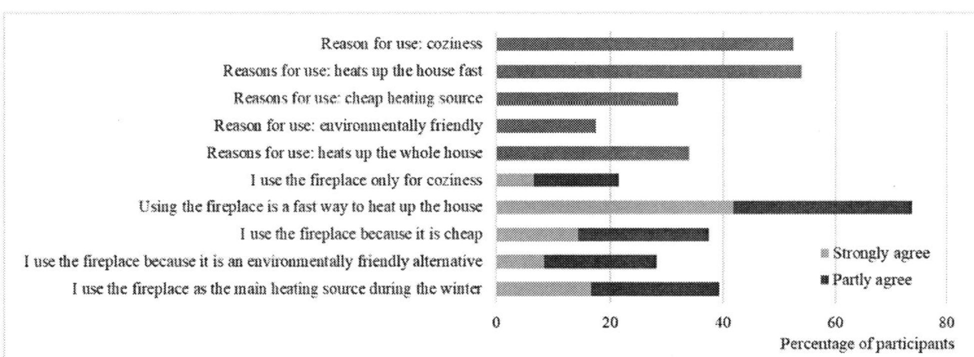

Figure 4. Percentage of participants that answered positive (i.e. ticked off an answer or agreed to a statement) to questions regarding use of the fireplace or wood stove.

3.2. Electrical energy savings

This section analyzes the effect of different time schedules for the stove operation on the space-heating demand. Other energy-related user behaviors are kept unchanged, such as the start set-point temperature and the door opening. Figure 5 shows the results of the space heating demand in kWh/m^2 per year (heating season). The total space heating demand increased when the wood stove was used, but electrical energy savings were achieved (see Table 3). The space heating demand covered by the wood stove and the achieved electrical savings were similar in the patterns 'cozy', 'cheap' and 'fast' and were similar in the patterns 'environment' and 'main heat'. The potential electrical energy savings [%] are calculated as [11]

$$E_{\text{saved}} = 1 - \frac{Q_{\text{SH,wood stove}} + Q_{\text{SH,electric}}}{Q_{\text{SH,ref,total}}} \tag{1}$$

where Q_{SH} is the energy demand for space heating. The savings presented in Table 3 are in comparison to the reference case (i.e. no wood stove) for each retrofitting package. Using a wood stove can result in up to 32% savings of the electrical space heating demand. Pattern 'environment' resulted in most savings (21-32%) while pattern 'cozy' resulted in the lowest amount of savings (6-11%). The highest reductions were achieved in the moderate retrofitting package (R2), closely followed by retrofitting package R3. Figure 5 and Table 3 show that the relative space heating demand covered by a wood stove increased after renovation, though the absolute values decreased. This is because less energy is needed to heat up the house after retrofitting and the indoor temperature drops less often below the set-point for space heating.

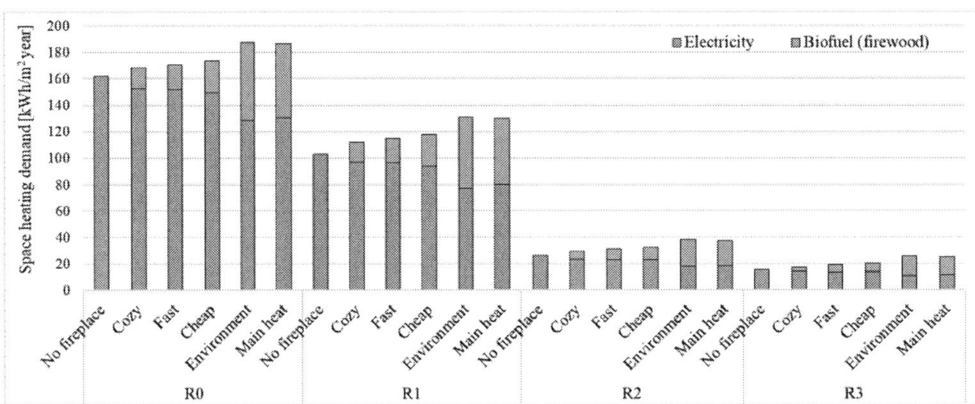

Figure 5. Space heating demand covered by electricity and biofuel in the four retrofitting packages and for the fireplace behaviour patterns.

Table 3. Electrical energy savings achieved in comparison to the reference case (i.e. no wood stove) within each retrofitting package.

	R0 - no renovation	R1 - minimum	R2 - moderate	R3 - major
Cozy	5.9%	6.0%	11.1%	10.4%
Fast	6.2%	6.2%	12.7%	14.6%
Cheap	7.9%	9.0%	13.5%	11.5%
Environment	20.8%	25.2%	32.3%	31.9%
Main heat	19.4%	22.5%	30.6%	29.6%

3.3. Overheating

The thermal comfort was assessed in terms of occupied overheating hours. Figure 6 shows the number of occupied hours in the living room (where the wood stove is located) when the temperate exceeds 24, 25 and 26°C, respectively. The percentage of occupied hours over 25°C decreased after full retrofitting, which is in agreement to results from Skare [11]. After retrofitting, the number of occupied hours over 24°C increased when no wood stove was used, though it stayed the same in 'cozy', 'cheap' and 'fast' and decreased in 'environment' and 'main heat'. Figure 7 shows the relation between the number of hours above 26°C and the wood stove use (i.e. space heating demand covered by firewood). There is a clear positive correlation between the two, indicating that the number of overheating hours increases when the wood stove is used more often. It also shows that when the space heating demand is lower (after retrofitting), it results in less overheating hours. The control strategy of the wood stove starts a combustion cycle when the pattern is active and when the temperature drops below 22.1°. As explained by Skare [11], after increasing building insulation levels, the living room temperature drops less often below the start set-point temperature and the wood stove is used less often. The results also showed that due to local temperature peaks when using the wood stove, the humidity levels dropped, as pointed out by Hamon et al. [8]. When the wood stove was not in use, the humidity levels were significantly more stable.

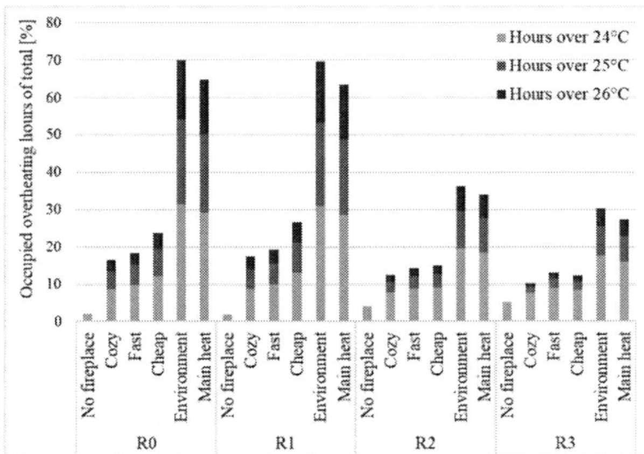

Figure 6. Number of occupied overheating hours over 24, 25 and 26°C in the living room for the five fireplace behaviour patterns.

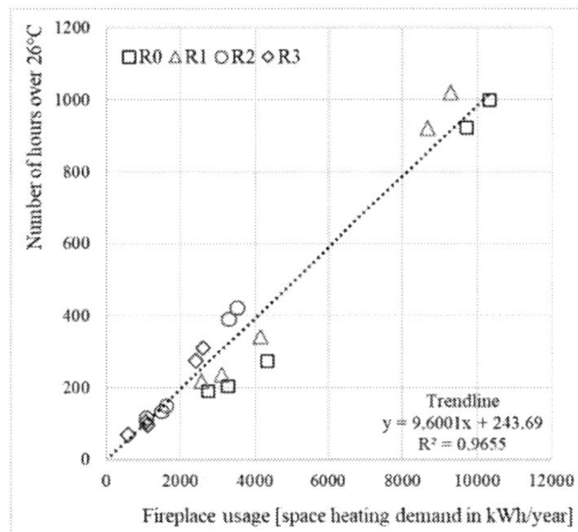

Figure 7. Number of hours over 26°C as a function of wood stove usage (space heating demand covered by firewood)

4. Limitations and future work

The survey sample in this study was not random because of how the survey was distributed. Therefore, results cannot be assumed representative for all of Norway. It was not possible to identify how often the defined patterns are used from the collected data. After completing the data analysis, it was found that some questions may not have been precise enough or that more information should have been asked. For example, clear definitions should be provided for the pattern variables. It could also be interesting to investigate how much fuel people typically use for one combustion cycle and how often they refuel. For future work it is suggested that the survey is revised and randomly distributed.

For the sake of simplicity, this work modeled the fireplace as an electric radiator implemented in a single-family house located in Trondheim. Other types of fireplaces, housing types and climate zones were outside the scope of the present study. In the future, a more realistic model of the stove heat emission can be considered. Indoor comfort was only assessed by evaluating overheating hours, but other thermal comfort criteria could be used. Future work

IOP Conf. Series: Earth and Environmental Science **352** (2019) 012022 doi:10.1088/1755-1315/352/1/012022

may include assessment of humidity levels and indoor air quality. The standard room air model in building performance simulation packages, such as IDA ICE, assumes well-mixed air with uniform temperature and concentration of pollutants and does not compute the airflow inside the room. The limitations of this modeling approach to assess the indoor thermal environment of building heated by wood stove has been investigated by Georges et al. [10] using field measurements. It is shown that the thermal stratification can significantly impact thermal comfort. To investigate this effect, alternative room air models should be considered such as zonal models [22] or CFD [23]. The retrofitting packages were not optimized for the housing type and no alternative combinations of heating sources were included. The nominal power of the stove and its thermal storage should also be adapted to the building thermal properties, such as the insulation level and thermal mass [12]. Savings in terms of fuel cost and CO_2 emissions could be investigated as well.

5. Conclusions

Using firewood in addition to electricity is a common space heating solution for residential buildings in Norway. A survey was designed and sent out to investigate how and why fireplaces and wood stoves are used. Five fireplace behaviour patterns were defined from the survey data and were implemented in a simulation model. Depending of the building insulation level and fireplace behavior patterns, simulation results showed that using a wood stove can result in up to 32% savings on the electrical space heating demand. Pattern 'environment' resulted in most savings (21-32%) while pattern 'cozy' resulted in the lowest amount of savings (6-11%). The highest relative reductions were achieved in the moderate retrofitting package, closely followed by the major retrofitting package. Results showed that the number of overheating hours increased when the wood stove covered more of the space heating demand and that the number of overheating hours decreased after full renovation of the building envelope. Active use of the wood stove is effective as retrofitting measure when the aim is to save fossil fuels or electricity. However, the nominal power of the wood stove and its thermal storage should be adapted to the building thermal properties in order to maintain a comfortable temperature in the room.

Acknowledgments

The authors would like to acknowledge the Norwegian University of Science and Technology, NTNU, for all financial support via the strategic research program ENERSENSE. Martin Thalfeldt and Laurent Georges would like to thank the Research Council of Norway and a number of industrial partners through the project WoodCFD. The authors would also like to thank everyone that filled out the survey.

References

[1] European Commission Energy performance of buildings accessed: 2019-03-31 URL https://bit.ly/2UncsxY
[2] Economidou M, Atanasiu B, Despret C, Maio J, Nolte I and Rapf O 2011 *Buildings Performance Institute Europe (BPIE)* 35–36
[3] Statistics Norway 2019 Statistikkbanken, 06266 accessed: 2019-04-26 URL https://bit.ly/2PtiyYK
[4] Statistics Norway 2012 Statistikkbanken, 10576 accessed: 2019-04-26 URL https://bit.ly/2VqhRpC
[5] Olje- og energidepartementet 2008 Strategi for økt utbygging av bioenergi accessed: 2019-04-29 URL https://bit.ly/2XUpjWK
[6] 1969 Byggeforskrift 1969 accessed: 2019-04-26 URL https://bit.ly/2DDrSoh
[7] Direktoratet for Byggkvalitet 2017 Byggteknisk forskrift (tek 17)
[8] Hamon M, Cao G, Skreiberg Ø, Georges L, Seljeskog M, Khalil R, Sevault A and Mathisen H M 2018 Assessment of the effects of using wood stoves on indoor air quality in two types of norwegian houses *Cold Climate HVAC Conference* (Springer) pp 887–897
[9] Georges L, Skreiberg Ø and Novakovic V 2014 *Energy and Buildings* **72** 87–95
[10] Georges L and Skreiberg Ø 2016 *Journal of Building Performance Simulation* **9** 663–679

[11] Skare A B 2018 *Validation of a method to select the optimal nominal power of a wood stove in residential buildings* Master's thesis NTNU
[12] Thalfeldt Marting S A, Georges L and Skreiberg Ø 2019 Simplified power sizing method for the correct building integration of wood stoves *13th REHVA World Congress, CLIMA conference 2019* (REHVA)
[13] Georges L, Skreiberg Ø and Novakovic V 2013 *Energy and Buildings* **59** 203–213
[14] Finstad A, Flugsrud K, Haakonsen G and Aasestad K 2004 *Vedforbruk, fyringsvaner og svevestv.* (Statistics Norway)
[15] Kipping A and Trømborg E 2015 *Energy* **93** 655–671
[16] IBM 2017 Ibm spss statistics 25
[17] Thyholt M, Pettersen T D, Haavik T and Wachenfeldt B J 2009 *Energy Analysis of the Norwegian Dwelling Stock* vol 37
[18] EQUA Solutions AB 2018 Ida indoor climate and energy (version 4.8)
[19] Standard Norge 2016 *NS 3031 Energy Performance of Buildings: Calculation of Energy Needs and Energy Supply* (Standard Norge)
[20] Hagen H 1990 *Byggforsk 552.103 Oppvarming av boliger. Energiforbruk og kostnader.* (SINTEF Byggfors)
[21] Standard Norge 2013 Ns 3700 criteria for passive houses and low energy buildings: Residential buildings
[22] Georges L, Thalfeldt M, Skreiberg Ø and Fornari W 2019 *Building and Environment* **149** 169–181
[23] Georges L and Novakovic V 2012 On the integration of wood stoves for teh space-heating of passive houses: assessment using dynamic simulation *First Building Simulation and Optimization Conference, BSO12* (IBPSA England)

IOP Conf. Series: Earth and Environmental Science **352** (2019) 012023 doi:10.1088/1755-1315/352/1/012023

Refurbishment of historic buildings at a district scale: *Enhancement of cultural value and emissions reduction potential*

A Loli[1], C Bertolin[1] and T Kleiven[1]

[1] Department of Architecture and Technology, Norwegian University of Science and Technology, 7491 Trondheim, Norway

Abstract. The historic buildings have a significant value in providing a sense of identity to the cities and the community. On the other hand, due to their age, they show the highest ratio of living discomfort and energy consumption. Therefore, their refurbishment is a very important process because, if done right, it will not only reduce their energy demand and increase the living comfort but will also strengthen the social and cultural benefits through leisure and tourism.

In the city of Trondheim, as in many other European cities, the historic buildings have been erected in different architectural periods, which manifest diverse historic and technical features. A categorisation of the wall sections of historic buildings has been done for each city's development period regarding their construction material and technique, building functionality and protection status.

The scope of the article is to estimate the potential for reduction of greenhouse gas emissions at a street/neighbourhood/city level prior to applying large-scale intervention measures. This can be achieved by proposing refurbishment alternatives for wall and window sections that preserve the historic value and at the same time, approach or even meet the actual technical standards. Afterwards, the carbon footprint of the refurbishment action itself and the environmental benefits after the refurbishment (operational phase) is estimated for each category of wall sections. The environmental results, multiplied with the total surface of sections carrying the same attributes, give the overall potential of reduction for the entire group of buildings. Based on this, the on-site renewable energy that would lead to achieving zero-emission targets can be calculated. The framework is also important because it does not treat each building separately, but it suggests refurbishment scenarios for specific categories of buildings built in different historical periods.

1. Introduction

By definition, a historic building is a single manifestation of immovable tangible cultural heritage in the form of an existing building [1]. In addition, a historic building manifests significant cultural, artistic, and social values which gives a specific scope to its refurbishment process. For spreading the refurbishment of historic buildings at an urban scale (street, neighbourhood or city), the definition should encompass not only the very valuable historic buildings (the so-called monuments) but also a large number of buildings in European towns and cities which are far less important from a historical and architectural point of view but, taken as a whole, represent an important part of heritage [2]. By doing so, the intervention process at district scale will incorporate residential buildings or relatively recent buildings, which have protection status from governmental institutions.

IOP Conf. Series: Earth and Environmental Science **352** (2019) 012023 doi:10.1088/1755-1315/352/1/012023

However, due to the wide range of required expertise and the complexity of the interdisciplinary research, the overall intervention process in historic buildings is a complex issue that needs to be effectively resolved. Generally, the affected groups of stakeholders attempt to make decisions in their favour and sometimes the collaboration among different categories of stakeholders seems complicated for lack of an integrated framework and interconnected vision covering the different perspectives. Moreover, an extended literature review concluded that sustainable improvements are based on the operational phase i.e., after the conclusion of interventions and no method considers the environmental impact of the refurbishment process itself [3].

The European projects related to maintenance of historic buildings are primarily focused on energy efficiency solutions [4, 5], as suggested from the directive of the European Parliament [6]. Regarding environmental impact, a framework that considers the carbon emissions in the selection of interventions has been proposed for the maintenance of masonry building [7]. This paper discusses further the importance of the environmental approach in the decision-making process, in order to select the most suitable adaptation interventions which can be applied to diverse categories of buildings. Indeed, the built environment has different attributes depending on the location, materials, history, time of construction, esthetical features, etc. Systematic and clear categorisation of the built environment according to the construction techniques, architectural values, buildings maintenance and performance conditions might be a helpful step to insert the process of refurbishment in a standardized path that is understandable to each practitioner.

To achieve satisfactory results from plan to practice, it is required strong cooperation from specialists of different fields such as urban planners, architects, engineers, researchers, building conservators, buildings owners and others involved in heritage management [8]. This article aims to group the main attributes, meaningful for historic buildings, into three main categories: technical, heritage significance and environmental impact related. The combination of the scale of material decay and its historic value would lead to the suggestion of appropriate refurbishment works that retain significance while the environmental impact of the work during and after refurbishment would be the driver for selection of sustainable scenarios.

2. Materials and methods

As the name "sustainable refurbishment of historic buildings" indicates, the involved groups of interest can be grouped in three main categories: the materials' science community (service life and risk assessment specialists), cultural heritage specialists (historians, architects, governmental institutions) and the sustainability community (with a special focus on environmental impact). Each of the main communities evaluates the built environment in specific aspects and the results are commonly expressed in grading systems regarding respectively the level of decay, historic value classification and amount of environmental impact.

2.1. Decay assessment

Buildings are prone to changing requirements over long periods and it is very important to work with the time rather than against it. For this reason, service life prediction (SLP) is an important process prior to interventions that should be applied to existing buildings [9]. There are many ways for calculating the SLP but according to the report "Performance-based methods for service life prediction" [10], the methods have been divided into two major groups: factor methods and engineering design methods. The application of the methods estimates the service life for a particular component or assembly (in years) by considering specific technical conditions. According to the factor method, the service life (in years) is calculated on the basis of the following equation:

$$ESL = RSL \cdot f_A \cdot f_B \cdot f_C \cdot f_D \cdot f_E \cdot f_F \cdot f_G \quad (1)$$

where ESL = estimated service life; RSL = reference service life; f_A = factor A: quality of components; f_B = factor B: design level; f_C = factor C: work execution level; f_D = factor D: indoor environment; f_E = factor E: outdoor environment; f_F = factor F: in-use conditions; and f_G = factor G: maintenance level.

The service life prediction is strictly connected with the decay assessment process because the indoor and outdoor environment can create a proper state for decay. The level of decay (in %) is linked with the value of the SLP (in years) through polynomial functions as shown graphically in Figure 1.

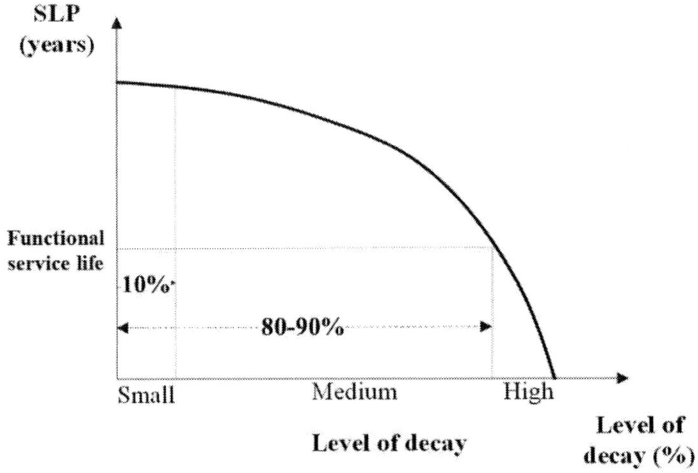

Figure 1. The link between the service life prediction (SLP) and the level of decay of the materials.

The prediction of the decay rate on building components is a necessary step to determine the time when a refurbishment process should be conducted before the end of technical/functional service life is reached i.e., before the component is rendered obsolete. The categorisation of decay results during the decay assessment (small, medium and high) would serve as the starting point in suggesting refurbishment interventions and their frequency of application.

For building components which are estimated to live for decades, the value of SLP needs to be corrected by introducing a new correction factor that considers the effects of the climate change in construction materials. Moreover, when the interventions are applied to historic buildings which hold in addition cultural and social values; the location, aspects of use, and the economic cost should be assessed in the final calculation.

2.2. Historic value assessment

Although a historic building does not necessarily have to be a heritage-designated building, its expected standard lifespan is considered at least twice longer than the lifespans of a building with no or low significance value. For this reason, they require high-quality interventions to ensure long term performance of the building, secure convey of their values and fulfilment of sustainability-driven criteria.

Based on the value that they represent, historic buildings have different protection statuses which can be of an international, national or local level. The international or national institutions (UNESCO, ICOMOS, Riksantikvaren in Norway, etc.) deal mainly with the conservation of the outstanding buildings (the so-called monuments) which have specific protection status and big restriction for change. The protection status of this group of buildings is regulated mainly from municipalities following the restrictions and requirements during the intervention processes.

In the city of Trondheim, apart from the listed buildings (Fredet), there is a large number of protected buildings (Vernet) which are categorised in three main groups (A, B and C) according to the value that they represent [11]. Further interdisciplinary developments (e.g. DIVE method [12]) have been applied in some case studies to these historic buildings in order to categorise them according to the value and giving instructions about the type of the intervention for building component [13]. The recommendations and limitations of the groups of historians, architects and others involved in cultural heritage preservation need to be included in the suggestion of the refurbishment scenarios that need to be applied in protected buildings.

2.3. Environmental assessment

The interventions in historic buildings need to follow the sustainability principles and to adopt the minimal technical interventions through principles of compatibility, reversibility and retreat-ability. The environmental impact of the intervention works can be assessed by applying the Life Cycle Assessment (LCA) analysis [14, 15] to the materials and processes used during the adaptation procedure. According to the EN 15978:2011 standard [16], the intervention works refer to the modules B2 until B5, respectively Maintenance (B2), Repair (B3), Replacement (B4), and Refurbishment (B5).

3. Results

The historic and cultural assessment of buildings indicates the issues to deal with regarding their protection class, rather than dealing with exact instructions on how these problems should be solved [12]. On the other hand, prediction of the materials durability for different building components tells when and where the refurbishment intervention should be done, but not what type of intervention is needed. For this reason, the results of these two independent assessments should be merged prior to deciding the type, the time and the location of the intervention. Both the results of the decay and historic assessments are obtained from classifications which are independent of each other and can be shown in a 2-D chart (Figure 2).

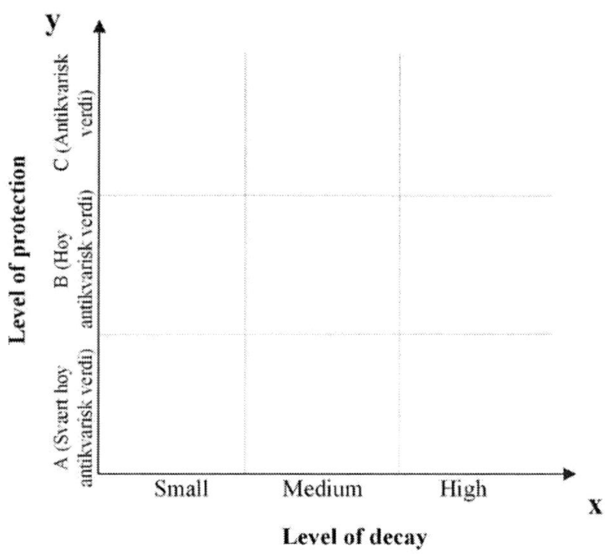

Figure 2. Regrouping of building components according to their decay status and protection level.

The service life is predicted for building components or assemblies' level while the categorisation according to the historic value is performed at building scale with different suggestions about interventions that are allowed to different components.

Each of the cells in Figure 2 contains information about the durability and performance of the constituting component/assembly (level of decay – x-axis) and the scale of allowed intervention in accordance to their protection status (y-axis). For each cell, a refurbishment intervention technique that fulfils both the requests of reducing the decay and keeping unchanged or enhancing the value is provided.

In most of the cases, especially in components with high capacity to change, different scenarios of interventions achieve this dual goal. Therefore, the most appropriate intervention scenario is selected in accordance with its environmental impact. Given the fact that the proposed refurbishment scenarios approach or even meet the actual technical standards [17] (e.g. thermal insulation), then the environmental impact of how it is achieved should be considered. Moreover, applying the environmental

IOP Conf. Series: Earth and Environmental Science **352** (2019) 012023 doi:10.1088/1755-1315/352/1/012023

impact of the refurbishment works as a driver for scenario selection, would lead to the principles of maintenance, repair and reuse as much as possible from the original fabric over the replacing procedures, which is in the same line with the standard approach used by conservators. The new variable transforms the diagram of Figure 2 from a 2-D chart into a 3-D chart (Figure 3) with embodied emissions of the intervention works (modules B2-B5 regarding EN 15978:2011) as the third constituent of it.

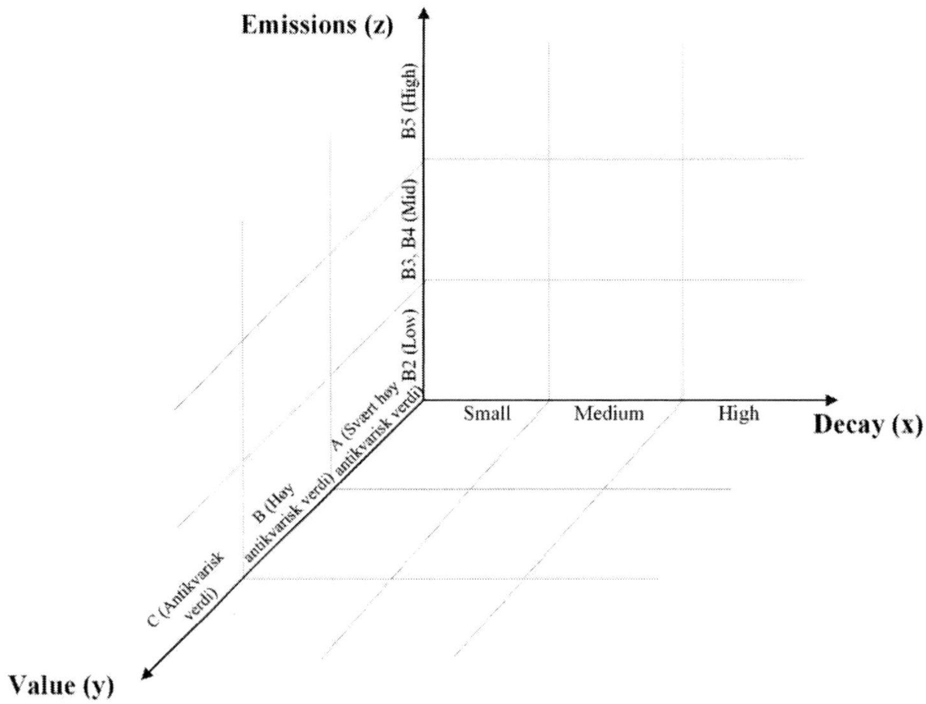

Figure 3. Inclusion of the environmental cost (embodied emissions) of the intervention works in the decision-making process.

However, in common practice, especially in public procurement works, the variable that mostly drives the decision of following a specific scenario rather than others, is the economic value of the work. In the new diagram, the monetary aspect is partially considered looking at the prediction of the level of decay and therefore the SLP (x-axis) and together with the law restrictions established to retain building historic, cultural and social values (y-axis) and the environmental impact in the z-axis, they compound the three pillars of the sustainability (environmental, social and economic) in the decision-making process.

Ideally, a good refurbishment would not only keep the same level of decay and protection class but would help to decrease the decay category and enhance the historical value of the building. Therefore, the next refurbishment rounds and the building itself would result in less environmental cost because the demand for the next interventions would be of a smaller scale and with an increased time of intervention intervals.

As our goal is the sustainable refurbishment of historic buildings, the main focus will be given to the components connected directly to the energy efficiency and GHG emissions, i.e. the external components such as outer walls, windows and roof. The environmental impact should be firstly estimated for the intervention processes itself (modules B2-B5 regarding EN 15978:2011) and later, the emissions during the operational phase B6 should be calculated in order to evaluate the payback energy to be generated by renewables. The emissions for each intervention scenario suggested for each cell of Figure 3 (for each component with a certain level of decay and historic value class), are calculated at the unit level which can be for $1m^2$ for walls and roofs or unit for windows.

The aim of the framework is to estimate the potential of the reduction of the emissions of buildings with similar typology and technique of constructions in the street, neighbourhood or city scale. Depending on the typology of constructions, this can be achieved in two ways:

- When the buildings are very similar to each other (e.g. surface and number of stories), the results of emissions from the refurbishment of an unit scale ($1m^2$) can be multiplied with surface of walls of a building to estimate the results at building scale and then multiplied per number of similar buildings to achieve the carbon footprint of the specific work at district scale. A sum of the emissions of each "winning" refurbishment work corresponding to each cell of Figure 2, would give the total impact of the entire refurbishment of the district.
- When the neighbourhood is heterogeneous regarding surface and height of buildings, the emissions' results from the refurbishment in unit scale can be multiplied directly with the surface of walls at district scale with similar shared criteria (the same value in x and y-axis) to estimate the environmental cost of the similar work and then, of the whole neighbourhood.

3.1. Example of application

The method described above has been applied to two buildings with different status of decay and level of protection. The application of the framework in an entire district is an ongoing process that will be published on a later stage of the study.

The buildings are located in Møllenberg area in Trondheim as it is shown in Figure 4. Because of the esthetical attributes that they carry, the municipality of Trondheim has classified Building 1 with the level of protection C and Building 2 with the level of protection B.

Figure 4. Photos of the Building 1 (left) and Building 2 (right) situated in Wessels Gate in Trondheim.

The service life prediction has been estimated for both wall sections using the factor method. The lifespan of the wood products in construction is over 30 years, but for some components, it reaches more than 100 years [18]. For typical timber log wall, the reference service life (RSL) has been assumed 50 years. Factor values for each component of the equation (1) have been selected according to the recommendations of ISO standards [19, 20] as given in Table 1.

Table 1. The factor values for the estimation of the service life.

Factor	Building 1		Building 2	
	Condition	**Factor**	**Condition**	**Factor**
A – Inherent performance level	Good quality	1.2	Good quality	1.2
B – Design level	Normal design	1.0	Good design	1.2
C – Work execution level	Normal	1.0	Normal	1.0
D – Indoor environment	Average risk	0.9	Average risk	0.9
E – Outdoor environment	Frequent risk	0.8	Frequent risk	0.8
F – Usage conditions	Residential use	1.0	Residential use	1.0
G – Maintenance level	Reduced maintenance	0.8	Good maintenance	1.2

The estimated service life ESL of the walls of the building 1 and 2 are:

$$ESL_1 = 50 \times 1.2 \times 1.0 \times 1.0 \times 0.9 \times 0.8 \times 1.0 \times 0.8 = 34.56 years \quad (2)$$

$$ESL_2 = 50 \times 1.2 \times 1.2 \times 1.0 \times 0.9 \times 0.8 \times 1.0 \times 1.2 = 62.21 years \quad (3)$$

The ESL of the external wall of Building 1 is approximately 70% of the RSL which corresponds to medium level of decay. In Building 2, the ESL is higher than the RSL due to good cyclic maintenance which makes the level of decay of a small scale.
The buildings have been constructed at the same time and their typical log sections (Figure 5) are similar for both buildings. Both the wall sections, after additional insulations works performed in the '80s, have a thermal value U = 0.32 W/(m^2K) while according to the actual Norwegian TEK17 standard, this value for outer walls should be below 0.18 W/(m^2K) [17]. Walls have vapour barrier layers on both sides, recommended for cold climate, to avoid the risk of condensation. However, due to different levels of protection and levels of decay, different intervention scenarios will be suggested to them in order to reduce the decay and improve thermal performance.

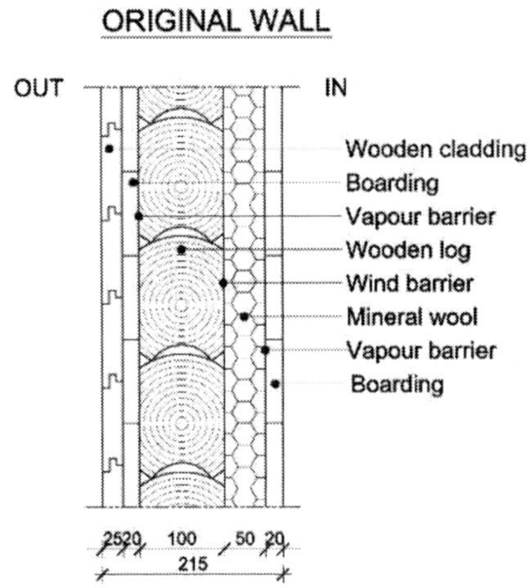

Figure 5. The original section of the wall sections

In Building 1, with C level of protection, it is allowable to have intervention works from outside. This type of intervention is also recommended as the thermal insulation is spread uniformly through the wall, minimizing cold bridges. To reach the desired U-value and reduce the decay, it is recommended to add 50 mm of insulation from outside and replace the original external cladding. This is the only scenario suggested for this type of wall (Figure 6). Another scenario would be the implementation of insulation works from inside, but this option is excluded because, as the external cladding needs to be replaced, it gives the possibility to work from the outer side.

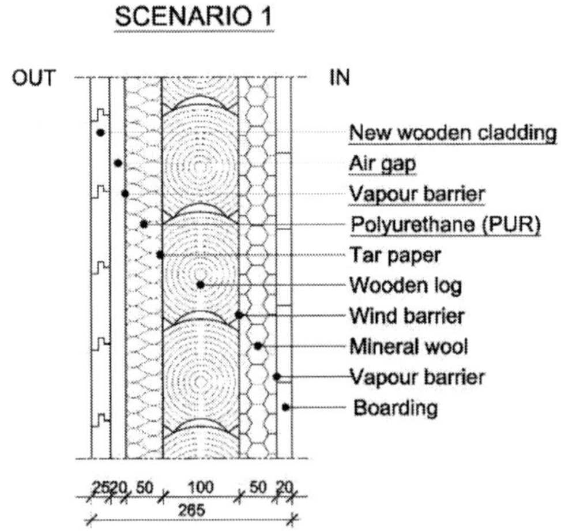

Figure 6. The proposed scenario for the outer walls of Building 1.

New paint of the façade is also included in the intervention works. The carbon footprint is calculated for middle change (Repair, Replacement) and it consists of the emissions for the production and transportation of new materials, the emissions during the installation works and the emissions of the end of live processes for the old materials:

$$E = E_{(p)} + E_{(t)} + E_{(i)} + E_{(eol)} = 19.17 + 0.64 + 0.23 + 2.45 = 22.49 kgCO_{2eq} / m^2 \quad (4)$$

For building 2, the intervention works will be executed from inside due to the protection status of the building. The wall is in generally in good conditions, but its thermal insulation needs to be improved. For this, two intervention scenarios are suggested: additional insulation from inside or replacement of the original insulation as shown in Figure 7.

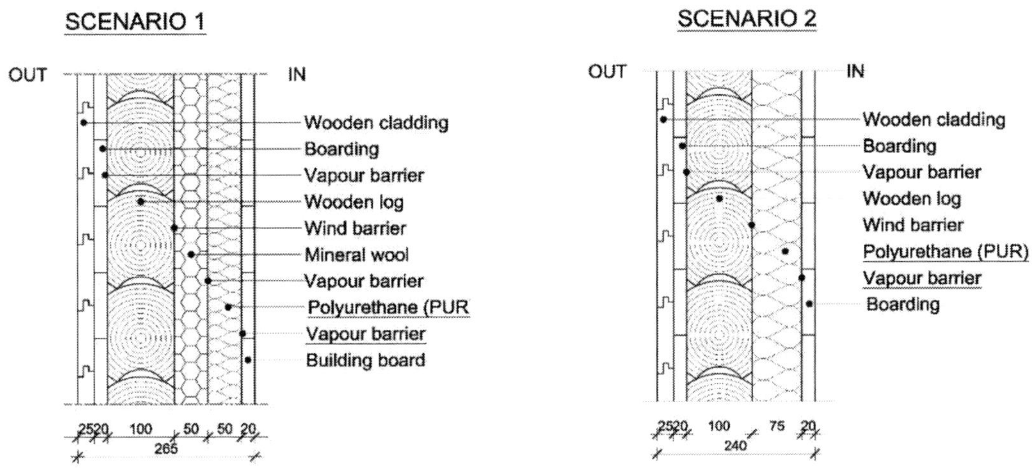

Figure 7. The proposed scenarios for the outer walls of Building 2.

Both scenarios reach the recommended U-value, therefore it will be the carbon footprint of the works which will be the driver for the selection. The environmental cost of the scenarios is calculated still for middle changes as follow:

$$E_1 = \left[E_{(p)} + E_{(t)} + E_{(i)} + \cancel{E_{(eol)}} \right]_1 = 5.70 + 0.14 + 0.08 = 5.92 kgCO_{2eq} / m^2 \quad (5)$$

$$E_2 = \left[E_{(p)} + E_{(t)} + E_{(i)} + E_{(eol)} \right]_2 = 8.65 + 0.14 + 0.12 + 3.55 = 12.46 kgCO_{2eq} / m^2 \quad (6)$$

From the comparison, the first scenario is the most environmentally friendly.

The 3-D chart that connects the level of decay, level of protection and level of embodied emissions of the interventions is given in Figure 8 for both buildings. The carbon footprint of the repair/replacement interventions for 1m^2, multiplied with the total area of walls with similar features in the entire neighbourhood, generates the environmental cost of each work at the district level. The same procedure can be applied to different combinations in the 3-D chart, which encompasses all possible cases that can be met in the district.

The intervention works aim simultaneously the reduction of the level of decay, minimisation of the carbon footprint of the actual/ future interventions and whenever it is possible, the increase of the protection class of the building. Graphically speaking, the objective of the intervention works is to shift the entire set of dots in the 3-D chart as close as possible to the origin of the coordinates.

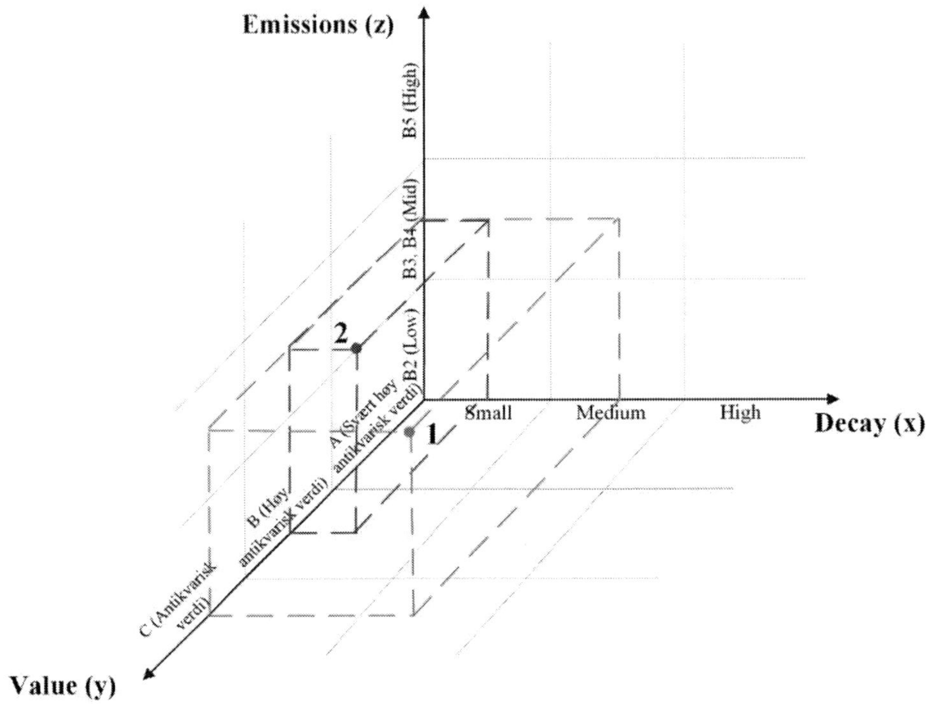

Figure 8. Position of the interventions in buildings in the 3-D chart (cyan dashed line for Building 1 and red dashed line for Building 2).

4. Discussions and Conclusions

The historic value of the building is the only non-renewable asset among the other assets and it should be preserved at any cost during the decision-making process. For this reason, it should be the primary driving factor during refurbishment works. Exceptions to this judgement can be done if the stability of the structure is compromised (due to hazardous events or long abandonment) and the proposed measurements may exceed the allowed level of intervention. After the historic and cultural values are ensured, their link with the level of decay of the material would suggest suitable refurbishment interventions. Considering the growing demands for minimizing the environmental impact, an intervention with satisfactory emission results during and after refurbishment can be preferred. By applying this framework, the results of each community remain independent and the conclusions of each component are taken into account.

The target of preserving the cultural heritage faces challenges in privately owned buildings, especially when the protection class is low. The proposed method does not treat the buildings individually, but it gives suggestions on a neighbourhood scale, thus enabling higher financing opportunities for private owners through collaboration with governmental institutions and involvement in larger projects.

The 3-D diagram presented in Figure 3 suggests a method to analyse, at once, the existing constrains and guidelines when planning an intervention to historical buildings, e.g. for improving the energy performance [21]. The technical compatibility, economic viability, use of the building and the indoor/outdoor environments are considered during the assessment of the level of decay strictly linked to the SLP; the heritage significance is ensured from the historic value assessment and law regulations; while the energy performance is the driver considered during the environmental assessment in a life-cycle approach.

The enlargement of the historic buildings' concept to a street, district or urban level reflects the refurbishment needs for specific categories of buildings, usually private residential houses, which were neglected in the previous interventions as they were not holding memorable historic values. The proposed method makes the path for sorting and assessing the limitations and possibilities of these buildings regarding their historic values, physical characteristics or decay rate, easier and more flexible. Moreover, the assessment and then, the choice of a sustainable solution for groups of buildings that hold the same significance, preservation status and condition of the decay, would result in time and money-saving rather than treating each building separately. These results, incorporated with the energy efficiency reduction due to refurbishment works, help to estimate the operational energy needed from the neighbourhood which serves as a base for the calculation of energy produced from the renewables, thus facilitating the move towards Zero Emission Neighbourhoods in historic urban areas.

5. References

1. European Committee for Standardization (CEN), *EN 15898:2011 - Conservation of cultural property - Main general terms and definitions.* 2011: Brussels, Belgium.
2. Hofmann, F.G. and Aachener Institut für Bauschadensforschung und Angewandte Bauphysik, *Urban Heritage - Building Maintenance: COST Action C5 ; Final Report.* 2002: Office for Office Publ. of the Europ. Communities.
3. Bertolin, C. and Loli, A., *Sustainable interventions in historic buildings: A developing decision making tool.* Journal of Cultural Heritage, 2018. **34**: p. 291-302.
4. Berg, F., Genova, E., and Rroström, T. *Interdisciplinary building categorization - A method to support sustainable energy efficiency strategies in historic districts.* in *Central Europe Towards Sustainable Building 2016: Innovations for Sustainable Future, CESB 2016.* 2016. Grada Publishing.
5. Troi, A. and Bastian, Z., *Energy efficiency solutions for historic buildings: A handbook.* 2015: Birkhauser Verlag AG. 1-336.
6. European Parlament, *Directive 2012/27/EU of the European Parliament and of the Council of 25 October 2012 on energy efficiency, amending Directives 2009/125/EC and 2010/30/EU and repealing Directives 2004/8/EC and 2006/32/EC (Text with EEA relevance).* Official Journal of the European Union, 2012.
7. Forster, A.M., et al., *Green maintenance for historic masonry buildings: an emerging concept.* Building Research & Information, 2011. **39**(6): p. 654-664.
8. Loli, A. and Bertolin, C., *Towards Zero-Emission Refurbishment of Historic Buildings: A Literature Review.* Buildings, 2018. **8**(2): p. 22.
9. Brand, S., *How buildings learn: What happens after they're built.* 1995: Penguin.
10. Hovde, P.J. and Moser, K., *Performance based methods for service life prediction.* State of the art reports, CIB Report: Publication, 2004. **294**.
11. Trondheim Kommune. *Aktsomhetskart kulturminner.* 2019; Available from: http://www.trondheim.kommune.no/byantikvaren/kartinfo/.
12. Reinar, D.A., Westerlind, A.M., and Riksantikvaren, *Urban heritage analysis: A handbook about DIVE.* 2010, Riksantikvaren: Oslo, Norway.
13. Byplankontoret - Trondheim Kommune, *Kulturmiljøet Kjøpmannsgata - En kulturhistorisk stedsanalyse av bryggerekken og tilhørende områder.* 2016, Trondheim kommune.
14. International Organization for Standardization (ISO), *ISO 14040:2006 - Environmental management - Life cycle assessment - Principles and framework.* 2006: Geneva, Switzerland. p. 20.
15. International Organization for Standardization (ISO), *ISO 14044:2006 - Environmental management - Life cycle assessment - Requirements and guidelines.* 2006: Geneva, Switzerland. p. 46.

IOP Conf. Series: Earth and Environmental Science **352** (2019) 012023 doi:10.1088/1755-1315/352/1/012023

16. European Committee for Standardization (CEN), *EN 15978:2011 - Sustainability of construction works. Assessment of environmental performance of buildings. Calculation method.* 2011: Brussels, Belgium.
17. KRD (Ministry of Local Government and Regional Development), *Forskrift om tekniske krav til byggverk (Byggteknisk forskrift TEK17) - Regulations on technical requirements for construction works.* 2017: Oslo, Norway.
18. Pearson, T., Swails, E., and Brown, S., *Wood product accounting and climate change mitigation projects involving tropical timber.* 2012, Winrock International report to the International Tropical Timber Organization.
19. International Organization for Standardization (ISO), *ISO 15686-8:2008 - Buildings and constructed assets - Service-life planning - Part 8: Reference service life and service-life estimation.* 2008: Geneva, Switzerland. p. 36.
20. International Organization for Standardization (ISO), *ISO 15686-1:2011 - Buildings and constructed assets - Service life planning - Part 1: General principles and framework.* 2011: Geneva, Switzerland. p. 21.
21. European Committee for Standardization (CEN), *EN 16883:2017 - Conservation of cultural heritage - Guidelines for improving the energy performance of historic buildings.* 2017: Brussels, Belgium.

IOP Conf. Series: Earth and Environmental Science **352** (2019) 012024 doi:10.1088/1755-1315/352/1/012024

A parametric study of the energy performance and carbon footprint of super-insulation in terrace constructions

Christofer Skaar, Jørn Emil Gaarder, Malin Sletnes
SINTEF Community

Corresponding author: christofer.skaar@sintef.no

Abstract. Energy requirements for buildings are continually tightened, as seen in the ambitions to introduce near zero-energy building (nZEB) requirements in Norwegian and European building codes from 2020. One consequence of this is an increased use of insulation. However, standard insulation may cause challenges in many circumstances, for example where increased wall dimensions lead to reduced daylight levels or where increased insulation leads to increased floor height. Super-insulation materials are a possible solution to these challenges. Although several super-insulation products exist on the market, there is still a need for proven system solutions that provide the required level of insulation, along with reduced thickness in the constructions. An additional challenge is that these solutions should also be cost-effective and carbon-effective. The economic benefits should outweigh the costs and the carbon footprint should ideally be reduced, but at least not significantly increased.

To analyse the potential of super-insulation, we have performed a parametric case study of terrace constructions based on super-insulation and compared these with a baseline solution. The terrace construction uses vacuum insulation panels (VIP) as the main insulation. The top plate insulation is tapered mineral wool, aerogel is used in the edges and on top of the construction there are wood tiles. The parameters that have been varied are i) terrace dimensions, ii) width of the edge with non-combustible aerogel, iii) the thickness of the VIP layer, iii) the slope of the tapering, and iv) the heat conductivity of the VIP panels.

To evaluate the benefits of the super-insulation an analysis of energy performance in the use phase has been done. As the energy efficiency of the super-insulation solution is improved, this gain can be used either to reduce thickness or to increase energy performance. Both these will have an impact on the costs. To evaluate the environmental performance of the solution a screening LCA has been performed, with focus on the carbon footprint. The results of the case study show under which circumstances the super-insulation solution has better performance than the baseline, and vice versa. Key parameters that drive energy performance and carbon footprint are identified, providing suggestions for further research.

IOP Conf. Series: Earth and Environmental Science **352** (2019) 012024 doi:10.1088/1755-1315/352/1/012024

1. Introduction

Improved energy performance has been a target ever since the national building act was introduced in Norway in 1965. This has resulted in a significantly reduced energy use in new buildings compared to the average [1] and the Norwegian and European ambition level is to require that all new buildings are nearly zero-energy buildings (nZEB) from 2020 [2]. One consequence of this is that there has been an increased use of insulation in buildings. However, standard insulation may cause challenges in many circumstances, for example where increased wall dimensions lead to reduced daylight levels or where increased insulation leads to increased floor height. Super-insulation materials are a possible solution to these challenges. Although several super-insulation products exist on the market, there is still a need for proven system solutions that provide the required level of insulation, along with reduced thickness in the constructions and with the additional challenges that these solutions should also be cost-effective and carbon-effective [3-5]. The economic benefits should outweigh the costs and the carbon footprint should be reduced both for the construction itself and for the building in a life cycle perspective.

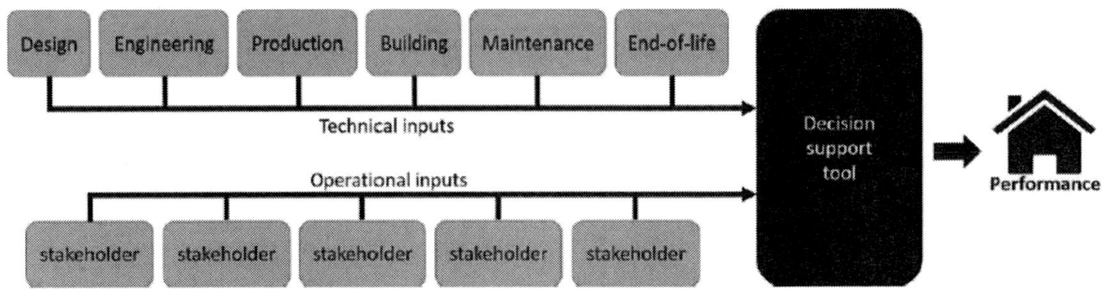

Fig. 1. An extended scope for performance-based design as a framework for a decision support tool
Figure 1: Decision support tool in a building context [6]

To reduce the carbon footprint, it is necessary to see the construction in a building context and with a life cycle perspective already from the design stage [7-11]. Figure 1 shows how performance-based design can be used as a framework for a decision support tool. This framework can be used from design stage to end-of-life stage, and it takes into account both the technical performance of the construction (e.g. U-value, load bearing capacity, etc.) and the operational performance (e.g. carbon footprint, life cycle cost, etc.) [6]. The main challenge is to provide reliable decision support early in the design process, where there is still flexibility to choose between different constructions, materials and processes. As there is a large range of options available, a parametric decision support tool will make it feasible to analyse several options. Such a tool can either be used to explore options manually or it could be integrated with optimisation tools [12]. Here we present the results from a case study of super-insulation in a terrace construction. The goal of the case study is to identify the key parameters driving energy performance and carbon footprint for the construction itself. The constructions are seen in a building context, but the scope here is limited to energy consumption in the use stage. Other potential consequences at the building level, such as reduced building height, increased daylight, are not addressed here.

2. Methods

The methodology in this case study is based on a combination of carbon footprint analysis and energy performance analysis. The calculation of the carbon footprint is based on life cycle assessment (LCA) and the calculation of the energy performance is based on calculating the U-value of the construction, shown in Figure 2 and Figure 3 respectively.

IOP Conf. Series: Earth and Environmental Science **352** (2019) 012024 doi:10.1088/1755-1315/352/1/012024

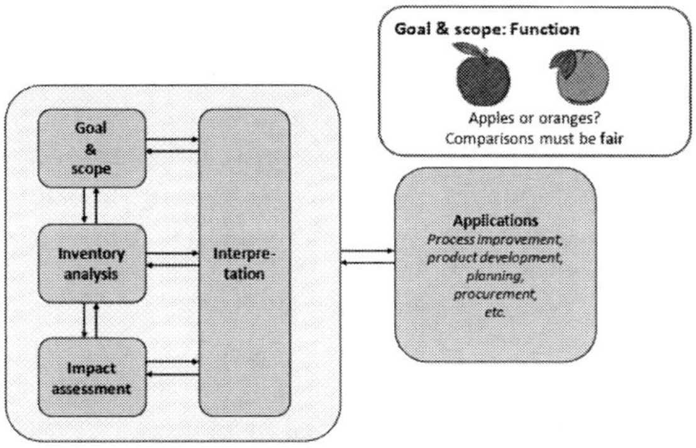

Figure 2: Life cycle assessment [13]

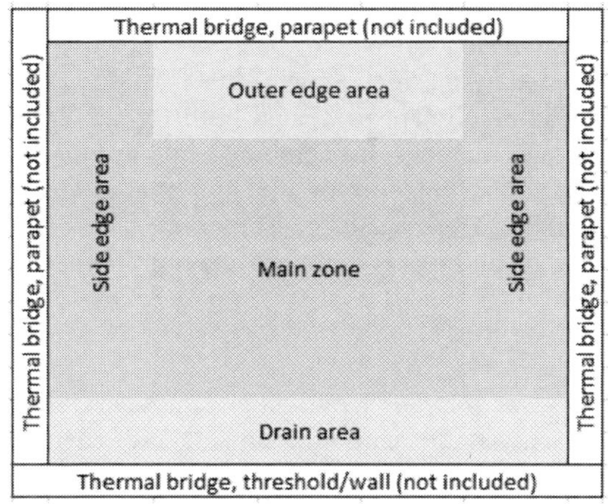

Figure 3: Calculation of U-value

There are four stages in an LCA: i) goal and scope definition, ii) inventory analysis, iii) impact assessment, and iv) interpretation [14, 15]. The first stage of an LCA defines the goal and scope of the study, including defining the purpose of the study, identifying the function of the system and defining the system boundaries. When comparing different solutions, it is necessary to ensure that the product systems meet the same functional requirements to ensure a fair comparison. This is based on first identifying the function and then defining a quantified functional unit. The system boundaries are defined using the modular approach as defined in EN 15978 for buildings [16] and EN 15804 for construction products [17]. The system boundaries are shown in Figure 4, with the included modules highlighted.

The second stage of LCA is to develop the life cycle inventory (LCI). To perform the analysis an adaptation of the tool developed in the ZEB research centre has been used [18]. The foreground inventory is based on work in the ongoing SuperIsol project [19]. The background inventory is based on a combination of data from ecoinvent 3.1. (using the system model *recycled content*) [20], scientific literature, and Environmental Product Declarations (EPD) [21-23].

The third stage of LCA is the impact assessment. Here this is limited to carbon footprint, quantified as global warming potential using the CML impact assessment methodology and in accordance with EN 15804 [17]. The fourth stage is the interpretation of the results. As LCA is an iterative process, this is done throughout the study. Finally, it should be noted that this analysis is not a full LCA, but it is a simplified approach based on the LCA methodology.

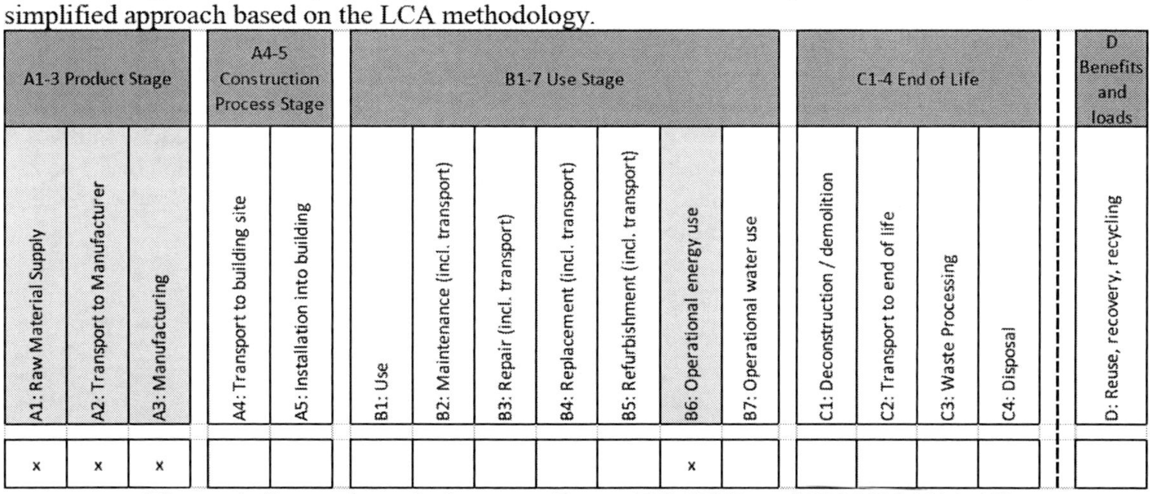

Figure 4: System boundaries according to EN 15978 and EN 15804 [16, 17]

IOP Conf. Series: Earth and Environmental Science **352** (2019) 012024 doi:10.1088/1755-1315/352/1/012024

The energy performance analysis is calculated for a building located in Oslo and based on the annual heat loss of the construction, with direct electric heating. Figure 3 illustrates the elements included in the calculation of the U-value for the constructions.

The results are presented both for the carbon footprint and for the energy performance analysis. The two methods are combined by applying a carbon footprint factor (kg CO_2-eq. per kWh) for the energy consumption in the use phase, and a simplified assumption that energy consumption is identical to electricity consumption. The choice of carbon footprint factor is an ongoing debate in Norway, and this is therefore a factor that has been included in the sensitivity analysis [24-26].

3. Case study

3.1. Functional unit, system and system boundaries

To analyse the carbon footprint and energy performance potential of super-insulation, we have performed a parametric case study of terrace constructions. The constructions using super-insulation are compared with a baseline solution using mineral wool. The purpose of the comparison is to analyse the scale of potential differences; it is not intended as a conclusive comparison between different materials. The functional unit is 1 terrace construction meeting the requirements in the Norwegian building code, over a period of 30 years.

The system boundaries are shown in Figure 4, showing that the included modules are the product stage (A1-A3) and the operational energy in the use stage (B6). The modules A4-A5, B1-B6, B8, C1-C4 and D are outside the system boundaries. This is a simplification based on two assumptions. The first is the assumption that the environmental impact of repair and maintenance are due to material consumption, and that these will correlate with the results for A1-A3. The second assumption is that the end-of-life stage is less significant when the impact assessment is limited to carbon footprint.

3.2. Terrace construction

Figure 5 shows a general illustration of the terrace construction using vacuum insulation panels (VIP) as the main insulation. The top plate insulation is mineral wool, the bottom plate insulation is tapered mineral wool, aerogel is used in the edges and a thin concrete plate constitutes the top layer. In the case study, only the insulation materials (VIP, aerogel and mineral wool) are included. This is because there is little variation for the other materials in the variants.

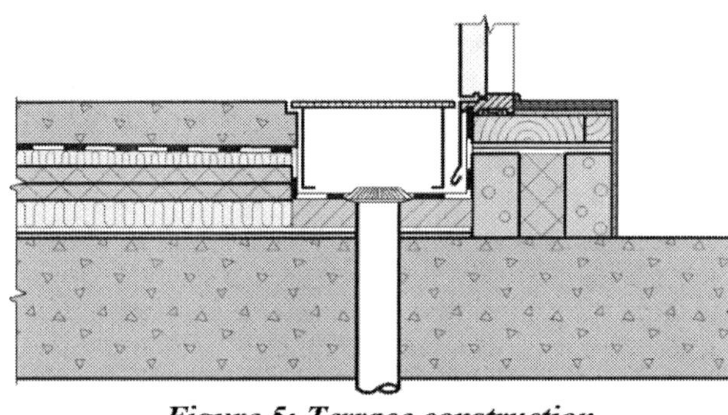

Figure 5: Terrace construction

The minimum performance requirements for the constructions is that they satisfy the requirements in the Norwegian building code. Examples of performance requirements are thermal conductivity, load bearing and safety in case of fire. For load bearing, it can furthermore be relevant to distinguish between uniformly distributed loads, line loads and concentrated loads, as each material will have different properties in this regard.

It should be noted that there is a significant difference in height between the mineral wool and the base case super-insulation construction is 183 mm. This difference can have consequences at the building level, e.g. reduced floor height.

3.3. Parametric variation

The selected parameters are those that typically can be varied when designing a terrace construction. In addition, the heat conductivity of the VIP panels is a significant parameter to include, to address the possibility of degradation of the VIP panel over time (e.g. due to aging effects or damage). The parameters are shown in the list below, with values for each variant shown in Table 1. The standard

IOP Conf. Series: Earth and Environmental Science **352** (2019) 012024 doi:10.1088/1755-1315/352/1/012024

VIP solution is shown in the column labelled *base case*. The VIP solutions are compared with a traditional construction using mineral wool. Here the parameters are identical to the base case with VIP, but the volume of materials are different. The parameters that have been varied, are:

i) terrace dimensions
ii) width of the edge with non-combustible aerogel
iii) the thickness of the VIP layer
iv) the slope of the tapering
v) the heat conductivity of the VIP panels

Table 1: Parametric variation of terrace construction and corresponding material volumes (parameter variation is shown in green).

		Mineral wool	*Base case VIP*	*A*	*B*	*C*	*D*	*E*	*F*	*G*	*H*	*I*	*J*
Parameter	Dimensions [m x m]	**4x4**	**4x4**	3x3	6x6	4x4	4x4	4x4	4x4	4x4	4x4	4x4	4x4
	Edge [mm]	-	**600**	600	600	300	900	600	600	600	600	600	600
	VIP thickness [mm]	-	**60**	60	60	60	60	30	90	60	60	60	60
	Slope [-]	**1:40**	**1:40**	1:40	1:40	1:40	1:40	1:40	1:40	0	1:100	1:40	1:40
	VIP heat conductitity [W/mK]	-	**0.007**	0.007	0.007	0.007	0.007	0.007	0.007	0.007	0.007	0.005	0.011
Volume	VIP [m3]	-	**0.52**	0.23	1.47	0.69	0.20	0.2604	0.7812	0.52	0.52	0.52	0.52
	Aerogel [m3]	-	**0.44**	0.31	0.69	0.33	0.76	0.2196	0.6588	0.44	0.44	0.44	0.44
	Mineral wool [m3]	**5.02**	**1.42**	0.68	4.15	1.42	1.42	1.4245	1.4245	0.74	1.0138	1.42	1.42

3.4. Energy performance

The energy performance is here defined to be identical to the heat loss of the construction, and any other energy consumption is not considered here. The annual heat loss for each variant is calculated based on the location of Oslo, with a constant indoor temperature of 20 °C and a heating demand in October—April. The heat is supplied with direct electric heating, where 1 kWh of heat loss is equal to 1 kWh supplied electricity. Table 2 shows the energy performance of all variants, as well as the base case VIP and the traditional construction using mineral wool.

Table 2: Parametric variation of terrace construction and corresponding energy performance

	Mineral wool	Base case VIP	A	B	C	D	E	F	G	H	I	J
U-Value, VIP [W/m2K]	0.119	0.119	0.132	0.103	0.109	0.128	0.171	0.116	0.139	0.130	0.107	0.138
Heat loss (Oct-Apr) [kWh]	192	120	120	373	176	206	275	187	224	209	172	222

3.5. Carbon footprint

The carbon footprint has been calculated for each of the variants of the construction, taking into account the amount of key materials to build the terrace and the energy performance of the construction in a 30–year time perspective. 30 years has been selected as an estimate of the reference service life of the terrace. Only the materials VIP, aerogel and mineral wool have been considered in these solutions. For energy, two variants have been included.

The sources for the GWP values have been found in EPDs and scientific literature. It should be noted that there is a high level of uncertainty for the carbon footprint calculations for VIP and aerogel. This is due to the low number of life cycle assessments and Environmental Product Declarations (EPD) for these materials, as they are novel compared to mineral wool. This is a key uncertainty and is addressed in the following section.

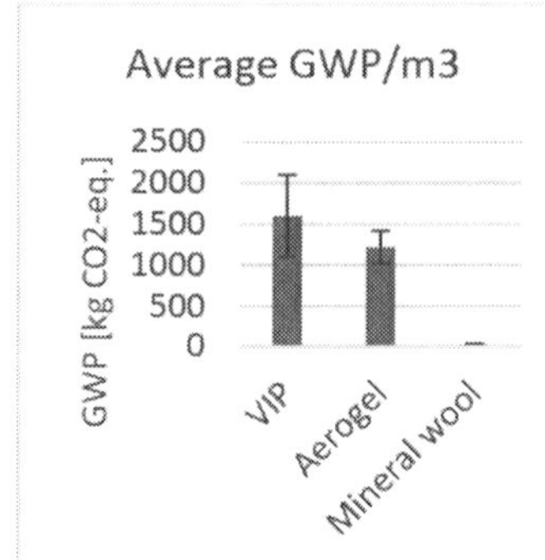

Figure 6: Average GWP per m³, with 95 % confidence limits (A1-A3)

3.6. Sensitivity and uncertainty analysis

The sensitivity of the design is covered through the parametric variation described above. The GWP factor for electricity is a value-based choice and is often debated. The sensitivity of the results related to the choice of GWP factor for electricity is therefore evaluated, with two different factors. The first factor is for the current Norwegian physical grid mix and the second is the based on a scenario analysis (the ZEB factor, [25]). The factors used here are 24 g CO_2-eq. per kWh and 130 g CO_2-eq. per kWh, respectively.

There are several potential suppliers for each of the materials that make up the constructions. The uncertainty of supplier selection is estimated using for each material with values from relevant EPDs or from scientific literature. Based on this, the average value and the 95 % confidence interval are calculated. This provides an average result for each construction, with an estimate of the uncertainty. However, it should be noted that there are three challenges with this approach. The first is that it assumes a normal distribution, the second is that the sample size is small (it is limited to relevant products with an EPD or documented in scientific literature), and the third is that there is a risk of sampling bias as it can be assumed that it is more likely for producers with a good environmental performance to have an EPD.

4. Results and discussion

To evaluate the benefits of the super-insulation an analysis of energy performance in the use phase has been done. As the energy efficiency of the super-insulation solution is improved, this gain can be used either to reduce thickness or to increase energy performance. Both these will have an impact on the costs. To evaluate the environmental performance of the solution a screening LCA has been performed, with focus on the carbon footprint. The results of the case study show under which circumstances the super-insulation solution has better performance than the baseline, and vice versa. Key parameters that drive energy performance and carbon footprint are identified, providing suggestions for further research.

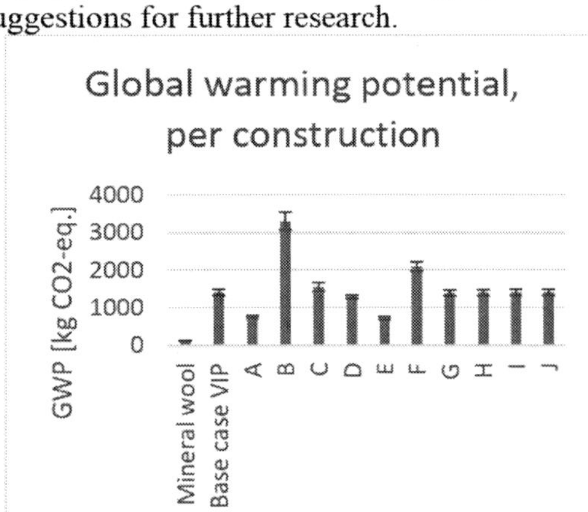

Figure 7: GWP per construction, with 95 % confidence limits (A1-A3)

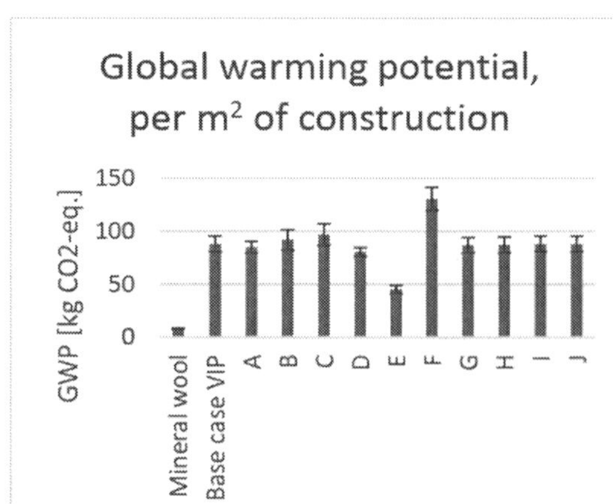

Figure 8: GWP per m² of construction, with 95 % confidence limits (A1-A3)

Figure 7 shows the total carbon footprint (cradle-to-gate, A1-A3) for each variant of the terrace construction and Figure 8 shows the carbon footprint normalised per m^2 of terrace construction. The first figure shows that the most significant parameter for the total carbon footprint is here—as expected—the dimensions of the terrace. However, even when we normalise the results per m^2 of terrace—as shown in Figure 8—an increase in the terrace dimension will only lead to a slight increase of the carbon footprint per m^2. The reason for this increase is that edge will have a relatively lower share of the area. This is also reflected in variants C and D, where the edge is decreased and increased. For variants I and J there are no differences, as these solutions have the same amount of materials as the base case. The change in these two variants is in the heat conductivity of the VIP panels, which is not reflected in the cradle-to-gate footprint.

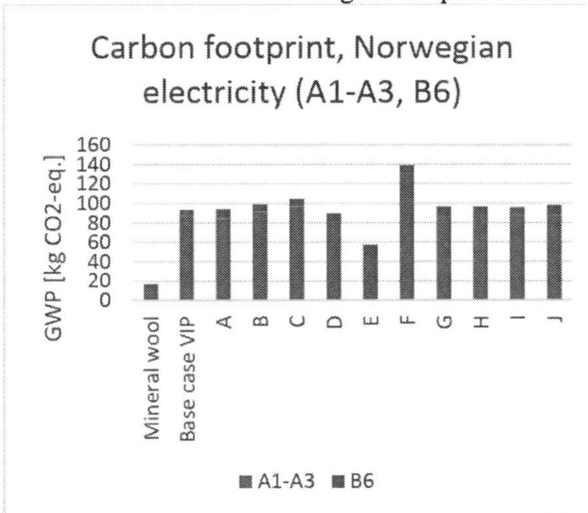

Figure 9: GWP per m2, Norwegian electricity (A1-A3 + B6)

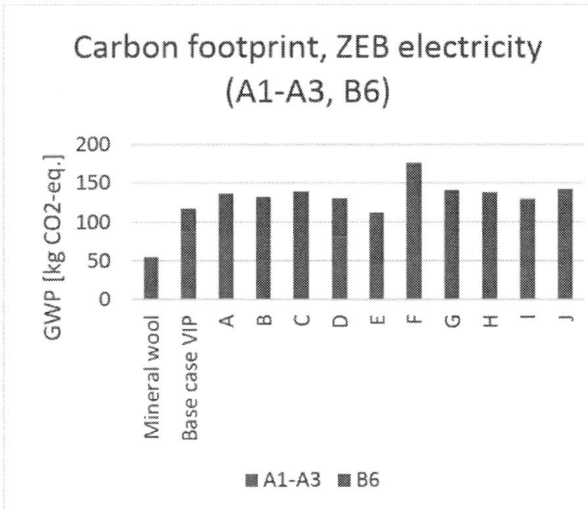

Figure 10: GWP per m2, ZEB electricity (A1-A3 + B6)

If we include the electricity consumption in the use phase, the different energy performance of each construction can be analysed. Figure 9 and Figure 10 shows the carbon footprint both for the production of the materials (A1-A3) and the electricity consumption in the use phase (B6). Here two different CO_2-factors for electricity have been used, first with today's Norwegian physical mix and the second with the ZEB factor. Both figures show that it is the carbon footprint of the materials that has the main contribution to the total carbon footprint, especially for the scenario with the Norwegian electricity. However, we also see the importance of having a life cycle perspective, e.g. when we compare A and B (differences in heat loss changes the conclusion when we include the use phase) and I and J (damaged VIP panels will here lead to 25–30 % increased heat loss). For A, the reduced carbon footprint comes at the expense of worse energy performance. The improved energy performance of B far outweighs the difference in carbon footprint for the materials. For the VIP solutions, only the solution E (with decreased VIP thickness) has a better performance than the base case. The solution with the highest carbon footprint is F, where the VIP thickness is increased from 60 to 90 mm. None of the VIP solutions outperform mineral wool, when we look at the terrace construction alone. However, one argument for using super-insulation is that it may have advantages in a building context. E.g. reduced thickness can lead to reduced floor height in the building or the possibility to add an extra floor within regulatory restrictions.

To provide an estimate of how large these advantages would need to be in a building context, we can compare the increased carbon footprint from the terrace construction (A1-A3 + B6) with the carbon footprint for low-carbon concrete. The maximum carbon footprint for low-carbon class A concrete (B30, M60) is 200 kg CO_2-eq. per m^3 concrete [27]. For the scenario with Norwegian

IOP Conf. Series: Earth and Environmental Science **352** (2019) 012024 doi:10.1088/1755-1315/352/1/012024

electricity mix, a 4x4m terrace would need to provide a saving of 0.65 m3 low-carbon concrete. For the scenario with the ZEB electricity factor, the saving would need to be in the magnitude of 3–10 m^3 of low-carbon concrete. When we know the difference in height between the mineral wool and base case super-insulation is 183 mm, it is likely that a saving can be achieved at the building level.

These results indicate that super-insulation in terrace constructions can lead to a reduction in carbon footprint, but only if the advantages at the building level outweigh the higher carbon footprint of the construction itself. It should be noted that there is high uncertainty in the GWP calculations for VIP and aerogel, as these are relatively novel materials with less available documentation than mineral wool.

5. Conclusions

The results of the case study of a terrace construction show that if super-insulation simply replaces traditional insulation it will likely lead to a higher carbon footprint. However, if there are additional benefits in a building context (e.g. reduced floor height), super-insulation may lead to reduced carbon footprint for the whole building. The most significant parameters for the VIP solution are terrace dimensions, thickness of VIP, and increased heat conductivity due to damage. A key recommendation for further work is to investigate the effect at the building level of using VIP in constructions.

6. Acknowledgements and declaration of interests

The project SuperIsol is funded by the Research Council of Norway under the large-scale programme for energy research (ENERGIX), project number 282389 [19].

References

[1] Byggforskserien, "473.003 Energieffektive bygninger. Begreper og definisjoner," ed, 2015.

[2] *Directive 2010/31/EU of the European Parliament and of the Council of 19 May 2010 on the energy performance of buildings*, 2010.

[3] N. Lolli and A. G. Hestnes, "Greenhouse gas analysis of insulation options in residential energy retrofitting," presented at the PassiveHus Norden 2013, Goteborg, 2013.

[4] R. D. Schlanbusch, S. M. Fufa, K. Sørnes, and T. Kristjansdottir, "Energi- og klimagassanalyse av isolasjonsmaterialer," SINTEF akademisk forlag, Rapport2014.

[5] N. Lolli and I. Andresen, "Aerogel vs. argon insulation in windows: A greenhouse gas emissions analysis," *Building and Environment,* vol. 101, pp. 64-76, 2016/05/15/ 2016.

[6] N. Labonnote, C. Skaar, and P. Rüther, "The potential of decision support systems for more sustainable and intelligent constructions: a short overview," *Procedia Manufacturing,* vol. 12, pp. 33-41, 2017.

[7] A. Hollberg and J. Ruth, "LCA in architectural design—a parametric approach," *The International Journal of Life Cycle Assessment,* vol. 21, pp. 943-960, 2016.

[8] S. M. Fufa, C. Skaar, K. Gradeci, and N. Labonnote, "Assessment of greenhouse gas emissions of ventilated timber wall constructions based on parametric LCA," *Journal of Cleaner Production,* vol. 197, pp. 34-46, 2018.

[9] G. Lobaccaro, A. H. Wiberg, G. Ceci, M. Manni, N. Lolli, and U. Berardi, "Parametric design to minimize the embodied GHG emissions in a ZEB," *Energy and Buildings,* 2018.

[10] B. Wrålsen, R. O'Born, and C. Skaar, "Life cycle assessment of an ambitious renovation of a Norwegian apartment building to nZEB standard," *Energy and Buildings,* vol. 177, pp. 197-206, 2018.

[11] B. Soust-Verdaguer, C. Llatas, and A. Garcia-Martinez, "Simplification in life cycle assessment of single-family houses: A review of recent developments," *Building and Environment,* vol. 103, pp. 215-227, Jul 2016.

[12] C. Skaar, N. Labonnote, and K. Gradeci, "From Zero Emission Buildings (ZEB) to Zero Emission Neighbourhoods (ZEN): A Mapping Review of Algorithm-Based LCA," *Sustainability,* vol. 10, p. 2405, 2018.

[13] C. Skaar, M. Bergem, and P. D. Jakobsen, "Environmental impact - TBM versus D&B," in *Tekna TBM applications II*, ed. Bergen, 2011.

[14] ISO, *Miljøstyring, livsløpsvurdering, prinsipper og rammeverk (ISO 14040:2006) = Environmental management, life cycle assessment, principles and framework (ISO 14040:2006)* vol. NS-EN ISO 14040. Lysaker: Standard Norge, 2006.

[15] ISO, *Miljøstyring, livsløpsvurdering, krav og retningslinjer (ISO 14044:2006) = Environmental management, life cycle assessment, requirements and guidelines (ISO 14044:2006)* vol. NS-EN ISO 14044. Oslo: Standard Norge, 2006.

[16] CEN, *NS-EN 15978: Bærekraftige byggverk : vurdering av bygningers miljøpåvirkning : beregningsmetode = sustainability of construction works : assessment of environmental performance of buildings : calculation method* vol. NS-EN 15978:2011. Oslo, Norway: Standard Norge, 2011.

[17] CEN, "EN 15804:2012+A1:2013: Sustainability of construction works - Environmental product declarations - Core rules for the product category of construction products," ed. Geneva, Switzerland: International Organization for Standardization, 2013.

[18] R. Woods and M. Samdal, "ZEB Final Report 2009-2017," The Research Centre on Zero Emission Buildings (ZEB), Trondheim, Norway2017.

[19] SuperIsol. (2019, 2019-05-08). *Nye superisolasjonsløsninger og -produkter for norske bygg*. Available: https://prosjektbanken.forskningsradet.no/#/project/NFR/282389

[20] ecoinvent. (2018, 2018-05-30). *ecoinvent*. Available: http://www.ecoinvent.org

[21] IBU. (2016). *IBU*. Available: http://www.epd-online.com

[22] EPD-Norway. (2016). *EPD-Norway*. Available: http://www.epd-norge.no

[23] Environdec. (2016). *Environdec*. Available: http://www.environdec.com

[24] N. Lolli and A. G. Hestnes, "The influence of different electricity-to-emissions conversion factors on the choice of insulation materials," *Energy and Buildings,* vol. 85, pp. 362-373, 2014/12/01/ 2014.

[25] M. K. Wiik, S. M. Fufa, T. Kristjansdottir, and I. Andresen, "Lessons learnt from embodied GHG emission calculations in zero emission buildings (ZEBs) from the Norwegian ZEB research centre," *Energy and Buildings,* vol. 165, pp. 25-34, 2018.

[26] Standard Norge, *NS 3720:2018 Metode for klimagassberegninger for bygninger* vol. NS 3720:2018. Oslo: Standard Norge, 2018.

[27] Norsk Betongforening, "Publikasjon nr. 37: Lavkarbonbetong," Norsk Betongforening2015.

IOP Conf. Series: Earth and Environmental Science **352** (2019) 012025 doi:10.1088/1755-1315/352/1/012025

Designing a ZEN Campus: An exploration of ambition levels and system boundaries

Christofer Skaar[1], Christian Solli[2] and Jonas Vevatne[3]

1) SINTEF Community
2) Norwegian University of Science and Technology (NTNU)
3) Norwegian Directorate of Public Construction and Property (Statsbygg)

Corresponding author: christofer.skaar@sintef.no

Abstract. The Norwegian University of Science and Technology (NTNU) is gathering from dispersed locations to one central campus at Gløshaugen, requiring an estimated 92 000 m^2 new buildings and upgrading 45 000 m^2 of existing buildings. NTNU has high environmental ambitions for the new campus, including zero-emission ambitions. This paper explores system boundary definitions and ambition levels in a Zero Emission Neighbourhood (ZEN) context. A key element is a plus energy campus that provides a surplus of renewable energy in the operational phase, that can compensate the carbon footprint of buildings, infrastructure and mobility. Preliminary energy and carbon analyses of the campus have been performed A key result is an overview of design choices and methodology choices for concept stage calculations for a zero-emission campus. Six system boundaries have been defined, with the production to consumption ratio varying from 19 % to 132 %. The lowest includes all buildings, the highest includes production from all buildings, but consumption only from new and renovated buildings. The main finding is that it is possible to realise a plus energy campus for new and renovated buildings, but not including non-renovated buildings. A plus energy campus requires a combination of PV and seasonal energy storage.

IOP Conf. Series: Earth and Environmental Science **352** (2019) 012025 doi:10.1088/1755-1315/352/1/012025

1. Introduction

The Norwegian University of Science and Technology (NTNU) is gathering the campus in Trondheim from dispersed locations around the city to one central campus at Gløshaugen. This is a significant building project both in the context of the university and in the context of Trondheim municipality. The project requires a substantial development of the existing campus at Gløshaugen; there is an estimated need for 92 000 m^2 of new buildings and upgrading of 45 000 m^2 of existing buildings. This means that the renovation will affect almost 15 % of the existing approximately 320 000 m^2 of buildings at NTNU, in addition to expanding the campus by almost 30 %. NTNU is partner in the Zero Emission Neighbourhood (ZEN) centre [1] and has high environmental ambitions for the new campus, including zero-emission ambitions e.g. as defined by ZEN. The ZEN concept of zero emissions builds on the work from the Research Centre on Zero Emission Buildings (ZEB) [2], which developed methodology for how to design and build zero-emission buildings. A ZEB building is a building where the production and export of renewable energy is exported, which offsets the carbon footprint of the building in a life cycle perspective.

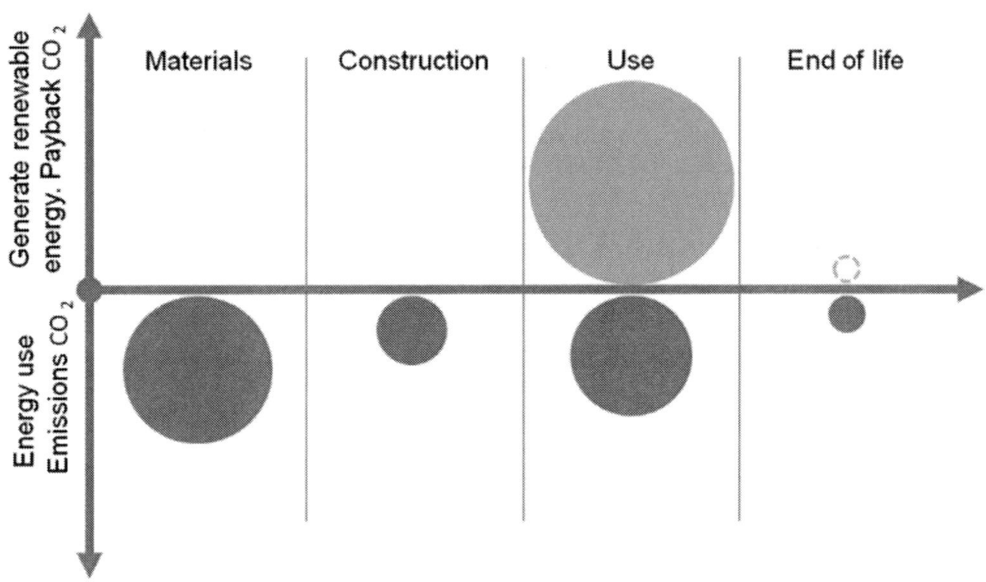

Figure 1: The Zero Emission Building framework [3, 4]

The Research Centre on Zero Emission Neighbourhoods builds on the ZEB concept and expands it both by lifting it up to a neighbourhood perspective and expanding the scope from carbon footprint to a total of eight aspects, as shown in Figure 2. Here the ambition level is linked to performance compared against a reference scenario for the neighbourhood. ZEN also includes mobility in the use phase (e.g. transport of students and employees to and from the university), and this requires significant further development of the methodology [5-7]. But achieving zero emissions is also in ZEN based on export of new renewable energy. The purpose of this paper is to analyse if and how the NTNU campus at Gløshaugen can generate more new renewable energy than is consumed by the campus itself and to explore how the definition of the system boundaries influence this. The results are a synthesis that is based on work performed in the campus development project at Gløshaugen.

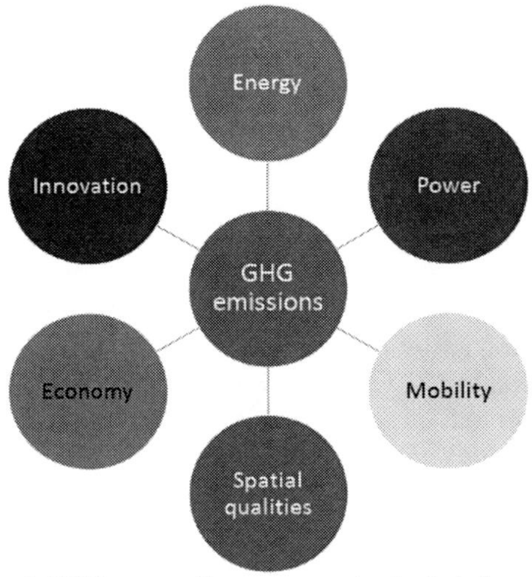

Figure 2: ZEN aspects (from presentation by I. Andresen)

Although the main purpose of this paper is on the energy surplus from the campus, it should be recognised that this is only one aspect of many that are important for a ZEN pilot project. ZEN has developed key performance indicators (KPI) for all the eight aspects shown in Figure 2, and these aspects should not be seen in isolation as there is a complex interaction between them.

2. Method

This paper explores system boundary definitions and ambition levels for a zero-emission campus in a ZEN context, outlining key design choices both for the technical system and for the definition of the system boundary, which all have carbon footprint impacts. To reach the carbon ambitions in a neighbourhood perspective it is necessary to reduce the carbon footprint of the buildings (including their infrastructure and foundations), to build and renovate energy-efficiently, and to produce new renewable energy onsite. A key element is a plus energy campus that can provide a surplus of renewable energy in the operational phase that can be used to compensate for the carbon footprint of buildings, infrastructure and mobility. Renewable onsite energy production onsite is photovoltaics (PV), other technologies were considered but not included (e.g. building integrated wind).

This paper presents a synthesis of the system boundary discussion, based on work performed in relation to the campus development project at Gløshaugen. Preliminary energy and carbon analyses of the campus have been performed by several consultancy companies through the pre-project period, and these identify both energy flows and the potential carbon footprint of the planned campus [8, 9]. In addition, a document outlining the environmental premises for the campus project has been developed in the pre-project stage by NTNU and Statsbygg, with contribution from ZEN [10]. These outline the technical aspects of energy production and generation, and this is interpreted in light of the zero-emission ambition using the ZEN framework. A limitation of this work is the exclusion of the quick clay challenge related to infrastructure and foundations, which will require in-situ improvement before they can be used (e.g. buildings or seasonal thermal energy storage). Here the focus is primarily on how the campus can provide a surplus of new renewable energy. This paper systematises the work that has been performed in the pre-project and analyses the implications for the energy production and consumption at the campus.

3. System boundary choices

3.1. Introduction

The system boundaries of ZEN cover three categories: buildings, infrastructure and mobility. Although it is possible to have new renewable energy production from mobility, it is the infrastructure and buildings that may have the most significant contribution in the context of the campus. For this reason, the discussion here does not include mobility. For buildings, the system boundaries in ZEN build on the work from the Research Centre on Zero Emission Buildings (FME ZEB) [3]. However, when expanding to a neighbourhood perspective, this is more complex than adding up the sum of the individual buildings [5, 6]. Particularly, it is of interest to discuss how the system boundaries should relate to the existing buildings on campus that are outside of the project and how they should relate to the surrounding energy system (both heat and electricity).

3.2. Energy consumption

The energy consumption at Gløshaugen consists of heating, cooling and electricity. Electricity is today primarily supplied from the electrical grid. For heating the main supply is from the municipal district heating system, supplemented with internal surplus heat e.g. from high performance computers. There are considerable seasonal differences in consumption patterns, both due to the physical climate and to the characteristics of a university campus with significantly reduced activity in the summer.

3.3. Energy production

Multiple studies of the energy system at Gløshaugen were performed as part of the pre-project [8, 9], investigating the potential for on-site production of new renewable energy and for short term and seasonal energy storage .

3.3.1. Electricity from photovoltaics

An analysis of the potential for producing electricity from photovoltaics has shown that it is possible to generate 2.2 GWh from new buildings (plans are not yet fixed for the buildings, so this is based on an average estimate for the campus development project), 3.6 GWh from existing buildings (where 0.5 GWh will be from renovated buildings). All figures depend on a high degree of utilisation of roofs and facades.

3.3.2. Heat from heat pumps

Heat pumps have been a focus area at NTNU for many years and is also a technology that will be used in the campus project. There has been a continued effort to use heat pumps to increase energy efficiency [11], currently the consumption of district heating from the city grid is approximately 20 GWh [8]. Although the energy consumption has been reduced, there is still the challenge of high peak loads.

3.3.3. Other energy technologies

Other energy technologies—e.g. wind turbines, solar thermal energy collectors, combined heat and power from biomass—have been considered in the preliminary stages, but these technologies have not been analysed in detail. The main reason for this is that the initial results show that photovoltaics and heat pumps currently have the highest production potential.

3.4. Energy optimisation

Optimising the energy flows in the short and long term can provide additional benefits. Two key approaches have been investigated; short-term and long-term storage of energy. Short-term energy storage can provide benefits on energy and power demand, e.g. through load shifting and peak shaving. Long-term energy storage can provide benefits of reduced waste heat and reduced use of peak load, which is often based on non-renewable energy. For the long-term energy storage, NTNU has performed a concept study of a system termed Energihub Gløshaugen. The concept study Energihub Gløshaugen is a cooperation between NTNU, Statkraft Varme and Trønderenergi, investigating the potential role of Gløshaugen in relation to the surrounding energy infrastructure. Two key challenges addressed are short-term energy storage and long-term (seasonal) energy storage. In a ZEN context, the long-term storage of energy can have a high impact on the energy surplus, and this is therefore the main focus in here. This storage is a borehole thermal energy storage (BTES) consisting of approximately 300 boreholes of 120 meters depth.

3.5. Buildings

There are currently around 320 000 square meters of floor area in the existing buildings at Gløshaugen. The campus project requires a substantial additional development, with an estimated need for 92 000 square meters of new buildings and upgrading of 45 000 square meters of existing buildings. The type and age of buildings vary considerable at Gløshaugen; from the main building (*Hovedbygningen*) in 1910 to the ongoing ZEB Flexible Lab that is currently being constructed.

In terms of system boundaries, there is a main distinction between buildings that will be directly affected by the campus project (e.g. new and renovated buildings) and existing building that will not undergo renovation. The energy consumption in the existing buildings is high, both due to the age of the buildings and due to a high number of energy-intensive activities such as laboratories and through-the-clock activities.

4. Results and discussion

A key result from the synthesis of the energy studies and the campus development plans is an overview of design choices and methodology choices for concept stage calculations for a zero-emission campus, as shown in Table 1. The system boundary consists of three building categories: new buildings, renovated buildings, and non-renovated buildings. Six system boundaries have been defined and evaluated in terms of energy production and consumption ratio in the operational phase. Depending on the system boundary definition, the production to consumption ratio is 19 % to 132 %. The lowest ratio is when all buildings are included, including non-renovated (F in Table 1). The highest ratio is when production from all buildings is included, but consumption only includes new and renovated buildings (E in Table 1). It should be noted that the system boundaries are inconsistent in variant E, where production is included for non-renovated buildings but not consumption.

Table 1: System boundaries of production/consumption and coverage ratio [10]

	Production			**Consumption**				
	New buildings	*Renovated buildings*	*Non-renovated buildings*	*New buildings*	*Renovated buildings*	*Non-renovated buildings*	*EnergyHub*	***Coverage ratio***
A	X			X				**30 %**
B	X	X		X	X			**23 %**
C	X	X	X	X	X			**49 %**
D	X	X		X	X		X	**107 %**
E	X	X	X	X	X		X	**132 %**
F	X	X	X	X	X	X	X	**19 %**

As we can see from the column *Coverage ratio* in Table 1, only two of the six alternative system boundaries will lead to an energy surplus with the defined system boundaries. These are options D and E, where the EnergyHub contributes to the increased coverage ratio.

Option A and B provide the most clear-cut system boundary options. Here only affected buildings are included, and the difference is whether renovated buildings are included or not. The main drawback of this option is that it does not provide a full picture of the extent of the campus project, thus providing simple system boundaries at the cost of losing system complexity and interplay.

Option C expands the system boundaries to include the production from all buildings at Gløshaugen, but only the consumption of the new and renovated buildings is included.

Option D and E both include the effect of seasonal storage, through the EnergiHUB Gløshaugen project. Option E is the same as option C plus seasonal storage. These two options are the only ones that can provide an energy surplus, which shows that without seasonal storage it will be difficult to achieve the zero-emission ambition.

Finally, option F encompasses all of Gløshaugen. This option does not differentiate between existing buildings, renovated buildings and new buildings. A key strength of this option is that it is the entire campus that is included within the neighbourhood system. However, it also means that the campus project not only must offset its own carbon footprint, it must also atone for past building traditions.

A key challenge that is immediate from Table 1, is that it is not obvious how the system boundaries should be defined when we are dealing with a complex development project within a neighbourhood. We see that there can be challenges in how existing buildings should be included, in particular if the buildings are only partially renovated. For example, how should we define the system boundaries if we install PV on the roof of an existing building, but without making any other renovation? Here there is no clear-cut approach that can be recommended, but there are three options that could be further explored. The first is to leave out the building entirely, also excluding the PV production. The challenge in this case is that if it is the campus project that is the cause of the PV, then it will have the expenses without any of the benefits. The second option is to include the building entirely. The challenge in this case is that the energy performance of the existing building will also be included, without the campus project having any influence on this. A third option can be to allocate energy production from such buildings based on the influence the campus project has had. E.g. we can have a situation where the PV installation is initially planned to happen during a major renovation 20 years in the future, but the campus project makes it happen in 5 years. In this case, it could be argued that the PV production is within the system boundaries in the 15 years between planned and actual instalment.

This illustrates the challenges with the system boundaries. If there were multiple partners, economic boundaries could be used. E.g. if neighbourhood developer x pays building owner y to install PV, the share could be considered "nearby production" and the allocation be based on the business model between x and y. For the campus, there are few such distinctions, as there is one owner.

These initial results also highlight the need for accommodating for different time horizons and the need for different time resolutions. The campus is a continually evolving entity, and this should be reflected when analysing it as a zero-emission neighbourhood e.g. in defining the system boundaries or reference neighbourhood. For energy, the short-term and long-term analyses depend on having a relevant time resolution. This can for example by the hour or minute for electricity, whereas district heating might only need weekly or monthly data.

This study has focused on the system boundaries from the perspectives of energy and carbon footprint. This means that cost barriers are not explicitly addressed here, although some of the underlying studies have included aspects of cost (e.g. when selecting energy technologies to include in the analyses). Other barriers that should be addressed in future work include the technological, organisational and economic integration between stakeholders.

5. Conclusions

A plus energy campus is a basis for a zero-emission campus. The main finding is that it is not possible to realise a plus energy campus for the entire Gløshaugen campus. This is based on the initial analyses—the volume of existing buildings and the energy use in these are currently just too high. However, it is possible to reach a plus energy campus for the campus project itself (i.e. the buildings affected by the campus development plans). In this case it is dependent on a combination of PV and seasonal energy storage. This stresses the importance of seeing the zero-emission neighbourhood in the context of the surrounding energy system. A significant future challenge is the energy consumption of existing buildings not included in the campus development.

Acknowledgements

This article has been written within the Research Centre on Zero Emission Neighbourhoods in Smart Cities (FME ZEN, project number 257660). The authors gratefully acknowledge the support from the ZEN partners and the Research Council of Norway.

IOP Conf. Series: Earth and Environmental Science **352** (2019) 012025 doi:10.1088/1755-1315/352/1/012025

References

[1] FME ZEN. (2018, 2018-05-30). *FME ZEN*. Available: http://fmezen.com/

[2] M. R. K. Wiik, S. M. Fufa, J. Krogstie, D. Ahlers, A. Wyckmans, P. A. Driscoll, *et al.*, "Zero Emission Neighbourhoods in Smart Cities: Definition, Key Performance Indicators and Assessment Criteria: Version 1.0. ZEN report no. 7. Billingual version. Nullutslippsområder i smarte byer: Definisjon, nøkkelindikatører og vurderingskriterier," SINTEF akademisk forlag, Rapport2018.

[3] ZEB. (2018, 2018-05-30). *The Research Centre on Zero Emission Buildings*. Available: http://www.zeb.no

[4] M. K. Wiik, S. M. Fufa, T. Kristjansdottir, and I. Andresen, "Lessons learnt from embodied GHG emission calculations in zero emission buildings (ZEBs) from the Norwegian ZEB research centre," *Energy and Buildings,* vol. 165, pp. 25-34, 2018.

[5] C. Lausselet, V. Borgnes, and H. Brattebø, "LCA modelling for Zero Emission Neighbourhoods in early stage planning," *Building and Environment,* vol. 149, pp. 379-389, 2019.

[6] C. Skaar, N. Labonnote, and K. Gradeci, "From Zero Emission Buildings (ZEB) to Zero Emission Neighbourhoods (ZEN): A Mapping Review of Algorithm-Based LCA," *Sustainability,* vol. 10, p. 2405, 2018.

[7] C. Lausselet, V. Borgnes, L. A.-W. Ellingsen, A. H. Strømman, and H. Brattebø, "Life-cycle assessment methodology to assess Zero Emission Neighbourhood concept. A novel model," SINTEF akademisk forlag, Rapport2019.

[8] Fjerdrumsmoen et al., "Energiproduksjon - Tomtevurdering Nye Campus Grov estimat eksisterende Campus," Multiconsult ASA 10205530-RIEn -NOT-001 / 10205530-RIEn - NOT-002, 2018.

[9] A. Borg, P. Bernhard, H. Holmberg, and J. Tønnesen, "Potensial for lagring og styring av energiproduksjon og -forbruk - Campus Gløshaugen. Rapport delprosjekt 3.," Asplan Viak AS 618681-02, 2018.

[10] C. Solli, J. Vevatne, C. Skaar, I. Andresen, and T. Sørgjerd, "Premissdokument miljø - Vedlegg F.1 til OFP-rapport," Statsbygg2018.

[11] A. H. Collett, "NTNU har halvert bruk av fjernvarme," in *Universitetsavisa*, ed. Trondheim, 2014.

IOP Conf. Series: Earth and Environmental Science **352** (2019) 012026 doi:10.1088/1755-1315/352/1/012026

The Norwegian ZEB definition and lessons learnt from nine pilot zero emission building projects

Inger Andresen[a], Marianne Kjendseth Wiik[b], Selamawit Mamo Fufa[b], Arild Gustavsen[a]

[a] Norwegian University of Science and Technology (NTNU), 7491 Trondheim, Norway
[b] SINTEF Building and Infrastructure, Børrestuveien 3, 0373 Oslo, Norway

inger.andresen@ntnu.no

Abstract. In the frame of the Norwegian Research Centre on Zero Emission Buildings, nine real-life zero emission pilot building projects were initiated during the period of 2009 to 2017. Eight of the pilot building projects have been constructed and are currently in use, whilst one of them is still in the planning phase. All the pilot projects have been followed up by researchers during the design, construction, and operation phases, by e.g. analyses of construction documents, and post-occupancy surveys. The main lessons learnt from this work are presented in this paper. They include the importance of an integrated design process, having clear goals and associated assessment methods, following a strategy based on 'trias energetica', as well as choosing locally sourced materials with low embodied carbon and long service lives, and increasing the focus on the users.

1. Introduction

The Norwegian Research Centre on Zero Emission Buildings (the ZEB Centre) was established in 2009 with the aim to develop materials and solutions for new and existing buildings with net zero greenhouse gas (GHG) emissions over their lifetime (www.zeb.no). The centre was led by the Norwegian University of Science and Technology (NTNU) and SINTEF as research partners and encompassed 21 public and private partners from the Norwegian building and construction industry.

During the ZEB Centre, nine real-life zero emission pilot building projects were initiated in the period from 2009 to 2017. The pilot projects are listed in Table 1. Eight of the pilot building projects have been constructed and are currently in use, whilst one of them (Zero Village Bergen) is still in the planning phase. A summary of key data is also given in Table 1, and an illustration of each project is shown in Figure 1.

The pilot projects have been followed by researchers from the ZEB centre during the design, construction, and operation phases. Information has been gathered through participation in design workshops, analyses of construction documents, measurements, and interviews with participants in the design, construction, and operations phases.

This paper presents a summary of the main lessons learnt from the design, construction and operation of these zero emission pilot building projects. It will also discuss the definition of Zero Emission Buildings (ZEB), how the pilot projects have responded to the definition, and how the definition may be further applied in Norwegian building regulations.

Table 1. Key data from the zero emission pilot building projects.

HFA: Heated Floor Area
T_{amb}: annual mean ambient temperature, I_{hor}: annual total solar radiation on the horizontal
*: These projects did not compensate for all material emissions.

Name of building	Type of building and heated floor area	Location and weather data	ZEB ambition level (see section 2)
Powerhouse Kjørbo	Office building, renovation HFA: 5000 m^2	Sandvika, Norway T_{amb}: 5.9°C I_{hor}: 960 kWh/m^2	ZEB-COM÷EQ
ZEB House Multikomfort	Single family house, demonstration building HFA: 200 m^2	Larvik, Norway T_{amb}: 7.6°C I_{hor}: 974 kWh/m^2	ZEB-OM
Visund Haakonsvern	Office building HFA: 2000 m^2	Bergen, Norway T_{amb}: 7.5°C I_{hor}: 764 kWh/m^2	ZEB-O÷EQ
Skarpnes	5 single family houses HFA: 770 m^2 (total)	Arendal, Norway T_{amb}: 7.5°C I_{hor}: 934 kWh/m^2	ZEB-O
ZEB Living Lab	Single family house, research test facility	Trondheim, Norway T_{amb}: 5.1°C I_{hor}: 890 kWh/m^2	ZEB-OM*
Powerhouse Brattørkaia	Office building HFA: 14000 m^2	Trondheim, Norway T_{amb}: 5.1°C I_{hor}: 890 kWh/m^2	ZEB-COM÷EQ
Heimdal School	High school HFA: 25000 m^2	Trondheim, Norway T_{amb}: 5.1°C I_{hor}: 890 kWh/m^2	ZEB-OM*
Campus Evenstad	Educational building HFA: 1100 m^2	Evenstad, Norway T_{amb}: 4.7°C I_{hor}: 663 kWh/m^2	ZEB-COM
Zero Village Bergen	800 dwellings, terraced houses and blocks HFA: 80 000 m^2	Bergen, Norway T_{amb}: 7.5°C I_{hor}: 764 kWh/m^2	ZEB-O

2. The ZEB Definition

The ZEB definition builds upon the framework for net zero energy buildings defined by Sartori et al. [1]. In this definition, the term *Net ZEB* is used to refer to buildings that are connected to the energy utility infrastructure, and the wording 'Net' underlines the fact that there is a balance between energy taken from and supplied back to the energy grid over time. The Net ZEB balance is calculated as in Eq. (1):

$$\text{Net ZEB balance: } |\text{weighted supply}| - |\text{weighted demand}| = 0 \quad (1)$$

In a (net) zero *emission* building (ZEB), as defined by the Norwegian Research Centre on Zero Emission Buildings, the balance is measured in terms of greenhouse gas equivalent emissions during the lifetime of a building, instead of energy demand and generation. The greenhouse gas (GHG) emissions are calculated using CO_2 equivalent (CO_{2eq}) conversion factors for each energy carrier

($kgCO_{2eq}/kWh$) or building materials ($kgCO_{2eq}/m$, $kgCO_{2eq}/m^2$, $kgCO_{2eq}/m^3$, $kgCO_{2eq}/kg$ or $kgCO_{2eq}/pc$). According to the definition, a ZEB can be achieved by offsetting greenhouse gas (GHG) emissions from the entire life cycle of the building through the generation of onsite renewable energy [2]. The ZEB Centre's definition is very ambitious; therefore, a stepwise approach with different ambition levels has been developed to allow flexibility for different types of buildings and local boundary conditions. The main ambition levels applied for the pilot projects in the ZEB Centre are as follows, starting with the lowest [2]:

ZEB-O÷EQ: Emissions relating to all energy use for operation "O", except energy use for appliances/equipment (EQ), shall be compensated for with renewable energy generation.

ZEB-O: Emissions relating to all operational energy "O" shall be compensated for with renewable energy generation.

ZEB-OM: Emissions relating to all operational energy "O" plus embodied emissions from materials "M" shall be compensated for with renewable energy generation.

ZEB-COM: This is the same as ZEB-OM, but it also considers emissions relating to the construction "C" phase. The additional phases included are transport of materials and products to the building site and the construction installation processes.

Some of the zero emission pilot building projects have used a variation of these ambition levels, e.g. the two Powerhouse projects have included emissions from construction, but not emissions from the energy use of electrical appliances during the operation phase. The ambition level for the Powerhouse projects has therefore been set to ZEB-COM÷EQ.

Figure 1. The ZEB Pilot building projects. From upper left to lower right: Powerhouse Kjørbo, ZEB house Multikomfort, Visund Haakonsvern, Skarpnes, ZEB Living Lab, Powerhouse Brattørkaia, Heimdal School, Campus Evenstad, Zero Village Bergen.

3. Lessons learnt from the design process

3.1. Integrated design

All the zero emission pilot building projects have endeavoured to use an integrated design process, as described in [3]. Such a process involves establishing clear goals, employing multi-disciplinary cooperation from the early design stages, implementing a high level of energy integration and synergy of systems, and using modern performance prediction tools throughout the process to improve the environmental performance of a building.

Experiences show that having a well-defined goal for the environmental ambition is crucial for the success of zero emission building projects [3]. Moum et al [4] emphasise the importance of strong collaboration and involvement of stakeholders, through management style and the establishment of good meeting arenas. In the evaluated pilot projects, project participants seemed to have a shared understanding and acceptance of the purpose and need of a ZEB definition. The actors saw the possibility of learning and gaining new knowledge as highly useful for their own career and company competitiveness. There was a special focus on enabling good interdisciplinary collaboration for reaching the ZEB ambition.

To achieve a well-functioning ZEB, the different design concerns must be viewed as integral parts of the overall concept with all its differing requirements, e.g. minimized GHG emissions, excellent energy performance, high quality indoor and outdoor environments, and good architectural qualities. Each of the various technologies or strategies applied do not necessarily have to be new or innovative, but the whole concept will be. This was expressed as followed by one of the participants in the design of Powerhouse Kjørbo [5]:

It's not that we've invented new solutions; we've used known solutions [...]. It's mostly known solutions that have been combined and that have been dimensioned optimally. Instead of choosing installed systems in the design based on isolated performance criteria in each specialist area (e.g., HVAC, acoustics, electric installations), the approach emphasized the function of each part in view of the whole.

The underlying design strategy is based on the fact that passive energy design has a significant importance in achieving Zero Emission Buildings. A way of illustrating such a design strategy is through the *Trias Energetica* pyramid [6]. Trias Energetica is a three-step approach that gives priorities for realising an integrated sustainable energy solution, and consists of the following steps:

1) Reduce energy demand and related GHG emissions by applying energy reducing measures such as efficient building form, plan, orientation and facade layout, good thermal insulation, air tightness, and heat recovery. In a ZEB process, this step also includes designing for low GHG emissions from the construction and use of building materials by selecting local materials with low embodied emissions and long services lives. 2) Employ technologies for utilising local renewable energy sources, such as solar thermal systems and photovoltaics, heat pumps, biomass, waste and wind. In a ZEB process, this also includes properly integrating building services and renewable energy sources to avoid a double up of building materials, for example building integrated photovoltaics. 3) Finally, if there still is a need for some auxiliary energy, this takes place at the third level, using the least polluting fossil fuels in the most efficient way. In a ZEB process, one would of course seek to avoid the use of fossil fuels entirely.

The researchers of the ZEB Centre have developed a more detailed model consisting of 9 steps: 1) Location, orientation, and form, 2) daylight and sun, 3) material choices, 4) the building envelope – insulation and air tightness, 5) energy efficient lights and appliances, 6) efficient heating, ventilation, and air conditioning (HVAC) systems, 7) measurement and control, 8) renewable thermal energy, and 9) renewable electricity. However, the process of going through these steps should not be interpreted as a linear one. This is because the design process is so highly integrated. It is necessary to see the interrelations between the different passive and active measures and optimise the whole building concept. In the following, four of these aspects are addressed in light of the pilot building projects, namely energy reduction, efficient HVAC systems, local renewable energy production, and material use.

3.2. Reduce first

Across all of the zero emission pilot buildings, measures have been taken to reduce energy demand as much as possible. This involves implementing a well-insulated, air-tight building envelope, passive solar strategies, utilising daylight, and demand control. All of the pilot buildings are designed in accordance with the Norwegian passive house standard [7,8] or better. U-values for exterior walls are between 0.10 and 0.15 W/(m^{2}K), whilst roofs have typically lower U-values. U-values for windows are lower than 0.8 W/(m^{2}K). There has also been a strong emphasis on achieving continuous air and moisture barriers in the building envelopes. In places where the air and moisture barriers are perforated by ducts or pipes, cuffs and/or tape are used to avoid leakages. The air leakage number, n_{50}, has been designed to be below 0.6 air changes per hour (ACH) for all of the pilot buildings. Measurements with blower doors tests show that all projects except one (Campus Evenstad), achieved this goal. One project (Visund Haakonsvern) even obtained a value of 0.1 ACH, which is extremely low.

3.3. Efficient HVAC systems

All projects have focused on reducing the installation and running costs of HVAC systems, by employing efficient energy solutions. This is made possible by utilising passive and active measures to reduce the need for heating and cooling, such as solar shading, thermal mass, demand control, and installing glazing with low U-values. The heat distribution system may be much more compact than in a conventional building if the facades are very well insulated and the heating load is low. For example, in the ZEB residential buildings, there is only one radiator per floor, centrally located near the technical room, which allows for shorter distribution pipes and less technical installations. The energy use for fans is minimised through designing systems with low pressure drops, and includes demand control, short and spacious distribution channels, partly relying on using secondary spaces such as corridors as return air channels and taking advantage of thermal buoyancy when extracting air. Ventilation air volumes are reduced by activating thermal mass to avoid overheating, and by using low-emitting materials to minimise indoor air pollutants. Specific fan power (SFP) factors vary between 0.5 and 1.5 kW/(m^{3}s) as design values across the ZEB pilot projects.

3.4. Utilising local renewable thermal energy and electricity

A wide range of on-site renewable energy systems were considered for the pilot projects, including different kinds of heat pumps, solar thermal collector systems, bio-energy systems, photovoltaic systems and wind power installations. In most cases, heat pumps with ground or sea water source systems were selected to cover 80-100% of the heating and cooling loads, due to their high cost-efficiency. Seven of the ZEB pilot buildings have heat pump systems. Two of the pilot projects, the ZEB House Multikomfort and ZEB Living Lab, have solar thermal systems in addition to heat pumps, mainly for demonstration and testing purposes. The building at Campus Evenstad is connected to a combined heat and power (CHP) plant based on gasification of wood chips, while Heimdal school uses a CHP system based on biogas. The ZEB House Multikomfort also has a grey water heat recovery system. Where district heating is available, this is used for auxiliary heating. Wind power was not selected in any of the projects because of low energy yields due to insufficient wind speeds on site. In all except one of the pilot projects (Campus Evenstad), photovoltaic systems have been chosen. All are based on crystalline silicon technologies with module efficiencies ranging from 16 to 21%. All are grid connected, and one has battery storage for electric car charging (ZEB House Multikomfort). The annual output of the PV systems has been calculated to range from 113 to149 kWh per m^2 of PV module area [9].

3.5. Materials matter

Research from the ZEB Centre shows that the relative share of embodied GHG emissions in construction materials and technical installations may be significant for buildings where the operational energy demand has already been minimised [10], and that the production and replacement of building materials may contribute between 55-85% of total emissions cross the Norwegian ZEB pilot projects. In the zero emission pilot building projects, there have been efforts to reduce embodied GHG emissions in construction materials and in the construction phase.

The most efficient design strategies and material choices for achieving low embodied emissions identified through the pilot projects were; area and material reduction, application of reused and recycled materials, using materials with low embodied carbon, sourcing local materials, and adopting materials with high durability and a long service life [11, 12]. The ZEB case studies highlight that it may be challenging to keep up the focus on embodied emission reduction during a complex project process. This is because decisions regarding design and material alternatives are based on many criteria including technical properties such as load bearing capacity, fire safety, durability and sound proofing properties; as well as data availability, cost and time issues [12]. Challenges during the project also included unforeseen changes in the design and construction phases, such as unexpected ground conditions or new design requirements.

4. Lessons learnt from the construction process

Several of the issues addressed in section 3.1 on the integrated design process, may also be transferred to the construction process. Throndsen et al [5] conclude that in building projects with a high degree of innovation and development, it is extra important to focus on good communication between the design team and the executing part on the building site. In the ZEB pilot projects, there was often a need to adjust and improve solutions directly on the building site. This required an understanding of the intentions behind construction drawings. One criterion for success was found to be having enough time to go through all the challenges and issues with the construction manager and sub-contractors [4].

Participants in the pilot projects also agreed that "closeness", in terms of frequent communication both face-to-face and by other means, between the central and defining actors in the project team was a key to success [4]. Failures increased once the distances between actors increased. A problem that appeared was a disbelief shared by the subcontractors when they first became acquainted with the functional demands and the specifications from the design process - a process they knew little about. According to respondents, one important role for a process manager in the later stages could have been to liaise between the project management level (construction manager and contractor) and the subcontractors.

Some of the participants in the pilot projects also mentioned that contract frameworks could be a contributing source of errors, as turnkey contract frameworks typically dis-incentivised subcontractors from making order changes, also after they were clearly deemed necessary by consulting engineers [4]. Moum et al [4] suggest that a more collaborative framework could have been applied to allow subcontractors to identify and share the risk taking with the main contractors and project owner. They further propose that one way to incentivise sub-contractors to join in more time-consuming contract arrangements could be to highlight the skills and knowledge development benefits available for participants in such projects.

One of the pilot projects, Campus Evenstad, was closely followed up by the researchers with respect to documentation of GHG emission in the construction phase [13]. Following up on this, ZEB researchers have investigated these themes further to ascertain measures for emission reduction. Some of these measures include using biofuels instead of fossil fuels in construction machinery, avoiding foundations and groundworks during the cold season, utilising district heating for drying and heating of construction works, and electrifying construction site machinery that would otherwise use fossil fuels [12,13].

5. Lessons learnt from the operation phase

5.1. Energy performance – measurement versus predicted

Three of the pilot building projects have been in operation long enough to achieve at least one year of actual energy measurements, namely; Powerhouse Kjørbo, Visund, and Skarpnes. The measurements show that overall, the total energy performance of the buildings is harmonious with the predicted performance [14,15,16,17]. However, in the office building projects, the measured energy use for lighting was found to be somewhat higher than predicted values, whilst in the residential buildings, the heating energy use is higher than calculated. On the other hand, the measured output from the PV systems was higher than predicted across all three projects.

5.2. Indoor environment
The above-mentioned pilot buildings have also been evaluated with respect to indoor environment. After some running-in problems of some technical systems were resolved, the user evaluations showed that the buildings performed very well [4,5,14,15,16,17]. Also, physical measurements of indoor air temperatures and CO_2-levels, showed values that were within established comfort standards [14-17].

5.3. Don't forget the users
Berker [18] identifies three main practical/user issues that are central to address in ZEBs: 1) The level of end-user control, 2) the level of complexity of systems, and 3) the need for information about correct use. Users are much less satisfied when they cannot understand how things work or are unable to control temperature and ventilation. Moreover, the perception of personal control over an environment increases satisfaction. How much control the user gets over his or her environment is much more relevant in a building that is finely tuned to achieve an ambitious overall energy use. Furthermore, very energy efficient buildings tend to be more complex than traditional ones, and this makes it more difficult for users to make informed decisions about the operation. Thus, there is a need for good information about the use of the building. Such information, Berker claims, should ideally be given repeatedly and adapted to the users' needs.

6. Conclusions
The ZEB Centre's experiences with the pilot buildings show that ZEBs can be realised in Norway for different building typologies and locations. The ZEB definition may advance a Norwegian ZEB standard, where common traits include low-energy, well insulated, timber-framed buildings utilizing passive strategies with high-efficient HVAC systems, and local renewable energy systems. Realizing such buildings does not require new or unknown technologies, but the combination of known strategies, materials and technologies in new and smarter ways. Also, it requires increased investment in time and skills in the early design phases, new contracts for sharing the risk, and more consideration of users.

When considering Norwegian building regulations, a range of statutory frameworks and tools exist for aiding building professionals in achieving a ZEB-O ambition level. For example, the Norwegian technical specification SN/TS 3031 [19] provides energy performance calculation methodologies, whilst the Norwegian passive house standards NS 3700 [7] and NS 3701 [8] provide criteria to ensure buildings with low energy demand. However, more focus should be given to the other ZEB ambition levels, such as the materials in ZEB-OM and the construction processes in ZEB-COM. The new Norwegian standard for greenhouse gas calculations of buildings, NS 3720 [20], is a good starting point for evaluating the effect of material choices. Also, guidelines for zero emission construction sites have been developed based partly on the experiences from the ZEB pilot projects [21]. All together, these experiences and guidelines pave the way for the advancement of Zero Emission Buildings.

References
[1] Sartori I, Napolitano A, Voss K 2012 Net Zero Energy Buildings: A consistent definition framework *Energ and Buildings* **48** pp 220-32
[2] Fufa SM Schlanbusch RD Sørnes K Inman M Andresen I 2016 A Norwegian ZEB Definition Guideline *ZEB report 29* (Trondheim: SINTEF Academic Press)
[3] Andresen I, Hegli T 2017 Integrated Energy Design, in *Zero Emission Buildings*, ed AG Hestnes and NL Eik-Nes (Bergen: Fagbokforlaget)
[4] Moum A, Hauge LÅ, Thomsen J 2017 Four Norwegian Zero Emission Pilot Buildings – Building Process and User Evaluation *ZEB report 30* (Trondheim: SINTEF Academic Press)

[5] Throndsen W, Berker T, Knoll E 2016 Powerhouse Kjørbo. Evaluation of Construction Process and Early Use Phase *ZEB report 25* (Trondheim: SINTEF Academic Press)
[6] Lysen, EH 1996 The Trias Energetica – Solar energy strategies for developing countries *Proceedings of the Eurosun'96 conference* (München: DGS-Sonnenenergie)
[7] NS 3700: 2013 Criteria for passive houses and low energy buildings - Residential buildings (in Norwegian) (Oslo: Standards Norway)
[8] NS 3701: 2012 Criteria for passive houses and low energy buildings - Non-residential buildings (in Norwegian) (Oslo: Standards Norway)
[9] Andresen I 2017 The ZEB pilot buildings, in *Zero Emission Buildings,* ed AG Hestnes and NL Eik-Nes (Bergen: Fagbokforlaget)
[10] Wiik MK, Fufa SM, Kristjansdottir T, Andresen I 2018 Lessons learnt from embodied GHG emission calculations in zero emission buildings (ZEBs) from the Norwegian ZEB research centre *Energ and Buildings* **165** pp 25–34
[11] Fufa SM et al. 2017 The influence of estimated service life on the embodied emissions of zero emission buildings (ZEBs) when choosing low-carbon building products *Proceedings of XIV DBMC Conference* (Rilem: Ghent) p 365
[12] Wiik M.K et al. 2018 Design Strategies For Low Embodied Carbon In Building Materials. In: Pomoponi, F et al (eds) *Embodied Carbon in Buildings: Measurement, Management, and Mitigation* (Springer, Cham) pp 323-39
[13] Wiik MK et al 2017 ZEB Pilot Campus Evenstad, administration and educational building. As-built report. *ZEB report 36* (Trondheim: SINTEF Academic Press)
[14] Sørensen Å, Mysen M, Andresen I, Harklau B, Lunde A 2017 The Visund office building, Haakonsvern, Bergen. As Built Report. *ZEB report 38* (Trondheim: SINTEF Academic Press)
[15] Sørensen Å, Mysen M, Andresen I, Harklau B, Lunde A 2017 Zero emission office building in Bergen: Experiences from first year of operation *Energy Procedia* **132** pp 180-85
[16] Sørensen Å, Andresen I 2017. Powerhouse Kjørbo. As Built Report *ZEB report 35* (Trondheim: SINTEF Academic Press)
[17] Sørensen Å, Imenes AG, Grynning S, Dokka TH 2017 Energy measurements at Skarpnes zero energy homes in Southern Norway: Do the loads match up with the on-site energy production? *Energy Procedia* **132** pp 567-73
[18] Berker T 2018 From potential to performance – people matter, in *Zero Emission Buildings*, ed AG Hestnes and NL Eik-Nes (Bergen: Fagbokforlaget)
[19] SN/TS 3031: 2016 Energy Performance of buildings – Calculation of energy demand and supply (in Norwegian) (Oslo: Standards Norway)
[20] NS 3720:20018 Methods for greenhouse gas calculations for buildings (Oslo: Standards Norway)
[21] Fufa, SM Emission free construction sites. State-of-the-art. Guideline for innovative procurement processes (in Norwegian) *SINTEF Fag 49* (Oslo: SINTEF Building and Infrastructure).

Acknowledgements
This research has received funding from the Research Centre for Zero Emission Buildings. The authors are thankful for the support from ZEB partners and the Research Council of Norway.

IOP Conf. Series: Earth and Environmental Science **352** (2019) 012027 doi:10.1088/1755-1315/352/1/012027

Defrosting and Frost Protection Measurements of Heat Recovery in Cold Climate

D Johansson, Å Wahlström, M Dahlblom

Building Services Engineering, Lund University, Lund Sweden
dennis.johansson@hvac.lth.se, asa.wahlstrom@hvac.lth.se

Abstract. The installation of technical solutions for heat recovery from the ventilation air is a very important step, both with respect to improving the indoor climate and as an energy-saving measure. Ventilation without heat recovery use a substantial part of an entire building's heating use, especially if the building is a low energy building. To realize a large implementation of heat recovery property owners must increase their knowledge about how these systems work and they must be able to rely on them functioning properly. The function and performance of a ventilation system when it comes to heat recovery can be a problem as soon as temperatures drop far below zero. When freezing occurs some form of defrosting is required and this can lead to indoor climate problems and to little or no heat recovery being obtained, which will reduce the energy efficiency and, at the same time, may compromise moisture safety. The goal of this project is to increase knowledge about how heat recovery systems for residential buildings with respect to defrosting and frost protection measures in heat exchangers. The knowledge can thereafter be used to develop effective defrosting strategies and frost protection measures so that the energy use and power requirements are minimized. Measurements were performed for heat recovery systems placed in dwellings in north Sweden to investigate occurrence of frost formation. The paper will present these measurements and some comparing simulations which can serve as a basis for further development of good defrosting strategies.

1. Introduction

Ventilation without heat recovery use a substantial part of an entire building's heating use, especially if the building is a low energy building. Typical ventilation losses without heat recovery are at least 43 kWh/(m^2·year) in Malmö, southern Sweden, and 78 kWh/(m^2·year) in Karesuando, northern Sweden. Therefore, heat recovery ventilation is a well-known, economical technology whereby heat can be recovered from the extract air [1,2]. In practice, an efficiency of the magnitude of 80 % is used, although it can be questioned whether this is an optimal figure. Mechanical extract-only ventilation systems are common in residential buildings and, even if it were possible, by the addition of a heat pump to recover their heat energy, the problems with these systems include draughts, uncontrolled air movement patterns and electricity-dependent heat recovery units [2,3]. Heat recovery, by installing mechanical supply and extract ventilation systems, is therefore advantageous from many perspectives.

Heat recovery in ventilation systems is usually accomplished by installing either a plate heat exchanger with a cross- or counter-flow set up, or a thermal rotary wheel. Both types have been used extensively and successfully in office buildings for decades but in residential buildings, as mentioned above, this is a rather new phenomenon. Non-residential buildings normally have a higher air change rate and moisture is seldom a problem, as is often the case in dwellings where specific processes like showering and cooking generate a lot of moisture [4]. This means that non-residential buildings do not

IOP Conf. Series: Earth and Environmental Science **352** (2019) 012027 doi:10.1088/1755-1315/352/1/012027

have the same problems with frost and there is little experience of the problems that can occur due to freezing.

A 'public procurement for innovation' process for systems for heat recovery from ventilation air in existing apartment blocks has recently been carried out in Sweden [5]. The procurement was successful in as much as it has placed great focus on the issue and raised awareness in the property industry with respect to the large opportunities that arise by installing heat recovery systems. There was, however, a problem with the Swedish 'public procurement for innovation' process as it could not show results that came anywhere near to the theoretical figures, which to a great extent depended on imbalances in airflows and inappropriate defrosting measures. In order to attain the desired energy savings and avoid problems with frost, a greater understanding of how the whole ventilation system works together with the building and its occupants is required. Even if programs exist to simulate buildings and installation systems, methods must be developed to properly include defrosting and frost in combination with user behaviour.

Present methods and tools have not been adapted to actual conditions prevailing in buildings and this can lead to sub-optimisation. Nor are there any guidelines about how control and regulation should be carried out so that ventilation units can be designed with high energy efficiencies and offer good indoor comfort. If there were such a method development, it would be possible to more precisely predict the energy efficiency and this is important input data for energy calculations and, in turn, it would make it possible for the design energy use to correspond with the measured quantities. The same applies to design and actual indoor environments, and to moisture safety issues.

Temperature efficiency is negatively influenced by frost and, as frost occurs when it is cold outdoors, this leads to a poorly functioning heat recovery process when it is needed the most. This means that a considerably lower degree of energy efficiency is obtained as well as a higher energy use at the same time as the indoor climate is affected negatively.

The Swedish 'public procurement for innovation' process has shown that it is desirable to continue to develop heat recovery technology so that post-heaters can be avoided in residential buildings, at least in the south of Sweden. This will mean cheaper installations, not least important for renovation work, but defrosting and the avoidance of frost will be the deciding factors for making progress.

Correspondence between methods used in practice and existing freezing protection measures, and an optimal way to handle freezing issues seems to be totally lacking. Svensk Ventilation (The Swedish Association of Air Handling Industries) has formulated a definition of energy efficiency so that it can be calculated while taking defrosting into account [6]. However, a more holistic method is required that can handle a building as a system and make it possible to create better and more predictable designs, as well as innovations and developments that will improve energy efficiency and ensure acceptable indoor climates. With regard to temperature efficiency, thermal wheels have higher efficiencies than plate heat exchanger units of the same volume, especially at conditions where freezing occurs, and they are easily adjustable, but do incorporate more technology and moving parts.

In practice, freezing is dealt with in some more or less effective ways. Sometimes, in small units with thermal wheels, it is not dealt with at all, resulting in problems. Frost protection or defrosting strategies include:

- Blocking or turning off the supply air. This will melt the ice formation, which means that supply air will enter through leaks, which causes draughts and poor indoor climate and, in low-energy and air-tight housing, this is a non-working solution.
- Bypassing the supply air on the outside of the heat exchanger in the heat recovery unit. This gives low supply air temperatures and draughts, and consequently poor indoor climate.
- Pre-heating supply air or heating the extract air.
- Switching airflow direction or switching between serial heat exchangers. This has been tested but has been difficult to make reliable [7].

Using the first three methods, the degree of heat recovery is low as well as having the mentioned problems. As the manufacturer often has poor knowledge about how and when problems arise, defrosting is often done unnecessarily and with a too great a margin of safety. More knowledge is

IOP Conf. Series: Earth and Environmental Science **352** (2019) 012027 doi:10.1088/1755-1315/352/1/012027

required here regarding under which conditions condensation or freezing occurs and how these processes could be practically controlled, for example, by measuring the moisture content or pressure drop in the air handling unit or pre heat from ground loops etc. In order to take this into account and design a suitable method for defrosting, methods are needed that can determine when defrosting is necessary with respect to different moisture contents and temperature loads caused by the occupants. It is important to gain this knowledge so that problems can be anticipated and avoided.

At present there is no method that describes how the defrosting function or frost protection measure should be designed to provide the best possible energy recovery at the same time as moisture safety and indoor environment are taken into account.

1.1. Aim

This paper presents measurements done in 6 air handling units in residential buildings in cold climates as a part of the bigger project to increase knowledge about how heat recovery systems for residential buildings can be improved with respect to defrosting and frost protection measures in heat exchangers, so that the energy use and power requirements are minimized. The measurements show temperature efficiencies obtained in reality and try to explain those. In addition, the measurements were put in the perspective of simulations based on a typical residential building, occupants and units presents the theoretical influence from frosting in the heat recovery of the air handling unit.

2. Methods

Table 1 shows the nomenclature used in this paper. Measurements were done in 6 different air handling units in six residential buildings to cover a few different types of air handling units. The measurements were done in northern Sweden to increase the number of hours with cold climate. Figure 1 shows the measured outdoor temperature frequencies at the measurement locations together with standard outdoor climate [8]. The measured units are given in Table 2.

Table 1. Nomenclature used.

Abr.	Quantity	Unit	Abr.	Quantity	Unit
Tout	Temp outdoor	°C	Tex	Exhaust air temp	°C
Tvvx	Temp after heat recovery	°C	ηs	Temp eff of sup of heat rec	
Ts	Supply air temp	°C	ηe	Temp eff of ex of heat rec	
Tet	Extract air temp	°C	Texs	Sat temp of exhaust air	°C

Table 2. Measured heat recovery units.

Abr	City	Heat exchanger	Size	Notes
A	Kiruna	Coil	Apartment block	
B	Kiruna	Rotor	Apartment	Not possible to measure Tvvx
C	Kiruna	Plate	Apartment block	
D	Malmberget	Rotor	4 apartments	
E	Umeå	Plate	Apartment block	
F	Umeå	Plate	Apartment block	

The measurements were done with loggers for temperature and relative humidity with an accuracy on temperature of ±0.25°C and relative humidity of ±2.5% [9]. Temperatures before and after the air handling unit in both exhaust and supply stream, and the point between the heat recovery and auxiliary heat coil were measured every 10 minutes. The measurements were done during the winter of 2012 in Umeå and 2018 and 2019 in Kiruna and Malmberget. When the exhaust temperature decrease to the freezing point and condensation occurs, the problem with frost starts. If it is a rotor, this can be melted or sublimates when the rotor part is turned to warmer air meaning that the rotor handles frosting at lower exhaust temperatures. The temperature efficiency is defined for supply and exhaust airstreams respectively according to Equation 1 and 2.

$$\eta s = \frac{Tvvx-Tout}{Tet-Tout} \quad (1)$$

$$\eta e = \frac{Tet-Tex}{Tet-Tout} \quad (2)$$

The simulations were performed based on a constant Tsa of 20°C, a constant Tet of 22°C, corresponding to a room temperature close to 22°C, and different constant moisture supplies, which is defined as the difference between indoor and outdoor vapour content of 1, 2 and 4 g/m³. These figures were chosen since previous measurements in 325 apartments had an average moisture supply of approximately 2.2 g/m³ in wintertime [10,11,12,13,14,15]. Two typical defrosting strategies were checked for, Tex<0°C and the combination of Tex<0°C and a condensing situation in the exhaust air, which means Tex<Texs, with a continuously controllable heat recovery. It was assumed that the temperature over the heat recovery unit outlet was constant. Number of defrosting hours and number of degree hours for heating the supply air to Ts were simulated as a function of $\eta = \eta_s = \eta_e$. Energy for heating the supply air is the degree hours multiplied by airflow, air density and air specific heat capacity.

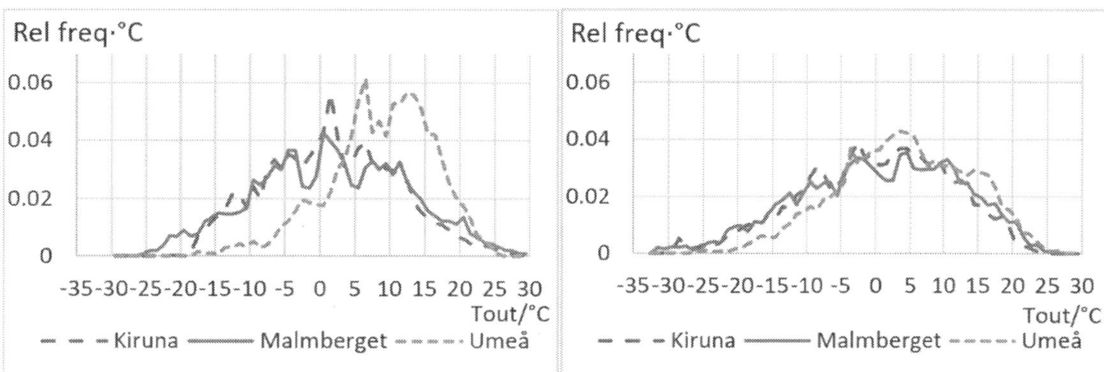

Figure 1. Relative frequency of the outdoor climate during the measurements to the left and during a normal year to the right.

3. Results and Discussion

Figure 2 shows all temperature efficiencies as function of time. For unit B, η_s cannot be calculated. For unit A there was a breakdown of the loggers in late spring 2018 and a breakdown of probably the pump of the coil system in autumn 2018. In summer time the weather was extraordinary warm and with those temperatures, the denominator in the equation 1 and 2 can be very close to zero. For unit B there was a summer period where it lowers the temperature efficiency accordingly to control the supply air temperature. And in the end of the measurement period there was a frost protection induced decrease in efficiency. Unit C had a much lower η_e compared to η_s, which depends on a much higher exhaust than supply airflow. Freezing was not indicated. This design with much larger exhaust flows was also used in the public procurement, Exhaust air was collected also from the basement while air was only supplied into the apartments. The design has the benefit that freezing is avoided but the total energy efficiency will be less since the supply air to the basement is not using any heat recovery.

Unit D had reasonably high temperature efficiency and has a reasonable balance between supply and exhaust airflow. However, it is important that there is an under pressure in the building for moisture safety reasons. Here the rotor can be controlled continuously to achieve the desired supply air temperature. In the end of the period there was a decrease in efficiency due to frost, and from the landlord it was clear that frost protection did not work. They had to switch it off and heat the unit by an electrical heater to melt it and restart it repeatedly. The two units in Umeå, E and F, showed no

IOP Conf. Series: Earth and Environmental Science **352** (2019) 012027 doi:10.1088/1755-1315/352/1/012027

frosting indication, however there was only a few hour with low outdoor temperatures. The moisture supply as the indoor vapour content minus outdoor vapour content was also low in buildings of units C and D.

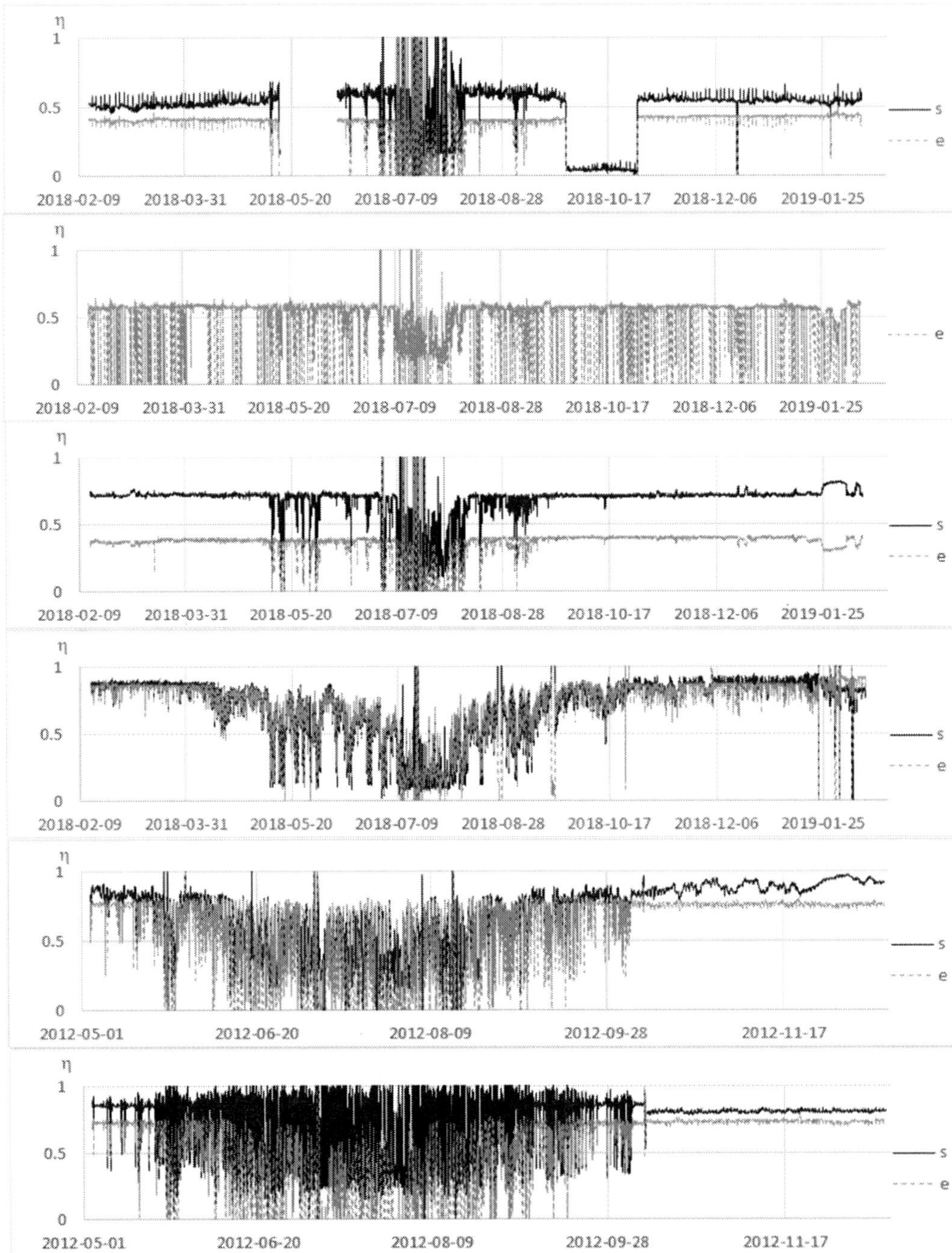

Figure 2. Heat recovery temperature efficiencies of the 6 units from A above to F below as a function of time.

Figure 3 shows η_e as a function of Tout for the measured units. It can be expected that if frosting occurs, whether there is a working frost protection or not, the temperature efficiency would decrease, particularly based on the exhaust side. If the frost protection works by decreasing the supply air flow in any way, it would not be possible to see it on ηs, but still on η_e. In these graphs, every measurement point is a dot, and it is not possible to distinguish many dots at the same location in the graph, so they do not show frequencies. Unit A and C had low temperature efficiencies and the decrease at low temperatures for Unit C has its explanation in a higher efficiency in ηs meaning that the airflow imbalance is even higher for some reason. For Unit D there was a small tendency of decreasing efficiencies at low outdoor temperatures. Comparisons between ηs and ηe as a function of exhaust temperature should also indicate a lower efficiency at a lower temperature. This can only to a small amount be indicated for Unit D despite it has a rotor. Even if Unit B has a number of points at low efficiencies, that number of points is not very high. To be able to give a more quantitative result, Figure 4 shows the duration of the temperature efficiencies together with the needed added heating to the supply air between Tvvx and Ts, noted dt. Low temperature efficiencies occur but that is when it is summer climate and heat recovery is not needed.

Figure 5 shows duration of the exhaust temperature and exhaust saturation temperature which indicates that the exhaust temperature is below freezing very seldom in the case of plate exchanger or coil units, A, C, E and F. In those, the saturation temperature is lower than the exhaust temperature indicating that both saturation temperature and subzero temperatures should be good to detect. The constant η_e (see Figure 4) indicates that condensation and freezing is a rare situation in the plate exchanger units. In the rotary units, with higher η, freezing will occur at lower Tex than 0°C

As a comparison with balanced airflows in both air streams, the simulation results are shown in Figure 6. An occupation with a moisture supply of 4 g/m³ will require a control strategy for detecting defrosting need close to just measure sub-zero temperatures in the exhaust air compared to detecting actual frost from condensation. Still 29 k°C·h is needed to heat the air with η close to 100% compared to 174 k°C·h without heat recovery. 1 g/m³ will clearly lower the influence from frost. No added moisture at all would not give any defrosting need.

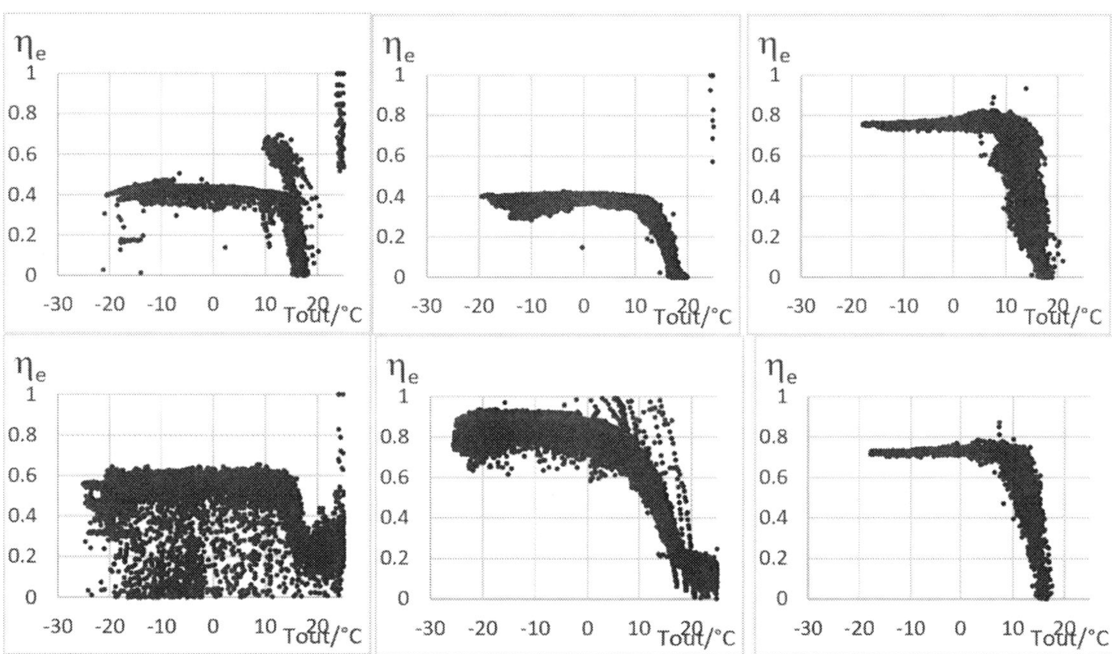

Figure 3. Exhaust based temperature efficiency as a function of outdoor temperature for the units A, C, E in the upper row and B, D, F in the lower row.

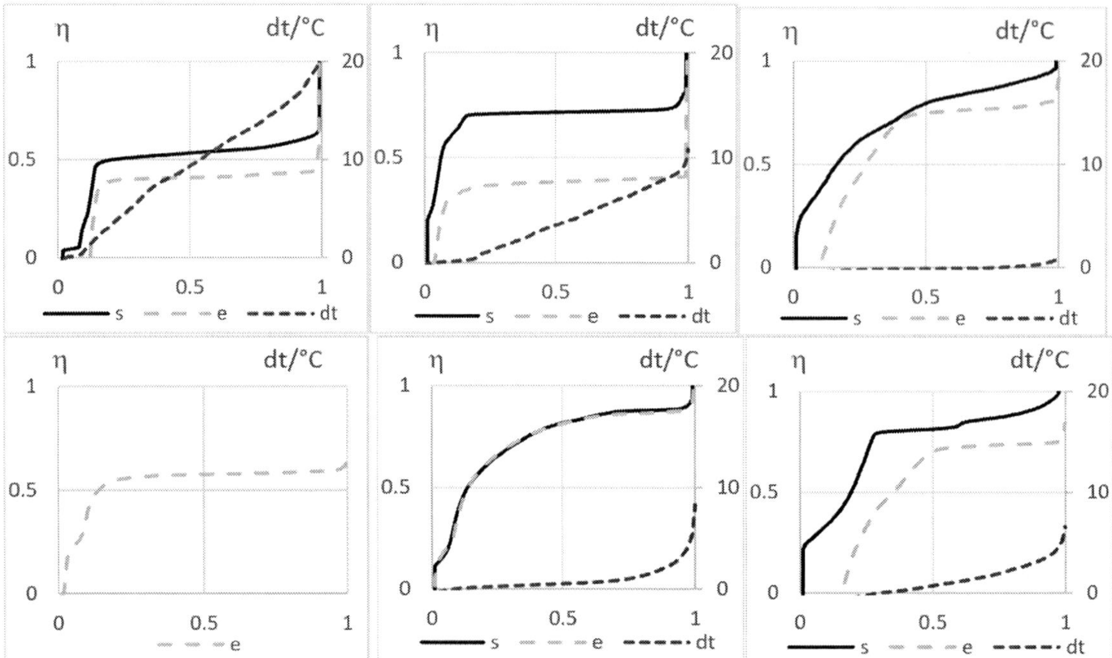

Figure 4. Duration of temperature efficiencies and added temperature, dt = ts – tvvx, for the units A, C, E in the upper row and B, D, F in the lower row. All x-axes show ratio of time below value.

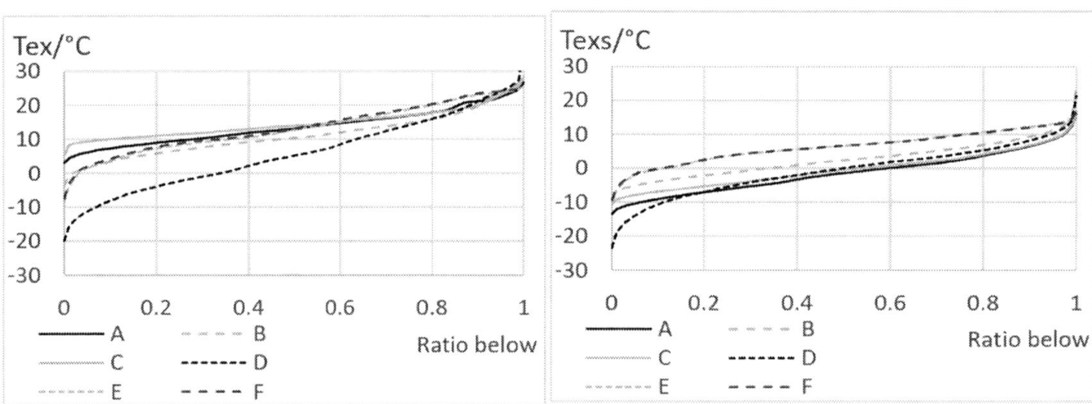

Figure 5. Duration of exhaust temperature, left, and exhaust saturation temperature, right.

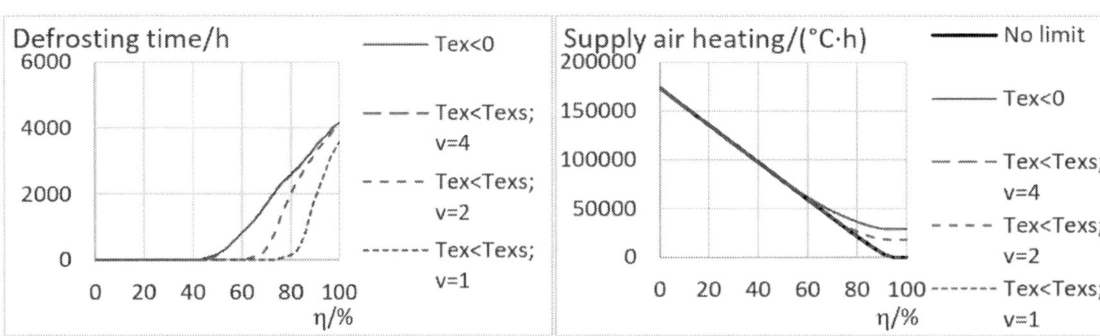

Figure 6. Simulated time of defrosting need and supply heat to reach Ts = 20° at Tex = 22°C if defrosting is made when Tex<0°C, or (Tex<Texs)&(Tex<0°C) respectively. v is moisture supply = indoor vapour content minus outdoor vapour content in the unit g/m³.

IOP Conf. Series: Earth and Environmental Science **352** (2019) 012027 doi:10.1088/1755-1315/352/1/012027

4. Conclusion

The analyzed temperature efficiencies in the measured, only a few years old, units were found to be rather low, particularly for non rotary heat exchangers. Imbalance or elsewise low efficiencies led to the fact that not much frosting conditions occurred in the measurements, despite one of the coldest Swedish cities. It means that sorting out other problems than frosting may be first priority before dealing with frosting.

There was frosting problems measured in the rotary units, and part of the reason is higher temperature efficiency and therefore lower exhaust temperatures, but also that they served fewer dwellings meaning that the variation in moisture supply most probably were higher.

The simulations show the potential for frosting problems if the balance of the airflows is more proper and the temperature efficiency is high. If the moisture supply is high, defrosting can be detected by only measuring the exhaust temperature, but future simulations with varying moisture supply over time including varying temperatures over the heat recovery will give more accurate limit values.

Future research will use these measurements to describe moisture loads and indoor temperatures from residential buildings and based on that test and simulate air handling units with high temperature efficiency set up properly so exhaust air is freezing. Also, measurements will be performed on new, high performing units to help with future recommendations on defrosting strategies that is optimal taking into account both energy and economy.

References

[1] Nilsson, P. E., editor 2003, Achieving the desired indoor climate, Studentlitteratur, Lund, Sweden

[2] Johansson, D. 2005, Modelling Life Cycle Cost for Indoor Climate Systems, Lund University, TVBH-1014, Doctoral thesis

[3] Engdahl, F. 2002, Air - for Health and Comfort – An anlysis of HVAC Systems' Performance in Theory and Practice, Lund University, Sweden, Report TVBH-1013, Doctoral thesis

[4] Nevander, L.E., Elmarsson, B. 1994, Fukthandbok – Praktik och teori, Svensk Byggtjänst, Stockholm, Sweden, in Swedish.

[5] Wahlström, Å. 2014, Teknikupphandling av värmeåtervinningssystem i befintliga flerbostadshus - utvärdering, BeBo-rapport, www.bebostad.se

[6] Svensk Ventilation, 2013, Årsverkningsgrad för värmeåtervinning med luft-luftvärmeväxlare. Riktlinjer för redovisning av produktdata.

[7] Vladykova, P. 2011, An energy efficient building for the Arctic climate Is a passive house sensible solution for Greenland?, doctoral thesis, DTU, Copenhagen, Denmark, R-243

[8] Meteotest 2019, Meteonorm handbook, manual and theoretical background, Switzerland, http://www.meteonorm.com/

[9] ONSET, 2018. HOBO ® MX2101 Data Logger, https://www.onsetcomp.com/

[10] Bagge, H. 2011, Building Performance – Methods for improved prediction and verification of energy use and indoor climate, Lund University, Sweden, TVBH—1019, Doctoral thesis

[11] Bagge, H., Johansson, D., Lindstrii, L. 2012, Measured Indoor Hygrothermal Conditions and Occupancy Levels in an Arctic Swedish Multi-family Building, Proceedings of Cold Climate HVAC conference 2012, Calgary

[12] Bagge, H., Johansson, D., Lindstrii, L., 2010, Indoor Hygrothermal Conditions in Multi-family Dwellings – Measurements and Analysis, Proc. of Building Envelope, Clearwater Beach

[13] Bagge, H., Johansson, D. 2013, Indoor Moisture Conditions in Multifamily Dwellings: Measurements and Analysis, ASHRAE Indoor Air Conference, Vancouver

[14] Johansson, D., Bagge, H., Lindstrii, L. 2012, Measured indoor temperature, moisture supply occupancy and mixed gas concentrations in apartment buildings – correlation between parameters, proceedings of Healthy Building Conference, Brisbane

[15] Johansson, D., Bagge, H., Lindstrii, L. 2013, Indoor Temperature – Analysis of Two Years of Measurements in 1300 Apartments in Sweden, Clima Conference, Prague

IOP Conf. Series: Earth and Environmental Science **352** (2019) 012028 doi:10.1088/1755-1315/352/1/012028

Towards low carbon economy – green bond and asset development

S Nenonen[1], A Koski[2], AP Lassila[2] and S Lehikoinen[2]

[1]Faculty of Built Environment, Tampere University, Finland,
[2]University Properties of Finland Ltd, Finland

suvi.nenonen@tuni.fi

Abstract. Green Bonds are an instrument for driving the environmentally friendly and low-carbon economy. Green Bonds are bonds whose proceeds are earmarked for and transparently channelled to environmentally-friendly projects and activities. The real estate industry has a multi-decade track record of addressing environmental impacts through the use of rating systems certified by independent third parties. Green building certification systems address multiple environmental impacts and measure outcomes across all asset lifecycle phases. Using bonds for such investments is not new but in an effort to improve transparency and increase opportunities both for issuers and investors it can take a more active role in combating climate change. The goal of this paper is to describe how property developer can use Green Bond as one instrument in sustainable life cycle management and continuous development of properties. The method used is a case study of Finnish property owner company, which commits to invest the funds raised in certified, environmentally responsible and energy-efficient projects. The single case study method employed in this study captured the process of case organization towards Green Bond initiative. More precisely the data was gathered by qualitative document analysis (QDA). The results show that company begin the process with focusing on environmental sustainability especially putting the effort in the first phase to energy efficiency. The Green Bond initiative provided a new avenue towards economic sustainability. Additionally, issues like shared use of facilities was discussed from social sustainability perspective. The results are interesting for property owners who are interested in systematic development towards regenerative built environment.

1. Introduction

Institutions of higher education have long-served society as leaders and innovators in research, discovery, and social responsibility. Now, they are assuming a leadership role in environmental responsibility, sustainability, and campus sustainability in particular. [1] The studies about universities' sustainability have indicated that there are many factors, such as psychological needs, physical facilities, personal motivations, public perception, price mechanisms and policies, for engaging the campus community in sustainability. [2] Campus asset management intends to use the proceeds of the bond to develop campus in a way that delivers positive environmental and social outcomes.

In more general level the underlying economic conditions and the need for growth, due to the growing population, have to include environmentally sustainable policies in order to address the problem in accordance with a healthy environment. In an effort to face the needs of our society we

IOP Conf. Series: Earth and Environmental Science **352** (2019) 012028 doi:10.1088/1755-1315/352/1/012028

have to move from the idea of circular economy towards regenerative economy. [5] The climate change and the irreversible consequences it brings with, become a matter that concerns all the economic players e.g. property owners of university campus

This papers aims to describe a process towards a regenerative economy by presenting the Green Bond case of university property owner in Finland. The question responded is how property developer of university campus can use Green Bond as one instrument in sustainable life cycle management and continuous development of properties towards stronger sustainability.

2. Sustainable campus

In terms of university campus sustainability, the College Sustainability Report Card is a useful operational list of criteria. It is an initiative of the Sustainable Endowments Institute in Cambridge, Massachusetts, USA. The criteria are organized by the four types or categories of sustainability [1] (Table 1).

Ecological criteria	Economic criteria	Institutional criteria	Energetic criteria
Food and recycling: dining services policies, including recycling and composting programs	Endowment transparency: accessibility to endowment investment information and shareholder proxy voting records	Administration: sustainability policies and commitments by school administrators and trustees	Climate change and energy: energy efficiency, conservation commitment to emissions reductions, use of renewable energy on campus
Green building: campus-wide green building guidelines and green building design for new and existing buildings	Investment priorities: prioritization of return on investment, investment in renewable energy funds, and investment in community development loan funds	Student involvement: student participation in sustainability initiatives and support for these activities by school administrators	
Transportation: alternative transportation for students, faculty, and staff; alternative fuel or hybrid technology for campus fleets			

Different Green Lease and Green Bonds are part of the economic criteria. However, such decisions and actions require the green buildings in campus portfolio. This is supported e.g. by environmental certification of building projects and e.g. by the Better Buildings Partnership (BBP), a collaboration of property owners working together to improve the sustainability of existing building stock [3]. Green bonds can mobilize resources from domestic and international capital markets for climate change adaptation, renewables and other environment-friendly projects. Such economic approaches enhance all stakeholders to include sustainability issues at a different level in economic considerations. [4] In Figure 1 the path from more conventional to regenerative economy is presented, showing that regeneration goes far beyond sustainability. [5]

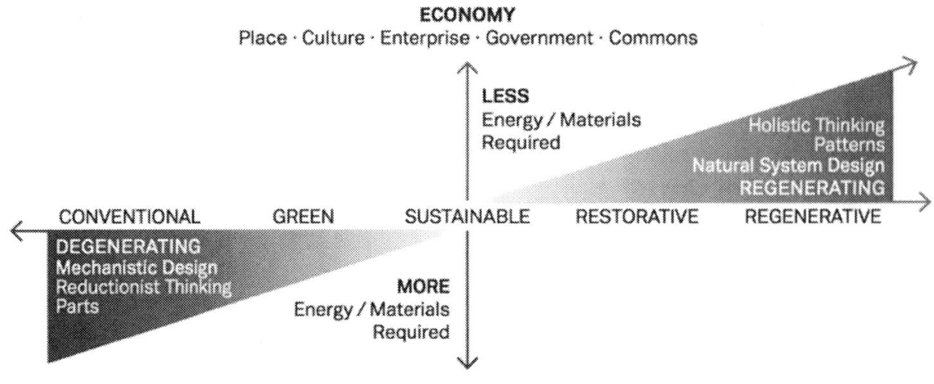

Figure 1. Towards regenerative economy - From Business as Usual to Regenerative Sustainability [5]

3. Green Bond

The term green bonds refer to bonds aimed at financing investments with an environmental benefit or a focus on reducing vulnerability to environmental changes. The focus is on investments relating to mitigating or adapting to climate change. This enables to broaden funding base by gaining access responsible investors, who include a broad range of non-economic criteria, including social, environmental and governance, in their investment choices. Green Bonds are fixed income securities which finance investments with environmental or climate-related benefits. Green Bonds are an integral component of "green finance" more generally, which aims to "internalize environmental externalities and adjust risk perceptions" for the sake of increasing environmentally friendly investments. [6;7] The Green Bond market has grown exponentially over the past few years. It is often seen as a financial instrument that may help overcome the low-carbon investment challenge. Green Bonds can mobilize resources from domestic and international capital markets for climate change adaptation, renewables and other environment-friendly projects. They require that the proceeds will be invested in projects that generate environmental benefits. In its simplest form, a bond issuer will raise a fixed amount of capital, repaying the capital (principal) and accrued interest (coupon) over a set period of time. The issuer will need to generate sufficient cash flows to repay interest and capital. This is how United Nations include Green Bonds in their development program. [7].

Green Bonds can help investors implement their long-term climate strategies and enable responsible investors to have alternatives to broaden their delete portfolios. In turn, Green Bonds can help bond issuers communicate their sustainability strategies, create internal synergies between financial and sustainability departments, and expand and improve relationships of borrowers with debt providers. Additionally, Green Bonds can support the implementation of national climate policies – through improved awareness and more efficient capital allocation, especially in the perspective of redirecting capital towards low-carbon and climate resilient projects. The investors are looking at corporates' sustainability/ESG (Environmental, Social and Governance) performance to better understand which ones are well positioned to transition to a green economy. Corporates with higher levels of sustainability performance and with a lower policy risk are attractive to investors. [11]

Investors are often looking for more transparency when gathering the details of the bond. This includes use of proceeds, project evaluation criteria, management of proceeds and issuers reporting on the Green Bond. A second-party opinion by a trusted and a respected evaluation firm provides a clear assessment of the issuer's Green Bond framework. [8] As an example of such stakeholder CICERO is a climate research institute based in Oslo and the leading provider of second opinions. It evaluates the issuer's framework for both project selection and investment. [9]

Financing of investments through Green Bonds has grown rapidly in recent years. But definitions of what makes a bond "green" vary. Various certification mechanisms have evolved to allow more granularity as well as continuity in assessment [10]. While the Green Bond is still the phenomena in development there are some challenges: question of environmental integrity of Green Bonds is related to definitions and procedures: how to define "greenness", which ultimately depends on the objectives of the use of Green Bonds. The second challenge is how market-driven and public support measures may help increase the tangible financial contribution of Green Bonds to the low-carbon transition. The Green Bond market is increasingly seen as having important potential to contribute to the systematic labelling of financial assets financing LCCR (low-carbon, climate-resilient) investments. Enhanced transparency of information provided by Green Bonds can unlock a number of benefits for issuers, investors and policymakers supporting the growth of the market. While there is an increasing consensus that this additional transparency brings added value, there are however neither harmonized definitions and taxonomies, nor a common reporting framework for labelled Green Bonds. To ensure its meaningful contribution to the low-carbon transition through improved transparency of information, public and private market actors will need to address these challenges and guarantee the environmental integrity of Green Bonds and improve climate-related disclosures for other financial products. [11] It is important to learn about the process of applying the Green Bond framework in different asset segments and in different cultures. The following case study focus on university properties and Green Bond.

4. Case Study

4.1. Company description

The case company, University Properties of Finland Ltd (SYK) is a nationwide owner and developer of higher education campuses outside of the Helsinki metropolitan area. SYK's portfolio of properties include buildings used for various purposes and are categorized as office and educational buildings. Altogether SYK owns 1.1 million square meters of space and yearly turnover is about 150 million euros. According to organization's vision it aims to be the most valued campus developer in Europe. Organization is owned by the Government of Finland and nine Finnish universities.

4.2. Methods

The single case study method employed in this study captured the process of case organization towards Green Bond initiative. More precisely the data was gathered by qualitative document analysis (QDA). It is a research method for rigorously and systematically analyzing written materials. The material before (2013-2018) and after (2019) the Green Bond initiative was analyzed by researcher. Material included meeting memos, organizational policies, sustainability reports and strategy documents. This was compared with Regenerative Economy (Figure 1) framework. Although documents alone can by no means give a complete picture of an organization's approach and activity, they are strongly indicative of where efforts and aims are focused.

The workshops for the internal group (Green Bond Committee) of the Green Bond preparation (3 workshops) and implication (2 workshops) was conducted and followed by author in order to understand the specifications. The workshop participants included both asset management and property management experts from the case company. The secondary data included the interviews with stakeholders in the Green Bond preparation phase. The interviews were supplementary data for Green Bond decision making of the case company.

4.3. Analysis of the phase 1 Steps towards from green to sustainability 2013-2018

The case company has paid attention to environmental issues since its founding in 2009. It covered sustainable development as a part of its strategy and documented its long-term sustainability vision (until 2030). Company signed 2011 the TETS agreement, Finnish Property and Building Sector Energy Efficiency Voluntary Agreement, that municipalities or private companies can sign to support meeting national commitments.

2013 the company made a decision to aim to certificate the largest construction projects by using BREEAM-certification (Building Research Establishment Environmental Assessment Method). Concretely, the first BREEAM certification achieved with Kampus-areena in Tampere, which is a 15 000 gross square meter building and was finished on August 2015. The eight floor building consists of bigger three floor base and five floor tower. 60% of the premises are used by the university and 40% is rented to an office and business premises. It is a platform for students, researchers and industry professionals to collaborate.

From 2013 to 2018, case company has reduced its CO2 emissions from energy by 12,9 %. Until year 2018 the company has succeeded on planning and constructing nine buildings with a grade of "Very Good". On June 2017 the case company was selected by the Ministry of the Environment as the best commercial property owner in Finland regarding completed energy savings during the years 2010 – 2016. This was also the year when the company established an internal Green Bond Committee to prepare the Green Bond initiative. The Green Bond Committee consisted of four members: Chief Financial Officer, Director of Project Management, Environmental Engineer and Controller. The phases are summarized in Table 2.

Table 2. From energy efficiency to wider scale of ecology.

2009	2011	2013	2015	2017	2018
Setting goals	Signing voluntary Finnish Property and Building Sector Energy Efficiency Agreement (TETS)	Making decision of all major construction being BREEAM certified	The first BREEAM certification achieved with Kampus-areena building in Tampere	Green bond preparation started	Nine BREEAM certifications achieved First Green Bond released in November

The company reports on responsibility in accordance with the GRI (Global Reporting initiative) principles and they publish an integrated report in the GRI database.

This phase from 2009 to 2018 included many green initiatives, which improve the energy efficiency in building portfolio. This was ensured by using environmental certification systems. After the first success in environmental certifications the asset management initiatives for economic sustainability was the interest of the property owner and the first Green Bond was released in 2018.

4.4. Eligible projects

In order to achieve Green Bonds to be used to finance new Eligible Projects the company had to fit into one of the following five categories:

- BREEAM certification with a certification level of at least Very Good and an energy use that is at least 15% lower than that required by the Finnish national building code
- Existing buildings with BREEAM In-use certification (two parts out of three: Asset and Building Management) with a certification level of at least Very Good
- Energy efficiency projects leading to a reduction of energy use of at least 25 per cent.
- Technical solutions utilizing renewable energy of solar, geothermal or wind power to satisfy the energy needs of properties.
- Clean transportation infrastructure for electric vehicles.

In addition, case company's Green Bonds will not finance fossil fuel generation projects.

Eligible Projects were evaluated and selected by companies Green Bond Committee. The Green Bond Committee approved Eligible Projects in consensus.

The division of the allocation of Green Bond proceeds between new projects and refinancing will be included in company's annual Green Bond Investor Letter including: a list of the all projects financed including allocated amount, a brief description and expected impact and information about the division of the allocation of Green Bond proceeds between new projects and refinancing

The project 1 was Medisiina D in Turku, Finland. It is a multi-user building and hub of medical development that serves the needs of the Hospital District of Southwest Finland, the University of Turku and Turku University of Applied Sciences. In addition to each of the three having premises of their own in the building, there are shared information, restaurant, teaching and meeting spaces as well as some shared laboratory facilities. Equipment supply premise have been consolidated in a single unit, compared to 15 locations previously. Shared use reduces the need for space and resources and creates cost savings.

The project 2 was Aurum Building also in Turku, Finland. It is still under construction. Aurum´s main users will the Faculty of Science and Engineering of Åbo Akademi University and the Department of Chemistry of University of Turku. Other users include the Biodiversity Unit and the Library of the University of Turku. The spatial solutions support joint activities and shared services. This contributes to a high rate of utilization and the economical use of space.

The third project is Verso in Lappeenranta, Finland. The objective of Verso project in Lappeenranta was the increased sharing of space between Lappeenranta University of Technology and Saimaa University of Applied Sciences. The Verso project will involve upgrading the university´s main building and two laboratories, adding more premises for businesses and modernizing building services. Efficiency will be improved by more streamlined operations and higher rates of utilization. The projects are described in Table 3.

Table 3. Eligible Projects.

	Project	**Location**	**Project status**	**Project finalized**	**Total estimated project costs, €m**	**Allocated proceed 31.12.2018, €m**
New projects	Medisiina D	Turku, Finland	In use	13.3.2018	72	36
	Verso Project	Lappeenranta. Finland	In use	20.10.2018	40	40
	Aurum Building	Turku, Finland	Under construction		80	7
Total					192	83

The case company established a dedicated account for the net proceeds of issued Green Bonds. The allocation of funds from the proceeds will be monitored by company treasury and verified by external auditor annually. If the account has a positive balance, funds will be deducted and added to company's lending pool in an amount equal to all disbursements made from that pool during the quarter in respect to Eligible Projects. Unallocated funds may be invested in liquidity reserves and managed accordingly. If, for any reason, a financed Eligible Project no longer meets the eligibility criteria, the project will be removed from the eligible pool of projects financed with Green Bond proceeds.

4.5. Phase 2 Steps from sustainability towards regenerative approach

The company reports on its green bonds through its annual Green Bond Investor Letter. The letter will cover a list of all projects financed, including their allocated amount, a brief description and expected impact. Reporting includes also the division of allocation of financing between new and refinanced projects. The company aims to report on the type and level of certification, energy use and CO_2 emissions when relevant and potentially other environmental benefits. The Annual Investor Letter will make use of the continual monitoring that company performs on its consumption and emissions and report on the actual ex-post impacts when feasible. Emissions are based on the relevant emissions conversion factors from the local grid in question. It is recognized that it is difficult to define quantitative indicators for old buildings with a wide range of uses. The annual letter, together with the statement of external verification, will be made publicly available online on company's website.

Transparency, reporting, and verification of impacts are key to enable investors to follow the implementation of Green Bond programs. Procedures for reporting and disclosure of Green Bond investments are also vital to build confidence that Green Bonds are contributing towards a sustainable and climate-friendly future, both among investors and in society. Company´s Green Bond Framework, together with its Sustainability Framework, Energy Management Framework, Environmental review and guidelines of company, Sustainability Vision 2030 and its approved Energy Efficiency, Carbon Neutrality and BREEAM Policy, aims to support the transition to low carbon and climate resilient growth.

Company aims to reduce energy consumption with 7.5% in 2025 compared to 2016, with an interim milestone set for 4% reduction by 2020. Company has the following comprehensive set of energy efficiency and emissions targets for new and retrofitting building construction:

1) energy efficiency rating E-number of best A class (which requires 20% higher energy efficiency than under Finnish regulations);
2) less than 150 kWh/sqm/a purchased energy;
3) at least 75 % of energy used in building coming from renewable energy sources; and
4) less than 25 kg CO_2/sqm/a emissions.

It will fulfill BREEAM certification grade of "Very Good" for new construction as well as for major renovations but has ambitions for achieving the grade of "Excellent" for new buildings. The intention is to put effort also to BREEAM-In-Use certifications in various university campus. Similarly, they have set the aim for carbon neutrality by 2030. These future visions have been summarized in Table 4.

Table 4. Future development from sustainability towards regeneration.

2020	2021	2023	2025	2030
Carbon neutral goals by compensation	Biodiversity principles for design and maintenance	Potential of Social Bond	Updating Energy Efficiency targets based on Finnish Property and Building Sector Energy Efficiency Agreement (TETS)	Carbon neutral goals without compensation

The company has defined the most important sustainability topics which support the business and strategy in company level. Additionally, it is currently working to define more precise metrics to indicate the development. This future path from 2020 have the following regenerative initiatives: goals for carbon neutrality and social bonds. The property and asset management portfolio is more interlinked. When the goals are achieved they provide more good to build environment instead of only improving it.

5. Conclusions

The results show that company begin the process with focusing on environmental sustainability especially putting effort in the first phase to energy efficiency of property portfolio. The second phase included already more conscious and systematic intention to achieve the interlinkage between the property and asset management by Green Bonds. Additionally, issues like shared use of facilities was discussed from social sustainability perspective. The future visions include more ambitious goals towards regeneration. This makes the company also more dependent on the ecosystem around it.

University campus as part of sustainable and even regenerative society is a central matter. In the case study Green Bonds along reasonable trajectories avoids CO2-emissions corresponding to the total emissions in national level. This is part of societal impact of university in general. Also education and research can be beneficiaries in various ways. The benefits of the use of Green bonds are multiple. The common purpose to help the environment or to invest in a positive social outcome brings in an outstanding cohesiveness to the stakeholder relationship.

The Green Bonds market is expected to progress over the coming years on a large-scale. In order to make this shift happen, there must be some changes inside the market. The requirements of Green Bond projects will be higher and more aligned. Transparency and common frameworks are essential in the future.

The transition to a sustainable global economy requires scaling up the financing of investments that provide environmental and social benefits. Next to Green bonds e.g. Social Bonds can play an essential role also in the case study organization. This paper is providing one example of steps towards regeneration by using economic and ecological goal settings. The next step is to identify the potential of Social Bonds and provide new perspective to responsible property owner and regenerative actions.

References

[1] Emanuel R and Adams JN 2011 College students' perceptions of campus sustainability *The International J. of Sustainability in Higher Education* **12** pp 79-92

[2] Too L and Bajracharya B 2015 Sustainable campus: engaging the community in sustainability, *International J. of Sustainability in Higher Education* **16** pp 57-71

[3] Bugden K, Botten C, Staheli J, Cross, S and Highmore S 2013 Green Lease Toolkit. In Centre T (Ed.). London: *The Better Buildings*

[4] Peretti G et al. (Eds.) (2019) Regenerative construction and operatio. COST Action CA16114 RESTORE *Working group three report: Regenerative Construction and Operation*

[5] Brown M, Haselsteiner E, Apró D, Kopeva D, Luca E, Pulkkinen K and Vula Rizvanolli B (Eds) 2018 Sustainability, Restorative to Regenerative COST Action CA16114 RESTOR, *Working Group One Report: Restorative Sustainability*

[6] Fullerton J (2015) *Regenerative Capitalism - How Universal Principles and Patterns will Shape our New Economy* Capital Institute

[7] Weidmann J 2017 *Green bond issuance and other forms of low-carbon finance* Global Public Investor Symposium on Green bond issuance and other forms of low-carbon finance Frankfurt am Main

[8] Pereira da Silva L 2017 *Green finance: can it help combat climate change?* Global Public Investor Symposium Green bond issuance and other forms of low-carbon finance Frankfurt am Main

[9] Glomsrød S and Taoyuan W 2016 Business as unusual - the implications of fossil divestment and green bonds for financial flows, economic growth and engery market. *Working paper 2016:01 Technical report* CICERO - Center for International Climate and Environmental Research

[10] Ehlers T and Packer F 2017 Green Bond Finance and Certification *BIS Quarterly Review* September 2017 Available at SSRN: https://ssrn.com/abstract=3042378

[11] Shishlov I, Nicol M and Cochran I 2018 *Environmental integrity of green bonds: stakes, status and next steps* Green Bonds Research Program Work Package 2 (INIS-FR-19-0222) France

IOP Conf. Series: Earth and Environmental Science **352** (2019) 012029 doi:10.1088/1755-1315/352/1/012029

Web tool for quantitative sustainability evaluation of hygro-thermally optimized insulation solutions

S Walther[1], M Birkved[2], T H Perkov[1] and S P Bjarløv[1]

[1] DTU Civil Engineering, Department of Civil Engineering, Technical University of Denmark (DTU), Brovej Building 118, DK-2800 Kgs. Lyngby, Denmark

[2] SDU Life Cycle Engineering, Department of Chemical Engineering, Biotechnology, and Environmental Technology, University of Southern Denmark (SDU), Campusvej 55, DK-5230 Odense-M, Denmark

E-mail: walthed@gmail.com

Abstract. This study is part of the EU funded project RiBuild and aims to create a web tool for supporting decisions relating to energy retrofits of historic buildings. The web tool provides a quantitative sustainability evaluation of hygro-thermally optimized insulation solutions through a greenhouse gas (GHG) emissions indicator. The study presents LCAs of five insulation systems for installation in historic buildings. The web tool accommodates 160 locations in seven countries and generates 210,180 impact profiles by combining insulation system, installation location, and heating systems. In 2 % of the scenarios, the induced impacts exceed the avoided impacts. Across all analyses, the mineral wool insulation system is the least GHG intensive solution. This study illuminates that the environmental justification of internal insulation depends on the geographical location of installation and the installed heating system in the retrofitted building. To further improve the sustainability performance of energy retrofits in historic buildings, the implementation of insulation recycling solutions and measures for reducing production impacts of insulation systems are needed.

1. Introduction

The potential consequences of climate change and the concomitant rise in global mean temperatures has received global attention and priority. The European Council [1] has set a target of 40 % reduction in GHG emissions by 2030 compared to 1990 levels. As the existing building stock accounts for 40 % of the energy use and 36 % of the GHG emissions [2, 3], energy retrofits for improvement of the energy efficiency of existing buildings offer a great potential to reduce the European GHG emission levels.

Historic buildings are part of the existing building stock and represent architectural, as well as historical and cultural values, that need to be preserved. Consequently, energy retrofits of historic buildings pose different constraints and challenges than traditional insulation retrofits, as their visual character and building fabric must be considered [4, 5]. External insulation retrofit measures impact the aesthetic appearance of a building, why these solutions are not always a possible option when renovating historic buildings. In this light, different approaches such as internal insulation retrofits, must be considered. However, these solution options are less favorable from a hygro-thermal point of view, as the temperature of the external wall is decreased when applying insulation internally. The subsequent lowered dew point and increase in relative humidity between insulation and the external wall aggravate the risk of condensation and mold growth [6].

The Horizon 2020 project *"Robust Internal Thermal Insulation of Historic Buildings"* (RiBuild) is a research project which aims to develop comprehensive guidelines for improving energy performance of historic buildings while maintaining a minimal risk of mold growth through hygro-thermal optimization of internal insulation retrofits. RiBuild is a collaboration between research institutions across seven European countries: Belgium, Denmark, Germany, Italy, Latvia, Sweden, and Switzerland. In the RiBuild project, historic buildings are defined as all protected and non-protected buildings constructed before 1945 [6].

1.1. Objective and scope

With this study, a quantitative sustainability assessment web tool is created for the RiBuild project to assess the climate burden impact of insulation systems for energy retrofits of historic buildings. The sustainability assessment provides an additional angle to the hygro-thermal optimization to support decision-making of the most appropriate internal insulation solution in a renovation scenario. This study will provide a quantitative life cycle assessment (LCA) of hygro-thermally optimized insulation solutions for historic buildings. In practice, the web tool accommodates a range of unit climate burdens (kg CO_2-eq) for a range of unit operations on induced impacts (i.e. from production, transport, installation, and disposal of the insulation systems) and avoided impacts (i.e. from saved energy consumption), which through the user interface can be combined with a multitude of application scenarios (geographic locations, heating systems, and wall types in the existing building). The study covers five project tailored insulation systems with various thickness (consisting of the insulation materials calcium silicate, polyurethane, phenolic foam, mineral foam, and mineral wool) for installation in 160 locations across the seven participating countries with a lifespan of 30 years from 2020-2050.

2. Method and materials

The LCA of the insulation systems in this study follows the ISO14040 and ISO14044 standards [7, 8], and covers the following steps: *goal and scope definition*, *life cycle inventory analysis*, *life cycle impact assessment*, and *interpretation*.

The intended application of this study is a comparative LCA of five project tailored insulation systems selected on their hygro-thermal performance for an internal insulation retrofit of historic buildings.

The system boundaries of the insulation systems covered, are in accordance with EN15804:2013 [9]: the production stage (A1-A3), the assembly stage (A4-A5), the operation energy use stage (B6), and the end-of-life stage (C1-C4), thus, making this a 'cradle-to-grave' simplified LCA. The impacts from the life cycle stages A1-A5 and C1-C4 are considered induced impacts, whereas the impacts from the life cycle stage B6 are considered avoided impacts via saved energy consumption for space heating. The saved energy consumption is calculated using the Heating Degree Days (HDD) method. only transmission losses are considered, and infiltration losses are disregarded. Maintenance (B2) and replacement (B4) have been excluded from this LCA, as the projected service lives of the materials for the insulation systems are longer than the life cycle defined in this assessment. The life cycle inventories (LCI) are modeled as attributional with allocation at the point of substitution, as this assessment is a comparison of specific goods which is in accordance with the ILCD guidelines [10]. LCIs have been modeled using the *SimaPro 8.5.2.0* software. The functional unit (F.U.) for the web tool is defined in alignment with the intended application of the study as:

1 m^2 insulation system with the equivalent U-value installed as a part of an energy retrofit of a historic building with the lifespan of 30 years from 2020 to 2050.

IOP Conf. Series: Earth and Environmental Science **352** (2019) 012029 doi:10.1088/1755-1315/352/1/012029

The LCIA method used is CML-IA baseline 3.05 with normalization factor EU25+ 2000.

2.1. Life cycle inventory

The five insulation systems and their variations for assessment in the web tool were chosen by the RiBuild research team based on their hygro-thermal properties (Table 1).

Table 1: Overview of material properties of the insulation systems included in the web tool. Insulation system S1 has four thickness variations, S3 has six, while S1, S4, and S5 have three each

System ID	Thickness [mm]	Material	λ [W/mK]	ρ [kg/m^3]	Manufacturer
	10	Lime Plaster	0.68	1480	Calsitherm
S1	25, 30, 50, 80	Climate Board	0.059	187	Calsitherm
	5	Glue Mortar	0.6	1410	Calsitherm
	10	iQ-Top	0.111	700	Remmers
S2	30, 50, 80	iQ-Therm	0.031	45	Remmers
	8	iQ-Fix	0.497	1500	Remmers
S3	20, 45, 65, 80, 90, 110	Phenolic Foam	0.018	40	Kingspan
	12.5	Gypsum Board	0.25	712	Knauf
	10	Climate Plaster	0.68	1480	Calsitherm
S4	60, 80, 100	Mineral Foam	0.045	115	Xella
	5	Glue Mortar	0.6	1410	Calsitherm
	0.22	Vapor Barrier	-	363.63	Saint-Gobain Isover
S5	70, 95, 120	Mineral Wool	0.032	33	Saint-Gobain Isover
	22	Air	-	-	-
	12.5	Gypsum Board	0.25	712	Knauf

The study seeks to assess product-specific systems by using country and manufacturer-specific data through environmental product declarations (EPDs) and data sheets when possible. Background data and additional data was derived from the *ecoinvent 3.4* database. The inventory modeling of phenolic foam insulation was based on Densley Tingley et al. [11]. No information was found on the Climate Plaster and the Glue Mortar, both in S4, and these system parts were assumed to be identical with the Calsitherm Lime Plaster and Glue Mortar in S1. The remaining production processes and background processes were modeled using existing *ecoinvent 3.4* processes. In the end-of-life stage, all materials were assumed to be landfilled.

The reduced transmission losses occurring as a result of an internal insulation retrofit lead to savings in energy consumption and consequently avoided GHG emissions. For each country, the three most common heating systems for residential space heating have been chosen based on data from the Heat Roadmap Europe and the Swiss Energy Department [12, 13] (Table A1). For the biomass furnace, country-specific modeling was only possible for Switzerland due to limitations of *ecoinvent 3.4*. For the remaining countries, Europe was chosen as representation. The energy carrier composition for district heating and electricity were based on country specific mixes. The remaining heating systems were modeled using country-specific processes from *ecoinvent 3.4*. The web tool accomodates five choices of existing wall types for calculating reduced transmission losses (Table A2).

Weather data for calculating HDD profiles for retrofit locations for the period 2020-2050 were obtained from the European research project Climate for Culture. These weather data were derived from climatic simulations based on the Intergovernmental Panel on Climate Change (IPCC) AR_4 A_1B climate change scenario and was subsequently adjusted for these projections of future GHG emissions scenarios [14, 15].

3. Results

The sustainability web tool provides a climate burden indicator to assist selection of the most appropriate insulation system for retrofits of historic building. In the user interface part, the user is requested to define the renovation scenario through specification of the renovation location, the heating system of the building, and the composition of the existing wall (Figure 1). Based on these choices, the web tool generates the climate burden indicator for the insulation systems in terms of induced impacts from 1 m^2 insulation systems and avoided impacts via the saved energy consumption for 1 m^2 wall section. For the specific scenario defined by the user, the results for all insulation solutions options are displayed. A mock-up of the web tool has been developed using Microsoft Excel and generates 213,180 possible impact profiles for application solutions.

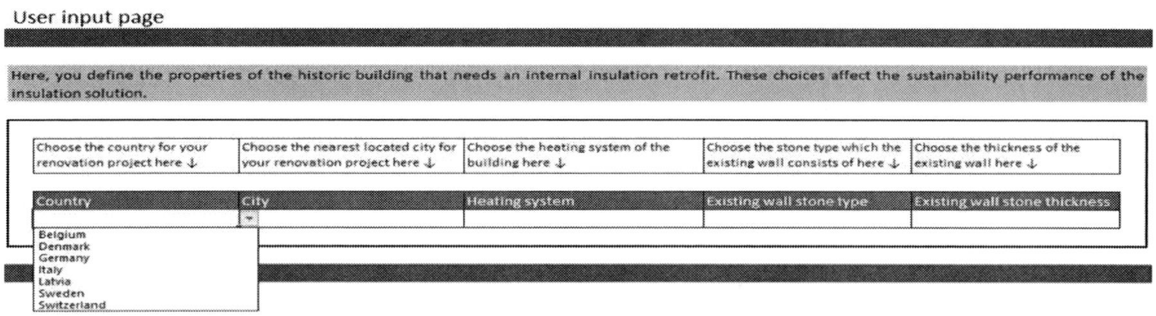

Figure 1: Screenshot of the user input page of the web tool

In order to determine the most appropriate insulation system for a renovation project, the insulation systems are compared based on their induced impact, the avoided impacts and their net impact profiles. For equal comparison of the insulation systems, the F.U. has been adjusted to:

1 m^2 insulation system with U-value 0,5 W/m^2K installed as a part of an energy retrofit of a historic building with an existing wall of Weinerberger brick with thickness 400 mm and λ = 0,801 W/mK and a lifespan of 30 years from 2020 to 2050.

For all application scenarios, S5 (mineral wool) has the smallest induced impact profile and S1 (calcium silicate board) has the biggest (Figure 2).

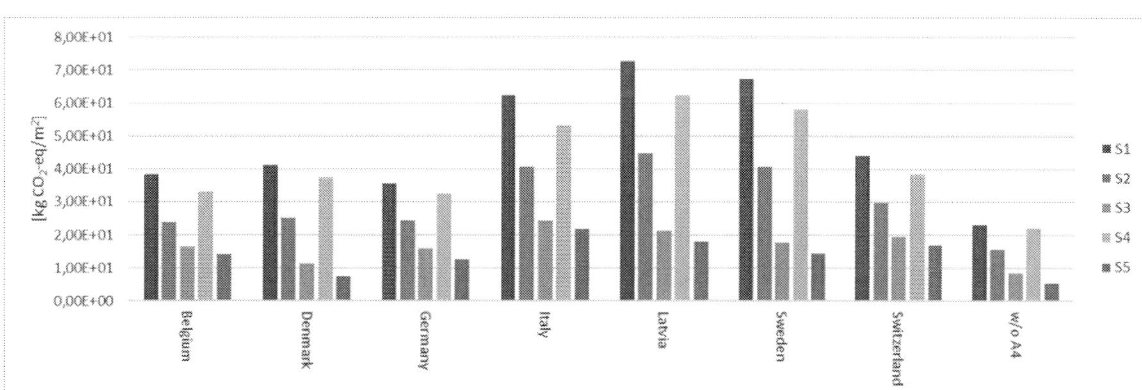

Figure 2: Induced impact profiles (A1-A5 and C1-C4) for all insulation systems for each country and excluding the transportation stage A4

IOP Conf. Series: Earth and Environmental Science **352** (2019) 012029 doi:10.1088/1755-1315/352/1/012029

The transportation impact can potentially double the induced impacts depending on the distance. Most of the materials contained in S5 are produced in Denmark resulting in a small contribution to the total induced impact when comparing with the scenario excluding the transportation stage. For S5 installed in Italy, the transportation contributes with approximately 75 % of the total induced impact of S5.

The avoided impact from saved energy consumption consists of the avoided transmission loss through the wall section with and without a retrofit, the climatic profile of the location, and the GHG emission intensity of the heating system. The comparison of the avoided impacts is based on the average HDD profile for each country (Figure 3).

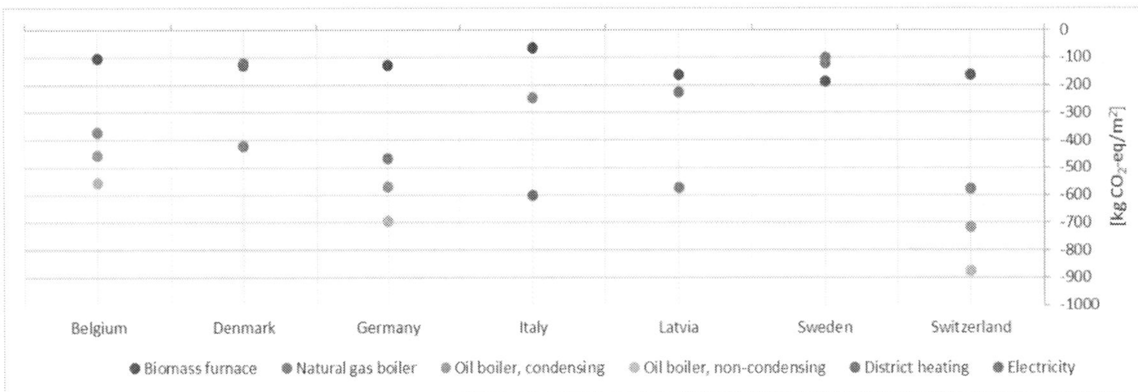

Figure 3: Avoided impacts (in kg CO_2-eq/F.U.) for each heating systems within each country for the entire period 2020-2050

The avoided impacts based on the saved energy consumption for space heating for each country changes with the type of heating system used. For all countries except Sweden and Denmark, the biomass furnace results in the smallest avoided impact. For these two countries, the smallest avoided impact came from district heating. For Sweden, the biomass furnace resulted in the largest avoided impact, highlighting the low GHG emission intensity of the Swedish district heating and electricity production. For the countries with oil boilers as an option (Belgium, Germany, and Switzerland), the non-condensing oil boiler results the largest avoided impact. For Denmark and Latvia it is the natural gas boiler which results in the largest avoided impact. Electricity results in the largest avoided impact in Italy which highlights the high GHG emission intensity of the Italian electricity production.

The insulation systems' net impact profiles follow the trend of the avoided impacts (Figure 4). All net impact profiles show negative values which indicates that the GHG emissions savings from saved energy consumption are greater than the induced impacts from the insulation systems throughout the life cycle. The heating systems with the highest GHG emission intensity result in the biggest GHG emissions savings for the insulation systems. For all application scenarios S5 (mineral wool) performs best and S1 (calcium silicate board) performs worst.

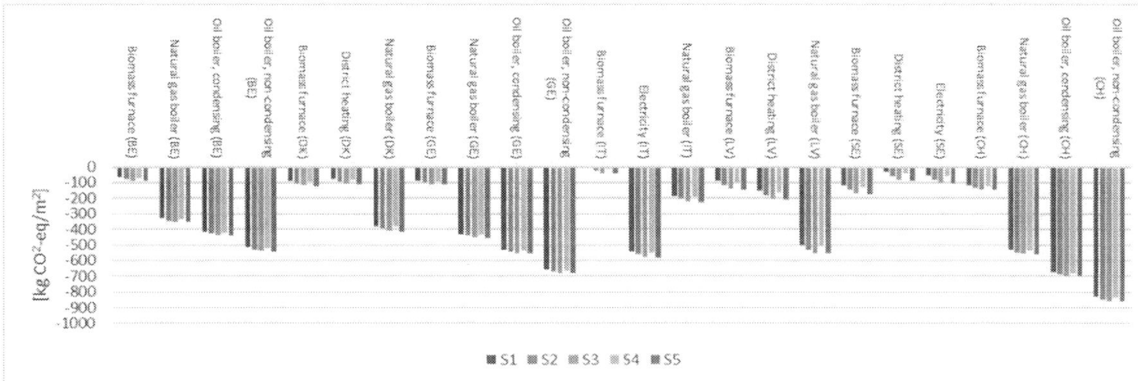

Figure 4: Net impact profiles for all insulation systems for each country specific heating system

3.1. All net impact profiles generated in the web tool

Some of the 213,210 net impact profiles generated from the application scenarios in the web tool result in positive net impact profiles (Table 2). Positive net impact profiles occur when the insulation systems impact with a higher GHG emission intensity than what is avoided due to saved energy consumption over the 30-year lifespan. The frequency of application scenarios with a positive net impact is most pronounced for Italy, followed by Sweden. Italy has the smallest average saved energy consumption and Sweden has the least GHG emission intensive heating systems (Figure 3). No scenarios produced a positive net impact in Latvia.

Table 2: Number of net impact results for each country and the number and percentage within these with a positive net impact

Country	n	n > 0	% > 0
Belgium	6080	8	0.13
Denmark	4560	1	0.02
Germany	91200	7	0.01
Italy	54720	4121	7.53
Latvia	2280	0	0
Sweden	28500	141	0.49
Switzerland	25840	16	0.06
Total	213180	4294	2.01

4. Discussion

The comparison of the insulation systems' induced impacts for the seven countries did not yield differences in the ranking of the most appropriate insulation solution. However, the transportation distance's contribution changes the performance of the individual insulation system when comparing between the countries. Thus, when deciding on materials for an internal insulation retrofit, the location of the production site should be taken into consideration.

For the avoided impacts, a large avoided impact was associated with either a cold climate, a GHG emission intensive heating system or a combination of the two. Sweden yielded the largest average saved energy consumption, but the smallest distribution range values of avoided impact for the heating systems, showing that the energy grid supplying heating has a considerable influence on the sustainability benefit of a retrofit in a renovation scenario. A study conducted by Ghose et al. [16] concluded that maximum benefits only occur if the post-retrofit energy grid is based on high coal generation, supporting the findings of this study.

2 % of all net impact results were positive (i.e. an excess of induced impacts). In these scenarios insulation retrofit measures would not be sustainable solution as the avoided impacts associated with energy consumption savings do not cover the induced impacts associated with the insulation systems. The occurrence of positive impact profiles was most pronounced for Italy, followed by Sweden, and shows that in general, the energy retrofits are most sustainable for colder climates with heating systems of high GHG emission intensity. Dylewski & Adamczyk

IOP Conf. Series: Earth and Environmental Science **352** (2019) 012029 doi:10.1088/1755-1315/352/1/012029

[17] find that the most ecological cost-effectiveness was found for the scenario of worst performing heat source in the coldest climate zone and the least ecological cost-effective solution was found for the scenario including the best performing heat source in the warmest climate zone. In addition, Oregi et al. [18] (cited in Karimpour et al. [19]) note that the induced impact from embodied energy of building materials can account for up to 35 % of future emission targets of a building constructed in a mild climate.

With the development towards more sustainable and less GHG emission intensive energy grids, the avoided impacts are diminished and the induced impacts become more influential of the insulation systems' overall performance. However, a decreased avoided impact implies a less GHG intensive energy grid for heating which in itself is a more sustainable scenario. Sohn et al. [20] emphasized that integrating a dynamic energy mix in building LCAs is preferable to a static energy mix, as future projections are available and this provides an indication of the building's overall environmental performance against possible future energy mix scenarios. When taking this development towards more sustainable technology into account, measures towards reducing induced impacts from production are needed to improve the overall sustainability performance of the insulation systems [21, 22]. Another approach would be introducing recycling solutions, either through the use of recycled materials or improving the recycling potential when carrying out energy retrofits. Ingrao et al. [23] highlighted the improvement of insulation systems' environmental performance by using recycled insulation materials compared to conventional (i.e. straight from the plant) insulation materials.

In this study, the best performing insulation system consisted of mineral wool as insulation material (S5). As the only system, S5 needs a vapour barrier which implies a risk of condensation from cooling of the existing wall. Therefore hygro-thermal simulations are essential when choosing an internal insulation solution for energy retrofits of historic buildings. The hygro-thermal performance and the climate burden impact profile might potentially result in opposing recommendations for decisions regarding the most appropriate choice of insulation system.

5. Conclusion

A web tool for assessing the most appropriate insulation solution for energy retrofits of historic buildings based on their climate burden impact has been created. The web tool generates a total of 213,210 impact profiles. This study compares the life cycle performance of the five project tailored insulation systems included in the web tool based on the same U-value of 0,5 W/m^2K.

Of the five insulation systems, S5 performed best across all analyses. However, as this insulation system needs a vapour barrier, it might not be the most appropriate insulation system when taking the hygro-thermal performance into account. For decision-makers who commission the LCA of insulation systems for energy retrofits of historic buildings, these conditions must be carefully considered as they will affect recommendations regarding the most appropriate solution for renovation of historic buildings.

Analysis of the avoided impacts showed that the combination of a large saved energy consumption and low GHG intensity heating system resulted in minor avoided impacts, whereas the combination of small saved energy consumption and high GHG intensity heating system resulted in a large avoided impact.

In some scenarios, insulation retrofit measures did not provide a sustainable solution as the induced impacts exceeded the avoided impacts. This situation was most pronounced for Italy, followed by Sweden, and showed that in general, energy retrofits are most sustainable for the combination of colder climates and GHG intensive heating systems. In combination with the projected development towards more sustainable energy grids, these findings elucidate that the environmental justification of internal insulation in energy retrofits depends on the geographical location of installation and the type of heating system supplying the building being retrofitted. Therefore, the implementation of recycling solutions and measures for reducing production impacts are needed to further improve the sustainability performance of energy retrofits in historic buildings.

Acknowledgement

This study is part of the RiBuild project, which has received funding from the European Research Council under the European Union's Horizon 2020 research and innovation program, grant agreement No. 637268.

Appendix A.

Table A1: Heating systems for space heating

Heating system	Countries
Biomass furnace	All countries
Natural gas boiler	CH, BE, DK, GE, IT, LV
Oil boiler, cond.	CH, BE, GE
Oil boiler, non-cond.	CH, BE, GE
District heating	DK, LV, SE
Electricity	IT, SE

Table A2: Stone types and thickness variations for the existing walls and their thermal conductivity coefficient

Stone type	Thickness [mm]	λ [W/mK]
Brick Weinerberger	300, 400, 500, 600	0.801
Lime Sand Brick	300, 400, 500, 600	1.0
Sandstone Monte Merlo	300, 400, 500, 600	0.956
Tuff Ettringen	300, 400, 500, 600	0.481
Granite	300, 400, 500, 600	1.7

References

[1] General Secretariat of the European Council 2014 *COM(2014) 0015* 1–15 URL `http://www.consilium.europa.eu/en/meetings/european-council/2014/10/23-24/`
[2] Vilches A, Garcia-Martinez A and Sanchez-Montañes B 2017 *Energy and Buildings* **135** 286–301
[3] Energy Performance of Building Directive 2018 The Energy Performance of Buildings Directive: Fact sheet URL `https://ec.europa.eu/energy/sites/ener/files/documents/buildings`$_{p}erformance_{f}actsheet.pdf$
[4] Magrini A and Franco G 2016 *Journal of Cultural Heritage* **21** 834–841
[5] Webb A L 2017 *Renewable and Sustainable Energy Reviews* **77** 748–759
[6] Blumberga A *et al.* 2016 *State of the art on historic building insulation materials and retrofit strategies* 1st ed
[7] International Organization for Standardization 1997 *ISO 14040: Environmental Management - Life Cycle Assessment - Principles and Framework* (ISO)
[8] International Organization for Standardization 2006 *ISO 14044: Environmental Management - Life Cycle Assessment - Requirements and Guidelines* (ISO)
[9] Dansk Standard DS/EN15804:2013 2013
[10] Wolf M A, Chomkhamsri K, Brandao M, Pant R, Ardente F, Pennington D W, Manfredi S, de Camillis C and Goralczyk M 2010 ILCD handbook - general guide for life cycle assessment - detailed guidance Tech. rep. European Commission Joint Research Centre JRC Via E. Fermi, 2749 - 21027 Ispra (VA) Italy
[11] Densley Tingley D, Hathway A, Davison B and Allwood D 2017 *Proceedings of Institution of Civil Engineers: Construction Materials* **170** 91–103
[12] Heat Roadmap Europe 2019 Profiles of heating and cooling energy demands in 2015 for EU28 countries URL `https://heatroadmap.eu/heating-and-cooling-energy-demand-profiles/`
[13] Prognos AG 2014 *Bundesamt für Energie BFE, Prognos AG (2012)*
[14] Leissner J and Kilian R 2014 *Climate for Culture: Built cultural heritage in times of climate change* URL `https://www.climateforculture.eu/index.php?inhalt=furtherresources.projectresults`
[15] Alley R B and Berntsen T 2000 *A report of Working Group I of the Intergovernmental Panel on Climate Change* URL `http://pure.iiasa.ac.at/id/eprint/6101/2/sres-en.pdf`
[16] Ghose A, McLaren S, Dowdell D and Phipps R 2017 *Building and Environment* **117** 274–287
[17] Dylewski R and Adamczyk J 2016 *Journal of Cleaner Production* **133** 467–478
[18] Oregi X, Hernandez P, Gazulla C and Isasa M 2015 *Buildings* **5** 354–380
[19] Karimpour M, Belusko M, Xing K and Bruno F 2014 *Building and Environment* **73** 106 – 114
[20] Sohn J, Kalbar P and Birkved M 2017 *Procedia Environmental Sciences* **38** 737–743
[21] Ylmén P, Mjörnell K, Berlin J and Arfvidsson J 2017 *Building and Environment* **118** 174–183
[22] Sohn J L, Kalbar P P, Banta G T and Birkved M 2017 *Journal of Cleaner Production* **142** 3243–3253
[23] Ingrao C, Scrucca F, Tricase C and Asdrubali F 2016 *Journal of Cleaner Production* **124** 283–298

IOP Conf. Series: Earth and Environmental Science **352** (2019) 012030 doi:10.1088/1755-1315/352/1/012030

A Norwegian zero emission neighbourhood (ZEN) definition and a ZEN key performance indicator (KPI) tool

Marianne Kjendseth Wiik[a], Selamawit Mamo Fufa[a], Inger Andresen[b], Helge Brattebø[b], Arild Gustavsen[b]

[a] SINTEF Building and Infrastructure, Børrestuveien 3, 0373 Oslo, Norway
[b] Norwegian University of Science and Technology (NTNU), 7491 Trondheim, Norway

marianne.wiik@sintef.no

Abstract. Within the Norwegian Research Centre for Zero Emission Neighbourhoods (ZEN) in smart cities, a definition for achieving zero emission neighbourhoods will be developed and tested against nine pilot areas. The ZEN definition considers a series of assessment criteria and key performance indicators (KPI) under seven categories; GHG emissions, energy, power/load, mobility, economy, spatial qualities and innovation. This paper presents the first draft of the ZEN definition, and discusses some of the opportunities and challenges in implementing assessment criteria and KPIs (ZEN metrics) into a ZEN KPI tool. This paper briefly presents the ZEN pilot areas and maps out existing tools used by ZEN stakeholders for the documentation of ZEN metrics. The paper goes further by presenting a ZEN KPI tool conceptual framework for the implementation of the ZEN definition in ZEN pilot areas and outlines a specification for the future development of a ZEN KPI tool. Finally, the paper presents further work for the development of a ZEN KPI tool.

1. Introduction

The United Nation's sustainable development goals (UN SDG) focus on a range of issues to be addressed by society; including healthy living and well-being (SDG 3); affordable, reliable and sustainable energy (SDG 7); sustainable economic growth and productive employment (SDG 8); resilient infrastructure (SDG 9); inclusive, safe, resilient and sustainable cities (SDG 11); suitable consumption and production patterns (SDG 12); combat climate change (SDG 13); protect restore and promote sustainable use of terrestrial ecosystems (SDG 15) and revitalise global partnership for sustainable development (SDG 17) [1]. These are some of the issues addressed in the Norwegian Research Centre for Zero Emission Neighbourhood (ZEN) in smart cities programme [2]. In the ZEN research centre, a definition for zero emission neighbourhoods in smart cities is under development to reduce direct and indirect greenhouse gas (GHG) emissions towards zero [3, 4]. The definition is a working draft that draws on several existing sources and professional expertise to form a comprehensive neighbourhood assessment tool.

The aim of this paper is to outline the background material paramount to the development of a ZEN definition, as well as to outline the development of a key performance indicator (KPI) tool that the ZEN definition will be tested and developed against. This paper briefly presents the ZEN pilot areas and maps out existing tools used by ZEN stakeholders for the documentation and implementation of assessment criteria and KPIs (ZEN metrics). The paper presents a ZEN KPI tool conceptual framework for the implementation of the ZEN definition in the ZEN pilot areas and outlines a specification for the future development of a ZEN KPI tool. It also discusses some of the opportunities and challenges in implementing ZEN metrics into a ZEN KPI tool. Finally, the paper presents further work for the development of a ZEN KPI tool.

2. Background

Several existing neighbourhood assessment tools and research projects have informed the development of the first draft of the ZEN definition, as well as professional expertise input from over fifty stakeholders. Such tools and projects include the Research Centre on Zero Emission Buildings methodology for measuring and reporting GHG emissions for the whole life cycle of a building [5-9], the PI-SEC – Planning Instruments for Smart Energy Communities research project which will provide a tool to resolve both municipal planning (top down) and project planning and construction (bottom-up) needs [10], the Horizon 2020 Smart Cities and Communities programme for secure, clean and efficient energy and focuses on efficient and user-friendly technologies and services within energy, transport and ICT [11, 12], the definition of positive energy blocks (PEB) in Horizon 2020 [13], the BREEAM Communities methodology [14], and the CITYKeys research project that supports the development of smart city solutions and services to address societal challenges [15]. More details about these tools and research projects can be found in [3, 4].

In addition, through a series of ZEN partner workshops on design and planning, energy supply, and buildings and materials, over fifty professional experts have given input to the ZEN criteria, and identified national and international standards relevant for the development of a ZEN definition. More specific details on the outcome of these workshops can be found in [4]. The standards identified include NS-EN 15978: 2011 Sustainability of Construction Works – Assessment of Environmental Performance of Buildings – Calculation Method [16], NS 3720: 2018 Method for Greenhouse Gas Calculations for Buildings [17], NS 3457-3: 2013. Classification of Construction Works – Part 3 Building Types [18], NS 3451: 2009: Table of Building Elements [19], ISO 52000: 2017. Energy performance of buildings - Overarching EPB assessment - Part 1: General framework and procedures [20], SN/TS 3031: 2007. Calculation of energy performance of buildings - Method and data [21], NS 3454: 2013. Life cycle costs for construction works - Principles and classification [22], NS-EN 16627: 2015. Sustainability of construction works - Assessment of economic performance of buildings - Calculation methods [23], ISO 15686-5: 2017. Building and construction assets - service life planning. Part 5: Life-cycle costing [24], and NS-EN 16258: 2012. Methodology for calculation and declaration of energy consumption and GHG emissions of transport services (freight and passengers) [25].

2.1. The ZEN definition

In the ZEN research centre, a neighbourhood is defined as a group of interconnected buildings with associated infrastructure, located within a confined geographical area. A zero emission neighbourhood aims to reduce its direct and indirect greenhouse gas (GHG) emissions towards zero over the analysis period, in line with a chosen ambition level with respect to which life cycle modules and building and infrastructure elements to include. The neighbourhood should focus on the following, whereby the first four points have direct consequences for energy and emissions:

a) Plan, design and operate buildings and associated infrastructure components towards zero life cycle GHG emissions.
b) Become highly energy efficient and powered by a high share of new renewable energy in the neighbourhood energy supply system.
c) Manage energy flows (within and between buildings) and exchanges with the surrounding energy system in a smart and flexible way. Flexibility should facilitate the transition to a decarbonised energy system and reduction of power and heat capacity requirements.
d) Promote sustainable transport patterns and smart mobility systems.

e) Plan, design and operate with respect to economic sustainability, by minimising total life cycle costs.
f) Plan and locate amenities in the neighbourhood to provide good spatial qualities and stimulate sustainable behaviour.
g) Development of the area is characterised by innovative processes based on new forms of cooperation between the involved partners leading to innovative solutions.

In addition, seven categories and accompanying assessment criteria and key performance indicators (KPI) have been identified by expert stakeholders as important for a Norwegian zero emission neighbourhood in smart cities definition (i.e. GHG emissions, energy, power/load, mobility, economy, spatial qualities and innovation). A full overview of the ZEN assessment criteria and KPIs can be found in [13] and are summarised in Table 1. These assessment criteria and KPIs utilise existing policies, frameworks, standards and references that architect, engineer and constructor (AEC) professionals are already familiar with. KPI results will be used to track, understand and validate the progress and performance of the ZEN pilot areas. Some assessment criteria and KPIs are valid for a building (B) system boundary, neighbourhood (N) system boundary or both (BN). In addition, the assessment criteria and KPIs are to be reported across a range of project phases, namely; the strategic planning phase, the implementation phase and the operational phase. KPIs for the spatial qualities and innovation categories still need to be developed.

Table 1. ZEN assessment criteria and key performance indicators [3, 4].

Category and Assessment Criteria	KPIs	Unit	System boundary	References
GHG Emissions				
Lifecycle GHG emissions				[16-19]
	- per building	$kgCO_{2eq}/m^2$ heated floor area (BRA)/yr	B	
	- per infrastructure	$kgCO_{2eq}/m^2$ outdoor space (BAU)/yr	N	
	- per user	$kgCO_{2eq}$/capita	BN	
	- total life cycle GHG emissions	tCO_{2eq}	BN	
GHG emission reduction				
	- reduction compared to a base case	%	BN	
Energy				
Energy efficiency in buildings				[2, 20, 21, 26]
	- net energy need	kWh/m² heated floor area (BRA)/yr	B	
	- gross energy need	kWh/m² heated floor area (BRA)/yr	B	
	- total energy need	kWh/m² heated floor area (BRA)/yr	B	
Energy carriers	- energy use	kWh/yr	BN	
	- energy generation	kWh/yr	BN	
	- delivered energy	kWh/yr	BN	
	- exported energy	kWh/yr	BN	
	- self consumption	%	BN	
	- self generation	%	BN	
	- colour coded carpet plot	kWh/yr	BN	
Power / Load				
Power/load performance				[2, 27]
	- yearly net load profile	kW	BN	
	- net load duration curve	kW	BN	
	- peak load	kW	BN	
	- peak export	kW	BN	
	- utilisation factor	%	BN	
Power load flexibility				
	- daily net load profile	kW	BN	
Mobility				

Mode of transport	- per mode of transport	%	N	[14, 15, 17, 25]
Access	- to public transport	Metres and frequency	N	
Economy				
Life cycle costs (LCC)	- per building - per infrastructure - per user - total LCC	NOK/m^2 heated floor area (BRA)/yr NOK/m^2 outdoor space (BAU)/yr NOK/capita NOK	B N BN BN	[19, 22-24, 28]
Spatial Qualities				
Demographic needs and consultation plan	- to be developed	- to be developed	BN	[14]
Public space	- to be developed	- to be developed	N	

2.2. *The ZEN pilot areas*
Altogether, the ZEN pilot areas encompass one million square meters of development for more than 30,000 inhabitants. The pilot areas include Campus Evenstad, Ydalir in Elverum, Furuset in Oslo, Fornebu in Bærum, Zero Village Bergen, NTNU Campus and Sluppen in Trondheim, the NRK site in Steinkjer and NyBy in Bodø. A mapping of the ZEN pilot area's characteristics regarding location, stakeholders, goals, measures, status and challenges can be found in [29].

3. Methodology
First steps involve mapping existing tools used by ZEN stakeholders to measure ZEN metrics. This is followed by discussing the purpose and user needs of a ZEN KPI tool. This information is used to develop a ZEN KPI conceptual framework that can then be used to theoretically test the ZEN definition with real data from the ZEN pilot areas. The testing will be a continuous, iterative process that goes beyond the scope of this paper. Part of this process will involve ascertaining suitable minimum requirements, ambition levels, reference values, threshold values, weighting and benchmarking for the successful implementation of a ZEN KPI tool.

4. Results and Discussion

4.1. *Existing tools*
The ZEN research centre stakeholders use a plethora of tools to document the assessment criteria and KPIs as outlined in Table 1. A mapping exercise has been carried out, and the following tools have been documented for each category: GHG emissions (e.g. ZEB emission tool, OneClick LCA, bLCAd-tool, områdeLCA, EFFEKT, PI-SEC indicator tool, Simapro Arda, vegLCA), energy (e.g. IDA-ICE, ZEB energy tool, PI-SEC indicator tool, Energyplus, SIMIEN), power/load (e.g. eTransport), mobility (e.g. PI-SEC), economy (e.g. Bygganalyse), spatial qualities (e.g. BREEAM Communities, PI-SEC planning wheel, Urbanetic Fabric, GIS, CITYBES, SSB data) and innovation (to be mapped). The list of tools that may be used in ZEN is by no means exhaustive. However, there is a distinct lack of harmonisation between these tools, which leads to difficulties in harmonising the inputs and outputs of a potential ZEN KPI tool. Some of the main issues include differences in system boundaries, methodology and data(bases) used, as well as differences in data resolution used dependent on the phase of the project (i.e. typically a low data resolution is used in the early planning phases and a high data resolution in the latter project phases). In addition, there is an increase in data complexity when one shifts the focus from an individual building or infrastructure to the whole neighbourhood level. The various tools identified in the mapping exercise use a range of different computing formats. A ZEN KPI tool would need to be flexible and compatible with these different formats (e.g. csv, xml, json, sql and html) [30].

4.2. The ZEN definition and ZEN KPI tool specification

The main purpose of the ZEN KPI tool is to operationalise the ZEN definition, and aid ZEN stakeholders through the planning, design, construction and operation of zero emission buildings, infrastructure and neighbourhoods. As a result, the ZEN KPI tool should be flexible for a range of conditions, easy to understand and clearly communicate the output of results. In addition, the ZEN KPI tool should be transparent and facilitate for comparisons between components, buildings, infrastructure and neighbourhoods. The ZEN KPI tool will need to process large amounts of data, as well as harmonise various standards, methods, data(bases), tools, system boundaries and interpret quantitative and qualitative measures. The ZEN KPI tool should be made available in Norwegian and English since a range of stakeholders will be using it. The ZEN KPI tool will also need to be robust and withstand future developments and changes, for example the future development of CO_2 emission requirements and tightening of energy requirements in Norwegian building codes [31].

Some of the challenges identified by stakeholders in the ZEN pilot areas include project organisation and management, lack of knowledge, legislation, goal conflicts, time and cost pressures, new energy technologies, system boundaries, risks and uncertainties, flexibility, transferability, complexity and scale [29]. A guideline for the implementation of the ZEN definition has identified a range of contextual parameters that affect the assessment of a pilot area [4]. These include the various project phases a pilot area progresses through (i.e. strategic planning phase, implementation phase and operational phase), the scope of assessment (i.e. material/component, building/infrastructure, neighbourhood and city) and which stakeholders are involved during the various project phases (i.e. planners, developers, architects, engineers, utility companies and citizens etc.).

However, one of the main challenges with the ZEN definition is that it is an early, first version that has not yet been tested against ZEN pilot areas or through the proposed ZEN KPI tool. As a result, it will be necessary to revise parts of the ZEN definition to take into consideration the practicalities of implementing various KPIs. For example, the access to public transport KPI under the mobility category is likely to be expanded to include access to other amenities, green spaces, city centres and local shopping centres. Similarly, the scope of the spatial qualities' category will be expanded to cover user qualities and contain more KPIs to reflect this. For example, the demographic needs and consultation plan assessment criteria may be expanded to include KPIs that consider gender, age, minority background, disabilities, population and resident density, occupation, total number of users and ZEN residents. These are useful indicators for other assessment criteria within the ZEN definition. For example, the total number of users can be used in calculating life cycle GHG emissions per user and life cycle costs per user. This shows that a future ZEN KPI tool will need to be dynamic and take into consideration these interconnections.

4.3. The ZEN KPI tool conceptual framework

A conceptual framework has been developed for the ZEN KPI tool from the information and experiences gathered on the ZEN definition, ZEN pilot areas, ZEN stakeholders and existing tools used to measure ZEN metrics, see Figure 1. It builds upon the initial ZEN toolbox framework developed by Houlihan Wiberg and Baer in [32]. The framework takes into consideration the various contextual parameters that will affect input data (grey boxes on the left-hand side of the figure), such as project phase, scope and stakeholders. It also maps out the various existing tools used by ZEN stakeholders (red box) to assess the various ZEN metrics outlined in the ZEN definition (yellow box) and shows how the results from these tools will feed into the ZEN KPI tool, structured according to the seven ZEN definition categories. Once the input data has been gathered an iterative process will take place within the ZEN KPI tool (green box) which involves obtaining and setting reference values, threshold values, weighting and benchmarking. Reference values will be set according to current standard practice (e.g. net energy need values set according to requirements in the Norwegian building regulations (TEK17) for new buildings), whilst the threshold, weighting and benchmarking values will be decided upon through a series of expert workshops with ZEN stakeholders, and by testing proxy data gathered from the ZEN pilot areas in the ZEN KPI tool. Ambition levels and goals can then be set according to the benchmark values. Finally, the results can be visualised in a range of ways, for example, in terms of a spider diagram, traffic light system, or mapped onto a city information model (CIM) of the neighbourhood. The results from the ZEN KPI tool can also be used in a series of visualisation tools (blue box) developed by the ZEN research centre, such as a dashboard of results, GIS, virtual reality and augmented reality.

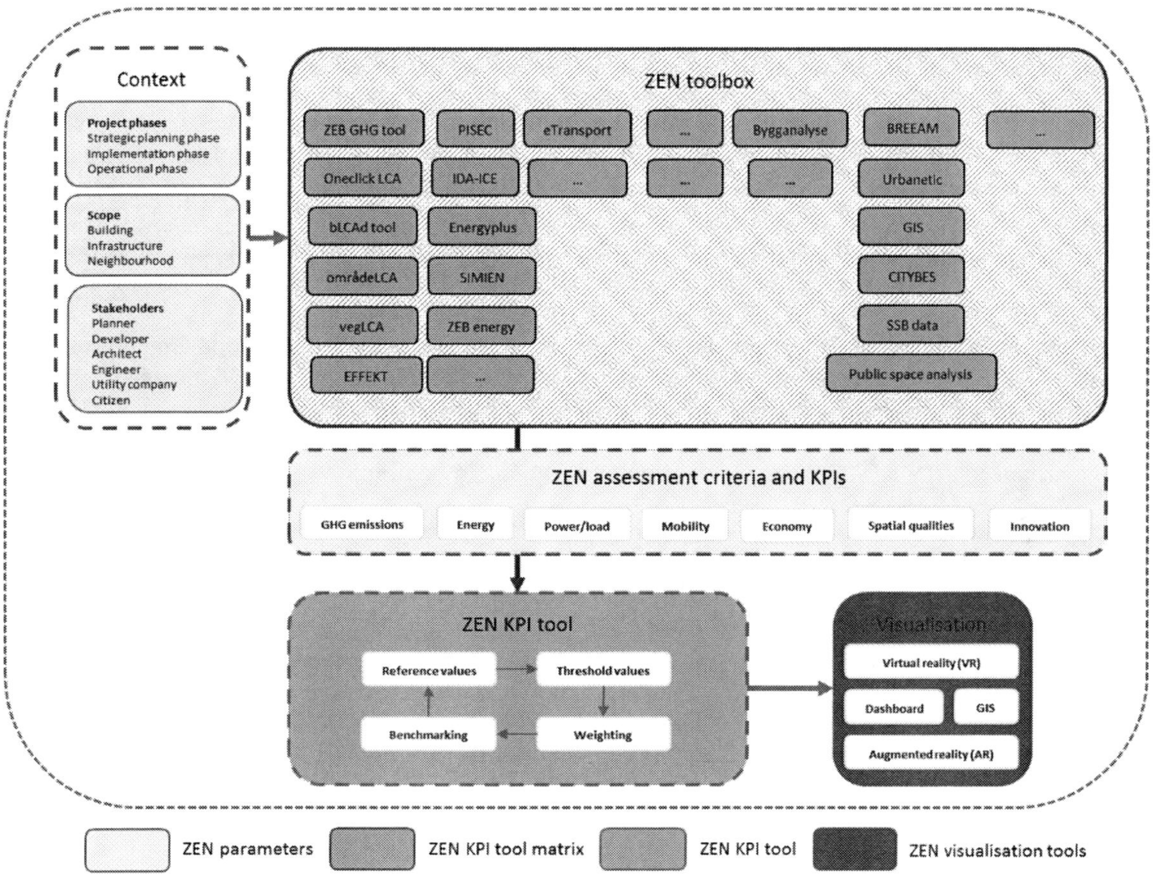

Figure 1. ZEN KPI tool conceptual framework developed from [32].

It is acknowledged that not every KPI is suitable for weighting or benchmarking in the ZEN KPI tool, but are useful information to document and use in some of the other KPIs. For example, it may be difficult to weight and benchmark a net load duration curve in the power/load category but is useful information for weighting and benchmarking the utilisation factor. The authors acknowledge that it may also be difficult to implement ambition levels and goals in accordance with the ZEN definition when many of the ZEN pilot areas are already underway and have already set their own goals for reaching a ZEN. Efforts will need to be made to harmonise existing goals and ambitions with that of the ZEN research centre's definition. In other cases, some KPIs are important to document in order to identify areas for improvement within the neighbourhood. For example, the GHG emission result and net energy need per building indicators help identify which buildings lead to higher GHG emissions or higher energy use in an early planning phase and can then be used to target areas for improvement in the implementation and operational phases.

Different types of multicriteria decision management (MCDM) and multicriteria decision analysis (MCDA) will be considered. It is important to find a method that makes it possible to assess and range design alternatives so that they can be compared to quickly ascertain the best course of action. When it comes to weighting, the aggregation model defines how weights and value functions for each attribute are synthesised in order to find an overall value for each category. The simplest one, and by far the most frequently used, is the simple additive weighting (SAW) model, whereby:

$$v(x) = \sum_{i=1}^{n} w_i v_i (x_i)$$

v is the overall value of the evaluation object x,
xi is the measurement of object x on attribute i,
vi is the single-attribute value function,
wi is the weight of attribute i,
n is the number of attributes.

Another issue that needs to be resolved within the ZEN KPI tool is the requirements for data resolution. For example, is a detailed life cycle GHG emission assessment at the material component level required at an early strategic planning phase? Or is it enough to simplify the process and calculate GHG emissions based on rules of thumb for the whole neighbourhood? For example, one could multiply empirical data from previous projects with building quantities in the new project (e.g. X $kgCO_{2eq}/m^2/yr$ per m^2 of floor, wall and roof etc. for each building within the development) and then carry out more detailed calculations during the implementation and operational phases when more project specific data is available.

Finally, the authors have identified three ZEN pilot areas for the further development and testing of the ZEN KPI tool, namely; Fornebu which is in the strategic planning phase, Ydalir which is in the implementation phase and Campus Evenstad which is in the operational phase. Future work will involve gathering data from these pilot areas to help test and simulate a first version of the ZEN KPI tool based on the proposed ZEN KPI tool conceptual framework and ZEN definition.

5. Conclusion

This paper has presented the current working draft of a Norwegian definition for zero emission neighbourhoods, briefly outlined the ZEN pilot areas and mapped out existing tools used by ZEN stakeholders to document ZEN metrics. This paper has used this information together with experiences and feedback from stakeholders to present a ZEN KPI tool conceptual framework for the implementation of the ZEN definition into ZEN pilot areas and has outlined a specification for the future development of a ZEN KPI tool. It has also discussed some of the opportunities and challenges in implementing ZEN metrics into a ZEN KPI tool. Finally, this paper has presented further work for the development of a ZEN KPI tool.

References

[1] United Nations General Assembly 2015 *Transforming our world: the 2030 agenda for sustainable development* [Available from: http://www.un.org/ga/search/view_doc.asp?symbol=A/RES/70/1&Lang=E]

[2] FME ZEN 2016 *The research centre on zero emission neighbourhoods in smart cities* [Available from: http://fmezen.com/]

[3] Wiik M et al 2018 *Zero emission neighbourhoods in smart cities: Definition, key performance indicators and assessment criteria: version 1.0. billingual version. Nullutslippsområder i smarte byer: definisjon, nøkkelindikatører og vurderingskriterier* Oslo: ZEN Project Report (7) SINTEF Academic Press ISBN: 978-82-536-1593-6

[4] Wiik M, Fufa S, Bær D, Sartori I, Andresen I 2019 *A ZEN guideline for the ZEN pilot areas. version 1.0.* Oslo: ZEN Project Report (11) SINTEF Academic Press ISBN: 978-82-536-1608-7

[5] Fufa S, Schlanbusch R, Sørnes K, Inman M, Andresen I. 2016 *A Norwegian ZEB definition guideline* Oslo: ZEB Project Report (29) SINTEF Academic Press

[6] Kristjansdottir T et al. 2014 *A Norwegian ZEB definition: embodied emissions* Oslo: ZEB Project Report (17) SINTEF Academic Press

[7] Wiik M, Fufa S, Kristjansdottir T, Andresen I. 2018 *Lessons learnt from embodied GHG emission calculations in zero emission buildings (ZEBs) from the Norwegian ZEB research centre* Energy Build **165** 25-34 https://doi.org/10.1016/j.enbuild.2018.01.025

[8] ZEB 2019 *The research centre on zero emission buildings* [Available from: http://www.zeb.no/]

[9] Schlanbusch R, Segtnan I. 2018 *bks 473.010. nullutslippsbygninger (ZEB) retningslinjer og beregningsmetoder* Oslo: SINTEF
[10] Walnum H, Sørnes K, Mysen M, Sørensen Å, Almås A. 2017 *Preliminary toolkit for goals and KPIs* Oslo
[11] European Commission 2017 *Smart cities & communities* [Available from: https://ec.europa.eu/inea/en/horizon-2020/smart-cities-communities]
[12] RemoUrban 2019 *Smartcities network* [Available from: http://www.remourban.eu/smartcities-network/title.kl]
[13] European Commission 2017 *Horizon 2020 work programme 2018-2020. 10. secure, clean and efficient energy Contract No.: European Commission Decision C(2017)7124 of 27 October 2017*
[14] BREEAM 2012 *The BREEAM communities technical manual* BRE [Available from: https://www.breeam.com/communitiesmanual/]
[15] Bosch P, Jongeneel S, Rovers V, Neumann H, Airaksinen M, Huovila A. 2017 *CITYkeys indicators for smart city projects and smart cities*
[16] EN 15978 2011 *Sustainability of construction works - assessment of environmental performance of buildings - calculation method* European Committee for Standardization, Brussels, Belgium
[17] NS 3720 2018 *Method for greenhouse gas calculations for buildings* Oslo: Standard Norge
[18] NS 3457-3 2013 *Klassifikasjon av byggverk – del 3 bygningstyper* Oslo: Standard Norge
[19] NS 3451 2009 *Bygningsdelstabell / table of building elements* Oslo: Standard Norge
[20] ISO 52000-1 2017 *Energy performance of buildings -- overarching EPB assessment -- part 1: general framework and procedures* International Organization for Standardization
[21] NS 3031 2007 *Calculation of energy performance of buildings - method and data* Oslo: Standard Norge
[22] NS3454 2013 *Life cycle costs for construction works - Principles and classification* Oslo: Standard Norge
[23] NS-EN 16627 2015 *Sustainability of construction works - assessment of economic performance of buildings - calculation methods* Oslo: Standard Norge
[24] ISO 15686-5 2017 *Building and construction assets - service life planning. part 5: life-cycle costing* Switzerland: International Standard Organisation
[25] EN 16258 2012 *Methodology for calculation and declaration of energy consumption and GHG emissions of transport services (freight and passengers)* Oslo: Standard Norge
[26] IEA EBC 2014 *IEA solar heating and cooling (task 40) EBC annex 52 towards net zero energy solar buildings* http://www.iea-ebc.org/projects/completed-projects/ebc-annex-52/
[27] IEA EBC 2014 *EBC Annex 67 energy flexible buildings* [Available from: http://www.iea-ebc.org/projects/ongoing-projects/ebc-annex-67/ http://www.annex67.org/
[28] Norconsult Informasjonssystemer 2017 *Norsk prisbok* Norconsult Informasjonssystemer AS
[29] Baer D, Andresen I 2018 *ZEN pilot projects: mapping of the pilot projects within the research centre on zero emission neighbourhoods in smart cities* Oslo: ZEN Project Report (10) SINTEF Academic Press
[30] Resch E, Andresen I 2018 *A database tool for systematic analysis of embodied emissions in buildings and neighbourhoods* Buildings 8(106):16
[31] Direktoratet for byggkvalitet (DIBK) 2018 *Utredning av livsløpsbaserte miljøkrav i TEK* Oslo
[32] Wiberg A, Baer D 2018 *ZEN Toolbox. First concept for the ZEN toolbox for use in the development of Zero Emission Neighbourhoods. Version 1.0.* Oslo

Acknowledgements
This research has received funding from the research centre for zero emission neighbourhoods in smart cities (FME ZEN). The authors are thankful for the support from ZEN partners and the Norwegian Research Council.

IOP Conf. Series: Earth and Environmental Science **352** (2019) 012031 doi:10.1088/1755-1315/352/1/012031

Energy analysis and energy planning for kindergartens based on data analysis

Yiyu Ding*, **Helge Brattebø**, **Natasa Nord**

Department of Energy and Process Engineering, Norwegian University of Science and Technology, NO-7491 Trondheim, Norway

e-mail: yiyu.ding@ntnu.no

Abstract. The aim of the study was to utilize different building data for prediction of development in energy use of a typical building type. In this study, energy use and its future development for kindergartens in Trondheim, Norway, were analyzed. The energy use data were retrieved from the energy monitoring platform of Trondheim Municipality. The total area of all the kindergartens was about 76 000 m^2, where the area of each kindergarten was ranging from 100 – 4 471 m^2. Firstly, typical heat and electricity duration curves per m^2 of kindergartens in Trondheim within six years were identified. Secondly, the kindergartens were divided into two cohorts based on their connection to district heating (DH). The average total annual energy use was 177 kWh/m^2 for kindergartens without DH, and 168 kWh/m^2 for kindergartens connected to DH. The peak load values were similar for both cohorts, about 140 W/m^2. Analysis of the duration curves showed a bigger electricity load variation for the kindergartens without DH connection. Within the building cohort with DH, three cases were found depending on the energy share from DH; i.e. DH high share, DH average share, and DH low share. By following different background data for CO_2 factors of electricity and local DH, the kindergarten with DH high share had almost the lowest annual CO_2 emission. Contrarily, the annual CO_2 emission of a kindergarten with lower share of DH, or without DH, usually had a wider range of emissions due to its dependence of the electricity production mix. Finally, a prediction was made by assuming 14.2 % growth rate of kindergartens on the ground of the average six-year total kindergarten area. The result showed that if more than 50- 67 % of the new building area would be connected to DH, a smaller increase of CO_2 emission from the projected area could be achieved, depending on the relevant CO_2 factors. This proved that buildings with DH were more robust than the one without DH concerning CO_2 emission. The suggested analysis method and identified duration curves could be used to as a reference example for defining energy profiles of other building types. These profiles are necessary for diversifying and upgrading local energy supply pathways, infrastructure sizing, and improving urban energy planning.

1. Background

Approximately 36-40 % of energy is consumed in building service around the world each year, and it is responsible for nearly 40 % of direct and indirect CO_2 emissions [1]. Therefore, urban building stocks are expected to make high contribution for low energy use and reduction of greenhouse gas emissions. In Norway, due to cheap and green electricity power from the abundant hydro-power, coverage rate of district heating (DH) system is small. DH only contributes approximately 11 % of total heating demand

IOP Conf. Series: Earth and Environmental Science **352** (2019) 012031 doi:10.1088/1755-1315/352/1/012031

in Norway [2]. Norwegian residential and service buildings are highly reliant on electricity for space heating (SH) and domestic hot water (DHW). Whereas, driven by the motivation of economic and environmental benefits of DH, relevant regulations and investment subsidies have been introduced to expand the build-up of DH system in Norway. As the third largest city in Norway, Trondheim municipality has been committed improving urban plans for better living environment under the pressure of urbanization, population growth, and mitigation of anthropological carbon footprint [3].

The aim of this article was to identify energy profiles of one typical building type in Trondheim. Typical profiles of energy use can be used as input to building simulations and model calibration. The historical energy use data of kindergartens from 2013 to 2018 was retrieved from the energy monitoring platform of Trondheim Municipality [4]. The outdoor weather data and energy use was given in hourly resolution. Besides kindergarten, school, heath/nursing center, sports center and others are also monitored.

2. Methods

2.1. *Building general information*

During the six years, numbers of total kindergartens have been increased from 83 to 99. Based on the connection to DH, the kindergartens were divided into two cohorts, Cohort 1 and Cohort 2. In Cohort 1, the buildings are not connected to DH, and supplied by electricity only, and in Cohort 2, the buildings are connected to DH. The yearly building numbers and building area of the two cohorts were compared in Table 1. In total, there were 559 hourly files of kindergartens being used in the analysis.

Table 1. Building numbers and area of Cohort 1 and Cohort 2.

		2013	2014	2015	2016	2017	2018
Building numbers (-)	Cohort 1	66	66	68	68	71	71
	Cohort 2	21	23	26	27	28	28
	Total	83	89	94	95	99	99
Building area (m^2)	Cohort 1	36979	38855	40890	40890	43259	43259
	Cohort 2	24623	26317	30105	31766	32768	32768
	Total	61602	65172	70995	72656	76027	76027

It shows that generally the share of Cohort 2 is smaller than Cohort 1 but growing, especially when it comes to the building area. As shown in Figure 1, the blue square stands for Cohort 1 and the red for Cohort 2, and the green line demonstrates the percentage of Cohort 2, Cohort 2 covers around 43 % of total building area till 2018. This can be explained in Figure 2 by plotting the relation between building area and weekly-based load needs. Most of the kindergartens in Cohort 1 were built within small to medium size (in blue stars), while kindergartens in Cohort 2 were within medium to large size (red circles). The area of each kindergarten varies largely from 100 to 4471 m^2.

2.2. *Energy duration curve per m^2*

There is a big variety of the building area of each kindergarten, hence, the load duration curves were analyzed based on energy demand per m^2. For buildings in Cohort 1, the duration curves were made only by electricity use. For buildings in Cohort 2, the duration curves of electricity and DH were analyzed separately. Yearly duration curve of each building was obtained by sorting annual load hourly profile from highest to lowest values, and average duration curve was made by the mean values of all the curves. From the average energy use under its outdoor temperature, energy signature was established to imply the relation between energy demand per m^2 and outdoor temperature. MATLAB was used for energy data analysis.

IOP Conf. Series: Earth and Environmental Science **352** (2019) 012031 doi:10.1088/1755-1315/352/1/012031

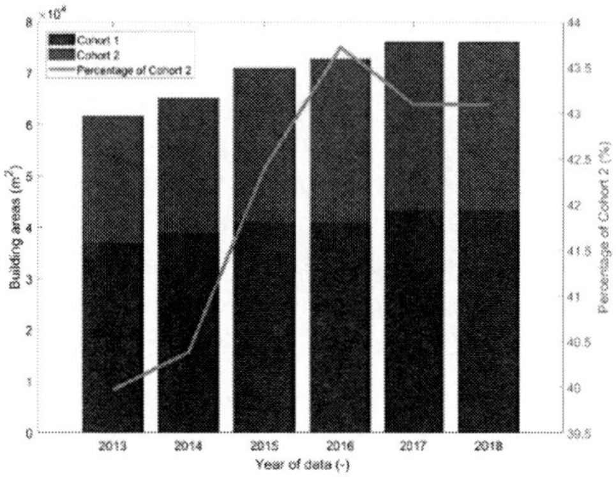

Figure 1. Building area comparison of Cohort 1 and Cohort 2.

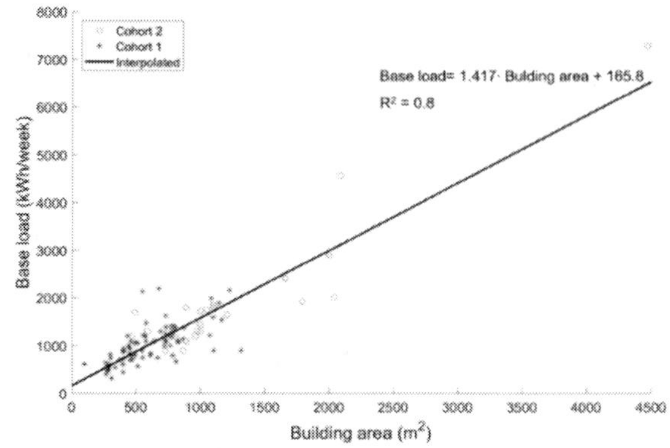

Figure 2. Building area vs Building weekly base load of Cohort 1 and Cohort 2.

2.3. *Energy coverage rate in Cohort 2*

In Cohort 2, heating demand was provided by DH and the other energy demand by electricity. In order to see the contribution from the two energy supply ways, Figure 3 demonstrates the energy coverage rates from DH and electricity in Cohort 2. In Figure 3, DH was marked in red and electricity in blue, each bar stands for the average energy use situation of one kindergarten from 2013 to 2018, and all the 28 kindergartens were included. From the bar chart, three cases were defined, they were named as DH average share, DH high share and DH low share. In the case of DH high share, nearly 76.9 % of total energy use comes from DH, 31.4 % higher than the case of DH low share. On average, DH supports 60.0 % of total energy use, as listed in Table 2.

Table 2. Energy coverage rate of three cases in Cohort 2.

	From DH (%)	From electricity (%)
DH average share	60.0	40.0
DH high share	76.9	23.1
DH low share	45.5	54.5

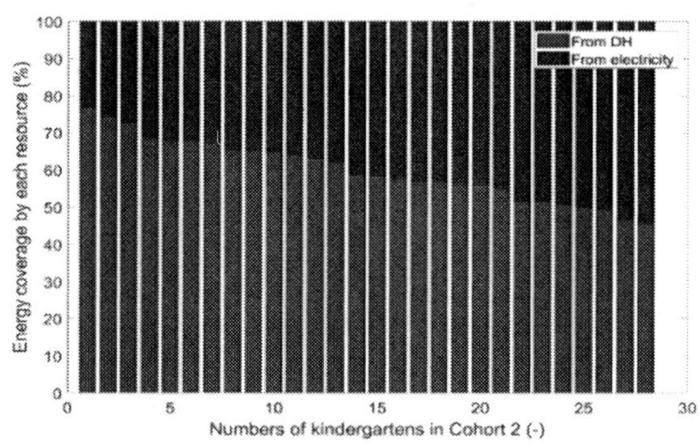

Figure 3. Energy coverage rates of DH and electricity in Cohort 2.

2.4. *CO_2 factors of electricity and DH production*

Benefitting from the modern transmission technology and the characteristic of electricity, electricity is capable of long-distance transmission with less than 5 % of loss. Norway is connected in the Nordic power grid and further expanded into the wider European grid, and electricity is traded in the free market. Within the Norwegian border, CO_2 factor of electricity can be as low as 10 gCO_2/kWh (named as $CO_{2\text{-}EL1}$), which is mainly contributed by the abundant hydro-power, however this factor can be high up to 110 gCO_2/kWh ($CO_{2\text{-}EL2}$) in the Nordic region since fossil fuels are involved in the electricity production mix. Distinguished from electricity, the transmission loss of heating can be quite high, which makes DH not suitable for long-distance transport. Therefore, the equivalent energy and environmental factors of DH is mostly locally specified. From the information of Norsk Fjernvarme, during 2010 to 2018 most of the DH in Trondheim has been provided by waste incineration, followed by fossil gas with the

IOP Conf. Series: Earth and Environmental Science **352** (2019) 012031 doi:10.1088/1755-1315/352/1/012031

contribution of around 10 %, and the small rest comes from flexible electricity, bio-energy, ambient heat, and fossil oil [5]. In Norway, in accordance to NS 3720-2018, the CO_2 emission from waste-to-energy for energy production (electricity and DH) has been allocated to the sector of waste management instead of energy sector [6]. The CO_2 factors of DH production in Trondheim were calculated based on the annual production composition of energy sources. Three typical CO_2 factors of DH were found, they are the average value from 2010 to 2018, value of 2015 as the 9-year lowest, and value of 2010 as the 9-year highest. These factors were used as background data for the assessment of CO_2 emission, respectively. The CO_2 factors of DH production in Trondheim are listed in Table 3, and the CO_2 data of fossil gas, bio-energy and fossil oil can be found in Norsk Energi [7].

Table 3. CO_2 factors of DH production in Trondheim.

		2010-2018: $CO_{2\text{-DH1}}$	2015: $CO_{2\text{-DH2}}$	2010: $CO_{2\text{-DH3}}$
Composition of energy sources (%)	Waste incineration	74.0	83.1	61
	Fossil gas	10.8	5.9	20
	Flexible electricity	8.5	5.0	6
	Bio-energy	4.0	4.0	5
	Ambient heat	0.8	1.0	1
	Fossil oil	1.9	1.0	7
CO_2 factors (gCO_2/kWh)		41.66	23.5	76.3

2.5. Annual CO_2 emission of one typical kindergarten and future prediction

From Figure 2, a typical kindergarten in Trondheim was determined at 700 m^2 concerning the main size ranges the two cohorts. For the buildings in Cohort 1, as addressed above regarding the difficulty of splitting energy share from heating and electricity, therefore, the annual CO_2 emission comparison of one typical kindergarten between Cohort 1 and Cohort 2 was made based on the annual average energy demand of Cohort 1. For Cohort 2, the three cases regarding DH shares were considered separately.

After the annual CO_2 emission calculation of one typical kindergarten was made and compared, the impact of new building area was predicted. In this article, 10 000 m^2 of new building area of kindergarten (A_{new}) was assumed to be added in Trondheim. The building area growth rate (r) was defined as the ratio between (A_{new}) and the annual average total building areas of kindergarten throughout the six years, which is 70 413 m^2. The increasing building area rate is 14.2 %. This growth rate was used as the reference line, and compared with the CO_2 growth rate based on different background data by varying the percentage of new building area connected to DH (x). For simplicity, the annual CO_2 emission was calculated based on the CO_2 factor of Nordic electricity ($CO_{2\text{-EL2}}$) and the three DH production factors. Meanwhile, for the new area connected to DH, the case of DH average share was used. In Function (1), as the denominator, the average annual average CO_2 emission of all the kindergartens ($\overline{CO_2}$) was calculated from the annual average energy use of Cohort 1 and Cohort 2 within the six years. The comparison between growth rates of building area and CO_2 can be explained as:

$$r - \frac{CO_{2-added}}{\overline{CO_2}} \cdot 100\% \tag{1}$$

$$CO_{2-added} = [A_{new} \cdot (1-x) \cdot E_{EL} + A_{new} \cdot x \cdot E_{DH-EL}] \cdot CO_{2-EL2} + A_{new} \cdot x \cdot E_{DH-DH} \cdot CO_{2-DHi}$$
$$(i = 1, 2, 3)$$

When Function (1) = 0, there is a break-even point that the increasing rates of CO_2 emission and new building area are same. When Function (1) < 0, it means if increasing new building area by 14.2 %, more than 14.2 % more CO_2 emission would be produced. On the contrary, when Function (1) > 0, it implies that slower CO_2 emission growth could be achieved.

3. Results

3.1. *Results of energy duration curve and Energy signature per m^2*

The annual average duration curves were presented in Figure 4, Figure 5, and Figure 6, and the annual energy demand were summarized in Table 4. Average duration curves were plotted in black thick lines. The peak load for the two cohorts are almost same. The maximum deviation from the average curves are 27.2 % in Cohort 1 and 24.3 % in Cohort 2. The deviation considers 0- 4000 hour in the duration curve. Energy loads during the last 4760 hours are small, and have minor influence of the grid and plant sizing. Moreover, peak load for Cohort 1 can only expect from electricity; while the peak load for Cohort 2 can be satisfied by DH and electricity, it releases the maximum demand of power grid. Although electricity use in Cohort 2 has weak relation with outdoor temperature, the duration curves of six years have similar pattern except higher use in 2013. It may be explained that fewer kindergartens were used for the analysis, and it caused the large deviation. The detailed annual duration curves can be found in Appendix Figure A 1 to Figure A 6, and there were several unknown high peak loads in Cohort 1.

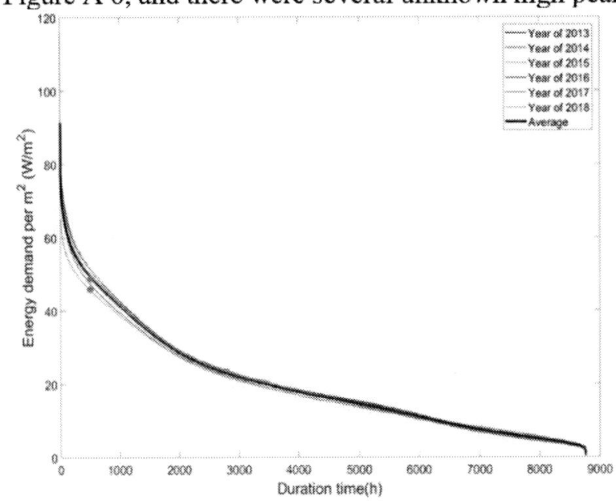

Figure 4. Average total energy duration curves of Cohort 1.

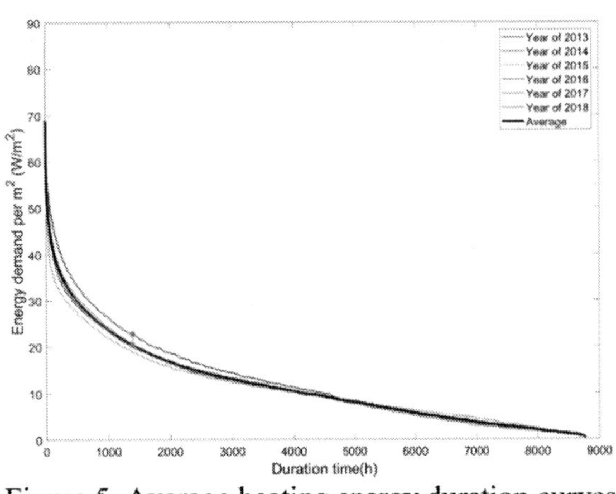

Figure 5. Average heating energy duration curves of Cohort 2.

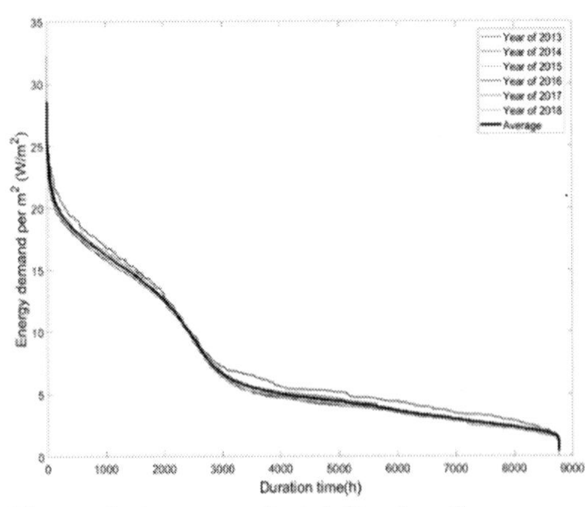

Figure 6. Average electricity duration curves of Cohort 2.

Table 4. Average annual energy use of Cohort 1 and Cohort 2.

		2013	2014	2015	2016	2017	2018	Average
Cohort 1	E_{EL} (kWh/yr)	182.3	169.6	169.6	180.9	180.8	179.8	177.2
Cohort 2	E_{DH-DH} (kWh/yr)	111.6	100.9	98.8	102.2	101.6	102.9	103.0
	E_{DH-EL} (kWh/yr)	69.9	65.4	64.6	62.9	62.5	63.1	64.7

Moreover, to see if the energy use followed the outdoor temperature (t_{od}), heating degree days (HDD)/ heating degree hours (HDH) and energy signature were adopted as rough measurements.

Firstly, heating degree days (HDD)/ heating degree hours (HDH) is the integral of difference between indoor and outdoor temperatures, and is robust tool of predicting space heating. 12- 18 °C are commonly used as the effective indoor temperature to avoid oversizing of heating plants [8]. In this article 18 °C was chosen to roughly estimate the colder and milder weather conditions. The HDH of the six years can be found in Table 5. The average annual heating use of Cohort 2 (E_{DH-DH}) better followed the outdoor temperature (t_{od}) than the average annual energy use of Cohort 1 (E_{EL}).

Table 5. Heating degree hours of six years.

	2013	2014	2015	2016	2017	2018
°C·h	107562.4	94982.4	99146.4	106567.2	105487.2	106156.8

Secondly, energy signature curve can be used as a function of t_{od} to describe and predict heating energy demand [8] [9]. Figure 7 and Figure 9 were made by average hourly energy demand of six years (105 168 hourly data). For buildings in Cohort 1, it is rather difficult to draw one interpolation curve to describe the relation between energy demand $P(t_{od})$ and (t_{od}) in the whole temperature range. There was a break around 5 °C, and energy demand turning back and forth with t_{od}. The appearance of break has been discussed before in caused by changing of heating equipment under different t_{od} [10]. In this article, it can be explained that some electric heating equipment may be shut down during off- work hours in Cohort 1. For example, electric resistant heater has little thermal inertia, which makes it unnecessary to keep on with non-appearance of occupants. Since electricity is used both for heating and other electric appliances, it is not easy to make accurate calculation of energy consumption share for heating and other electric uses. To know the daily operation routine of these buildings is needed. For buildings in Cohort 2 of hydronic DH system, SH and DHW are measured in one meter. The DHW use of one kindergarten was assumed as constant as its six-year average use, and its annual use followed the Norwegian statistic data [12], which is around 9 kWh/m^2 in most of Norwegian kindergarten. Figure 8 presents the distribution of the ratio between the annual hot water use and total heating needs within the six years. Clearly, DHW accounted for less than 9 % of total heating demand in most of the kindergartens and had a small influence in the whole picture. In this article, to describe the relation between SH and t_{od} more accurately, DHW use was deducted from the total DH needs. DHW use profile was roughly assumed as the DH use when t_{od} higher than 18 °C (the effective indoor temperature) in May, June, and August (kindergartens are mostly closed in July). For weekends, coefficient of 0.2 was considered. As shown in Figure 9, it is relatively easy to establish the energy demand function of t_{od} in polynomials through the entire outdoor temperature range. The function was written as:

$$P(t_{od}) = p_1 {t_{od}}^{i} + p_2 {t_{od}}^{i-1} + p_3 {t_{od}}^{i-2} + p_4 \qquad (2)$$

$$(i = 1, 2, 3. \, If \; i - 2 < 0, p_3, p_4 = 0; \; if \; i - 2 = 0, p_4 = 0)$$

IOP Conf. Series: Earth and Environmental Science **352** (2019) 012031 doi:10.1088/1755-1315/352/1/012031

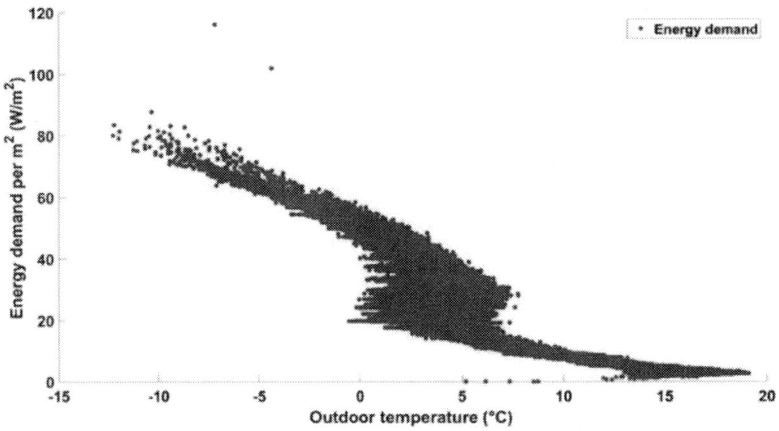

Figure 7. Energy demand vs Outdoor temperature of Cohort 1.

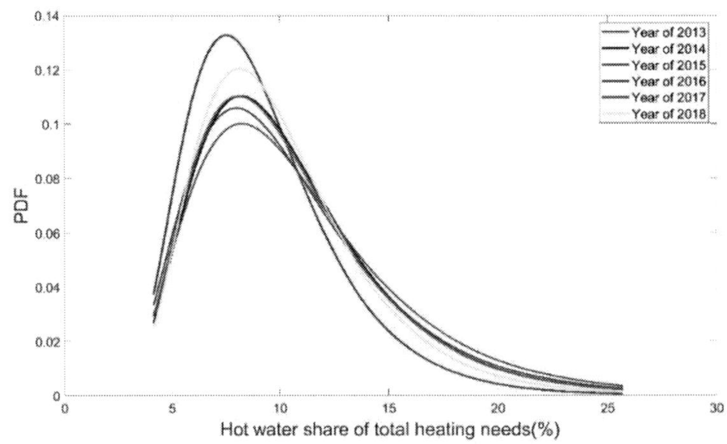

Figure 8. Distribution of hot water use in total DH needs of six years.

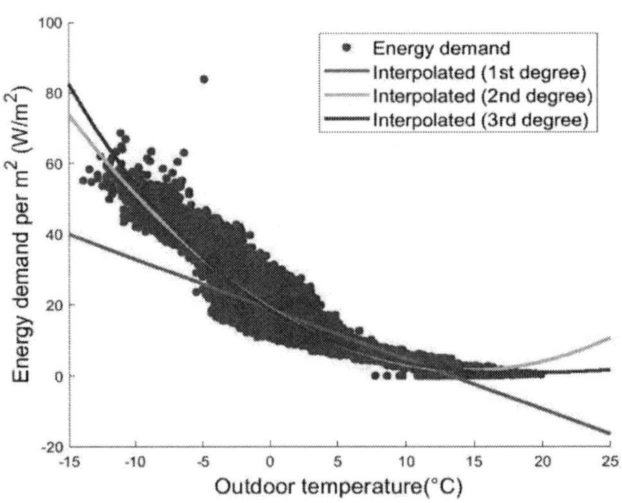

Figure 9. Energy signature curve of DH demand of Cohort 2 under 1st degree, 2nd degree, and 3rd polynomial.

To make sure of the goodness-of-fit of the model, the coefficients of determination R^2 was used. The value of R^2 should not be less than 0.75 as a rule of thumb in the analysis of building energy [12]. The coefficients of Function (2) and R^2 of each polynomial were shown in Table 6. It can be seen that even the simplest 1st degree polynomials satisfies the requirement of R^2 and fulfil the prediction of energy demand. This can be used to predict hourly heating load in the accordance with reference weather year, which is developed based on decades of weather data and can be found in database library [13]. The load profile can be used as input to energy system modelling, such as EnergyPLAN [14].

Table 6. Coefficients of Function (2) and R^2

	p_1	p_1	p_3	p_4	R^2
1st degree	-1.414	18.82	/	/	0.7927
2nd degree	0.08309	-2.412	19.02	/	0.8977
3rd degree	-0.001359	0.1027	-2.403	18.72	0.8995

3.2. Calculation of CO_2 emission of one typical kindergarten

In Figure 10, the stand-alone two bars at right side represent the building without DH. The annual CO_2 emission can be hugely increased from 1.2 tCO_2/yr to 13.6 tCO_2/yr when CO_2 factors of electricity changed from 10 to 110 gCO_2/kWh by making it the worst case. In the green square, three cases of different DH shares were compared, and their combinations regarding CO_2 factors were made as: blue bars of Norwegian electricity ($CO_{2\text{-EL1}}$) with average DH production ($CO_{2\text{-DH1}}$), orange bars of Nordic electricity ($CO_{2\text{-EL2}}$) with average DH production ($CO_{2\text{-DH1}}$), yellow bars of $CO_{2\text{-EL2}}$ with DH production of 2015 ($CO_{2\text{-DH2}}$), and purple bars of $CO_{2\text{-EL2}}$ with DH production of 2010 ($CO_{2\text{-DH3}}$). All the blue bars still gave the smallest values in each case since CO_2 factor electricity was 10 gCO_2/kWh. From the results, it can be seen that if electricity shoulders more energy supply, the total annual CO_2 emission can be varied a lot depending on the CO_2 factor of electricity. While in the case of DH high share, the variation of CO_2 emission under different background data was relatively small. Generally speaking, in the comparison of building with and without DH by using the same total energy demand, even in the case of DH low share under the highest DH production factor ($CO_{2\text{-DH3}}$), the total annual CO_2 emission (11.7 tCO_2/yr) can still be lower than the case without DH (13.6 tCO_2/yr) by 14 %.

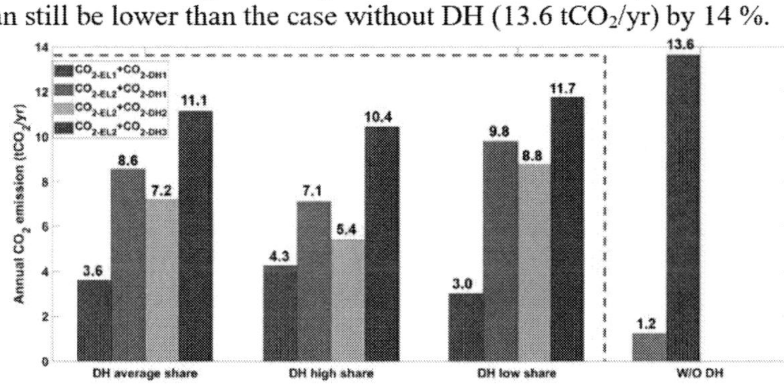

Figure 10. Annual CO_2 emission of one kindergarten of 700 m^2.

3.3. Assessment of CO_2 impact of new building area

After 10 000 m^2 of new building area of kindergartens was assumed to be built in Trondheim, the calculation of annual CO_2 emission regarding the new area was made. Through changing the penetration rates of new building area supplied by DH (x) between 0 % and 100 %, three kinds of growing trends of added annual CO_2 emission were calculated by following each DH production factor. As plotted in Figure 11, when all new buildings had only electricity, the added annual CO_2 emission would be 194.9

tCO_2/yr, and this is same for the three growing trends. When half of the new building area being connected to DH system, annual CO_2 reduction would be between 22.5 and 49.7 tCO_2/yr. Since it was predicted to follow linear CO_2 reduction with variation of DH penetration, the annual CO_2 emission would be double if all the new building area being connected to DH. The orange line represents the best case since DH production factor in 2015 was smallest, while the yellow line has mildest reduction slope due to the choice of highest DH production factor, and the blue line of the average DH production factor is in between.

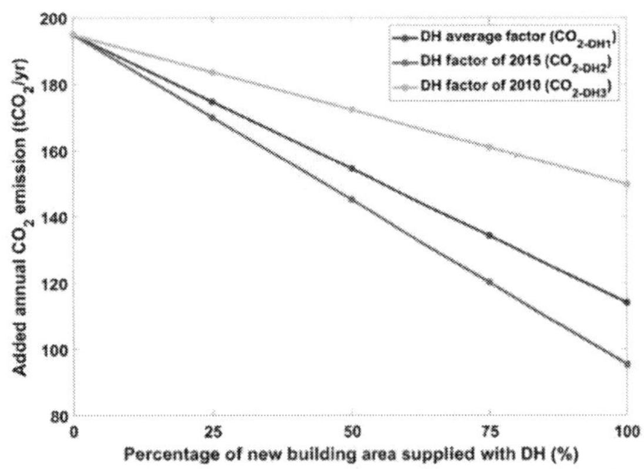

Figure 11. Annual CO_2 addition of 10 000m^2 new building area.

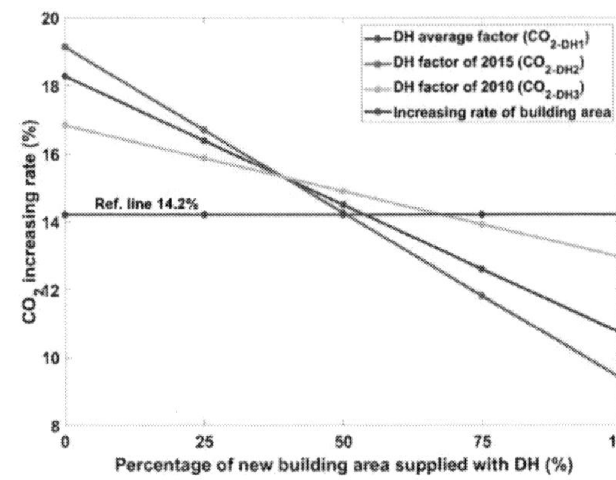

Figure 12. CO_2 increasing rate of 10 000m^2 new building area.

On the ground of the 6-year average annual area, the growth rate of building area, 14.2 %, was shown as the purple reference line in Figure 12. The region above the horizontal line had higher increasing rate of CO_2 than that of building area. It means if 14.2 % more building area being introduced, more than 14.2 % more CO_2 would be emitted; while the region below the line had smaller CO_2 increasing rate than the building area increasing rate, and this is what is expected to happen in the future to slower carbon footprint growth. The orange line representing the smallest DH production factor ($CO_{2\text{-}DH2}$) had the steepest slope. After more than half of new building area using DH, slower CO_2 increasing rate can be realized. When using the highest DH production factor ($CO_{2\text{-}DH3}$), the break-even point can reach at 67 % as shown in the yellow line of the mildest slope. Therefore, the breaking point located between 50 % and 67 % of new building area connected to DH under different CO_2 background data.

4. Summary and Future work

In this article, a typical energy profile of kindergarten in Trondheim was identified. The energy use data was retrieved from energy monitoring platform of Trondheim Municipality in total 559 hourly files. Two cohorts, namely Cohort 1 (not connected to DH) and Cohort 2 (connected to DH) were analyzed and compared. Under various building areas of the kindergartens, energy profile per m^2 of all kindergartens from 2013 to 2018 was defined and the average profile of each cohort was obtained. For Cohort 1, it is difficult to draw a robust energy signature regarding the energy demand and outdoor temperature, other issues and scheduling may be considered. While for Cohort 2, hot water use can be estimated as the only DH use in summer period and deducted from the total heating needs, in order to establish energy signature more accurately. Within the six- year duration curves, the annual average energy use of Cohort 1 was 177.2 kWh/(m^2.yr), and annual average electricity and heating of Cohort 2 was 64.7 kWh/(m^2.yr) and 103.0 kWh/(m^2.yr), respectively. Within Cohort 2, there were three cases depending on the energy contribution from DH and electricity, from DH high share, DH average to DH low share. 700m^2 was chosen as the representative building area of one kindergarten. Its annual CO_2 was compared between with and without DH based on the same total annual energy use. For the background data of electricity, two CO_2 emission were used. The one within Norwegian border gave the best results in all cases; when extended the border to the Nordic region, CO_2 emission jumped to higher level. For the CO_2 factors of DH production in Trondheim, the average factor from 2010 to 2018, the

IOP Conf. Series: Earth and Environmental Science **352** (2019) 012031 doi:10.1088/1755-1315/352/1/012031

factor in 2015 as lowest, and the factor in 2010 as highest, were used. The kindergarten with DH high share had lowest annual CO_2 emission and smaller CO_2 variation. By using the DH factor in 2015, it supported the lowest emission. For the kindergarten had low share of DH or even without DH, the CO_2 emission had a wider range. This is mainly caused by their higher dependence of the electricity production mix since electricity can be traded in the free market. Moreover, the building with only electricity is more likely to have unknown high peak load. As a mild prediction, 10 000 m^2 was assumed to be built in Trondheim. On the ground of average total kindergarten area within six- year, the growth rate of building area, 14.2 %, was used as the reference line. The growth rate of CO_2 emission could be slower than that of the building area, if more than 50 % and 67 % of new building area would be connected to DH. The break-point locates depending on the energy sources of local DH production, which determines the CO_2 factor.

The results of this article showed that building connected to DH system was more competent than the building of only- electricity concerning the CO_2 emission, and its energy demand easier to be predicted. In the future work, energy data and profiles of other building types and reference weather data in Trondheim shall be defined and analyzed. These profiles can be used to diversify and upgrade energy supply ways and improve urban energy planning.

Appendices

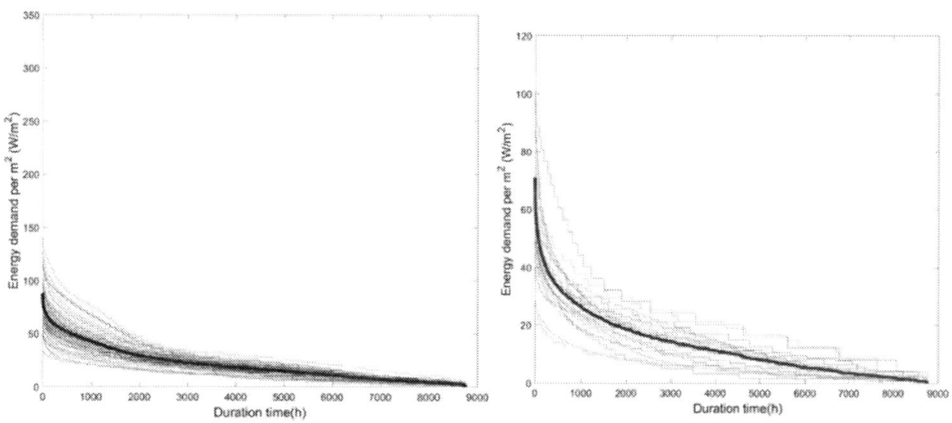

Figure A 1. Annual energy duration curves of Cohort 1 and Cohort 2 (only DH) in 2013.

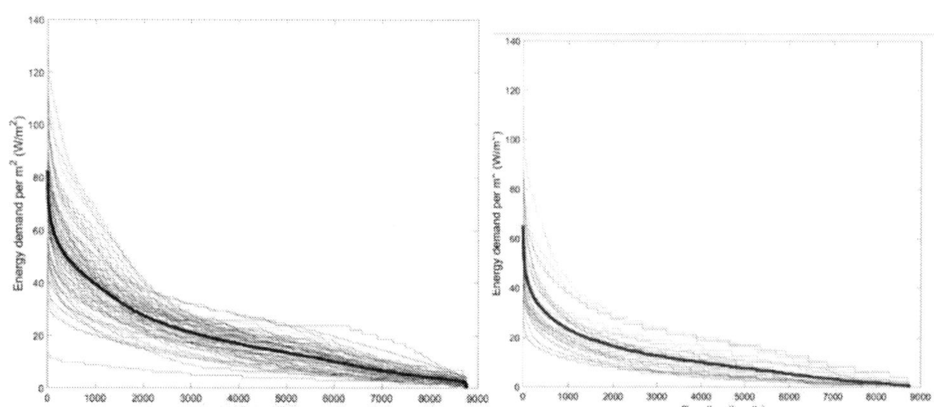

Figure A 2. Annual energy duration curves of Cohort 1 and Cohort 2 (only DH) in 2014.

IOP Conf. Series: Earth and Environmental Science **352** (2019) 012031 doi:10.1088/1755-1315/352/1/012031

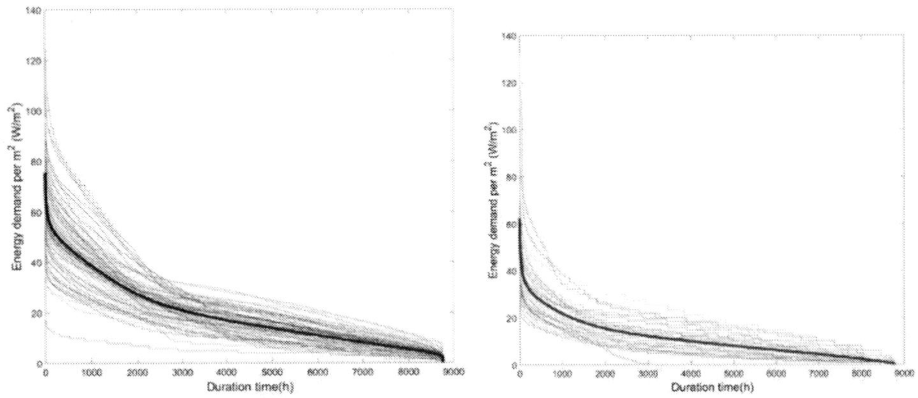

Figure A 3. Annual energy duration curves of Cohort 1 and Cohort 2 (only DH) in 2015.

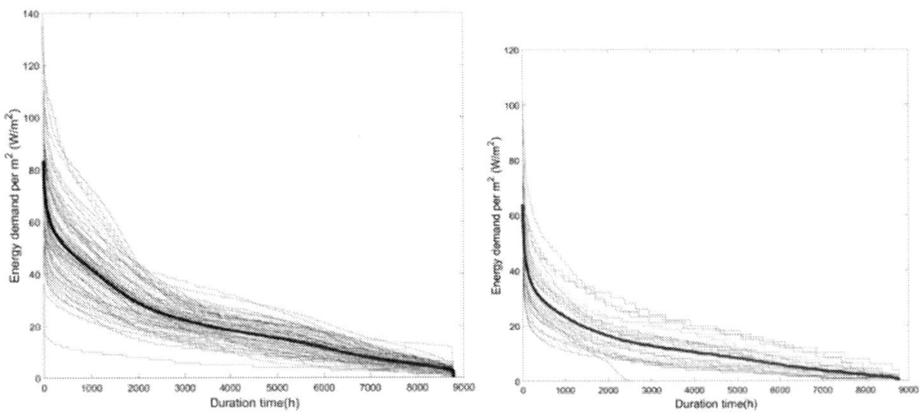

Figure A 4. Annual energy duration curves of Cohort 1 and Cohort 2 (only DH) in 2016.

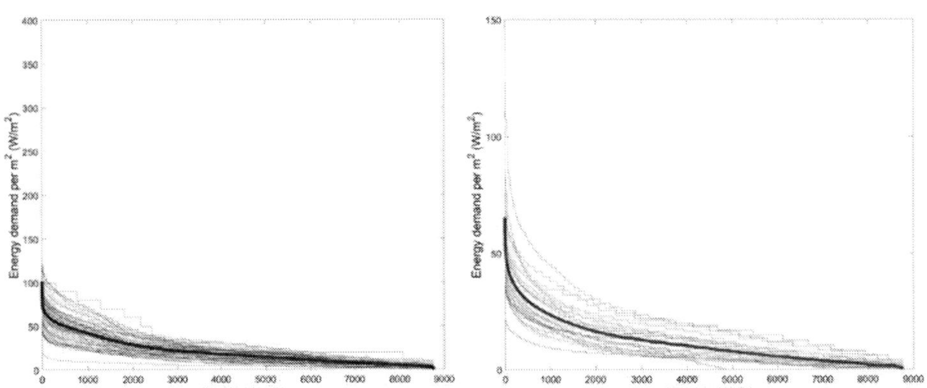

Figure A 5. Annual energy duration curves of Cohort 1 and Cohort 2 (only DH) in 2017.

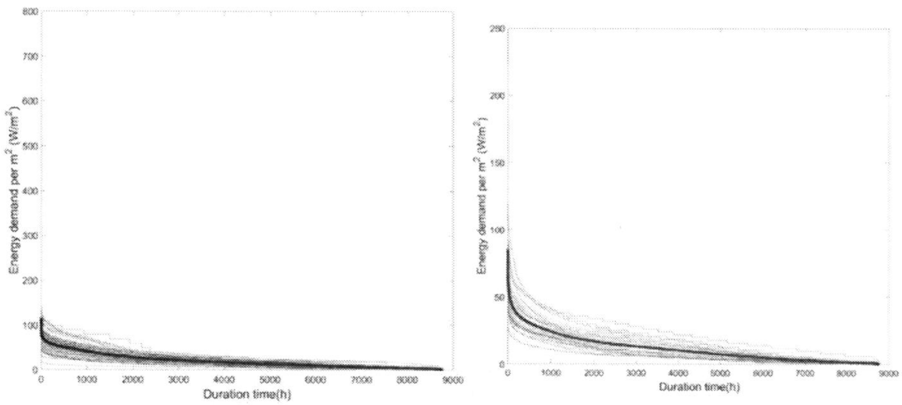

Figure A 6. Annual energy duration curves of Cohort 1 and Cohort 2 (only DH) in 2018.

References

[1] International Energy Agency. Energy Efficiency: Buildings.
[2] Patronen J, Kaura E and Torvestad C 2017 Nordic heating and cooling.
[3] SSB Statistics Norway 2018 Population projections.
[4] Energy monitoring platform of Trondheim Municipality. http://www2.esave.no.
[5] Norsk Fjernvarme. http://www.fjernkontrollen.no/.
[6] Climate Agency. NS 3720: 2018- Method for greenhouse gas calculations for buildings.
[7] Norsk Energi 2014 Klimaregnskap for Fjernvarme.
[8] Frederiksen S and Werner S 2014 *District heating and cooling* (Lund: Studentlitteratur) chapter 4.
[9] Hitchin R and Knight I 2016 Daily energy consumption signatures and control charts for air-conditioned buildings. *Energy and Buildings*, 112.
[10] Djuric N and Novakovic V 2010 Correlation between standards and the lifetime commissioning. *Energy and Buildings*, 42(4).
[11] Norwegian Water Resources and Energy Directorate (NVE) 2014 Analyse av energibruk i undervisningsbygg, RAPPORT.
[12] Menard S 2000. Coefficients of determination for multiple logistic regression analysis. *The American Statistician 54*: 17-24.
[13] Hensen J and Lamberts R 2011 *Building Performance Simulation for Design and Operation* (Spon Press) chapter 3.
[14] Sustainable Energy Planning Research Group. EnergyPLAN-Advanced Energy Systems Analysis Computer Model Documentation Version 14.

IOP Conf. Series: Earth and Environmental Science **352** (2019) 012032 doi:10.1088/1755-1315/352/1/012032

Using residential buildings to manage flexibility in the district heating network: perspectives and future visions from sector professionals

P V K Andersen[1], S Georg[2], K Gram-Hanssen[3], P K Heiselberg[4], A Horsbøl[1], K Johansen[3], H Johra[4], A Marszal-Pomianowska[4] and E S Møller[1]

[1] Department of Communication and Psychology, Aalborg University, Rendsburggade 14, DK-9000 Aalborg Øst, Denmark
[2] Department of Planning, Aalborg University, A.C. Meyers Vænge 15, 2450 Copenhagen SV, Denmark
[3] Danish Building Research Institute, Aalborg University, A.C. Meyers Vænge 15, 2450 København SV, Denmark
[4] Department of Civil Engineering, Aalborg University, Thomas Manns Vej 23, DK-9220, Aalborg Øst, Denmark

Corresponding author: Katinka Johansen, kaj@sbi.aau.dk

Abstract. Intermittent energy resources challenge the ways in which the existing energy system operates. Studies suggest that residential buildings can provide a flexibility service for district heating (DH) systems. This technique involves load shifting by heating buildings to higher temperatures at times when energy is more readily available, thus diminishing heating needs at times of peak demand or when energy is scarce. Based on three Future Workshops (FWs) where DH professionals and other relevant DH stakeholders participated and discussed this topic, this paper reports on the extent to which these actors see energy flexibility as a realistic future development, and on what they see as key potentials and challenges in that regard. Preliminary results indicate that the mix of the actors and the specific local context greatly influence how this topic is understood, emphasizing the importance of including local context in investigations of energy flexibility. FW participants included representatives from DH companies, municipalities, building associations, technology developers, etc. The FWs were conducted at three different localities of Denmark: Copenhagen, Aalborg and Sønderborg, i.e. the national capital, a regional capital and a smaller city, respectively.

1. Introduction

The relative instability in energy supplies and the comparative stability in energy consumption patterns, with fluctuations during the day and the year, is a key challenge as the energy systems include more renewable energy resources (RES), for example wind and solar [1–3]. Studies suggest that residential buildings may contribute to the management of such flexibility in the district heating (DH) sector as buildings can be used for short term heat storage. Realizing this potential entails increasing the temperature in buildings when renewable energy is more readily available, thus diminishing heating demands during the times of the day when energy is in peak use [4].

However, to assess the feasibility of such proactive use of residential buildings to manage instability in heating related energy consumption patterns, more in-depth understandings of the agency and behaviour of the actors using those buildings is necessary. Obtaining such knowledge entails gaining greater insights into everyday residential household practices and potential for change, i.e. insights into the patterns of heating related energy consumption among residents living in different types of households. The question is if the technical potential for using buildings to manage flexibility in the DH system is both practically realizable *and* compatible with patterns of energy consumption in households, and in this regard if the relevant actors are willing to - or are indeed practically able to – change and adapt their heating related energy consumption. Questions of comfort, consumer sovereignty and top-down control versus individual freedom are at the fore.

1.1 The DH socio-technical system, interdisciplinary competencies and collaboration

Understanding the full potentiality of this future vision for integrating buildings actively into the energy system requires cross-sectorial and technical integration. It requires involving the multiple sector-professionals that work at all levels of the energy production and distribution system [5], and it requires assessments of the potential for change of practices in the residential buildings that will potentially be used more proactively for heat storage. In other words, understanding what is currently at stake for DH, and in this light what the potentials for change are, requires both insights and understanding of the dynamics in the complete socio-technical systems that ultimately deliver DH to the end-users.

The complexity of such analysis calls for the integration of perspectives and analysis in the complete DH socio-technical system, ranging from micro-level perspectives (e.g. end-user perspectives or specialist technical knowledge) to macro-level perspectives (societal trends and tendencies and political decision-making), and this calls for interdisciplinary collaboration and research. To facilitate the first steps of such interdisciplinary collaboration, the Future Workshops (FWs) as a method promotes discussions of future perspectives/imagined scenarios regarding certain topics [6], typically with the participation of relevant sector professionals, members of the public and other stakeholders (see section 3).

This paper presents the preliminary results of three Future Workshops where DH professionals and other relevant DH stakeholders participated and discussed the potential use of residential buildings to manage flexibility in the district heating systems. Specifically, this paper reports on the extent to which these actors see energy flexibility as a realistic future development, and on what they see as key potentials and challenges in that regard.

The paper is organized in the following way: Section 2 introduces the relevant technical perspectives of the case for using residential buildings to manage flexibility in the DH system; section 3 describes the FWs as a method; the preliminary findings/results from the FWs are presented in section 4. Finally, the implications for practice and future research are discussed in section 5.

2. Technical perspective: the potential for using residential houses to manage DH flexibility

The increasing share of renewable energy resources (RES) in the energy system can induce a mismatch between energy use and production, thus creating a challenge for the stability of energy grids. To solve this problem, increasing efforts have been made to develop smart technologies [3] that are able to manage a RES-dominated and thus more intermittent energy network [2]. Coupling this total network (including electrical, thermal and gas grids, various forms of energy storage, electric transportations, local industries and buildings) with smart metering and control networks can account for the actions of all the actors in the network. This will enable more reliable management of energy supplies, increasingly based on the more intermittent RES energy production.

In this Smart Energy Grid context, a new paradigm has emerged when considering the building stock: Buildings are no longer considered as passive energy end-users. On the contrary, as an important and potentially active element of the energy grid, buildings can not only modulate their energy use, but also store, produce, self-consume or re-inject a significant share of energy. Thus, buildings may help reduce the mismatch between instantaneous energy use and energy consumption, in this way enhancing the integration of RES in the grid. Such demand side management techniques, also known as building energy flexibility strategies, can help to solve local or global congestion issues occurring at critical periods in the weak points of the bottleneck of a grid.

IOP Conf. Series: Earth and Environmental Science **352** (2019) 012032 doi:10.1088/1755-1315/352/1/012032

Within this network, the same technique can be applied to district heating specifically. Here, the energy flexible buildings can use what is known as penalty signals[1] as control information from the DH system to modulate the heat energy demand time profile. In this way, energy use can be shifted in time. This technique is also known as load shifting. This means that energy use can be shifted from critical periods, for example when heat demand is high, heat energy production is costly or has a high CO_2 release intensity, and towards periods of time when there are more abundant supplies of RES production, CO_2 release intensity is less and the production cost is also lower (see *Figure 1*). Additionally, maximum heat power peak demand often drives the limiting sizing factors of DH production and distribution systems. High peak power demands induce oversizing of the DH production units and distribution infrastructures, which are only rarely used at full capacity. In other words, the use of building energy flexibility strategies can reduce investment and operation costs for the DH systems by performing what has also been known as peak shaving and valley filling (see *Figure 1*). There are different perspectives to take into account when considering the use of buildings for managing heating energy flexibility: the amount of heat shifted in time, the duration of this flexible window, the direction of the change (increase or decrease compared to non-flexible case), and the costs of activating this heating energy flexibility.

Currently, there is no scientific agreement on the definition and assessment of (heating) energy flexibility. However, the latter is usually defined as "*the ability for a building to adapt its profile of energy use to the requirements of the grid (penalty signal) without jeopardizing technical and comfort constraints*" [7]. Most of the current research focuses on the heating energy flexibility of single buildings. However, in the case of a large scale implementation of demand side management, heating energy flexibility should be analyzed at a larger scale, in this way considering the aggregated flexibility potential of building clusters, neighborhoods or even entire cities. This will also account for the significant time constant of the DH or cooling networks.

The building stock has potential for thermal storage in hot water tanks (heating system and/or domestic hot water tank) [8] and in the thermal mass of its structure and indoor environment [9]. The latter can be used as a cost-effective thermal storage solution when employing indoor temperature set-point modulation [10], and excess RES production can thus be stored in the built environment. This allows for significant space heating energy shifting over a few hours for poorly insulated buildings, and for more than 24 hours for well-insulated buildings with large effective thermal inertia [11]. This also allows for heat power demand reductions in peak power periods, especially in the morning.

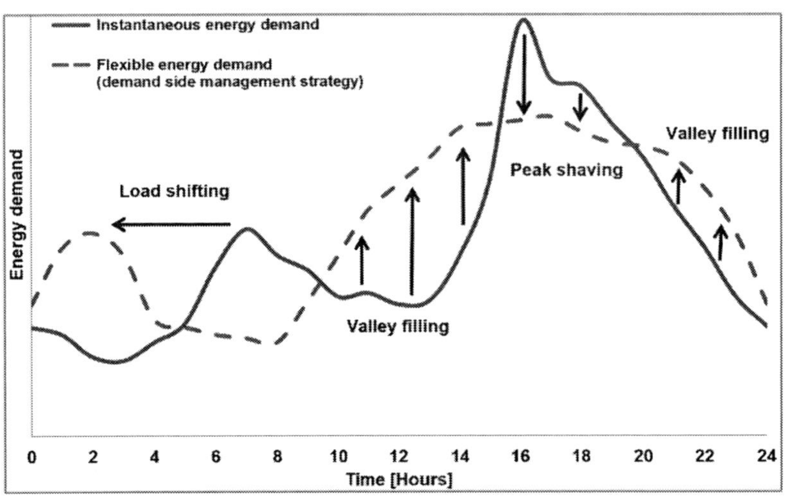

Figure 1: Example of demand side management / energy flexibility strategy with peak shaving, valley filling and load shifting.

[1] Heating energy price, CO_2 intensity of current heating energy production, local state of the DH distribution network, marginal heat production cost, etc.

3. The Future Workshops as a method

Since the potential DH transition processes require rethinking well-established structures, services and DH business models, it can be assumed that disparate parties will have to work together in novel ways for such a system to become a reality. However, in order to develop new ways of working across professional boundaries, the many different professional perspectives on such transition processes must be identified. The FW method is well suited for this.

The Future Workshop is a method developed by Robert Jungk and Norbert Müllert in 1987. The method was originally developed to support democratic, participative and collective decision making between community groups on a shared future, and for many years FWs have proven valuable as instruments for communication about planning and the creation of future visions and scenarios for certain topics. FW have been particularly prevalent within the domains of education, software design, and health [12,13].

3.1. The Future Workshops in practise

For this purpose, the future workshop method was inspired by the generic structure of the so-called ACTVOD-method [14], and thus the FWs were specifically adjusted to facilitate the inter-professional sharing of experiences and perspectives on a future intermittent energy system. More specifically, the overall aim of these FWs was to facilitate discussions between sector professionals on the potentials of and challenges for using residential buildings as flexibility generators in the district heating system, and thus the FW design also aimed for the discovery of collaborative and efficient ways to explore the potential for the integration of buildings to provide flexibility in the DH system.

Three FWs were organized at different locations in Denmark (Sønderborg, Aalborg and Copenhagen), and these covered both rural and urban areas. For each workshop, key local professionals were invited. These included representatives from municipalities, utilities, housing associations, private companies, and technology developers. The participants were divided into groups of 5-6, comprising a mix of different professionals. A total of 36 participants divided into 7 groups took part in the FWs. All three workshop were started with a short introduction and organized in three main phases: These phases included I) mapping, identification and prioritization of main actors and infrastructures relevant for the realization of an intermittent district heating system and the challenges associated with this; II) visualization of visions for solving the identified challenges; and III) critical discussions of the feasibility of these visions.

The FWs were carried out with the help of game boards with concentric circles in various colors. On these game boards, FW participants indicated their chosen prioritizations of given topics via notes/post-its (*see Figure 2*)

Figure 2: Example of Future Workshops game boards.

During the first phase (I) of the FWs, the participants were individually asked to identify different actors of relevance for realizing a transition to a more flexible district heating system. The identified actors were written down on separate cards, and then prioritized by the groups according to their importance. For this purpose, each group was asked to place and to prioritize the different actors according to their importance on the concentric circles, with the inner circle representing the highest priority and the outer circle representing the lowest priority. Participants were then asked to attach small notes for each of the actors, explaining what they saw as a central challenge for the specific actor in relation to the topic. The process was then repeated, but this time focusing on the identification and prioritization of technologies and infrastructures deemed important for achieving a more flexible DH system. Inspired by the previous exercises, in the second phase (II) the participants were asked to identify and describe what they saw as the two most important challenges for realizing the DH transformation. Based on these choices, the

IOP Conf. Series: Earth and Environmental Science **352** (2019) 012032 doi:10.1088/1755-1315/352/1/012032

groups were asked to visualize their ideal visions for a future that addressed these challenges. Finally, in the third phase (III) the participants took part in a common discussion on the feasibility of the realization of the presented visions for the future of DH. In other words, leaning on the expertise and insights of these sector professionals, the FWs created an opportunity to glimpse into future visions and scenarios for the DH systems based on the different types of sector expertise represented in the FWs. These all reflected different perspectives and analysis' of the DH system and future potential.

4. The Future Workshops: preliminary insights

In the following sections, preliminary findings/results from the FWs are presented. All of these findings are based on video-documentation from the FWs, on the collected notes and drawings from the FWs and on the summary report authored by Pernille V. K. Andersen, Anders Horsbøl and Emil Styrbæk Møller (see [15]). Here, the selected key findings from the FWs are presented in the same order as phases I, II and III of the workshops described in section 3.1.[2]

4.1. DH transitions: key actors and technologies as anticipated by FW participants

In groups, the FWs participants were asked to prioritize what they saw as among the most important actors for the integration of households in the future flexible DH systems. The groups could choose various actors as among the most important. 6 of the total 7 FW groups prioritized national politicians/authorities and utilities, 4 of the 7 groups prioritized housing residents, their organizations and different types of landlords, and 2 FW groups prioritized municipalities as among the most important.

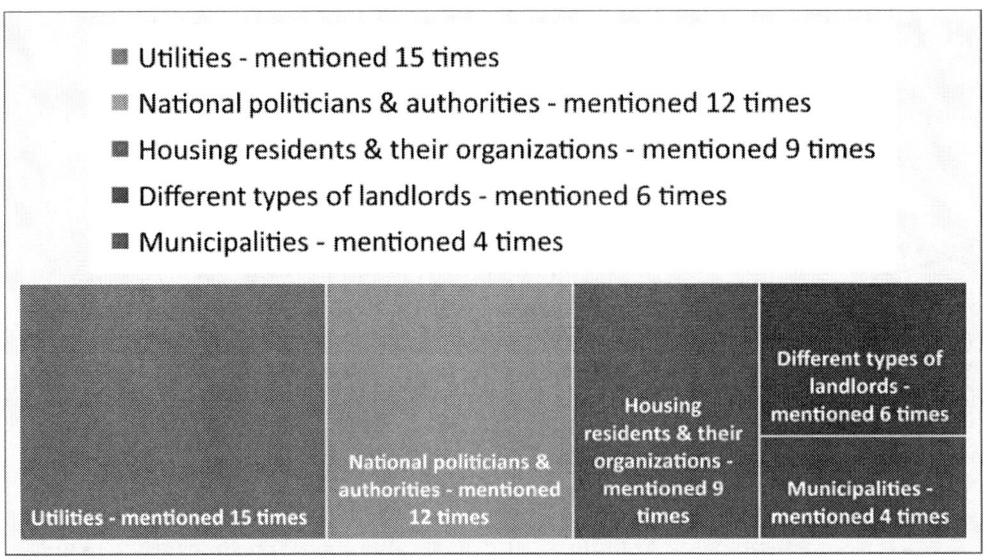

Figure 3: This graph shows the amount of times these key actors were mentioned when FW participants answered the question: What are/should be the most important actors for the integration of households in the future flexible DH systems? The size of the boxes represent the relative amount of times the actors were mentioned.

[2] The findings presented here are a part of a much larger dataset. This dataset will be analysed further to understand the professionals' different stakes in, attitudes towards and perspectives on the energy system more in-depth.

IOP Conf. Series: Earth and Environmental Science **352** (2019) 012032 doi:10.1088/1755-1315/352/1/012032

Evidently, there were very different perspectives on the role of residency-owners among the FW participants. This is underlined in the following quotes from the FWs:

Private company employee: *"We have this experience with the public that (...) they [the end users] just expect that there is heating in the radiator* [Utility employee: *Yeah, and that it's environmentally friendly how it's done] and there is water in the taps."* Social housing representative: *"Actually, I don't really agree with you there. I think that the end-users have a lot to say." (...)* Energy consultant: *"I think that they [the residents] are pretty significant. [As a resident in my] apartment I want to be able to control what heating I have. And I meet a lot of people who want to sleep in a very cold bedroom. [And] if I look at the fluctuations, I think ... well maybe it's during the night that you want to turn the heating on a bit when the wind turbines are producing, like ..."* Utility employee: *"That's when you want it cold."* Energy consultant: *"I know a lot of people who (...) want to control it themselves. (...) Where are we at if, as a consumer, you can't even control how the heating is on...?"* [3]

Overall, some FW participants emphasized the end-user preference of individual choice, while others argued that more top-down control would be necessary if these techniques are to work. These perspectives represent different extremes in views on individual freedom/consumer choice versus top-down controlled planning. Buildings and building design overall was mentioned as a part of the DH system transition by 5 out of the 7 groups in total. Moreover, the degree to which this was considered a viable and realizable option seemed to vary considerably from region to region. Actors judged as second-most important for the DH transitions overall were consultants, researchers, relevant tech-industries, janitors and other sector-related craftsmen.

When asked what they saw as the main and important technologies and techniques in achieving the potential transformation of the DH sector (I), 6 out of the 7 workgroups mentioned the word "control" 11 times in total, and 5 of the groups prioritized this concept as most important. To what extent the groups were thinking in terms of control/management at end-user level was not always quite clear, however. While 3 of the 7 groups mentioned several variations of the notion of data collection as an important technology for flexibility in the DH sector, this was never prioritized most. Words also mentioned, albeit less frequently, were "surveillance", "sensors" and "smart technologies." Incentives were also discussed, but these were never highly prioritized by the workshop participants.

4.2. DH transitions: key future challenges as anticipated by FW participants

Based on the reflections among FWs participants regarding key technologies and actors in the DH transition processes, groups were also asked to specify what they considered the most critical future challenges (and potentials) for integrating residential buildings more proactively into the DH system for managing flexibility (II & III).

The groups summarized the key challenges for such a potential DH transition via the following themes: A) The current political/governance framework, here particularly the challenges in creating political systems facilitating long-term strategizing and thinking in energy systems integration, data-sharing and system compatibility. B) Systems flexibility, here the need for managing daily flexibility and seasonal flexibility in supply, and in this context also the need to view the energy systems holistically, i.e. including electricity, gas, excess heating, geothermal systems etc. The importance of mentioning and *defining* the different types of flexibility was also highlighted. C) Challenges in identifying viable businesses cases and business models, here including long-term investment security comprising the potential for integrating novel systems innovations. D) Insufficient documentation of the potentials for flexibility in the DH system, too few demonstration projects and questions of data accessibility. E) Another key challenge and question concerned the management / facilitation of the DH transition process, for example, which actors or types of actors within the DH system will be in charge? F) As a future potential, pricing structures as motivation were also mentioned, here comprising thoughts

[3] These quotes are from the Future Workshop in Copenhagen. The quotes are translated from Danish to English by the authors.

on, for example, enhancing a better understanding of pricing structures among the end-users, rewarding end-users for doing 'the right thing' (incentives) and seeking to influence consumption choices.

4.3. DH transitions: on intra-sectorial differences and perceptions of these among FW participants

While the technical potential for actively integrating buildings into the energy system (and particularly the DH system) is there (see section 2), multiple uncertainties prevail: Not only is the collaboration of actors that use buildings a prerequisite for the success of this technique, as the quotes below underscore, there are also multiple structural differences and challenges within the sector itself:

Private company employee 2: "*Well ... I certainly know district heating companies that complain that their production price is expensive in these times of peak demand. (...) But I think it is very different from district heating plant to district heating plant what sort of challenges they have with this sort of stuff.*" *(...)* Private company employee 1: *"and there isn't really any ... if the users are ready to receive, I mean, if the houses are designed so that they are ready to receive these kinds of flexible solutions ... "* [4]

Evidently, some actors from utilities and district heating plants saw the potential of using buildings more proactively in the management of the DH system, and in different ways, some utilities and companies are already experimenting with some of these techniques. Other actors and utilities see less (or no) potential in this regard. Preliminary research evidence suggests that the diverse nature of the utilities may inform future outlooks for DH among the sector professionals: For example, the size and technical infrastructure of utilities themselves may hold different (immediate) potentials for the use of buildings for flexibility management. This may inform the future visions and imagined prospective potentials for, in this case, the DH sector of the actors involved in those places [16]. Overall, via the FWs it is difficult to precisely trace the extent to which different sector professionals are thinking in terms of integrating buildings in the DH sector more proactively, in what way they may be doing so, and what the nature of the differences in these future outlooks among different sector professionals may practically imply.

5. The Future Workshops: implications for practice, policy and research

5.1 Methodological reflections

In this research, the FWs were developed to facilitate inter-professional sharing of experiences and perspectives, including discussions on the potentials, challenges and visions for a future intermittent energy system.

What these particular FWs also underlined is the importance of who participates; i.e. what background do the participants have, and in what way are they associated with the FW topic. People who are less knowledgeable of, for example, the DH infrastructure, practice and relevant governance framework may see and imagine novel and creative systems solutions, while people who work within the sector may be more pragmatic in their approach. Potentially, this synergy may be both constructive and innovative (as was the original intention of the FWs). Importantly, FW group dynamics, for example power-plays among the FW participants, will inform the nature of the conversation as the workshop proceeds. A less fortunate consequence of inviting very heterogeneous participants could be that the conversation is difficult because people are just too 'far apart' in their approaches to and perspectives on the questions. While different types of insights are hopefully produced via FWs that include a broad panel of relevant stakeholders, overall critically 'thinking through' the potential impact of different 'levels' of systems expertise and/or systems insights for dialogue among the FWs participants is crucial.

5.2 Implications for practice and research

The FWs were held at three different locations in Denmark. Interestingly, the FWs held at these different Danish locations, and thus with invited DH stakeholders and sector professionals that mostly referred to the local DH plants, had quite different outcomes. This emphasizes the importance of regional variations in the DH sector in several ways: a) The technical infrastructure of specific DH plants matter for the

[4] These quotes from the Future Workshop in Aalborg.

challenges that the different sector professionals face and the way they think, here in terms of integrating buildings in the DH sector more proactively. Accordingly, b) local contexts matter for what future visions and perspectives are imagined among FW participants, just as differences among sector professionals themselves matter.

Considering the diversity among DH plants in Denmark alone [17], the degree of variation which is considered viable and realizable, here regarding residential building energy flexibility in DH systems, is hardly surprising. Practically, however, this also suggests that the potential integration of this strategy for DH flexibility management may have very local variations.

While the technical potential for integrating residential households more pro-actively in the DH systems for managing flexibility is there [4], the extent to which this is imagined as possible depends on the nature of the local DH socio-technical systems. Crucially, it also depends on the household residents, and on the extent to which they are willing to - and indeed able to – adapt their heating related energy consumption patterns to meet the needs for this potential transition.

Perhaps the observed diversity in the FWs among the multiple energy system stakeholders, with different stakes in, attitudes towards and perspectives on the energy system (and on energy transformations overall), should be expected. Accordingly, perhaps some level of friction, deriving from value differences and priorities amongst the diverse actors and stakeholders throughout the entire DH socio-technical system, should also be expected. Overall, more research is needed to understand and trace the dynamic interaction of these divergent views on and assessments of future potentials and challenges for the DH system among the stakeholders involved, and accordingly what this implies for energy transitions related practice and policy.

6. Conclusion

As a result of societal transitions to renewable energy resources, sometimes there are mismatches between the levels of energy production and energy consumption. In the case of district heating, research suggests that buildings/residential houses may be used proactively to manage DH related flexibility in energy use and consumption.

In this research, FWs, with various DH sector stakeholders, were designed to focus on visions and predictions for the potentials and challenges of more proactively integrating residential houses in the DH sector to manage energy flexibility. According to the FW participants, some of the challenges for achieving this DH systems transition revolved around: the political/governance framework, including current policy challenges of doing long-term investments, lack of demonstration projects, challenges of data accessibility, and the overarching question of who manages such transitions. The FWs also revealed significant regional differences in views and perspectives on this possible DH systems change.

While the technical potential for using residential buildings to manage DH systems flexibility is there, certain questions linger. For example: is this compatible with household energy consumption patterns? Are household residents able to – or willing to – change/adapt their energy consumption patterns? To fully understand the divergent views on the potential for DH systems transitions among the diverse sector stakeholders, the importance of the DH local systems context, and what this all implies for DH related future practice and policy, more research would be beneficial. To this end, the FWs as a method may prove useful for future research enquiry.

Acknowledgements

This work was financed by the InterHUB project, funded by Aalborg University' strategic funding for interdisciplinary research. InterHUB: Intermittent energy – Integrating Households, Utilities and Buildings. For more information on the project see: www.interhub.aau.dk

References

[1] IEA. International Energy Agency 2017 *Energy Policies of IEA Countries - Denmark 2017 Review*

[2] Mathiesen B V., Lund H, Connolly D, Wenzel H, Ostergaard P A, Möller B, Nielsen S, Ridjan I, KarnOe P, Sperling K and Hvelplund F K 2015 Smart Energy Systems for coherent 100% renewable energy and transport solutions *Appl. Energy* **145** 139–54

[3] European Commission 2006 *European SmartGrids Technology Platform Vision and strategy for Europe's Electricity Networks of the Future* vol 19 (Brussels)

[4] Lund H, Werner S, Wiltshire R, Svendsen S, Thorsen J E, Hvelplund F and Mathiesen B V 2014 4th Generation District Heating (4GDH) *Energy* **68** 1–11

[5] Lund H, Østergaard P A, Connolly D and Mathiesen B V 2017 Smart energy and smart energy systems *Energy* **137** 556–65

[6] Jungk R and Müllert N 1987 *Future workshops: How to create desirable futures* (London: Institute for Social Inventions)

[7] Reynders G, Amaral Lopes R, Marszal-Pomianowska A, Aelenei D, Martins J and Saelens D 2018 Energy flexible buildings: An evaluation of definitions and quantification methodologies applied to thermal storage *Energy Build.* **166** 372–90

[8] Arteconi A, Hewitt N J and Polonara F 2013 Domestic demand-side management (DSM): Role of heat pumps and thermal energy storage (TES) systems *Appl. Therm. Eng.* **51** 155–65

[9] Johra H and Heiselberg P 2017 Influence of internal thermal mass on the indoor thermal dynamics and integration of phase change materials in furniture for building energy storage: A review *Renew. Sustain. Energy Rev.* **69** 19–32

[10] Hedegaard K, Mathiesen B V, Lund H and Heiselberg P 2012 Wind power integration using individual heat pumps - Analysis of different heat storage options *Energy* **47** 284–93

[11] Dréau H J P H J Le 2019 Influence of envelope; structural thermal mass and indoor content on the building heating energy flexibility *Energy Build.* **183** 325–39

[12] M. G J . K 1991 *Design at Work - Cooperative Design of Computer Systems* ed J Greenbaum and M Kyng (New Jersey, USA: Lawrence Erlbaum Associates Publishers)

[13] Müller M and Druin J 2012 Participatory Design. The Third space *The Human - Computer Interaction Handbook* ed J A Jacko (CRC Press)

[14] Lauttamäki V 2016 ACTVOD-futures workshop – a generic structure for a one-day futures workshop *Foresight* **18** 156–71

[15] Andersen P V K, Horsbøl A and Møller E S 2019 *Interhub Future Workshops: sammenfatning af indsigter* (Allborg)

[16] Upham P, Bögel P M and Johansen K 2019 *Energy Transitions and Social Psychology: A Sociotechnical Perspective* (Abingdon & New York: Routledge)

[17] Danish Energy Agency 2017 *Regulation and planning of district heating in Denmark*

IOP Conf. Series: Earth and Environmental Science **352** (2019) 012033 doi:10.1088/1755-1315/352/1/012033

Efficiency of BIPV system – Field study in Norwegian climate

Lars Gullbrekken[a], Nora Schjøth Bunkholt[b], Steinar Grynning[a], Martin Bellman[c] and Tore Kvande[b]

a SINTEF Building and Infrastructure, Department of Architecture, Building materials and Construction, Høgskoleringen 7b, 7491 Trondheim, Norway

b Department of Civil and Transport Engineering, Norwegian University of Science and Technology (NTNU), NO-7491 Trondheim

c SINTEF Industry, Department of Sustainable Energy Technology, Trondheim

* corresponding author: email: lars.gullbrekken@sintef.no.

Abstract. Use of solar cells (PV) and solar collectors are key remedies in buildings where a large part of the energy supply should be based on renewable energy. The aim of this work has been to evaluate calculated and measured solar production of two identical BIPV roofs located at the ZEB Living Laboratory situated at NTNU-campus in Trondheim. Temperature, irradiance and wind speed and direction at the rooftop of the building have been monitored since the construction of the house. There was found a large difference in energy production of the northern roof section and the southern. One possible explanation is shading of the northern roof because of low solar azimuth during the measuring period. In order to avoid such disadvantages, design of the PV-roofs should be considered early in the design phase of the building project. A small difference was found between the hourly measured and the calculated values of the PV performance based on the monitored local climate data. Use of generic climate data expect to cause a larger difference between measured and simulated energy performance due to lack of consideration to local conditions.

1. Introduction

Use of solar cells (PV) and solar collectors are key remedies in buildings where a large part of the energy supply should be based on renewable energy. PVs are for example used in the pilot buildings of the ZEB-research project [1]. However, in Nordic climate, the use of solar collectors and PV can be challenging due to snow, wind and temperatures below zero. Characteristics for Norway are the varied climate and the rugged topography with corresponding large local differences over short distances and extreme seasonal variations. The climate is putting a great strain on the building envelopes of our buildings. The building envelope and the roof in particular may be exposed to severe wind, snow loads, precipitation, freeze thaw cycles and large temperature fluctuations.

Interest in building integration of photovoltaics and solar collectors, which means that PVs and solar collectors become an integral part of the building, is growing worldwide. Photovoltaics integrated in the building envelope, such as the roof or the façade, is referred to as Building Integrated Photovoltaics (BIPV). According to Ceron [2] almost 50 % of the BIPV market is roof applications. The primary function of the roof as a climate screen must still be fulfilled with BIPV-systems.

To maximize energy yield, to lower degradation processes, and from a building physical point of view, ventilation below BIPV is necessary. Ventilation may reduce the temperature of the solar cells, which is dependent on the heat gains from the sun and the heat losses to the surroundings. The efficiency of the PV-panels is strongly dependent on the temperature of the PV-panel [3,4,5]. Assuming still air and PV attached to a building with good rear ventilation Häberlin [6] suggests a 30°C temperature increase relative to ambient temperature at 1 kW/m^2 insolation. For crystalline silicon modules this will give an efficiency decrease of 15 % [7].

The performance of three different commercially available PV modules in outdoor conditions in Southern Norway was compared by Midtgard et al. [8]. As expected, it was found that the amorphous module had the lowest overall efficiency and the monocrystalline the highest, also when considering gross efficiency at all irradiances over a complete year. A good correspondence between the measurements and the calculations of the energy yield was found. The calculations were performed with the software PV Potential Estimation Utility. Another study by Imenes [9] showed a relatively large difference in the energy yield of 13 BIPV systems situated at different locations in Norway. This was among others due to shading challenges and non-optimal site conditions. The importance of including knowledge about the PV planning into the early stages of a building project was pointed out as a central measure in order to avoid shading problems and unforeseen added costs during the installation phase.

In order find the best cooling strategy for BIPV panels Misiopecki et.al [10] performed a numerical study (CFD calculations). A comparative analysis of different air gap widths for cooling of the roof was conducted. The results showed that an air gap height of 50 mm was not sufficient to provide cooling of a 70 m long BIPV roof. The developed model was considered a good starting point for further investigations. However, to the authors knowledge such investigations in Nordic climate has not been performed.

The aim of this work has been to evaluate calculated and measured solar production of two BIPV roofs with identical PV installations. The roofs are located at the ZEB Living Laboratory situated at the NTNU-campus in Trondheim. This has been done through an extensive measurements campaign where analysis of solar production, weather conditions and temperature on PV modules and in the air cavity beneath the modules has been carried out.

2. Method

2.1. Test site

The ZEB Living Laboratory is a demonstration building that is representative, as a typology, of the most common Norwegian dwelling – the single-family house – to show that CO_2-neutral constructions (ZEB-OM ambition level) can be realized in the Norwegian climate with todays' technologies.

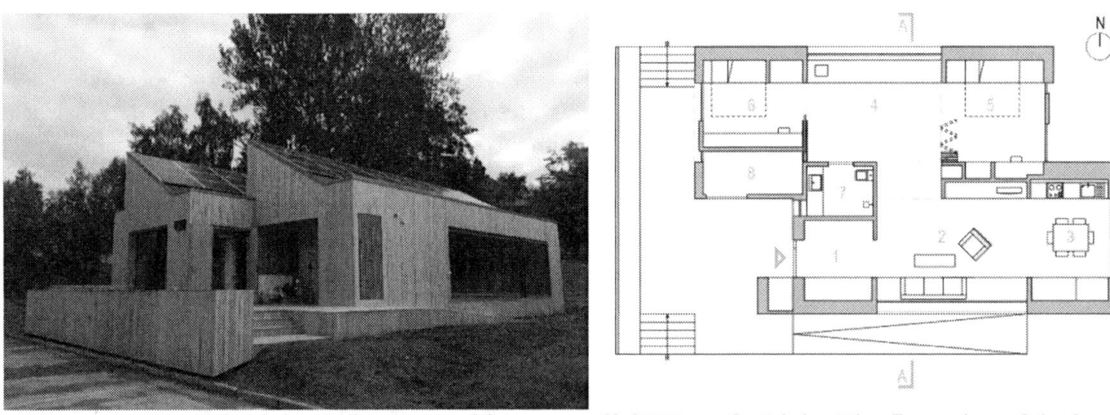

Figure 1. Left: Zeb Living Laboratory with two parallel PV-roofs. Right: The floor plan of the house consists of combined living room and kitchen and two bedrooms. Illustration: Gabrielle Lobacarro

2.2. PV Roof

The modules on each roof is arranged in one upper and one lower string of 12 modules (the middle row of modules is split between the upper and the lower string), resulting in a total of 39.6 m^2 PV on each roof. The mounting system is InterSole from Renusol (Figure 2). The PV modules are mounted on aluminium profiles, which are fastened on top of a HDPE membrane. The air gap beneath the PV modules is 49 mm. The lower part of both roofs is covered by asphalt roofing. Since the PV system does not replace any other part of the building envelope (the roof would have been weather-proof without it) the system can only be referred to as semi-integrated. The PV system replaces other roofing materials, but it is the membrane and not the PV modules that act as the climate shield. The specification of the roofs is given in Table 1. Data is collected using the Labview system.

Temperatures are measured on the rear face of the PV-panel and on the lower face of the air cavity by thermocouples of type T.

Figure 2. Closer view of the two parallel PV-roofs. Photo: Clara Good

Table 1. Specification of the PV roofs

Roof tilt	30°
PV	poly-Si REC 260PE
Efficiency	15.8 %
Area (gross modules/total)	1.65/79.2 m^2
Total installed power	12.48 kWp
Inverter power per roof	5.25 kW
Weight	10.9 kg/m^2

2.3. Calculations

In order to calculate the energy production, the dynamic simulation program PVSol with 3D visualization and detailed shading analysis of photovoltaic systems was used. The simulations included the set up at the ZEB Livinglab and were performed using measured climate data on an hourly basis in the proximity of the Livinglab, including irradiance onto horizontal plane, outside temperature, humidity and wind speed.

IOP Conf. Series: Earth and Environmental Science **352** (2019) 012033 doi:10.1088/1755-1315/352/1/012033

3. Results

3.1. Energy production, solar radiation and PV temperature

Measured energy production and solar radiation as a function of time in October 2018 is given in Figure 3. Figure 4 shows energy production as a function of solar radiation for the two PV roofs. The relation between energy production and temperature on the rear surface of the PV panels at variable solar radiation is presented in Figure 5.

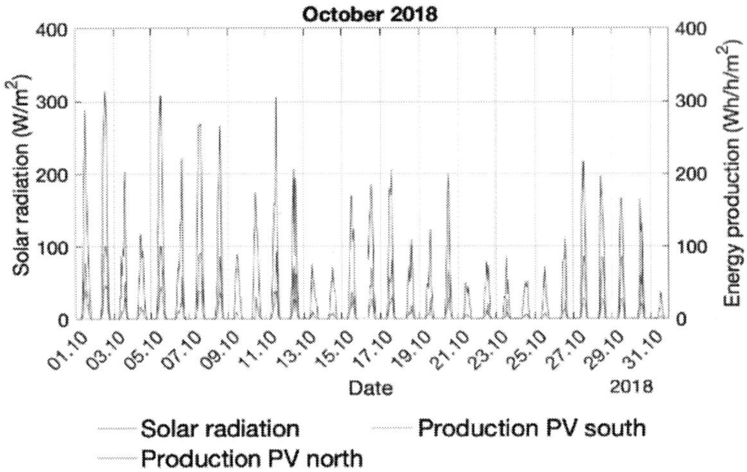

Figure 3. Energy production and solar radiation as a function of time in October 2018.

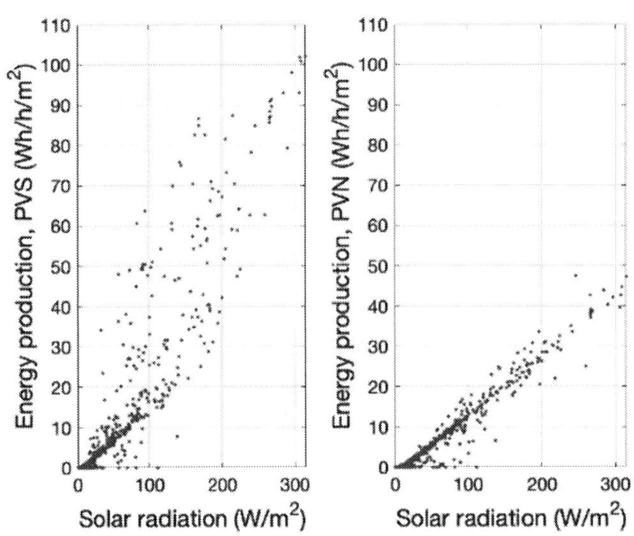

Figure 4. Energy production as a function of solar radiation in the horizontal plane. The southernmost PV-roof is given in the left diagram. The northernmost PV-roof is given in the right diagram.

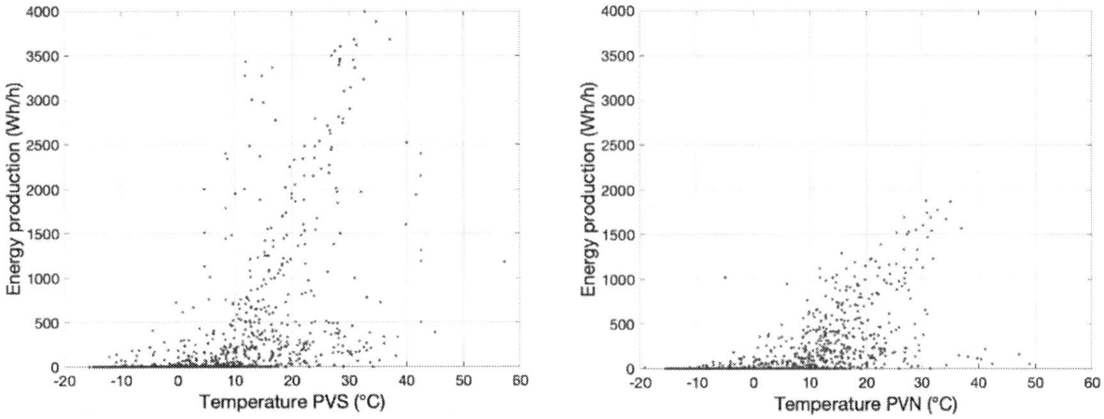

Figure 5. Relation between temperature on the rear surface of the PV-panels and energy production. The southernmost PV-roof is given in the left diagram. The northernmost PV-roof is given in the right diagram.

3.2. Calculations and measurements

Measured and calculated energy production of the two roofs are given in Figure 6-9.

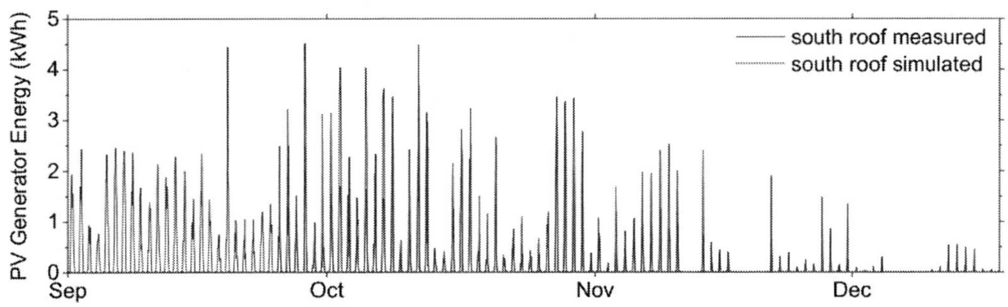

Figure 6. Calculations and measurement of the energy production for the south roof.

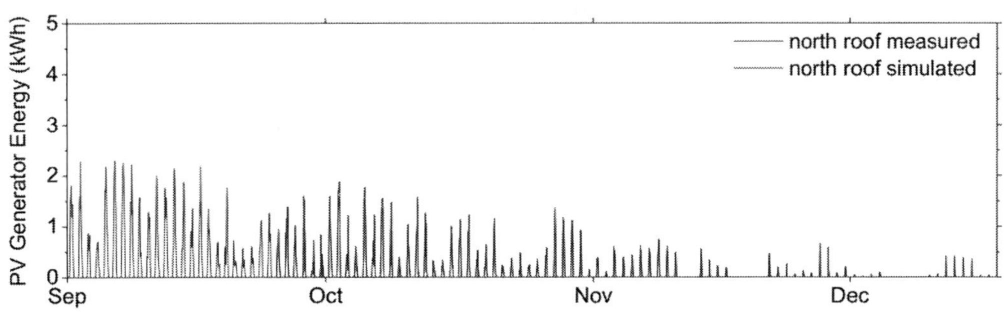

Figure 7. Calculations and measurement of the energy production for the north roof.

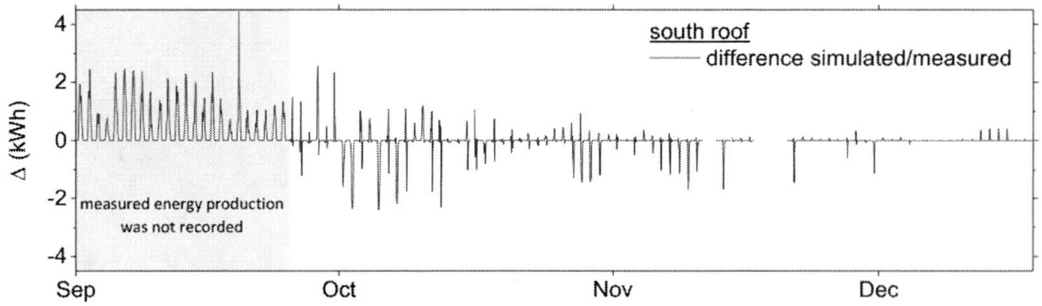

Figure 8. Difference between the simulated and measured energy production

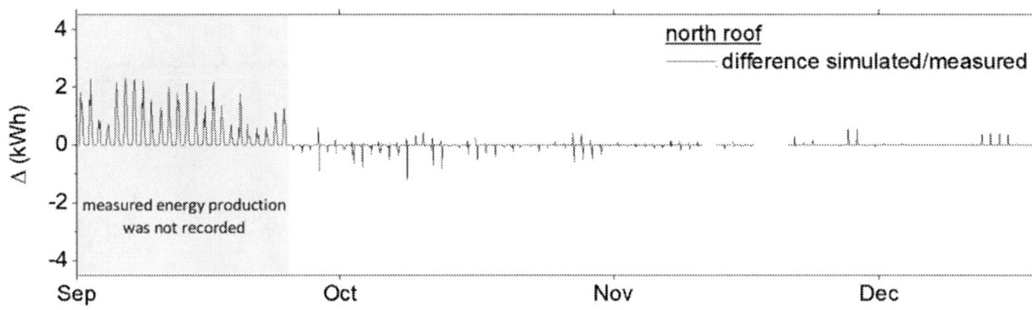

Figure 9. Difference between the simulated and measured energy production.

Figure 10. ZEB Living Laboratory during December 2018 with snow on the PV roofs.

4. Discussion

4.1. Energy production and PV temperature

As presented in Figure 5, the temperature conditions on the two roofs were approximately equal. The results show that increased energy production was associated with increased temperature on the PV panels. This is due to the fact that both energy production and temperature on the PV-panels increase with increased solar radiation.

According to the specification of the PV modules the efficiency is given as 15.8%. This value corresponds to the measured results in Figure 4 which gives the energy production as a function of solar radiation in the horizontal plane.

4.2. Calculations and measurements

A large difference in energy production between the north roof and the south roof was found, as presented in Figure 6 and 7. One possible explanation is shading of the north roof because of low solar azimuth during the measuring period. In order to avoid such problems, design of the PV-roofs should be considered early in the design phase of the building project. This has also been pointed out by [9]. Shading should be avoided in order to ensure a high performance of the PV-systems.

Figure 8 and 9 shows that for most cases the energy production is underestimated in calculations. The north roof shows a small difference between measurements and calculations. The difference is larger for the south roof. This could be due to shading effects from nearby buildings and trees not taken into account in the simulation program. However, the results imply that the simulation program can calculate the effect of the shading of the north section by the south roof in a good way. A high correspondence between measurements and calculations requires correct input data. In this case the input data were measured temperatures, irradiance as well as wind speed and direction at the rooftop of the building. As can be expected, use of generic climate data will cause a larger difference between measured and simulated energy performance.

In the period from 5th-20th of December the PV roof was covered by snow, see Figure 10. As Figure 8 shows there were some periods with sun during this period. This causes the calculated energy production to be overestimated during this period. However, the solar radiation is low in winter and hence it is a very limited amount of energy production that is lost due to the fact that the PV-panels were covered by snow. The situation in November and December can be compared to January and February conditions concerning solar radiation and snow cover. However, during March and April the solar irradiance will increase. The loss in production due to snow cover can be assumed to be larger in these periods.

5. Conclusion

The temperature conditions on the north and south roof were measured to be approximately equal. Still, a large difference in energy production between the two roof sections was found. A possible explanation is shading of the north section by the south roof because of low solar azimuth during the measuring period. In order to avoid such problems, design of PV roofs should be considered early in the design phase of the building project. Shading should be avoided in order to ensure a high performance of PV-systems.

The results show that the difference between the hourly measured and calculated values of energy production of the PV-roofs is small. In this case the input data for the energy production calculation were measured temperatures, irradiance and wind speed and direction at the rooftop of the building. Use of generic climate data will most likely cause a larger difference between measured and simulated energy performance.

References

[1] Dokka T, Wiberg A, Georges L, Mellegård S, Time B, Haase M, Maltha M, Lien A; *A zero emission consept analysis of a single family house*; ZEB Project report 9; SINTEF Byggforsk; 2013.

[2] Ceron I, Caamano-Martin E, Neila FJ; 'State-of-the-art' of building integrated photovoltaic products; *Renewable Energy*; 2013;58:127-133.

[3] Wilson M, Paul M; Effect of mounting geometry on convection occurring under a photovoltaic panel and the corresponding efficiency using CFD; Solar Energy, 2011;85,10:2540-2550.

[4] Mattei M, Notton G, Cristofari C, Muselli M, Poggi P; Calculation of the polycrystalline PV module temperature using a simple method of energy balance; *Renewable Energy*, 2006;3,14:553-567.

[5] Norton B, Eames P, Mallick T, Huang M, Mccormack S, Mondol J, Yohanis Y; Enhancing the performance of building integrated photovoltaics; *Solar Energy*, 2011;8,58:1629-1664.

[6] Häberlin H; *Photovoltaics System Design and Practice*; Wiley; 2012.

[7] Radziemska E,Klugmann E; Photovoltaic maximum power point varying with illumination and temperature; *Journal of solar energy engineering*, 2006;12,81:34-39.

[8] Midtgard OM, Sætre TO , Yordanov G, Imenes AG, Nge CL A qualitative examination of performance and energy yield of photovoltaic modules in southern Norway *Renewable Energy,* 2010; 35, 6: 1266-1274

[9] Imenes A. G., Performance of BIPV and BAPV installations in Norway, *2016 IEEE 43rd Photovoltaic Specialists Conference (PVSC)*, Portland, OR, 2016: 3147-3152.

[10] Misiopecki C, Gustavsen A, Time B; *Cooling of PV panels by natural convection*; ZEB Project report 6; SINTEF Byggforsk; 2012.

IOP Conf. Series: Earth and Environmental Science **352** (2019) 012034 doi:10.1088/1755-1315/352/1/012034

Relation between daylight availability and electric lighting in a single-family house

G Lobaccaro[1*], M Savarino[2], F Goia[1], E Fabrizio[2]

[1] Department of Architecture and Technology, Norwegian University of Science and Technology, Trondheim, Norway.
[2]Department of Energy, Politecnico di Torino, Torino, Italy

*gabriele.lobaccaro@ntnu.no

Abstract. Daylight availability is an important aspect that can potentially improve both the quality and the energy performance of buildings. However, it is not always straightforward easy to assure that an increase in the daylight availability leads to a reduction of electric energy use for artificial lighting. In this study, experimental measurements and numerical simulations were conducted to analyse the relation between the uses of artificial light and the daylighting availability for different groups of users who lived for one month each in a Zero Emission Building single-family house located in Trondheim, Norway. The use of electric lighting and the outdoor environment conditions (irradiance and illuminance on the horizontal plan) were recorded through advanced daylighting simulations, carried out with *DIVA-for-Rhino*, the daylighting availability during the periods of occupancy was then reconstructed, using as input data the outdoor environmental variable recorded during the experimental analysis. The results show that the coefficient of correlation between daylight availability and the artificial light is in general low and the use of artificial lighting seems to be largely independent from the availability of natural light.

1. Introduction

A homogenous daylight distribution has relevant benefits on the building's energy saving, on the human health and on the occupants' well-being [1]. In order to maximize and optimize the quality of daylighting, the building surrounding and the climatic boundary conditions should be taken into account to while designing a building. Maximizing the sunlight penetrating in the interiors also allows a reduction of electric energy use for artificial lighting [2]. A relationship between increased daylighting and reduced use of electric lighting can be found for office buildings: a decrease in the electric lighting use occurs when a proper range of illuminances is provided on the horizontal work plane [3]. However, it can be more challenging to find such a relationship in the context of residential building, probably because of the different user behavior in domestic setting [4]. There, the users' habits and preferences can turn out to be highly unpredictable and cause relevant impact on the buildings' energy use. The study presented in this paper originates from this challenge and aims at investigating how the users interact with electric lighting in a domestic environment, and more precisely at analyzing if the availability of natural light is correlated with the use of artificial lighting by real users.

2. Method and materials

This work is the follow up of a previous study where the relationship between electric lighting and daylight availability in a residential context [5]. The previous study aimed to evaluate the correlation between the electric energy use for lighting and the average indoor illuminance on the horizontal plane placed at 0.85 m from the floor level, equal to 100 m^2, in a residential building in Nordic climate. In the previous study, the correlation was assessed considering the whole floor area of the house without any distinction between the different rooms. The findings demonstrated that, in residential context, it is rather difficult to find a strong inverse correlation between the daylight availability and the use of artificial lighting. This means, in practice, that users generally do not switch off or dim the light when the internal illuminance owing to natural light increases. However, such a missing correlation might be linked to the very heterogeneous conditions in the different rooms of the building. In this study, the level of detail is then increased, and the relationship between daylighting and energy use for artificial lighting is broken down to the single rooms of the building.

2.1. *The experimental facility*

The Zero Emission Building (ZEB) Living Laboratory a single-family house located in Trondheim (Norway, latitude 63°25' N and longitude 10°27' E) (Figure 1) has been used to collect data on user occupancy, daylighting, and artificial lighting use. The ZEB Living Laboratory was designed to be representative of the Norwegian residential building stock for detached house typology. This research facility was built to carry out experimental investigations at different levels [6]. It is equipped with sensors that measure the indoor environmental quantities, such as temperature, CO_2 concentration and relative humidity. The monitoring system keeps also track of the electric energy use in the building, with a degree of detail down to the individual power line, light source, and appliances. This information allows one to know the actual energy use of the house, when it is occupied. Outdoor sensors are also installed to measure the climatic boundary conditions of the system [7].

Figure 1. The ZEB Living Laboratory and its surrounding.

2.2. *Structure of the methodology*

The methodology is structured in two parts (Figure 2). One part is related to the calculation, starting from experimental data, of the electric energy use for lighting in the building down to each individual room (Figure 2 on the left). The other part is related to the estimation of the daylighting availability (Figure 2 on the right) for the periods of occupancy. The data collection includes indoor data for electrical use and outdoor data for solar radiation and definition of indoor illuminance level respectively. To collect data of illuminance in the indoor environment, illuminance sensors, with an accuracy of ±10%, placed on the ceiling of all the living areas were used. Electric energy meter for lighting and dimmer status of each light source were used for the characterization of the lighting system in each room. Regarding the outdoor environment, a pyranometer with accuracy of ±3% was used to measure global solar irradiance on the horizontal plane, while the direct measurement of the different components of solar radiation (direct and global diffuse) was not implemented, but reconstructed from experimental data. Finally, the statistic tool of Pearson Correlation Coefficient (r) has been used to assess the correlation between the two variables: (i) the electric use for lighting, evaluated in a single room and (ii) the average illuminance on the horizontal plane placed at 0.85 m from the floor level, in each analyzed room.

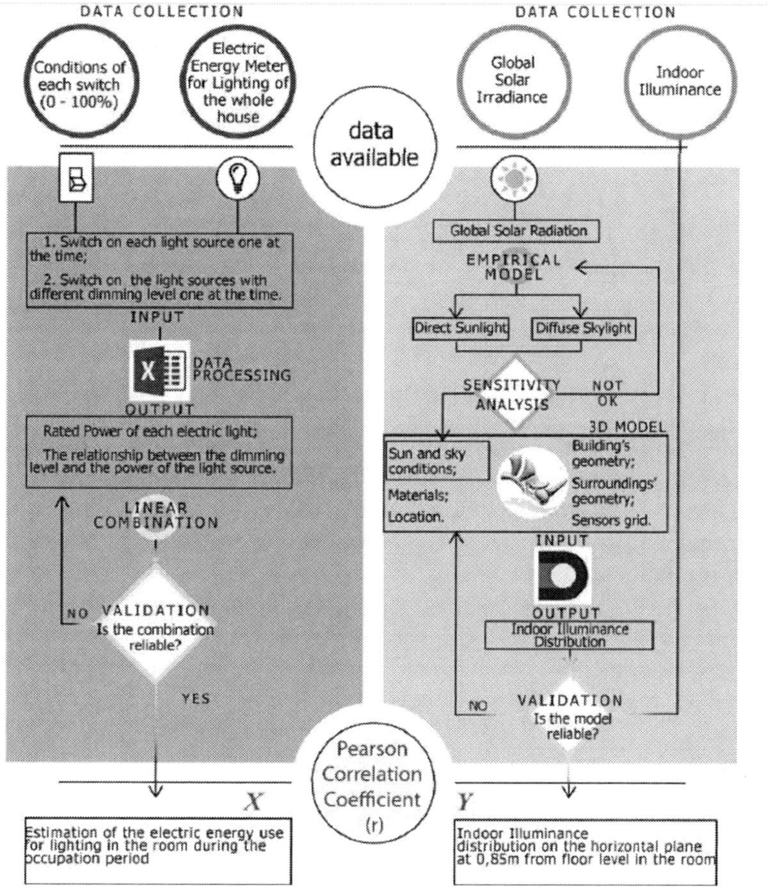

Figure 2. Flow chart of the methodology.

2.2.1. *Monitoring experiment at ZEB Living Laboratory*

This study is included in a wider qualitative and quantitative monitoring experiment which took place in the ZEB Living Laboratory between October 2015 and April 2016. During this period, six groups of users moved in the ZEB Living Laboratory and they used it as their own home for twenty-five days each. The groups, composed by two or four people, were chosen to be representative of the three main demographic categories, such as: young students, families with children, and elderly couples [8]. The users were invited to continue with their routines and habits by avoiding any unusual behavior. Furthermore, in order to observe the adaptation of the users to the house systems, very basic information was provided about the building operation. In order to give the users the time to get familiar with the system, only the data of the last week of the occupational period of each group was considered in this study. This strategy allowed one minimizing any uncertainty due to the unawareness of the users of the building operation system and interference. The experiment was conducted according to the best practice and Norwegian regulations and the consent to use personally non-identifiable data for research activities was released by the Norwegian centre for research data (NSD, Norges Samfunnsvitenskapelig Datatjeneste).

2.2.2. *Daylight simulations and model validation*

In order to reconstruct the inner spaces of the ZEB Living Laboratory (Figure 3a), a 3D model was built in *Rhinoceros* environment. The urban surrounding constitutes by the nearby buildings and terrain profile were also modelled to reproduce the geometry of the urban surrounding. Indoor illuminance levels for the analyzed rooms (the bedrooms west and east, the kitchen and the living area (Figure 3b)) were recreated through climate-based simulations using *DIVA-for-Rhino*, a *Radiance*-based software by customizing Perez sky model as input date and time as well as measured irradiances data (section 2.3.3).

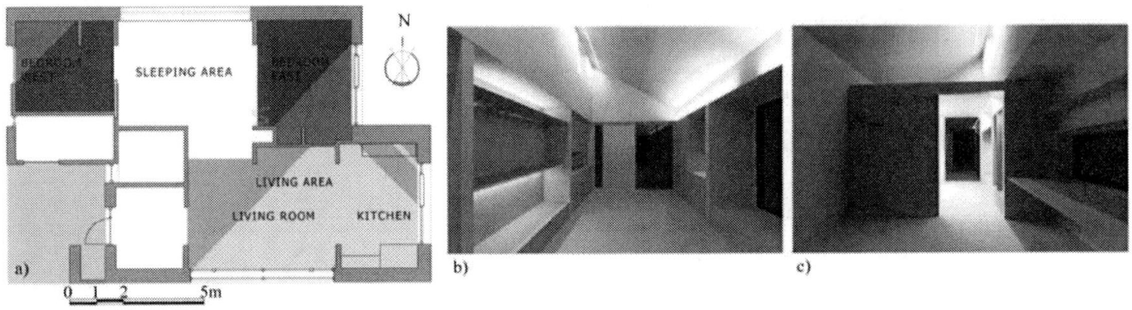

Figure 3. (a) Plan of the ZEB Living Lab: highlighted the analysed areas; (b) Rendering of the living room and (c) inside the bedroom east setting the Radiance's simulation parameters (Table 2).

By referring to a similar example in literature [9], *Radiance* simulation parameters (Table 1) and primitives (Table 2) were set to simulate the indoor materials of the ZEB Living Laboratory. The values of the natural illuminance on the horizontal plane positioned at 0.85 m above the floor level were extracted from the simulations.

Table 1. Radiance's simulation parameters.

ambient bounces (ab)	ambient divisions (ad)	ambient supersamples (as)	ambient resolution (ar)	ambient accuracy (aa)
5	1024	16	256	0.10

Table 2. Radiance's simulation parameters.

Description	Material/Colors	Radiance material	RGB	Specularity	Roughness
Ceiling	Opaque	WoodenCeiling	0.6/0.4/0.3	0	0
Wall	Opaque	WoodenInteriorWall	0.6/0.4/0.3	0	0
Floor	Opaque	WoodenFloor	0.5/0.3/0.2	0	0.02
Furniture	Opaque	WoodenFurniture	0.5/0.3/0.2	0	0
Single Glazing	Translucent	Glazing_SinglePane_88	0.96/0.96/ 0.96		
Triple Glazing	Translucent	Glazing_TriplePane_Krypton	0.51/0.51/0.51		
Mullions	Opaque/ dark grey	MullionsSheetMetalmatted	0.1/0.1/0.1	0.8	0
Outside Wood	Opaque	OutsideWood	0.5/0.3/0.2	0	0

2.2.3. *Model validation and sensitivity analysis*

This *Radiance* engine allows one obtaining climate-based daylighting metrics [10] by using typical weather data for a specific geographical location. In this case, the International Weather for Energy Calculations (IWEC) converted in energy plus weather (.epw) data file of Trondheim was used [11]. The original file was updated in order to include the actual measured data values of the global solar radiation recorded by the pyranometer at the period of the experiments. In fact, climate-based simulations require data of the direct normal and diffuse horizontal radiation to recreate the luminous distribution of the sky vault. However, the pyranometers installed in the ZEB Living Laboratory senses only the global solar irradiance on the horizontal plane but does not provide information about the direct and diffuse radiation. Therefore, to obtain values of solar radiation in direct normal and diffuse horizontal components (as required by the .epw data file), the following empirical model was considered:

$$Global\ Radiation = Diffuse\ Radiation + Direct\ Radiation\ \left[\frac{Wh}{m^2}\right]$$

Where, the global solar radiation, measured by the pyranometer, was distributed in its two components according to the following criteria:

- When the measured global solar radiation was lower than 100 W/m^2, it was assumed to be only diffuse solar radiation component;
- When the measured global solar radiation was higher than 100 W/m^2, it was considered diffuse solar radiation until 100 W/m^2, and the excess part was equivalently divided in direct (50%) and diffuse (50%) solar radiation.

In order to assess the reliability of this simple empirical model, a sensitivity analysis was conducted through different sets of analysis by setting the following inputs:

- In the first set of simulations, the direct solar radiation component in the original .epw data file was increased by 20% and the diffuse solar radiation component was reduced accordingly to reach the total value of the global radiation measured by the pyranometers;
- In the second set of simulations, the direct solar radiation component in the .epw file was decreased by 20% and the diffuse solar radiation component was increased accordingly to reach the total value of the global radiation measured by the pyranometers.

The sensitivity analysis showed that the empirical model is not very sensitive to the differences between direct and diffuse solar radiation components replaced in the .epw data file at the same level of global solar radiation (Figure 4).

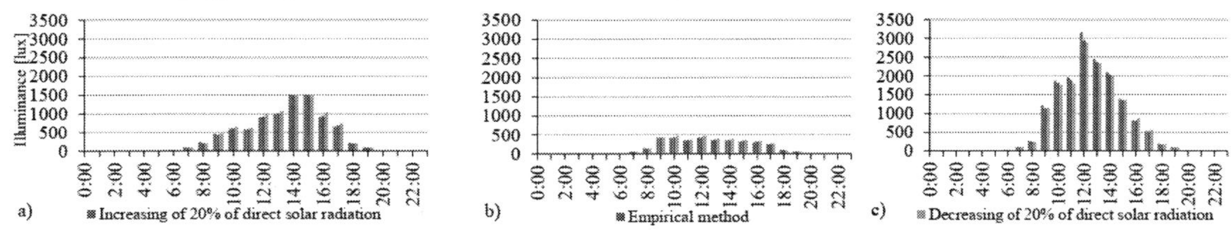

Figure 4. Sensitivity analysis: comparison between the illuminance values carried out from the analysis in *DIVA-for-Rhino* of daylight autonomy conducted in the (a) bedroom west, (b) bedroom east and (c) living area and the kitchen together on the 13th of April.

The proposed empirical model was used to split the global solar radiation measured by the pyranometers into direct and diffuse components and replace them as inputs data in the .epw data file. Daylight Autonomy simulations were run for the analyzed rooms and the outputs were compared with the collected data in order to assess the reliability of the empirical model. The data of indoor illuminance sensed by the indoor sensors facing downwards placed at the ceiling height of the analyzed rooms were compared against the simulated indoor illuminance values calculated at ceiling height. The data collection was carried out in three days of May (Figure 5), characterized by a) overcast (1st of May), b) intermediate (7th of May) and c) clear (9th of May) sky conditions. The analyses showed that the simulated values qualitatively and quantitatively approximate the real behavior of the natural light.

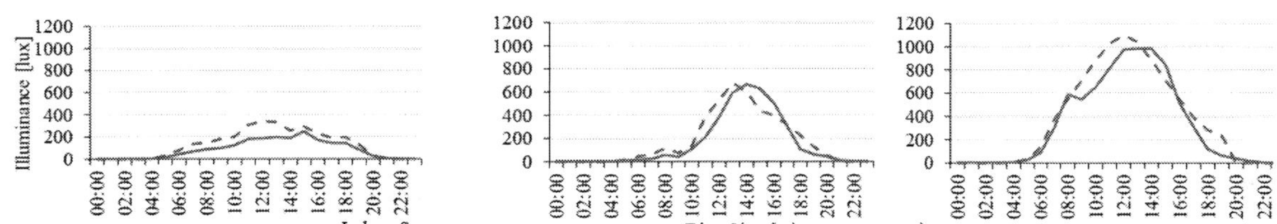

Figure 5. Comparison between the illuminance values carried out from the analysis of daylight autonomy and the values recorded by sensors installed on the ceiling of the living room in a) overcast (1st of May), b) intermediate (7th of May) and c) clear (9th of May) sky conditions.

2.2.4. *Characterization of the electric lighting*

The lighting fixtures of the ZEB Living Laboratory are LED strips that operate with a potential difference of 12V. A power transformer, always active for the conversion from 240V to 12 V, draws a permanent base-load of 33W. This base-load must be subtracted to the overall electric energy use measured every hour in order to be able to assess the energy use down to each individual room and lighting fixture. The characterization of the lighting system was carried out during January and February 2018. In particular, the following features were analysed:

- Energy usage of each lighting at the highest of its power, which means with no dimming (this condition will be indicated as Dimmer status 100%);
- The relationship between the dimming level and the energy use of the single lighting source, analysed only for the most used lighting sources.

Experimental data were collected and elaborated to classify the components of the lighting system. In the first step of the electric lighting characterization, the light sources were switched on for a sufficient amount of time, and one at the time, to allow the acquisition data system to record the electric energy used only by the active light. Then, the electric energy use in that time interval was divided by the time interval to find the power of the involved electric light, and then the base-load was subtracted. Below, a calculation example is shown for one light source. This operation was repeated for all the lights.

$$BaseLoad = 33\ W$$

$$E_{analysed\ light} = 86\ Wh\ ; t = 0.52\ h; P_{analysed\ light} = \frac{86\ Wh}{0.52\ h} = 170\ W$$

$$Nominal\ Power_{analysed\ light} = 170W - 33\ W = 137\ W$$

The second step aimed at assessing whether or not a relationship between the light dimming and the power of the light source could be found. This procedure was necessary to estimate the contribution of each lighting fixture to the overall electric energy meter using the available data of the condition of each switch (percentage of dimming). This part was conducted for the light sources placed on the ceiling of each room. After this, it was possible to assess the electric use in every single room by linear superposition, knowing the condition of each physical switch recorded by the monitoring system during the occupation period and its nominal drawn power at different dimming states (Figure 6).

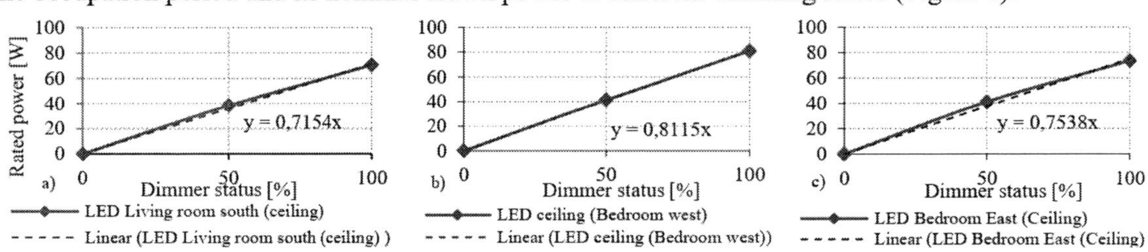

Figure 6. Plot of the Dimmer status-Rated power relationship for the LED installed on the ceiling of a) living room, b) bedroom west and c) bedroom east.

3. Results and discussion

In the Table 3 the results carried out from the calculation of the Pearson Correlation Coefficient (r) show the differences in the groups' behavior. Generally, the groups belonging to the same category tend to behave in the same way towards artificial lighting. A peculiar case is the group of students that is the only one using actively the sleeping area during the day. The Day 7 constitutes a particular case. In contrast with the other cases of non-applicable (N/A) results, in this case the electric consumption stays constant, despite the increasing or decreasing of the daylight availability. It means that the use of artificial lighting is totally independent of the daylight availability. This may also be due to the use of curtains, which is not recorded, or the need for a higher level of illuminance on a specific area of the room or simply that the user forgot to switch off the light when left the room. This day also shows the difficulty in defining preferences and assessing universally accepted level of indoor illuminance in a residential context. In this case, the different pattern and preferences are shown by the use of artificial light in the bedrooms. Even if in the bedroom facing west a level of illuminance higher than 100 lux is reached during the day (>), this is not related to a decreasing of the electric energy use for lighting

(Figure 7 a and b). On the other hand, a strong negative correlation was found for the bedroom facing east, despite the low level of illuminance (Figure 7 c and d).

Table 3. The results of the Pearson Correlation Coefficient (r) for the five groups of users monitored.

Day	Year Period Users	Pearson Correlation Coefficient r				Year Period Users	Pearson Correlation Coefficient r			
		Liv. room	Bed. west	Bed. east	Prev. Study		Liv. room	Bed. west	Bed. east	Prev. Study
Day 1		N/A*	N/A	N/A	0,097	2016	N/A	N/A	N/A	-0,367
Day 2	2015	0,324	**0,565**	-0,040	0,282		N/A	N/A	N/A	0,331
Day 3		0,300	0,385	**0,673**	0,388	18-24.01	N/A	N/A	N/A	0,008
Day 4	27.11/04.12	-0,392	-0,431	-0,420	-0,398	Family	N/A	N/A	N/A	0,380
Day 5	Two students	-0,271	-0,302	-0,279	-0,279	with two	N/A	N/A	N/A	0,290
Day 6		**0,021**	-0,066	-0,147	0,325	children	**0,599**	N/A	N/A	**0,603**
Day 7		-0,473	N/A	**-0,770**	-0,470		-0,485	N/A	N/A	-0,490
Day 1		**-0,574**	N/A	N/A	-0.531	2016	-0,392	-0,460	N/A	0,376
Day 2	2016	**-0,589**	N/A	**-0,573**	**-0.589**		0,192	N/A	N/A	0,172
Day 3		**-0,575**	N/A	N/A	0.013	12-18.03	-0,435	-0,473	-0,348	-0,474
Day 4	09-15.02	**-0,761**	N/A	N/A	**-0.786**	Family	-0,356	-0,503	N/A	0,002
Day 5	Retired couple	**-0,771**	N/A	-0,588	**-0.796**	with two	-0,473	-0,477	-0,429	-0,490
Day 6		**-0,732**	N/A	N/A	**-0.769**	children	-0,367	-0,415	-0,248	-0,249
Day 7		-0,380	N/A	-0,274	-0.476		0,128	-0,192	0,033	0,170
Day 1		N/A	N/A	0,455	0.079					
Day 2	2016	-0,329	N/A	N/A	-0.085					
Day 3		**-0,629**	N/A	N/A	-0.441					
Day 4	11-17.04	-0,467	N/A	N/A	-0.650					
Day 5	Retired couple	-0,052	N/A	N/A	-0.439					
Day 6		-0,282	**0,556**	N/A	-0.348					
Day 7		N/A	N/A	0,455	**0.726**					

In bold are highlighted all the strong correlations.

* N/A (not-applicable): represent the cases in which there is no electric energy use for lighting (lights switched off) in the room during the daylight hours of the day, therefore it is not possible to assess the correlation.

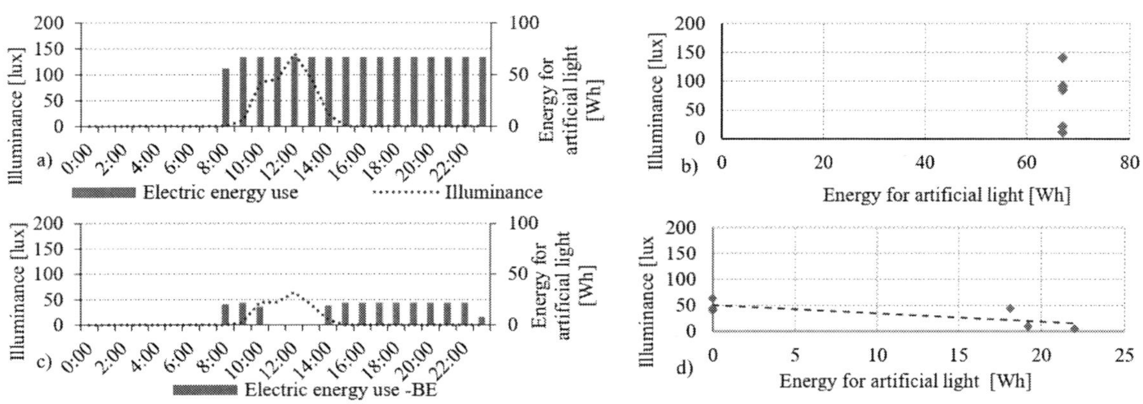

Figure 7. The hourly illuminance values and energy for artificial light and the correlation between illuminance level and energy for artificial light in bedroom facing west (a and b) and east (c and d).

The analysis related to the family with children that experienced the limited daylight hours in January, reveals that there was no the energy use for indoor lighting during the working hours. This underlines completely different habits from the group of students. Moreover, on day 6 a strong direct correlation in the living room demonstrates that even in this case the indoor illuminance level was, apparently, not enough for visual task. In fact, the positive value indicates that the energy use for electric lighting increases as the indoor daylight illuminance increases. For the second family with children, the Pearson's coefficient takes a medium value most of the days. The electric lights in the bedrooms are used only in the early morning and late afternoon. Even in this case, the use of electric lights is more dependent on the occupancy schedule than on the availability of daylight. Furthermore, the results show that the period 09-15 February 2016 (retired couple), counts more values in the range of strong correlation ($\pm 0.5 \leq r \leq 1.0$). In this period, there are also values close to -1 especially in the living room.

IOP Conf. Series: Earth and Environmental Science **352** (2019) 012034 doi:10.1088/1755-1315/352/1/012034

It means that the relationship between the daylight availability and the electric energy use for lighting is strong, which reflects that the electric lighting energy use decreases as the daylight availability increases (i.e. the users tended to switch off the lights in the room when the daylight provision of the space increased). Finally, not-applicable cases is the most common result in the bedroom (64% in the bedroom facing west and 58% in the bedroom facing east) while only 7% of the cases in the living room. This latter data might prove the user's carefulness in switching off the lights when leaving the room.

4. Conclusions

This study confirmed the main findings of the previous work [5]. Despite the level of detail for the calculation of daylight availability increased to the individual room. These are:

- It cannot be found a strong correlation between daylight availability and energy for lighting;
- It is difficult to obtain a robust correlation in the domestic context of residential buildings due to the fact that the users' behaviour is often unpredictable when they interact with artificial lighting system in their everyday life.
- The use of electric lights is more dependent on the occupancy schedule, the users' behaviours, the culture, the habits and the psychological aspects rather than on the availability of daylight.

Acknowledgments

Part of the data presented in this paper were collected in the framework of The Norwegian Research Centre for Environment friendly Energy Research (FME) on Zero Emission Buildings, which is gratefully acknowledged. The authors wish to thank the researchers of the Research Centre on ZEB which were involved in the qualitative experiment (Dr. Thomas Berker, Dr. Ruth Wood, and Dr. Marius Korsnes), and the users which volunteered to be part of the experiment in the ZEB Living Laboratory.

References

[1] J. Mardaljevic, "Rethinking daylighting and compliance," *SDAR* Journal of Sustainable Design & Applied Research,* vol. 1, no. 3, pp. 1-9, 2013.

[2] O. Walkenhorst, J. Luther, C. Reinhart and J. Timmer, "Dynamic annual daylight simulations based on one-hour and one-minute means of irradiance data," *Solar Energy,* vol. 72, no. 5, pp. 385-395, 2002.

[3] L. Bellia, F. Fragliasso and E. Stefanizzi, "Daylit offices: A comparison between measured parameters assessing light quality and users' opinions," *Building and Environment,* vol. 113, pp. 92-106, 2017.

[4] J. Mardaljevic, M. Andersen, N. Roy and J. Christoffersen, "Daylighting metrics for Residential Buildings," in *CIE 27th Session, July 10th -15th,* Sun City, 2011.

[5] G. Lobaccaro, S. Esposito, F. Goia and M. Perino, "Daylighting availability in a living laboratory single family house and implication on electric lighting energy demand," *Energy Procedia,* vol. 122, pp. 601-606, 2017.

[6] L. Finocchiaro, F. Goia, S. Grynning and A. Gustavsen, "The ZEB Living Lab: a multi-purpose experimental facility," in *Gent Expert Meeting, April 14-16th,* Ghent University – Belgium, 2014.

[7] F. Goia, L. Finocchiaro and A. Gustavsen, "The ZEB Living Laboratory at the Norwegian University of Science and Techonology: a zero emission house for engineering and social science experiments," in *Proceedings of 7PHN Sustainable Cities and Buildings*, Copenhagen, 2015.

[8] R. Woods, T. Berker and M. S. Korsnes, "Making a home in Living Lab: the limitations and potentials associated with living in a research laboratory," in *DEMAND Centre Conference, April 13-15th,* Lancaster, 2016.

[9] A. J. Jakubiec and C. F. Reinhart, "DIVA 2.0: Integrating daylight and thermal simulations using Rhinoceros 3D, Daysim and EnergyPlus," in *Proceedings of Building Simulation: 12th Conference of International Building Performance Simulation Association*, Sydney, 2001.

[10] J. Mardaljevic, "Simulation of annual daylighting profiles for internal illuminance," *Lighting Research & Technology,* vol. 32, no. 3, pp. 111-118, 2000.

[11] US Department of Energy, "Weather Data," EnergyPlus U.S. Department of Energy's (DOE) Building Technologies Office (BTO), the National Renewable Energy Laboratory (NREL), 2016. [Online]. Available: https://energyplus.net/weather. [Accessed 15 May 2016].

IOP Conf. Series: Earth and Environmental Science **352** (2019) 012035 doi:10.1088/1755-1315/352/1/012035

Defining a niche for Facilities Management in Smart Cities

[1]Carmel Lindkvist, [2]Alenka Temeljotov-Salaj, [1]Dave Collins and [2]Svein Bjørberg

[1]Department of Architecture and Planning, NTNU, Trondheim, Norway
[2]Department of Civil and Environmental Engineering, NTNU, Trondheim, Norway

Abstract. A wide variety of disciplines engage in Smart Cities as the scope and breadth of it is broad. Within the built environment, there is much discussion on how planning and construction phases influence Smart Cities. There is little discussion on the role of Facilities Management (FM). However, there has been much work on the influence FM has on design phases to ensure functional and usable buildings. Neglecting to scale up the influence of FM from individual buildings to city scale may have long term consequences on sustainability of cities. Taking a mindset of including an FM perspective early in the development of a Smart City is considered here in terms of social aspects of required services. The work draws on ideas of Urban FM to operationalize local needs which also responds to the need to link to broader city sustainable strategies. The starting point is from the development of two districts in Trondheim, Norway. In a three-day workshop, students interpret the needs of the area based on their own knowledge-based perspective, guidance from tutors of Urban FM and through engagement with local users of the area. The work highlights the potential of including an FM perspective in the development of cities.

1. Introduction

Facilities management (FM) perspective of Smart Cities is necessary to consider as buildings become more and more connected to wider districts. Smart Cities and ZEN mentality argues away from the individualistic approach of how buildings are used and focuses more on relationships between buildings, this introduces further complexity on how to maintain and manage FM services in buildings, neighbourhoods and city scales. To date, most work in these areas are based on planning and design. There is very little understanding on how buildings are managed from a use stage which is where FM takes on the responsibility. This knowledge gap highlights the contradiction within sustainability angle of ZEN and Smart Cities as sustainability is for the long term, FM manages the long-term realities that come from ideas in planning and design stages and ignoring the role of FM may have long term consequences on how realities live up to ideas.

Facility managers as the custodians of buildings are in a prime position to ensure buildings can be adapted to meet owners and user's needs as well as societal requirements. The connectivity of buildings within a district is apparent particularly within the idea of Plus Energy Buildings (PEB). These buildings produce more energy than they consume and therefore supply energy back to the grid or to surrounding buildings. In this way, these building create a community of energy users. Energy users are not easy to predict as their behavior is linked to function and use – two core concepts of FM. Lifecycle Assessment (LCA) tools often do not reflect the impact of user's behavior on energy load, indeed the electricity consumption by the office equipment is generally excluded from the plus-energy calculations (Fouquet et al., 2014; ISO/FDIS 52000-1:2016 Energy performance of buildings). A study in Vienna on a so-called 'plus plus' energy buildings found that appliances and powerful scientific desktop work stations which were not included in the original plans for the building combined with unintuitive lighting user interface resulting directly in excessive internal energy loads which both increased energy needs for cooling, but also the heating energy needs due to opened windows (David et al., 2017). In this study the goal of energy consumption in usage was not reached which impacted the ability for the building to provide plus energy to the surrounding area. However, while this study considered the impact of user behavior in the singular PEB, it did not consider the co-dependence of user behavior within the surrounding buildings. It is this co-dependence on buildings with disconnected organizations increases the complication in understanding how buildings are being used within and in the surrounding areas and how this impact on the district as a community. The implications of this is that in order to develop areas that will be developed in sustainable ways, it is important to understand their function and use or how users wish to use the area. This paper aims to build this understanding from a bottom up approach in first identifying what are the function and use of an area.

2. FM within urban development

FM is regarded as a relative newcomer to the real estate and construction sectors being formalized in the late 1970s. It is a profession that encompasses multiple disciplines, which include architects, economists and engineers, to ensure functionality of the built environment by integrating people, place, process and technology managing both soft services such as catering and hard services such as building maintenance (Atkin and Brooks, 2015). The broad nature of the profession makes it challenging to define, but what it ultimate seeks to do is to support non-core needs of an organization that are defined at the strategical level of management, planned at the tactical level and realized at the operational level. In the planning of cities, processes are in place to design cities to meet citizens future needs. However, the future is uncertain and therefore it is necessary to consider how cities that are developed are adaptable for changing citizen demands as well as regional and national goals. The challenges of linking strategic goals at the city level to operational at the local and district level is made clear within the planning literature. There are many diverse service needs within a city as neighborhoods are made up of different people who assign different degrees of importance to an area and it is unclear how these local agendas become integrated at the policy level (Turcu, 2013). In this way, there is no clear connection between the strategic level of political and municipal goals and local operational level of the citizens working and living in the neighborhood.

Urban FM is a concept which connects the knowledge of facilities management to that of urban planning providing a platform for public and private sectors to come together for the benefit of the community (Roberts, 2004). Alexander and Brown (2006:255) state three key areas of how FM can develop corporate social responsibility within communities:

- ensuring that the physical environment does not deteriorate
- creating a platform for skilled employment opportunities
- the opportunity for genuine involvement of the communities in the design and management of services and the urban environment

Clearly even in buildings that aspire to exemplary performance, if the basic functioning of building systems are not correct, fine-tuning individual component efficiencies and other details within the design will not result in desired performance outcomes (David et. al., 2017; Fedoruk et al., 2015). The main barriers, in practice, to achieving ambitious design goals regarding energy performance are not economic or technical, but institutional (Lindkvist, 2018; Fedoruk et al., 2015). Institutional in the sense the various stages of the building life cycle are specified, contracted and implemented without room for the necessary overlap of practices within these stages. There are key attributes of green building performance typically included in most assessment tools are shown resource use, emissions/ waste, and health and comfort issues (Cole, 2012). Comfort provisioning dominates in user expectation for energy services (Dalamagkidis et al., 2007). In this way to ensure energy targets are met being able to understand use and comfort from a user perspective key. There are various studies within the building context where users are excluded from the rationality of sustainable technical implementation resulting in unintended consequences that may run counter to the sustainable goals. One example is where users manipulate the sustainable aspects for their own comfort needs (Petersen

et al., 2017). The exclusion of users on a city scale indicates that such an affect can result in a counter effect on sustainable goals. On the other hand, districts acting individually will also be detrimental to sustainable goals as sustainable success calls for holistic approaches and partnerships. In this way, the inclusion of the top down approach within the district level is important. As Turcu (2013, pp.711) states *"...a community goal may not always be to reach a defined (policy) target/indicator), but to respond to local condition(s) which impact or influence that certain target/indicator"*. Cities can become "sustainable heroes" by focusing on sustainable integration from both top down and bottom up approaches, accounting for both the strategic and operational outlooks of the city. Cities reflect a complexity of a "co-evolutionary and non-linear nature of change which incorporates a range of actors and networks operating over long time-scales" (Dixon, Eames et al. 2014). The context of cities and the multiple dimensional aspects of sustainability necessarily means that the a cross disciplinary approach is necessary in order to account for organizational and structural conditions as well as the broad range of stakeholders with vested interests (Schweber and Leiringer, 2012). The planning of the regeneration of a neighborhood in a city environment is not just about the specific location of that space. It is also about the connectivity to it's wider environment and this is often lacking in the planning of city and built environment dimensions (Dixon, Eames et al. 2014). However, we see taking an urban FM perspective will facilitate connectivity.
The critical challenge for contemporary urbanism is to understand how to develop the knowledge, capacity and capability for public agencies, the private sector and multiple users in city regions systemically to re-engineer their built environment and urban infrastructure in response to climate change and resource constraints (Eames, Marvin et al. 2014). In terms of a neighborhood scale – the end-user is not just one user with diverse demands but a number of diverse end-users with diverse demands and therefore present challenges to combine diverse understandings on the same dwelling. Stakeholders view a building with different lenses depending on their role in the neighborhood (Karlsson, Lindkvist et al., 2013). There are not just diverse agendas making development of a holistic social perspective of a neighborhood challenging, there are also practical problems of developing sustainable neighborhoods. As neighborhoods are made up of diverse groups of people with different interests and the focus of sustainable neighborhoods is to make them more livable. We examine services in terms of what services are required to make the neighborhood livable. With the increased use of terms such as 'smart' and 'intelligent' to describe infrastructure, it is important to remember that the true 'intelligence' of a facility comes from its designed ability to serve specific functions in a useful and convenient way (Chen, 2010). Engaging users to understand future function and use of space creates increased opportunity of developing districts that will enable sustainable behavior.

3. Method
In February and March 2019, three-day workshops were held with students of a refurbishment, technology and management course. The workshops occurred over a week period where the student worked inbetween each one. The objective of the workshops was to gain student perspectives of further development and transformation of real case areas in two districts, Sluppen and Brattøra in Trondheim. The two area aim to be Positive Energy Blocks. The two districts are quite different as Brattøra has undergone a more recent transformation than Sluppen which is still being developed. In the course, students examined the concept of Urban FM in early lectures and were also given a presentation on FM in a Smart City vision. They were divided into two groups, given a practical exercise to draw on some Urban FM ideas and apply them to the two districts. Prior to the workshops, a survey was conducted by the workshop coordinators with a residents or citizens living or working in the districts in order to gain an understanding of what services were needed to develop these areas. The survey was conducted over one week and the response rate was small. The survey was given to one housing association in one neighborhood (known as "Borettslag" in Norwegian) and residents living close to the area known by the researchers. The survey was also given to a business to be disseminated to employees working in the area. The students were given survey results – 19 survey respondents in Sluppen and 5 survey respondents in Brattøra. It is not known what was the entire number of the population sample was given the survey in the end, but a response rate of 24 was considered low given it's close location to the city of Trondheim. Students also did on-site visits to Brattøra and Sluppen where they conducted

IOP Conf. Series: Earth and Environmental Science **352** (2019) 012035 doi:10.1088/1755-1315/352/1/012035

observations of the area and spoke to people who were in the area on the day of 26 February between the hours of 1400-1700. The students in the Brattøra area did not use the survey in their presentation because of the low response rate. The weakness of the methods used here is that the survey was disseminated within a short amount of time and only to a limited number of citizen living or working close or in the areas. Having more time to work with more housing associations in the areas as well as linking in with businesses to disseminate results may have increased the number of respondents. The methods were also a learning process for the students as part of a wider course on refurbishment. The objective of the class was to enable students to increase their understanding of the impact services as well as use participatory process of engaging in observation and talking to people who live and work in an area under transformation. The students were successful in the learning goal, but it did impact on the quality of data received during site visits and adhoc interviews with peope they spoke to. The two student groups had very different districts in terms of development, there was some commonalities, where they focused on mobility and demography in the area, seeing that as core to creating connectivity to the Trondheim city culture. Below is an outline of key findings from these workshops.

4. Findings for Sluppen

Sluppen is 4km from Trondheim city center and has primarily office buildings. It is located near a large road, the E6 which leads directly to the city center. It has poor public transport links with the need to cross the E6 to access the majority of buses from the core Sluppen business area. It is currently being developed and the population of this area is expected to increase.

Service and comfort needs were assessed in responses of 9 people who participated in the survey and who either worked or lived within a range of 3km distance of Sluppen. It was primarily residents with some workers included in results. The age range was primarily between 36 and 55 years – indicating a wide demographic who have an interest in the area. The focus of the survey focused on what respondents would like from the area and what they do not like. Table 1 illustrate the results.

Table 1 Results of respondents likes and dislikes of Sluppen

Likes	Dislikes
Quiet areas	Lack of green areas
To live close to the city center	Noise
Live close to school	Poor public transport Poor
Live close to work	maintenance of the area
Live close to the river	
Feeling of a nice neighborhood with good communication with each other	

The above indicates that people either living or/and working in the area seek good public transport connecting where they work and live. They seem to like the quietness and nature of the area and want a community atmosphere. While they wish to live close to the city, they do not want noise in the area. The likes that the respondents chose from primarily link to comfort issues while the dislikes link to aesthetic issues in the area.

The students identify services that they deemed would improve the area. The students linked to health through identifying the need to develop hiking areas and improving the area for pedestrians and cyclists. This may have been linked to Sluppen being near a nature area but also identifies how respondents wish to reduce their usage of car and the current public transport not being efficient to do so. From the strategic perspective in the city, the students believed they may have to decide whether there were short term solutions to respond to the current needs in the area that would impact on the sustainable aspects by meeting people need's for comfort and healthy areas as well as develop long-term planning of the area.

4.1 Student observations

The students in their observations focused on the attractiveness and access to social facilities of the area. They referred to the area as a 'Car city' and linked to this evidence of there being a lot of car parks in the area as well as a petrol station referring. There is a lack of green infrastructure in the area and existing activities could be improved in terms of their visibility. They considered the demographic of the area in terms of families – "is there enough schools? Kindergartens?" and students "Can students go out and be social?". They believed the E6 impact decreased the attractiveness of the area as it was a source of pollution and noise from traffic.

While the students were quite critical of the area – they saw potential in the area. They focused on increasing the residents in Sluppen but being inclusive referring to a "Sluppen for all ages". They believed there is a need to go from a commercial area to a recreational area – which deflects away from an image of being a 'car city' and more toward the identified need of healthy living. The students also saw a need for green and blue areas. Also, as they viewed some of the buildings as empty – they saw a need to consider multipurpose buildings. The concluding thoughts were to stop Sluppen from being a "ghost city" after working hours and weekends and create better connectivity to the town center. Overall, the students believed in order to increase and optimize on the area, they believed the area to be more adaptable for not just the core functions in which buildings were developed, but view buildings through diverse functionality, to reduce it being empty when all the workers have gone home for the evening.

5. Findings for Brattøra

The students described Brattøra as an area in Trondheim which is an artificial island. While it is part of Trondheim – it was quiet and was not a busy area during their observation which was on a Tuesday between 1500-1700. The municipality's objective that this area should be known as Trondheim's most technological developed area obscures the socializing aspects of the area. They see PowerHouse Brattørakaia as a statement building for the water front. The students found that people only go to the area if they have business or a need to come to the area. They do not tend to come to the area to be social or to enjoy the views of the fjord. The current buildings being built are very architectural attractive but are quite functional to cater for the needs of their occupants who are primarily office workers or students – but if an area is to impact on a district level – there needs to be consideration for outside the walls of buildings. In this area, while the needs of the primary users seemed to be met, the potential for the wider public use of the area were unclear to the students other than enjoying the views that the existing nature from the fjord provides.

5.1 Student interviews with pedestrians in Brattøra

In order to find out more beyond their observations and document search, the students asked people on the street what they thought of the area. The table here presents a general perspective of people they spoke to in terms of their likes and dislikes.

Table 2 Likes and dislikes based on student observations and interviews they met on the street

Likes	Dislikes
Nice architecture	Barriers is the railway – walking over the bridge at the train station may be problematic for people who bike
Not many cars as construction area	Most people who work there seem to use the car
Bike and walking paths good	The bus transport is not great for parents who use it for young children*
Good waste management in the area	They thought there was a lot of asphalt in the area – "asphalt jungle"
	The area is very quiet after working hours and this is reflected by the food store which closes at 8pm

IOP Conf. Series: Earth and Environmental Science **352** (2019) 012035 doi:10.1088/1755-1315/352/1/012035

	People would like more cafés and nice places to stay there in the evening There is no community life The coast is not taken advantage of in terms of social activities and accessing views of the fjord

*The public transport has radically changed since the study was conducted

The like focus on very technical aspects of the area in that is has good architecture, good bike and walking paths as well as good waste management, however, the social aspects appear to be lacking where there are limitations of possibilities beyond buildings functionalities in the area. This is alludes to how the building cater for the individual need of their occupants, but lack social connectivity which may lead to the perception of no community life.

5.2 Student observations

The students believe the areas could become more social rather than an area that is functional to the workers of the businesses. They wish Brattøra to be a place to socialize and increase in it being more welcoming place for Trondheim. They did not observe housing in the current situation which they viewed as a problem as Brattøra appears "deserted at night". They believe housing could make it more lively and they compared it to another part of Trondheim which had a waterfront with residential house as well as bars and restaurants. In terms of development – they believed both increased housing and development of bars and restaurants need to happen at the same time. While Brattøra can be reached by bus or car, people feel the area is too far from city center and it is not easy to reach by public transport. They proposed to keep the area free from bus and use self-driving automobiles which fits with the theme of it being a technological area. They propose these self-driving cars to service the whole area as well as roof top buses that can move around north of the area. They also propose E-scooters (electric scooters) which are currently being used in other cities and provide different stations for electric-scooters in and outside the area. Further potential is to motivate people to come to Brattøra and enjoy the beautiful views of the fjord through boat trips rather than go for just working or studying. Thought to improve the area by having a swimming area in the sea as well as increasing the number of food places by or on the sea through use of floating restaurants and a park for kids to play. The students believed Brattøra is an attractive area but is very quiet in the evening, they believed that it needs more integration with the city center. While there is a lot already in the area with innovative buildings, hotel, swimming pool and offices, the main task in Brattøra is to attract people to come to the area and to invite people to spend more time in the area, particularly after working hours. The students acknowledge the area was quite successful in economic and environmental aspect but lacked social aspect and they questioned the sustainability of the area with social aspects not being clearly obvious based on their observations. In considering, this they focused on how different facilities could improve the area, while also considering the wider long-term development.

6. Discussion and conclusion

The students of this workshop had very limited time to develop their ideas, but they were quite good at indicating how the built environment sets limitations on use and functionality of an area. "Smart" is not just about technology, it must link to function by also examining the space between buildings as social arenas that build communities and create connectivity between districts. As Chen (2010) claims true 'intelligence' of a facility comes from its designed ability to serve specific functions in a useful and convenient way (Chen, 2010). Urban FM is an intermediary to examining links to function and use in the development of districts. It has the potential to develop business opportunities where areas lack services, identify ways to build communities and serve needs of areas to be healthy linking cities ability to be 'Smart' in meeting sustainable goals.

One of the three areas identified by Alexander and Brown (2006) on how FM can develop corporate social responsibility within communities was primarily considered by the students which was *the opportunity for genuine involvement of the communities in the design and management of services and the urban environment.* While the students identified that there are diverse service needs within a city as neighborhoods are made up of different people who assign different degrees of importance to an area, they also indicated that the areas did not reflect how local agendas become integrated at the policy level (Turcu, 2013). While Sluppen is an area planned to go through radical redevelopment, there were a lot of needs and lack of facilities in the area that the students identified. Knowing what these needs are could potentially help decision makers to develop plans to meet needs that require a short-term investment in improving the sustainability of the area and may also fit in with the long-term plan. This could be potentially done on what comfort issues impact on the sustainable use of the area an what will therefore impact on carbon emissions and energy use in these districts. If comfort provisioning dominates in user expectation for energy services and impact on their ability to behave in sustainable ways (Dalamagkidis, Kolokotsa, Kalaitzakis, & Stavrakakis, 2007, Petersen et al., 2017), these expectations should also be considered on current and future uses of an area. The critical challenge is to develop the knowledge, capacity and capability for public agencies, the private sector and multiple users in city regions systemically to re-engineer their built environment and urban infrastructure in response to climate change and resource constraints (Eames, Marvin et al. 2014). One way of doing this, is by considering the development of services from an Urban FM perspective by considering how services can impact on user behavior as well as meet the use, functionality and comfort provisioning of an area for citizens.

References

[1] Alexander, K. and Brown, M. (2006), "Community-based facilities management", *Facilities*, Vol. 24 Nos 7/8, pp. 250-68.

[2] Atkins, B.L. and Brooks, A. (2015) *Total facilities Management* Oxford: Blackwell Science

[3] Chen, Z. (2010) 2Facilities intelligence and evaluation: A multicriteria assessment approach", *Energy and Buildings*, Vol. 42, no.5, pp.728–734.

[4] Cole, R. (2012) "Transitioning from green to regenerative design". *Building Research & Information,* 40, 39-53

[5] Dalamagkidis, K., Kolokotsa, D., Kalaitzakis, K. and Stavrakakis, G.S. (2007) "Reinforcement learning for energy conservation and comfort in buildings" *Building and Environment,* 42, 2686-2698.

[6] David, A., Leeb, M. AND Bednar T. (2017) "Comparison of the planned and the real energy consumption of the world's first (Plus)- Plus Energy Office High-Rise Building" *Energy Procedia,* Vol. 132, pp.543-548.

[7] Dixon T., M. Eames, J. Britnell, G. B. Watson and M. Hunt (2014). "Urban retrofitting: Identifying disruptive and sustain technologies using performative and foresight techniques." *Technological Forecasting and Social Change,* Vol. 89 pp.131-144

[8] Eames, M., S. Marvin, M. Hodson, T. Dixon, P. Guthrie and S. Lannon (2014) "Briefing: Re-Engineering the City 2020-2050-Urban Foresight and Transition Management", *Urban Design and Planning: Proceeding of the Institution of Civil Engineers*

[9] Fedoruk, L.E. Cole, R.J., Robinson, J.B. and Cayuela, A. (2015) "Learning from failure: understanding the anticipated–achieved building energy performance gap", Building Research & Information, 43:6, 750-763

[10] Fouquet, M., Lebert,A., Lasvaux, S., Peuportier, B., Roux, C., et al. (2014) "Illustration of methodological challenges in energy and environmental assessment of buildings" SB 14 Barcelona, Oct 2014, Barcelona, Spain. World SB14 Barcelona conference proceedings

[11] ISO/FDIS 52000-1:2016 Energy performance of buildings

[12] Karlsson, A., Lindkvist, C., Wojtczak, E.; Stachurska, K., Holm, D., Sørnes, K., Schneuwly, P., Tellado, N. and Rodriguez, F. (2013) "Common barriers and challenges in current nZEB practice in Europe: D.1.1 Report" *In:* Karlsson, A., Lindkvist, C. (eds) ZenN project report

[13] Lindkvist, C. (2018) *Utopia for whom? Project and operational perspectives of energy efficient buildings.* EuroFM. 2018. ISBN 978-94-90694-09-8. Research Papers For the 17th EUROFM Research Symposium EFMC 2018, 5-8 June in Sofia Bulgaria, Tucker, M. (ed.) (17).

[14] Pettersen, I.N., Verhulst, E., Kinloch, R.V. and Junghans, A. (2017) "Ambitions at work: Professional practices and the energy performance of non-residential building in Norway" *Energy Research and Social Science* Vol.32, pp.112-120

[15] Roberts, P. (2004) "FM: new urban and community alignments", *Facilities*, Vol. 22 Nos 13/14, pp. 349-52.

[16] Schweber, L. and R. Leiringer (2012) "Beyond the technical: a snapshot of energy and buildings research" *Building Research & Information* Vol., 40, no.4 pp.481-492

[17] Turcu, C. (2013) "Rethinking sustainability indicators: local perspectives of urban sustainability" *Journal of Environmental Planning and Management* Vol. 56, no.5, pp.695-719

Acknowledgments

Arthurs would acknowledge the role of students involved in developing this work from Refurbishment, Technology and Management course, Norwegian University of Science and Technology.

IOP Conf. Series: Earth and Environmental Science **352** (2019) 012036 doi:10.1088/1755-1315/352/1/012036

Local energy markets as a solution for increased energy efficiency and flexibility

I Ilieva[1], **B Bremdal**[1,2], **A A S de la Nieta Lopez**[3], **S H Simonsen**[4]

[1] Smart Innovation Norway, Håkon Melbergs vei 16, 1783 Halden, Norway
[2] University of Tromsø, Hansine Hansens veg 18, 9019 Tromsø, Norway
[3] Utrecht University, Vening Meineszgebouw A, Princetonlaan 8a, 3584 CB Utrecht, The Netherlands
[4] Skagerak Energi, Floodeløkka 1, 3915 Porsgrunn, Norway

Iliana.ilieva@smartinnovationnorway.com

Abstract. With increasing share of distributed renewable energy resources in the grid and arising energy consumer awareness on environmental challenges new market models are sought where energy can be traded in an efficient and end-user centric way. This trend, together with the increasing consciousness on the benefits of local consumption and production has given rise to an increased focus on local energy market structures. Within the E-REGIO project, funded through the ERA-Net Smart Grid Plus initiative, local energy markets have been paid particular attention. This paper discusses opportunities associated with local energy trading, as verified through the E-REGIO local energy system pilot - Skagerak EnergiLab in Norway. Embracing, among others, local loads, energy storage system, PV generation and a large consumer (stadium facility), the pilot-based simulations have produced some useful insides on the implementation of local energy markets and have helped collect learnings that can be of benefit for future local energy market establishments.

1. Introduction

Local energy markets can be considered as a good solution to pending challenges in the local grid. Resent research indicates increased interest towards local markets, giving visibility to the ambition to move electricity trade closer to the end-user [1], [2], [3]. Utilizing a trading platform that can match supply and demand on a local level, thus creating a marketplace for flexible loads, distributed generation and storage, local markets can effectively alleviate problems in the local grid, at the same time as they facilitate a more efficient use of local energy resources. Not less important is the fact that local markets often engage market participants through innovative business models, encouraging them to install technology solutions for local production, distributed storage and demand response [4].

As "local" markets can be considered markets that facilitate trade within a smaller town or a neighbourhood. A precondition for the realization of local trade is the existence of demand that needs to be covered by some form of supply. With the accelerating spread out of distributed energy resources (DER) in the grid, it becomes increasingly attractive to utilize locally produced or stored energy.

The local market envisioned in this work incorporates two types of trade – in energy and in flexibility – which are facilitated by means of a market pool manager. The use of a market platform to facilitate the exchange among independent local suppliers (providers of energy and flexibility) and buyers (in demand of energy and flexibility) can, from economic viewpoint, bring optimal resource distribution and remuneration for trading parties. Yet, in the context of existing grid constraints, some major benefits can be associated with improved operation of the distribution grid (e.g., better quality of supply and

IOP Conf. Series: Earth and Environmental Science **352** (2019) 012036 doi:10.1088/1755-1315/352/1/012036

avoidance of local congestions) and increased self-consumption on an individual (prosumer) or collective level. In particular, from a grid operator point of view, local markets for energy and flexibility can have a highly positive effect due to the possibility to effectively manage and aggregate local assets by means of market rules. Thus, grid challenges associated with rapidly increasing capacity needs (e.g., as related to higher penetration of distributed generation and electrical vehicles) can be to a large extent mitigated.

Further, the current work considers the presence of distributed storage as highly beneficial for a well-functioning local market structure and this attitude is well documented within the following chapters. More specifically, the availability of distributed storage can be of particular importance as it can be utilized as an energy bank for both commercial and technical purposes.

The above topics are being met with growing interest as they concern a wide range of actors in the electricity value chain – end users (consumers/prosumers), distribution system operators (DSOs), energy retailers, transmission system operators TSOs), regulators, technology providers and others. Providing reference towards a unique market design and supporting it by specific real-case implementation with a consequent simulation results, the current scientific contribution helps fill in the research gap associated with the variety of possible cases for operational local market establishments.

To start with, this research work represents the local energy market design as envisioned in the E-REGIO project and reflects on the specific case-based implementation, as valid for the Skagerak pilot. The discussion continues with a description of the pilot site and how it is used to test new and innovative energy-solutions for local energy balance and flexibility that benefit both electricity end users and the grid. Thus, the possible contributions of local market structures to improved grid operations, as experienced by the grid operator Skagerak Nett, are exposed. Finally, this paper provides a conclusive part that reflects on the experiences gained so far from the local energy market implementation and builds the basis for the creation of guidelines directed towards best practices and regulatory recommendations as associated with local energy trading.

2. E-REGIO's local energy market design

The E-REGIO project explores the role of local system operator (LSO) as a facilitator of grid and community services based on an innovative platform that manages local trade [5]. The platform considers the available energy and flexibility resources in the grid to find an optimal market-based match that satisfies the ambitions of local prosumers, consumers and storage owners at the same time as it supports the local grid in its operation.

2.1. Overview

The local market defined within E-REGIO is considered attractive for community members due to its ability to cater for more efficient utilization of renewable and local energy, facilitate smart mobility services and contribute to service diversity and economic benefits for the end users. The design of the local market considers roles, responsibilities and rights associated with local market participation, as well as such that relate to the activities which can be carried out in the market. The activities are to deliver values to different customers of the LSO and within the E-REGIO project this value is investigated through different business cases. The discussion provided in this paper is centred around a business case for the DSO, where increased flexibility utilization by means of local market mechanism enables a more efficient distribution grid operation at the same time as benefits (both direct economic and from value added services) are brought to the end users (providers of flexibility).

From a technology perspective, the efficient local market trading is facilitated by means of an ICT platform that exploits advanced algorithms to find optimal utilization of local energy resources considering both the grid's needs and the preferences of local market participants. Thus, the digitization uptake associate with the utilization of big data, machine learning, and artificial intelligence is well presented into the local market processes. This also helps to accommodate new or amplified business opportunities for the local market participants, bringing increased interest in attaining market membership.

2.2. Actors and roles

As a facilitator of the local trading (by means of the platform functionalities), the LSO is a most central local market actor. Within the E-REGIO project the LSO's roles are defined in more detail as: aggregator, market maker, intermediary (between the external electricity markets and the local market), local energy retailer (which caters for attractive, clear and concise energy contracts), energy storage manager, EV operator and local service provider. Considering the preferences of its customers, the LSO carries out a joint optimization of the flexibility loads. In the process the delivery of either grid or community service, or both can be prioritized. Finally, the contracts offered by the LSO can be referred to as flexibility contracts, energy contracts and contracts for various services (e.g., related to community storage, smart mobility or complementary ones).

When it comes to the specific LSO customers, these have been divided into two groups. The first group encompasses the end users which are considered the smaller (internal) customers of the LSO. Thus, the end users' group contains the local consumers, prosumers, producers, as well as storage owners and neighbourhood managers, the last being defined as legal entities that manage the neighbourhood community's energy resources. In the second group of LSO customers are classified the external (middle voltage (MV) or high-voltage (HV) level) actors such as DSOs, TSOs, balancing responsible parties (BRPs) or distribute generation facilities of a larger scale (as compared to the local prosumer/producer).

In order to maintain a reliable power supply and ensure grid balancing, external LSO customers may be willing to buy flexibility services. The scope of flexibility services that the LSO can offer is substantiated by its own portfolio of end users. The overview of trading actors and contracting regimes within the E-REGIO project has been provided previously in [6]. While the LSO plays the role of market pool manager, DSO, TSO, energy retailers and producers and the existing electricity market are considered predominantly buyers of energy/flexibility services. These parties are bounded with the LSO through the so called "external contracts". On the other hand, "internal contracts" are meant to define the trading premises of smaller-scale energy and flexibility traders. These are consumers, prosumers, ESS and local producers, as representative for the LSO's market portfolio.

3. The Skagerak pilot

The Skagerak pilot (Skagerak EnergiLab) is owned by Skagerak Nett which is one of the largest distribution grid owners in Norway and a part of Skagerak Energi Group. Skagerak Energi Group also includes Skagerak Kraft (dealing with power generation), Skagerak Varme (dealing with district heating) and parts of other businesses such as natural gas and electricity retail. Skagerak Energi has a partnership with Odds ballklubb, Norways oldest and most environmentally friendly soccer club. Their home stadium, which has the name Skagerak Arena, is used as a pilot site to test new and innovative energy solutions for local energy balance and flexibility that benefits both the customers and the grid.

More specifically, the Skagerak pilot, as shown in figure 1, consists of three large PV installations covering the east, south and west roofs of the tribunes adding up to approximately 5000m2 in addition to a large energy storage battery. The PV panels and the battery are connected to the local grid supplying the stadium in two different locations. The pilot aims to demonstrate how the PV generation, different usage of the battery and the local consumption can efficiently interact.

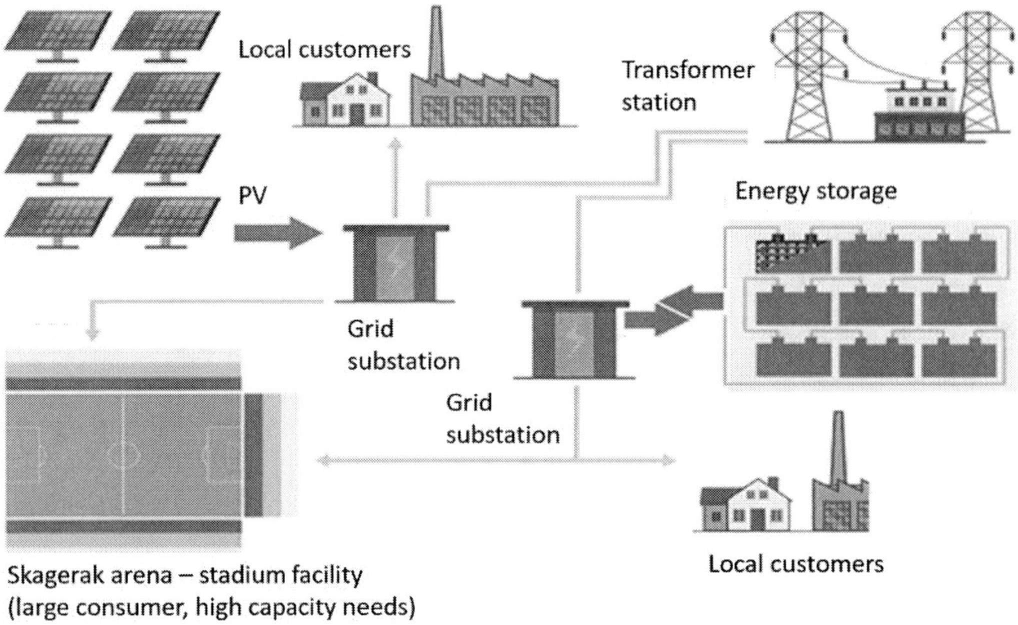

Figure 1. Overview of the E-REGIO pilot at Skagerak, Norway.

The pilot's PV and energy storage installations will be used to test different use cases for the three actors, subgroups of the described above LSO customer types – end users, power generator and DSO. The energy storage is used by all three actors for each use case isolated but also in combination. In this regard the pilot's ambition is to demonstrate how surplus storage, peak shaving, congestion management, voltage support, frequency support and arbitrage can be successfully achieved through the E-REGIO platform for local trade.

4. Simulation-based studies

As a background to the simulation studies performed, it should be clarified that the optimal usage of PV and energy storage system (ESS) is assumed achievable by means of local market trade. Thus, the simulation results presented hereby indicate the impact that efficient trade with local energy and flexibility will have on the overall local system operation, with associated profits and costs.

In order to evaluate the E-REGIO concept within a real environment, a simulation effort has been carried out with focus on the above described Skagerak pilot. The simulations address two grid services relevant for the local grid area – dealing with energy surplus and peak shaving. These two services are facilitated based on the local resources available at the Skagerak pilot. In particular, the Skagerak pilot is equipped with a local PV power plant of 800kWp and a local energy storage system of 1000kWh/800kW. Further, the pilot is characterized by a certain local consumption and a connection to the distribution network.

To show the local profiles of PV and load, figures 2 and 3 portray the local aggregate daily consumption and the aggregate daily PV production. Considering the exhibited consumption and production profiles, utilization of a local storage facility becomes of high relevance.

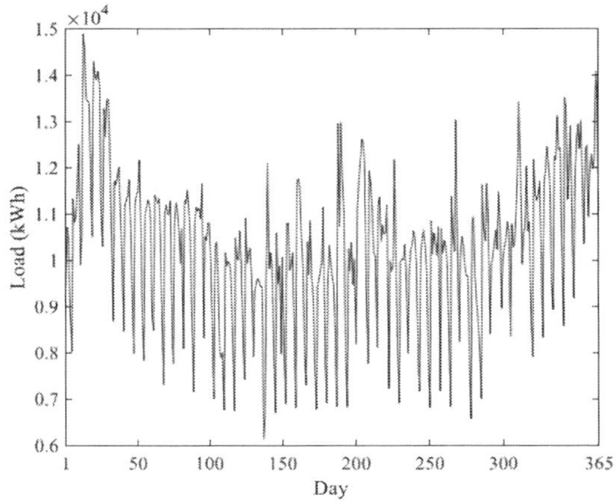

Figure 2. Local aggregate daily consumption.

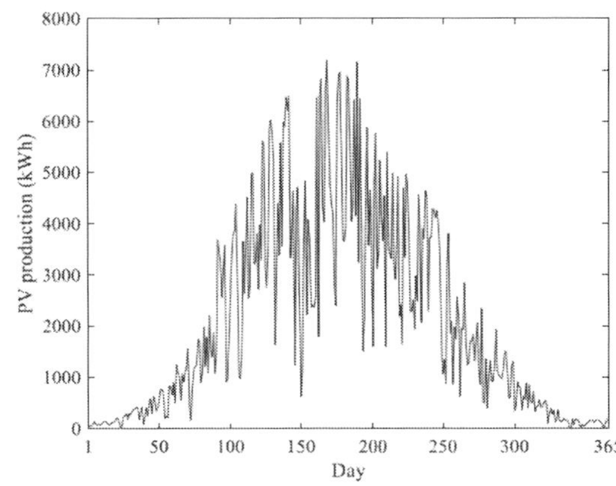

Figure 3. Local aggregate daily PV production.

4.1. Description of the simulations

From a conceptual point of view the achievement of maximum self-consumption, from which the grid also benefits, has a first priority. However, within the simulation effort described below some assumptions are made to simplify the modelling work and which may to some extent be limiting the wider range of possibilities for local market trading. Yet, the E-REGIO local market platform to be running towards the project end will cater for increased number of options, as associated with e.g., the ownership of local resources, order of priorities in the optimization process and others.

To develop the simulations, the LSO is considered as the owner of the local resources such as the PV and the ESS. The LSO also manages the resources, maximizing profits but subject to power limits. The simulations reflect a local market concept, but as previously mentioned, involve some model-specific assumptions that are further referred to in the following paragraphs.

Within the simulations, optimization is related to maximizing incomes from injecting energy into the network, minimizing costs of the energy delivered through the distribution network, and minimizing the maximum power limit (kWh/h) per month. The mathematical model allows modelling the services of the Skagerak pilot. The objective of the mathematical model, as mentioned earlier, is to maximize the profits of the local system operator. These profits come from managing all the resources involved in the local energy system as well as from minimizing the energy delivered per hour from the distribution network (peak shaving). The mathematical model maximizes the profits because the local market has an income and costs. The income comes from selling the locally produced surplus energy, while the costs are associated with all resources utilized to satisfy the local consumption needs. The end users are the ones to pay the costs for the local market's operation. For this reason, the local market's optimization should aim to reduce those costs as well as the power limit cost. In parallel, the local market helps with congestion management through the limit of maximum power from the distribution network.

To understand how the local energy system behind the customer's smart meter has been modelled, figure 4 shows an illustration of its different participants. The participants are the network (distribution network), PV (PV production), ESS (energy storage system), and local load (the local consumption). As it was mentioned above, the network, PV and ESS are managed by the LSO, which is, in these specific simulations, considered as their owner. Moreover, the LSO tries to reduce the costs and consume locally the local resources. The local use of the resources is supported by the utilization of PV and ESS, which is considered free of cost. An exception arises in the case when the ESS is charged with energy generated outside the local system.

To model the local market, energy flows among participants are needed. Four local market participants with energy flows between them have been considered, as shown in table 1. Network-delivered energy can be delegated to the ESS and local load (NET^{ESS} and NET^{LOAD}), the PV has energy flows to the three other components ($PV^{ESS}, PV^{NET}, PV^{LOAD}$), the ESS can charge from the network and PV (NET^{ESS} and PV^{ESS}) and discharge to the local load and network (ESS^{LOAD} and ESS^{NET}), while the local load can only receive energy ($ESS^{LOAD}, PV^{LOAD}, NET^{LOAD}$).

Table 1. Energy flows (start/finish) between participants.

Energy flows (start/finish)	NET	PV	ESS	LOAD
NET	No	No	Yes	Yes
PV	Yes	No	Yes	Yes
ESS	Yes	No	No	Yes
LOAD	No	No	No	No

Once the energy flows in the local system are introduced, the costs are only for $C^{NET^{LOAD}} = C^{NET^{ESS}} = 0{,}02048$ €/kWh from the 9th month to 4th month and 0,01858 €/kWh from the 5th month to the 8th month (the one-year period September 2017 to August 2018 has been considered for simulation purposes). The tariff for PV^{NET} and ESS^{NET} as a prosumer is 0,0039 €/kWh. Besides, the energy per hour imported from the network is a cost equal to 5,54 €/kWh/h per month.

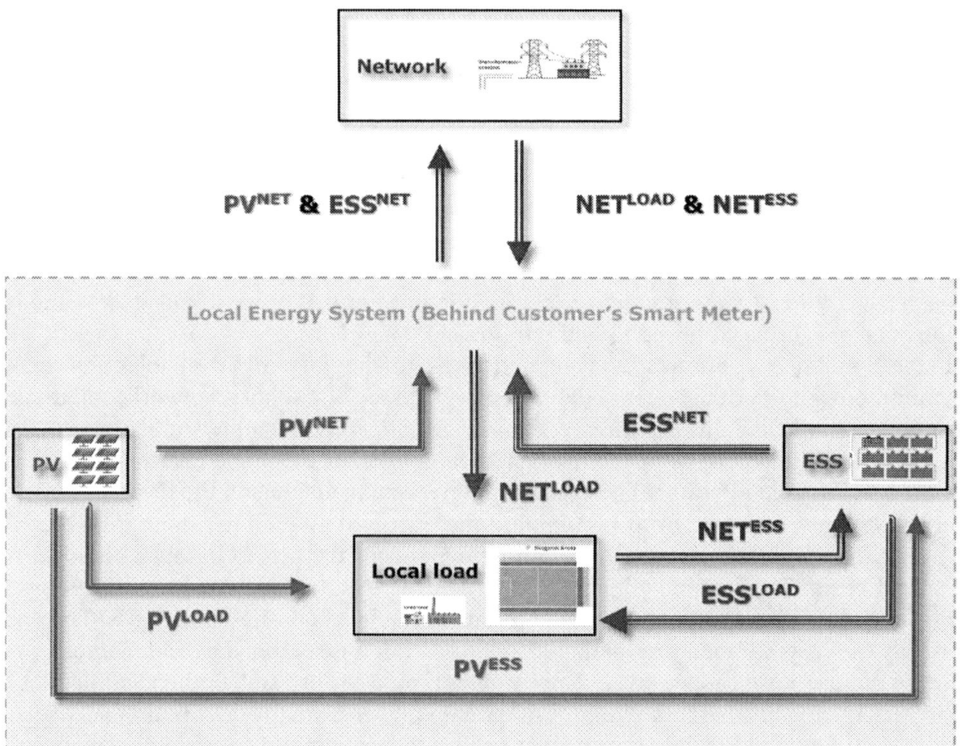

Figure 4. Illustration of the local energy system and its energy flows.

4.2. Results

The volatility of the local PV production and consumption as shown in figures 2 and 3 plays an essential role for the flexibility utilization and for how the local system can mitigate this volatility. In this relation, figure 5 displays the results for the optimal profile of the total energy flow from the network, the limit of peak shaving that is the maximum energy per hour kWh/h from the network and the load. Figure 6 shows how the total energy flow from the network is divided into the ESS (NET^{ESS}), and the load (NET^{LOAD}). As it can be observed in figure 5, the energy from the network is lower than the load and the flexibility stemming from the local energy system is higher. Thus, the more local resources as PV and ESS are being utilized, the higher the efficiency of local flexibility in matching the local power balance. This higher rate of flexibility utilization is associated with the increase in local resources to satisfy the local load. In particular, the local system can exploit the ESS for balancing purposes after integrating the PV production locally.

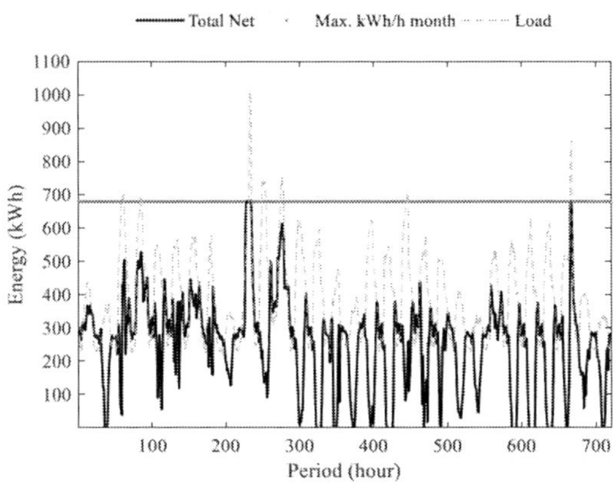

Figure 5. Total energy imported from the network (Total Net), maximum kWh/h per month and the load.

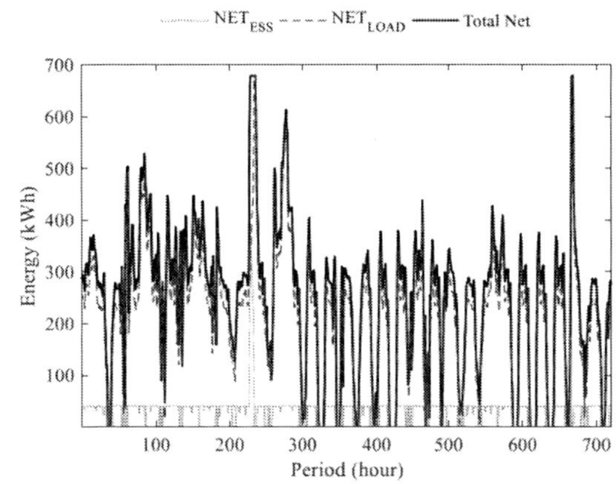

Figure 6. Energy flow from the network to the ESS (NET_{ESS}), from the network to the load (NET_{LOAD}), and the total energy imported from the network (Total Net).

Table 2 shows the annual profits of the local energy system behind the customer's smart meter and maximum kWh/h for two local market cases: i) a system without PV and ESS, and ii) a system that includes the local resources PV and ESS, with a round trip efficiency of the ESS equal to 80 % during a one-year period. Thus, table 2 indicates that case ii) has more flexibility because maximum kWh/h is reduced by 24,3 %. This lower maximum kWh/h (peak shaving) reduces the cost (negative profit) of the local system. The energy imported from the DSO network goes from 3.703.454,03 kWh without PV and ESS to 3.313.817,76 kWh with PV and ESS for the whole year.

Table 2. Annual profits, maximum kWh/h from the network and differences between cases i) and ii).

Case	Profits (€)	Maximum (kWh/h) per month from the network
1. Without PV and ESS (i)	-135.306,0	11.133,3
2. With PV and ESS with an RTE 80 % (ii)	-113.044,2	8.431,3
3. Δ (1. - 2.)	-22.261,8	2.702,0
4. Δ (1. - 2.) %	16,5	24,3

As indicated by the simulation results, utilization of the local flexibility resources will have a positive impact on the local system operation. In E-REGIO this is to be achieved through well-designed flexibility contracts that ensure the optimal solution is attained. The local market trading facilitated within E-REGIO caters for reduced energy import and for mitigated local congestions. In addition, accommodating for the utilization of local PV generation and storage, the local market contributes towards increased local system efficiency and decreased CO_2 emissions. More specifically, the simulation results point out that the local energy market would be able to efficiently reduce the CO_2 footprint with approximately 9.351.270 gCO2eq, given that the ESS losses are equal to 20 %.

5. Conclusions

This paper addresses central topics in the E-REGIO project related to the creation of local markets. A prerequisite for any form of trade is that there is a demand that seeks some form of supply to cover the request for energy. With the advent of energy prosumers and energy production within or close to the vicinity of where it will be consumed a business case for "short travelled" energy arises. Provided that separate economies are involved the prerequisites for commercially based exchanges arise. Within E-

REGIO, the practical implementation of such exchanges is to be carried by means of an advanced ICT platform for local trade.

Two basic forms of exchanges are possible - energy trade and flexibility trade. The former can be in the form of direct peer-to-peer exchanges or through entity in the middle, a broker or pool manager, which in the case of E-REGIO refers to the role of the Local System Operator (LSO). The latter accounts for up and down regulations to maintain local energy balance.

Local energy balance can be attractive for several parties as illustrated by the pilot case of Skagerak and the simulations. First of all, it can be an instrument to maintain positive and negative peak loads within certain upper and lower limits. As shown this can be highly beneficial for the local grid owner, which wants to maintain high quality of supply in its grid and to avoid local congestions. At the same time, it creates an impetus for prosumers to increase self-consumption on an individual or collective level. As illustrated by the simulations, an ESS will be quite essential in order to maximize self-consumption and the economic yield. In the case of Skagerak this is currently possible since the pilot case represents a simplified model of a local market with a minimum of economic entities within the market boundary. A practical consequence is that the ESS can be operated "behind the meter". The meter represents a single point of coupling and defines a separate economic zone for the PV and ESS owner as well as the load. The local market can then exploit the differences in tariffs and prices to achieve desired economic ends.

However, with the existing regulations, it is generally not possible to place multiple prosumers and multiple consumers within a single economic zone and to achieve a form of internal trade with a common ESS and increase self-consumption as each entity of this kind would require a meter. Thus, each new entity would appear in the form described here. But the relevance of the presented approach could still be relevant for multiple entities under prevailing regulatory conditions. That is, for buildings and areas where there is one meter and multiple economies (such as in the case where there are several tenants within the realm of one property). As supported by the ERA-Net community the E-Regio project is obliged to contribute to new standards and policies. In this role the project has made recommendations in line with the knowledge gained and the work that has been done and remains to be undertaken. In particular, the parties involved in the project have actively provided input to the EU Commission regarding the forthcoming directives on energy communities.

Acknowledgments
The E-REGIO project (2017-2020) has received funding from the ERA-Net Smart Grids Plus initiative, with support from the European Union's Horizon 2020 research and innovation program. The authors are grateful for working on the important topic of local energy markets, as part of the project's scope.

References

[1] Minniti S, Haque N, Ngyen P and Pemen G 2018 Local markets for flexibility trading: key stages and enablers *Energies* 2018, 11(11), 3074

[2] Ramos A, De Jonghe C, Gómez V and Belmans R 2014 Reaching the smart grid's potential: defining local markets for flexibility *IAEE Conference 2014*

[3] Teotia F and Bhakar R 2016 Local energy markets: Concept, design and operation *2016 National Power Systems Conference (NPSC), Bhubaneswar* pp. 1-6

[4] Bremdal BA, Olivella-Rosell P, Rajasekharan J and Ilieva I 2017 Creating a local energy market *CIRED Conference 2017* Glasgow

[5] Ilieva I, Wang X, Rajasekharan J 2018 The innovative role of local system operator as a facilitator of grid and community services in the local electricity market *CIRED Workshop 2018*, Lubljana

[6] Ilieva I and Gramme E 2019 DSOs as beneficiaries of innovative contracts and services, facilitated through local electricity market structures *25th International Conference on Electricity Distribution* Madrid 2019

IOP Conf. Series: Earth and Environmental Science **352** (2019) 012037 doi:10.1088/1755-1315/352/1/012037

Future energy pathways for a university campus considering possibilities for energy efficiency improvements

Natasa Nord, Nina Holck Sandberg, Huy Ngo, Eirik Nesgård, Aleksandra Woszczek, Tymofii Tereshchenko, Jan Sandstad Næss, Helge Brattebø

Norwegian University of Science and Technology, Kolbjørn Hejes v 1B, Trondheim, NO-7491, Norway

Abstract. The study aimed to show in a systematic way possible energy efficiency measures that would decrease the total energy use at the university campus in Trondheim, Norway. The entire study was developed in close collaboration with the NTNU Property and Technical Management divisions, meaning that suggested energy efficiency scenarios and other assumptions were highly relevant. Currently, the campus floor area is about 300 000 m^2 and consists of buildings combining offices, lecturing halls, study halls, and laboratories. The campus building stock has been built from 1910 to 2002. To perform this study, building performance simulation and the dynamic segmented modeling were combined. A dynamic neighborhood building stock model was utilized to aggregate the outputs from the building simulation and evaluate global effects of energy efficiency measures. Reference building models for each university cohort were developed based on the methodology for defining the reference buildings. The results of the single reference building analyses showed that a decrease of up to 50% in heating energy use might be achieved by increasing efficiency of the ventilation system and by decreasing the temperature of the heating system. The results showed that in spite of building stock growth, the estimated energy use would decrease from 2017 to 2050 by 10% for the standard renovation, and by 26% for the combination of ambitious renovation and technical improvements.

1. Introduction

Energy planning of the building stock is a highly important topic and highly relevant for energy policy, requirements, and standards development. Regardless of its importance, this research topic is still in its infancy. Different tools have been developed, but they have only partially succeeded due to sectional and particular interests when developing the tools [1, 2]. There are several reasons for this, such as a fragmented building industry, complex building ownership, divergent interests regarding energy use and supply.

Different methods are suggested to model energy use and emissions of the building stocks. The extrapolation method considering the occupant behavior is done by many researchers in Japan. Energy Solar Planning (ESP) tool is a simple tool for municipalities' district planning based on a steady-state monthly energy balance method [1]. Monte Carlo simulation has been used to predict space heating energy use of housing stocks in a bottom-up approach [3], but the model shows to have uncertainties in prediction. A software to deal with this topic is the Sustainable Urban Neighborhood modeling tool (SUNtool) and its successor CitySim. Even though these tools are dealing with occupant behavior, they are not treating energy storage associated with buildings or district energy systems. In no of the over mentioned tools, the dynamic segmented modeling of the building stocks over many years has not been treated. The dynamic segmented modeling of the building stocks is highly necessary tool to estimate development of the building stocks in the future and thereby development of global emission in the future [4].

The aim of the study was to develop energy pathways for a university campus in Trondheim, Norway. This campus will undergo various developments and for this purpose detailed building simulation and the dynamic segmented modeling were combined. Building models covering different building codes were developed in IDA ICE.

The paper is organized as follows. Methodology is introduced in brief by presenting the main calculation steps and data collection process. In the result section, energy use of the university campus buildings at different standard are firstly introduced. Based on the dynamic segmented modeling of the building stock, total energy demand until 2050 was presented. Due to effectiveness of the paper, only the main results are introduced.

2. Methodology

In this study, the methodology was including the following steps: 1) data collection about building properties and energy use at the university campus, 2) definition of a typical university building with typical energy efficiency measures, 3) detail building simulation of a typical university building, and 4) energy use aggregation and projection by using the dynamic segmented modeling. The collected data about the university were analyzed and organized to develop a typical university building for further analysis in the dynamic segmented modeling for the building stocks. The mathematical background for the dynamic segmented modeling of the building stock was implemented in a tool called Dynamic Zero Emission Neighbourhood (ZEN) model [5]. Therefore, in the text below, the details about that tool are explained.

2.1. Dynamic Zero Emission Neighbourhood model

The ZEN model developed by Næss et al. [5] investigates the development of a neighbourhood building stock over time in the context of its size, composition, energy use and greenhouse gas emissions associated with energy use at neighbourhood level. The model is generic and can be used for any type of neighbourhood (residential, service or mixed).

The ZEN model is dynamic and allows studying the long-term development of a neighbourhood building stock. At the start of the modelling period, the model uses a detailed description of the initial stock as well as given or assumed plans for future construction. In addition, demolition and renovation activity can be based on concrete plans or modeled by use of probability functions. Various floor area types in the buildings are grouped together in floor area classes. The building stock is segmented in archetypes defined by floor area class, cohort, and renovation stated. The model calculates the heated floor area for each archetype for all years of simulations. Buildings can move between archetypes after renovation. Furthermore, the ZEN model allows for detailed long-term energy analyses of a neighbourhood building stock. The model is described in detail in the report written by Næss et al. [5].

In this study, archetype-specific hourly energy load profiles for heating and electricity were used as the input. These hourly profiles were developed based on the detail simulation of the typical university building. In this study, the archetypes of the university building development were developed based on the building year. Detail description about the archetypes and the university development is given in the subsection below.

2.2. Data collection

The energy use data of Gløshugen campus were collected from the energy monitoring system. The heat energy use, electricity use, and other relevant indicators were obtained and organized. In order to develop sophisticated model of the campus buildings, several data sources have been employed. For building envelope and geometry, data from energy certification documents issued by Norwegian Water Resources and Energy Directorate (NVE) [6] were used. The classification of campus area use was received from the NTNU Technical Management Division. In order to define technical systems and their functions, energy certifications were used together with the standard, NS3031 [7], and the national building code, TEK17 [8]. In addition, maintenance personnel contributed with experiences and data understanding.

2.3. Reference model

The analyzed NTNU campus consists of 46 buildings, excluding NINA building, Norwegian Institute for Nature Research, and ZEB laboratories. The total gross area is about 300 000 m^2. The campus has its own district heating (DH) ring supplied by the DH utility company in Trondheim. In addition, waste heat recycled from an IT data center was also utilized. The reference model of the campus building should be based on the most common building types in the analyzed area. It was important to choose buildings from the same construction period, location, and building category.

First, a statistical analysis was made to identify number of buildings and their related properties. A previous detail statistical analysis of the energy use at the NTNU campus gives relevant inputs [9]. The statistical analysis of the university campus showed that the largest number of the buildings was built in the period of 1951-1970. This corresponds to 26 of all the buildings in the campus area. Further, it was necessary to define building geometry and building envelope. After evaluation of all the criteria, 18 buildings were selected as reference buildings. The analysis of these 18 buildings showed that the highest electricity use was 197 kWh/m^2, while highest heating use was 510 kWh/m^2. At the same time the most energy effective buildings showed that the total specific energy use was 121 kWh/m^2. This shows that variation in energy use is rather big at the campus level. Finally, the energy use was defined as the average and constituted 133 kWh/m^2 for electricity use and 140 kWh/m^2 for the heating use.

After evaluation of the building content and construction year for all the available buildings, the cohorts were introduced as shown in Table 1.

Table 1. Overview of cohorts

Cohort	Model
Before 1950 – C1	B1
1951-1970 – C2	B2
1971-1999 – C3	B3
2000-2010 – C4	B4
2017 – after – C5	B5

Based on the information from Technical Management Division about the area classification, it was found that the total area was divided into 140 rooms and 18 zones. Finally, all the zones were combined to form the nine most representative. The statistical analysis of the rooms and zones at the campus is given in Figure 1.

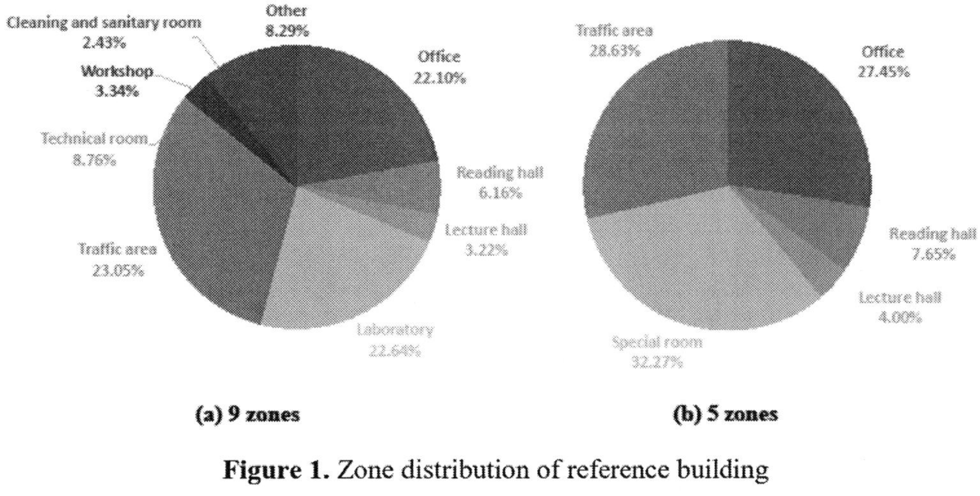

Figure 1. Zone distribution of reference building

As given in Figure 1, the content of the university reference building was defined. Figure 1 shows that office, traffic area, and laboratory accounted for the largest share of 67.79 % in total. Due to similar functionality of some zones, it was decided to combine them. Laboratories, workshop rooms, and various rooms requiring a lot of electricity (server room, supercomputer room, refrigeration room, etc.) were merged into a single zone. This zone is referred to as Special room. Technical rooms, laundry and sanitary facilities, and others were neglected, because they occupied a small area and had small contribution to energy use. The final zone distribution implemented in the study is shown in Figure 1b. Finally, the geometry and size were selected for the reference building. The most relevant information for the model development are summarized in Table 2.

Table 2. Reference model building areas

Building's geometry	Parameter	Reference model
General	Total area [m^2]	7220.00
	Heated are gross [m^2]	7159.20
	Floor area [m^2]	1805.00
	Number of floors	4
Total zone area/ per floor area	Office [m^2]	1967.60 / 491.90
	Library [m^2]	545.20 / 136.30
	Educational facilities [m^2]	282.00 / 70.50
	Special room [m^2]	2321.20 / 580.30
	Traffic area [m^2]	2043.20 / 510.80

Based on the geometry and building envelope parameters, the model was built in IDA-ICE simulation software. The simulation model and the floor area distribution are shown in Figure 2.

(a) Simulation model (b) Floor area

Figure 2. Simulation model developed in IDA-ICE

2.4. Establishment of energy efficiency measures

To find solutions for reduction of energy use at the university campus, the energy efficiency measures were introduced. This step aimed to show potential of energy efficiency measures for educational buildings based on currently technical possible solutions. To efficiently post-process building simulation outputs in the dynamic segmented model for the building stock, the suggested energy efficiency measures were combined into four packages of the measures: 1) standard renovation of building envelope, 2) ambitious renovation of building envelope, 3) technical and operational improvements, and 4) combination of the last two packages. The summary of established package measures is shown in Table 3.

Table 3. Establishment of Energy efficiency measures

Package		Building envelope	Energy efficiency measures
P1: Standard package		Outer walls 1	Insulation with 50mm mineral wool
		Roof	Insulation with 50mm mineral wool
		Windows 1	TEK17 level (U-value 0.8 $W/(m^2K)$)
		Air tightness	Improvement of leakage rate to 1.5 l/h
		Thermal bridge	Improvement of thermal bridge to 0.06 $W/(m^2K)$
P4= P2+P3	P2: Ambitious package	Outer walls 2	Insulation with 100mm mineral wool
		Roof	Insulation with 50mm mineral wool
		Windows 2	Ambitious level (U-value 0.6 $W/(m^2K)$)
		Air tightness	Improvement of leakage rate to 1.5 l/h
		Thermal bridge	Improvement of thermal bridge to 0.06 $W/(m^2K)$
	P3: Technical package	Heat recovery ventilation	Replacement of heat recovery with 80%
		Low temperature heating system	Switch from 80/60°C to 60/40°C

2.5. *Development of future energy use*

The reduction of energy use is one of the objectives that our university, NTNU, put forward within their energy strategy. In order to investigate possible future energy development pathways, a neighbourhood building stock energy model was developed. It was founded on dynamic material flow analysis principles. The detailed model description and background can be found in the report [5]. Two scenarios were introduced to analyze possible energy efficiency developments: baseline scenario and advanced renovation scenario.

Baseline scenario considered that the future development of the existing and new buildings would follow current trends. This means that the renovation activities and new buildings would happen in compliance with present policy and regulations. The existing building stock was assumed to undergo standard renovation in a 40-year renovation cycle. The new buildings were expected to be built according to passive house requirements.

Advanced renovation prioritized increased energy efficiency of the building stock. This meant that the existing buildings were expected to undergo advanced renovation, whereas the new buildings are presumed to be built according to passive house requirements. Energy supply systems are assumed to be the same as in *Baseline scenario*.

3. Results

The most relevant results for the campus development and future energy use are presented below. Duration curves for heating use after the energy efficiency implementation are presented too.

3.1. *Building stock development*

Figure 3 shows the building stock development at the university campus in the period 2017-2050. Figure 3 shows also the change in the total floor area for various cohorts. Please note that the cohorts are marked with "C", while the meaning and the content of each cohort is given in Table 1.

The building stock model considered plans about new construction at the campus towards 2025. Therefore, an increase in stock size occurred in this period, see Figure 3 at about 2025. Later on, it was assumed that some buildings would be demolished, as they reach their estimated end of life. The building stock model did no add new construction to replace the demolished buildings, and therefore the simulated stock seemed to be decreasing after 2025. This decrease was a result of the demolition of buildings from the cohort group 1951-1970.

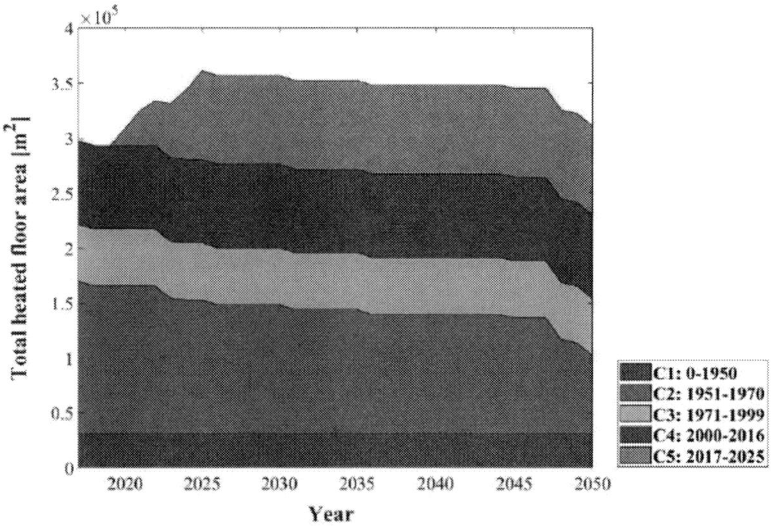

Figure 3. Neighbourhood building stock development at Gløshaugen campus 2017-2050. Total stock size and distribution to cohorts

3.2. Heating demand after implementation of energy efficiency measures

Taking in account that the building type 2, B2, see Table 1, included the most buildings at the university campus, the results are given for this cohort. Specifically, the focus was on the heating use, because the biggest potential for savings were identified there by implementing technically available measures. In addition, the building category 4, B4, with the newest building requirements and implemented measures is presented. Figure 5 and Figure 6 show the heat duration curves with the introduced renovation packages, while Table 4 and Table 5 show specific DH energy use and savings.

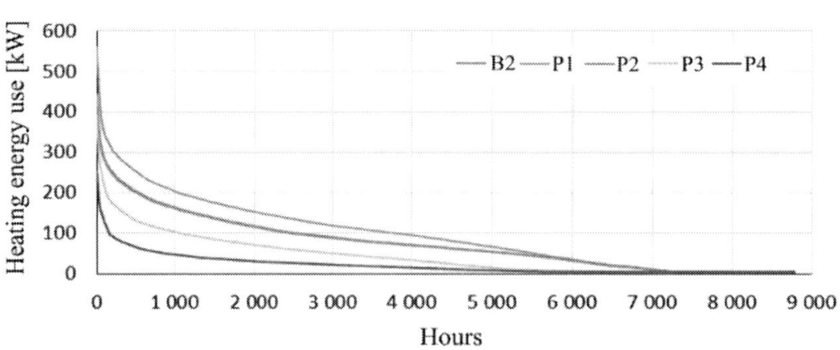

Figure 4. Heat duration curve for B2 model and corresponding renovation packages

Table 4. Specific heating energy use for B2 model with introduced energy efficiency measures

	B2	P1	P2	P3	P4
DH (kWh/m²)	119.6	95.4	93.2	53.9	27
Savings (kWh/m²)		24.2	26.4	65.7	92.6
Savings (%)		20	22	55	77

As it can be seen from Figure 5 and Table 4, big savings could be achieved with the façade renovation packages. The results showed values in the range of 20-22%, while the technical package, P3, yielded a saving of 55%.

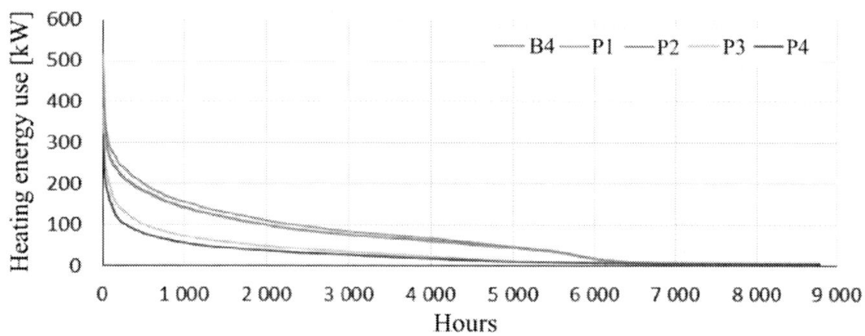

Figure 5. Heat duration curve for B4 model and corresponding renovation packages

Table 5. Specific heating energy use for B4 model with introduced energy efficiency measures

	B4	P1	P2	P3	P4
DH (kWh/m^2)	83.8	81.2	79.7	40.4	32.1
Savings (kWh/m^2)		7.1	8.6	47.9	56.2
Savings (%)		8	10	54	64

Figure 6 and Table 5 show that not as much savings for the building type B4 was achieved in comparison to B2 model. However, the technical packages, P3, P4, still provide high savings. Further, for both B2 and B4 models, it can be noticed that the heat duration curves had different shapes. The reason for this was that they belong to different construction year with different technical requirements. Regarding the specific energy use, there was no increase due to implementation of ambitious renovation in combination with technical improvements, the technical package P4, because, the indoor temperature was always satisfied during the working hours regardless of the cohorts and the technical packages.

3.3. Results on campus energy use development towards 2017-2050

Figure 7 shows the total energy demand with respect to cohorts and for two development scenarios.

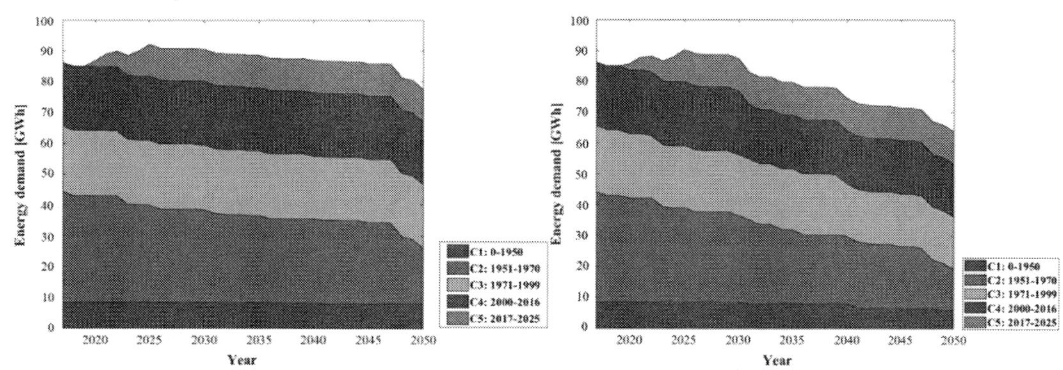

(a) Standard renovation (b) Ambitious renovation

Figure 6. Energy demand with respect to cohort group

Figure 7a shows the results with respect to *Baseline scenario*. The results shows that energy demand of Cohort 1, 3, and 4 (C1, C3, and C4) decreased slightly as a result of renovation activity. For the newest cohort, C5, after the completion of the construction, the total energy demand would remain the same. It can be noted that Cohort 2, C2, marked red, diminishes substantially over the modelling period. The reason for this is demolition and renovation activities that would occur in this cohort group.

Figure 7b shows development with respect to *Advanced renovation*. At the beginning of the simulation, the total energy demand increased due to the emergence of new construction reaching a maximum of 90 GWh in the year 2025. The total energy demand decreased constantly due to renovation and demolition. Compared to the 2017 level, this means that the total energy demand of the campus in 2050 would be expected to be 26 % lower. The drastic decline in the total energy demand (by almost 60% compared to the 2017 level) occurred in Cohort 2 as a result of demolition and renovation activities. The energy demand of Cohort 1, 3, and 4 diminished over the simulation period and the total energy demand of the newest cohort remained unchanged after 2025 until the end of the modelling period.

Table 6 shows the results on development of the energy use in the university campus introduced in Section 2.5. The results show a percentage change in the value of energy demand at year 2050 with regard to the 2017 level.

Table 6. Energy demand and GHG emissions in all scenarios in 2050 compared to 2017 level

	Baseline	Advanced renovation
Electricity demand	+1.4%	-0.5%
Heat demand	-28.7%	-66.7%

The analysis on the energy use revealed that despite the stock growth, the total energy demand would decrease. The main reason for this was due to a substantial decrease in heat demand due to introduction of energy efficiency measures and construction of new low energy and passive standard buildings.

4. Conclusions

The aim of the study was to develop energy pathways for the university campus in Trondheim, towards 2050, because reduction of energy use is one of the our university objectives. The study showed that most of the buildings at the campus were built between 1951-1970. Four energy efficiency packages were introduced for reduction of energy use. Energy efficiency packages were mainly focused on heating saving potentials. It was found that saving potentials were highly dependent on the construction period of the buildings. Further, the technical package P4, ambitious renovation in combination with technical improvements, showed the greatest improvement in terms of energy efficiency. However, a substantial heating energy could be saved by implementation of simple technical measures. Specifically, improvement in the ventilation system gave the best results.

Detail analysis of the electricity use is required to introduce measures that would decrease electricity use. All energy efficiency measures should be analyzed with consideration of cost estimates. Real savings could be higher than the simulated results depending on energy price models. In addition, since the expensive peak load was reduced due to implementation of some energy efficiency measures, the economic benefits could also be higher than simulated.

Due to the planned relocation of the campuses, the campus building stock is expected to grow substantially until the year 2025 as a consequence of the construction activity. After 2025 the stock is estimated to gradually decrease as a result of the demolition of buildings from the cohort group 1951-1970.

Finally, the study demonstrated that the ZEN model for the building stock development was reliable for the future analyses of energy demand for a neighborhood like the NTNU campus. The conclusions from this study were used as the part of the investigation study for the campus development.

References

[1] Robinson D. Integrated resource flow modelling of the urban built environment. Building Performance Simulating for Design and Operation2012. p. 441-68.

[2] Chen Y, Hong T, Piette MA. Automatic generation and simulation of urban building energy models based on city datasets for city-scale building retrofit analysis. Applied Energy. 2017;205:323-35.

[3] Kavgic M, Summerfield A, Mumovic D, Stevanovic Z. Application of a Monte Carlo model to predict space heating energy use of Belgrade's housing stock. Journal of Building Performance Simulation. 2014.
[4] Sandberg NH, Sartori I, Brattebø H. Using a dynamic segmented model to examine future renovation activities in the Norwegian dwelling stock. Energy and Buildings. 2014;82:287-95.
[5] Næss JS, Sandberg NH, Nord N, Vestrum MI, Lausselet C, Woszczek A, et al. Neighbourhood building stock model for long-term dynamic analyses of energy demand and GHG emissions. 2018.
[6] Directorate NWRaE. Energy use. 2011.
[7] NS 3031, Calculation of energy performances of buildings, Method and data. In: Norway S, editor.2007.
[8] Regulations on technical requirements for construction works. Norwegian Building Authority. 2017.
[9] Guan J, Nord N, Chen S. Energy planning of university campus building complex: Energy usage and coincidental analysis of individual buildings with a case study. Energy and Buildings. 2016;124(Supplement C):99-111.

IOP Conf. Series: Earth and Environmental Science **352** (2019) 012038 doi:10.1088/1755-1315/352/1/012038

Influence of emerging technologies deployment in residential built stock on electric energy cost and grid load

Ruslan Zhuravchak, Natasa Nord and Helge Brattebø
Norwegian University of Science and Technology
Kolbjørn Hejes vei 1B, Trondheim, NO-7491, Norway
E-mail: ruslan.zhuravchak@ntnu.no

Abstract. High penetration rates of novel building energy technologies has prompted a growing concern about their microeconomic effect and grid influence. Deployment of photovoltaic (PV), solar water heating (SWH) systems and energy storage solutions, in addition to the growth of electric vehicles (EV) fleet, are reshaping the structure of built stock and lead to the changes in its electric energy demand profile. Long-term forecasting of such structural changes is necessary to guide the decision-making process that would satisfy the needs of both, energy consumers and the suppliers. Whereas electric energy price model is one of the key influencing factors of technologies acceptance for households, peak loads and grid feed-in determine the needed capacity of power grids. The objective of this study was to assess both, the aggregated cost of energy and the changes in cumulative load profiles that are excepted by 2050 for one of residential building typologies in Norway. Methodologically, it was achieved with descriptive statistics, stochastic forecasting and detailed energy performance simulation. Annual electric energy cost for consumers were evaluated under six pricing models. The results suggested that time-of-use and variable maximum power extraction models represent the lowest and the highest extremes in energy cost. At the aggregated level, peak load will decrease in range 1% to 13% compared to current level. Peak PV feed-in will reach up to 40% of peak load by 2050.

1. Introduction

Renewable energy technologies and energy conservation measures together with the emerging electric mobility solutions are penetrating the residential built stock rapidly. The motives and the consequences of such shift have to be studied in a systemic manner [1] to ensure the compliance to long-term strategic development plans for communities.

The benefits for consumers typically involve achieving the economic [2] and possibly environmental objectives [3] that may result from the reduced energy use and from power feed-in on the long term. More detailed analysis, however, suggests that these technologies may cause even further inconsistency in grid interaction [4]. Hence, their deployment at a large scale may require added power generation capacity and grid reinforcement. For the Norwegian power system this issue becomes even more challenging given that electricity is the major source of energy for space heating. Thus, Norwegian power grid development and maintenance for the period 2016 - 2025 will be supported by NOK 140[1] billion of investments [5].

Under such investment strategies, the expected increase in electric energy nominal price for households is 30% by 2025. In addition, a number of changes in electric energy pricing model will be introduced [5] by the Norwegian Water Resources and Energy Directorate (NVE). This brings to the necessity to evaluate the future costs of electric energy under various pricing methods.

Forecasting the future state of built stock is crucial for developing further power grid investment plans and energy pricing methods. Realistic long-term, large-scale energy forecasting models need to account for: 1) Detailed information on how certain technologies affect the energy performance at a single building level; 2) Typological complexity, variability, and dynamic nature of built stock; and 3)Uncertainty in consumers' decision to accept a certain novel technology.

"White-box" energy performance simulation is considered as an effective tool used to evaluate the results of technological interventions applied at a building level. Large-scale energy planning, however, is hindered by additional challenges, associated with the heterogeneous structure of building stock. Figure 1 illustrates the variability of energy use for residential buildings in Norway. A number of factors lead to significant variance, even within the same building type [6]. It can be attributed to the properties of envelope, energy supply system, appliances and occupant behaviour. Appropriate statistical methods that quantify this variability should be used at the model development and validation steps to minimise the error [7].

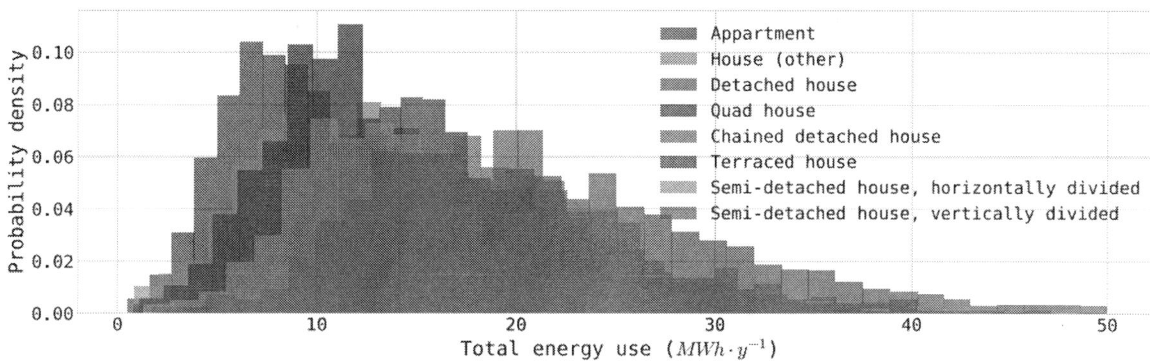

Figure 1. Energy use in the Norwegian residential built stock - univariate distribution

The deployment of novel technologies in buildings has a large number of underlying sources of variability that cannot be accommodated by deterministic modelling principles. Social and economic environment, amongst the other factors that affect the decision-makers, can be considered as random variables. Studying the dynamic evolution of such random phenomenas has to involve stochastic forecasting with the appropriate tools from probability theory [8]. This enables to account for a full spectrum of likely future outcomes, to quantify the uncertainties and thus, serves as a basis for well-informed strategic planning. Consequently, stock-wide influence of novel technologies on future electric energy prices and on power grid needs to be estimated in a quantitative, probabilistic terms. Developing the feasible methodological approach to achieve such objective is the key reasoning behind this study. It is exemplified with one residential building typology - semi-detached house divided vertically. The approach involves typological sampling, detailed building energy performance simulation, and stochastic forecasting. For the exemplification purposes, four technologies were considered: EV, PV, domestic hot water tank (DHWT) and SWH system.

[1] As of May 2019, 1 NOK = 0.1018 EUR.

2. Methodology

The suggested methodology consisted of five steps:

- Descriptive statistics - estimating central tendency, dispersion and distribution of variables within the typology to establish key properties of the representative building;
- Detailed energy performance simulation of the representative building without and with the technologies under the study and all possible combinations of those;
- Statistical simulation of stock-wide acceptance of technologies;
- Aggregation of the resulting cumulative energy costs under the available pricing methods;
- Aggregation of grid load and grid feed-in.

For detailed energy performance simulation, IDA-ICE software was used. Analytical tasks were carried out in Python programming language using the following libraries: 1) NumPy, SciPy, Statsmodels, and Networkx for numerical computing and simulation; 2) Pandas for data wrangling; 3) Matplotlib and Seaborn for data visualisation.

2.1. Descriptive statistics

The analysis of Norwegian Energy Performance Certificates (EPC) was carried out at this step. The background related to the dataset, as a component of EPBD [9] implementation in Norway, are available in source [10]. Amongst 18100 records for all the residential building categories listed in Figure 1 except apartments, semi-detached house divided vertically reached 9.7% by records count, 9.0% by heated floor area and 9.3% by energy use. The distribution of samples based on heated floor area (m^2) is illustrated in Figure 2.

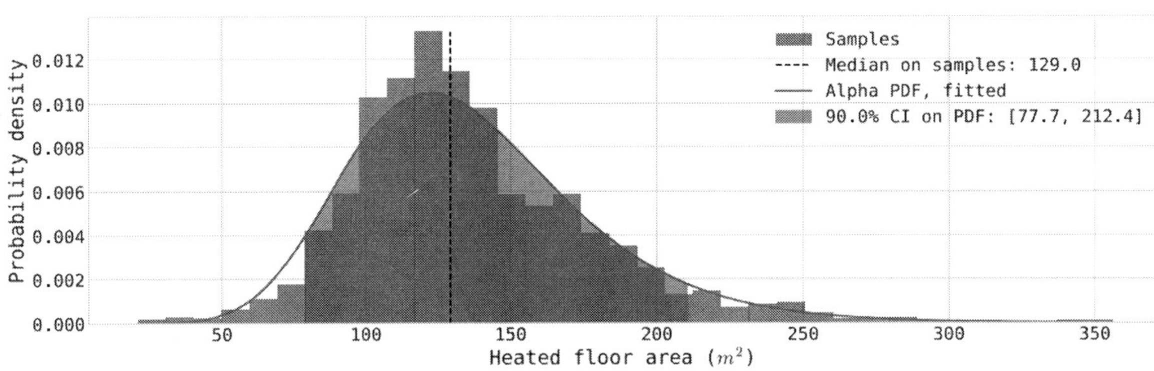

Figure 2. Heated floor area - univariate distribution (data-specific and theoretical best fit)

It was found that Alpha Probability Density Function (PDF), compared to other theoretical PDFs, describes this variable in the dataset best. For this type of PDF with its specific parameters, 10th and 90th percentiles, or confidence interval (CI) is [77.7, 212.4]. Subsequent steps in the methodology focused on this range only.

Further analysis of the EPC dataset revealed that 60% of buildings within this typology were constructed before 1990, as concluded earlier in source [11]. An additional step was aimed at describing the energy intensity within the building typology. Figure 3 illustrates the linear relationship between the age of buildings and the energy intensity. The figure indicates that the energy performance of buildings constructed before 1990 remains relatively poor. They contribute significantly to the stock-level energy use and their refurbishment should be of priority. The overall linear trend was further used for model validation, as elaborated in section 2.2.

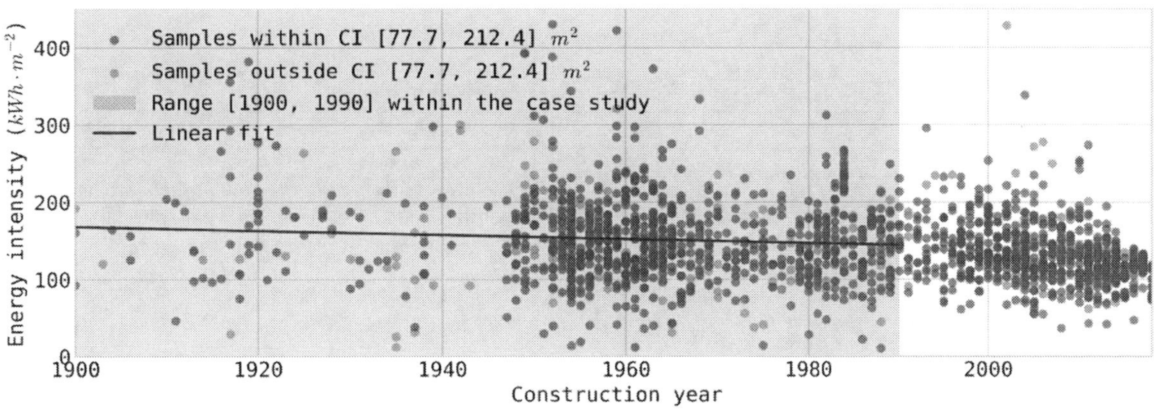

Figure 3. The relationship between building's age and energy intensity

2.2. Energy performance simulation

Based on the information from previous section, a selected representative building was two-storey, three-bedroom, single-family house built under the standards available before 1990. Heated floor area of the building modelled (122.2 m^2) close to the medium value for this typology (129.0 m^2 as specified in Figure 2). The unit was located at the end of a terraced building to represent the worst-case scenario. It was assumed that electricity is the only source of energy for space heating and domestic hot water (DHW) supply. A multi-zone model of the building was developed in IDA-ICE [12] with climate data for Bergen, Norway. Based on NS3031 [13], the heating system was designed for a desired temperature of 21/19 $°C$ in occupied/non-occupied hours during the coldest period of the year. The internal gains, electrical appliances, and DHW use were modelled in detail to take into consideration occupant behaviour as shown in [14].

The outputs of energy performance simulation for this (seed) model are load and energy use profiles on hourly basis over one year of operation. Simulated annual specific energy use was compared against standards (178 kWh/m^2 [15]), other source [16] and the EPC dataset (Figure 3) as a model calibration step.

A calibrated seed model was further extended with all unique combinations of four technologies: EV, PV, DHWS and SWH (Table 1).

Electric load profiles for EV defined based on the data for the most common commercial solutions [17]. Model inputs related to technical specifications for PV systems derived from manufacturers and distributors [18, 19]. A list of unique combinations of technologies in Table 1 accounted for the technical relationship between SWH system and the DHW tank - DHWT with the demand management mechanism is an independent component whereas SWH does requires DHWT for its operation.

The annual load and building energy use profiles were simulated for all the combinations of technologies listed in Table 1. Load profiles over the three coldest days are illustrated in Figure 4 for the simplest (seed) and the most complex models. As a convention, a reversed PV power profile displayed here. This subset of the load profiles, when the demand for space heating is highest, demonstrates the key differences between the models:

- PV power, if self-consumed, enables to decrease grid load substantially. The mismatch between PV generation and peak load, however, makes such benefit insignificant for grid stability;
- SWH system in combination with DHW tank are instrumental in reducing the total load

IOP Conf. Series: Earth and Environmental Science **352** (2019) 012038 doi:10.1088/1755-1315/352/1/012038

Table 1. Unique combinations of technologies

Index	EV	PV	DHWT	SWH	Comment
1					Seed model (none of the technologies accepted).
2	■				Battery capacity 22 kWh. 50% charging daily.
3		■			PV area 16 m^2. 4000 kWp installed capacity.
4			■		Volume 200 l. Heater capacity 2 kW.
5			■	■	Collector area 16 m^2. 200 l storage tank.
6	■	■			
7	■		■		
8		■	■		
9	■	■	■		
10	■		■	■	
11		■	■	■	
12	■	■	■	■	All technologies accepted.

during the peak periods. Their performance varies strongly in response weather conditions;

- EV may contribute substantially to the afternoon peak if no advanced charging control/scheduling used.

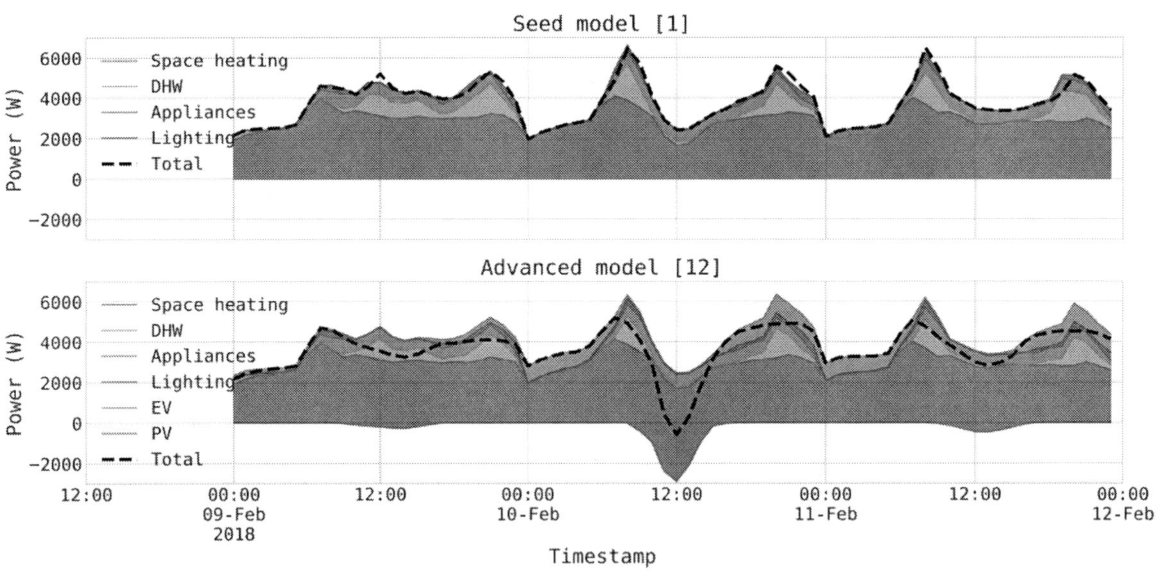

Figure 4. Load/feed-in profiles for model [1] (top) and [12] (bottom)

2.3. Stochastic model

This section is concerned with developing a stochastic model used to predict a stock-wide penetration level of novel technologies. The key concept is stochastic process of building's evolution. It defines the sequence and the time steps with which the technologies can be accepted in one building. The acceptance of any technology in a building is denoted by a discrete (binary) random variable X:

$$X = \begin{cases} 0, & \text{if technology is } \textbf{not accepted} \\ 1, & \text{if technology is } \textbf{accepted} \end{cases} \tag{1}$$

Variable X is characterised by the synthetic parameter - technological acceptance rate (TAR). TAR reflects the probability P with which any of the available technologies will be accepted ($X = 1$) at the time step t, such that:

$$P\{X = 0\} + P\{X = 1\} = 1 \tag{2}$$

Under the absence of perfect knowledge, TAR is meant to encapsulate all kinds of judgments and influencing factors that drive the decision to accept any technology. Through social advertisement, economic incentives and other energy-related programs at the municipal or national scale, this parameter can be influenced.

Given that at some time step t, variable X yields an acceptance, the exact technology is drawn from a random, uniformly distributed variable denoted as U which is one realization of an exhaustive list of the available technologies $u = \{EV, PV, DHWT, SWH\}$. Thus, when the first acceptance occurs:

$$P\{U = EV\} = P\{U = PV\} = P\{U = DHWT\} = P\{U = SWH\} = \frac{1}{4} \tag{3}$$

A general form of Equation 3 for the acceptance n is:

$$\forall n \in [1,\ 4], \quad \forall U \in u \qquad P\{U = u\} = \frac{1}{5 - n} \tag{4}$$

The overall process resembles a random walk over the discrete filed with inconsistent time step. Figure 5 illustrates one sample path with TAR = 0.07 as a network graph. Current state of the building always corresponds to the seed model. The final state, however, depends on if/how many acceptances occurred and which technologies were drawn, but always has a reference model in Table 1.

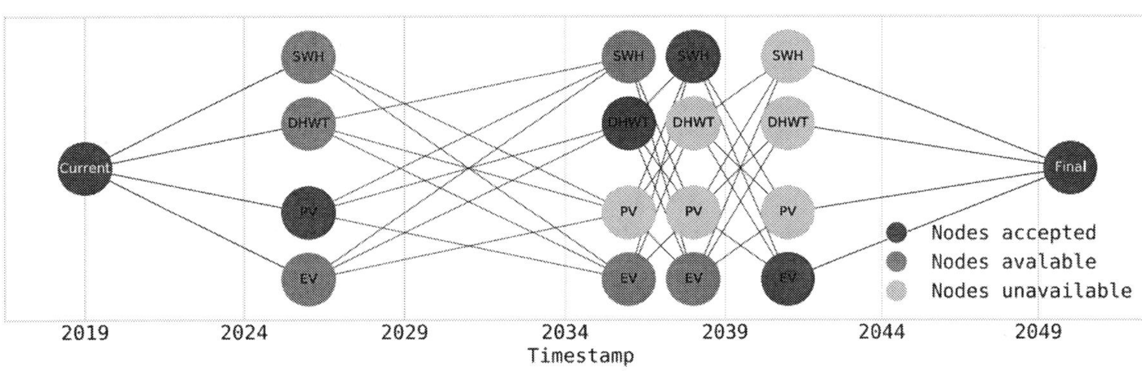

Figure 5. Sample path - acceptance of technologies in the building

Such simulations were carried out for each out of 1000 individual buildings within the synthetic built stock. Its final state can be presented as a bar chart in Figure 6, illustrating a number of times when each unique combination of technologies can be observed in 2050 under TAR = 0.07. The results in Figure 6 represent one possible outcome (trial) of the built stock evolution. To address the objectives of the study, 1000 trials were simulated, where TAR takes a random value from continuous normal (Gaussian) PDF with mean value $\mu = 0.05$ and the standard deviation $\sigma = 0.015$.

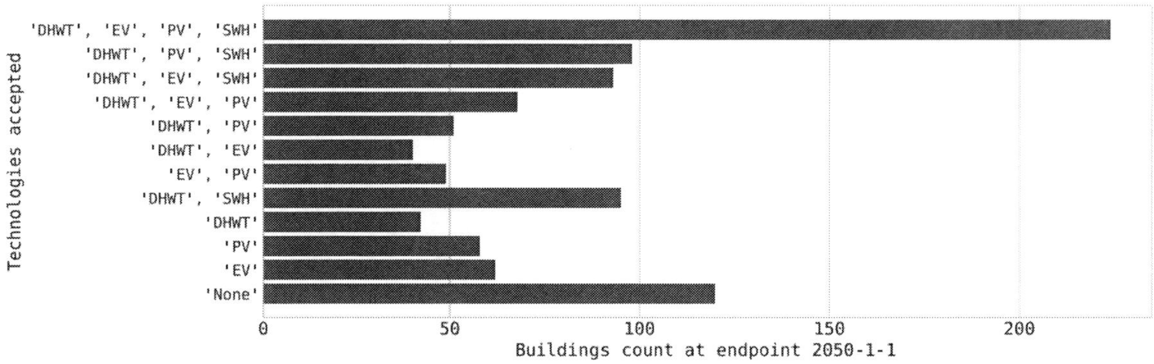

Figure 6. Number of buildings per combination of technologies accepted

Given that simulated final states of built stock are available, the corresponding data on grid interaction from the reference models and the energy price can be aggregated to the stock level.

2.4. Energy price calculation methods

Currently in Norway, the electricity bill for residential building owners consists of two elements: energy and electricity grid fee. Further, the electricity grid fee consists of energy part and a constant part. This constant part should reflect the power demand of a user, but in the current grid fee model it is not considered. Finally, this means that the electricity grid fee is not reflecting the real electricity grid cost caused by the power demand of residential buildings.

In this study, the energy cost was based on the Nord Pool hourly data for 2017. The energy cost was calculated by using hourly electricity demand and the hourly electricity cost from the Nord Pool market. The grid fee part may be defined in different ways depending on the model. In this study, the six models were analysed: 1) the current spot price model, 2) the current period model, 3) the maximum measured power model with the constant coefficients over the year, 4) the maximum measured power model with the variable coefficients over the year, 5) the subscribed power model with 4 kW subscribed power, and 6) the time of use model. The specific values for each of these models were provided from Haugaland Kraft. In the case of the electricity export, the feed-in tariff, when the building installed PV, a possible income for the building was calculated. This income was calculated in the same way as the grid fee cost for each model, except that each element was weighted 80% and 25% was taken for the taxation. An incentive to motivate building owners to install PVs would be to decrease the taxation, but this was not analysed in this study. For each of the scenarios in Table 1, monthly and annual electricity cost for each of the six pricing models were calculated. Due to effectiveness of the paper, the pricing models are not introduced in detail mathematically. Currently, in Norway there are some suggestions for the feed-in tariff, but it is not yet widely used. The values used in this paper were discussed with the company Haugaland Kraft.

3. Results

Given the simulation procedures and assessment methods discussed above, the results represent the likely levels of penetration of technologies under the study. The associated grid interaction and the prices for energy are elaborated here.

3.1. Aggregated grid interaction

In Figure 7 illustrates 1000 load duration curves aggregated to the synthetic stock level. Each curve reflects an evolved unique stock, composed of buildings with various combination of technologies (Table 1) in 2050. For comparative purposes, aggregated duration curve from seed model is displayed also.

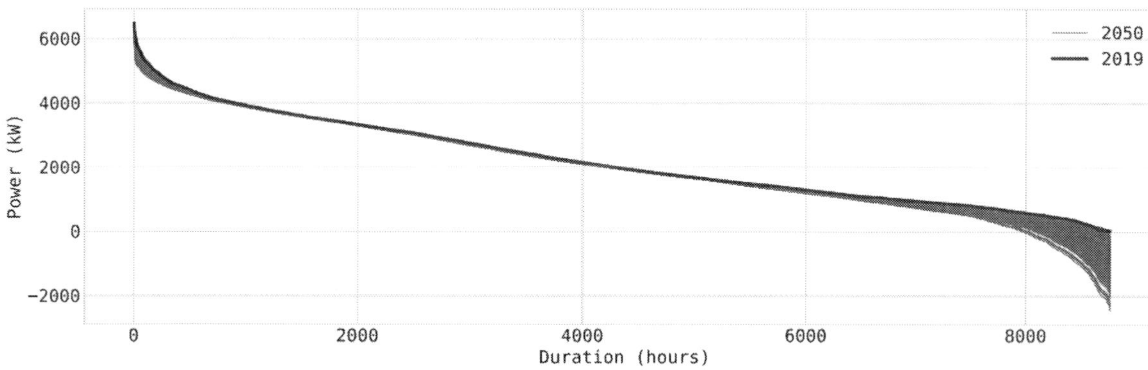

Figure 7. Aggregated load duration curves

It can be observed that the middle part of the curves remains unchanged. Substantial changes, however, are expected in peak loads and at grid feed in. Peak load observed on the annual basis is expected to decrease between 1% and 13%. The running time with low loads will change, which can be attributed to higher levels of self-sufficiency. The figure indicates that under this case study, extra peak power generation will not be necessary since additional load caused by the the EVs will be compensated by PV and SWH systems. Penetration of PV systems would yield significant levels of peak grid feed-in which can reach up to 40% of peak load.

3.2. Cumulative electricity cost

Cumulative annual cost for electricity, according to six pricing methods discussed in Section 2.4 and under various TAR, are illustrated in Figure 8.

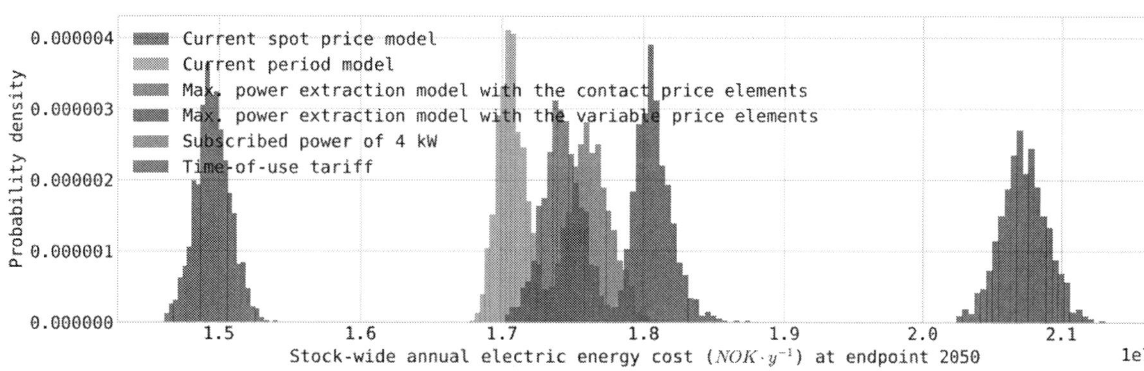

Figure 8. Aggregated electric energy costs per pricing model

It is evident from the Figure 8, that time-of-use tariff and variable maximum power extraction models refer to the lowest and the highest electric energy price under any acceptance of novel technologies. Overlapping variance makes it more challenging to establish preferences between the other four models. Thus, under high TAR (left part of each histogram), cost for energy with the spot price model will be lower compared to subscribed power (or contract maximum power) model with low acceptance of novel technologies.

4. Discussion

A methodological procedure, proposed in this study, accounts detailed information at the building level and random causality behind the acceptance of novel technologies. Modelling principles (and the results) are not idealised by deterministic scenario modelling, but benefit from probabilistic approaches instead. The likely developments of grid interaction and energy price are evaluated, providing the necessary background for informed energy planning.

As elaborated in Section 2.1, a dynamic stochastic process relies strongly on synthetic TAR. More detailed analysis of consumers' willingness to deploy [20] each particular technology, their economic, environmental and social motives, is needed to produce more accurate results.

The dataset and the pricing methods used to assess the stock-wide cost for energy are limited by those currently available. The results are sensitive to any future changes in e.g. taxation mechanisms, incentives for energy efficiency measures and/or low energy use. With methodological and instrumental toolset used in this study, the feasibility of such strategic initiatives can be assessed.

5. Conclusions

Meeting future strategic energy- and environment-related plans for cities and communities requires a consideration of structural changes in the built environment. These changes are, to the large extent, shaped by the penetration of novel technologies, particularly in the residential buildings. This study evaluated two key aspects of such changes in the future grid stability and cost for energy. The scope covered EV, PV, DHWT, and SWH systems deployment. These objectives were achieved through a comprehensive methodological approach that involves descriptive statistical analysis, building energy performance simulation with model calibration step and stochastic forecasting of built stock evolution. It is exemplified with one residential building typology in Norway. The results suggested that for the given case study, additional power demand is not likely to occur, regardless of increased use of EVs. PV feed-in, however, would reach substantial levels (up to 40% of peak load), depending on the penetration rates for the PV systems. Considering different pricing methods, those based on the pricing of power extraction would cover better electricity grid cost, while they would result in highest electricity cost for the building users. The limitations of the study are associated with narrow scope and data scarcity. However, it was shown that the methodology is applicable for the analysis of future developments of built stock and answering the given research questions. Large scale energy planning and policy making, therefore, may benefit from the approaches and methods provided in study.

Acknowledgments

The authors gratefully acknowledge the support from the Research Council of Norway through the research project Methods for Transparent Energy Planning of Urban Building Stocks - ExPOSe (project number 268248) under EnergiX program.

References

[1] Nykamp H 2017 *Environmental Innovation and Societal Transitions* **24** 83 – 93
[2] Becchio C, Bottero M C, Corgnati S P and Dell'Anna F 2018 *Land Use Policy* **78** 803 – 817
[3] Lausselet C, Borgnes V and Brattebø H 2019 *Building and Environment* **149** 379 – 389

[4] Sørnes K, Fredriksen E, Tunheim K and Sartori I 2017 *Energy Procedia* **132** 610 – 615 11th Nordic Symposium on Building Physics, NSB2017, 11-14 June 2017, Trondheim, Norway

[5] Hansen H, Jonassen T, Løchen K and Mook V 2017 Forslag til endring i forskrift om kontroll av nettvirksomhet - Utforming av uttakstariffer i distribusjonsnettet Tech. Rep. 5 Norges vassdrags- og energidirektorat

[6] Yoshino H, Hong T and Nord N 2017 *Energy and Buildings* **152** 124 – 136

[7] Fonseca J and Panão M J O 2017 *Energy and Buildings* **152** 503 – 515

[8] Moral P and Penev S 2017 *Stochastic Processes: From Applications to Theory* Chapman & Hall/CRC Texts in Statistical Science (CRC Press)

[9] Directive (EU) 2018/844 of the European Parliament and of the Council of 30 May 2018 amending Directive 2010/31/EU on the energy performance of buildings and Directive 2012/27/EU on energy efficiency Official Journal of the European Union

[10] Brekke T, Isachsen O K and Strand M 2018 EPBD implementation in Norway. Status in December 2016 Tech. rep. Enova, Norwegian Water Resources and Energy Directorate (NVE) and Norwegian Building Authority (DIBK)

[11] Statistisk sentralbyrå (SSB) / Statistics Norway 2013 Statistisk årbok 2013

[12] EQUA Simulation AB 2019 IDA indoor climate and energy - a new generation building performance simulation software

[13] Standard Norge 2014 NS-3031: Beregning av bygningers energiytelse - Metode og data

[14] Nord N, Tereshchenko T, Qvistgaard L H and Tryggestad I S 2018 *Energy and Buildings* **159** 75 – 88

[15] Statistisk sentralbyrå (SSB) / Statistics Norway 2011 Energy consumption in households, 2009

[16] Enova SF 2017 Enovas byggstatistikk 2017

[17] Sørensen Å L, Jiang S, Torsæter B N and Völler S 2018 Smart ev charging systems for zero emission neighbourhoods - a state-of-the-art study for norway. zen report no. 5 Tech. rep. SINTEF and NTNU

[18] Norsk solenergiforening (NSF) 2019 Produser din egen strøm med solcellepanel

[19] FME SUSOLTECH-IFE, BIPVNO-SINTEF, BIPVNO-NTNU, Solenergiklyngen and Solenergiforeningen 2018 Muligheter og utfordringer knyttet til bygningsintegrerte solceller (BIPV) i norge 2018

[20] Zheng D, Yu L, Wang L and Tao J 2019 *Sustainable Cities and Society* **44** 291 – 309

IOP Conf. Series: Earth and Environmental Science **352** (2019) 012039 doi:10.1088/1755-1315/352/1/012039

Opportunities for Local Energy Supply in Norway: A Case Study of a University Campus Site

S Backe[1], Å L Sørensen[2], D Pinel[3], J Clauß[4] and C Lausselet[4]

[1] Norwegian University of Science and Technology, Department of Industrial Economics and Technology Management, Trondheim, Norway

[2] SINTEF Building and Infrastructure, Oslo, Norway

[3] Norwegian University of Science and Technology, Department of Electric Power Engineering, Trondheim, Norway

[4] Norwegian University of Science and Technology, Department of Energy and Process Engineering, Trondheim, Norway

E-mail: stian.backe@ntnu.no

Abstract. Neighbourhoods can contribute to climate change mitigation by supplying and/or facilitating renewable energy sources (RES). In this context, we evaluate opportunities related to the energy system of a Norwegian university campus, Campus Evenstad, by quantifying monetary value of local energy resources like solar photovoltaics (PV) and a bio-based combined heat and power (CHP) plant in a cost analysis. Environmental value is discussed regarding operational control of energy units to minimize emissions. Using mixed integer linear programming (MILP), we also present results from an investment analysis for local thermal and electric energy system to achieve different levels of emission compensation. Results show that local electricity supply generates most monetary value through saved costs related to reduced power grid import, and that solar PV is the most cost-efficient resource to achieve compensation of GHG emissions.

1. Introduction

Within the Research Centre on Zero Emission Neighbourhoods in Smart Cities (FME ZEN), the definition of a 'Zero Emission Neighbourhood' (ZEN) [1] highlights measures related to energy efficiency, renewable energy, flexible operations and economic sustainability.

A method for emission compensation for buildings has been developed in the Norwegian Zero Emission Building (ZEB) Research Centre [2]. A similar method is now used in a new Norwegian standard NS 3720:2018 [3]. The focus was on nZEBs [4], which are buildings with a low energy requirement. In addition, the delivered energy to nZEBs should to a significant extent be covered by local renewable energy sources (RES). This includes electricity and heat produced and delivered inside or nearby the building. The idea is to compensate for the total life-cycle greenhouse gas (GHG) emission measured in CO2 equivalents by producing more on-site energy than needed for self-consumption. The locally produced energy is based on RES, and the emission credits gained by feeding the grid with surplus energy lead to emission credits by using a marginal approach. A modular LCA approach inspired by the standard NS3720:2018 has recently been proposed for the neighbourhood level [5].

This paper presents a case study of Campus Evenstad which has a total floor area of 10,000 m^2, and consists of several buildings, local energy production and local energy storage. We

(a) Photo by Arne Nyaas (Fjellfolk Media)

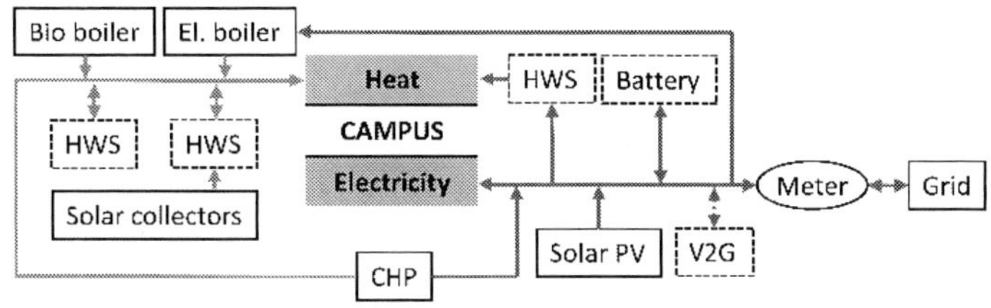

(b) Sketch of energy system at Campus Evenstad.

Figure 1: Campus Evenstad (1a) and the connection between energy system units (1b). Blue arrows shows possible flow of electricity, while orange arrows shows possible flow of heat in the heating grid. Dotted boxes symbolize storage units. Some hot water storage (HWS) tanks are fueled by electricity, while others are connected to the heating grid. Vehicle-to-grid (V2G) is still under development.

investigate the following questions: (1) What economic and environmental value does the local energy production and storage represent at Campus Evenstad? and (2) Which investments in local energy production and storage is required to compensate for GHG emissions related to energy use at Campus Evenstad, in a cost-efficient manner?

The structure of the paper is as follows: Section 2 gives an overview of our case study with estimated and measured energy data. Section 3 quantifies monetary and environmental value of these energy assets. Section 4 presents our investment analysis of how to achieve a net zero emission balance at Campus Evenstad. Finally, Section 5 concludes the paper and suggests future work.

2. Background

Campus Evenstad is a pilot project in FME ZEN with a low-carbon energy system (see Figure 1). The campus was also a pilot in FME ZEB, where the new administration building became Norway's most ambitious ZEB pilot classified as ZEB-COM [6]. The end-users at Campus Evenstad include around 70 employees (academic employees, operators and administrative staff) and 250 students. The buildings range from the energy efficient ZEB-COM building and the passive house classified student dorms to buildings from the 1800s, 1970s and 1980s consuming large amounts of energy both for heating and ventilation. The site is connected to the power grid and has a local heat distribution grid.

Table 1 presents an overview of on-site energy supply assets (excluding the power grid). The annual production is estimated for the CHP and the solar collectors and is based on full-year measurements from 2016 for the bio boiler, the electric boiler and the solar PV. The numbers are preliminary since there have been changes since 2016 with more buildings connected to the local heating grid. In addition to the generation units in Table 1, there are eleven hot water storage (HWS) tanks of varying size and temperature-dependent storage capacity. There was installed a li-ion battery in 2018 that can store 204 kWh and charge/discharge at 120 kW. It can perform roughly two hours of island operation in winter times and can enable start-up of the CHP independent of grid import. Four electric vehicle (EV) charging stations are installed at the campus: One charging point delivers 20 kW and the three remaining deliver 10 kW each. A newly planned charging station will be able to draw energy from the EVs (vehicle-to-grid, V2G) turning them into mobile batteries.

Campus Evenstad consumes around 1, 100, 000 kWh electricity including locally produced electricity annually (see Table 2). Based on Table 1, current local supply can deliver 20% of

IOP Conf. Series: Earth and Environmental Science **352** (2019) 012039 doi:10.1088/1755-1315/352/1/012039

Table 1: Generation of heat and electricity from local units at Campus Evenstad in 2016. Estimates are marked with *.

Generator	Capacity, thermal	Capacity, electricity	Annual generation
CHP, Thermal	100 kW	-	400,000* kWh
CHP, Electricity	-	40 kW	160,000* kWh
Boiler, Bio	350 kW	-	300,000 kWh
Boiler, Electric	315 kW	-	275,000 kWh
Solar collectors	100 m^2	-	40,000* kWh
Solar PV	-	60 kW	62,000 kWh

annual electricity consumption. Electric import is volatile and can at times reach over 400 kWh/h. The building with the highest consumption of electricity makes up around 26% of the total electricity consumption of the campus. Table 2 contains more details on the interaction between the campus and the power grid between 2015 and 2017.

Table 2: Key performance indicators (KPIs) for electricity at Campus Evenstad (the grid utilisation factor is the annual average load divided by the annual peak load, self-consumption is the share of locally produced energy that is consumed on-site and self-generation is the share of total consumption that is produced on-site [7]). Estimates are marked with *.

KPI, electricity	2015	2016	2017†
Net import [kWh]	1,012,941	1,058,962	906,955
Max import power [kWh/h]	436	479	468
Average power [kWh/h]	116	121	104
Grid utilisation factor [%]	27	25	24
Export [kWh]	0	158	70
Delivered electricity PV [kWh]	62,454	61,960	62,000*
Delivered electricity CHP [kWh]	-	160,000*	160,000*
Self-consumption [%]	100	99.93	99.97
Self-generation [%]	6*	21*	24*

†*Excluding January*

Energy used for heating purposes adds up to around 620, 000 kWh. Comparing this to the heat production listed in Table 1 and electricity import listed in Table 2, the CHP plant has not yet been utilized to its full potential (4,000 hours of operation) during the first years in operation. The thermal demand varies between 250 and 350 kWh/h throughout the year, depending on both space heating (SH) and domestic hot water (DHW) demand. Most thermal demand is distributed through the heating grid, and the remaining demand is supplied by electric heaters in buildings. The CHP is first priority supply for the heating grid. The CHP unit alone cannot always meet heat demand and can neither ensure fully reliable heat supply, so the bio boiler is second priority. Table 1 shows that the CHP and the bio boiler could meet current heat demand when combined. When bio-based supply is unavailable, the electric boiler supplies heat demand (third priority). When heating demand is met by the CHP, electricity is produced (CHP produces 0,4 kWh of electricity per kWh of thermal energy). Avoiding use of electric boiler can thus (1) save imports from the grid and (2) facilitate more electricity generation by the CHP.

The large HWS tanks are connected to the local heating network with some flexibility to match heating demand. The flexible operation of the heating system is dependent on (a) the dimension of the heat supply system, (b) the volume of the water storage tank, (c) the placement of the temperature sensors in the water tanks (if temperature-based), (d) the hysteresis set-points and (e) the layout of the storage tanks in combination with the heat distribution system.

3. Value of energy assets

3.1. Monetary value

Campus Evenstad has a prosumer agreement facilitating cost savings through self-consumption of electricity and revenue from selling surplus electricity not exceeding 100 kWh/h. It is based on net metering of delivered energy (import) from the grid (consumption minus production). Electricity produced at Campus Evenstad has additional value through a Norwegian-Swedish system for market-based electricity certificates. This income is independent of whether the produced electricity is self-consumed or exported. To allow for higher export, Campus Evenstad could apply for concession to become an electricity producer subject to fixed and variable grid feed-in costs.

The billing contract with the retailer follows the daily varying spot price of the power market in addition to a fixed monthly price. The grid tariff has three parts: (1) a fixed annual part, (2) an energy part related to transportation losses, and (3) a power part based on the highest hourly average power from the last 12 months.

Based on cost data from the energy retailer (Ishavskraft and spot prices from Nord Pool) and the distribution system operator (Eidsiva Nett), we present estimates of the power purchase rates at Campus Evenstad in Table 3.

Table 3: Estimate of power purchase rates for Campus Evenstad based on open sources. The energy part of the energy rate is linked to the wholesale spot price in Nord Pool. The energy dependent tax charges include an excise tax for electricity and a charge for electricity certificates.

Cost category	Fixed part	Energy part	Power part
Energy rate (ex. VAT)	588 NOK/yr	0.06-2.07 NOK/kWh	-
Grid tariff (ex. VAT)	13,200 NOK/yr	0.04 NOK/kWh	432 NOK/kWh/h
Tax charge (ex. VAT)	800 NOK/yr	0.1658+0.02 NOK/kWh	-

Assuming an average spot price of 0.50 NOK/kWh, total energy savings (including savings on VAT and grid charges and tax charges based on Table 3) from consuming locally produced energy at Campus Evenstad can be estimated to 0.91 NOK/kWh plus 0.02 NOK/kWh from electricity certificates. Annual savings based on the numbers from Table 1 can be estimated to:

$$(62,000 \text{ kWh} + 160,000 \text{ kWh})(0.91 + 0.02 \text{ NOK/kWh}) = 206,460 \text{ NOK} \quad (1)$$

Maximum annual savings could thus amount to 17-18 % of the total electricity bill with currently installed assets.

The normal payment for prosumer export is spot price plus a positive or negative charge (+/- 0.05 NOK/kWh) to compensate for local grid losses. If we assume the local grid charge is 0 NOK/kWh, energy export is only worth the spot price plus electricity certificates (0.52

IOP Conf. Series: Earth and Environmental Science 352 (2019) 012039 doi:10.1088/1755-1315/352/1/012039

NOK/kWh). Compared to self-consumption (0.93 NOK/kWh), and depending on the spot price of electricity, this is about half as profitable. Campus Evenstad currently has very high self-consumption (see Table 2).

The deployment of demand side flexibility by activating thermal and electric storages is remunerated under current agreements if they can be utilized to (1) maintain a high degree of self-consumption (which is more valuable than export) and (2) reduce the maximum peak load over 12 months (very demanding task). This requires a reliable data logging system and advanced processing of data to plan and react to high loads. The battery, with a potential to deliver 120 kW for 2 hours, could contribute to peak shaving. Other flexible resources, like V2G and flexible energy loads at campus, could also contribute. The challenge is to ensure reliability of the flexible resources, so that they are prepared to respond when the load is peaking.

3.2. Environmental value

It is possible to distinguish emissions related to electricity from the grid and to the local energy system at Campus Evenstad through an estimation of a CO2eq. intensity [8]. Furthermore, it is possible to define the emission-optimal operation for Campus Evenstad using electricity either from the grid or from the local energy system depending on what leads to the lowest overall emissions [9].

In Norway, electricity is mainly produced from hydropower with a low average CO2eq. intensity. The CO2eq. intensity in the Norwegian grid is strongly dependent on electricity exchanges with neighbouring countries relying on more carbon-intensive fuels, such as gas or coal. Norway usually imports electricity during the night when electricity is cheap and exports during the day when electricity is expensive. This leads to higher average CO2eq. intensities during the nights and lower average CO2eq. intensities during the day (see Figure 2).

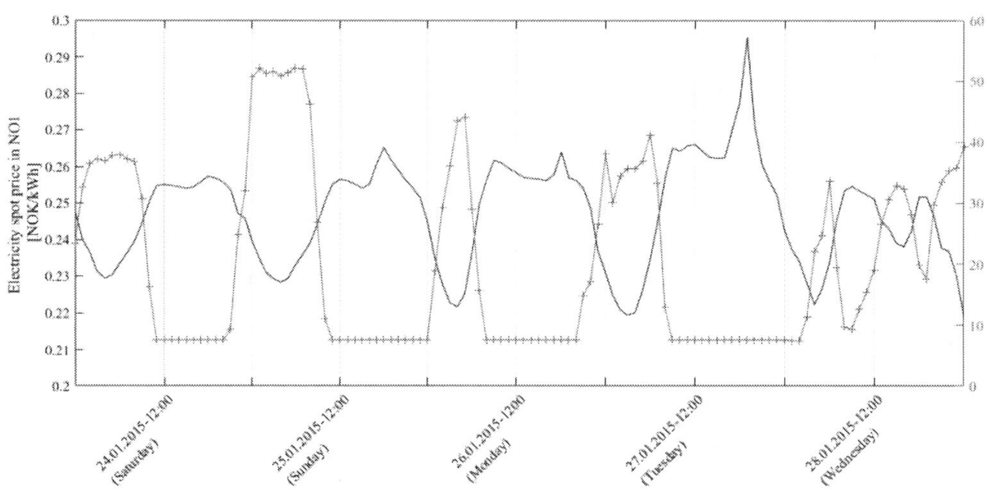

Figure 2: Electricity spot price and CO2eq. intensity in NO1 during 5 days in January 2015. Reference: [8]

If controlling the electricity import according to an average CO2eq. intensity at Campus Evenstad, electricity import will increase during peak hours. Already existing peak loads may be amplified, potentially leading to higher electricity prices and grid congestions. Furthermore, Figure 2 shows that a price-based control and a CO2-based control would lead to contradictory operation periods, meaning that minimizing CO2 emissions at the same time as minimizing costs would be challenging. Maximizing self-consumption of local electricity production from solar PV and bio-based CHP will guarantee both: (1) low emissions and (2) minimize energy transport losses.

Applying the CO2eq. intensity as a control signal for the energy system operation aims at supporting emission reduction goals, but reducing peak load is also a relevant goal at Campus Evenstad. Blackouts have been a problem at Campus Evenstad which increases the motivation for procuring and operating energy flexible assets to increase reliability of supply.

4. Investments to achieve a 'zero-emission balance' for the neighbourhood

The following section presents an investment analysis to achieve a Zero Emission target for Campus Evenstad. We apply the optimization model ZENIT (Zero Emission Neighbourhood Investment Tool) [10], which is a mathematical programming model to minimize the the cost of investing and operating the energy system of a ZEN during its lifetime.

It considers the investment cost of energy supply and storage technologies, as well as the operation and maintenance costs. We simulate one single year with hourly timesteps and scale the costs to the lifetime of the ZEN to limit the computation time. The lifetime of the system is assumed to be 60 years. We require emissions to be compensated through clean electricity exported from the neighbourhood assuming it reduces some of the more carbon-intensive production in other parts of the power system. We refer to this relation of equality between emission and compensation as the 'zero-emission balance', and it is modeled as a constraint in the optimization by using the product of CO2 factors (gCO2eq/kWh) and consumption (kWh) of the different fuels. Other constraints include load balancing of supply and demand for electricity and heating, as well as state-of-charge constraints for storage units. Technologies can be invested in at the building level and at the neighborhood level with different costs and characteristics. A more detailed description of the model can be found in [10].

We investigate fulfilling the 'zero-emission balance' both annually and seasonally. The analysis was conducted both with a grid connection of 800 kW and the limitation of 100 kW on exports. The revenue from export of electricity was based on historic spot prices.

We estimated the hourly energy load in the building stock as described in [11]. The CO2eq. factors used were 17 gCO2eq/kWh for electricity (inspired by average hourly emission factor in price zone NO1 [8]) and 7 gCO2eq/kWh for wood chips [12]. The emission factor for electricity from the CHP is 28 gCO2eq/kWh assuming a 25% efficiency of electricity generation from the CHP plant. Thermal efficiency of the CHP is assumed to be 40%. Emissions from the life-cycle stages other than the operational phase were not included, so we assume solar panels and solar thermal on-site emits 0 gCO2eq/kWh. We also performed an analysis using *asymmetrical* CO2eq. factors for the annual balance, meaning there is a different weighting on imported CO2eq. and exported (compensated) CO2eq. For the asymmetrical CO2eq. factors, we assumed import of electricity remained at 17 gCO2eq/kWh but were set at 136 gCO2eq/kWh for exports from renewable on-site sources inspired by the Norwegian standard NS 3720:2018.

The results are displayed in Figure 3. The striped part of each bar represents the existing energy system at Campus Evenstad, while the non-striped part represent additional capacity (left axis). For the storage technologies, the squares represent existing capacity while the circles represent additional capacity (right axis).

With annual asymmetric CO2eq. factors (Figure 3, far left), some additional PV (around 10 kW) and investment in a heat pump (around 80 kW) were obtained as a result. This highlights the impact of the choice of CO2eq. factors in resulting designs, i.e. emission compensation through export of electricity is very dependent on the estimation of CO2eq. factor and whether it is a marginal or an average estimation.

With an annual symmetrical CO2eq. factor (Figure 3, middle left), the results indicate that Campus Evenstad would need a large investment in PV and a small investment in heat pump.

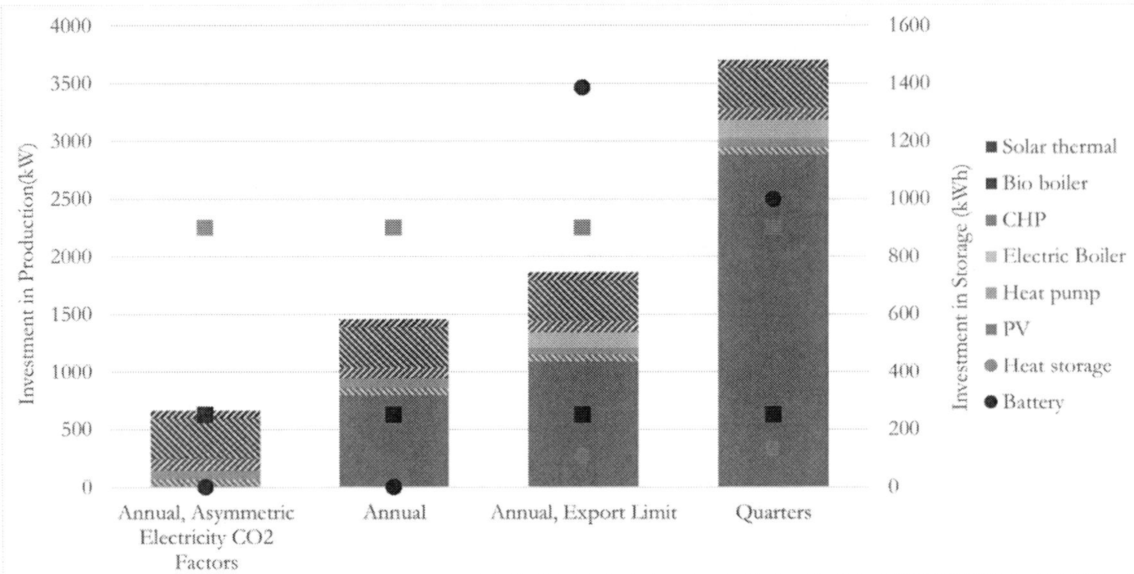

Figure 3: Results from four instances in ZENIT. From the left: (1) annual asymmetrical CO2eq. factor, (2) annual symmetrical CO2eq. factor, (3) annual symmetrical CO2eq. factor with 100 kW export limit and (4) seasonal symmetrical CO2eq. factor.

These investments would come in addition to the already existing system of bio boiler, CHP and solar thermal. Note that this is under the assumption that wood chips have a CO2eq. factor of 7 gCO2eq/kWh [12]. No additional investment in heat storage or battery would be required. The PV area would be around 4,000 m^2 (about 10 times the current area).

With an annual symmetrical CO2eq. factor with the 100 kW export limit (Figure 3, middle right), the results suggest a slight increase in the PV investment, investment in an electric boiler and a large investment in batteries. This is because the PV production that can no longer be sent to the grid directly, but is still needed to compensate for emissions. The peak import remains unchanged, but the self-consumption of electricity is greatly increased. Compared to the same case without the export limit, total energy system costs increase with 41%.

With a seasonal symmetrical CO2eq. factor (Figure 3, far right), massive investments in PV are chosen. A large battery is also chosen by the investment model. In the case of seasonal balances, the compensation far outweighs the emissions. Indeed, the amount of PV resulting of the optimization represent what is necessary to fulfill the balance in the highest demanding season, which is during the winter. In other quarters, this amount of PV is over-dimensioned and results in a lot of electricity production which results in high exports (high compensation) and low import (low emissions). Compared to the case with annual symmetrical CO2eq. factor, total energy system costs increase with 112%.

The model suggests investment in heat pumps and PV in all instances. However, a heat pump will increase the electricity demand (since heat supply is mostly based on bio) and more PV is not likely to increase reliability as the production follows solar conditions. Refurbishment of older buildings would also be a way to reduce energy demand and increase efficiency of buildings, which would in turn reduce the amount of PV needed to fulfill the 'zero-emission balance'. Using a planning tool from the project PI-SEC [13], we find that an upgrade to TEK 10 standard for older buildings achieves energy savings in the range of 30%.

5. Conclusion

Self-consumption of generated electricity has higher economic value than export under current agreements. Self-consumption will contribute positively to emission reductions when based on renewable sources. The objective of operational control of electricity import through battery storage and demand response must be chosen carefully to balance conflicting objectives of monetary cost minimization and CO2eq. minimization. Alternative billing agreements based on power rather than energy will make remuneration related to operational control more feasible.

Across different ambitions and assumptions to achieve a 'zero-emission balance' at Campus Evenstad, investment in PV is a preferred strategy by the ZENIT optimization model when minimizing costs. Depending on the 'zero-emission' ambition and constraints related to power exports, battery storage is also key to achieving the target.

Further research should consider a more detailed mapping of the energy efficiency solutions, and it should separate between electric-specific and thermal energy efficiency. Embodied emissions should be considered for the energy infrastructure (PV, CHP, batteries, etc) and other materials. Ensuring reliable electricity supply or self-consumption should be further investigated as Campus Evenstad is prone to blackouts.

6. Acknowledgements

This article has been written within the Research Centre on Zero Emission Neighbourhoods in Smart Cities (FME ZEN). The authors gratefully acknowledge the support from the ZEN partners and the Research Council of Norway. A special thanks to Zdena Cervenka who has initiated and motivated this work on behalf of Statsbygg.

References

[1] M. R. Kjendseth Wiik, S. Fufa, J. Krogstie, D. Ahlers, A. Wyckmans, P. Driscoll, H. Bratteb, and A. Gustavsen. Zero Emission Neighbourhoods In Smart Cities: Definition, key performance indicators and assessment criteria: Version 1.0. Bilingual version, 12 2018.

[2] S. M. Fufa, R. D. Schlanbusch, K. Sørnes, M. R. Inman, and I. Andresen. A Norwegian ZEB definition guideline. SINTEF Academic Press, 2016.

[3] NS 3720:2018 - Method for greenhouse gas calculations for buildings. Norsk Standard, 2018.

[4] M. Hamdy, A. Hasan, and K. Siren. A multi-stage optimization method for cost-optimal and nearly-zero-energy building solutions in line with the EPBD-recast 2010. *Energy and Buildings*, 56:189–203, 2013.

[5] C. Lausselet, V. Borgnes, and H. Brattebø. LCA modelling for Zero Emission Neighbourhoods in early stage planning. *Building and Environment*, 149:379–389, 2019.

[6] M. R. K. Wiik, Å. L. Sørensen, E. Selvig, Z. Cervenka, S. M. Fufa, and I. Andresen. ZEB Pilot Campus Evenstad. Administration and educational building. As-built report. *ZEB Project report*, 2017.

[7] I. Sartori, A. Napolitano, and K. Voss. Net zero energy buildings: A consistent definition framework. *Energy and buildings*, 48:220–232, 2012.

[8] J. Clauß, S. Stinner, C. Solli, K. B Lindberg, H. Madsen, and L. Georges. A generic methodology to evaluate hourly average CO2eq. intensities of the electricity mix to deploy the energy flexibility potential of Norwegian buildings. In *10th International Conference on System Simulation in buildings*, 2018.

[9] J. Clauß, S. Stinner, I. Sartori, and L. Georges. Predictive rule-based control to activate the energy flexibility of Norwegian residential buildings: Case of an air-source heat pump and direct electric heating. *Applied Energy*, 237:500–518, 2019.

[10] D. Pinel, M. Korpås, and K. B Lindberg. Cost Optimal Design of Zero Emission Neighborhoods' (ZENs) Energy System: Model Presentation and Case Study on Evenstad. In *International Symposium on Energy System Optimization (ISESO), Karlsruhe*, 2019. Pre-print available at `https://arxiv.org/abs/1903.07978`.

[11] K. B. Lindberg, G. Doorman, J. E. Chacon, and D. Fischer. Hourly electricity load modelling of non-residential passive buildings in a nordic climate. In *2015 IEEE Eindhoven PowerTech*, pages 1–6. IEEE, 2015.

[12] T. H. Dokka, I. Sartori, M. Thyholt, K. Lien, and K. B. Lindberg. A Norwegian zero emission building definition. Passivhus Norden, 2013.

[13] H. T. Walnum, K. Sørnes, M. Mysen, Å. L. Sørensen, and A.-J. Almås. Preliminary toolkit for goals and KPIs. PI-SEC; Planning Instruments for Smart Energy Communities, 2017.

IOP Conf. Series: Earth and Environmental Science **352** (2019) 012040 doi:10.1088/1755-1315/352/1/012040

Geo-Solar District Heating – Lessons Learnt from a not Realised Project

Dietrich Schmidt[1]and Kilian Stroh[1]
[1]Fraunhofer Institute for Energy Economics and Energy System Technology IEE
Königstor 59, DE-34119 Kassel, Germany

Email: dietrich.schmidt@iee.fraunhofer.de

Abstract. For the new housing area with approx. 130 low energy residential houses, located in Kassel, Germany, various supply concepts are investigated. Main objective of the project is the development of an innovative and optimised heat supply concept based on renewable energies and a low temperature district heating for later realisation.
During the course of the project several heat supply concepts have been investigated and compared to give an input to the decision and realisation process. A promising heat supply concept is based on a central ground source heat pump in combination with a low temperature district heating (40°C supply temperature) for space heating and decentralised solar thermal systems for domestic hot water preparation. The advantages of this supply variant are comparable low annual heating costs and about 60% lower CO_2-emissions in comparison to the reference system (decentralized air / water heat pump with solar thermal system).
Due to time constrains it was decided not to realise the above mentioned energy concept. However, the research project took up the task to investigate the transferability of the concept in technological and economical perspectives to other locations and boundary conditions. From the results of the study the importance of a very high connection rate can be highlighted. Also the question of the particular business models for such capital demanding projects (high investment costs for the heating grid vs. low energy demand/sales) has been regarded.

1. Introduction

During the planning phase of new residential areas, the investigation of suitable energy sources and supply strategies is crucial. The use of renewable energy sources (e.g. solar energy and geothermal energy) offers great potential for a sustainable and efficient supply of heat. However for optimized usage of these resources, it is necessary to identify appropriate technologies to ensure efficient supply of the new housing area.

In order to identify the best possible system solution for the named housing area, different supply strategies have been investigated and compared. The main objective is the development of an innovative heat supply concept based on renewable energies and an optimised supply concept. Central challenges in achieving these objectives is the identification of the most promising and efficient technical solutions for practical implementation. Furthermore aspects of future network management as well as business models for distribution and operation have been considered. As a result, the focus of this investigation was on an assessment of suitable centralised or decentralised supply concepts for the new residential area using renewable energy sources in a cost efficient way.

The development of the energy concept has been carried out in two project phases. The first project phase consisted of a pre-study in order to identify the most efficient and economical heat supply concept.

In the course of the second project phase a detailed concept was elaborated which included selecting, dimensioning and detailed cost determination of the various components.

2. Description of the housing area

The planning area in focus is located in the city of Kassel (Germany). The area is surrounded by existing buildings of the district and is located in an urban ventilation path. For that reason combustion of oil or wood (fine dust emissions) should be avoided. Due to the location of the area a connection to the existing district heating network of Kassel is not feasible because of logistical and economic reasons. Instead a local district scheme was in focus for the planning. The concept involves principally the use of renewable energy sources (RES) such as geothermal and solar energy for low temperature district heating supply. The additional use of a combined heat and power (CHP) plant has also been investigated. Furthermore the implementation of intelligent storage systems and thermal load shifting concepts has been envisaged [1].

Figure 1. Map of the investigated area. Source: City of Kassel/Germany [2]

Table 1. Assumptions for the planned buildings in the housing area [2], [3]

Total number of buildings	131
Dwelling units	253
Persons per dwelling unit	2.3
Total net dwelling area	28,975 m^2
Roof shape	SFH, SDH and TH = Gable roof, MFH = flat roof
Heat emission system	surface heating

The new housing estate will be characterized by a very compact and south oriented construction; 1-2 storey detached and semi-detached houses in the north, two-storey terraced houses in the centre and large three-storey apartment buildings in the south. All buildings have specific heat demands of 45 kWh/m^2a and a specific domestic hot water (DHW) demand of 730 kWh/pers·a. Thus, the demand is significantly below the maximum energy demand for new buildings (< 50 kWh/m^2a) according to the current German energy saving ordinance EnEV [4].

3. Development of innovative heat supply concepts

For the identification of a possible supply concept in a first step DHW, heating and cooling demands have been calculated. Based on the resulting energy demand suitable supply concepts have been elaborated.

3.1. Determination of heat/ cool demand and DHW demand

The monthly heat and cool demand as well as DHW demand was calculated corresponding to the monthly balance method of DIN V 18599 [5] including part 2 [6], part 4 [7], part 5 [8], part 8 [9] and part 10 [10].

The total annual energy demand according to the monthly balance method DIN V 18599 [4] was calculated for a heating demand of Q_{hd} = 1.539 MWh/a, for cooling demand Q_{cd} = 319 MWh/a and for domestic hot water demand Q_{DHW} = 382 MWh/a. These values correspond to a very good insulation standard of the buildings. From detailed simulations a heating load of 927 kW has been estimated [3].

3.2. Selection of supply variants

Based on the low determined monthly energy demand different supply options have been examined. Since it has been decided to supply the district preferably with renewable energy sources the utilization of near-surface geothermal energy by means of borehole heat exchangers (BHE), the use of solar energy or the installation of a bio-methane powered CHP have been found to be eligible. For economic reasons, the use of a natural gas powered CHP was examined as well. In general two decentralised (D1 and D2) supply concepts and two centralised supply variants (C1 and C2) have been developed and compared.

4. Outline of assessed decentralised heat supply variants

As part of the first decentralized supply concept (D1) a decentralised air / water heat pump system was investigated. During the summer months, the domestic hot water is prepared by solar thermal energy. The disadvantage of this concept is the high initial cost and the noise emissions of distributed generation units. As part of the second decentralized supply concept (D2) a decentralized gas fired condensing boiler with solar DHW supply was investigated. The first supply concept D1 served as a reference case for the more detailed assessment (see also [3]).

Table 2. Overview of the investigated decentralised supply variants [1], [3]

Decentralized heat supply variants	
Decentralized (D1):	Decentralized air / water heat pump with solar DHW support
Decentralized (D2):	Decentralized gas fired condensing boiler with solar DHW support

IOP Conf. Series: Earth and Environmental Science **352** (2019) 012040 doi:10.1088/1755-1315/352/1/012040

5. Centralised heat supply variants

Table 3. Overview of the investigated centralised supply variants [1], [3]

Centralised heat supply variants	
Centralized (C1):	Centralized geothermal powered HP for district heating supply at low temperature level of 40°C. DHW preparation by solar/ heating rod
Centralized (C2):	Centralized natural gas powered CHP for district heating supply in combination with demand-oriented power generation, a large heat storage and HP.

5.1. Centralised heat supply variant (C1): heat pump and borehole heat exchangers

The first supply variant consists of a centralised heat pump connected to borehole heat exchangers (BHE). Depending on the operation, the ground acts as heat source and/or thermal storage. For the thermal regeneration of the ground unglazed solar collectors (swimming pool absorbers as a low-cost option) are intended to be used (see Figure 2). It is conceivable to use the district heating during heating period to provide (low temperature) heat.

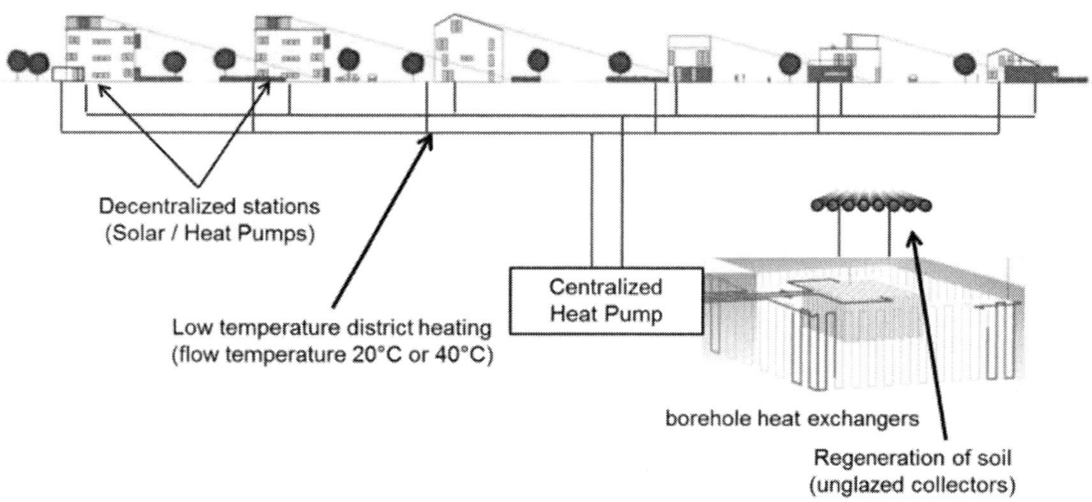

Figure 2. General description of hydraulic of centralised geothermal powered HP for district heating supply C1, [1] and [3]

In supply variant (C1) a centralised ground coupled heat pump feeds the district grid at a temperature level of 40°C. The heat for space heating is supplied directly by the district heating network through the use of heat exchangers. For preparation of DHW different variants are possible. In case of separated domestic hot water preparation thermal solar collectors (e.g. flat-plate collectors) or a heating rod (with a preheating via district heating) could be used. The solar panels could be installed on the roof or on the planned carports.

The advantage of this supply variant is the direct use of heat for space heating from the grid (substation required). No decentralized heat pumps must be installed and thus the investments costs are significantly lower [3].

The DHW is stored in buffer tanks which are intended to be located in the thermal envelope of the buildings.

IOP Conf. Series: Earth and Environmental Science **352** (2019) 012040 doi:10.1088/1755-1315/352/1/012040

As storage option for heat for space heating the BHE field is also available. This field is planned at the border of the development area (see figure 1). The thermal regeneration of the ground takes place via unglazed solar absorbers. The absorber technology is very cheap and effective (approx. 1 €ct / kWh of heat in an installation area of approximately 1.400 m^2 [3]).

5.2. Centralised supply variant (C2): heat pump and CHP plant

In the second centralized supply variant a combination of a CHP plant (140kWel / 207kWth) and a ground coupled heat pump with 535 kW heating capacity has been investigated [1]. According to manufacturer's instructions the heat pump can achieve a COP greater than 4.0 even if the supply temperature is higher than 35°C. As fuel for the CHP plant natural gas is foreseen, renewable energy sources bio methane is also eligible but not calculated for the cost assessment. To increase the system efficiency a solar collector field (vacuum tube collector) is considered. Additionally a gas fired backup boiler is planned. The network operation temperature for space heating is 35°C - 50°C. For domestic hot water preparation the grid temperature will be increased to 70°C 1-2 times a day for ~ 1.5 h.

As heat storage option for heat for space heating the BHE field is available, see variant C1. For the storage of high temperature heat of 70°C-80°C (variant C2), two buffers with a volume of 25 m^3 are foreseen. It has been determined that this size is sufficient for the sole heating of all decentralized DHW storage tanks located in the buildings including heating of the district heating grid.

Three control strategies for the operation of CHP and heat pumps are eligible. The first possible strategy is the usage of the CHP plant in case of base load and the heat pump for covering the peak load.

The second possible strategy is to run CHP and heat pump in parallel. In this variant the CHP plant charges the storage tank with high temperature heat.

The third strategy is to run the CHP plant during peak load and the heat pump during base load. The heat pump supplies the heat generated to the network or to the storage and thus covers the entire base load heat demand.

6. Foreseen network design and heat losses

Based on the investigated supply options (refer to Table 2 and Table 3), the selection of suitable district heating pipes has been made. Within the project the installation of PEX twin pipes (PN 6/10 bar) is considered. These pipes are cost effective and are characterized by low heat losses. The heat loss through the district heating network are estimated to be ~ 40 MWh/a, corresponding to ~ 2.5% related to the useful heat of 1.564 MWh/a [1].

7. Assessment and selection of supply the concept for the housing area

For assessment and selection of suitable supply variants different parameters and characteristics have been considered: The parameters are classified into "technical", "economic" and "soft" factors; for each case a subdivision is made. The "technical" parameters include the primary energy demand, CO_2 emissions and space requirement for installed supply technology. Depending on the supply concept other parameters have been taken into account indirectly. For the HP the seasonal performance factor (SPF), for CHP the coefficient of performance (COP) and for the use of solar systems the efficiency of the absorbers, solar yield and solar fraction has been considered. The "economic" parameters included the annual heating costs (including investment and operating costs, maintenance costs and arising costs for the consumer) as well as price stability. Moreover the determination of the amortization period and the expected return on invest was of importance.

IOP Conf. Series: Earth and Environmental Science **352** (2019) 012040 doi:10.1088/1755-1315/352/1/012040

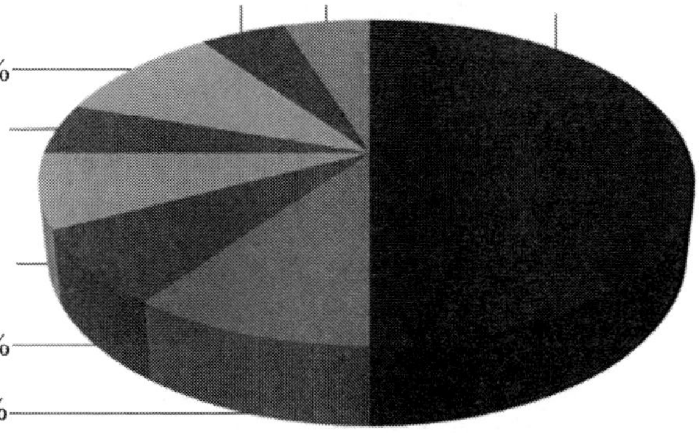

Figure 3. Weighting factors for assessment of supply concepts as discussed with the partners [1]

The "soft" factors included the consideration of the level of innovation and user-friendliness. Furthermore customer satisfaction and customer loyalty are factors that have to be considered, because they also potentially contribute to the successful implementation of the project. Both, the assessment parameters and the weighting factors have been discussed and decided by the project partners.

The key evaluation parameters with their weighting factors, which have been used for comparison of the variants, are shown in Figure 3. According to the weighting factors it turned out that the variants C1 (Centralized HP, decentralized DHW preparation by solar/ heating rod – 40°C) and C2 (Centralized fossil powered CHP/HP) are the most suitable and selected solutions. Since the variant C2 integrates the combustion of fossil fuels, it has not been chosen for further investigation. In comparison to reference variant D2 both centralised variants are advantageous due to environmental issues (CO_2 emissions, primary energy demand). The supply concepts differ essentially with regard to the dependency on price of electricity (C1) and natural gas (C2). Finally the heat supply concept C1 has been chosen for a more detailed analysis based on the results from the first project phase [1]. The second phase consists of elaboration of the favoured supply variant and development of a detailed concept.

8. Discussion of the results

In particular renewable energy sources offer great potential for sustainable and efficient heat supply of buildings. However, for optimized utilization of these resources it is necessary to identify appropriate technologies to ensure efficient supply for the new housing area. For this reason, extensive studies have been carried out to identify suitable supply options. Since, in addition to the technical aspects economic aspects play a crucial role, it turned out that the centralized heat supply variants integrating a district heating grid (C1) seemed to be more favourable in comparison to the analysed variants [1]. For this supply variant a detailed feasibility study and a detailed estimation of costs has been conducted. Subsequently the development of a business model has been carried out.

The results of the detailed analysis [3] show that the besides of environmental reasons (e.g. lower CO_2 emission) the variant C1 is economical competitive when external funding (e.g. via the German National funding scheme for innovative district heating systems “Modellvorhaben Wärmenetze 4.0”) could be integrated. In this case the utility company is able to make a “competitive” offer to the end consumer. Otherwise the reference supply systems D1 turned out to have lower total heating costs. The reason turned out to be higher investment costs/capital costs for the BHE and the heating grid in variant C1. Furthermore, a high connection rate is crucial for a suitable business concept. The detailed techno-economic analysis shows that the connection rate is a crucial parameter with a very high sensitivity on the resulting annual heating costs, e.g. a 20% reduced connection rate of 80% results already in 7% higher estimated annual heating costs for the end user. In comparison to that the sensitivity of a better building standard (lower heat demand) has a much lower impact on the resulting estimated heating costs.

Although the technical/environmental as well as the economic parameter turned out to be favourable the decision has been taken not to realise the proposed and analysed supply system variant. The main reason for this were time constrains. The long time needed to get a permit to start the drilling for the BHE and utilise the ground together with delays in the planning were the reasons.

9. Conclusions and outlook

In order to identify the best possible system solution for a new housing area planned in Kassel (Germany), different supply strategies were assessed and compared. Main objective is the development of an innovative and optimised heat supply concept based on renewable energies and low temperature district heating. Central challenge in achieving this objective is the identification of the most promising and efficient technical solutions for practical implementation.

As a result from a first project phase [1] an innovative centralized heat supply concept (C1) has been developed and selected in a political process for realization. In the course of the second phase detailed analyses of the possible system design and operation are conducted. Furthermore aspects of future network management as well as business models for distribution and operation have been considered in a detailed study [3].

Mainly because of time constrains the decision has been taken not to realise the identified concept; the planned core drillings and Enhanced Geothermal Response Tests (EGRT) have not been carried out.

The project has shown that not only good system design integration innovative technologies in combination with a good economical parameters and a suitable business concept are important for a project realisation. Furthermore, these kind of innovative projects have special boundary conditions. There is a good chance that these kinds of projects are not going into realisation when common expectation on a design time frame and on the profitability is applied. Moreover, a good communication built on the trust between the project partners is a must.

Acknowledgement

The content of the paper is an outcome of the collaboration of the Fraunhofer Institute for Energy Economics and Energy System Technology (IEE) and Kassel University in a close cooperation with the City of Kassel and the local utility company Staedtische Werke Kassel AG. The financial support given to the project by the City of Kassel and the Staedtische Werke Kassel AG is greatly acknowledged by the project partners. Furthermore, the financial support given by the German Federal Ministry of Economic Affairs and Energy under support code: 03ET1336 A, B, C & E to conduct the second phase of the project is highly acknowledged by all involved project partners. The authors are responsible for the content of this paper.

References

[1] Schmidt, D.; Kallert, A.; Orozaliev, J.; Best, I.; Vajen, K.; Reul, O.; Bennewitz, J.; Gerhold, P. Development of an innovative low temperature heat supply concept for a new housing area. In: Proceedings of the 15th International Symposium on District Heating and Cooling, September 4-7, 2016, Seoul, Korea.

[2] Development plan Nr. IV/65 „Zum Feldlager“; Architektur+Städtebau Bankert, Linker & Hupfeld, City of Kassel- Urban planning, building supervision and monument conservation; 02 February 2013

[3] Schmidt, D.; Kallert, A.; Orozaliev, J.; Best, I.; Vajen, K.; Reul, O.; Bennewitz, J.; Gerhold, P. et al.. Geosolare Nahwärmeversorgung für die Neubausiedlung “Zum Feldlager”. Final report of the BMWi research project 03ET1336. Fraunhofer IEE, Univeristy of Kassel, German Heat & Power association AGFW, Staedtische Werke Kassel AG, City of Kassel, 2018.

[4] EnEV2014 - Regulation on energy-saving insulation and energy-saving systems engineering for buildings (Energy Saving Ordinance - EnEV); 01. Mai 2014

[5] DIN V 18599-1: 2011-08. Energy efficiency of buildings – Calculation of the net, final and primary energy demand for heating, cooling, ventilation, domestic hot water and lighting

[6] DIN V 18599-2: 2011-08. Energy efficiency of buildings – Calculation of the net, final and primary energy demand for heating, cooling, ventilation, domestic hot water and lighting – Part 2: Net energy demand for heating and cooling of building zones

[7] DIN V 18599-4: 2011-08. Energy efficiency of buildings – Calculation of the net, final and primary energy demand for heating, cooling, ventilation, domestic hot water and lighting – Part 4: Net and final energy demand for lighting

[8] DIN V 18599-5: 2011-08. Energy efficiency of buildings – Calculation of the net, final and primary energy demand for heating, cooling, ventilation, domestic hot water and lighting – Part 5: Final energy demand of heating systems

[9] DIN V 18599-8: 2011-08. Energy efficiency of buildings – Calculation of the net, final and primary energy demand for heating, cooling, ventilation, domestic hot water and lighting Part 8: Net and final energy demand of domestic hot water systems

[10] DIN V 18599-10: 2011-08. Energy efficiency of buildings – Calculation of the net, final and primary energy demand for heating, cooling, ventilation, domestic hot water and lighting - Part 10: Boundary conditions of use, climatic data

IOP Conf. Series: Earth and Environmental Science **352** (2019) 012041 doi:10.1088/1755-1315/352/1/012041

OmrådeLCA, assessment of area development: Case study of the Zero-Emission Neighbourhood Ydalir

V L Yttersian[1*], M Fuglseth[2], C Lausselet[1] and H Brattebø[1]

[1] Norwegian University of Science and Technology, NTNU, Trondheim, Norway
[2] Asplan Viak AS, Sandvika, Norway

[*]vidar_ly@outlook.com

Abstract. The built environment is a major contributor to global greenhouse gas (GHG) emissions. There is a growing need to quantifying GHG emissions, and LCA tools can be used to help address them and find the key contributors. In order to compare different solutions of development of the built environment in the early stage planning process, the OmrådeLCA tool has been developed. OmrådeLCA has the unique advantage, compared to other similar tools, of using the system expansion approach, which allows for comparing different scenarios based on the same functional unit. We applied OmrådeLCA on Ydalir (Norway), a zero emission neighbourhood in an early stage planning process. The results show how significant the share of emissions from transportation are, contributing to more than 60 % to the total GHG emissions. Sensitivity analysis shows that choices made, and data used in modelling transportation significantly impact resulting GHG emissions. Thus, conducting a thorough analysis of factors affecting transportation is important for obtaining representative results when using OmrådeLCA. The results from the assessment of Ydalir with OmrådeLCA have been compared with results from the same case assessed with a tool developed by NTNU. The comparison shows relatively small differences in calculated results. This small degree of variation between the two tools in calculated results, demonstrates that OmrådeLCA can provide good estimates even at an early stage. This gives the tool great utility value because it is in the early stages of a project major actions can be performed and decisions made that will affect the emissions the most.

1. Introduction

The built environment, referring to the building and transportation sector, is a major hotspot of environmental impacts, accounting for 55 % of total global greenhouse gas (GHG) emissions [1]. To reduce the impact the built environment has on the climate, several different measures and actions can be implemented [2]. Zero Emission Buildings (ZEB), compensate for total greenhouse gas emissions emitted over the building lifecycle by producing local renewable energy. The construction of ZEBs and Zero Emission Neighbourhoods (ZEN) [3] are measures which can contribute to minimising GHG emissions from new construction. Conducting GHG emission calculations is essential in order to document and assess the impact from the built environment and to compare different solutions and alternatives. LCA is a methodological framework used to analyse the environmental impacts of goods, services, materials, etc. The ISO standards ISO 14044 [4] applies for how to perform an LCA. In Norway, LCA methodology for assessing buildings has been standardised in NS 3720 [5].

A study by Wiik et al. [6] on lessons learnt from embodied GHG emission calculations in zero emission buildings from the Norwegian ZEB research centre found that the main contributors to total GHG emissions for building components are the building envelope (ca. 65%), and for the life cycle stages, the production and replacement of materials (ca. 55–87%). The study by Wiik et al. considered buildings adhering to a high building standard, resulting in a reduced need for operational energy. Dahlstrøm et al. [7] compared GHG emissions from a single-family dwelling built according to conventional building standards with a higher energy (passive house) standard, also considering the impact of the choice of emission intensity for electricity. The study highlights the increased significant the emissions from operational energy use got when shifting from both a higher building standard to a conventional one, and an electricity production mix with a low emission intensity to one with a higher intensity.

When expanding the scope and system boundaries to be applicable on a neighbourhood scale such as ZEN, transportation must also be accounted for. Lotteau et al. [1] carried out a study on 21 existing case studies on the neighbourhood scale. Although they found significant differences both in the type of projects, the scope of the studies, functional units and system boundaries, some general trends were identified. The buildings (construction and use of materials over the life cycle) were found to be the main contributors to GHG emissions, followed by transportation. However, Norman et al. [8] found transportation to be the largest contributor to total GHG emissions, contributing to 40-60 % of the total.

There is growing interest in the field of urban sustainability assessment on the neighbourhood scale, as mention by Lotteau et al. [1]. When developing areas according to the ZEN principle, robust quantitative estimation methods for GHG emissions applicable on the neighbourhood scale and at different stages of the planning process are needed. These methods and tools must meet the best practice for assessing GHG emissions and be in line with current frameworks and standards.

OmrådeLCA is an assessment tool for GHG emissions from areas, developed by Asplan Viak AS. The tool is based on the LCA framework, and considers emissions arising from use of building materials, operational energy use in buildings, and transportation of inhabitants on the neighbourhood scale.

The motivation for conducting the study described in this paper has been to further develop OmrådeLCA from its original version, as a high-quality tool for assessing GHG emissions from areas that in a transparent way can show results and importance of different choices taken and crucial factors for the development of new neighbourhoods. To assess the quality of the updated version of OmrådeLCA, the tool has been applied to the specific case Ydalir (Norway) and results have been compared with results from an assessment of the same case using a similar tool developed by NTNU. The study also considered the underlying methodology of OmrådeLCA in comparison with the NTNU ZEN-tool.

2. Method

2.1. Model

OmrådeLCA (which can be translated into English as AreaLCA or NeighbourhoodLCA) is a calculation tool developed by Asplan Viak in order to assess accumulated GHG emissions of area development. The underlying methodology of OmrådeLCA aims to shed light on the question; What is the best development path regarding GHG emissions for a given project [9, 10]?

The tool is intended mainly for use in an early planning stage, often before a regulation plan is approved, to compare different scenarios for development of an area or neighbourhood. Choices under consideration may be a different mix of building functions (such as apartments, single-unit dwellings, kindergarten and offices), and energy- and transportation solutions. To be comparable, these different scenarios must be assessed using the same functional unit, as stated in the LCA standards ISO 14040 and ISO 14044. This is done by using a methodological approach in LCA known as system expansion. Each scenario is assessed on the basis of the same overall functional mix of built area or number of inhabitants, per building category. The functional unit is set according to the scenario providing the largest built area or number of inhabitants, and for the other scenarios, it is assumed that any area that is not provided within the given scenario must be provided (built) in another (alternative) location. An illustration of this approach is given in Figure 1. The functional mix illustrated in this example is a mix of apartments, offices and a kindergarten. The difference between scenario 1 and 2 is the built square metres of housing and kindergarten that needs to be compensated for in the "alternative fulfilling of the function". As most other assessment tools use an attributional LCA approach, where only the environmental impact of each scenario as it is defined is assessed separately, this system expansion approach distinguishes OmrådeLCA from other tools [9, 10].

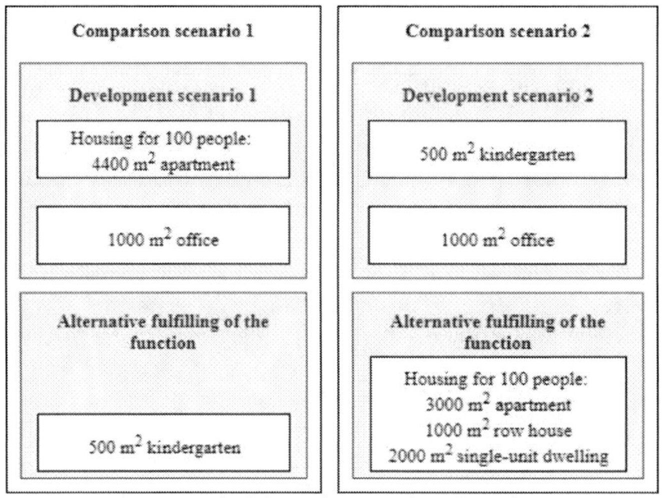

Figure 1. Methodological approach in OmrådeLCA.

The tool calculates emissions based on built area (BRA) and divides the calculations into three primary modules. These are materials, energy use, and transportation. The modules included in the tool compared to the NS3720's modular structure can be seen in Figure 2. OmrådeLCA uses data from national and local statistics, travel habit surveys, LCA databases and LCA studies. These generic data are combined with case-specific data which can be entered in the tool. Various modelling choices and scenario selections can also be made within the tool in order to tailor the analysis to a specific area development case.

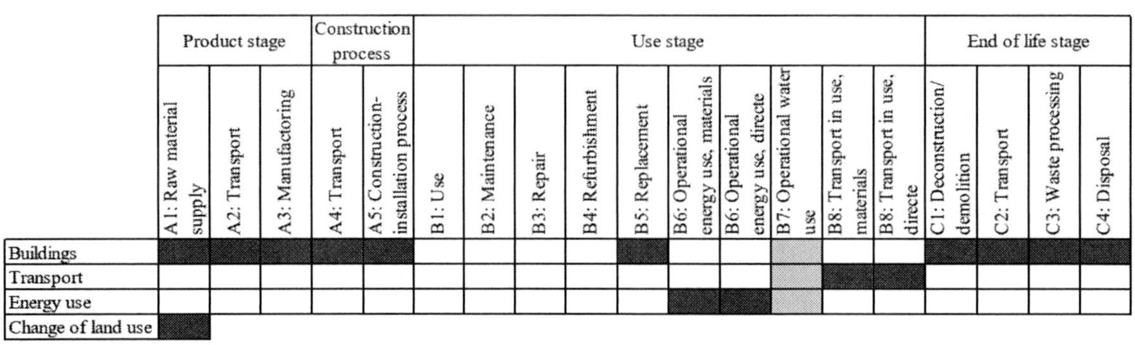

Figure 2. Modules included in OmrådeLCA according to NS 3720.

Through the described study, OmrådeLCA has been further developed from the original version made by Asplan Viak. The most important changes are as follows: emissions from use of construction materials have been subdivided according to the different modules of NS 3720; local energy production can be implemented easily; use of specific data for energy mix, development pattern and energy standards has been enabled. Also, result calculations have been made more transparent. Results are expressed in different functional units, and given according to the different modules specified in NS 3720, as well as cording to the different building categories considered. In addition, sensitivity analysis of different assumptions has been implemented.

2.2. Ydalir, a ZEN case study

Ydalir is an area located 1,5 km outside the Norwegian city of Elverum. It is to be developed as a residential area over the next 20 years, consisting of approximately 100 000 m^2 of residential buildings, a kindergarten and a school [11]. Ydalir is one of nine pilot projects within FME ZEN, the Research Centre on Zero Emission Neighbourhoods. A neighbourhood within FME ZEN must aim to reduce its GHG emissions towards zero over its life cycle [12].

In the case of Ydalir, several measures for reducing GHG emissions have been implemented. Transportation by car has been limited by considerably lowering the share of parking lots per household, where the parking lots in addition are located in a separate shared parking house prepared for charging of electric cars. Ydalir is in close location to the city-centre where transportation by walking and cycling are given priority in infrastructure development, together with preparations for better public transportation opportunities. Other measures include local energy production by solar panels and CHP (combined heat and power) using biofuel as energy carrier, use of materials with low associated GHG emissions in buildings, and implementation of passive house standard or better in buildings [11].

These measures have as far as possible been implemented in OmrådeLCA. Improved energy standards and use of local energy production are both easily implemented. Transportation is however not as easily implemented in the tool. Travel survey (RVU) data for cities in Norway of the approximately same size as Elverum has been used as a basis. The data have further been adjusted according to a study [13] on the impact of transportation changes to account for the expected reduction in car usage at Ydalir. The measure of using materials with low associated GHG emissions in buildings is not used in OmrådeLCA for this case. The function exists in OmrådeLCA but is not used since there are no values for how low the emissions are estimated to be compared to "normal" or ZEB buildings. Therefore, due to the lack of data, a conservative approach has been chosen, where GHG emissions from materials are assumed to correspond to other ZEB projects or passive houses.

The system boundary for the assessment is set to include materials for buildings and local energy production, operational energy use in buildings and the transportation of residents. A 60 year period of analysis has been used.

3. Results and discussion

3.1. Comparison of results and methodology of OmrådeLCA and the ZEN-tool

Results from the OmrådeLCA assessment of Ydalir can be seen in Figure 3, per main category; materials (including materials for local energy production), operational energy use and transportation. Corresponding results for a reference scenario representing development according to today's building standard in an average location in Elverum are also included for comparison.

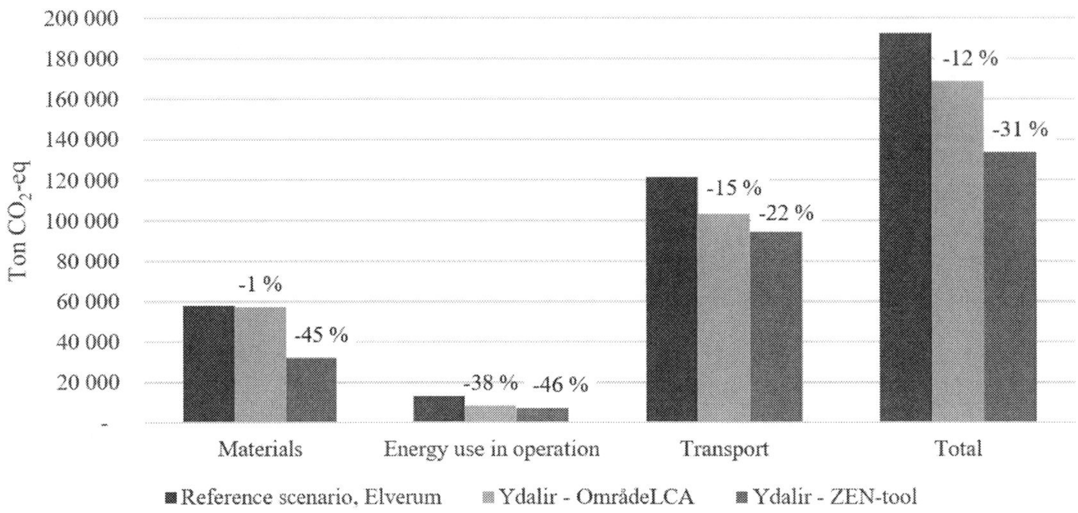

Figure 3. Total GHG emissions from the assessment of Ydalir with OmrådeLCA and the ZEN-tool.

The percentages presented in the figure represent the difference in emissions assessed with the two tools compared to the reference scenario. The difference from the ZEN-tool compared to OmrådeLCA is a reduction of 44 %, 14 % and 8 % respectively for the categories materials, energy use and transportation, and 21 % reduction comparing the total emissions. The difference in calculated GHG emissions from materials is mainly due to quite significant differences in emission intensities per m^2 used in the two tools. OmrådeLCA, intended for use in an early planning stage, uses average values for emissions per square meter for the given building category and building standard. The ZEN-tool, on the other hand, uses specific data from LCAs performed on the planned buildings at Ydalir. The resulting emission intensities per m^2 in the ZEN-tool are lower than corresponding calculations for other ZEB buildings. The ZEN-tool, using actual numbers causes the emissions from materials in the ZEN-tool to likely be more correct for this specific case. The results, 44 % decrease in emissions from materials assessed with the ZEN-tool compared to OmrådeLCA, thus show that by using average values in OmrådeLCA, results can be quite inaccurate. However, by using an additional reduction/increase in the emission intensity in OmrådeLCA in cases where the emission intensity is estimated to be lower or higher than the average, results in OmrådeLCA can be much more accurate.

The two tools give quite similar answers regarding emissions from operational energy use. Both tools calculate energy demand according to the Passive house standard. The calculations of emissions from local energy production are quite similar too. They both assign net negative emissions to electricity produced with solar panels (due to replacing grid electricity), and assumes the CHP replacing district heating or grid electricity. The difference in the results lies in some small variation of the data used for emission intensities and production data for solar panels, together with some differences in how estimations and allocations have been performed for the on-site energy production by the CHP. The difference (14 %) in the results regarding emissions from energy use for the two tools has less of a significance in view of the small share of the total emissions coming from energy use (4,9 % in the OmrådeLCA assessment of Ydalir).

IOP Conf. Series: Earth and Environmental Science **352** (2019) 012041 doi:10.1088/1755-1315/352/1/012041

Although calculations of transportation demand are based on the same transport survey data in both tools, there are still some significant differences (8 %) in calculated transportation emissions. This is caused by differences in the background on emissions from vehicle production and direct emission from vehicle operation.

Figure 3 shows results of an assessment where the same elements are included in both tools. However, there are some system boundary variations between the tools in this respect. The ZEN-tool includes local infrastructure, meaning roads, sidewalks and outside parking lots. This is a strength of the ZEN-tool compared to OmrådeLCA. However, OmrådeLCA is intended at an early stage where information on the size and amount of such infrastructure will rarely be available. In addition, if it was to be included in OmrådeLCA, it must also be accounted for in the alternative location of built area in order to fulfil the functional mix, which will be challenging to estimate with sufficient accuracy. However, as emissions from the infrastructure included in the ZEN-tool account for only 6 % of total emissions in the Ydalir case, this is not considered to be a very significant weakness in OmrådeLCA. OmrådeLCA, include opportunities to account for emissions from additional building components, soil stabilization and change in land area. OmrådeLCA also provides greater flexibility in adapting model parameters than the ZEN-tool. However, as the case-specific data needed to apply these features in the Ydalir case was not available, the potential impact of these additional features was not quantified.

The main difference in methodology for the two tools lies in which project stage they are intended to be used. OmrådeLCA is an early stage tool, meaning that generic data are used more extensively, with the opportunity of replacing these with specific data where available, while the ZEN-tool relies mostly on specific data. Another key difference is the possibility in OmrådeLCA to compare results towards a reference scenario or several other alternatives using the same functional mix of built area. This makes it possible to find not only the alternative with lowest associated GHG emissions for a given area, but also the alternative causing the lowest amount of GHG emissions for a given function, e.g. 10 000 inhabitants.

By comparing with a reference, it is easier to evaluate the environmental performance of the Ydalir development. As can be seen in Figure 3, the assessment of Ydalir with OmrådeLCA causes 12 % less GHG emissions than the reference scenario. The use of solar panels and better building standards reduces emissions from operational energy use by 38 %, but as energy use accounts for only a small part (4,9 %) of total emissions, this does not greatly affect total emissions. The impact of solar panels is also diminished because of the low associated emissions per kWh from the Norwegian electricity grid mix causing more emissions from materials than gained during energy production, compared to just using electricity from the grid. Calculated emissions from use of building materials are almost identical as in the reference case despite the fact that the emissions intensity per m^2 used for ZEB/passive houses is higher than in the reference case, where TEK 17 standard is used. The reason for this is a high area efficiency in the Ydalir case, causing less m^2 required to accommodate the given number of inhabitants than required in the reference case.

The reduction in emissions from transportation of 15 % must be regarded as a conservative estimate. As mentioned previously, several ambitious measures to reduce car travel are planned at Ydalir. However, due to the lack of RVU data to model the planned effects, conservative estimates have been used. It is therefore reasonable to assume that the actual reductions in emissions from transportation will be greater than 15 %.

3.2. Importance of transportation

Travel habit data vary significantly between different locations, and it can be challenging to assess which dataset most accurately corresponds to the location of a specific project or alternative location. OmrådeLCA therefore includes the possibility to compare results for different choices regarding travel data selected, as well as using project specific data. Figure 4 illustrates how much calculated emissions related to transportation vary depending on the chosen RVU dataset for the Ydalir case, using datasets for areas in the Oslo and Akershus areas as examples.

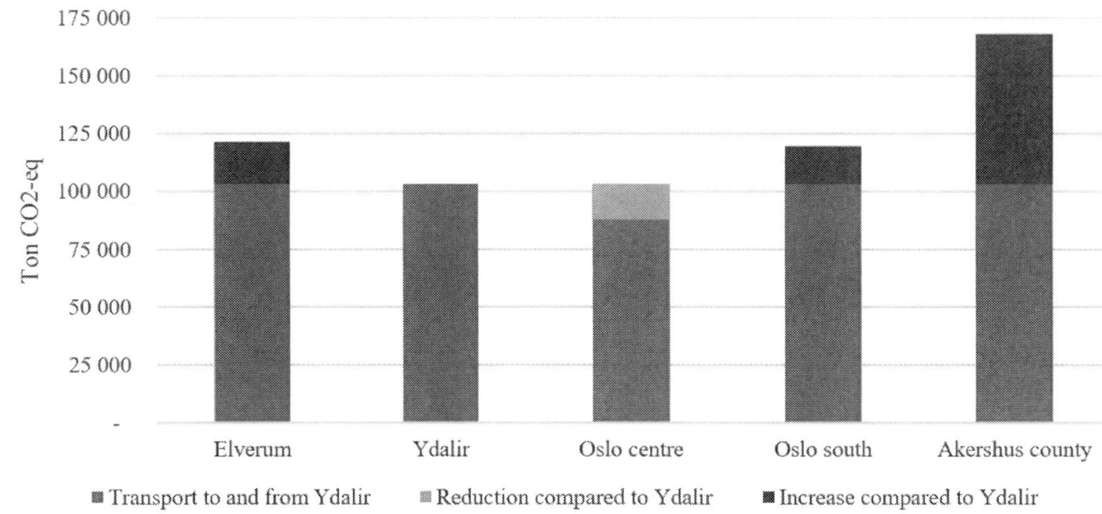

Figure 4. Emissions from transportation when using different RVUs.

Emissions varies from a 15 % decrease by choosing Oslo centre to a 63 % increase by choosing Akershus county when comparing with the Ydalir RVU. This is a large difference, showing how important it is to do thorough analyses of transportation. This is supported by differences shown in Figure 3 for emissions from transportation. When performing a sensitivity analysis for some key parameters, shown in Figure 5, it is also evident that transportation affect the results greatly. Each parameter of the analysis is in turn altered with a 25 % increase, and by doing so for the average travel distance, the total emissions from Ydalir increase by 15,3 %. This underscores the importance of performing thorough analyses of transportation when conducting an LCA on the area/neighbourhood scale.

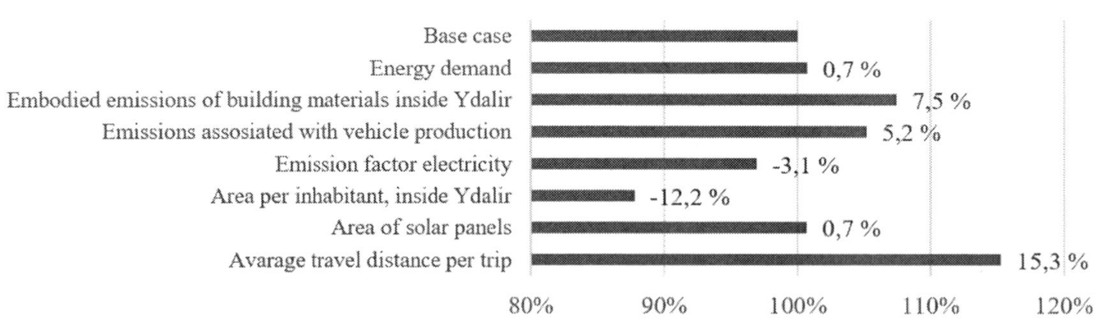

Figure 5. Sensitivity analyses of emissions from Ydalir, where each parameter is increased by a factor of 25 %.

The decrease in emissions of Figure 5 when increasing the area per inhabitant is because of a lower total amount of inhabitants of Ydalir, leading to diminished total transportation demand. This can lead to the misinterpretation that fewer people per square meter is better. If the functional unit was not total emissions from the neighbourhood, but emissions per inhabitant, the result would be the opposite.

IOP Conf. Series: Earth and Environmental Science **352** (2019) 012041 doi:10.1088/1755-1315/352/1/012041

4. Conclusion

Calculations of GHG emissions and reduction measures are a vital aid in the effort to reduce greenhouse gas emissions from the built environment. OmrådeLCA is a calculation tool that assesses GHG emissions of the built environment on a neighbourhood scale in an early stage planning. It calculates GHG emissions in three main parts, materials, operational energy use and transportation. OmrådeLCA have in this study been further developed from its original form, and been tested on the case of Ydalir, a ZEN consisting of both residential houses, a kindergarten and a school. To verify the results and methodology of OmrådeLCA the results from the same case of a tool made by NTNU have been compared to OmrådeLCA's. The two tools generally calculate emissions in a similar way, and thus yield quite similar results. The main differences in the calculation results are due to differences in the background data used. OmrådeLCA, intended for an early stage of a project, uses mostly generic data, while the ZEN-tool, intended for use at a later stage when data are available, uses mostly specific data. The small degree of variation between the two tools in calculated results, demonstrates that OmrådeLCA can provide good estimates even at an early stage. This gives the tool great utility value because it is in the early stages of a project major actions can be done and decisions made, that will affect the emissions the most. OmrådeLCA's system expansion approach is a feature that make OmrådeLCA different from a lot of other tools, and makes it well suited to assess the overall consequences for emissions from area development.

Transportation is the category contributing to the largest share of total GHG emissions in both tools. Sensitivity analyses performed with OmrådeLCA also show that emissions from transportation depend greatly on the choice of travel survey data used and assumptions regarding travel distances. In order to obtain representative and robust results, it is therefore important to conduct thorough analysrs and interpretation of data concerning transportation, and also to show the effect of different measures. This study shows that OmrådeLCA has the functionalities needed to make good and robust analyses, and the tool is therefore well suited to be used as a decision tool in case of area development and minimization of greenhouse gas emissions.

References

[1] Lotteau M, Loubet P, Pousse M, Dufrasnes E, Sonnemann G. Critical review of life cycle assessment (LCA) for the built environment at the neighborhood scale. 2015.

[2] Agency EE. Trends and projections in Europe 2018, Tracking progress towards Europe's climate and energy targets. 2018.

[3] ZEB. ZEBs definisjoner 2018 [cited 2018 19.12.]. Available from: https://www.zeb.no/index.php/no/om-zeb/zeb-definisjoner.

[4] Miljøstyring, Livsløpsvurderinger, Prinsipper og rammeverk, (2006).

[5] Metode for klimagassberegninger for bygninger, (2018).

[6] Wiik MK, Fufa SM, Kristjansdottir T, Andresen I. Lessons learnt from embodied GHG emission calculations in zero emission buildings (ZEBs) from the Norwegian ZEB research centre. Energy and Buildings. 2018.

[7] Dahlstrøm O, Sørnes K, Eriksen ST, Hertwich EG. Life cycle assessment of a single-family residence built to either conventional- or passive house standard Energy and Buildings. 2012.

[8] Norman J, MacLean H, Kennedy C. Comparing High and Low Residential Density: Life-Cycle Analysis of Energy Use and Greenhouse Gas Emissions. 2006.

[9] Borg A, Fuglseth M. Klimagassberegninger områdeutvikling Hopsfossen. 2018.

[10] Dahlstrøm O, Solli C. Klimagassberegning av utviklingsscenarioer. 2018.

[11] Vekst E. Ydalir Masterplan del 1 - versjon 2. 2017.

[12] Centre ZR. Sustainable neighbourhoods with zero greenhouse gas emissions [cited 2019 26.04.]. Available from: https://fmezen.no/.

[13] Christiansen P, Engebretsen Ø, Hanssen JU. Parkeringstilbud ved bolig og arbeidsplass. 2015.

IOP Conf. Series: Earth and Environmental Science **352** (2019) 012042 doi:10.1088/1755-1315/352/1/012042

Latent heat storage for centralized heating system in a ZEB living laboratory: integration and design

Alexis SEVAULT[1], Fabian BØHMER[2], Erling NÆSS[2], Liang WANG[1]

[1] SINTEF Energy Research, P.O. Box 4761 Torgarden, NO-7465 Trondheim, Norway
[2] NTNU Department of Energy and Process Engineering, Trondheim, 7491, Norway

Alexis.Sevault@sintef.no

Abstract. The ZEB Lab project, coordinated by SINTEF and NTNU, aims at building a ZEB (Zero Emission Building) in Trondheim (Norway) in 2019, to be used both as office building and living laboratory. An innovative latent heat storage (LHS) unit using phase change material (PCM) will be integrated in the centralized heating system. The LHS unit will be able to store excess heat from various heat sources connected to the heating system, when they are not required for space heating. One challenge is to make use of the full potential of the PCM latent heat to have a compact and effective unit, while the unit itself should have a low associated CO_2-footprint. The LHS system consists of two units designed for a total heat storage capacity of 0.6 MWh, corresponding to the heat needed on top of the heat pump to cover for up to 3 consecutive days in the coldest period of the year, with a maximum combined effect of 26 kW. A bio-based wax is used as PCM with melting temperature 37 °C and measured latent heat 198 kJ/kg. Dynamic system modelling is used to support the design of the LHS unit and ensure sufficiently high heat transfer rates.

Keywords. Phase Change Materials, PCM, Thermal Energy Storage, Heat Storage, Building, Centralized Heating system.

1. Introduction

1.1. Background

The ZEB Lab project (www.zeblab.no) , coordinated by SINTEF and NTNU, aims at building a ZEB (Zero Emission Building) in Trondheim (Norway) starting in 2019, to be used both as office building and living laboratory [1]. The building will rely on innovative technologies, both regarding building materials and energy system. One technology to be implemented and tested in the centralized heating system of the building will be a latent heat storage (LHS) unit using a phase change material (PCM).
The LHS unit will be integrated in the low-temperature heating system centred around a heat pump providing hot water for space heating. An essential asset of the LHS unit is to be able to provide high heat storage capacity and heat effect within the narrow temperature range offered by low-temperature heating. During hours of low heat demand, the LHS unit will be able to store excess heat. During high demand, the stored heat will be released either to provide heat directly to the heating circuit or to support the heat pump by compensating for a drop of return temperature below the optimal intake temperature. Since such LHS technology is not commercially available yet, an experimental unit has to be custom-designed following the best integration path in the heating system.

1.2. Objective
The main objectives of the present study are: (1) to assess one scenario for full integration of the LHS unit in the centralised heating system of the ZEB Flexible Laboratory; (2) to dimension the LHS unit according the building energy demand, while selecting the most appropriate PCM; and (3) to model the dynamic thermal performance of the designed LHS unit to validate its general design.

1.3. Literature review
A well-known challenge with using PCM for thermal energy storage is the poor thermal conductivity for available PCMs, limiting heat transfer rates [2, 3]. Comprehensive work has been done to increase the heat transfer rates within LHS systems by utilizing heat transfer enhancement techniques in numerical investigations and in experimental setups. However, only a few full-scale active LHS systems are in operation, making it challenging to document the potential upsides of coupling a LHS system to a heat pump for peak shaving and heating purposes.

Hirmiz et al. [4] studied the integration of LHS systems into heat pump systems to improve the demand side flexibility and, ideally, the strategy for the LHS system to cover the complete heat demand during peak periods. By utilizing a TRNSYS numerical model, it was concluded that a LHS system can completely offset peak heat demand periods within 2 to 6 hours, reducing peaks in the power grid. Through modelling and measurement data analysis, Jokiel et al. [5] evaluated a LHS system installed to reduce the required chiller capacity for three ammonia chillers/heat pumps covering the base load for heating/cooling at the Bergen University College (Norway). A dynamic system model was developed using Modelica [6] to better understand the dynamics of melting and solidification of the PCM. The model proved to correctly predict the measured data, within an acceptable accuracy, especially regarding the accumulated values of absorbed and released heat.

Bonamente et al. [7] studied the potential for system optimization in an existing ground-source heat pump heating system by implementing a TES unit. Computational fluid dynamic calculations were carried out and validated against measured data using two TES solutions: one using water as storage medium, and the other using PCM. Results showed that the COP of the system was increased from 2.9 to 3.2 and 3.4 for, respectively, heating and cooling modes when using water as TES medium. By using a PCM, the system COP was increased to 4.13 and 5.89 for, respectively, heating and cooling modes. In addition, the total volume of the PCM thermal storage was 10 times more compact compared to the water tank system making it more suitable for indoor installation and use.

Shifting the cooling load during simulated summer conditions was experimentally tested by Moreno et al. [8] by coupling a TES system to a heat pump. Thermal behaviour for the TES system was evaluated for cold storage and for space cooling. Two different TES configurations were tested, one using water and the other using PCM. The latter configuration utilized macro-encapsulated PCM with a phase change temperature of 10 °C. It was concluded that PCM storage is favourable to water storage. With identical volumes, the PCM tank was able to store 35.5 % more cold energy on average compared to the water storage tank. Other results indicated that by increasing the heat transfer rate for the PCM storage, it could store 14.5 % more cold energy, while delivering an acceptable indoor temperature for a 20.65 % longer duration compared to the water storage.

2. Integration of LHS unit

2.1. Centralized heating system
The building heating system is centralized and originally relies on a 25-kW heat pump system and several heat sources and heat sinks along the hot water heating circuit. The selected heat pump system is designed to provide a temperature lift on the hot side from 35 °C to 40 °C. Besides the heat pump, heat from the local district heating is used as heat source for the building, providing hot water at 45 °C. Throughout the heating circuit, preheating of domestic hot water, room radiators and heat exchangers providing heated air for ventilation are used as heat sinks to heat up the building. Additional components enabling research experiments in the different rooms of the buildings are also

IOP Conf. Series: Earth and Environmental Science 352 (2019) 012042 doi:10.1088/1755-1315/352/1/012042

planned but constitute minor heat sinks and heat sources on the heating circuit and thus are out of the scope of the present study. Without the LHS unit, the heat pump is meant to cover the maximum heat demand of the building, calculated to ca. 26 kW, necessary to maintain all rooms in the building at a comfortable temperature on the coldest days of the year in Trondheim (Norway). Using the LHS unit to support peak heating demands, the size and nominal effect of the heat pump can be significantly reduced, so that it operates more effectively. Combining a constant heat pump effect down to 12 kW with a charged LHS unit of 0.6 MWh would allow for a total heating effect of 26 kW for up to 42 h.

2.2. LHS unit integration

Among the possible scenarios to integrate the LHS unit in the centralized heating system, the integration enabling thermal buffering to support the heat pump was selected (see Figure 1). Depending on the heating demand in the building, the return temperature of the heating loop might be lower than 34 °C, and thus require additional power from the heat pump to sustain 40 °C as outlet temperature. Integrating the LHS unit downstream from the heat pump, with the option to circulate the return water through it or not, provides the opportunity to both charge and discharge the LHS unit, while smoothing the effect demand from the heat pump. Charging the LHS unit occurs when the heating demand is low, using 40 °C as inlet temperature, as it is generated by the heat pump. Using a PCM with phase change temperature within 34-37 °C, return water at lower temperature than 34 °C can circulate through the charged LHS unit and be heated up before entering the heat pump. Additionally, the LHS unit can be directly charged using the district heating loop providing hot water at 45 °C. Note that, as shown in Figure 1, two LHS units (LHS-1 and LHS-2) are integrated in the heating system, to ultimately allow for research experiments using various heat exchanger designs and test the thermal performance of several PCM. The present study focuses only on the design of LHS-1.

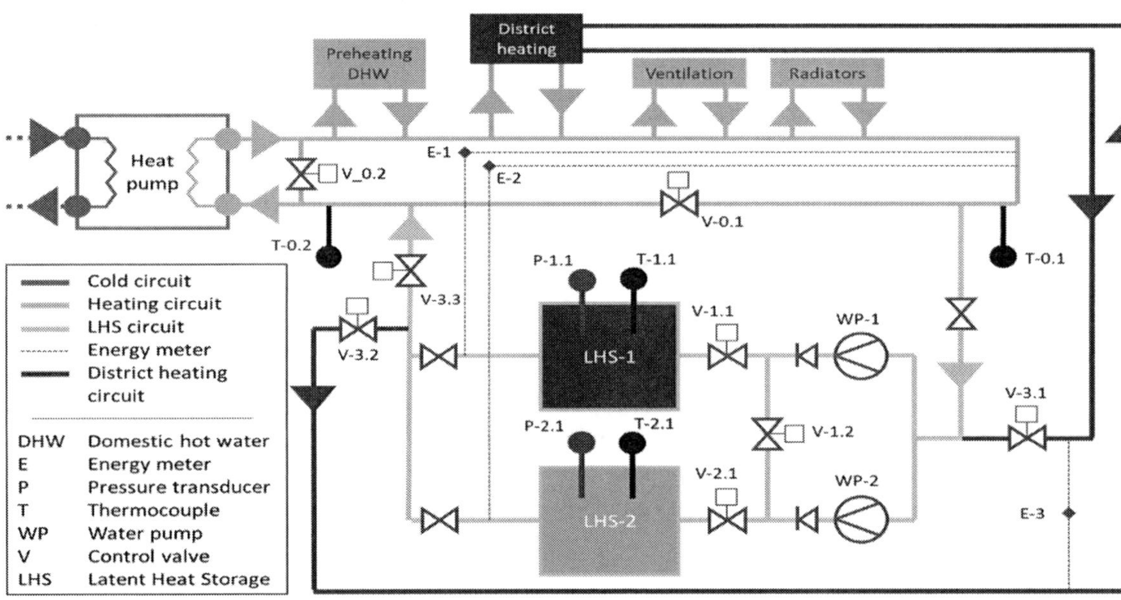

Figure 1. Process diagram of the centralized heating system showing the integration of the two LHS units. Only the instrumentation for control of the LHS units is shown for simplification.

Another feature available with this integration is the opportunity to use the LHS unit as a direct heat source in the building heating loop. This is meant to occur when the LHS unit is charged and the heating demand in the building is relatively low. Therefore, the heat pump can be bypassed, reducing significantly the energy use during these low-demand periods. This operational mode is especially interesting if energy price is integrated in the control system of the overall heating system.

IOP Conf. Series: Earth and Environmental Science **352** (2019) 012042 doi:10.1088/1755-1315/352/1/012042

2.3. Control strategy

As shown in Figure 1, the system allows for a variety of control strategies through a large number of control valves and two regulated water pumps. The selected strategy includes two levels: (1) a temperature-controlled strategy for charging and discharging using only the heat pump as heat source and heat sink; (2) a price-controlled strategy where the energy price is taken into account to decide when to harvest heat from the district heating network and when to use the LHS unit as direct heat sources for the building heating circuit. In both cases, the energy level of the LHS units is followed up using thermocouples located at various locations in the unit. Full charge is indicated by an average PCM temperature 4 K above its melting temperature range. Full discharge is indicated by an average PCM temperature 4 K below its solidification temperature range. In addition, three energy meters will enable to track the effect and accumulated transferred energy to follow up the thermal performance of the two units. The control system of the LHS system is to be fully integrated in the building control system, which will include a "researcher mode" to allow customizing and testing various control strategies.

3. Design and modelling

3.1. PCM selection and performance testing

The most suitable PCM for the LHS unit should primarily have a melting temperature within 34-37 °C, which yields only a limited range of commercially available PCMs. Table 1 lists a selection of commercial PCMs with melting temperatures ranging from 34 °C to 37 °C, as well as some of their thermodynamic properties given by the manufacturers.

Table 1. Selection of commercially available PCMs.

Product	*Type*	*Melting point [°C]*	*Heat of fusion [kJ/kg]*	*Thermal conductivity [W/(m.K)]*	*Manufacturer*
E34	Eutectic	34	240	0.54	PCM Products
A36	Organic	36	217	0.18	PCM Products
L36S	Salt hydrate	36	260	0.6	TEAP
E37	Eutectic	37	213	0.54	PCM Products
A37	Organic	36	235	0.18	PCM Products
CT37	Organic	37	202	0.24	CrodaTherm
PCM37	Organic	37	215	N/A	Microtek

After investigation of the pre-selected commercial PCMs listed in Table 1 for the melting temperature range 34-37 °C, the PCM CrodaTherm 37 (CT37) was selected. CT37 is a water-insoluble organic PCM, derived from plant-based feedstocks [9]. The PCM appears as a crystalline wax in solid state and oily liquid above melting temperature. The main arguments in favour of this PCM are its low degree of supercooling (cf. Table 2), its low-carbon footprint as well as its affordable cost. In addition, CT37 has low flammability, which is an essential criterion in buildings.

A sample of CT37 received by CrodaTherm was analysed by Thermogravimetric Analysis – Differential Scanning Calorimetry (TGA/DSC) at the SINTEF Energy Laboratory to evaluate the thermodynamic performance of the PCM. A measurement of 10 melting/solidification cycles was performed using a TGA/DSC SDT600 from TA Instruments, with controlled heating and cooling rates of 1 K/min ranging from 30 °C to 50 °C, in a nitrogen atmosphere. The results are shown in Figure 2. As indicated by the manufacturer, the first melting displays a significantly larger latent heat of fusion than the following melting/solidification cycles. Taking into account only the 9 following cycles, CT37 remains absolutely stable, yielding very similar heat flow patterns. The average latent heat of fusion is 198.6 kJ/kg (+/- 0.9 %) and the average latent heat of crystallisation is 196.4 kJ/kg (+/- 0.7 %). The average peak melting temperature peak is 36.5 °C (+/- 0.3 %) and the average solidification temperature peak is 34.5 °C (+/- 0.1 %). The weight loss is measured to 0.04 % along the first two cycles and then remains stable for the following 8 cycles. Note that thermodynamic property measurements might be variable from one device to another and is also known to depend on the sample mass and measurement procedure.

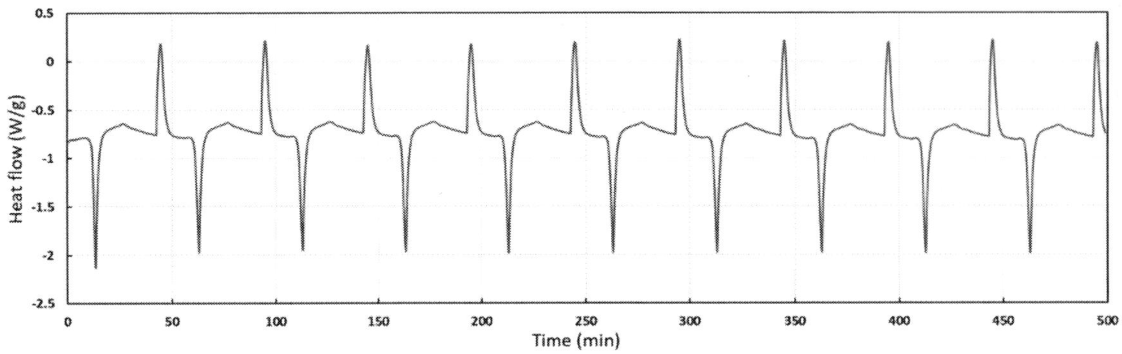

Figure 2. DSC measurements of heat flow absorbed and released by a sample of CT37 for 10 melting/solidification cycles.

3.2. General design

The general design parameter of LHS-1 are given in Table 2. The LHS unit dimensions are the first constraints to consider for the unit design due to the architecture of the building, limiting the access into the technical room through a 1.8-m wide corridor. This justifies the idea of having two LHS units whose dimensions allow to enter the building (see Table 2). The design of LHS-1, shown in Figure 3, is based on a fin-and-tube heat exchanger, filled with PCM. Water from the heating circuit circulates in the tubes. The design parameters of the fin-and-tubes heat exchanger are discussed in Section 0Headers at both ends of the unit enable a homogeneous distribution of the water across the tubes. A thick thermal insulation around the LHS-unit allows for a theoretical heat loss under 2 % per 24 h.

Table 2. General design parameters of LHS-1 using a fin-and-tube heat exchanger design.

Properties of LHS-1	*Values*
Dimensions of LHS-1 unit (height x width x length) *[m]*	1.5 – 1.4 – 2.5
Measured PCM melting temperature range and peak *[°C]*	35 – 39 (36.5)
Measured PCM solidification temperature range and peak *[°C]*	33 – 35.5 (34.5)
Measured PCM latent heat of fusion *[kJ/kg]*	198.6
Measured PCM latent heat of crystallisation *[kJ/kg]*	196.4
PCM density *[kg/m^3]*	957 (at 32 °C), 819 (at 75 °C)
PCM thermal conductivity *[W/(m.K)]*	0.24
PCM specific heat capacity (solid – liquid) *[kJ/(kg.K)]*	2.3 – 1.4
PCM degradation temperature *[°C]*	> 50
Total theoretical thermal storage capacity *[from 30 to 40 °C] [kWh]*	325
Ratio of latent heat to total heat storage capacity	90 %

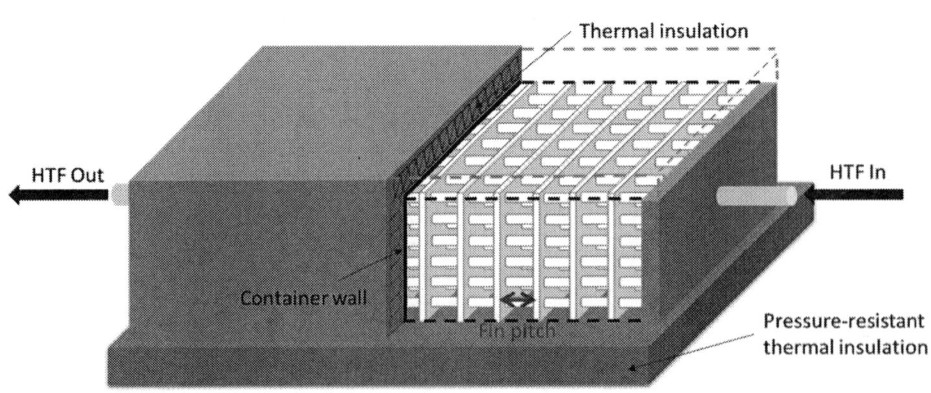

Figure 3. Simplified geometry of the LHS unit. The PCM (not shown here) will occupy the space between the plate fins.

4. Dynamic system modelling

4.1. Model description and assumptions

To investigate the transient nature of the charging and discharging processes of the LHS unit, a heat exchanger model using PCM for heat storage has been developed in the modelling and simulation software Dymola [10]. Dymola allows dynamic modelling of thermal systems with variable inputs. It is based on the open, object-oriented modelling language Modelica [6]. The specific heat exchanger model was developed using validated TIL libraries from expert thermodynamic model developers at TLK-Thermo GmbH [11]. The model is based on a solid-liquid fin-and-tube heat exchanger using PCM for TES custom-made by TLK-Thermo GmbH in 2017 and utilised in previous works [12]. This LHS model only has one mass flow, for the heat transfer fluid (HTF), water here. The model requires the following input parameters: heat exchanger basic geometry, fin geometry, HTF inlet temperature, HTF mass flow rate, PCM and material properties, as well as the initial temperature for respectively the HTF, PCM and heat exchanger. The LHS model consists of tubes and plate fins running through a tank filled with PCM. Heat transfer rates are calculated from the HTF to the PCM through tube walls and plate fins following a 1-dimension heat transfer model. The specific LHS model is still to be experimentally validated using PCM CT37, though it is assumed sufficient for a first assessment.

The dimensions of the tank are 1.5-m-high, 1.4-m-wide and 2.5-m-long. To be able to operate the LHS system during peak periods (e.g. cold days), the LHS unit should be fully charged in advance. The control strategy involves to fully charge the LHS unit during off-hours, typically 10 hours during the night or during warmer days since the storage capacity enable several worth of heat demand in the coldest days. The LHS model enables to evaluate the heat transfer area required to be able to fully charge the system in a predetermined number of hours, assuming a HTF inlet temperature of 40 °C as it can be delivered by the heat pump in off-peak hours. Note that this is the most critical case for charging processes since the connection to district heating enables quicker charge at 45 °C.

The following assumptions are made for the numerical simulations:

- Heat transfer coefficient between HTF and inner pipe wall are calculated using Dittus-Boelter correlation [13].
- Hysteresis effects are not accounted for in the PCM.
- No heat loss through tank wall.
- The PCM phase changes are complete during charging/discharging processes (no partial load).

Simulations are carried out for six different scenarios for both charging and discharging processes. Three scenarios are simulated with fin pitches of, respectively, 1, 2 and 4 cm, in the heat exchanger geometry for a constant HTF mass flow of 2 kg/s. The next three scenarios are simulated with a variable HTF mass flow of respectively 1, 2 and 3 kg/s, for a constant fin pitch of 2 cm. The heat exchanger consists of 225 pipes, arranged 15 x 15 across a transversal section, along the longitudinal axis, with 1 cm inner diameter. Fin thickness is set to 1 mm. The simulations are performed over a sufficiently long time period to allow for complete melting or solidification of the PCM. For the charging processes, the initial temperature is 30 °C in the whole LHS unit and the HTF inlet temperature is 40 °C. For the discharging processes, the initial temperature is 40 °C for the whole LHS unit and the HTF inlet temperature is 30 °C.

4.2. Results and discussion

Figure 4 shows the simulation results for one charging process with the tested fin pitches. Figure 4 (a) shows the average PCM temperature throughout the LHS unit. The charging or discharging time is defined here as the time required for the average temperature of the PCM in the LHS unit to reach the HTF inlet temperature. By reducing the fin pitch from 4 cm to 1 cm, the charging time is reduced from ca. 16 h to 9 h. This is mainly because of the increased surface heat transfer area from the increased quantity of fins, and because the shorter fin pitch also reduces the PCM volume in the LHS unit by 7

%. Figure 4 (b) shows the PCM liquid fraction and heat flow from the PCM to the HTF during the melting process. Heat transfer flows are initially high due to the larger temperature differences driving the heat transfer. They rapidly decrease within the sensible heat transfer region, until the heat flows stabilise when reaching the latent heat transfer region for fin pitches of 1 cm and 2 cm. The fin pitch of 4 cm does not yield such a sharp change during phase change.

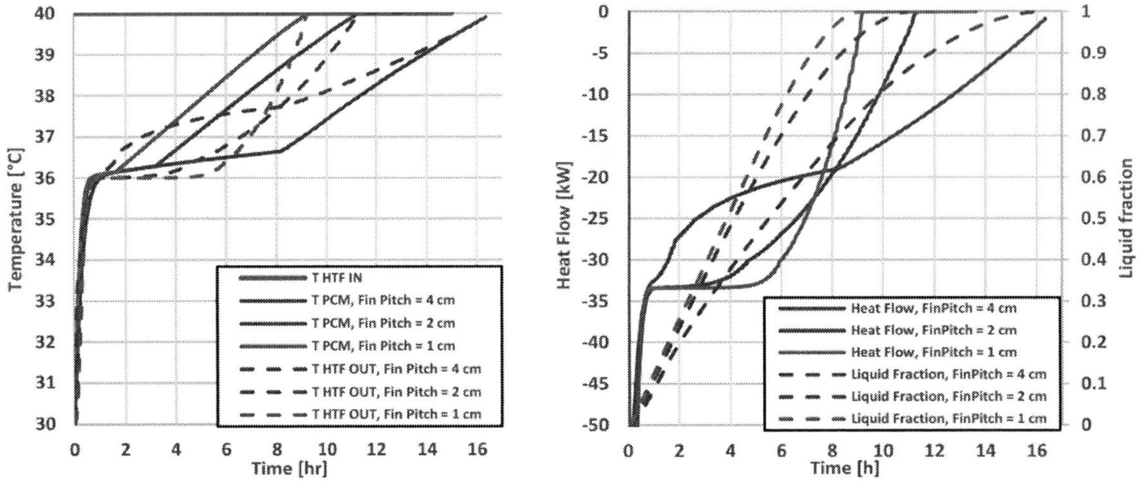

Figure 4. Charging process for various fin pitch with constant HTF mass flow of 2 kg/s, (Left) Temperature of HTF at inlet and outlet, and average PCM temperature; (Right) PCM liquid fraction and heat flow from PCM to HTF.

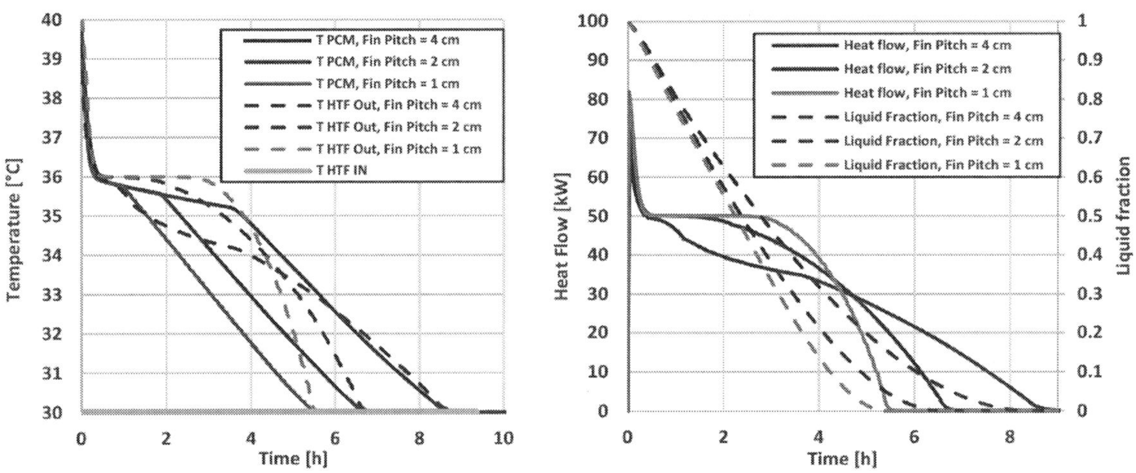

Figure 5. Discharging process for various fin pitches with constant HTF mass flow of 2 kg/s, (Left) Temperature of HTF at inlet and outlet, and average PCM temperature; (Right) PCM liquid fraction and heat flow from PCM to HTF.

Figure 5 shows the results of the discharging simulations for the tested fin pitches. As shown in Figure 5 (a), a fin pitch of 1 cm yields an HTF outlet temperature almost constant at 36 °C for 2 hours, which is optimal for the steady operation of the heat pump. Increasing the fin pitch results in steadily decreasing HTF outlet temperature during discharging. Figure 5 (b) shows that reducing the fin pitch provides higher and more constant heat flows in the latent heat transfer region, ultimately leading to shorter discharging times. According to the simulations, a fully charged system stores respectively 293 kWh, 286 kWh and 271 kWh for a fin pitch of 4 cm, 2 cm and 1 cm. This yields average heat transfer rates of ca. 35 kW, 44 kW and 50 kW for the various fin pitch distances during discharge.

These results validate the thermal performance for the proposed design of the LHS system, both in terms of charging times and heat transfer rates during discharge. Charging of the system may occur over night when the building is not in use, typically for a time period of maximum 10 hours. Figure 4

(a) shows that this can be achieve using a fin pitch distance of 1 or 2 cm. Figure 5 (a) confirms that the LHS unit can raise the return temperature of the heating circuit closer to design intake conditions for the heat pump (35 °C). Figure 5 (b) demonstrates that the LHS unit is able to deliver more than sufficient heat transfer rates, to meet the maximum heat demand in combination with the heat pump.

5. Conclusion

The aim of the present study is to evaluate the integration and design strategies for a LHS unit to be implemented in the centralized heating system of the ZEB Flexible Laboratory building in Trondheim (Norway). The LHS unit is designed based on a fin-and-tube heat exchanger filled with PCM whose phase change temperature is 35-37 °C. Using dynamical system simulations, the thermal performance of the proposed LHS design has been evaluated. The simulated LHS unit can store up to 325 kWh and simultaneously achieve sufficiently high heat transfer rates during discharge to successfully back up the heat pump during the coldest winter days.

Acknowledgements

The current study was carried out through the competence-building project PCM-Eff supported by SINTEF Energy Research using basic funding from the Research Council of Norway. The project aims at investigating novel phase change material solutions for efficient thermal energy storage at low and medium-high temperature (www.sintef.no/en/projects/pcm-eff/). The LHS unit is co-financed through the ZEB Flexible Laboratory by SINTEF, NTNU, Research Council of Norway and ENOVA.

References

[1] Time, B., et al. *ZEB Laboratories – Research Possibilities*. in *Proc. of NORDIC ZEB+ - 1st Nordic Conf. on Zero Emission and Plus Energy Buildings – 6-7 Nov. 2019, Trondheim (Norway)*.
[2] Shukla, A., D. Buddhi, and R.L. Sawhney, *Thermal cycling test of few selected inorganic and organic phase change materials*. Renewable Energy, 2008. **33**(12): p. 2606-2614.
[3] Sevault, A., et al., *Phase change materials for thermal energy storage in low- and high-temperature applications: a state-of-the-art, SINTEF Energy Research Report TR A7638 - ISBN 978-82-594-3684-9*. 2017.
[4] Hirmiz, R., et al., *Performance of heat pump integrated phase change material thermal storage for electric load shifting in building demand side management*. Energy and Buildings, 2019. **190**: p. 103-118.
[5] Jokiel, M., et al. *Phase change material thermal energy storage for a large scale ammonia chiller/heat pump system*. in *7th Conference on Ammonia and CO2 Refrigeration Technology. Proceedings: Ohrid, Macedonia, Mai 11-13, 2017*. 2017. International Institute of Refrigeration.
[6] *Modelica and the Modelica Association - Modelica*. 2019 [Visited 2019]; Available from: https://modelica.org/.
[7] Bonamente, E., A. Aquino, and F. Cotana, *A PCM Thermal Storage for GSHP: Simulating the System Performance via CFD Approach*. Energy Procedia, 2016. **101**: p. 1079-1086.
[8] Moreno, P., et al., *PCM thermal energy storage tanks in heat pump system for space cooling*. Energy and Buildings, 2014. **82**: p. 399-405.
[9] *Crodatherm 37 - Datasheet*, C.I. Plc, Editor. Consulted in May 2019.
[10] *Dassault Systèmes - DYMOLA Systems Engineering* 2019 [Visited 2019; Available from: https://www.3ds.com/products-services/catia/products/dymola/.
[11] *TLK-Thermo GmbH, TIL Suite Thermal Systems*. 2019 [Visited 2019; Available from: https://www.tlk-thermo.com/index.php/en/software-products/til-suite.
[12] Jokiel, M., et al., *Dynamic modelling of a refrigerated cabinet with integrated phase change material thermal storage*. Proceedings of the 25th International Congress of Refrigeration, August 24-30, Montreal (Canada), 2019.
[13] Dittus, F.W. and L.M.K. Boelter, *Heat transfer in automobile radiators of the tubular type*. International Communications in Heat and Mass Transfer, 1985. **12**(1): p. 3-22.

IOP Conf. Series: Earth and Environmental Science **352** (2019) 012043 doi:10.1088/1755-1315/352/1/012043

Influence of smart lighting control on the lifetime of high power LED luminaires

Janne Askola[1], **Hans Baumgartner**[1], **Tomi Pulli**[1], **Petri Kärhä**[1] and **Erkki Ikonen**[1,2]

[1] Metrology Research Institute, Aalto University, Espoo, Finland

[2] VTT MIKES, VTT Technical Research Centre of Finland Ltd, Espoo, Finland

E-mail: `janne.askola@aalto.fi`

Abstract. LED lighting and smart controlling can decrease the consumption of electric power in zero or positive energy buildings. This study demonstrates that cyclic dimming, which appears in smart controlling, e.g. through occupancy sensing, might have an effect on the luminaire lifetimes and failure rates. Two different types of LED luminaires were aged in 30-s cycled dimming and undimmed modes. The manufacturer-specified lifetimes for both luminaire types were 100 000 hours. For one of the luminaire types, four out of five cycled units failed before 30 000 operating hours. For the uncycled luminaires, the lifetimes estimated from the measurements were over 100 000 hours. For the other type, the estimated lifetimes were 75 000 hours and 64 000 hours for uncycled and cycled luminaires, respectively.

1. Introduction

The energy savings due to LED lighting products are remarkable when compared to the traditional incandescent and fluorescent lighting, making the LED products particularly attractive in zero or positive energy buildings. In many designs, additional energy savings can be achieved by adding smart control to the lighting, e.g. with occupancy or daylight sensing. The lights may be switches off completely when not needed, or dimmed to a lower level.

Typically, the luminaire manufacturers specify the lifetime of high power LED panels between 50 000 hours and 100 000 hours. According to the standard on LED luminaire lifetime projection, a minimum of 6 000 hours is needed for lifetime estimation and for a lifetime of 100 000 hours, measurement data of at least 17 000 hours of ageing should be in use [1].

Two different effects may influence the lifetime of LED luminaires of smart lighting. Cyclic dimming, which appears in smart controlling, effectively decreases the driving current of the LED chips which should result in extended lifetime. It is known that pulsing LEDs with different duty cycles effectively decreases the current and therefore decreases the junction temperature [2].

On the other hand, electronics hammer testing, where the temperature of the electronics is cycled between higher and lower levels, is one method used for accelerating ageing [3]. In a similar way, the smart lighting control may cycle the temperatures of the components of the system, thus introducing additional stress to the LED components and the driving electronics of the luminaires as compared to the traditional controlling, which may accelerate the ageing of the luminaire. The effect of constant thermal stress for LED lamps has been found to decrease the lamp lifetime [4, 5].

IOP Conf. Series: Earth and Environmental Science **352** (2019) 012043 doi:10.1088/1755-1315/352/1/012043

The combined effect of the decreased driving current and the thermal stress introduced by the cycling is of high interest when designing smart controlling level. In our research, we have studied the effect of cycled dimming on high power LED luminaires. Our measurement results show that the differences in the luminous flux values between the undimmed and periodically dimmed luminaires are clearly seen after 15 000 hours of operation.

2. Luminaire ageing and measurements

In total 20 luminaires from two different manufacturers were aged in room temperature conditions. Type A luminaires were built of 12 LED strips containing in total 38 LED chips and their controlling electronics. The LEDs were connected to the electronics in four parallel series of three strips. Type B luminaires consisted of two separate LED panels that have separate controlling electronics.

For both luminaire types, the specified lifetime expectancy set by the manufacturer was 100 000 hours. All luminaires were switched on for 9 hours followed by an off period of 3 hours simulating daytime use. During the on period, half of the luminaires were cycled between 100% and 20% electrical power at 30 second interval, later on referred to as cycled dimming. The luminaires were aged for 4.5 years, leading to an operating time, i.e., time when the lamps were switched on, of 28 000 hours.

Even though the manufacturers have designed the studied luminaires for outdoor usage, their LED components and driving electronics are comparable to the luminaires used in indoors. The values specified by the manufacturer for lifetime, luminous flux and electrical power of the studied luminaire types are presented in Table 1. The lifetimes have been specified at the ambient temperature of 25 ° C.

Table 1. Specified lifetimes, luminous flux and electrical power values for the studied luminaires.

Luminaire	Specified lifetime (h)	Specified lum. flux (lm)	Electrical power (W)
A	100 000	7 500	63
B	100 000	13 700	137

The luminaires were equipped with their own modular driving electronics that supply constant direct current to the LEDs. The dimming of the luminaires was controlled by restricting current through DALI connections included in the driving electronics. A regulated AC power source of 230 V was used to power the luminaires in measurements. The regular electricity grid was utilized in the ageing setup.

The luminaire temperatures during different stages of the operation cycles were measured to ensure that the temperature limits defined by the manufacturer are not exceeded. Temperatures of the LEDs were measured by setting a temperature measurement sensor as close as possible to the LED chip in such way that it was not in front of the LED. In Type B luminaire measurement, the sensor was placed under a separate, detachable lens array. No detachable lenses were used in Type A luminaires. The difference in the placement of the temperature sensor might explain some of the absolute temperature differences of the luminaires, but should not affect the temperature change due to the cycled dimming. The measurements were carried out at ambient temperature of 25 ° C.

The temperature profiles during ageing as measured by the temperature sensors near LED chips are presented in Figure 1. The temperatures are stated as temperature differences from the ambient temperature. For the Type A lamps, the average temperatures during the cycled

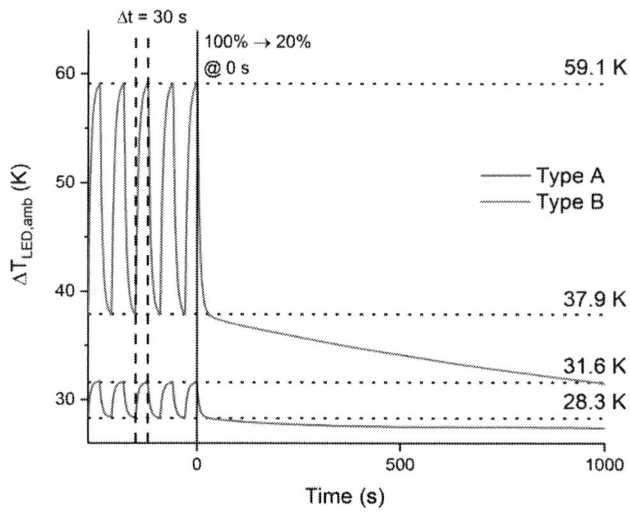

Figure 1. Temperature differences of the luminaires on the LED panel right next to LED chip compared to ambient temperature during the used 30 s dimming cycle and after setting the LED power from 100% to 20%.

dimming is 30 ° C above the ambient temperature, and the temperature varies by 3.3 K. For the Type B luminaire, the corresponding average temperature is 48.5 ° C above the ambient temperature and the temperature variation is 21.2 K. One difference influencing the temperatures in the two luminaire types is the design of enclosing. The enclosing in Type A luminaire acts also as the luminaire heat sink, whereas Type B has a separate heat sink inside the enclosing.

Using the cycled dimming, the temperatures at different stages of the dimming vary by 21.2 K and 3.3 K for Type B and Type A luminaires, respectively. In Figure 1 at the time of 0 s, the electrical power of the luminaires has been changed from 100% to 20%. It can be seen that from the time of 0 s until the time of 30 s, the temperature change rate is at largest, being approximately 0.71 ° C/s for Type B luminaire. In comparison, the temperature change rate after the time of 30 s is below 0.01 ° C/s and for Type A luminaire the change rate during the cycled dimming is 0.11 K. The temperature changes close to the LED chips are therefore maximized using the 30 s dimming cycle.

The luminaires were measured on average every 5 months using an absolute integrating sphere method for luminous flux, in total 12 times during 4.5 years. During the measurements, the electrical power consumption and the spectral radiant flux distributions of the luminaires were also measured. The electrical power of Type B luminaires was set to 70% using the DALI controlling to keep the temperature of the luminaire inside the closed integrating sphere below the manufacturer limit.

3. Results

Based on the measurement results, the luminaires with cycled dimming have lower lifetime expectancies as compared to the ones that were not dimmed. Four of the Type B dimmed luminaires broke down during the study, and only one operating normally. Two of the failed units broke down after 20 000 hours and the third failed before 28 000 hours. The fourth luminaire was visually observed to be broken after the last measurement round at 28 000 hours.

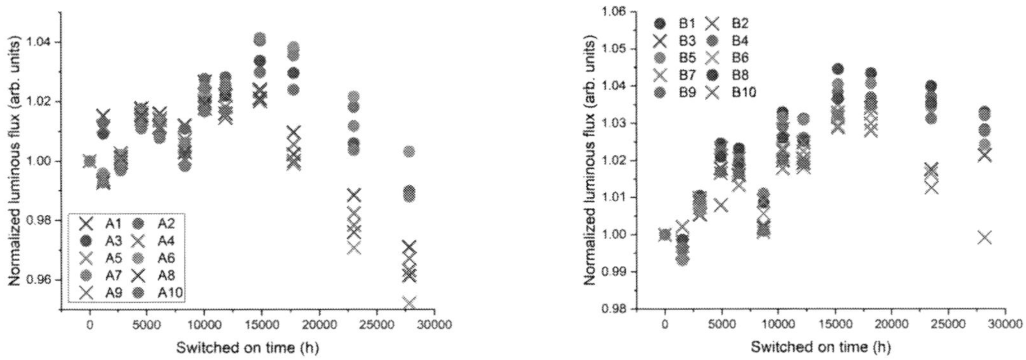

(a) Type A luminaires (b) Type B luminaires

Figure 2. Normalized luminous flux values of the 30-s cycled (cross) and naturally (circle) aged luminaires. The luminous flux of luminaires B2 and B6 has collapsed after 20 000 burning hours and luminaire B7 failed before 28 000 hours. Luminaire B3 was visually observed to be broken after the last measurement round.

The normalized luminous flux values of the two luminaire types are presented in Figure 2. At the measurement round of 28 000 hours, for both luminaire types the luminaires that were dimmed repeatedly and still operating normally had decreased on average 2–3% more in luminous flux than the ones that were not dimmed.

For all the studied luminaires, the luminous flux increases up to 5% during the first 15 000 hours of operation. In the results after 15 000 h, one can see that the luminous fluxes of the cycle dimmed luminaires are lower than those of the undimmed luminaires, indicating that the cycled dimming has a negative effect on the luminous flux. The standard deviation of the different luminaires of the same ageing group at each measurement round is within the uncertainty of the measurements. Estimating the lifetime after 15 000 operating hours with exponential decay shows that the undimmed Type A luminaires would be below 80% of their initial luminous flux levels after 75 000 hours. For the 30-s cycle dimmed luminaires, the corresponding lifetimes would be shorter, i.e., 64 000 h.

The luminous flux levels of the failed luminaires of Type B are well below 50% of the initial luminous flux values and both of the LED panels in the luminaires are still dimly lit. Based on this we can conclude that eight separate LED panels have failed in four different luminaires because of the cycled dimming.

Figure 3 presents the electrical power of Type B luminaires during the integrating sphere measurements. An increase of 10% in electrical power for the failed luminaires was measured on the last measurement round where the luminaires were still emitting light correctly. The electrical power of the Type B cycle dimmed luminaires gradually increases, and within 10 000 hours from the first observation of increased power, an average luminaire breaks down. No abnormalities in the measured electrical power values were noticed for Type A luminaires.

4. Conclusions and discussion

Based on our study, the adaptive control of smart lighting might have a negative impact on the lifetimes and failure rates of high power LED luminaires, resulting in higher maintenance rates and energy consumption than anticipated.

Using a 30-s dimming cycle, almost all the studied luminaires of Type B broke down before 1/3

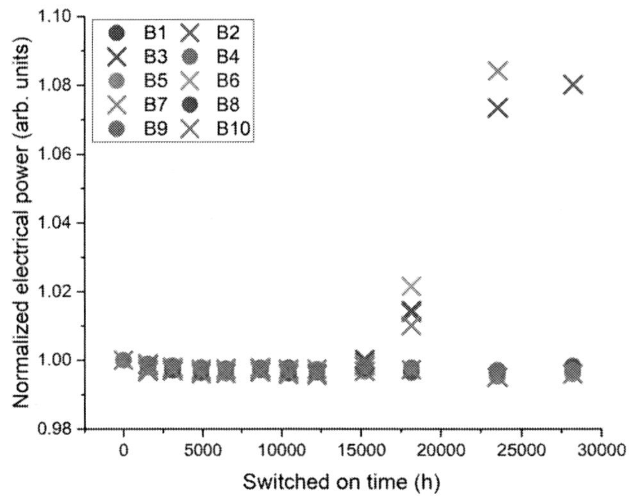

Figure 3. Normalized electrical power of Type B luminaires during the integrating sphere measurements. Luminaires B2 and B6 failed after 20 000 burning hours and luminaire B7 failed before the measurement round at 28 000 hours. As visually observed, luminaire B3 malfunctioned right after the last measurement round. Luminaires marked with crosses are aged with 30-s dimming cycle. The naturally aged luminaires are marked with circles.

of the specified lifetime. The cycle dimmed luminaires of Type A decreased in luminous flux more than the traditionally controlled luminaires, and the estimated lifetimes were approximately 15% lower for the cycle dimmed luminaires.

The lifetime expectancies of 50 000 to 100 000 hours of operation need in minimum approximately 8 000 to 17 000 hours of measured data by the standard of IES [1]. Our results show that the luminous flux for the studied luminaires increases initially, being at the maximum after 15 000 hours of operation, and the differences due to the cycled dimming are seen only after this point. To obtain more accurate lifetime expectations for the luminaires, an extended ageing period from 15 000 hours is needed.

The typical lifetime expectancies specified by the manufacturers for the luminaires in the market are approximately 20% lower at the moment, as compared to the luminaires used in the research. In addition to the lifetime expectancies, the manufacturers should specify the number of switching on times for their luminaires, which would imply the acceptable level of smart controlling.

The increasing electrical power of the luminaires before abrupt failure might be due to the failure of individual LED chips in the panels that results in increasing leakage current. In LED panels, the LED chips are typically connected in series, which results in failure of the whole panel if even one chip is constraining current. Similar failure mechanisms have been reported earlier for E27-base LED lamps [6].

Another possibility for the failure mechanism is that the thermal stress introduced by the 30-s dimming cycle wears out the driving electronics of the LED panels. The electronics are molded in epoxy and therefore finding the specific failing component is difficult.

We have reported that the short-term energy savings with smart control dimming might result in shorter lifetimes of the devices, and possibly in higher failure rates. Therefore, in smart controlled lighting, a compromise between the switched on time and the energy savings should be sought. Manufacturers should also specify maximum number of switched on times for their products to assist designing optimal smart controlling. The luminaire ageing study is still ongoing and further results are expected to be obtained.

References

[1] Illuminating Engineering Society of North America. Projecting Long Term Luminous Flux Maintenance of LED Lamps and Luminaires IES TM-28-14. New York: IESNA, 2014.
[2] Vaskuri, A., Kärhä, P., Baumgartner, H., Kantamaa, O., Pulli, T., Poikonen, T., & Ikonen, E (2018). Relationships between junction temperature, electroluminescence spectrum and ageing of light-emitting diodes. Metrologia, 55, S86–S95.
[3] U.S. Department of Energy (DOE) Report. Hammer testing findings for solid- state lighting luminaires `https://www1.eere.energy.gov/buildings/publications/pdfs/ssl/hammer-testing_Dec2013.pdf`. 2013.
[4] Baumgartner, H., Renoux, D., Kärhä, P., Poikonen, T., Pulli, T., & Ikonen, E. (2016). Natural and accelerated ageing of LED lamps. Lighting Research & Technology, 48(8), 930-942.
[5] Cai, M., Yang, D., Zheng, J., Mo, Y., Huang, J., Xu, J., Chen, W., Zhang, G., & Chen, X. (2016). Thermal degradation kinetics of LED lamps in step-up-stress and step-down-stress accelerated degradation testing. Applied Thermal Engineering, 107, 918-926.
[6] Vaskuri, A. (2018). Spectral Modelling of Light-Emitting Diodes and Atmospheric Ozone Absorption. Aalto University publication series Doctoral Dissertations; 202/2018. ISBN:978-952-60-8243-1

IOP Conf. Series: Earth and Environmental Science **352** (2019) 012044 doi:10.1088/1755-1315/352/1/012044

Exploring the tools and methods to evaluate influence of social groups on individual occupant behavior with impact on energy use

M K Annaqeeb[1,3], J W Dziedzic[1], D Yan[2] and V Novakovic[1]

[1]Department of Energy and Process Engineering, Norwegian University of Science and Technology, Trondheim Høgskoleringen 1, 7491, Norway

[2]School of Architecture, Tsinghua University, Beijing, 30 Shuangqing Rd, Haidian Qu, Beijing Shi, China, 100091

[3]Author to whom any correspondence should be addressed

masab.k.annaqeeb@ntnu.no

Abstract. One of the key elements in driving the energy performance of buildings has been recognized as occupant behavior (OB). However, available tools for assessing and simulating occupant behavior are based on fixed schedules and aggregated profiles, which fail to capture the diversity of OB. A significant aspect of OB is its relation to social groups and their influence and interdependence on each other. The data regarding the influence of social groups is important to achieve an effective model of OB as it accentuates the individual OB profile based on the influences it can have from the social groups they belong to. This added module is not present in traditional building simulation tools. This study aims to explore the tools and methods to evaluate the factors that are responsible for the influence of social groups on individuals' energy-related behavior. The paper investigates the kind of data sets needed for understanding this interdependence, including the occupant's social group, their standing in the group, and the intent behind different actions and its comparison to the actions the individual would take without any external influence. The results will be used to construct questionnaires, which can prove beneficial in developing social group profiles in OB models.

1. Introduction

Current statistics on worldwide energy consumption indicate that more than one-third of the expenditure comes from residential and commercial buildings [1]. Reportedly, this consumption would have been 40% higher if not for the implementation of energy-efficient technologies, but despite the technological advancement, the challenges in energy reduction are enormous, and these are compounded with the increasing global pressure and concerns to meet the worldwide goals for this reduction [1]. In addition, the disparity between the expected and the actual energy performance [2] [3]of buildings is another cause for concern, and it leads to a need of having a more innovative and holistic approach to bridge this gap and adapt to the energy demands in an efficient manner.

However, occupants are an integral part of the building systems, and due to the dynamic nature of occupant interactions with these systems, building sectors become much more complex in terms of their energy performance. While there can be several other factors involved in the contribution of the performance gap, most notably, faults in the mechanical and electrical systems, malfunctioning of equipment etc. [4], occupant behavior(OB) has been identified as one of the major drivers of this gap [5]. This trend has also been reflected in the fact that, over the last 30 years, the research community has shifted a lot of its focus towards OB [6]. This behavior, then becomes a deciding factor when it comes to the variability in the comfort settings and the implications it has on the energy usage.

In that context, there is a marked dissonance when it comes to the building design and performance simulations as well. Low-energy, or zero-energy buildings, are designed to comply with high-efficiency standards, which are vital for reducing carbon emissions, and these designs are dependent on an active occupant interaction with the building systems. However, the predictions of the energy performances result in an even larger error for these low-energy buildings [7]. This can be attributed to the actuality that the potential for achieving that high level of efficiency in these buildings is reliant on the building being operated in a specific way according to the design. Another side of this problem is also the rigidity in the building performance simulation (BPS) process, wherein OB is most often considered as a fixed or schedule-based model [8]. Earlier BPS tools utilized deterministic inputs to represent OB over the life cycle of the building, which took the form of daily, weekly, and monthly schedules. The main advantage of these inputs was the ease of use and simplicity of the models. However, more and more BPS tools have started incorporating stochastic models to capture the diversity and include the uncertainties which arise in OB [9].

The need to incorporate an interdisciplinary approach is based on this human-building dynamic, and this approach can be viewed at three different levels: the individual occupant, the group behaviors (building level), and the collective behaviors (district scale) [10]. In these terms, a multi-scalar approach, one that deals with different zones and interdisciplinary considerations of occupants' activities (habits, movement, presence, etc.) will be necessary for both the data collection and the simulation processes, and this will involve an amalgamation of engineers, architects, social scientists etc. Along these lines, Dziedzic et al. proposed a Building Occupant Transient – Agent-Based Model, which would contain several different modules for the different aspects of OB [11]. This would include distinct, individual modules in the simulations to account for the movement, habits, surroundings, and social structure.

This added modules for the social structure and influences is often neglected, since a lot of focus tends to be towards measuring the physical occupant interactions, but not the socio-psychological drivers of that behavior. Researchers have often been sensitive to the cost and technological availability, and largely ignored accompanying factors such as social equity, group status, peer influence etc. Sovacool et al. identified three main negative patterns from the energy research literature over the last 15 years and the first of them was that the social dimensions are under-examined, especially those pertaining to perceptions about energy use, and decision-making process among individuals, organizations, and communities [12].

2. Social group influences on individual occupant behavior

2.1 Occupant social groups and networks

Measurement of social influences is often conducted through the evaluation of social groups. These social networks, or groups, comprise of the group of people that occupants have to interact or share the space with. This is applicable to both residential and commercial buildings. It is an umbrella term which includes distinct nodes, where the nodes would represent the number of people in a particular group (size), the strength of the relationship between the members (connection), and in some cases, they represent events and ideas that affect the group as well [13]. These networks have been documented has having a significant effect on the energy-us behaviors of occupants [14] [15] [16]. While the framework of this paper has its focus on commercial buildings and workplaces, the intended scope is to recognize the kind of behaviors needed to be studied to collect the datasets, and that data collection can be adapted to residential buildings as well, by modifying the kinds of questions asked within the data collection surveys.

2.2 Theory of planned behavior

In order to quantify these influences and behaviors, literature in energy research and social science relies heavily on the *theory of planned behavior* [13] [18]. The theory states that there exist three factors behind the behavioral intentions and behaviors; attitudes and beliefs towards the action, beliefs about how significant others may respond to the action (*subjective norms*), and the *perceived behavioral control* (i.e. beliefs about the ease of performing the action). These three factors have been taken as the foundation in this study, to build the framework for the social structure module in OB modeling.

Abrahamse and Steg used the theory of planned behavior to explain subjective norms in residential households, that is, behaviors that would rely on the extent of importance other members of the household, and the social pressure to carry out the actions [19]. For instance, changing the thermostat settings as an energy saving measure might not be carried out if the family members disapproved of those settings. In addition, when it comes to workplaces, or commercial buildings, the utilization of energy is integrated with the group behaviors, since the equipment and facilities are often shared between co-workers. Chen and Knight used the same theory to examine the influence of colleagues among more than 500 employees in electric power companies in China, and the results indicated that the approval or disapproval from the colleagues was the major factor influencing the individual's energy saving actions [18].

Streamlining the study of these influences can also take the form of studying the social networks in relation to energy use. Anderson et al. used an agent-based model to investigate different types of social networks in buildings and the extent of their significance on administrative actions designed to reduce energy use [14]. The results indicated that accurate representations of the social networks, in relation to the historical data of the administrative actions, are vital in predicting energy use. Chen et al. also made use of agent-based modeling to build a network level computational model that simulates the decision-making process of individuals under different type of network configurations [15].

3. Framework for incorporating social influences in OB modeling

While quite a few studies have highlighted the impact of this social influence, not much research has been done on integrating this aspect in the OB simulation process. Anderson et al, did manage to conduct simulations based on occupant's susceptibility to social influences depending on their position in the social network, and the model also took into account the corresponding peer influence on their energy-use behavior [16]. However, this topic warrants a lot more study, especially in the wider context of how it fits into the overall OB profile.

3.1 General Principles

This framework takes into account all the above-mentioned factors that have been determined to be the drivers of the social influences in OB, combined with the principles of the theory of planned behavior, and dictates the datasets required from the framework's perspective. Since a major part of the topic has its focus on the social networks, this framework will involve the creation of a database that details the occupant's involvement and interaction with their respective networks. This would be a representation of the *subjective norms,* and will include a mapping of the network grid around the workplace, the number of people in the network that count as significant enough to influence the behavior, and the rank of the occupant in their networks. Considering the sharing of facilities in workplaces, the database also has to involve these interactions in designated areas, such as the common areas, lunchrooms/breakrooms, shared office spaces etc.

Regarding the *perceived behavioral control* (PBC) and the attitude/beliefs of the occupant towards different actions, gathered data has to include their perceptions about the energy-use, concerns for reducing that usage, and awareness of facilities or obstacles that accommodate a specific action. The collection of these datasets can then be utilized by connecting them to different energy-use behaviors such as thermal control, lighting control, ventilation etc.

3.2 Required Datasets to construct the module

As seen in figure 1, the datasets can be divided according to the principles, and the combined effect be correlated to the energy-related behavior using weighed factors that are derived on the basis of the extent of each parameter's perceived influence. For subjective norms, this influence would be the variability in the action, with and without the external influence. A general description of these is summarized as follows:

- Perceived behavioral control (PBC): Occupants perceptions about their own ability to conduct the action, awareness of the kinds of facilities that are available to ease the performance of the action, or obstacles that may hinder it. This could come in the form of technological aids present at the workplace, administrative actions that incentivize certain behaviors etc.
- Subjective norms: These factors will be tied in with each of the spaces that involve the interaction within the social network, and the factors are as mentioned in 3.1
- Beliefs/Attitudes: These would investigate the occupant's personal beliefs and perceptions about the action.

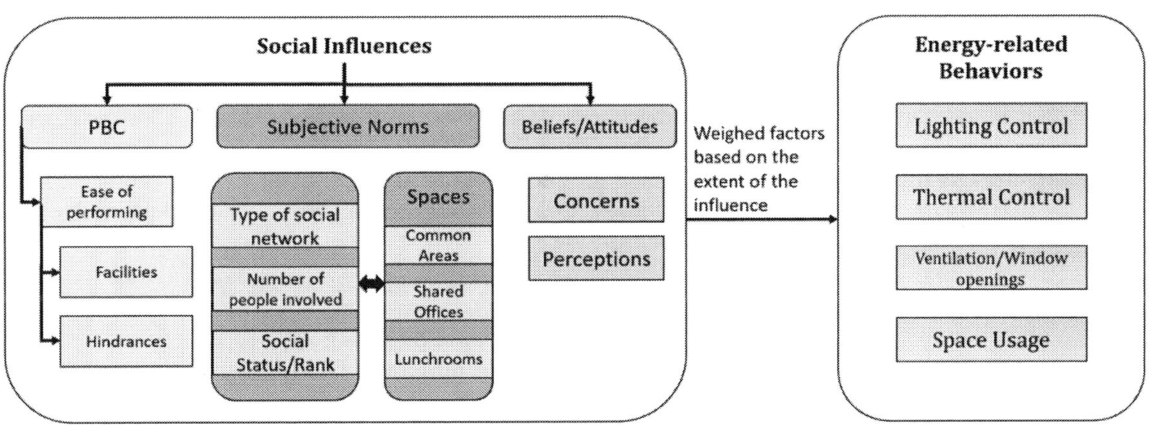

Figure 1. Schematic of datasets required.

3.3 Survey Methodology

There is a growing trend in adapting survey methodology to explore the nuances in OB. The reason behind this is that surveys still remain the most effective method for measuring variables that cannot be monitored or observed directly such as the intentions behind the energy-related behaviors, concerns, and perceived norms (latent variables) [20] [21]. However, surveys carry their own share of risks, and doubts regarding the validity, since discrepancies are often present between the reported and actual behaviors. To minimize these risks, several factors have to be taken into consideration during the design process of the survey/questionnaire. In addition, in order to incorporate the beliefs and motivations of the occupants' behind their actions, quantitative survey methods may not be sufficient [22], and qualitative interviews with the occupants might have to be combined with them to give a better understanding of the behaviour. Since the focus and scope of this work is towards OB modelling, the qualitative step will be taken into consideration after the initial results of the quantitative survey, in order to identify the factors that were insufficiently covered, and achieve better design of the process.

When measuring social and psychological parameters, it is necessary to avoid using absolute responses or brief scales, and instead adopt combined items and contextually large scales based on established theories. Vague quantifiers need to be avoided and equal number of choices should be provided on both the positive and negative sides. Apart from these, other general guidelines for designing questionnaires should be implemented as well, such as the use of simplified words over specialized ones, general aesthetic, and response that are mutually exclusive.

4. Discussions and conclusions

Energy research literature gives us definitive insights about the impact of social influences on occupant behavior, and the theory of planned behavior is an effective tool to quantify these influences. This paper brings out the necessity of considering these factors in conjunction with the energy-use behaviors to obtain a better approach to OB modeling. The guidelines and types of datasets outlined in this paper will be used to construct questionnaires and gather the required data. Results from the data collection will be vital for constructing the social modules in OB modeling, in addition to providing insights about general behavioral intentions and norms. However, most limitations here are due to the discrepancies in the reported behaviour. Special care has to be taken to avoid any kind of sampling errors, response bias, or reliance on occupant's estimations based on long-term memory.

References

[1] International Energy Agency, *World Energy Outlook* 2018 (IEA Publications)

[2] Azar E and Menassa C C 2012 A comprehensive analysis of the impact of occupancy parameters in energy simulation of office buildings *Energy and Buildings* **55** 841-53

[3] Menezes A C, Cripps A, Bouchlaghem D and Buswell R 2012 Predicted vs. actual energy performance of non-domestic buildings: Using post-occupancy evaluation data to reduce the performance gap *Applied energy* **97** 355-64

[4] Turner C and Frankel M 2008 *Energy performance of LEED for new construction buildings* (Washington DC: New Buildings Institute)

[5] Yan D, Hong T, Dong B, Mahdavi A, D'Oca S, Gaetani I and Feng X 2017 IEA EBC Annex 66: Definition and simulation of occupant behavior in buildings *Energy and Buildings* **156** 258-70

[6] Hong T, Taylor-Lange S C, D'Oca S, Yan D and Corgnati S P 2016 Advances in research and applications of energy-related occupant behavior in buildings *Energy and buildings* **116** 694-702

[7] Parker D, Mills E, Rainer L, Bourassa N and Homan G 2012 Accuracy of the home energy saver energy calculation methodology *Proc. of the 2012 ACEEE summer study on energy efficiency in buildings (Florida)* (Lawrence Berkeley National Lab Report)

[8] Hoes P, Hensen J L, Loomans M G, De Vries B and Bourgeois D 2009 User behavior in whole building simulation *Energy and buildings* **41(3)** 295-302

[9] Yoshino H, Hong T and Nord N 2017 IEA EBC annex 53: Total energy use in buildings—Analysis and evaluation methods *Energy and Buildings* **152** 124-36

[10] Hong T, Yan D, D'Oca S and Chen C F 2017 Ten questions concerning occupant behavior in buildings: The big picture *Building and Environment* **114** 518-30

[11] Dziedzic J, Yan D and Novakovic V 2017 Occupant migration monitoring in residential buildings with the use of a depth registration camera *Procedia Engineering: 10th Int. Symp. On heating, ventilation and air conditioning (Janin)* vol 205 (Janin: Elsevier) p 1193-200.

[12] Sovacool B K, Ryan S E, Stern P C, Janda K, Rochlin G, Spreng D, Pasqualetti M J, Wilhite H and Lutzenhiser L 2015 Integrating social science in energy research *Energy Research & Social Science* **6** 95-9

[13] Serrat O 2017 *Social network analysis*. In *Knowledge solutions* (Singapore: Springer) p 39-43

[14] Anderson K, Lee S and Menassa C 2012 Effect of social network type on building occupant energy use *Proc. of the Fourth ACM Workshop on Embedded Sensing Systems for Energy-Efficiency in Buildings (Toronto)* (New York: ACM Digital Library) p 17-24

[15] Chen J, Taylor J E, Wei H H 2012 Modeling building occupant network energy consumption decision-making: The interplay between network structure and conservation *Energy and Buildings* **47** 515-24

[16] Anderson K and Lee S 2013 Modeling occupant energy use interventions in evolving social networks *2013 Winter Simulations Conference (Washington DC)* (Washington DC: IEEE) p. 3051-305

[17] Scherbaum C A, Popovich P M and Finlinson S 2008 Exploring Individual-Level Factors Related to Employee Energy-Conservation Behaviors at Work *Journal of Applied Social Psychology* **38(3)** 818-35

[18] Chen C F and Knight K 2014 Energy at work: Social psychological factors affecting energy conservation intentions within Chinese electric power companies *Energy Research & Social Science* **4** 23-31

[19] Abrahamse W and Steg L 2011 Factors related to household energy use and intention to reduce it: The role of psychological and socio-demographic variables *Human Ecology Review* **18(1)** 30-40

[20] Holopainen R, Tuomaala P, Hernandez P, Häkkinen T, Piira K and Piippo J 2014 Comfort assessment in the context of sustainable buildings: Comparison of simplified and detailed human thermal sensation methods *Building and environment* **71** 60-70

[21] Steemers K and Manchanda S 2010 Energy efficient design and occupant well-being: Case studies in the UK and India. *Building and environment* **45(2)** 270-8'

[22] Bryman A 1988 *Quantity and quality in social research* (London: Unwin Hyman)

IOP Conf. Series: Earth and Environmental Science **352** (2019) 012045 doi:10.1088/1755-1315/352/1/012045

Reanalysis of occupant experiment in ZEB Living Lab

K S Skeie[1], M S Korsnes[2], R Woods[2]

[1] Department of architecture and technology, NTNU
[2] Department of Interdisciplinary Studies of Culture, NTNU

kristian.skeie@ntnu.no

Abstract. In 2015-2016, a six-month long experiment was carried out in the ZEB Living Lab, one of the first houses in Norway planned to reach net zero emission throughout its lifetime. By means of qualitative experiments, social scientists evaluated how six groups of occupants impacted the zero emission building, and how the zero emission building impacted its occupants in different ways. Data were collected through direct observation, diaries and interviews before, during and after the stay. In this paper, we take a closer look at the sensor data from the detailed monitoring system, generated during each group's 25-day long residing's. We find that despite the experiment being a controlled environment with many shared household characteristics, occupancy is determined by diverse factors, and the users' preferences and attitudes influence energy consumption. All major household appliances were the same, but the time of use, frequency, duration and type of use differed. The study contributes to increasing understanding of user behaviour and energy use in low energy houses. Providing occupancy profiles is valuable, in light of the increasing amount of local renewable energy production. These datasets can be used as input in energy simulation and planning of zero-emission neighbourhoods.

1. Introduction

The experiment with six occupant groups in the ZEB Living Lab was conducted October 2015 – April 2016. Data were collected by mixed-methods through direct observation, diaries and interviews before, during and after the respective stays of 25 days. Sensor data from the detailed monitoring system (temperature, humidity, CO2 levels, light levels, motion, energy use, airing), was presented to the users in interviews but has thus far not been analysed further.

In this paper, we revisit the written logbooks and analyse the sensor data using statistical techniques in order to gain better understandings of two issues: 1) Whether the self-logged activities correspond to the highly-energy consuming activities, and 2) whether there are significant differences between groups concerning energy use and peak power consumption. This second endeavour could shed light on the role of occupant life phase, age, and family situation, on the design of zero-emission arrangements.

Other research from the experiment studies domestication of the zero emission technologies [1, 2], and how a sense of home was established in the Living Lab [3, 4]. Feedback from the occupants has also been used to explore the concept that people living in the house will act as both producers and consumers of energy [5]. In these studies, opportunities and limitations of the living lab setting are also addressed. The qualitative data from interviews and observations are at the centre of analysis, leading to insights on the expected and unexpected uses of the technologies, but without further quantitative comparisons between groups. Some of the home living habits of the six groups concurred within the zero-emission ambitions, and some may lead to higher energy use than expected [2]. In this paper, we look into these previous findings by laying out the available sensor data.

However, the data gathered on energy use are not fully comparable between groups. Gaps frequently occur in the measurements, as indicated in the total equivalent days of 30-second monitoring data available per group (Table 1). The influence on heating need is not easily identified, as the outdoor temperature varies and the residencies are less than a month long. Moreover, the experiment was conducted right after construction. All technologies and sensors related to the heating system were not completely calibrated at the time of the experiment. Therefore, the analysis is limited to direct electricity use and domestic hot water use.

The six different user groups were composed as follows: two student groups, two families with small children and two elderly couples. Each group was asked to make the Living Lab their home, trying to use it as they would normally use their own houses. The average time spent at home differed between groups (and within the groups). Table 1 shows the average total hours per day (between Monday to Fridays) that no movement was detected in the house for an entire hour (see Figure 8).

Table 1. Overview of the resident groups, the data available for analysis and average time at home.

Group	Description	Analysis period	Unoccupied house
S1 Students	Male and female couple, 22 years old. Live in a 52 m^2 student apartment	18.5 weekdays[a]	5 hours per weekday
S2 Students	Two female friends, 20 and 21 years old. Live in a shared apartment together with three other girls, built 1905	12.3 weekdays[a]	8 hours per weekday
F1 Family	Mother 31 years old and father 36. Son 6 years old and daughter 2. Live in a row house of 185m^2, built 2007	18.3 weekdays[a]	8 hours per weekday
F2 Family	Mother 31 years old and father 37. Two daughters of 3 and 2 years old. Live in a detached house of 135 m^2	15.5 weekdays[a]	7 hours per weekday
E1 Elderly	Husband 81 and wife 68. Live in a detached house of 170 m^2	14.5 weekdays[a]	3 hours per weekday
E2 Elderly	Husband 61 and wife 56. Live in a detached house of 120 m^2	17.5 weekdays[a]	6 hours per weekday

[a] Total equivalent duration of monitoring data recorded at 30-second interval available for the analysis.

2. Method

The ZEB Living Lab is a single family house with a gross volume of approximately 500 m^3 and a heated floor area of approximately 100 m^2. The monitoring system is designed based on the experimental data needed to characterise the energy and environmental performance of the building fully. A description of the test facility highlighting the architectural features and technological aspects, including the specification and quality aspects of the measurement system is published in [6].

The analysis is concentrated on weekdays (Monday to Friday) to identify and classify variables related to occupancy profiles. Weekends are filtered out to look specifically at building occupancy and use patterns that emerge from daily activities. This choice also has a practical dimension as data acquisition losses occurred, most frequently on Fridays lasting until a restart on Monday mornings.

The variable categorisation that guides the analysis is taken from IEA EBC Annex 66 "Definition and Simulation of Occupant Behavior in Buildings" [7]. These reference procedures for obtaining occupancy profiles in residential buildings differentiates between variables that influence occupancy profiles such as socio-demographic driving-factors and variables influenced by occupancy impacting the use of equipment, building services and energy consumption [7].

First, data from the electricity meters are presented together with short descriptions from the diaries, outlining the routines in the house. We then investigate the variables influenced by occupancy and present them in two sections: The type of activity, the hours of presence in the different rooms. The use and description of lighting, appliances, cooking and demand for hot water.

3. Daily activities and energy use in the living lab

The occupants kept a logbook during their stay in order to get an overview of the routines in the house. From the logbooks, we recount (in Figure 2 – Figure 8) the type of activities undertaken, such as cooking, laundry, and dinners. The logged activities presented in the diagrams were selected to explain the events that can be seen in the hourly time of day electricity use. We found that occupants logged most activities, but not every single use such as dishwasher, or shower. I the figures below, some logbook entries are included of activities on days where there are gaps in the electricity measurements.

The guest visits occurred frequently among all the groups, and the time of day when the house was in use, cannot be presented here, but are discussed in the next section. During any given day, the number of visits and time of leaving, or home-comings of one or more building occupants, are too numerous to recount in these graphical representations.

On the left side a bar chart show the daily electricity consumption for different use. Then two rows show the hourly and 15-minute daily power load. Next to it is a heat map that show the hourly energy use horizontally day by day. A deep orange colour indicates higher energy consumption (>3 kWh/h). Finally, selected activities from the daily activity logs written by the occupants are listed to the left.

3.1. Students S1 & S2

Both student groups studied close by, on the same campus. The students found that the location of the living lab, on campus, altered how and where they spent their day. Both group S1 and S2 often ate lunch in the house, and S1 would spend more time at home than usual (19 hours per day on average). The activity diary and electricity use that occurs in the daytime show the variation between days. Friends would also come over frequently. S1 brought a vacuum cleaner and its energy use is visible on the time of use (Figure 1). The other groups used a smaller chargeable vacuum cleaner that was provided.

Figure 1. Energy use and daily activities of Student group 1.

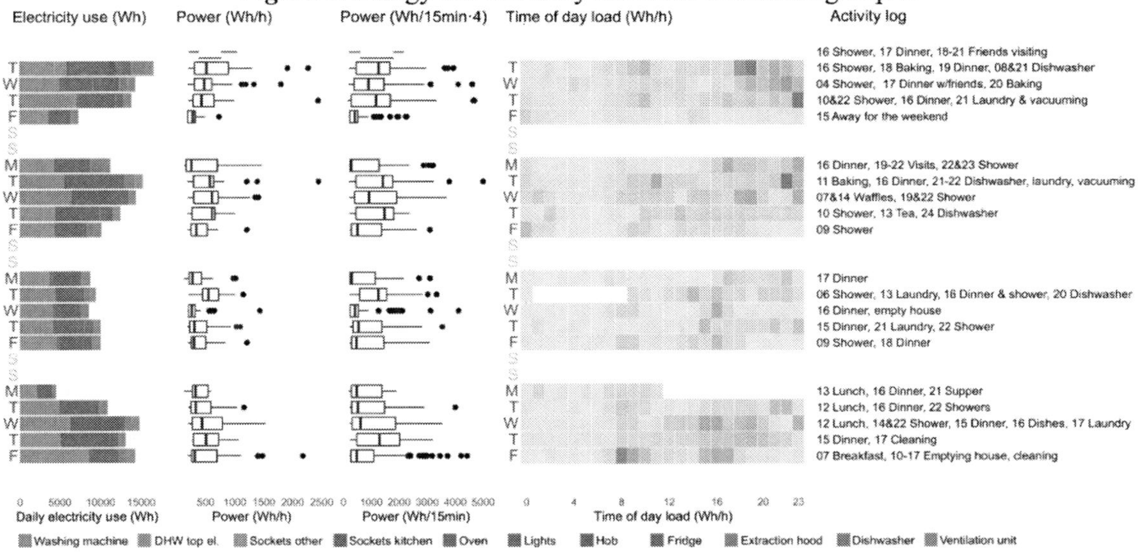

S2 Spent most of the daytime away from the house (8 hours unoccupied on average), but would sometimes come back for lunch or spend half the morning or afternoon there. The two students had some shared and many separate routines, such as eating and food preparation in the kitchen. Figure 2 shows the main energy use and power on an hourly basis to originate from showering. Unfortunately, many days are missing metering data, but the log provides an indication of the use on days without data.

Figure 2. Energy use and daily activities of Student group 2.

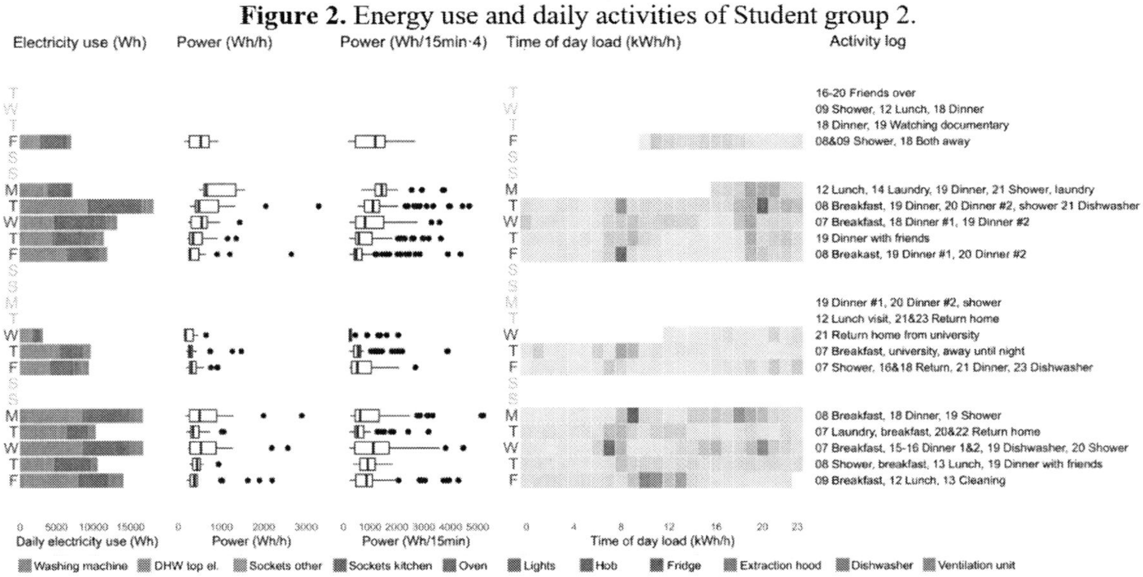

3.2. Families with young children F1 & F2

The two families have the most pronounced occupancy patterns, clearly indicating the time away to approximately 8 hours per work day and the electricity use for dinner preparation in the afternoons. Both families experienced sickness during their stay, which can be seen from the electricity use heat map. The first family F1 brought a portable electric radiator (1 kW), but for the sake of comparison with other groups, electricity consumed by this heater is subtracted from the data presented in Figure 4.

Figure 3. Energy use and daily activities of Family 1.

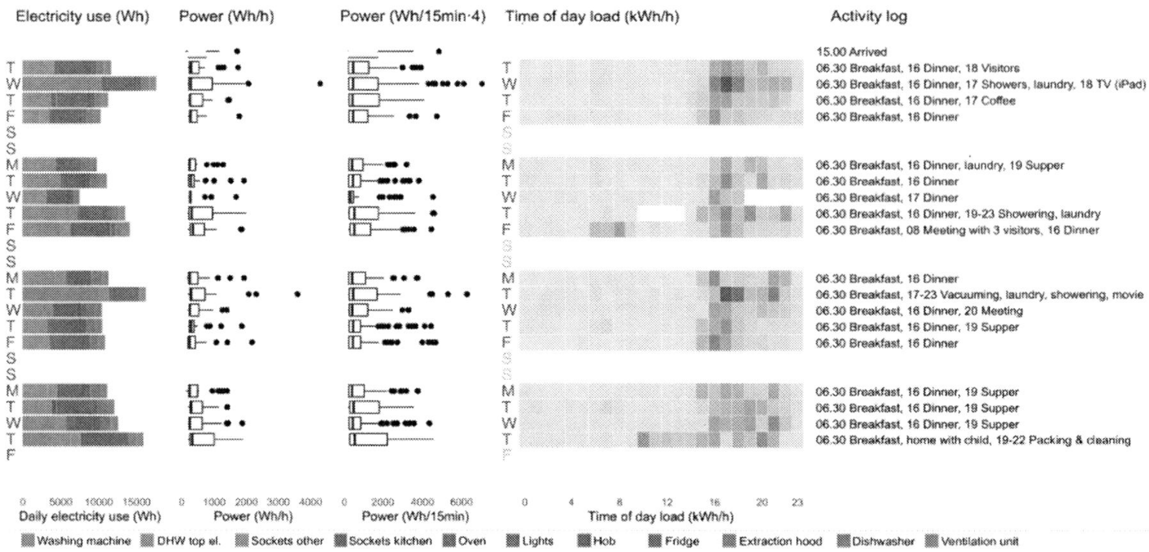

The second family, F2, had an electric car. The electric vehicle (EV) charger is connected to a metered socket in the technical room, drawing 2.15 kW in typical charging cycles lasting 6 hours. The car was left to charge after returning from work, or in the late evening (Figure 5). The bar chart in Figure 5, shows what the daily energy use for the electric car amounts to kWh/day and the box plots show loads.

Figure 4. Energy use and daily activities of Family 2 with EV-charger. Notice the bar plot scaling.

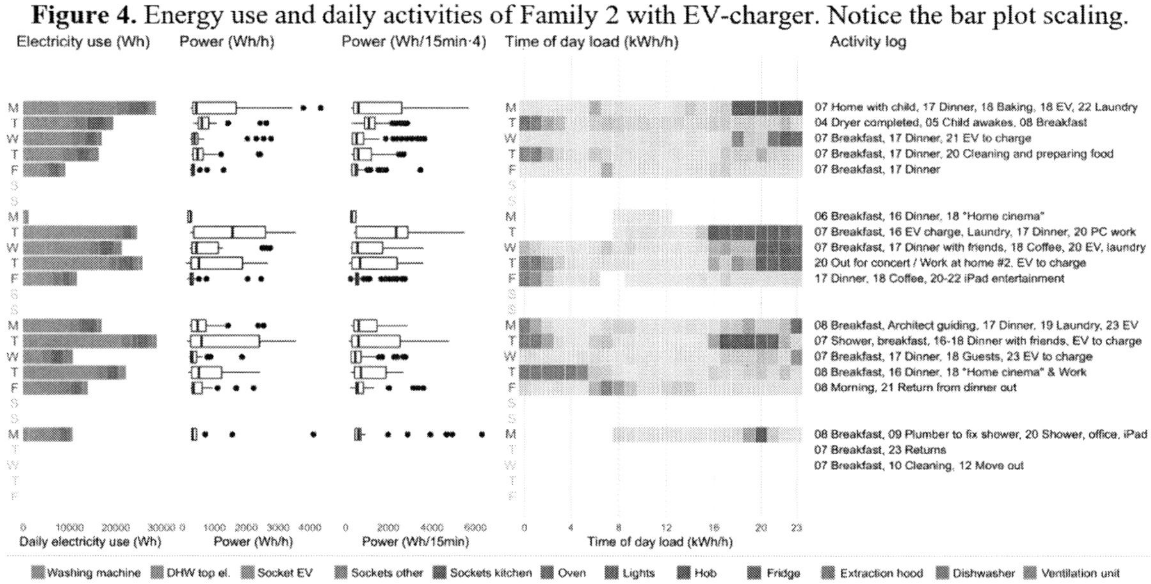

3.3. Elderly couples E1 & E2

The two older couples both held high levels of activity with many guests and family visits. One can see more electricity use in the daytime than for the families with children. E1 would bake frequently and spend more time in the kitchen. During their time in the living lab, the house was vacant approximately 3 hours per day on average, according to the presence detection sensors. Unfortunately, most of the data of week 4 is missing. Periods of the highest energy use originate from the kitchen.

IOP Conf. Series: Earth and Environmental Science **352** (2019) 012045 doi:10.1088/1755-1315/352/1/012045

Figure 5. Energy use and daily activities of Elderly couple 1.

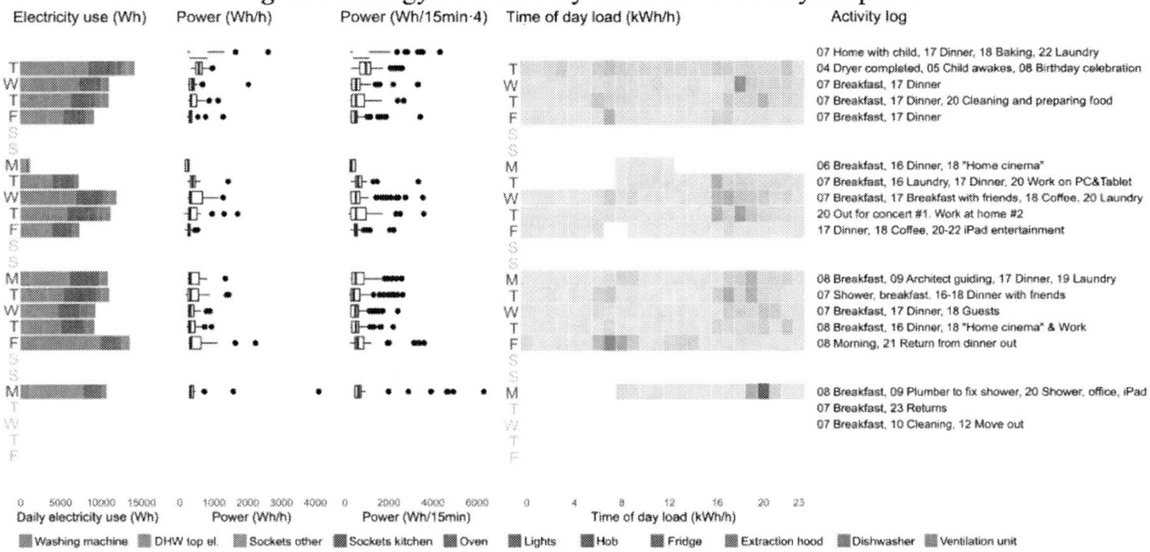

The other couple (E2) was away from the house about 6 hours on an average day. Some data are missing, including a one-hour power outage on Tuesday week 2. E2 was the only group to bring a TV.

Figure 6. Energy use and daily activities of Elderly couple 2.

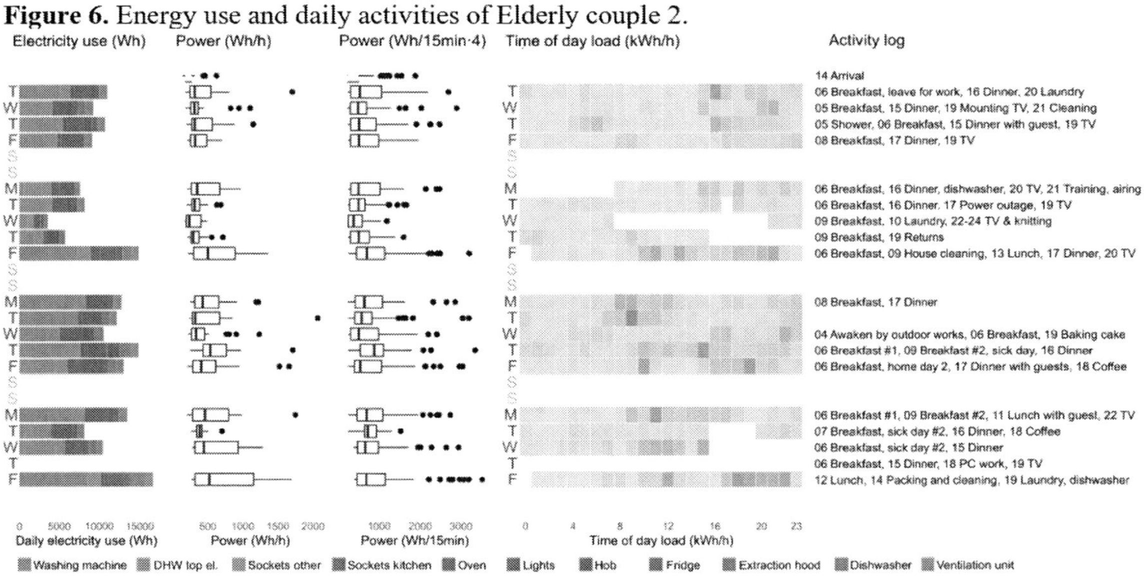

4. Amenities in the living lab

4.1. Electrical equipment

Participants were asked to bring with them what they regarded was necessary to create a home. This included electrical equipment. As larger appliances were in place (Table 3, Appendix), participants brought various smaller electric equipment from home, like a coffee grinder, kitchen machine, electric kettle, toaster, blender, waffle maker, stereo, hairdryer and chargers for tablets and phones. Simple plug-in energy meters revealed how much some of these devices were in use over the total stay (Table 5, Appendix). Sockets were also monitored on room level. The peak power registered for plug loads in each room, presented in Table 2, can largely be explained by the following observations. S1 brought a vacuum cleaner (moved around between sockets), F1 brought a radiator heater (moved between living S and living N sockets), F2 brought an iMac (on "Living N sockets" in Table 2), and E2 brought their TV (living N socket). The various kitchen equipment brought by the group all produces maximum loads of around 0.7 kW over 15 minutes or less over a full hour. The hourly peak value on each meter is well below 1 KW.

The major household appliances were all the same, selected to be energy efficient (Table 3, Appendix). In this case, the peak power between groups is similar. Except, the fridge has a rapid freeze cycle function likely used by the student groups (bringing maximum power to 2 kW over 15 minutes). It also appears the students did not use the oven on the weekdays. The induction hob has a rated power of 7.4 kW max, 2.3 kW per heating zone and 3.2 kW in boost mode (Table 3, Appendix). Table 2, show that over an hour, the peak power of the induction oven was well below 1 kW.

Table 2. Maximum peak power (kW) per 15 minutes and per hour on weekdays.

	Maximum power load in kW (kWh per 15 min · 4 / kWh per 1 hour)					
	S1	S2	F1	F2	E1	E2
Kitchen sockets	0.7 / 0.6	0.7 / 0.3	0.6 / 0.2	0.5 / 0.2	0.7 / 0.4	0.7 / 0.4
Living S sockets	0.6 / 0.1	0.1 / 0.1	1.4 / 1.0	0 / 0	0.9 / 0.3	0.8 / 0.2
Living N sockets	0.5 / 0.2	0 / 0	0.8 / 0.4	0.2 / 0.2	0.1 / 0.1	0.1 / 0.1
Bedrooms sockets	0.1 / 0	0.1 / 0.1	0.1 / 0.1	0 / 0	0.5 / 1	0 / 0
Bathroom sockets	0.2 / 0	0.3 / 0.1	0 / 0.1	0.2 / 0	0.7 / 0	0.2 / 0
EV-charge socket	-	-	-	2.2 / 2.1	-	-
Washer & dryer	2.1 / 1.2	2.0 / 0.6	2.1 / 1.4	2.1 / 0.9	1.7 / 0.6	2.1 / 1.4
Fridge & freezer	2.0 / 0.9	1.8 / 0.7	0.1 / 0.1	0.1 / 0.1	0.1 / 0.1	0.1 / 0.1
Induction hob	1.3 / 0.6	1.6 / 1.0	1.7 / 0.6	1.0 / 0.4	2.4 / 1.1	1.7 / 0.9
Oven	0 / 0	0 / 0	1.8 / 0.8	1.7 / 0.8	2.2 / 1.1	1.7 / 0.7
Dishwasher	1.7 / 0.8	1.6 / 0.6	2.1 / 1.1	1.7 / 0.6	2.1 / 0.6	1.9 / 0.7
Extraction hood	0.1 / 0.1	0.1 / 0	0.1 / 0.1	0.1 / 0	0.1 / 0	0.1 / 0.1
Indoor lighting	0.8 / 0.8	0.8 / 0.7	0.8 / 0.8	0.7 / 0.7	0.5 / 0.4	0.5 / 0.5

4.2. Electricity use for lighting

Earlier studies have not found clear correlations between daylight availability, control of shading devices and the artificial lighting [8, 9]. The LED lighting system has an installed power of 912 W, or 742 W installed indoor, 137 W outdoor and 33 W transformer loss (Table 4, Appendix). Figure 7 shows the average indoor lighting use per hour of the day. The profiles are affected by daily routines, time at home, dimming preferences, and reflect that occupants stayed in the house between October to April. Red coloured bars are main areas (living rooms, entryway & kitchen), yellow bedrooms, and green bathroom.

Figure 7. Mean indoor electricity used for lighting per hour (W), excluding standby power (30 W).

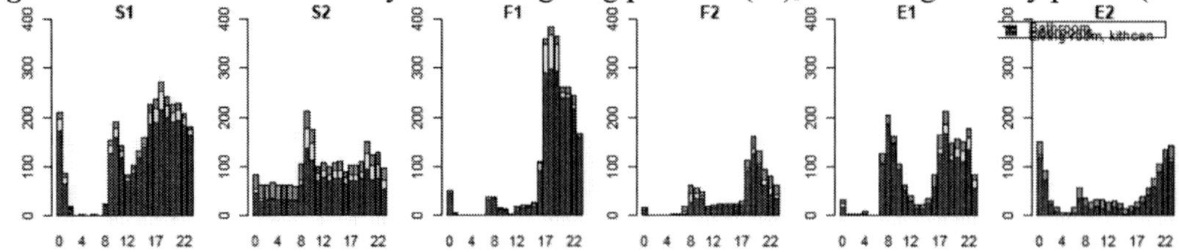

4.3. Domestic hot water use

The hot water accumulation tank was heated by a direct electric resistance heater of 2.7 kW to 70 degrees, which is a typical configuration in Norway. This electric hot water heating was included in the analysis. Due to the design of the heating system in the Living Lab, the total energy use for DHW heating could not be easily identified at the time of the experiment. The incoming tap water was preheated through a coil in the space heating tank which held a tank temperature of 35 to 40 degrees. This space heating tank can be heated by low-temperature sources; solar thermal panels, a ground source heat pump or direct electricity. Without the preheating coil, the electric heater in the DHW accumulation tank would need to operate for a longer time to lift the temperature for DHW. Therefore, the total energy need for hot water is higher than the electricity reported in this study, but the 2.7 kW electric power is representative of hot water tanks heated directly by electricity.

A split between the DHW used for personal hygiene, showers, or other domestic uses has not been attempted from the existing monitoring data, but it is safe to assume that showering is the greatest cause of energy use for DHW heating. Several groups mentioned that they showered longer in the lab because the shower head was nice and comfortable, also they did not pay for electricity.

4.4. Cooking

Table 1 gave an overview of total hours per day where the living lab is unoccupied. Figure 9 shows the day to day hourly variations when no movement is detected overlaid in grey colour. The orange colour in the heat maps shows when there is someone present in the kitchen. Together these two variables, kitchen presence and building vacancy, were found to correlate to the hourly load patterns. On days when the house was not left vacant in the daytime or evenings, more meals would be cooked.

Figure 8. Kitchen occupancy hour by hour on weekdays (presence over an entire hour in dark orange, lighter colour if vacant part of the hour). Hours of in-house movement not detected is overlaid in grey.

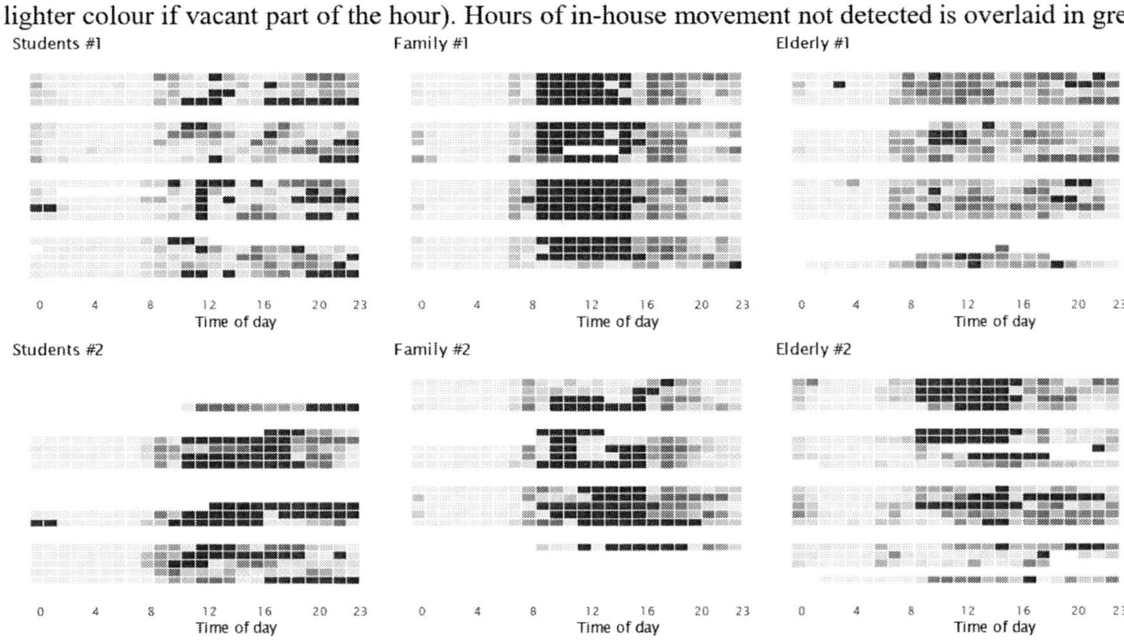

5. Conclusion

The analysis proves that all major explanatory variables are logged in the lab, given that the monitoring system operates. By laying out sensor data descriptively and combining it with the daily activity-logs, we created a starting point for other studies, i.e. to estimate occupancy from CO_2-levels or occupant habit effects on heating need using system identification techniques. The dataset also has the potential for occupancy detection and modelling, for example by using Markov-Chain or other techniques [7].

Despite that the experiment is in a controlled environment with many shared household characteristics, diverse factors determine occupancy, and the users' preferences and attitudes influence energy consumption. All major household appliances were the same, but the time of use, frequency, duration and type of use differed. Excluding space heating from the analysis, the most energy consuming activities are related to showering, cooking and EV-charging. Kitchen presence and building vacancy were found to correlate with electricity use, translating into the time spent at home and the meals cooked there, both varying on a day to day basis. Of the self-logged activities, cleaning also to some extent were of the more energy consuming activities, be it a combination hot water draws for handwashing dishes, vacuum cleaning or use of the dishwasher, laundry machine that according with the logbook coincided with food preparation or cleaning in the kitchen. Some patterns were expected, – families with small children and seniors had more stable routines. These patterns may have been more clear over longer periods and could be used to plan new experiments.

There are differences in energy use between the groups, but are they significant? The time of use on the course of the day differs significantly between the groups and from day to day, especially depending on whether someone is home in the daytime. This is an important consideration in the planning and operation of a ZEB building with solar PV aiming to optimize self-consumption, or a battery storage solution. The peak power per 15 minute or 1 hour; however, is not very different between groups, except in the case were additional space heaters, or electric cars are brought in. The magnitude of peaks, for most household appliances, are less than 1 kWh/h. Combined loads (represented by box plots) were below 4 kWh/h for all groups including EV charging and electric hot water heating (excluding space heating). Over an hour, this is also true for how the induction hob was used. Despite the rated maximum power of 7.4 kW, or 2.3 kW per heating zone, the highest load over an hour was 1.1 kWh/h. The study contributes to understanding user behaviour and energy use in low energy houses. In light of the increasing amount of local renewable energy production, providing occupancy profiles is valuable, to be used as input in energy simulation and planning of zero-emission neighbourhoods.

References

[1] Korsnes M S, Berker T and Woods R 2016 Compliance and Deviation: How occupants interact with a high performance zero emission building. In: *DEMAND Centre Conference,* (Lancaster

[2] Korsnes M S, Berker T and Woods R 2018 Domestication, acceptance and zero emission ambitions: Insights from a mixed method, experimental research design in a Norwegian Living Lab *Energy Research & Social Science* 39 226-33

[3] Woods R, Berker T and Korsnes M S 2016 Making a home in Living Lab: the limitations and potentials associated with living in a research laboratory. In: *DEMAND Centre Conference,* (Lancaster

[4] Woods R, Berker T and Korsnes M 2019 Homemaking in ZEB Living lab: Interpretations of a Zero Emission Housing Solution *Manuscript in preparation*

[5] Korsnes M 2017 Householders as co-producers: lessons learned from Trondheim's Living Lab. In: *ECEEE 2017 Summer Study on energy efficiency conference proceedings,* (Belambra Presqu'île de Giens, France: eceee)

[6] Goia F, Finocchiaro L and Gustavsen A 2015 The ZEB Living Laboratory at the Norwegian University of Science and Technology: a zero emission house for engineering and social science experiments. In: *7th Passivhus Norden - Sustainable Cities and Buildings,* (Copenhagen

[7] De Simone M, Carpino C, Mora D, Gauthier S, Aragon V and Harputlugil G U 2018 Reference procedures for obtaining occupancy profiles in residential buildings *Subtask A Deliverable IEA EBC Annex 66*

[8] Lobaccaro G, Esposito S, Goia F and Perino M 2017 Daylighting availability in a living laboratory single family house and implication on electric lighting energy demand *Energy Procedia* 122 601-6

[9] Savarino M 2018 Electric Lighting in a single family house in relation to daylight availability for real users. Politecnico di Torino)

(a) Appendix

Table 3. Overview of installed electrical appliances.

	Type of appliance	Energy label	Electrical power	Annual energy use[a]
Washing machine	Combined washing machine and dryer, 8 kg.	A+++ with heat pump	1000 W	177 kWh
Fridge / freezer	Combined fridge and freezer, 200/63 l.	A++	140 W	233 kWh
Hob	Induction top 4 zones, all with booster function	-	7.4 kW max, 2.3 kW per zone 3.2 kW boost	-
Oven	Combi-steam oven, 71 l.	A	3500 W	230 kWh
Dishwasher	Dishwasher, 60 cm.	A++	2200 W	241 kWh
Extr. hood	150-280 m^3/h, 70-125 W	-	195 W incl. lights	-

[a] Annual estimated energy use for appliances as specified by the manufacturer.

Table 4. Overview of installed LED lighting groups per room [9].

Location	Type of light	Electrical power	Dimming	Energy use per 1000h[a]
Technical room $5m^2$	Transformers	33 W (standby)	-	33 kWh
Entryway $8m^2$	Ceiling, Furniture	35+3 W	2 · 0-100	38 kWh
Living room S $25m^2$	Ceiling, Furniture, Floor lamps	70+22+19 W	3 · 0-100	111 kWh
Kitchen $13m^2$	Ceiling, Pendant lamp, Furniture	72+58+8 W	3 · 0-100	138 kWh
Living room N $20m^2$	Ceiling, Furniture	97+32 W	2 · 0-100	129 kWh
Mezzanine $12m^2$	Ceiling	44 W	1 · 0-100	44 kWh
Main bedroom E $16m^2$	Ceiling, Desk, Cabinet, Bed-light	76+2+2+22 W	4 · 0-100	102 kWh
Bedroom W $12m^2$	Ceiling, Desk, Cabinet, Bed-light	82+43+1+16 W	4 · 0-100	142 kWh
Bathroom $4.5m^2$	Ceiling	38 W	1 · 0-100	38 kWh
Outdoors	External lighting	137 W	1 · 0-100	137 kWh

[a] 1000 hours corresponds to about one year of usage according to EC consumer guidelines (EC, 2015).

Table 5. Observed plug loads and accumulated on plug-in el. meters (25-day periods incl. weekends).

	Students		Families		Elderly	
	S1	S2	F1	F2	E1	E2
Kitchen	Coffee grinder, el. kettle	El. Kettle: 3.6 kWh, blender & kitchen machine: 0.5 kWh	Blender, coffee maker, grinder, toaster: 5.6 kWh	Coffee maker with el. kettle: 3.4 kWh	Toaster: 2.7kWh, kitchen machine: 0.7kWh, waffle maker	Coffee maker, el. kettle, mixer & waffle maker: 7.5 kWh
Living S	Speaker	Stereo: 0.9 kWh	Additional space heater	0 kWh	Radio, tablet charger	Small heater, radio
Living N	Laptop, Tablet	Unknown	Battery radio, add. space heater	iMac: 32 kWh, chargers: 0.8 kWh	Unknown	TV, chargers for tablets & phones: 4.4 kWh
Bedrooms	Phone charger: 0.3 kWh	Unknown: 0.9 kWh	Tablet & chargers: 0.4 kWh	0 kWh	Unknown: 0.3 kWh	Unknown

Bathroom	Unknown: 0.1 kWh	Meter 2: 3.0 kWh	0 kWh	Hairdryer & shaver: 0.3 kWh	Unknown: 0.5 kWh	Hairdryer & shaver: 0.3 kWh

IOP Conf. Series: Earth and Environmental Science **352** (2019) 012046 doi:10.1088/1755-1315/352/1/012046

Thermal conductivity of high-performance insulation - a laboratory study. Realistic design values for use in energy-efficient buildings.

Malin Sletnes, Kristin Elvebakk, Jørn Emil Gaarder, Egil Rognvik, Steinar Grynning

SINTEF Community, Department of Architecture, Materials and Structures, NO-7465 Trondheim, Norway

malin.sletnes@sintef.no

Abstract. Aerogel insulation blankets could be the solution to some of the challenges encountered when designing (energy efficient) buildings. However, uncertainties regarding performance in different applications and environments may limit their use in cases where they could otherwise be employed to advantage. The purpose of this study was to examine how to best adapt European standards for the measurement of thermal insulation properties to aerogel insulation and highlight important aspects to be aware of when evaluating material parameters. We have systematically varied measurement parameters for thermal conductivity and compressive strength to map the impacts on reported material properties. Furthermore, we have compared the measurement conditions to actual conditions in selected building applications. In conclusion we propose that the European (EN) standard for measuring compression behavior for insulation materials should be revised and clarified with regards to materials that do not exhibit a clear elastic domain. We also suggest that the thermal conductivity of aerogel insulation blankets should be measured with a slight compressive load on the material, but that the calculation of the thermal conductivity should nevertheless be based on the measured sample thickness.

1. Introduction

The energy performance of buildings has increased significantly during the latter decades. One major contributor to this, in cold climates, has been to increase the thermal insulation levels of the buildings. This in term have led to thicker insulation layers and thicker walls and roofs. High performance insulation materials can contribute to maintaining high insulation levels with reduced wall and roof thicknesses compared to traditional materials. This will decrease the amounts of materials used in the structures and potentially increase sellable floor area of a building.

Although silica aerogels have been known for more than 80 years and have thermal conductivities approximately half those of mineral wool (in the range of 0.012 – 0.018 W/(mK)), the application of aerogel-based products as thermal insulation in buildings is not wide-spread. The main reasons for this are high material costs and dust formation. However, there are some cases where aerogel insulation could be a very good alternative despite the high costs[1-3], such as insulation of thermal bridges, retrofit insulation, and other applications where space restrictions are of essence. Nevertheless, aerogel insulation is rarely considered as an alternative by Norwegian developers or contractors. Part of the reason may be the lack of knowledge and experience on product properties and how to best implement these products into well-known building detail designs. Furthermore, lack of suitable product standards could be an obstacle. There is an ASTM standard for flexible aerogel insulation, and a CUAP [4]. However, there is currently no European EN or ISO standard for aerogel insulation materials, thus complicating the process of product documentation and certification in Europe.

Sometimes adapting existing measuring methods and conditions to new materials is not straight forward. For instance, the European standard for determining compression behaviour of thermal insulation materials, EN 826 [5], is clearly meant for materials which either fracture or weaken prior to

10 % strain or display a clear region of elastic deformation. In this paper, we present the issues that arise upon applying EN 826 standard to measure the compression behaviour of aerogel insulation blankets, which exhibit neither of these properties. Furthermore, most product standards for traditional thermal insulation materials specify that thermal conductivity (λ-value) should be determined directly at measured thickness. However, when this method was applied for aerogel insulation blankets in our previous work [6], we measured λ-values of about 0.017 – 0.018 W/(mK), about 20 % higher than the declared λ-value from the manufacturers (0.014 - 0.015 W/(mK)). In the succeeding discussion, the argument was made that it is important to apply a compressive pressure on the aerogel insulation during measurement of λ-values to avoid air gaps between the sample and instrument. The argument being that aerogel insulation is especially sensitive to air gaps because the thermal conductivity of aerogel insulation is lower than the thermal conductivity of air. Indeed, by compressing the samples during measurement, the measured λ-values improved from 0.018 to 0.015 W/(mK) for the same product [6]. However, the question remains whether measuring with or without compression gives the most realistic λ-value for use in real applications.

Thus, in this study we have focused on systematically changing the applied compressive stress on the sample to investigate the effect on the measured λ-value. The difference from previous work is that we are measuring both the pressure on the sample and sample thickness, whereas in previous work, only sample thickness was measured. Furthermore, we have measured the compressive strength of the insulation, and compared this to expected loads in some typical in-use conditions. In addition to discussing material properties and measurement methodology, we will give some examples which show the significance of the findings for building design with these materials and suggest improvements in measuring techniques which give more realistic design values.

2. Materials and methods

2.1. Materials

The laboratory measurements were performed on commercially available grey aerogel insulation blankets in the form of batts with a thickness of approximately 18 – 20 mm. The aerogel insulation blankets are composite materials of nanoporous silica aerogel in glass/polymer fibre mats. The grey colour is due to synthetic graphite, which has been added in order to reduce radiative heat transfer. Images of the aerogel insulation blanket at different magnifications are shown in Figure 1.

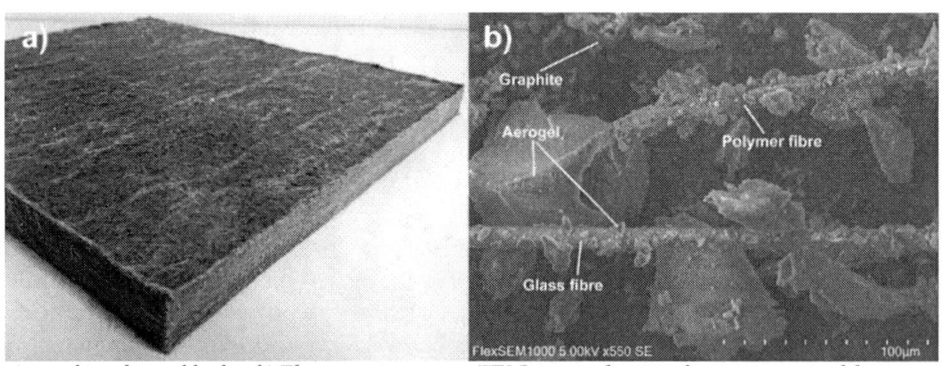

Figure 1 a) Aerogel insulation blanket. b) Electron microscopy (SEM) image showing the components of the aerogel insulation blanket.

2.2. Material characterization

2.2.1. Thermal Conductivity Apparent thermal conductivity λ (W/(mK)) was measured using a heat flow meter (HFM) apparatus in accordance with the governing standard EN 12667 [7]. The compressive stress on the samples during measurement was varied from 0.7 to 20 kPa.

2.2.2. Compressive stress Compressive stress-strain curves were recorded in accordance with EN 826 [7], using a preload of 250 Pa. The specimen size was 10 x 10 cm, and a total of 27 specimens from 4 different production batches were analysed. The sample thickness was measured for each sample prior to measurements. The average sample height was 18.5 mm, with a standard deviation of 0.4 mm.

3. Results and discussion

3.1. Compressive stress

Compressive stress versus strain curves for 27 samples of aerogel insulation blankets from 4 different production batches are shown in Figure 2. According EN 826, the compressive stress at 10 % strain (σ_{10}) should be calculated from the stress-strain curve in the following manner: A line is extended through the (steepest) straight portion of the curve, to the stress corresponding to the preload (250 Pa). The point of intersection is determined as the zero-deformation point. 10 % strain is then calculated starting from the zero-deformation point (i.e. zero-deformation point plus 10 % of the initially measured sample thickness), and σ_{10} is the compressive stress corresponding to this strain. However, EN 826 also states that if there is no distinct straight portion of the curve, this procedure shall not be used. In the alternative method, the zero-deformation point should be defined as the deformation corresponding to a stress of 250 Pa. The two methods are shown in Figure 3, and will be referred to as method A and B, respectively. For aerogel insulation blankets, using these two different approaches to determining the compressive stress give significantly different values, as presented in Table 1. Thus, the declared compressive stress at 10 % strain is very sensitive to the interpretation of the standard. Regardless of the chosen method, the standard deviation for σ_{10} is relatively high, but it is twice as high for Method A than for Method B (see Table 1), reflecting both the divergence in the measured stress-strain curves (Figure 2), and the added insecurity in locating the most linear portion of the curve for extrapolation.

Many commercial suppliers of aerogel insulation blankets do not state σ_{10}, however it appears that those who do use the interpretation that results in the highest declared value. For instance, the EPD of Aspen Spaceloft states $\sigma_{10} \geq 80$ kPa [8], which corresponds well with the values obtained via Method A. However, it is important to be aware that the actual deformation corresponding to a load of σ_{10} according to EN 826 could be significantly higher than 10 %. In average, when using Method A, the attained zero-deformation point already corresponded to 8.7 % actual reduction of the sample height. Adding 10 % strain to this results in an actual deformation of 18.7 % at σ_{10}.

3.2. Practical applications and load cases

For practical use in a building the aerogel blankets will be exposed to different loads. The insulation can be used in both walls, roofs and floors. In roof terrace constructions the aerogel blankets will be exposed to loads from snow, furniture and people, among other things. Examples of loads on the (membrane covered) insulation in a roof terrace construction with different top layers are shown in Table 2. We have tried to identify realistic loads which we assume will appear on the terraces, e.g. loads from heavy persons, furniture, and self-weight of the top layer. We have chosen to use the value for evenly distributed imposed load of 4.0 kN/m^2 from the Eurocode [9], because this is not easily defined in other ways. The Eurocodes are meant for calculation of the capacity of constructions, and these loads are therefore considered conservative compared to the loads the aerogel blankets are expected to be exposed to in real life. Evenly distributed imposed load could be e.g. load from people and furnishings. Snow loads in Norway vary depending on geographical location [10], however 3.8 kN/m^2 was chosen as a representative snow load on flat-roof terraces. Since the imposed load was marginally higher than the snow load, and it is assumed that the roof terrace will not be in use during the presence of heavy snow loads, only the imposed load was used in the calculations.

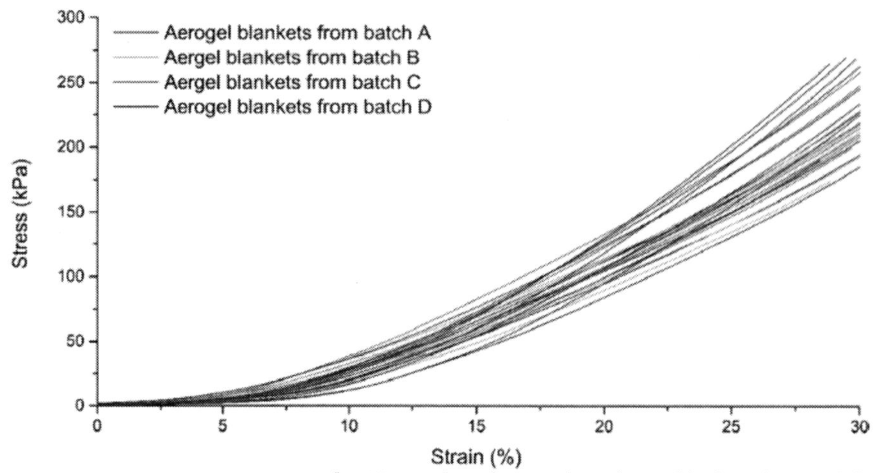

Figure 2 Compressive stress versus strain curves for 27 samples of aerogel insulation blankets from 4 different production batches. The assessment of whether these curves have a distinct straight portion decides the declared compressive stress at 10 % deformation according to EN 826.

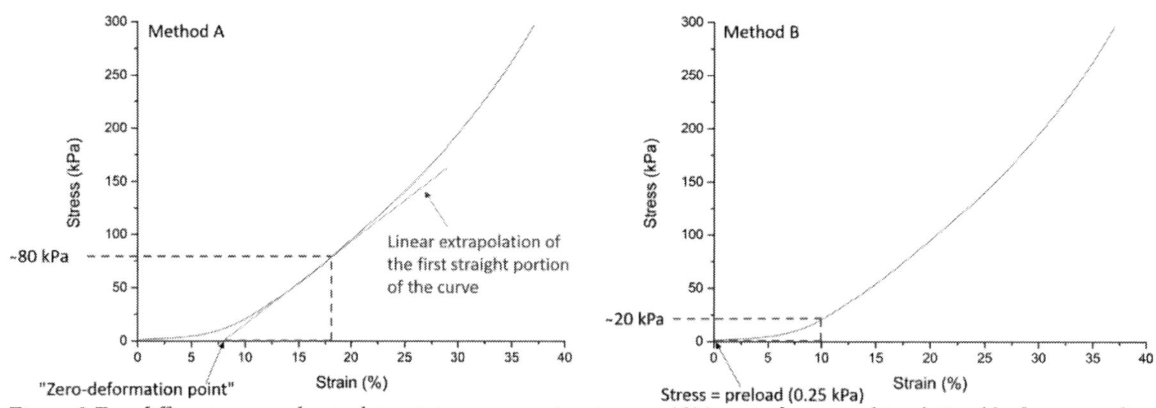

Figure 3 Two different approaches to determining compressive stress at 10% strain for aerogel insulation blankets according to EN 826.

Table 1 Compressive stress at 10% strain for aerogel insulation blankets according to EN 826, using 2 different approaches.

	Number of samples	Average compressive stress at 10 % strain, σ_{10}		Standard deviation	
		Method A [kPa]	Method B [kPa]	Meth. A	Meth. B
Batch A	8	115.0	24.2	15.4	7.8
Batch B	7	88.4	27.3	3.39	7.4
Batch C	6	94.0	27.8	2.84	6.6
Batch D	6	88.7	23.6	2.01	7.5
All samples	27	97.6	25.7	15.3	7.2

We assume that the weight from a heavy person (120 kg) standing on the top of the terrace will represent the most extreme concentrated load. Weight from heavy installations as jacuzzies are not considered relevant in this case, and weight from furniture etc. are considered as lower due to a bigger load distribution area. Considering the loads on top of the terrace construction, the load distribution of the top layer in the terrace will impact the resulting pressure on the aerogel blankets. The top layer will distribute the loads uniformly, as line loads or as concentrated loads, depending on how the top layers are connected to the terrace construction. In Table 2 we have included load calculations for a 70 mm concrete screed, 50 mm thick concrete tiles on pedestals and a wooden decking.

From Table 2 the most extreme loads will occur if the top layer consists of concrete tiles placed on small pedestals. The total load on the aerogel blankets from a person of 120 kg standing on the concrete tiles directly on top of a pedestal with a diameter of 10 cm will be more than 190 kPa. Increasing the pedestal diameter to 20 cm reduces the load significantly, to less than 45 kPa. As we can see, the dimension of the pedestals will be crucial. When comparing the loads in Table 2 to the σ_{10} for aerogel blankets Table 1, it is clear that all the concentrated loads are higher than the σ_{10} calculated using Method B, whereas most of the loads are lower than the σ_{10} calculated using Method A. Thus, the interpretation of the standard could be decisive for whether one evaluates the aerogel blanket compressive strength to be adequate for use in roof terraces. Therefore, knowledge of the material, and its compression behaviour beyond just the σ_{10} value is important in a design process.

Table 2 Examples of typical loads on the membrane covered thermal insulation in a roof-top terrace construction.

Type of top layer	Load distribution on membrane/ insulation [kPa]	Self-load of top layer [kPa]	Imposed load [kPa]		Resulting load on membrane/ insulation [kPa]
			Evenly distributed	Concentrated	
Concrete screed	Area load	1.7	4.0	-	**5.7**
Wooden decking[a]	Line loads	5.4	25.5	-	**29.9**
			-	49.8	**55.2**
Concrete tiles[b] (400 x 400 mm)	Point loads Ø20	5.1	20.4	-	**25.5**
			-	38.2	**43.3**
Concrete tiles[b] (300 x 300 mm)	Point loads Ø10	11.5	46.0	-	**57.5**
			-	152.9	**172.4**
Concrete tiles[b] (400 x 400 mm)	Point loads Ø10	20.4	81.6	-	**102.0**
				152.9	**192.8**

[a] 98 x 73 mm beams with a center-center distance of 60 cm
[b] on pedestals with a diameter of 10 or 20 cm

For applications of aerogel insulation blankets, the essential aspect is not the declared value as such, but the performance of the material with respect to the in-use conditions. Our results showed that the modulus (ratio of stress to strain) increases with applied load up to a pressure of at least 800 kPa, significantly higher than the expected loads in in-use conditions. Thus, the material will not fail or weaken if it is subjected to a load higher than the σ_{10}. Consequently, these aerogel insulation blankets could in many cases be considered compression proof, but one should take special care in cases where point and line loads may lead to local depressions and uneven surfaces.

3.3. Thermal conductivity

The λ-values as a function of the applied compressive stress during measurement for aerogel insulation from two different production batches are shown in Figure 4. It is evident that the λ-value decreases with applied compressive stress during measurement. Increasing the pressure from 0.7 kPa to 20 kPa resulted in a decrease in the λ-value of 0.82 ± 0.02 mW/(mK) for sample A and 0.95 ± 0.03 mW/(mK) for sample B compared to 0.7 kPa pressure, a 5.4 and 6.2 % decrease, respectively. However, the apparent improvement in λ-value is due to the decrease in sample thickness (d) during measurement, rather than an increase in thermal resistance. As shown in Figure 5, both the thermal resistance (R) and the sample thickness were reduced upon subjecting the sample to compressive stress during measurement. The λ-value was reduced because the effect of the sample thickness reduction was larger than the reduction in thermal resistance ($\lambda = d/R$). When negating the effect of the sample thickness reduction by calculating the λ-value using the measured sample thickness prior to compression for all measurements, the λ-value increased slightly, in accordance with the reduction in thermal resistance (Figure 6).

It is important to notice that the reduction in thermal resistance upon compression is very slight. Even under 20 kPa pressure, the thermal resistance was only reduced about 2 – 3 %. This is a potential advantage for aerogel insulation compared to other types of compressible insulation because the thermal resistance of the insulation layer can be considered intact even in applications where the insulation is subjected to compressive loads, such as in the roof-top terraces described above.

In our previous work, we measured λ-values as high as 0.017 – 0.018 W/(mK) for the same materials [6]. All measured values in this study were lower than this. The reason is that we used a preload of at least 0.7 kPa, thus insuring sufficient contact between the whole sample surface and the instrument plates. Sufficient contact is important for correct measurements of thermal conductivity; however, our results still show that it is possible to obtain artificially low λ-values by decreasing the sample thickness during measurement. Hence, in order to get the most realistic λ-values values for aerogel insulation, we propose that the materials should be measured with a compressive load of up to 1 kPa, but the declared lambda value should be calculated based on the nominal thickness.

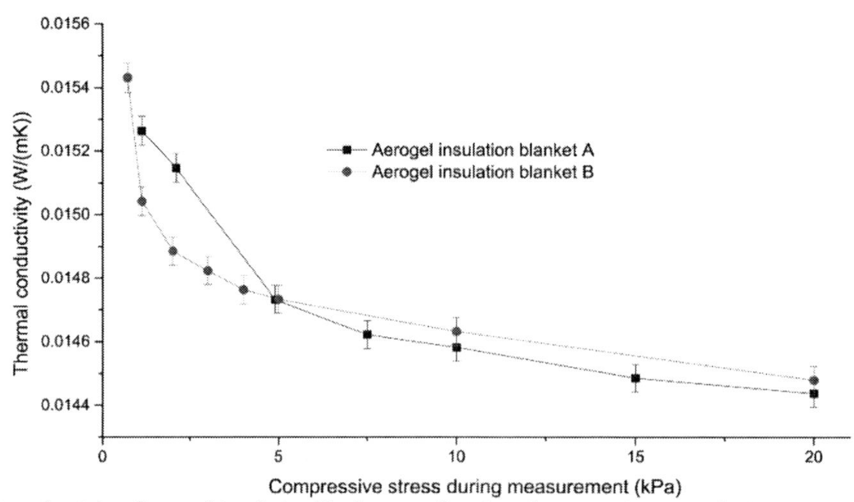

Figure 4 Thermal conductivity of aerogel insulation blankets as a function of compressive stress during measurment. The error bars indicate the expanded standard uncertainty (k=2) calculated from the repeatability of the measurements.

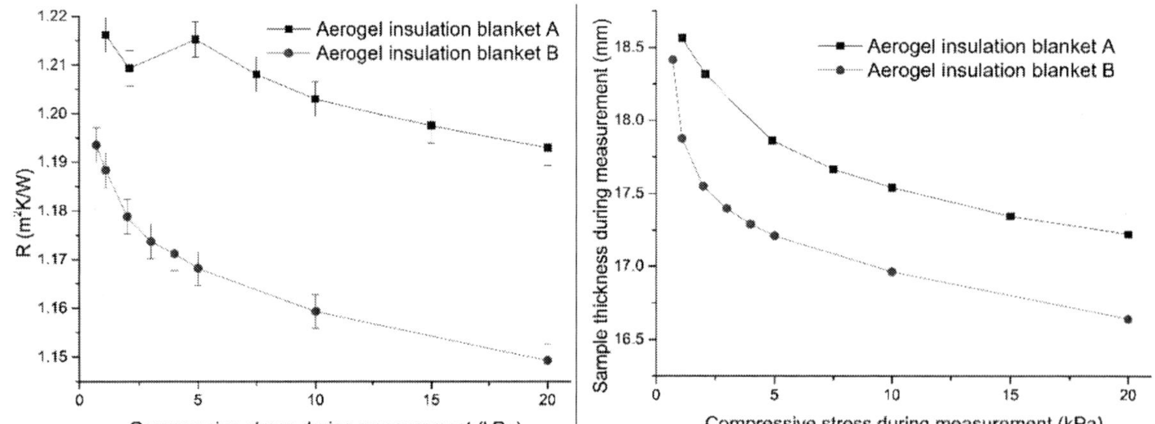

Figure 5 Thermal resistance (R-value) and sample thickness of aerogel insulation blankets as a function of compressive stress during measurement of thermal conductivity. The error bars indicate the expanded standard uncertainty (k=2) calculated from the repeatability of the measurements.

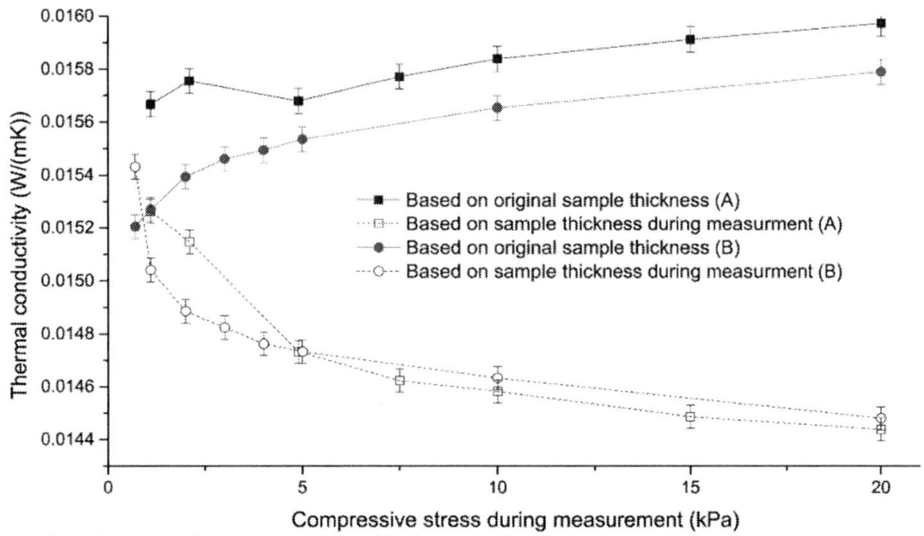

Figure 6 Thermal conductivity of aerogel insulation blankets as a function of compressive stress during measurement calculated based on sample thickness without compression and sample thickness during measurements.

Since the application of aerogel insulation is most relevant in restricted volumes, such as insulation of thermal bridges and as a supplement to traditional insulation materials the practical effect of variations in lambda value due to measurement techniques may in many cases be limited. An example of how variations in the lambda value could impact the heat loss from a thermal bridge at the leading edge of a floor slab (Figure 7) is shown in Figure 8.

An increase of heat conduction coefficient from 0,014 W/mK to 0,018 W/mK (29% increase) in the example will result in an increase of thermal bridge coefficient from 0,131 W/mK to 0,148 W/mK (13% increase). For a given 15 m x 15 m office building with 5 stories, the difference in the resulting heat loss from the thermal bridges in the example above is calculated to 5,1 W/K. This corresponds to an energy demand of approximately 514 kWh over a heating season from October to April in a typical Norwegian climate (using average temperatures for Oslo and a constant indoor temperature equal to 20 °C). Thus, the effect of different measurement conditions for thermal conductivity in this case is not large, but it is nevertheless important to use correct measurement methodology for obtaining declared values, also with regards to competitive conditions in the market.

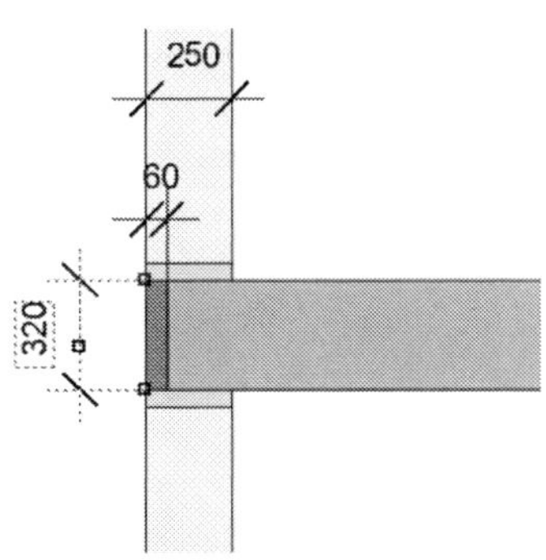

Figure 7 Aerogel insulated thermal bridge. The walls are constructed using insulated framework and the floor is a hollow-core concrete slab protruding 190 mm into the wall.

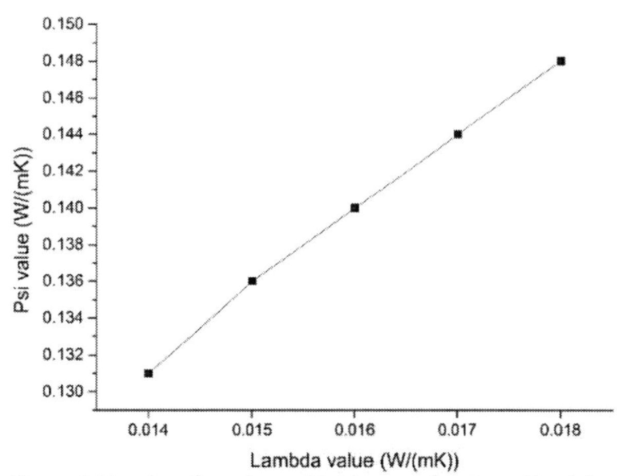

Figure 8 Heat loss from thermal bridge as a function of lambda value.

4. Conclusion and further work

This study has demonstrated some of the issues that arise upon applying European (EN) measurement standards for thermal insulation to measure the compression behaviour and thermal conductivity of aerogel insulation blankets. The methodology for calculating the compressive strength at 10 % strain in EN 826 is unclear due to the shape of the stress-strain curves for aerogel insulation. Applying two different methodologies gave average σ_{10} values of 97.6 kPa and 25.7 kPa, respectively. The lower value is not sufficient to withstand typically expected loads on a roof-top terrace, however the aerogel insulation will not fail or weaken if exposed to loads exceeding σ_{10}. Hence, those wishing to apply aerogel insulation in constructions where it will be subjected to compressive load should consult the stress-strain curve rather than simply relying on the reported value for σ_{10}. For the measurement of thermal conductivity of aerogel insulation blankets, we propose that the materials should be measured with a compressive load of up to 1 kPa, but the declared lambda value should be calculated based on the nominal thickness. This work highlights the need for a governing European product standard for aerogel insulation, and the adaptation of existing measurement standards to new types of thermal insulation.

5. Acknowledgements

This work was part of the IPN project SuperIsol, owned by Isola AS and funded by the Research Council of Norway under the large-scale programme for energy research (ENERGIX), project number 282389.

References

[1] U. Berardi and L. Ákos, "Thermal bridges of metal fasteners for aerogel-enhanced blankets," *Energy and Buildings,* vol. 185, pp. 307-315, 2019.

[2] U. Berardi, "Aerogel-enhanced systems for building energy retrofits: Insights from a case study," *Energy and Buildings,* vol. 159, pp. 370-381, 2018.

[3] E. Cuce, P. M. Cuce, C. J. Wood, and S. B. Riffat, "Optimizing insulation thickness and analysing environmental impacts of aerogel-based thermal superinsulation in buildings," *Energy and Buildings,* vol. 77, pp. 28-39, 2014.

[4] CUAP 12.01/36 Fibre Reinforced Silica Aerogel Thermal Insulation, 2011.

[5] EN 826:2013 Thermal insulating products for building applications - Determination of compression behaviour.

[6] M. Sletnes, B. P. Jelle, and B. Risholt, "Feasibility study of novel integrated aerogel solutions," Energy Procedia, vol. 132, pp. 327-332, 2017.

[7] EN 12667:2001 Thermal performance of building materials and products - Determination of thermal resistance by means of guarded hot plate and heat flow meter methods - Products of high and medium thermal resistance.

[8] "Environmental Product Declaration of Spaceloft Aerogel Insulation," ed: Aspen Aerogels Inc., 2015.

[9] NS-EN 1991-1-1:2002+NA:2019 Eurokode 1: Laster på konstruksjoner - Del 1-1: Allmenne laster - Tetthet, egenvekt og nyttelaster i bygninger

[10] NS-EN 1991-1-3:2003+A1:2015+NA:2018 Eurokode 1: Laster på konstruksjoner - Del 1-3: Allmenne laster – Snølaster

IOP Conf. Series: Earth and Environmental Science **352** (2019) 012047 doi:10.1088/1755-1315/352/1/012047

Process-related risks in refurbishment of dwellings using prefabricated wall elements with integrated PV and ventilation ducts

S B Holøs[1], V Lukina[2], Å. L. Hauge[1] and Kari Thunshelle[1]

[1]SINTEF Building and Infrastructure, 0314 OSLO, Norway
[2]Boligbygg Oslo KF

Sverreb.holos@sintef.no
Abstract. Prefabricated façade elements with integrated technical infrastructure is an attractive technology for refurbishment of existing dwellings. Heating and cooling demand can be reduced, local energy production introduced, and indoor air quality be improved, with disturbance to the tenants and building site being small and of short duration compared to more traditional building processes. On the other hand, unexpected events could largely reduce these benefits. Thus, risk management of the building process is of great importance. Focus group interviews and workshops were arranged before and after the building phase in a pilot project using such elements in Oslo, Norway. Representatives of building owner, design team and contractor contributed actively at the workshop. In a pre-building phase workshop, a range of hazards were identified and prioritized using a participative process facilitated by a neutral moderator. A large proportion of the prioritized risks in the building phase were connected to the renovated flats being occupied during the renovation. Other significant identified risks related to transport and logistics, and undetected challenges in the existing construction. Mitigation included prioritizing tenant information, including direct dialog, and increasing the presence of on-site workforce both for coordination with tenants and in order to respond quickly to unforeseen events. The participants emphasized that an open, cooperative processes with a high degree of trust and sense of a common goal had been important for the robust design that was developed prior to the workshop. During the retrospective evaluation, the participants concluded that the risk mitigation procedures had been successful in preventing some events as well as reducing the consequences of others. However, some of the measures to mitigate an identified risk of rain intrusion were inadequate, and it was acknowledged that the combination of bad weather and long working days could have identified this as preventable.

1. Introduction

In order to reduce emission of gases with global warming potential, reducing the energy demand of existing buildings is identified as an important goal. The European Union is committed[1] to increase energy efficiency of the building stock through increasing the rate of building renovation improving energy performance and indoor climate.

The project "Robust and Reliable technology concepts and business models for triggering deep Renovation of Residential buildings in EU" (4RinEU) has the overall objective of defining robust, cost-effective, tailorable, deep renovation technologies and usable methodologies, feeding into reliable

[1] Directive (EU) 2018/844

business models. The project was established on the premise that uncertainties in terms of costs and benefits in the life cycle is a barrier against deep renovation, and that more cost efficient and time saving renovation concepts can help overcoming these. Prefabricated façade elements including technical installations, including a Renewable Energy Source (RES), is one of the defined renovation technologies within 4RinEU.

It is postulated that this renovation technology can reduce heating and cooling demand, improve indoor climate and add local renewable energy production through a cost-efficient process with little negative impact on the building users and the environment, as it is possible to do the construction work without relocating inhabitants and with little need for rig area. On the other hand, the consequences of some undesirable events could be greater than in a traditional building process, exactly because the buildings are in use during the construction work.

Within the framework of the 4RinEU, a deep renovation demo project was developed together with Boligbygg Oslo KF – a municipal enterprise that owns, manages and lets social housing in Oslo. Before the deep renovation process, prefabricated elements with integrated technical components were not established on the market, a fact which added to the perceived project risks. For this Norwegian demo, integration of PV panels and ventilation ducts were chosen to be integrated in the elements.

1.1. Demo building

The building selected for deep renovation (Figure 1) is a two-storey timber-frame building from 1971, with only minor later upgrades, owned by Boligbygg Oslo KF. The building contains in total eight apartments, each with a floor area of approximately 42m^2, distributed around two staircases. The building is situated in a suburban area (Haugerud) in Oslo.
The apartments had electrical heating, one electrical heated boiler per apartment and natural (stack) ventilation.

Figure 1a and b. Building before (left) and after (right) renovation. Photo SINTEF (a) and Boligbygg KF (b)

1.2. Introductory condition assessment and premonitoring

An assessment of the technical condition of the building, with emphasis of energy performance, indoor air quality performance and technical renovation needs was performed. Moisture problems were detected, and further examined by partially destructive examination methods. Opening of the wall construction uncovered that the construction of the ground floor external walls differed from the original assumptions, as the main carrying construction was of aerated concrete blocks and not timber frame. Building air-tightness was tested by a blower-door test performed according to NS-EN ISO 9972.

Presence of materials of health or environmental concern was examined by a separate expert surveyor, and asbestos-containing sheets were detected in walls.

Energy performance reports and accounts of previous energy form local electricity provider were collected. Electricity and domestic hot-water usage, airing habits and indoor climate were monitored in four of the apartments for comparison with post-renovation conditions, results are not reported here.

As a supplement to the technical investigations, Boligbygg collected reports of user behaviour and preferences. These showed that tenants often closed the air inlets to avoid draft – often leading to bad air quality – and that preferred indoor temperatures were higher than standard values used in the calculation of building energy demand.

Furthermore, it was concluded that the building needed a renovation of the cladding, and extra insulation to reach a higher energy standard both for walls and roof. The technical systems were old, but functioning, and the electrical boilers were not in a condition where replacement was necessary.

1.3. Renovation concept design

A design team selected by the building owner were contracted at an early decision phase. Expertise on wooden elements and BIM skills were important selection criteria. in addition to architect, consultant engineers and contractor / manufacturer, experts from SINTEF and the specialized element manufacturer Gumpp & Maier were involved in the design process.

Improving energy performance and indoor climate and installing renewable energy, with as little tenant disturbance as possible were important goals for the building owner. Use of prefabricated façade elements with technical integration was decided from the start of the project. Due to the electrical boilers in each apartment and electrical heating, integrated PV panels in the south-facing façade were preferred over solar collectors and energy hub. The electricity from the PVs were primarily to be used for heating and lighting of staircases and operation of the balanced ventilation. The surplus will be sold to the grid because legal restrictions prevent use of produced energy within the apartments

Due to the low loadbearing capacity of aerated concrete, the original idea of wall mounting was not possible, and a new foundation for the elements was included in the concept. Prefabricated roof elements were chosen instead of insulating existing attic to ease moisture proofing of wall / roof all junctions. Existing insulation and cladding were kept, and wind and vapour barriers included in wall elements.

Balanced ventilation with heat recovery was considered a necessary part of the deep renovation to improve indoor air quality and improve energy performance. Several solutions for integration of ductwork were considered. Supply air ducts were integrated in north-facing wall element towards bedrooms, while exhaust ductwork was integrated underneath the stairs. A separate prefabricated element formed as a technical shaft with integrated air handling units (AHUs) was developed. Two apartments were connected to each AHU.

1.4. Contract model

Architect and several of the consulting engineers were selected from providers with framework agreement with Boligbygg Oslo KF. Considering the potential risks in the implementation phase, the building owner wanted to include element manufacturer and building contractor early in the design phase. At first, the search for such were unsuccessful, so more of the design work than intended was

IOP Conf. Series: Earth and Environmental Science **352** (2019) 012047 doi:10.1088/1755-1315/352/1/012047

performed before the element manufacturer, which also was able to undertake the building works, was contracted in an open-book participatory contract model, with a PV specialist producer as a subcontractor.

2. Methods

2.1. Pre-implementation workshop

Prior to the implementation, representatives of building owner (9), design team (4) and contractor (2) as well as three researchers following the process were invited by the client to a workshop. The process was facilitated by an external moderator with no prior knowledge of the process. After introductory reports of individual experiences from project participation, participants were asked to name important success criteria for the project. The participants were asked to describe "wishes" and "fears" for the building process individually, and then within groups of 4-7 people gather similar risks from the individuals. The grouped "fears" and "wishes" were then placed along two axes according to likelihood and magnitude (positive or negative) of consequences. Based on the results from all four groups, a plan was made in a plenary session to avoid the prioritized fears and realize the wishes. This plenary session was led by a representative of the building client.

2.2. Evaluation workshop and focus group interview

Following the conclusion of the building works, representatives from the design team, building contractor and client were again gathered to a workshop systematizing experience and identifying risks and undesired events, and how they were handled. The workshop was carried through as a focus-group interview. This method is well-suited for exploring attitudes and arguments, and frequently creates new insights when the participants react to each other's input [1]. A semi-structured interview guide that accommodated the participants' own inputs was used. The participants were asked to describe risk management according to phase. The interview was recorded but not fully transcribed. The analysis of the results was based on notes taken during the interview.

2.3. Monitoring and post-hoc experience

A temporary monitoring of indoor CO_2-levels, temperature and relative humidity in living rooms, temperature and relative humidity in sleeping room and supply air duct was set up for four apartments using TinyTag Ultra or Rotronic CP 11 data loggers in a period from December 15th to February, 10th. During the installation and collection of the data loggers, the tenants of these four apartments reported informally about their experience. Only preliminary results of the post-monitoring are reported here.

3. Results

3.1. Risk identification

The risks that were identified and classified into four groups during the pre-building workshop are reported in table 1.

Table 1. Identified wishes and fears, divided into four groups. The scenarios classified as "likely" were prioritized for further work.

	Unlikely	*Likely*
Desirable	Tenants serving cake	Positive press coverage
	Nice weather	Interest in the business
	Client adopts solution for all properties	Transfer of experience to other projects
	No surprises	Satisfied client
	Process according to plan (2)	Satisfied tenants (5)
	Everything fits	No injuries (3)
		No design-related defects
		Visually appealing building
		Delivery on time & budget
Undesirable	Montage failures	Elements not fitting (7)
	Technology not working	Delays (3)
	Failing target performance	Problems with ducts
	Injuries (tenants, workforce)	Insufficient control of existing cables and installations
	Damage to VPs	Unknown technical challenges in existing construction
	Injuries on site (5)	Unknown decay or asbestos
	Tenant locked in	Heavy rainfall during building
	Bad weather	Tenant denying access to apartment
	Operation problems	
	Connection problems to VPs	
	Negative press	

3.2. Risk management plan

A risk management plan (Table 2) to address the prioritized scenarios / targets identified) was produced and responsibility for implementation agreed, as a supplement for the different actors existing routines.

Table 2. Risk reducing actions identified during the pre-implementation workshop

Domain	*Scenario*	*Risk reducing action*
Logistic	Delay at subcontractor / supplier	
	Delay during transport	Sufficient material and manpower for temporary protection of construction. Spare capacity at factory to replace damaged element within set time limit
	Damage during transport	
	Damage to existing cables, pipes, etc.	Available contact info to local electrician plumber, etc. with short response time.
Health and safety	Need to deactivate fire-alarm	Fire guard in place
	Fire escape route blocked	Establish alternative escape route through window
Societal (tenants)	Failure to secure access to apartment	Involve tenants. Inform in relevant language. Information meeting with refreshments in addition to written.
	Annoyance with disturbed power, TV or other signal supply	Inform prior via tenant contact, and during process via signs.

IOP Conf. Series: Earth and Environmental Science **352** (2019) 012047 doi:10.1088/1755-1315/352/1/012047

	Injury to occupant	Establish good relations to occupants. Maximum continuity of presence by contractor.
Societal (other)	Neighbours gets hurt or interferes with building	Inform all tenants in neighbourhood about planned work
	Negative public impression of project	Information activities planned, using press and social media. Defined information responsibilities and strategy. Focus on effect and total economy – not costs.

3.3. *Preliminary monitoring and informal user feedback*

Preliminary monitoring results indicated that living-room temperatures were within 20-26 °C interval in 85 – 100 % of the monitored time period, while CO_2 levels were below 1000 ppm for 94-100 % of the time. Outdoor temperatures in the period varied between -13.4 and 10.1 °C, average 2.7 °C.

During the installation of logging equipment, two of the occupants reported improved thermal comfort, while one occupant complained that the supply air had to be blocked due to strong odour from tobacco smoke.

3.4. *Evaluation workshop*

The experiences form the installation phase from the evaluation workshop is summarized in table 3. Undesired events in the implementation phase included delayed supply of wooden cladding, water damage to one apartment due to heavy rainfall, delays of work due to presence of one occupant in an access-restricted area and need for mitigation of a damaged asbestos-containing board.

Table 3. Summary of experience as reported in the evaluation workshop

Positive experience	*Improvement potential*
Elements successfully installed	Massive time-use for detailing
Zero injuries	Delays at subcontractors and materials suppliers
Successful integration of PVs, including transport	Long and complicated list of owner demands
Efficient installation without scaffolding, building crane and minimal temporary installations.	Method for temporary moisture protection
Exemplary handling of elements, even with narrow construction site	Continuity on site
No problems with tolerances to existing walls, placing of windows and existing inlet air openings	On-site access to drawings (plan / details)
No problems with conjunctions to other elements, roof elements or elements with ductworks.	More attention to report on hazardous materials and substances, which described location of asbestos
Successful dismantling of existing windows, to maintain fire escape routes.	Time use for implementing health and safety routines by subcontractors
Minimal need to relocate tenants	(Location of) cables, pipes and ducts
High tenant satisfaction	
Short time for actual mounting of elements, acceptable overall progress.	
Good communication – common targets	

4. Discussion

4.1. Risk reducing activities prior to implementation

Several of the participants repeatedly emphasized that there was a high degree of openness in communication, trust and sense of common goals between the actors in the design and building phase of this project, and that this was an important for successfully tackling the design challenges. It is beyond the scope of this paper to discuss the importance of these factors for successfully reducing risk by achieving good and robust design, but previous research[2] have concluded that trust is important for group efficiency. Neither do we attempt to assign these *factors* to prior collaboration between participants in the design team [2], the type of contract [3], the fact that the project was perceived as ground-breaking, or any other possible causes suggested in the literature [4, 5].

Since the EU-project had quite specific targets for reducing energy demand and operational costs, a pre-monitoring process fed more information into the design team of the pre-renovation state of the buildings than what may be generally available. Also, cooperation with the researchers following the project may have contributed with risk-reducing information to the design process. As an example, researchers from SINTEF Building and Infrastructure were able to provide specific information on the interpretation of legislation on fire safety, thus reducing risks of not achieving necessary permits.

4.2. Risk identification process

The method of including both desirable and undesirable events in the risk identification process may have contributed to a broader perspective on what kind of events and consequences that needs consideration than a process only examining the unwanted. Statement of desires such as "winning a prize", "satisfied tenants" or more jocular "tenants serving cake" would quite likely have been missed if solely undesirable events were included and brings attention to positive aspects and end goals of the process. Also, the inclusion of tenant contacts and maintenance workers may have strengthened the focus on societal risks and the long-term perspective in the risk assessment.

4.3. Risk classification and adequacy of countermeasures

The method of classifying "fears" and "desires" into four categories could be regarded as a crude simplification of the method of assigning probabilities and consequences of all scenarios in quantitative risk assessment procedures[6]. All registered undesired events identified during the evaluation (section 3.4. were actually identified during the workshop. The risk of asbestos-containing material was not prioritized in the workshop, while the countermeasures against water-damage from rain and missing supply of components were prioritized, but the actions were inadequate to prevent consequences. The presence of an occupant in a restricted area had minimal consequences, as the person was detected, and proper adjustment of work progress could be made.

4.4. Cost and benefit of risk assessment and countermeasures

It has been described a number of psychological mechanisms influencing the perception of risk that in turn could lead to "wrong" allocation of resources for prevention, see [7] for a popular account and [8] for critical discussion. One such mechanism is that risks related to novelties and unusual circumstances may be given undue emphasis compared to well-known factors. A possible example from the described case may be that a high number of participants identified "elements not fitting" as a scenario of high concern, even if measurements in addition to the laser-scanning used for making the BIM were taken at least twice and plans to make smaller adaptations were in place. We have not attempted to specify costs for the risk identification process, the specific countermeasures or the actual costs of repairing the water-damage or mitigating the asbestos and are thus in no position to analyse cost and benefit in the actual case.

4.5. Recommendations for similar projects

To successfully manage the implementation-phase risk of deep renovation of dwellings without relocation of occupants using prefabricated building elements, we recommend workshops or similar processes involving design team, contractor, owner and end-user representatives in order to identify and manage any risks not being mitigated by the design and the operational procedures of the contractor. We further recommend that steps are taken to ensure that scenarios involving occupant or neighbour behaviour, logistical challenges and possible unknown building or installation properties are emphasized during the project. The process should ensure that not only peculiar features of the analysed project, but also well-known risks from other projects and activities are included. A question on "what problems have we experienced earlier in renovation projects?" could be beneficial. The authors stress that the recommended procedure does not replace the need for a thorough condition assessment prior to selection and design of the deep renovation method.

5. Conclusion and further work

The project was able to successfully demonstrate that the concept of renovating low-rise timber-based apartment buildings using prefabricated elements with integrated photovoltaic panels and ventilation ducts as well combined with heat-recovery mechanical ventilation was possible, and that installation was possible with minimal disturbance of the tenants.

The risk management procedures used in the building phase identified several realistic scenarios, and succeeded in reducing, but not avoiding, consequences of rain intrusion, interference with asbestos-containing materials and failure in delivery from materials suppliers. The inclusion of a broad range of participants in the process of identifying risks and opportunities resulted in a risk management plan adapted to the peculiarities of the actual project, with particular emphasis on societal risks and consequences.

Completion of the commissioning process and post-monitoring of energy use an indoor climatic quality will reveal whether the targets of reducing energy demand by 60 % and operational costs by 15 % were achieved, as well as any yet undetected functional flaws.

6. Acknowledgments

The work was funded by the European Commission, EC Grant Agreement Number: No 723829.

7. References

[1.] Morgan, D.L., *Focus Groups as Qualitative Researc*. 1997: Sage publications.
[2.] Buvik, M.P. and M. Rolfsen, *Prior ties and trust development in project teams - A case study from the construction industry.* International Journal of Project Management, 2015. **33**(7): p. 1484-1494.
[3.] Eriksson, P.E. and M. Westerberg, *Effects of cooperative procurement procedures on construction project performance: A conceptual framework.* International Journal of Project Management, 2011. **29**(2): p. 197-208.
[4.] Bygballe, L.E., M. Jahre, and A. Swärd, *Partnering relationships in construction: A literature review.* Journal of Purchasing and Supply Management, 2010. **16**(4): p. 239-253.
[5.] Hosseini, A., et al., *Project Partnering in Norwegian Construction Industry.* Energy Procedia, 2016. **96**: p. 241-252.
[6.] Kaplan, S. and B.J. Garrick, *On The Quantitative Definition of Risk.* Risk Analysis, 1981. **1**(1): p. 11-27.
[7.] Kahneman, D., *Thinking, fast and slow*. 2012, London: Penguin Books.
[8.] Aven, T., *On the allegations that small risks are treated out of proportion to their importance.* Reliability Engineering & System Safety, 2015. **140**: p. 116-121.

IOP Conf. Series: Earth and Environmental Science **352** (2019) 012048 doi:10.1088/1755-1315/352/1/012048

Review of HVAC strategies for energy renovation of detached houses towards nZEB in cold climates

V Heide[1], L Georges[1], A G Lien[2] and H M Mathisen[1]

[1]Department of Energy and Process Engineering, Faculty of Engineering, NTNU
[2]SINTEF Building and infrastructure, Architecture, Materials and Structures, Trondheim
vegard.heide@ntnu.no, laurent.georges@ntnu.no, anne.g.lien@sintef.no, hans.m.mathisen@ntnu.no

Abstract. This paper provides a review of heat and ventilation measures that can be applied to ambitious energy renovation of detached houses in Nordic countries. In this review, requirements for solutions are defined. Key technologies are described and analysed in the context of renovation. The review focuses on strategies that are simple, cost-effective and robust, and can be transposed to the Norwegian context. The review revealed that a wider range of concepts and strategies than commonly used in Norway seem to be relevant. No solution or system appears to be an obvious and universal choice. A number of very different system solutions, with their pros and cons, are relevant, depending on the individual house and situation. Some combined heat and ventilation systems include hydronic space heating. This is however not common in Norwegian houses, and installing this is a major cost and intervention. Wood stoves, on the other hand, are regular, and can be used for peak heating. These factors seem to be crucial for the choice of system. Improved airtightness after renovation makes systematic ventilation measures necessary. Assumptions for occupant preferences and behaviour also seem to be important for choice of system. There are also differences in the commonly used HVAC concepts and strategies for renovation between the Nordic countries. These differences do not seem to be explained by climate only, and differences in building code may be part of the reason. A number of demonstration projects on ambitious energy upgrading are completed, but few of them have been systematically monitored and evaluated.

1. Background

30% of Norwegian residential buildings are detached houses built between 1950 and 1990. This building stock is reaching a stage where major renovations are needed. Some of these houses are renovated step by step and some in one large operation. Unfortunately, only half of the renovations taking place in Norway include energy renovation. A recent study found that 60% of the Norwegian detached households made no changes to the ventilation system when performing an energy renovation, while only 9% installed balanced mechanical ventilation. Renovation usually includes a substantial improvement in air tightness. This reduces the air change due to infiltration, demanding systematic ventilation measures to provide sufficient indoor air quality (IAQ). The challenge is to develop ventilation concepts that provide good indoor environment and energy performance, and still are simple and affordable enough to actually be implemented.

Renovation requires solutions to a very complex situation. A wide range of heat, ventilation and air-conditioning (HVAC) strategies can be considered, not only the most common ones. Although heating and ventilation systems are interconnected, the main focus in this review is on the ventilation. Strategies

IOP Conf. Series: Earth and Environmental Science **352** (2019) 012048 doi:10.1088/1755-1315/352/1/012048

such as balanced, exhaust, decentralized and hybrid ventilation are key technologies considered in the review.

2. Method

In the review, research publications have been investigated according to several aspects: the systems overall complexity and robustness, its integration in the existing building structure, energy performance, indoor environment and the possibility to create significant thermal zoning, lifecycle costs, among others. The review focuses on measures that can be transposed to the Norwegian context. Several demonstration projects on ambitious energy upgrading are completed. Nevertheless, few of them have been systematically monitored and evaluated, and the number of studies on HVAC-systems in energy renovation of detached houses in a cold climate, are limited. Therefore, some studies on apartment buildings or milder climates are included. Not all technologies and strategies relevant for energy renovation are thoroughly investigated in a renovation context. Studies on the actual technologies or components themselves, even used in new buildings, can give useful insight. Therefore, some studies about new buildings are included. The reviews primary focus is deep energy renovation, or renovation towards nearly zero-energy buildings (nZEB). However, some less ambitious projects were included when relevant.

3. Requirements

3.1. Existing buildings and HVAC

Due to low electricity prices, very few of the Norwegian detached houses from 1950 – 1990 have hydronic heating [1]. The additional cost to install a hydronic system can strongly impact their cost-effectiveness compared to cheap direct electric heating. Strategies like ventilation radiators also need hydronic heating. Finding space for technical units, including ventilation channels, can be an issue in the small houses. In Norway, most houses have a wood stove that can provide useful peak heating on cold days, reducing the needed power output from the base heating system [1].

3.2. Indoor air quality (IAQ)

There have been some concerns about IAQ after deep energy renovation, due to reduced air change rates from infiltration with more airtight houses. Several studies report good IAQ and good user satisfaction when balanced ventilation is installed. Leivo [2] found slightly better IAQ and better occupant satisfaction after installing heat recovery into exhaust ventilation system in Finnish apartment buildings. Some studies also report good IAQ with exhaust ventilation. Wells [3] found no differences in IAQ between exhaust and balanced ventilation. However, occupants with exhaust ventilation reported their homes as more comfortable. Efficient filtering is an advantage for handling allergy and for polluted outdoor air. If filters are not changed regularly, they can degrade the IAQ.

3.3. Indoor thermal environment

In addition to comfortable operative temperature and moderate temperature stratification, the absence of cold draft is important.

3.4. Temperature zoning.

High insulation levels tend to reduce temperature differences inside the building. This enable to simplify the space-heating distribution using a fewer number of heat emitters [4, 5]. In highly insulated buildings however, many occupants report the bedrooms as too warm even though not equipped with heat emitters. One main reason seems to be that one-zone balanced cascade ventilation with efficient heat recovery moves heat from the warmest to the coldest rooms. High temperatures in bathrooms and living rooms are partly transferred to bedrooms. Many occupants open the windows in bedrooms to reduce the temperatures during winter time, thus increasing energy use for heating. This has been investigated for new buildings by Berge, Georges and Thomsen [4-6]. The same problems could occur in deep-renovated houses, with good insulation and balanced ventilation. Although this finding requires additional confirmation, Klinski [7]reported open bedroom windows in a renovated house where balanced ventilation has been installed. The desire for cool bedrooms is also reported in Denmark, typically 16°C in apartment bedroom [8]. Bjørneboe [9] reported that the occupants preferred a lower temperature in bedrooms also after renovation.

With increasing internal gains from appliances [10], more insulation and a warmer climate, overheating during summer needs to be addressed. Several studies also report floor heating in bathrooms working during summer, increasing the internal heat load [6, 7]. In some cases, better roof insulation can improve the thermal comfort. In one study [9], the house owners reported the house to be cooler in summer and warmer in winter after the renovation. In a Scandinavian climate, solar shading and window cross ventilation is an important strategy. The selected systems should be burglar-proof. Psomas et al [11] investigated automated roof windows to address the risk of overheating in energy renovated dwellings. They found that an automated window system with integrated heuristic passive cooling control strategies significantly decreases the thermal discomfort and overheating risk without any significant compromise on the indoor air quality.

3.5. Occupant sensitivity
Some strategies may fulfil the preference of specific occupants but not for the others. For instance, window ventilation is positive if cooler bedrooms are desired while balanced cascade ventilation favours warmer bedrooms [4, 5]. Some people are very sensitive to the fan sound in bedrooms but not everyone [12, 13]. Systems that can accommodate a very wide range of user preferences are less occupant sensitive. Some systems may also require specific knowledge or skills from the occupants.

3.6. Energy and environmental performance
In addition to a low energy consumption, strategies to minimize the peak power consumption, and adaption to variations in energy prices and supply, is interesting. The ability to store heat is a key factor. The ability to use low temperature heat sources can also be useful.

4. Ventilation

4.1. Mechanical ventilation with heat recovery (MVHR)
Balanced ventilation with heat recovery is one main ventilation strategy for energy efficient buildings. Efficient heat recovery ensures a low energy consumption, preheating of supply air gives good thermal comfort. Controlled ventilation rates and efficient filtering provide good IAQ. The challenges are the limitation for temperature zoning, the requirement to install ducts and the overall investment costs. Some solutions for integrating ducts are shown in Lien et al. [14]. Some other studies [15, 16] investigated a system with the heat exchanger and part of the ducts integrated into the façade. Installing MVHR in some rooms and leaving some rooms (typically bedrooms) outside the MVHR system, is also a way to reduce the duct work, although this is not described in field studies. In renovation projects, it is difficult to ensure low infiltration and this reduces the heat recovered by the heat exchanger compared to new buildings. Measured infiltration values before renovation are typically between 4 and 10 ach and are reduced to between 1.0 and 2.5 ach after renovation [9, 14, 17]. These studies refer to only one case each.

4.2. Real heat exchanger efficiency.
Several field studies show a lower heat recovery efficiency than the nominal value. Regarding rotating exchangers, Lassen [18] found 10 to 20% lower efficiency than the nominal value. Petersen et al. [19] measured 5 to 37% below nominal for five schools. Kamendere et al. [16] found a heat exchanger efficiency of 80% at an outdoor temperature of +15°C and 75% at -5°C. The efficiency of the whole ventilation system was lower due to temperature drop in ducts. Merzkirch et al. [20] found an average

IOP Conf. Series: Earth and Environmental Science **352** (2019) 012048 doi:10.1088/1755-1315/352/1/012048

efficiency of 65% for 20 centralized systems, and 70% for 80 decentralized systems. Østin [21] found an efficiency from 72 to 76 % when nominal value was 80% in a new single family house. In a renovated apartment La Fleur [22] reported on a measured recovery efficiency of 57.4% while the estimated value was 70% (based on the nominal value). This could indicate that the real energy performance of MVHR is slightly overrated, having consequences when comparing different strategies.

The possibility of temperature zoning is limited with regular balanced ventilation. However, some possible ways to achieve cool bedrooms without increasing energy consumption have been proposed, but not thoroughly investigated: reducing the efficiency of the heat exchanger [23], balanced supply and extract airflow rates also in bedrooms (meaning removing the cascade flow) [24], automatic stop (or reduction) of the air supply to bedroom when the window is open, or keeping naturally ventilated bedrooms outside the MVHR-system [25].

4.3. *Single room ventilation with heat recovery (SRVHR)*

Several different systems and sizes of decentralized ventilation with heat recovery can be found, from tiny regenerative units to larger counter-flow boxes. These can be completely integrated into the facade or mounted as a box inside the building envelope. Usually no internal ductwork is needed, resulting in lower pressure losses, lower fan consumption and easier installation. For renovation purpose this is a significant advantage, also affecting costs. Natural ventilation in bedrooms, combined with SRVHR in other room has also been used, but no study on this is found. In addition, SRVHR supports temperature zoning as they operate separately for the different rooms.

Single room units often have higher noise levels [20]. To achieve acceptable noise levels, users might reduce the airflow rate, resulting in unsatisfactory IAQ. Some devices have very simple filters, less efficient for pollen [26]. Due to smaller fans with low pressure increase, these systems are more sensitive to the dynamic pressure from the wind and the stack effect. This can cause unbalanced flows and thus lower heat recovery or increased frost risk. Dual units with axial fans are more affected [27, 28]. Nominal heat recovery efficiencies are in the same range as centralized heat exchangers (or only slightly below). Coydon [29] points out that the common methods to evaluate the energy performance of ventilation devices do not take into account all the important aspects, especially for façade integrated ventilation systems. He has developed a calorimetric evaluation method which seems to be more accurate for regenerative heat exchangers. Using this method on façade integrated heat exchangers, one regenerative with alternating flows gets an efficiency of 73,3%, and a counter flow 79,3%.

4.4. *Natural ventilation*

A number of detached houses are renovated without installing balanced ventilation. Keeping the original natural ventilation system is simple and cheap, without intervention. However, if the air change was based on general infiltration, this strategy collapses completely when the house gets airtight. Lack of ventilation strategy is not comparable to implementation of a natural ventilation strategy. Using natural ventilation requires a systematic approach, with careful design of air vents and stacks, to utilize and control wind- and buoyancy forces. The main challenge with natural ventilation is the lack of heat recovery, leading to colder supply air and a higher space-heating demand. Another issue is to control the ventilation airflow rate. With a purely manual control, the system is dependent on the occupant skills and behaviour. However, automatic window and vent control systems are possible. Sensor-assisted manual control (based on CO_2 or relative humidity) is so far not much investigated

Another issue is the pollution and sound levels outdoors. Some studies have shown low IAQ in naturally ventilated buildings, although this is not always the case. Kamendere [16] measured IAQ in two identical renovated apartment buildings, one with new MVHR and one with the original natural ventilation system. The building had CO_2-concentrations above 1000 ppm 68% of the time for MVHR and 40% using natural ventilation. The naturally ventilated building also had slightly higher temperatures and lower RH. From the CO_2-level measurements, it was concluded that the residents regularly ventilated by opening windows (partly compensating for variation in air change rates due to variable outdoor temperature and wind). There are also some reports of occupants blocking ventilation supply terminals in natural ventilated houses. This is usually explained as aiming to prevent draft (due to low temperature of the supply air and poor air mixing, resulting in downward movement of cold air), and a lack of understanding of the real need for sufficient air change. Partial blocking of supply air terminals is also reported in some cases with new MVHR [8]. This seems to depend on design of the inlet and localization of the draft, exposing people or not.

4.5. Ventilated windows (VW)

With exhaust- or natural ventilation, it is still possible to preheat the supply air to some extent by letting it pass through a culvert, or between the glass panes in a ventilated window (VW). The motivation is improving thermal comfort and reducing energy use. The air brings most of the heat loss through the window back in, and typical efficient U-values of 0,4 -0,5 W/m2*K [30]. A cooling effect can also be achieved by letting the air flow between the panes returns to the outside. Liu [31] has investigated the optimal configuration of the panes, for comfort and energy use. Simulations by Carlos et al [32] for Oslo-climate find a preheating of 2-5 degrees without sun, at typical airflow rates, most at the lowest outdoor temperatures. VW might result in more frosted windows on cold days, due to colder outer glass surface.

5. Combined systems (heat and ventilation)

Only technologies that need a specific discussion in the context of renovation are included in this section. Other technologies (like pellet boilers) needing no specific treatment in this context are therefore not discussed here.

5.1. Exhaust (extract) air heat pump (EAHP)

Mechanical exhaust ventilation has some advantages. It requires only one fan, a less complicated duct system and low pressure in the building minimizing the risk of condensation in the construction. The disadvantage is the lack of heat recovery and the risk of poor thermal comfort when cold air is brought directly through the building wall. Heat pumps using the extract air as the heat source can recover more energy than air-to-air heat exchangers. In addition, they can provide both space-heating and heating of domestic hot water (DHW). The maximum heat output is limited by the ventilation air flow, and this used to be limited to around 2 kW for a detached house [33]. However, 7-8 years ago, a new inverter-controlled heat pump that can cool the air to a lower temperature was introduced in the market, leading to 4-6 kW heating capacity. With laboratory testing, they found a typical seasonal coefficient of performance (SCOP) of 2.2 to 3.0 for these new heat pump types and 1.5 to 2.0 for the older type. This reported SCOP combined both space-heating and DHW.

Mikola [34] made field studies on EAHP into refurbished apartment buildings with exhaust ventilation in Estonia. He described several measures important for success: reducing thermal losses through the envelope and using the heat for both space heating and DHW. Also recommends using a low temperature heating graph, with a supply temperature from 55-60 to 40°C according to the outdoor temperature. However, lower radiator temperature reduces the ability to counteract down-draft from ventilation air, and he reported people reducing air change rates due to cold unheated supply air. This could indicate that the combination of exhaust ventilation and reduced temperatures in the existing radiators is not an optimal combination [22]. With floor heating (FH) or ventilation radiator (VR), the thermal comfort is reported to be better. With a ventilation radiator, outdoor air flows in through a duct in the wall and is heated by the radiator panels before entering the room (Figure 1). The higher air velocity and temperature difference increases the heat transfer and allows a lower water temperature. Several studies [35-37] show that VR provide a slightly better thermal comfort, although FH and conventional high-temperature radiators also proved to be satisfactory. When supplied by a heat pump, FH allowed a lower supply temperature and electricity use than VR and conventional radiators [35]. If used in highly-insulated houses, VR could cause problems with cold draft when the outdoor temperature is near the balance temperature of the house. As the radiators are not heating, supply air will not be

IOP Conf. Series: Earth and Environmental Science **352** (2019) 012048 doi:10.1088/1755-1315/352/1/012048

preheated [37]. Gustafsson also made an energy performance comparison of three HVAC systems for a renovated semi-detached detached house. With the insulation level of the EnerPHit-standard (renovation standard from Passivhaus Institute), in Stockholm climate, balanced ventilation with heat recovery and EAHP had the lowest energy consumption. Second best was exhaust ventilation with EAHP and VR. With slightly higher infiltrations (e.g. 1.5), or smaller ventilation rates (e.g. 0.35), the situation would change.

Despite similar climates, EAHP are frequent in Sweden and Finland but not in Norway. For many years EAHP were installed in 90% of all new built detached houses in Sweden[33]. Part of the reason could be the building code that favored EAHP. The Swedish code counts *delivered energy* and thus rewards the heat captured by the EAHP. In Norway, the requirements are only set for the *net energy demand (i.e. neglecting the heating system efficiency)*, so only part of the captured energy is accounted for. Another important aspect is probably the ventilation air flow requirements more strictly specified for different rooms in Norway.

A more precise term would actually be *extract* air heat pump, as most of these use the heat in the extract air. However, some compact units (see next subsection) include a heat pump that use the heat in the *exhaust* air (meaning after part of the heat is exploited by an air-to-air heat recovery unit). Østergaard [38] made a theoretical investigation on typical Danish detached houses from the 1990's. The study found that these houses can be heated with low-temperature heating with supply and return temperatures below 55°C/35°C for large parts of the year, with the existing radiators. When gone through reasonable energy renovation, these houses can be heated with a supply temperature below 50°C for more than 97% of the year. MVHR was assumed, thus it was a limited need to counteract the draft from the supply air.

5.2. Compact unit with heat pump
The standard compact unit consists of an air-to-air heat exchanger, a small exhaust air-to-water heat pump, a DHW tank and a supply air post-heater. Some of the tanks are equipped with coils for receiving solar heat (fig 2). The total power available from the ventilation air is only enough to heat highly-insulated buildings. Heating demand might be too large in a renovated house, also causing uncomfortably high air heating temperatures. However, the possibility of assistance from a drain-water heat exchanger (also called grey-water heat recovery) or a wood stove for peak heating could be investigated. Compact heat pump units that add ambient air when more energy is needed, are recently developed. These can definitely supply sufficient heat for renovated houses. Some of these also provide hydronic space-heating [39].

6. Conclusion

Requirements for HVAC-solutions for renovation of detached houses are suggested. Key technologies are described and analysed in the context of renovation. A main part of the contribution is a decision-making table. Some knowledge gaps have been identified. Some key conclusions are listed below:

- A number of demonstration projects on ambitious energy upgrading are completed. Nevertheless, few of them have been systematically monitored and evaluated.
- Improved airtightness after renovation makes systematic ventilation measures necessary
- A wider range of concepts and strategies than commonly used in Norway seem to be relevant.
- Whether or not the house has a hydronic space heating seem to be crucial for the choice of system. Most Norwegian detached houses do not have hydronic space heating. Installing this would be a major cost and intervention.
- Assumptions for occupant preferences and behaviour also seem to be important for choice of system.
- Between the Nordic countries, there are differences in commonly used HVAC concepts and strategies for renovation. They cannot be explained by climate differences only. Differences in building code, energy pricing and tradition may be part of the explanation.

IOP Conf. Series: Earth and Environmental Science **352** (2019) 012048 doi:10.1088/1755-1315/352/1/012048

Decision table **Qualitative assessment for planning process**	Energy-efficiency	Investment	Need hydronic	Space for technical	Need ducts	Noise	Filtering	Air quality	Temperature-Zoning	Cold draft	Occupant sensitive
Balanced ventilation	+ +	÷ ÷		÷	÷ ÷	?	+		÷		
Decentralized balanced ventilation	+	÷				÷ ÷	?		+		
Balanced.ventilation + window ventilation bedroom	+	÷		÷	÷		?		+		÷
Decentral.bal.ventilation + window ventilated bedroom	+	+					?		+		÷
Extract air heat pump, to DHW + ventilated window	+	÷			÷				+	÷	
Hybrid ventilation: balanced winter, natural summer	+ +	÷ ÷		÷	÷		?		?		
Natural ventilation	÷ ÷	+ +		+	+	?	÷	?	+	÷	÷
Extract air heat pump + ventilation radiator	+	÷	÷	÷	?				+		
Compact HVAC-unit	+ +	÷	?	÷	÷	?	+				
Ground source heat pump	+ +	÷ ÷	÷	÷							
Hydronic woodstove			÷	?							?
Solar thermal		÷	÷	?							
Air-air heat pump	+					÷					
Wood stove (not only peak)	+	+									÷
Pellet stove		÷									
Electric resistance heating	?	+ +		+							

Table1. Decision making table for qualitative assessment of HVAC-solutions for energy renovation of detached houses. Listed pros and cons estimated for different criteria.

References

[1] A. C. H. Bøeng, Bente and B. M. Larsen, "Kartlegging av oppvarmingsutstyr i husholdningene," NVE, SSB, 2014.

[2] V. Leivo, M. Turunen, A. Aaltonen, M. Kiviste, L. Du, and U. Haverinen-Shaughnessy, "Impacts of Energy Retrofits on Ventilation Rates, CO2-levels and Occupants' Satisfaction with Indoor Air Quality," *Energy Procedia,* vol. 96, pp. 260-265, 2016.

[3] E. M. Wells *et al.*, "Indoor air quality and occupant comfort in homes with deep versus conventional energy efficiency renovations," *Building and Environment,* vol. 93, pp. 331-338, 2015.

[4] L. Georges, K. Wen, M. J. Alonso, M. Berge, J. Thomsen, and R. Wang, "Simplified space-heating distribution using radiators in super-insulated apartment buildings," *Energy Procedia,* vol. 96, pp. 455-466, 2016.

[5] M. Berge, L. Georges, and H. M. Mathisen, "On the oversupply of heat to bedrooms during winter in highly insulated dwellings with heat recovery ventilation," *J Building Environment,* vol. 106, pp. 389-401, 2016.

[6] J. Thomsen, L. Gullbrekken, S. Grynning, and J. Holme, "Evaluering av boliger med lavt energibehov (EBLE)–samlerapport," 2017.

[7] M. Klinski, E. Fredriksen, K. Sørnes, and A. Svensson, "Etterevaluering av EKSBOpilotprosjekt. Enebolig i Asker," 2016, vol. SINTEF-notat 23.

[8] K. E. Thomsen *et al.*, "Energy consumption and indoor climate in a residential building before and after comprehensive energy retrofitting," *Energy and Buildings,* vol. 123, pp. 8-16, 2016.

[9] M. G. Bjørneboe, S. Svendsen, and A. Heller, "Evaluation of the renovation of a Danish detached house based on measurements," *Energy and Buildings,* vol. 150, pp. 189-199, 2017.

[10] A. Dugué, S. Raji, P. Bonnamy, and D. Bruneau, "E2VENT: An Energy Efficient Ventilated Façade Retrofitting System. Presentation of the Embedded LHTES System," *Procedia Environmental Sciences,* vol. 38, pp. 121-129, 2017.

[11] T. Psomas, P. Heiselberg, T. Lyme, and K. Duer, "Automated roof window control system to address overheating on renovated houses: Summertime assessment and intercomparison," *Energy and Buildings,* vol. 138, pp. 35-46, 2017.

[12] J. Kurnitski, L. Eskola, J. Palonen, and O. Seppänen, "Use of mechanical ventilation in Finnish houses," in *Proceedings of 2nd European BlowerDoor-symposium*, 2007, pp. 152-161.

[13] M. Baborska-Narozny and F. Stevenson, "Continuous Mechanical Ventilation in Housing □ Understanding the Gap between Intended and Actual Performance and Use," *Energy Procedia,* vol. 83, pp. 167-176, 2015.

[14] A. G. Lien, K. S. Skeie, E. O. Bjaanes, K. Hagen, and Y. Kvalø, "Oppgradering av et 60-tallshus og et 70-tallshus," 2017.

[15] G. Dermentzis, F. Ochs, D. Siegele, and W. Feist, "Renovation with an innovative compact heating and ventilation system integrated into the façade – An in-situ monitoring case study," *Energy and Buildings,* vol. 165, pp. 451-463, 2018.

[16] E. Kamendere, G. Zogla, A. Kamenders, J. Ikaunieks, and C. Rochas, "Analysis of Mechanical Ventilation System with Heat Recovery in Renovated Apartment Buildings," *Energy Procedia,* vol. 72, pp. 27-33, 2015.

[17] H. M. Tommerup, "Measurement results and experiences from an energy renovation of a typical Danish detached house," in *8th Symposium on Building Physics in Nordic Countries*, 2008: Danish Society of Engineers, IDA.

[18] N. Lassen, A. Fylling, M. Mysen, T. H. Dokka, and L. Bordewich, "Passivbygg som forskriftskrav i 2020," Report. 2009.

[19] A. J. B. Petersen, Ida, P. Schild, E. N. Haugen, G. Nilson, and K. Høydahl, "Forhold tilknyttet bruk av roterende gjenvinnere i skoler. ," 2009.

[20] A. Merzkirch, S. Maas, F. Scholzen, and D. Waldmann, "Field tests of centralized and decentralized ventilation units in residential buildings – Specific fan power, heat recovery efficiency, shortcuts and volume flow unbalances," *Energy and Buildings,* vol. 116, pp. 376-383, 2016.

[21] R. Östin, "Evaluation of a Single Family Low Energy Building in Cold Climate," *Energy Procedia,* vol. 132, pp. 9-14, 2017.

[22] L. La Fleur, B. Moshfegh, and P. Rohdin, "Measured and predicted energy use and indoor climate before and after a major renovation of an apartment building in Sweden," *Energy and Buildings,* vol. 146, pp. 98-110, 2017.

[23] L. Georges, F. Håheim, and M. J. Alonso, "Simplified Space-Heating Distribution using Radiators in Super-Insulated Terraced Houses," *Energy procedia,* vol. 132, pp. 604-609, 2017.

[24] E. Selvnes, "Thermal zoning during winter in super-insulated residential buildings," NTNU, 2017.

[25] V. Heide, "Energieffektivitet med åpent soveromsvindu i passivhus," presented at the Passivhus Norden 2013, Gøteborg, 2013.

[26] H. Manz *et al.*, "Performance of single room ventilation units with recuperative or regenerative heat recovery," vol. 31, no. 1, pp. 37-47, 2000.

[27] T.-A. Koiv, A. Mikola, and R. Simson, "Ventilation solutions in renovated apartment buildings in cold climate conditions," *J Engineering,* vol. 7, no. 03, p. 129, 2015.

[28] A. Mikola, R. Simson, and J. Kurnitski, "The Impact of Air Pressure Conditions on the Performance of Single Room Ventilation Units in Multi-Story Buildings," *Energies,* vol. 12, no. 13, p. 2633, 2019.

[29] F. Coydon, S. Herkel, T. Kuber, J. Pfafferott, and S. Himmelsbach, "Energy performance of façade integrated decentralised ventilation systems," *Energy and Buildings,* vol. 107, pp. 172-180, 2015.

[30] L. M. Raffnsøe, "Thermal Performance of Air Flow Windows–Master thesis," Technical University of Denmark, Department of Civil Engineering, 2007.

[31] M. Liu, P. K. Heiselberg, O. K. Larsen, L. Mortensen, and J. Rose, "Investigation of Different Configurations of a Ventilated Window to Optimize Both Energy Efficiency and Thermal Comfort," *Energy Procedia,* vol. 132, pp. 478-483, 2017.

[32] J. S. Carlos and H. Corvacho, "Ventilated double window for the preheating of the ventilation air comparison of its performance in a Northern and a Southern European climate," *Journal of Renewable Energy,* vol. 2013, 2013.

[33] M. Lindahl, C. Haglund Stignor, K. Andersson, and S. Thyberg, "Exhaust air heat pumps evaluated for Nordic circumstances," in *11th IEA HPP*, 2014.

[34] A. Mikola, "Problems with Using the Exhaust Air Heat Pump for Renovation of Ventilation Systems.pdf," 2015.

[35] A. Hesaraki, E. Bourdakis, A. Ploskić, and S. Holmberg, "Experimental study of energy performance in low-temperature hydronic heating systems," *Energy Buildings* vol. 109, pp. 108-114, 2015.

[36] Q. Jin, A. Simone, B. W. Olesen, S. Holmberg, and E. Bourdakis, "Laboratory study of subjective perceptions to low temperature heating systems with exhaust ventilation in Nordic countries," *Science and Technology for the Built Environment,* vol. 23, no. 3, pp. 457-468, 2016.

[37] M. Gustafsson *et al.*, "Energy performance comparison of three innovative HVAC systems for renovation through dynamic simulation," *Energy and Buildings,* vol. 82, pp. 512-519, 2014.

[38] D. S. Østergaard and S. Svendsen, "Theoretical overview of heating power and necessary heating supply temperatures in typical Danish detached houses from the 1900s," *Energy and Buildings,* vol. 126, pp. 375-383, 2016.

[39] Flexit, "https://www.flexit.no/globalassets/catalog/documents/26099_bro_116914no.pdf ".

IOP Conf. Series: Earth and Environmental Science **352** (2019) 012049 doi:10.1088/1755-1315/352/1/012049

On the relation of thermal comfort practice and the energy performance gap

Runa T Hellwig

Department Architecture, Design and Media Technology, Aalborg University, Rendsburggade 14, 9000 Aalborg, Denmark

rthe@create.aau.dk, ORCID: 0000-0002-5882-8033

Abstract. Recent research results from different countries show that although energy efficiency measures in buildings indeed led to lower energy use in buildings, there is a performance gap between the calculated energy use and the actual measured energy use in energy-efficient houses, leading to a higher energy use than predicted. Thermal comfort related behaviour is one out of manifold reasons contributing to this performance gap. Thermal comfort requirements are based on objectively measurable parameters. A number of contextual factors impact an individual's thermal comfort perception and preference. Technological opportunities and material arrangements offer several ways to conditioning indoor environments. Research shows that they shape the occupants' thermal comfort attitudes. Over time, technology as conditioning practice and insulation has led to different thermal comfort practice in buildings contributing to this performance gap. As humans show an excellent adaptation potential towards a wide range of temperature, enabling them to adapt to diverse climates but also seasonally, it follows also the adaptation process can work in the opposite direction. Hence, that with reduced exposure to outdoor weather and more narrow temperature ranges inside building humans might also adapt to indoor thermal conditions and get more sensitive to small indoor temperature changes, leading probably to higher indoor temperature over time ("indoor exposure rebound"). As our energy conservation efforts of the recent years show less effects than expected, it seems that the two mainly applied sustainability strategies efficiency and consistency have limited effects as they are affected by rebound phenomena. Sufficiency, as the third sustainability strategy, is not yet a generally accepted strategy. It refers to what has been described as "the right measure". The question of what would be "the right measure" of indoor thermal comfort, meaning what thermal conditions would be sufficient, can be raised. Based on a discussion of findings from literature, it will be concluded that there is a need for a new thermal comfort thinking in climates which have the need for seasonal or all year round active conditioning leading to a more sufficient conditioning practice.

1. Introduction

The need to adapting our built environment and the way we design and use it is a consequence of the progressing climate change. It demands us to rethinking our accustomed routines, behaviours and expectations. It is commonly agreed upon that we need sustainable ways to shape and use our built environment. Sustainability is defined via ecological, economical and socio-cultural quality criteria. In recent years, we have been rather successful in developing energy conservation strategies for new buildings as well as for energy renovation of the building stock. Quality management procedures and

IOP Conf. Series: Earth and Environmental Science **352** (2019) 012049 doi:10.1088/1755-1315/352/1/012049

measurements have been developed and applied in the building process in order to make sure that the planned energetical performance of a building comprising its envelope and building service systems can be realised in the expected way. Starting with the first low-energy houses in the 80ties and 90ties, we have been able to continuously lowering the energy demand for heating so that we now talk about nearly Zero Energy Buildings (Energy Performance of Buildings Directive EPBD 2010), Zero Energy Buildings or even Plus Energy Buildings – not on research project level or demonstration project level solely - but already on the market with pilot first users. Calculation procedures have become more sophisticated and the system boundaries have been extended from energy use to end energy, primary energy and the generation and use of energy on-site. The technological knowledge on how to design buildings for energy efficiency and the use of renewables gained in research and demonstration projects over a period of 30 years has also shaped our legislation (e.g. in Europe Energy Performance of Buildings Directive EPBD 2010, e.g. on national level in Germany "Wärmeschutzverordnung" WSVO 1995, "Energieeinsparverordnung" EnEV 2002 et seqq.) leading to an improved energy efficiency practice. With the first implementation of the EPBD ranges for comfortable indoor environments (thermal, air quality, acoustical and visual) have been defined in standards (EN 15251 [14] and its successor EN 16798 [15]) in order to ensure that energy conservation does not lead to unacceptable indoor environmental conditions. In recent studies, using data-sets on energy labelled residential buildings (e.g. [18]) it could be shown that there is an energy performance gap between the calculated energy demand and the actual energy used in the real building.

Besides the limitations and assumptions incorporated in energy demand prediction models, reasons for this performance gap are seen in the way buildings are used and users behave in these buildings (e.g. [21]). Thermal comfort practice is one important contributor as it seems to be subject to rebound effects[1]. The aim of this paper is to discuss the relation of thermal comfort and the energy performance gap; how the understanding of what thermal comfort means shapes the way buildings are planned and operated. It will be discussed how other sustainability strategies than energy efficiency can be used to unlock further saving potential. Sufficiency as a sustainability strategy addressing the relation between humans and their environment will be discussed as a key strategy leading, firstly, to a broader understanding of this relationship and, secondly, to developing more sufficient ways of how to design and operate indoor spaces comfortably. In this paper conditioning for thermal comfort in winter is the main focus.

2. Indoor temperature yesterday and today

Over the years, we have been observing an increase in room temperature in winter leading to increased energy use in buildings. Borsch-Laaks [6, p.27] describes the development of the room temperatures depending on the technologies available during certain periods, starting in the 17. and 18. Century, leading to *mean* room temperatures of about 10°C in kitchens or what today would be called the living room. In the early 20ies century all rooms had a heating device (stove) but bedrooms and side rooms normally were not heated. Grytli and Støa [20] point out the impact a central heating systems and electrical heating systems had on the development of more freely arranged layouts of the floor plans of residential buildings in Norway.

In 1858, Pettenkofer [28, p78] recorded an air temperature of 15°C in a classroom in January. Markham's (1947) findings (quoted in [1], p17) give us an impression of the increasing winter comfort temperatures in the UK: around 1900 it was 60°F (15.5°C) and in 1920 it was 64°F (17.8°C). The 1946 British Standards Code of Practice of Warmth in Houses gives a range of 60 to 68°F (15.5 to 20°C) for living rooms (from [1], Tab 2.1). According to a survey of Welch (1960, referenced in [1]), in new buildings with thermostats occupants opted for temperatures of 68 to 70°F (20 to 21.1°C).

[1] A gain in efficiency (here: energy efficiency) which does not lead to the same magnitude of reduction in use of a good (here: energy) is subject to a rebound effect.

IOP Conf. Series: Earth and Environmental Science **352** (2019) 012049 doi:10.1088/1755-1315/352/1/012049

In 1925, an instruction on how to operate a heating system in a German Ministry's building said that with permanent heating 18°C should not be exceeded as 19°C would be too warm for many persons (quoted in [7], p84/85). In 1935, the German Health Authority recommended winter temperatures between 17.5 and 18.5°C for decentralised or centralised heating and stated that a room with a temperature above 21°C is regarded as being "overheated" [7]. Rothfeld [29] stated 1916 that *more* than 20°C in classrooms would impair the children's learning performance. In East-German indoor climatic guidance values for residential indoor environments in the heating period were 18-23°C [2]. Today a minimum of 20°C is required in winter (Category II according to EN 15251 [14] and its successor [15]) most often far higher temperatures are measured.

From the UK we know about systematic analysis whether indoor temperatures have risen over the years [31] with the result that temperature in rented homes may have increased over time but not owner-occupied houses of the building stock. In the mid 80ies, temperature measurements in first German demonstration projects on energy efficient buildings ("Solarhäuser Landstuhl") showed a temperature practice between 17 to 23°C in the living rooms and 16°C to 21°C in the bedrooms [19]. Already back then, three types of users were defined: the energy saving user having a mean day-time temperature of 18°C and a night set-back temperature of 15°C, the normal user having a mean day-time temperature of 20°C and the energy wasting user with 22°C without having any night set-back. Eight years later, another low-energy house demonstration project ("Niedrigenergiehäuser Heidenheim") showed mean temperatures during the heating period between 17 and 21 °C [16]. In a Danish study it was found that users in houses with better energy labelling tend to have higher indoor temperatures compared with users of houses in buildings with less energy-efficient building label [21].

3. Energy performance gap

Based on 3400 German homes of the building stock, presumably not energy efficient homes, Sunikka-Blank and Galvin [33] found that the actual measured energy consumption is on average 30 percent points lower than the calculated energy demand. They call this the pre-bound effect.

Gram-Hanssen and Hansen [18] analysed data of actual measured energy consumption data of energy labelled detached houses from Denmark showing that for energy-*inefficient* houses the actual energy use is much *lower* than the calculated energy demand (Figure 1). Contrary, the energy-*efficient* houses show an actual energy use which is *higher* than the calculated energy use. They conclude that users adjust their behaviour in energy-*inefficient* houses, leading the lower energy use than expected.

4. Thermal comfort practice

Thermal comfort practice involves several parties, mainly the users, the building professionals (planners, operators) but also companies selling and advertising indoor climate conditioning equipment. In the following possible sources for thermal comfort related rebound effects are discussed.

4.1. Rebound effects related to thermal comfort practice

The performance gap between the calculated and the actual energy use of buildings can partly be explained with changed conditioning practice for thermal comfort. These are:

- extended availability of conditioning systems to more rooms, e.g. through the implementation of central heating systems (spatial rebound: entire unit vs selected spaces of a unit as described in section 2),
- changed conditioning schedules (temporal rebound: intermittent/night set-back or shut-off vs permanent, e.g. [19]),
- changed occupant behaviours (behavioural rebound, e.g. clothing habits, [21]), and
- changed temperature regime (extent rebound: e.g.[21]).

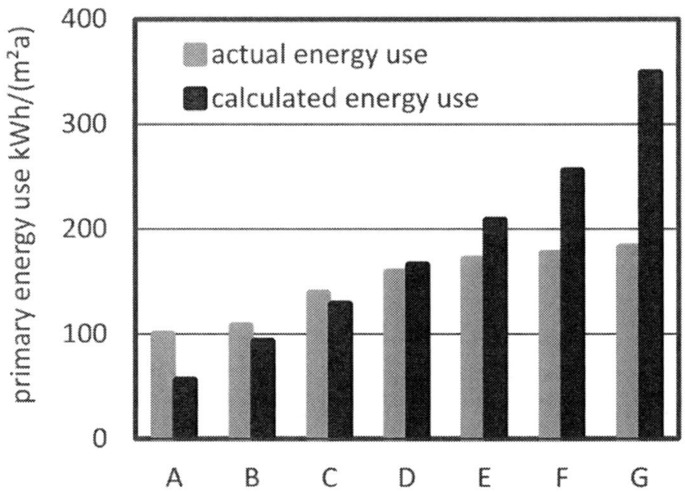

Figure 1: Energy performance gap – Example Danish data, N=135,311 homes, based on data from [18], mean values, variance not shown, A to G are energy efficiency labels.

4.2. Attitudes towards thermal comfort

In the 70ies thermal comfort was implemented as a product which can be sold. This leads to the expression of "providing thermal comfort". Building professional can provide a certain temperature range. But an occupant's comfort perception depends on a variety of other factors, which are often factors that cannot be measured or influenced e.g. by the facility manager. That these contextual factors (e.g. outdoor climate, behavioural factors, psychological factors as personal control) play a major role has been shown in numerous research results (e.g. [27], [3]; [10], [23]). De Dear et al. [10, p.3] asked whether building professionals can be seen as providers of comfort and occupants as passive recipients of comfort? Bordass and Leaman [7, p.192] suppose that the focus on comfort provision as a service of building professionals "...may deny occupants simple facilities for discomfort alleviation..." in the design process. Could it also add to an occupant's impression that the locus of indoor climate control was an external one [24]?[2] The often expressed opinion among building professionals that occupants exert unsuitable behaviour would also be supported by using the term "comfort provision". Could this attitude lead to occupants, e.g. in office buildings who would demand changes from the "comfort provider" instead of taking actions themselves, leading to a higher complaint rate? A facility manager or planner saying: ‚Occupants always open the windows which has a negative impact on the energy consumption; they better shouldn't have access to windows!‘ can cause occupants thinking they would not have the capability to open the window at a suitable time (negative social verbally persuasion[3]). From this argumentation it might follow that occupants may not feel responsible to *seek comfort* because comfort would be *provided* [24]. On the contrary, other research has shown, that occupants who are concerned about environmental issues react more relaxed

[2] The concept of locus of control has been used to describe generalised expectancies towards the belief of being in a position (internal locus) or not (external locus) to cause a change [30]. This concept was applied as one impact factor on the individual's level of control in a *conceptual approach of personal control in indoor environments* by [23].

[3] The concept of self-efficacy developed by Bandura [4] was applied as one impact factor on the individual's level of control in a conceptual approach of personal control in indoor environments by [23].

IOP Conf. Series: Earth and Environmental Science **352** (2019) 012049 doi:10.1088/1755-1315/352/1/012049

when in a green[4] building the thermal comfort conditions are somewhat outside the normally experienced ranges [26, 12, 13].

4.3. Comparable comfort conditions

The basis for energy efficiency comparison is the approach of "*equivalence of service*". The principle of *equivalence of service* (here thermal comfort) is what the European Energy Performance of Buildings Directive sets as precondition for energy efficiency comparison of different solutions. The energy performance can only be compared within the same building usage (e.g. residential) and with the same thermal comfort. Thermal comfort is expressed in this cas by the operative temperature the building is designed for and has the capability to deliver these conditions on demand. Shove [32] argues that this approach is one driver stabilising "...contemporary, but often recently established ideas, for instance about the meaning of comfort...", "...reinforcing the idea that such interpretations [here: thermal comfort requirements] are non-negotiable...". A planner might perceive a high pressure if he could not meet this *basic need*, followed by e.g. litigation issues. This is what actually happens in planning practice, that the planning does follow the explicit requirements laid down in a standard but seldom adopts the opportunity to fulfil the *goal* or *intention* of a standard in a *different way* than laid down in the standard.

4.4. Indoor exposure rebound

Thermal comfort has been described as a self-regulating system [27]. Based on the principles of human thermoregulation and the physical principles of heat exchange between humans and their environment, numerous studies have shown that there is more factors than pure thermoregulation that affect thermal comfort perception. The theory of adaptive thermal comfort mentions among others physiological adaptation, which stands for acclimatisation processes when exposed to different seasons or when moving from one climatic zone to another (e.g. [3], [34]). Also, psychological adaptation comprising expectations, learnt attitudes or individual concepts of what thermal comfort means play a role (e.g. [8]). These adaptive principles result into an excellent human adaptability to a wide range of temperatures. Thus if humans do rarely spent time outdoors they might adapt to *prevailing indoor temperatures*[5] instead. As we observing higher temperatures in energy-efficient homes (see section 2) this could also be seen as an effect of *indoor exposure adaptation*, hence it could be called an "*indoor exposure rebound*". It might be explained by the fact that exposure to more stable and narrow temperature ranges changes the expectations towards indoor temperatures and at the same time the human thermoregulatory system might not be "experienced" anymore with temperatures outside this narrow temperature range the body has adapted to. The human body may also get more sensitive to small temperature changes. Therefore the body might get more alert and therefore probably leading to discomfort perception followed by the user adjusting the thermostat to a new set-point.

5. Sustainability goals and strategies to get there

Sustainability rating systems aim at balancing the ecological, economical and socio-cultural aspects of our built environment. Thermal comfort is part of the socio-cultural set of sustainability criteria.

[4] Green stands here for all kind of buildings which's performance is described to be more sustainable than standard buildings.

[5] Prevailing indoor temperatures in the heating period are designed following the classic heat balance approach according to [14, 25]. The adaptive model [14] can be applied for periods of the year in which no active conditioning system is operating.

The assessment criteria are set according to a societies concepts about what sustainability means. To the authors individual perception, the definition of thermal comfort criteria in the frame of sustainability rating systems has lead to the above described reinforced attitude that comfort is not negotiable. Besides *criteria* for the assessment of the degree of sustainability of a building, there is also *strategies* to reach sustainability: efficiency, consistency and sufficiency. Efficiency (less resource use per unit of service) and consistency (recycling, renewable energy) have been widely accepted and have been already implemented in design, planning or operation procedures. They are known to be connected to rebound effects as descried earlier. The third strategy, sufficiency is less known and not yet common sense [17]. Sufficiency refers to the right measure. And is often associated to behavioural changes. Sufficiency can be influenced through – amongst other factors which have been described already [5] - changes in the socio-cultural column; here: *Thermal comfort.*

6. Towards sufficient thermal comfort - conclusions

Not getting disturbed thermally, hence reaching thermal comfort is what we are seeking for. However, since we have numerous technological means to make indoors spaces comfortable it seems that we got adapted to the indoor conditions we have created ourselves, still seeking to increase comfort which seems to lead to higher heating set-points, hereby not contributing to less energy use. At the same time we have learnt that a *maximised* comfort would be a *basic need.* While we are using all these technological means to *maximise* our comfort we seem to become more sensitive to temperature changes and seem to have forgotten about the excellent human adaptability to temperatures.

From the above discussed facts and perspectives we can conclude that the energy performance gap we have been observing cannot be overcome without addressing the user's role in thermal comfort perception. Furthermore, the focus in thermal comfort should be more on contextual indoor environmental factors than purely on temperature (and related factors as air velocity etc.). Instead, we should expect the user's to play an active role in this process, and support them in doing so by planning for e.g. an appropriate degree of personal control of the indoor environment or inform them about the intended functioning of a building as part of behavioural adaptation processes.

What is needed is a new understanding and thinking approach which addresses the question for the *right measure of thermal comfort* in climates which have the need for seasonal or all year round active conditioning leading to a more sufficient conditioning practice. The physical principles of heat exchange between humans and their environment and human thermoregulation are *one basis* of thermal comfort. However, the often neglected contextual factors should get more attention and become part of the planning practice.

There is already research results [e.g. 26, 12, 13] showing, that, provided the users are conscious about the "green" performance of their building and understand its importance, the controls are usable and they got factual information on how to make use of certain technological means to adapt [23], they will be able to use their building in the intended way. Shove [32, p.8] concludes that a solution would be to design buildings "...that do not meet present needs, and that do not deliver equivalent level of service, but that do enable and sustain much lower-carbon ways of life", hence with regard to the topic discussed here: more sufficient thermal comfort practice.

Acknowledgement
The author would like to thank the Obelske Familiefond, Denmark for supporting this work.
This work has been performed within the framework of the International Energy Agency - Energy in Buildings and Communities Program (IEA-EBC) Annex69 "Strategy and Practice of Adaptive Thermal Comfort in Low Energy Buildings". www.iea-ebc.org, www.annex69.org

This paper is based on previous talks/lectures of the author of this paper at the following events:
1. "Some thoughts …On thermal comfort and sustainability strategies in the built environment" Seminar at the Indoor Environmental Quality Laboratory, Sydney School of Architecture, Design and Planning, The University of Sydney, Australia 15. August 2018,
2. "On the relation of thermal comfort and sustainability strategies in the built environment" held at the "Thermal Comfort and Low Energy Cooling" Symposium organised by CEPT University, Ahmedabad, India on 31. October 2018.
3. "From energy efficiency towards sufficiency strategies for indoor environmental design" held as inauguration lecture at Department of Architecture, Design and Media Technology, Aalborg University, Denmark on 2. May 2019.

References

[1] Angus, T.C. (1968): The control of indoor climate. Pergamon Press, Oxford, London, Edinburgh, New York, Toronto, Sydney, Paris, Braunschweig.
[2] ARK Arbeitsgemeinschaft Raumklimatologie (1972): Empfehlung über raumklimatische Richtwerte für Wohnbauten vom 21. Januar 1972, Stadt- und Gebäudetechnik, 5, 121-126
[3] Auliciems, A. (1981) 'Towards a psychophysiological model of thermal perception', International Journal of Biometeorology, 25, pp. 109–122.
[4] Bandura, A. (1977): Self-efficacy: Toward a unifying theory of behavioral change. Psychological Review, 84, 2, 191-215.
[5] Bierwirth, A.; Thomas, S. (2019): Energy sufficiency in buildings. A concept paper. European Council for an energy efficient economy, 48 pp. www.energysufficiency.org, last accessed 15. May 2019.
[6] Borsch-Laaks, R. (2000): Wohnen ohne Feuchteschäden. Schwachstellen und ihre Vermeidung, Hinweise für Eigentümer und Mieter. Heidelberg, frankfurt, VWEE Energieverlag, 2nd Ed. ISBN: 3-87200-655-X
[7] Bradke, F.; Liese, W. (1952): Hilfsbuch für raum- und außenklimatische Messungen (Guidance to indoor and outdoor climate measurements). In German. 2nd ed., Springer. Berlin, Göttingen, Heidelberg
[8] Brager, G. S. and de Dear, R. J. (2003) 'Historical and cultural influences on comfort expectations', Buildings, culture and environment: Informing local and global practices. Blackwell Publishing Ltd, pp. 177–201.
[9] Bordass, W.; Leaman, A. (1997): Future buildings and their services. Strategic considerations for designers and clients. Buildings Research and Information, 25, 4, 190-195.
[10] de Dear, R. J., Brager, G. S. and Cooper, D. (1997) 'Developing an Adaptive Model of Thermal Comfort and Preference', in Final Report on ASHRAE Research Project 884. Macquarie University Sydney.
[11] de Wilde, P. (2014): The gap between predicted and measured energy performance of buildings: A framework for investigation. Automation in Construction, 41, pp 40-49 http://dx.doi.org/10.1016/j.autcon.2014.02.009
[12] Deuble, M. P., & de Dear, R. J. (2012). Green occupants for green buildings: The missing link? Building and Environment, 56, 21–27. https://doi.org/http://dx.doi.org/10.1016/j.buildenv.2012.02.029
[13] Deumling, R.; Poskanzer, D.; Meier, A. (2017): Who are the low energy users: Lessons for climate policy. Proceedings: eceee 2017 Summer Study on energy efficiency: Consumption, efficiency and limits paper 9-432-17 https://www.eceee.org/library/conference_proceedings/eceee_Summer_Studies/2017/9-consumption-and-behaviour/who-are-the-low-energy-users-lessons-for-climate-policy/, last accessed 22.01.2019
[14] EN 15251:2007: Indoor environmental input parameters for design and assessment of energy performance of buildings addressing indoor air quality, thermal environment, lighting and acoustics. Brussels: CEN (European Committee for Standardization)

[15] EN 16798:2019: Energy performance of buildings - Part 1: Indoor environmental input parameters for design and assessment of energy performance of buildings addressing indoor air quality, thermal environment, lighting and acoustics. (preliminary)
[16] Erhorn, H.; Reiß, J. (Ed.) (1994): Niedrigenergiehäuser: Zielsetzung, Konzepte, Entwicklung, Realisierung, Erkenntnisse. Fraunhofer-Institut für Bauphysik, Stuttgart. 1. Ed, 10/1994
[17] Fischer, C.; Grießhammer, R. (2013): When less is more. Sufficiency: Terminology, rationale and potentials. Oeko-Institute's Working paper 2/2013. Oeko-Institue e.V., Freiburg, Darmstadt, Berlin, October, 2013, www.oeko.de/oekodoc/1879/2013-007-en.pdf, last accessed 29.12.2017
[18] Gram Hanssen, K.; Hansen, A.R. (2016): Forskellen mellem målt og beregnet energiforbrug til opvarming af parcelhuse. SBI forlag. https://sbi.dk/Assets/Forskellen-mellem-maalt-og-beregnet-energiforbrug-til-opvarmning-af-parcelhuse/sbi-2016-09-1.pdf
[19] Gruber, E.; Erhorn, H.; Reichert, J. (1989): Cancen und Risiken der Solarchitekture: Solarhäuser Landstuhl. Köln, Verlag TÜV Rheinland. ISBN: 3-88585-676-X
[20] Grytli, E.; Støa, E. (Ed.) (1998): Fra Årestue til smarthus. Teknologien omformer boligen. Oslo, Norsk Arkitekturforlag, ISBN 82-7532-010-0
[21] Hansen A.R.; Gram-Hanssen, K.; Knudsen, H.N. (2018): How building design and technologies influence heat-related habits Building Research & Information, 46:1, 83-98, DOI: 10.1080/09613218.2017.1335477
[22] Hellwig R.T.; Teli, D.; Schweiker, M., Choi, J.-H.; Lee, J.M.C.; Mora, R.; Rawal, R.; Wang, Z, Al-Atrash, F. (2018): Applying adaptive principles – Developing guidance for planning practice. Proceedings Comfort at the Extremes Conference CATE 19, Heriot Watt University Campus Dubai, 10-12 April 2019
[23] Hellwig, R. T. (2015) 'Perceived control in indoor environments: a conceptual approach', Building Research & Information, 43(3), pp. 302–315. doi: 10.1080/09613218.2015.1004150.
[24] Hellwig, R.T. (2018). Revisiting overheating indoors. Proceedings of 10th Windsor Conference: Rethinking Comfort Cumberland Lodge, Windsor, UK, 12-15 April 2018. London: Network for Comfort and Energy Use in Buildings, http://nceub.org.uk, paper 0116, www.windsorconference.
[25] ISO 7730 (2006): Ergonomics of the thermal environment - Analytical determination and interpretation of thermal comfort using calculation of the PMV and PPD indices and local thermal comfort criteria (ISO 7730:2005); German version , May, 2006.
[26] Leaman, A., & Bordass, B. (2007). Are users more tolerant of 'green' buildings? Building Research and Information, 35(6), 662–673. doi:10.1080/09613210701529518
[27] Nicol, F. and Humphreys, M. A. (1973) 'Thermal comfort as part of a self-regulating system', Building Research and Practice, 1 (3), pp. 174–179.
[28] Pettenkofer, M. (1858): Über den Luftwechsel in Wohngebäuden. Münchem, Cotta'sche Buchhandlung
[29] Rothfeld M. (1916): Lüftung und Heizung im Schulgebäude (Ventilation and Heating in School Buildings) In German. Berlin, Julius Springer
[30] Rotter, J.B. (1966): Generalized expectancies for internal versus external control of reinforcement. Psychological Monographs, 80, 1, 1-28
[31] Shipworth, M. (2011): Thermostat settings in English houses: No evidence of change between 1984 and 2007, Building and Environment, 46, 3, 635-642 https://doi.org/10.1016/j.buildenv.2010.09.009
[32] Shove, E. (2017): What is wrong with energy efficiency? Building Research and Information. DOI: 10.1080/09613218.2017.1361746
[33] Sunikka-Blank, M.; Galvin, R. (2012) Introducing the prebound effect: the gap between performance and actual energy consumption, Building Research & Information, 40:3, 260-273, DOI: 10.1080/09613218.2012.690952
[34] Taylor, N.A.S. (2014): Human heat adaptation. Comprehensive Physiology. 4, 325-365. DOI: 10.1002/cphy.c130022

Energy Flexibility for an Institutional Building with Integrated Solar System: Case Study Analysis

Fatima Amara, Vasken Dermardiros, and Andreas K. Athienitis

Department of Building, Civil and Environmental Engineering, Concordia University Montreal, Québec H3G 1M8, Canada

E-mail: `fatima.amara@solarbuilding.ca`

Abstract. Buildings with renewable energy systems and sizeable energy storage capacity can provide significant energy flexibility. Defining appropriate load management strategies requires estimating this flexibility accurately. However, this is no easy task, since the energy flexibility depends on the building structure, heating/cooling system, meteorological conditions, occupant activities, demand response strategy, among other factors. This paper focuses on the thermal storage flexibility of a building component: a concrete floor with an embedded radiant hydronic system. An approach based on a combined adaptive Auto Regressive model with exogenous inputs (ARX) is used to quantify the thermal energy available in the concrete slab. The model describes the relationship between: (a) the geothermal pump system and weather variables and (b) the temperatures of the indoor air and the concrete surface slab. These temperatures are then used to assess the state of charge (SOC) of the slab. The SOC indicator is critical to reducing heating energy use during peak periods. Results show that the proposed model can accurately quantify the energy flexibility of the building through performance indicators. Finally, tests using real data confirm the validity of the model as a tool to estimate energy flexibility.

1. Introduction

Energy flexibility suggests solutions to reduce energy needs while facing weather and occupancy activities, facilitate demand response programs and real-time energy arbitration [1]. According to the upcoming challenges of energy demand-side management the increased uses of Renewable Energy Sources (RES) can significantly compromise the stability of electrical grids, caused by their intermittence of production. Similarly, the buildings without a large share of energy storage systems and batteries can also suffer from the instabilities of RES [2]. To tackle these challenges, energy flexibility of buildings can help to reduce the mismatch between instantaneous power demand and renewable energy production according to local climate conditions, user needs and grid requirements [3, 4]. In that context, buildings as commercial and institutional can supply flexibility strategies using different technologies, (e.g., pumps, natural ventilation, thermal storage in building structure, electricity storage in batteries of electric vehicles and electrical plug-load shifting) [5]. These various technologies could play a key role in reducing the pressure on the energy grids through strategies that combine measures for adjusting and shifting building power use [6, 7].

In 2018, the Canadian Energy Transition announced that electricity and heating in commercial and institutional buildings will be supplied by RES. This aims to increase electrical

Table 1: Nomenclature

Parameter	Definition
t	Discrete time
T_t^{slab}	Concrete slab surface temperature (C)
T_t^{int}	Indoor air temperature (C)
T_t^{out}	Outdoor temperature (C)
T_t^{sp}	Set-point temperature (C)
T^{max}	Maximum surface slab temperature (C)
Q_t^{slab}	Heat into the concrete slab (W)
Q_t^{hp}	Total overall heating (W)
Q_t^{AHU}	Air heating unit (W)
Q_t^{FCU}	Fan coil thermal unit (W)
Q_t^{stor}	Heat storage (W)
Q_t^{ref}	Theoretical limit of heat storage (W)
Q_t^{phs}	Potential heat storage (W)
P_t^{load}	Plug-loads (W)
G^{hi}	Global horizontal solar radiation (W/m^2)
e_t^{slab}, ε_t^{int}	Prediction error at time t of ARX models
n	Number of observations
$\mathbf{w}$	Coefficient vector of ARX model
$\mathbf{z}_t, \mathbf{x}_t$	Input vectors of ARX models
$\mathbf{K}$	Inverse autocorrelation matrix
μ	Forgetting factor
Acronyms	
ARX	Auto-regressive with eXogenous input
RLS	Recursive Least-Squares
MAE	Mean Absolute Error
SOC	State of charge
ED	Electricity demand

use in commercial buildings by at least 12.3% by 2030 [8]. For Quebec, commercial and institutional buildings represents approximately 18% of the total final consumption of energy in 2012 [9]. 37% of this energy comes from natural gas and 57% is electrical and principally produced by hydroelectric dams [9]. To reduce greenhouse gas emissions, building operators have gradually shifted from natural gas to electricity as their main energy source [10]. In this case, more electricity is consumed especially in winter because of heating needs. Accordingly, during this season two periods of peaks are observed during morning and evening which strains the grid. In a such situation, the general profile will vary depending on the category of building. For example, the residential building's profile fluctuation has exactly two peaks periods. The same profile is observed for institutional and commercial buildings, but with very big power demand. In addition, such buildings consume as much electricity as a dozen of residential buildings and hence offers greater potential for shifting or reducing electricity use from the grid [11]. So far, energy flexibility analysis in commercial and institutional buildings are limited in terms of developing flexibility strategies regarding energy demand profile during peak periods [6, 5]. Although, they must respond more and more flexible to the prerequisites in the energy system, orient themselves more towards energy production than demand. For this reason, Varennes library is presented in this paper as a case study to increase the knowledge on energy flexibility in institutional buildings. The library is in the city of Varennes near Montreal, it offers

IOP Conf. Series: Earth and Environmental Science **352** (2019) 012050 doi:10.1088/1755-1315/352/1/012050

an environment conducive to learning with unique and innovative features [12]. The library includes a building-integrated photovoltaic (BIPV) system, geothermal heat pumps as the main heating/cooling source, and hydronic radiant slab for thermal storage. Those components can be flexible and, in turn increase the energy flexibility of the building. In this paper we will focus our study to quantify energy flexibility of hydronic radiant slab component using two performance indicators: the available storage capacity and state of charge. The work intends to propose a time series analysis based on a combined Auto Regressive model with exogenous inputs (ARX) to estimate the concrete slab surface temperature. The estimated temperature is utilized to adaptively predict the indoor air temperature of the building. Subsequently, the modeling results are then explored for hydronic radiant slab energy flexibility evaluation.

Given the above discussion, the important contributions of this study can be categorized as follows:

(i) Developing an adaptive model capable of realizing a learning procedure by tracking the changes of building's air and concrete slab surface temperatures profiles.
(ii) Applying a transformation to the historical data is used to improve the proposed model performance by scaling data. In turn, the providing performance can estimate this flexibility accurately and makes its evaluation more trustworthy.

The remainder of the paper is organized as follows: Section 2 describes the details of the library building. The problem formulation is explained in Section 3 through proposing a time series analysis for the indoor air temperature and the radiant slab of the building during winter periods. Evaluation of energy flexibility of the concrete slab component using the combined ARX model results is developed in Section 4. This is followed by the conclusions in Section 5.

2. Case Study: Varennes library

The library used as a case study is representative of an institutional building category, shown in Fig. 1 (A). Built in 2014, and open to the public since 2015. The considered, very well-insulated building, is spread across two floors and covers a total area of 2100 m^2. The first radiant floor of the building has five zones and a heated slab space of 240.61 m^2. On the second radiant floor there is six zones with a total heated slab area of 435.62 m^2. The building is Leadership in Energy and Environment Design (LEED) certified and offering an environment conducive to learning due to the installed technologies of the thermal mass in floor slabs, geothermal, and BIPV (part BIPV/T–T is for thermal and signifies that there is heat recovery) [12].

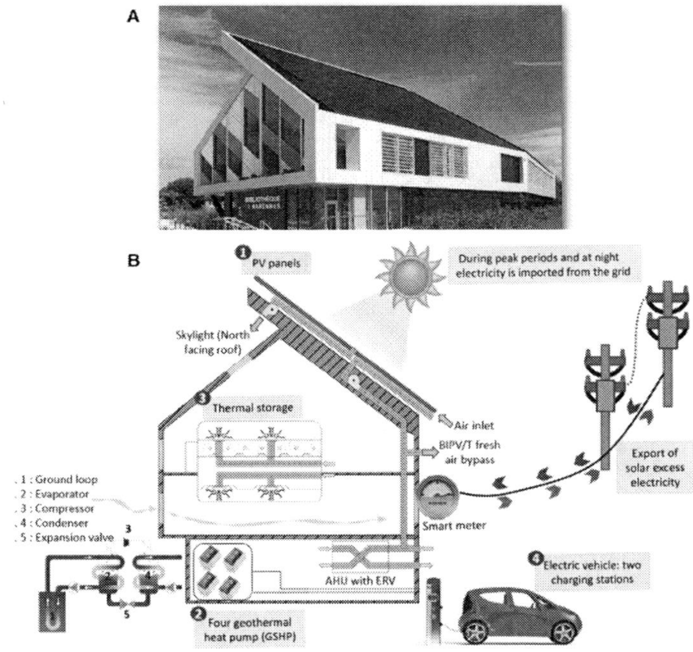

Fig. 1: Front view (A) and plan view (B) of the case study building.

The library heating and cooling is provided by four electric ground-source heat-pumps with a capacity of 105 kW, (see Fig. 1 (B)). The radiant hydronic slab system is installed on the concrete floor of the building. It is used for both heating and cooling. The slab surface is typically between $18 \longmapsto 29°C$. The library has a different operating schedule for weekdays and weekend, When occupied, air set-point temperatures are maintained between $21 \longmapsto 24°C$ in winter and summer, respectively, when unoccupied, temperatures are between $18 \longmapsto 28°C$. The building has two electrical vehicle stations that are accessible for public.

The paper proposes a methodology to obtain a model suitable to estimate the concrete slab surface and air temperatures, then be used to evaluate the energy flexibility of hydronic slab component using knowledge of building loads and weather conditions. The developed framework takes advantage of real-world data exploration measured by smart-meter technology. This data encompasses the information about outdoor temperature T^{out}, global horizontal solar radiation G^{hi}, heat pump load Q^{hp}, plug-loads electricity demand P^{load}, average air temperature T^{int}, and average concrete slab temperature T^{slab}, at 15 minute interval. Slab and air temperatures varies widely because they includes zones with different temperature set-points. These measured data were recorded on three years, but only the peak period in winter from January to February 2018 is considered in this study as is illustrated in Fig. 2.

3. Methodology

In this section, the description of the building modeling approach is provided to examine the relationship that geothermal pumps systems and weather conditions have with building's air and concrete slab temperature. The proposed approach is referred to the adaptive time series analysis on the basis of ARX model for the hydronic concrete slab and the building using real measurement data shown in Fig. 2. Two adaptive ARX models were applied and combined to describe the future of the indoor air temperature, and, in turn, offering an assessment of the energy flexibility of the building. The simulation period is two months interval where the first ARX model is used to model the hydronic radiant slab and estimate the concrete surface slab

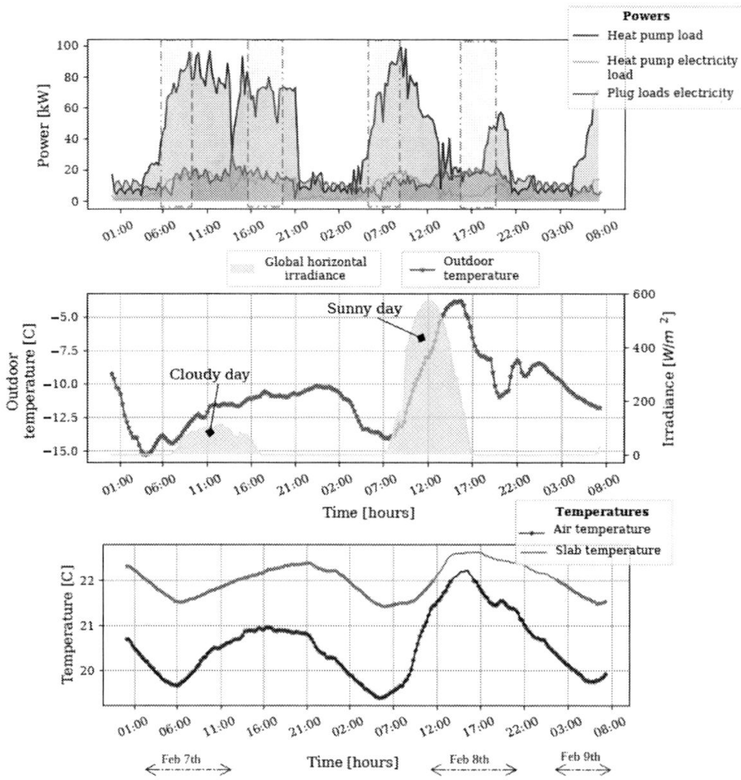

Fig. 2: Measurement data curves of cloudy and sunny days for Varennes library.

temperature T_t^{slab}. As a result, the estimated temperature is used as an input for the second ARX model to estimate the air temperature T_{t+1}^{int} for one-step ahead. The schematic diagram shown in the Fig. 3 depicts the structure of the combined models related to the hydronic concrete slab and the building modeling. In addition, the adaptive process is introduced in both models to track on-line the concrete slab and indoor air temperature changes.

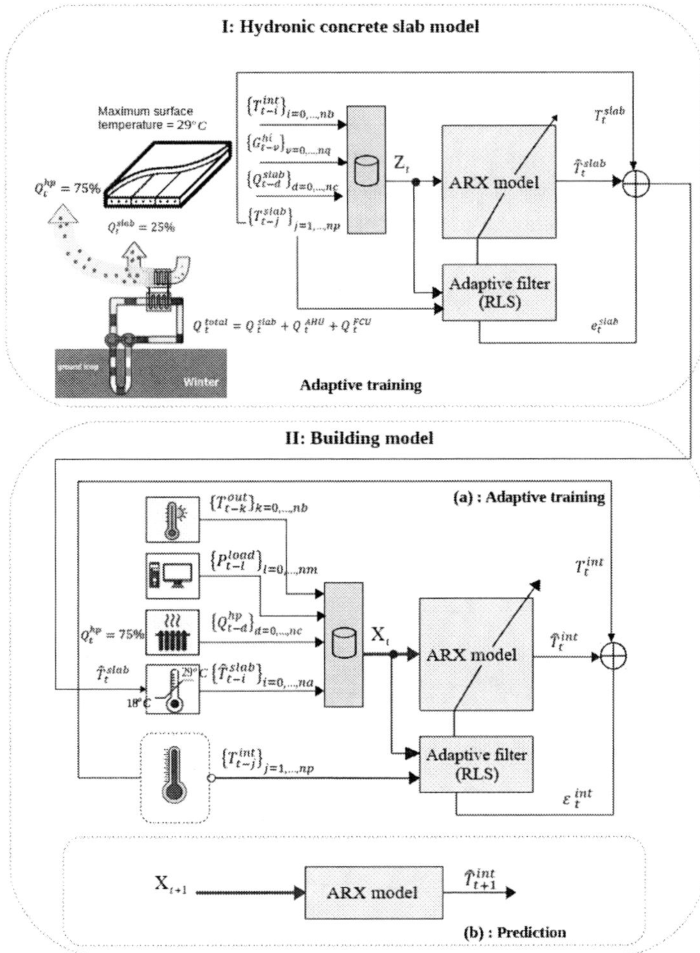

Fig. 3: Block diagram of the combined ARX models for modeling: (I) hydronic concrete slab, (II) building model

3.1. Hydronic radiant slab modeling

As a first step, the ARX aims to estimate the concrete slab surface temperature at current time step t. The model uses at each discrete time t, the measured air temperature, the global horizontal irradiance and the heat into the concrete slab denoted by Q_t^{slab}, as is given by (1).

$$T_t^{slab} = f(T_t^{int}, Q_t^{slab}, G_t^{hi}) + e_t^{slab} \tag{1}$$

where e_t^{slab} is the error which is independent of the input observations with $\mathbb{E}(e_t^{slab}) = 0$. In this model, it is assumed that the real measurements of total overall heating is the sum of heat into the concrete slab Q_t^{slab}, the heat to the air heating unit Q_t^{AHU}, and fan coil thermal unit Q_t^{FCU} amounts as is given by (2):

$$Q_t^{total} = Q_t^{slab} + Q_t^{AHU} + Q_t^{FCU} \tag{2}$$

The flow ratio slab, used to estimate T_t^{slab}, is approximately equal to 25% of Q_t^{total}. This value has been computed through a specific process based on the design flow-rate, described in the mechanical plans of the library. Therefore, 75% of total heat denoted by Q_t^{hp} goes into the air heating unit, and fan coil thermal unit. Note that, the adaptive ARX model utilized for hydronic concrete slab takes into account the logarithmic transformation of the heat into the slab and solar radiation data to estimate the temperature T_t^{slab}. In fact, the logarithmic transformation helps to stabilize the variances of the heat into the slab demand and irradiance, and makes their distributions symmetric when it shows different variations at distinct levels [1, 13, 14]. Additionally, it makes training phase less sensitive to the scale of heat into the slab's features, so we can better solve the coefficient vectors of the filter. We denote the original observations of the heat into the concrete slab and irradiation at time interval t as $Q_1^{slab}, \ldots, Q_n^{slab}$, and $G_1^{hi}, \ldots, G_n^{hi}$ their transformed observations as $\tilde{Q}_1^{slab}, \ldots, \tilde{Q}_n^{slab}$, and $\tilde{G}_1^{hi}, \ldots, \tilde{G}_n^{hi}$ which are the result of the equations $\tilde{Q}_t^{slab} = \log(Q_t^{slab} + 1)$, and $\tilde{G}_t^{hi} = \log(G_t^{hi} + 1)$ [14]. As a result, our ARX model is defined by (3):

$$T_t^{slab} = \sum_{j=1}^{p} \theta_j T_{t-j}^{slab} + \sum_{i=1}^{b} \varphi_i T_t^{int} + \sum_{d=1}^{c} \phi_d \tilde{Q}_t^{slab} + \sum_{v=1}^{q} \rho_v \tilde{G}_t^{hi} + e_t^{slab} \quad (3)$$

where, $\Theta_t = [\theta_{1t}, \theta_{2t}, ..., \theta_{pt}]^T$, $\Psi_t = [\varphi_{1t}, \varphi_{2t}, ..., \varphi_{bt}]^T$, $\Phi_t = [\phi_{1t}, \phi_{2t}, ..., \phi_{ct}]^T$, and $\Gamma_t = [\rho_{1t}, \rho_{2t}, ..., \rho_{qt}]^T$ are the coefficient vectors of T_t^{slab}, T_t^{int}, $\tilde{Q}_t^{slab}$, and $\tilde{G}_t^{hi}$ respectively. Accordingly, we have utilized an input vector $\mathbf{z}_t$ of the observations of the time series data for the ARX model given by (4):

$$\mathbf{z}_t = \left[T_t^{int}, T_{t-1}^{int}, ..., T_{t-b}^{int}, \tilde{Q}_t^{slab}, \tilde{Q}_{t-1}^{slab}, ..., \tilde{Q}_{t-c}^{slab}, \tilde{G}_t^{hi}, \tilde{G}_{t-1}^{hi}, ..., \tilde{G}_{t-q}^{hi}\right]^T \quad (4)$$

Three hours of past data was used for the output, the average air temperature, heat into the concrete slab and global horizontal irradiance have been assumed in this model. These data points time lags for 15 minutes interval are $p = b = c = q = 12$. Concurrently, to adjust for the changing slab conditions, the coefficients of the filter are adaptively updated using RLS adaptive algorithm, implemented according to [15]. The procedure for adaptation the RLS filter is shown in algorithm (1).

Algorithm 1: Summary of RLS algorithm

Input: $\mathbf{z}_t$
Output: T_t^{slab}
1 Initialize: $\hat{\mathbf{w}}(0) = 0$, $\mathbf{P}(0) = \delta^{-1}\mathbf{I}$, where $\mathbf{I}$ is identity matrix and δ is a normalizing constant
2 **for** $i = 1, ..., n$ **do**
3 Calculation of the adaptive filter output: $\hat{T}_t^{slab} = \mathbf{w}_t \mathbf{z}_t$, where
4 $\mathbf{w}_t = [\varphi_{1t}, \varphi_{2t}, ..., \varphi_{12t}, \phi_{1t}, \phi_{2t}, ..., \phi_{12t}, \rho_{1t}, \rho_{2t}, ..., \rho_{12t}]^T$
5 Estimation error: $e_t^{slab} = T_t^{slab} - \hat{T}_t^{slab}$
6 Calculation of Kalman gain: $\mathbf{K}_t = \dfrac{\mu^{-1}\mathbf{P}_{t-1}\mathbf{z}_t{}^T}{1 + \mu^{-1}\mathbf{z}_t \mathbf{P}_{t-1}\mathbf{z}_t{}^T}$
7 Update of the inverse correlation matrix: $\mathbf{P}_t = \mu^{-1}\left[\mathbf{P}_{t-1} - \mathbf{K}_t \mathbf{z}_t \mathbf{P}_{t-1}\right]$
8 Update of the filter coefficients: $\mathbf{w}_{t+1} = \mathbf{w}_k + \mathbf{K}_t e_t^{slab}$
9 $t + 2 = t + 1$, back to step 3

The forgetting factor value of RLS algorithm is $\mu = 0.84$, in accord with [15]. As a result, the estimated temperature $\hat{T}_t^{slab}$ at time instant t can be explored in order to predict the average temperature using the second ARX model as is illustrated in Fig. 3 (I).

The applied analysis of ARX model generates the estimated temperatures $\hat{T}_t^{slab}$ as is shown in Fig. 4 (A).

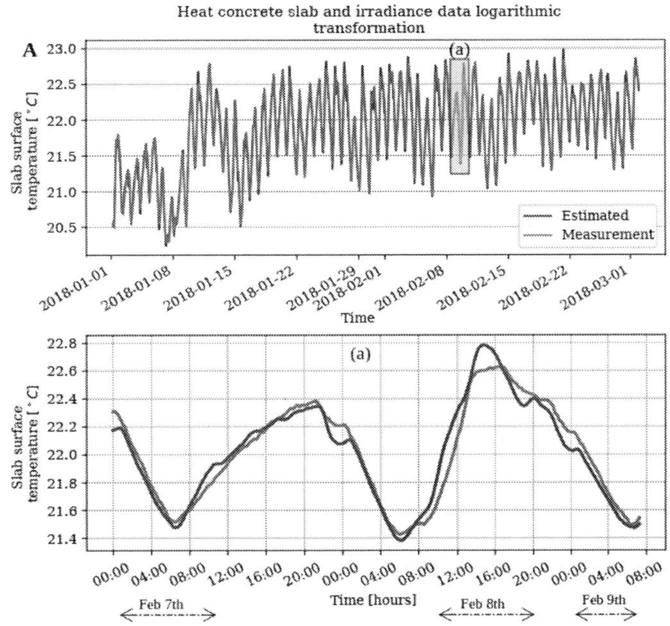

Fig. 4: Concrete surface slab temperature results, using logarithmic transformation of heat into the concrete slab and solar radiation.

Fig. 4 (A) shows the estimated slab temperature after the logarithmic transformation analysis. It can be seen that the logarithmic transformation applied on the heat into the slab demand and irradiance data can stabilize its variance variation and, in turn, makes the coefficients better condition for convergence. As a results, the logarithmic transformation allow to evaluate the output of the model better. Consequently, the estimated slab temperature of this case is used as an input for the modeling process of the average air temperature at one-step ahead. Notice that the logarithmic transformation works better for this case because the heat into the slab consumption and global horizontal solar radiation have an asymmetric distributions.

3.2. Building model

Compared to the hydronic concrete slab modeling approach, the building model estimates the average air temperature at one-step ahead, as is illustrated in Fig. 3 (II). The original data of outdoor temperature T_t^{out}, 75% of heat pump load Q_t^{hp}, plug-loads P_t^{load} and the estimated concrete slab temperature $\hat{T}_t^{slab}$ are used to predict the next value of the air temperature, as is given by (5) .

$$T_t^{int} = g(\hat{T}_t^{slab}, T_t^{out}, Q_t^{hp}, P_t^{load}) + \varepsilon_t^{int} \tag{5}$$

Since the real data of heat pump load does not have a symmetric distribution, the logarithmic transformation $\tilde{Q}_t^{hp} = \log(Q_t^{hp} + 1)$ is applied to stabilize the heat pump load variance before the modeling process, as is described in the Subsection 3.1. The second adaptive ARX model at each time instant t is expressed by (6):

$$T_t^{int} = \sum_{j=1}^{p} \alpha_j T_{t-j}^{int} + \sum_{i=1}^{a} \vartheta_i \hat{T}_t^{slab} + \sum_{k=1}^{b} \beta_k T_t^{out} + \sum_{d=1}^{c} \gamma_d \tilde{Q}_t^{hp} + \sum_{l=1}^{m} \psi_l P_t^{load} + \varepsilon_t^{int} \tag{6}$$

where ε_t^{int} is assumed to be a white noise with zero mean $\mathbb{E}(\varepsilon_t^{int}) = 0$. The variables $\alpha_{1t},...,\alpha_{pt}$, $\vartheta_{1t},...,\vartheta_{at}$, $\beta_{1t},...,\beta_{bt}$, $\gamma_{1t},...,\gamma_{ct}$ and $\psi_{1t},..., \psi_{mt}$ are coefficients that are estimated adaptively using

RLS filter procedure shown in algorithm (1). The RLS filter output is given by (7):

$$\hat{T}_t^{int} = \mathbf{w}_t \mathbf{x}_t \tag{7}$$

where $\mathbf{x}_t$ is the input vector for computing the coefficients vector $\mathbf{w}_t$ expressed by (8):

$$\mathbf{x}_t = \left[\hat{T}_t^{slab}, ..., \hat{T}_{t-a}^{slab}, T_t^{out}, ..., T_{t-b}^{out}, \tilde{Q}_t^{hp}, ..., \tilde{Q}_{t-c}^{hp}, P_t^{load}, ..., P_{t-m}^{load}\right]^T \tag{8}$$

The maximum number of the estimated slab temperature, outdoor temperature, heat pump load and plug-loads data time lags is assumed to be four hours, selecting an order $a = b = c = m = 16$, respectively. Note that for the air temperature order used during the training phase is also assumed to be $p = 16$. The forgetting factor value used in this model is $\mu = 0.80$. With regard to the on-line training phase shown in the semantic diagram of the Fig. 3 $(II \cdot a)$, the model utilizes T_t^{int} and $\mathbf{x}_t$ to update the ARX model and estimate the average temperature $\hat{T}_t^{int}$. Subsequently, the model exploits the next step information of $\mathbf{x}_t$ in order to predict the average temperature at one-step ahead as is illustrated in Fig. 3 $(II \cdot b)$. As a result, the predicted air temperature for this scenario is shown in Fig. 5 (B).

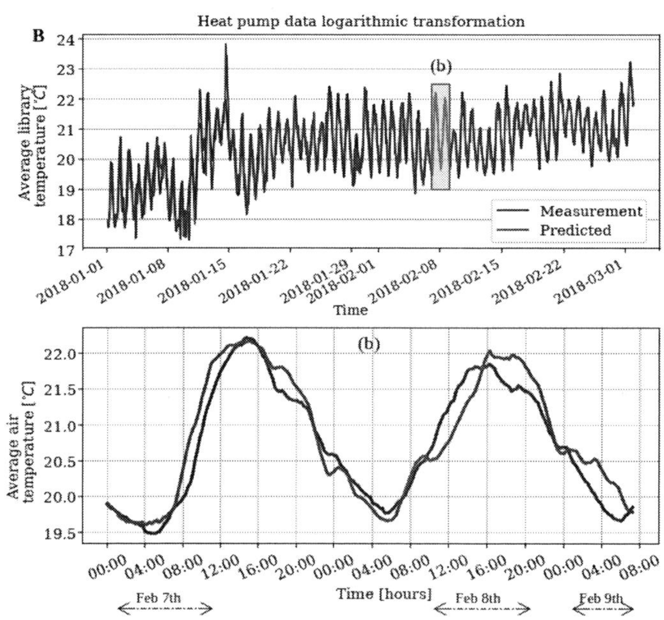

Fig. 5: The one step-ahead prediction of average air temperature using logarithmic transformation of heat pump load data.

Fig. 5 (B) demonstrates the logarithmic transformation analysis performance on the average air temperature prediction. There is a notable stabilization of the predicted temperature which can follow the air temperature profile and offer a realistic prediction. Consequently, the logarithm transformation of the raw data aids to improve the model performance, and in turn, will yield for more trustworthy evaluation of the energy flexibility.

4. Results and discussion

4.1. Energy flexibility evaluation

The improved estimation of slab surface and air temperatures from hydronic radiant slab modeling and building model are used to investigate the energy flexibility of the radiant

IOP Conf. Series: Earth and Environmental Science **352** (2019) 012050 doi:10.1088/1755-1315/352/1/012050

slab component through two performance indicators: available storage capacity and state of charge (SOC). These indicators can produce different interpretations depending on the purpose reference scenario for which it is used.

4.1.1. Heat storage in the concrete slab: To determine the available heat storage in the radiant slab, the averaged thermal capacitance of the concrete C^{th} is required. It is computed using the concrete thermally effective layers thickness of 125 mm for the first floor and 90 mm for the second floor, thermal conductivity and volume-related heat storage capacity. Afterwards, the available heat storage capacity Q_t^{stor} in (kWh) is calculated using (9) :

$$Q_t^{stor} = C^{th} \Delta T_t \tag{9}$$

where C^{th}= 33.86 $kWh/°C$, and ΔT_t is the temperature difference at time step t between the maximum slab surface temperature $T^{max} = 29$ $°C$, which is the upper limit of the concrete slab to maintain thermal comfort and the estimated slab temperature $\hat{T}_t^{slab}$ results obtained from case study (2), as is given by (10):

$$\Delta T_t = 29 - \hat{T}_t^{slab} \tag{10}$$

Therefore, the available stored energy Q_t^{stor} is divide by the averaged coefficient of performance (COP) to compute the available electricity demand which is the electricity due to heat pump denoted by ED_t. It is computed by the following equation and expressed in (kWh).

$$ED_t = \frac{Q_t^{stor}}{COP} \tag{11}$$

Notice that the COP is assumed has a constant value of 4, which is the average of the energy source used to heat the fluid circulating through the system's ground during winter.

4.1.2. Potential heat storage: The available heat storage capacity Q_t^{stor} indicator is quantified by analyzing its deviation from the reference scenario which is the maximum flexibility limit defined by $Q_t^{ref} = 304.6$ kWh. It is computed under the same conditions of maximum slab surface temperature T^{max} (see Fig. 6). The difference between Q_t^{ref} and the available storage capacity indicate the potential heat storage Q_t^{phs}. It is expressed in kWh as is given by (12):

$$Q_t^{phs} = \left| Q_t^{ref} - Q_t^{stor} \right| \tag{12}$$

On the other hand, the possible potential rate of heat storage in percentage is defined in (13). It is computed to quantify the rate of heat storage needed to achieve the reference scenario during morning peak period from (6-9 a.m.) and evening from (4-8 p.m.). We would like to note here that the $(1 - \eta) \times 100$ values describe the actual rate of heat storage capacity.

$$\eta(\%) = 100 \cdot \frac{Q_{i,j}^{phs}}{Q_{i,j}^{ref}} \quad i \in [6, 9\ a.m], j \in [4, 8\ p.m] \tag{13}$$

4.1.3. Concrete state of charge: The state of charge is defined as the percentage of the possible thermal charge that is present inside the concrete slab component. It involves the slab and indoor air temperatures estimation from hydronic radiant slab modeling and building model. It is measured by the following equation:

$$SOC_t = 100 \cdot \left| \frac{\hat{T}_t^{slab} - \hat{T}_t^{int}}{29 - \hat{T}_t^{int}} \right| \tag{14}$$

The results of the available heat storage, the electricity need, and SOC indicators captured using combined ARX models analysis during two months interval are illustrated in Fig. 6 (a). Fig. 6 (b) depicts the detailed results description during a cloudy day on February 7th, and sunny day on February 8th (occupied days time). Fig. 6 (c) represents the reference scenario related to the maximum flexibility, the potential heat storage capacity in kWh. The available and the potential rates of heat storage capacity computed during peak periods from 6 a.m. to 9 a.m and 4 p.m. to 8 p.m. considering a sunny day are illustrated in the Fig. 6 (d). These periods correspond to the highest electricity demand from the grid. Time plot of outdoor temperature and solar radiation data are shown in Fig. 6 (e).

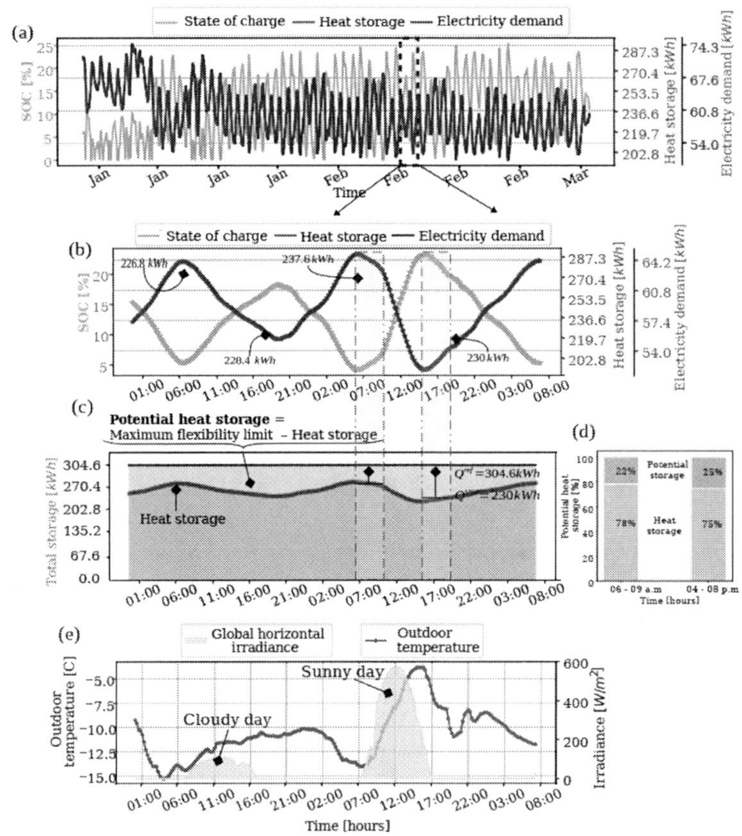

Fig. 6: Representation of the available heat storage capacity, electricity need, state of charge and the potential heat storage.

It can be seen in Fig. 6 (b) that there is an opposite behavior of flexibility between the available heat storage, electricity demand and SOC events, i.e. if in a certain upward event the heat storage is high, in the same moment, the SOC event will provide a lower efficiency and vice versa. When looking at the SOC indicator in sunny day, a considerable amount about 23 % is realized before evening peak because of the solar radiation. This rate is seven times high when loading with 16 hours of electricity of delay. In this case, the building can be beneficial of evening peak reduction of heating energy consumption, by preheating the library during day time, and storing it to the concrete slab of the library. This means we do not need to import the electricity from the grid during evening peak periods.

On the other side, based on the results of the actual heat storage capacity indicator during peak periods, the slab has the capacity to store about 226.8 kWh and 237.6 kWh during morning peaks and 228.4 kWh and 230 kWh during evening peaks, respectively. These capacities are compared then to the reference scenario associated to 304.6 kWh of the maximum flexibility limit as is discussed in the equation (13). Considering a sunny day, the Fig. 6 (d) depicts that the slab is capable to offer approximately 78% and 75% of flexibility to store heat during morning and evening periods peak respectively. Consequently, the possible potential rate of storage needed to reduce the deviation from the reference scenario is about 22% and 25% during morning and evening periods, respectively. This possible potential can be improved by increasing air or slab temperatures set-points. Although library air temperature goes rarely above to $21°C$, it is possible to increase the storage before peak times by changing the air set-point temperature variations up of $24°C$, without penalizing the thermal comfort.

5. Conclusion

This work has introduced an on-line combined time series model to evaluate the energy flexibility of Varennes library's hydronic concrete slab component. This model estimate on adaptive way the slab temperature. Moreover, the estimated temperature has been investigated to adaptively predict the indoor air temperature. The evaluation of the combined method using logarithmic transformation of raw data has demonstrated that this transformation is more favorable for improving estimation tasks of the proposed model. The performance of the approach is used to quantify the heat storage capacity and state of charge indicators. As a result, the available amount of heat storage offered by the slab can satisfactory be used to reduce evening peak because of solar radiation. In future work, the morning peak needs more investigation, with a possibility to preheating the building a day before. This peak has a broad complexity associated with other phenomena will be studied. This can help to develop a strategy able to displace the morning peak of the total overall heat consumption to improve energy flexibility quantification and its interaction with the grid.

5.1. Acknowledgments

The authors would like to thank Éric Dumont for his valuable suggestions to improve the quality of the paper. The authors would also thank Stéphane LaBarre for his cooperation to provide the real data of library electricity, as well as Régulvar for providing information and assistance on the library's controls.

References

[1] F. Amara, K. Agbossou, Y. Dubé, S. Kelouwani, A. Cardenas, and J. Bouchard, "Household Electricity Demand Forecasting Using Adaptive Conditional Density Estimation," *Energy and Buildings*, vol. 156, pp. 271–280, sep 2017.

[2] H. Johra, *Integration of a magnetocaloric heat pump in energy flexible buildings.* Denmark: Aalborg Universitetsforlag, 2018.

[3] Bart Bleys; Søren Østergaard Jensen; Anna marszal-Pomianowska, "Energy Flexible Buildings Annex 67," *IEA EBC Annex 67*, no. December, pp. 32–36, 2018. [Online]. Available: www.iea-ebc.org/projects/ongoing-projects/ebc-annex-67/

[4] A. Bálint, "Generalizability of Machine Learning Algorithms for Modelling and Control of Thermostatically Controlled Loads," Ph.D. dissertation, Universitat Politècnica de Catalunya, 2018.

[5] Joy Dalmacio Billanes, Zheng Ma, and Bo Nørregaard Jørgensen, "Consumer Central Energy Flexibility in Office Buildings," *Journal of Energy and Power Engineering*, vol. 11, no. 10, oct 2017.

[6] E. Mlecnik, "Integration von Energieflexibilität ein Erfahrungsbericht von der Technischen Universität Delft," 2018.

[7] M. Zheng, "User needs, motivation and barriers for application of Energy Flexibility in Buildings," IEA EBC Annex 67 Energy Flexible Buildings, Tech. Rep., 2017.

[8] Gourvernement du Québec, "Plan directeur en transition, innovation et efficacité énergétiques du Québec 2018 -2023," Tech. Rep., 2018.

IOP Conf. Series: Earth and Environmental Science **352** (2019) 012050 doi:10.1088/1755-1315/352/1/012050

[9] A. Clermont, "Contribution des bâtiments institutionnels dans l'atteinte de l'indépendance énergétique du québec et dans la lutte contre les changements climatiques," Ph.D. dissertation, Université de Sherbrooke, 2015.
[10] Natural Resources Canada, "Energy Efficiency Trends in Canada 1990 to 2013," Tech. Rep., 2016. [Online]. Available: https://www.nrcan.gc.ca/sites/www.nrcan.gc.ca/files/energy/pdf/trends2013.pdf
[11] A. Kathirgamanathan, K. Murphy, M. D. Rosa, E. Mangina, and D. P. Finn, "Aggregation of Energy Flexibility of Commercial Buildings," *eSim IPBSA Conference at Montreal, Canada*, pp. 173–182, 2018.
[12] V. Dermardiros, "Energy Performance, Comfort and Lessons Learned from an Institutional Building Designed for Net-Zero Energy," *ASHRAE Transactions*, pp. 1–29, 2019.
[13] J. W. Taylor and M. B. Roberts, "Forecasting Frequency-Corrected Electricity Demand to Support Frequency Control," *IEEE Trans. Power Syst.*, vol. 31, no. 3, pp. 1925–1932, 2016.
[14] R. J. Hyndman and S. Fan, "Density forecasting for long-term peak electricity demand," *IEEE Trans. Power Syst.*, vol. 25, no. 2, pp. 1142–1153, 2010.
[15] A. Sayed and T. Kailath, "Recursive Least-Squares Adaptive Filters," in *Digital Signal Processing Fundamentals*, nov 2009, pp. 1–40.

Case study of personal heaters in a Plus energy building – Simulations of potential energy savings and results from a field test

N. Lassen
Skanska Norge, Lakkegata 53. 0187 Oslo. Norge.

Abstract. Personal Comfort Systems (PCS) for heating, such as foot warmers, heated chairs and infrared heaters, can compensate for a lowering of up to 10°C in ambient temperatures in a heating situation. They are found to lead to a significant increase in occupant satisfaction with the thermal environment, as they enable for a personalized thermal environment. In this way, the use of PCS systems can ensure occupant satisfaction while widening the temperature dead-band, or difference between heating and cooling set-points in buildings. Several field studies from North America have indicated that the average dead-band between heating and cooling set-points is between 1 and 2 °C, leading to considerable amounts of energy used for over-heating and over-cooling. This kind of systems may therefore be an important contributor towards Plus energy buildings, but they are seldom used. The objectives of this study were (1) to test PCS heaters in a modern Norwegian office environment and see how they are appreciated by office workers and (2) to investigate how large energy savings such systems can contribute to in a Plus energy building. Powerhouse Kjørbo was selected as a case building for the study. Potential energy savings were calculated using energy simulations. The results in this case study were not able to confirm an increase in occupant thermal acceptability rate due to the use of a PCS heater. Interviews of occupants however suggest that PCS heaters are a good solution for improving the satisfaction of the limited number of occupants who have special needs, preferences or are located in a place with lower temperature. In buildings with an effective heating source, such as ground source heat pumps, PCS solutions are not likely to contribute to notable energy savings. In an indirect way, they may however still contribute toward realizing Plus energy buildings as PCS systems can help relax the demands set to other climate installations in the building. This again can allow the use of more environmentally friendly solutions such as utilization of thermal mass, temperature stratification and natural ventilation. They can also reduce installation costs by eliminating the need for more complicated and costly HVAC systems.

1. Introduction

Most office buildings in Europe and North America are controlled to a tight temperature dead-band between heating and cooling set-points, where the mean dead-band is found to be 1-2°C in North America [1,2] and 2°C in Norwegian buildings [3]. Tight temperature dead-bands come at a high cost in terms of both energy use, construction costs and system complexity. Therefore, solutions which make it possible to loosen the temperature dead-bands may be a significant contribution toward Plus energy buildings.

At the same time, the percentage of buildings in the ASHRAE Global Thermal Comfort Database II meeting the threshold of 80% satisfied occupants was found to be only 8%, if one includes votes from 0 to +3 ('neutral' to 'very satisfied'). In total 43% of the occupants are thermally dissatisfied, 19% neutral and 38% satisfied [4]. The reasons for this gap between the predicted and actual satisfaction is found to lie in personal differences between occupants, both clothing and metabolic rate

[5], as well as expectations and personality [6]. These studies conclude that the prediction accuracy of the PMV-PPD model is too low, and suggest that wider temperature dead-bands along with personal control would result in higher occupant satisfaction rates.

Personal Comfort Systems (PCS) is a collective term for systems that cool or heat individual occupants [7]. Several researchers see these systems as in important part of the solution for improving occupant satisfaction, while at the same time widening temperature dead-bands and saving energy. A review of scientific studies on PCS systems [7] has found that they can provide a corrective power (CP) of -1 - -6°C for cooling and 2 – 10 °C for heating. Corrective power is defined as the difference between two ambient temperatures at which equal thermal sensation is achieved - one with no PCS (the reference condition), and one with PCS in use. The same study also gathered information on occupant satisfaction from 13 of the studies, showing an impressive increase in occupant satisfaction for the PCS systems, also for ambient temperatures well outside of normal set-points. See illustration in Figure 1.

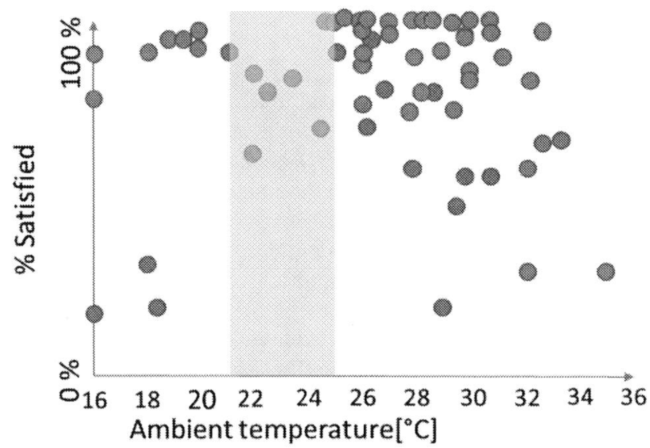

Figure 1. *Percent satisfied with thermal conditions in studies with PCS (red dots) and without PCS (blue dots). After Zhang, Arens & Zhai (2015)*

Aside from the benefits mentioned above, PCS can offer a wealth of data that can describe how individuals interact with heating/cooling devices in their own environment. This allows individuals' comfort and behaviour to be learned, and can inform centralized systems to provide "just the right" amount of conditioning to meet occupant needs [8].

Research on positive effects of widened temperature dead-bands have shown large potential effects. A North American parametric simulation study of seven climates and six model types found that reducing the heating set-point from 21.1 °C to 20°C saves an average of 34% of terminal heating energy. [9]. This result is however dependent on climate type, building envelope, heating system and heat source.

All the reviewed research suggests that PCS systems can have a strong positive effect on occupant thermal satisfaction. However, even though the idea of PCS systems has been a focus area of researchers since the nineties, there are still very few practical examples of organized use of PCS systems in commercial buildings. The objectives of this study are therefore (1) to test PCS heaters in a modern Norwegian office environment and see how they are appreciated by office workers and (2) to investigate how large energy savings can be anticipated in a ZEB building. Powerhouse Kjørbo has been selected as a case building for the study. Potential energy savings were calculated using energy simulations. A field test was conducted, where the occupants of 26 workplaces received web-based questionnaires with and without PCS heaters during a period of close to 4 months in the heating season.

IOP Conf. Series: Earth and Environmental Science **352** (2019) 012051 doi:10.1088/1755-1315/352/1/012051

2. Methods

2.1 Case Building

Both the field study and simulations are conducted with the plus energy building Powerhouse Kjørbo as a case example. Powerhouse Kjørbo is a complex of 5 cube-shaped office buildings which are connected by hallways. They are located at Kjørbotangen in Sandvika, west of Oslo, Norway. Only Unit 5 was used for energy simulations, while the field test was conducted in an open office landscape in the 1st floor of Unit 5. The building has an envelope heat loss coefficient of 0.25 W/m²K, is extremely air tight and has a high thermal mass. The ventilation system utilizes displacement ventilation and central air pre- heating and cooling with an efficient rotary heat exchanger. The heating system consists of centrally placed radiators and a ground source heat pump. The heated floor area of Unit 5 is approximately 1870 m².

2.2 Field study

The field study was conducted in the fall of 2018 in a part of the 1st floor. 26 workplaces were included in the test. It was known that several of the occupants in the room had previously complained about cold temperatures. The field study was part of a larger study where a complete system for continuous subjective occupant feedback was tested.

The test lasted for close to four months (Nov. 05 2018 to Feb 02, 2019), and web-based questionnaires were distributed to the occupants at 3 different instances. The first survey was done as a reference, before any equipment was installed. In Phase 2, a QR-code with an explaining text was attached to each desk. If scanned with a smartphone, the user would be taken to a webpage displaying four buttons for complaints regarding the indoor climate. No changes to the indoor climate were made when the buttons were pressed. The second survey was done when users had used this system for 11 days. In Phase 3, an infrared heater was installed under each desk. The heaters would warm the thighs and legs of each occupant when turned on. If a user scanned the QR code and pressed "Too cold", the heater would turn on for 30 minutes. An email was sent to the users notifying them that heaters had been installed. After 36 days, three randomly selected users were interviewed in an . After feedback from the users, the QR codes were removed and the users were offered a manual control possibility for controlling the heaters. The third survey was done when users had used this system for 12 days.

Table 1. Field study sequence and progress

	Date	Comment
Study start	05.11.2019	14 temperature sensors of type Disruptive technologies were installed in the space
First survey distributed	16.11.2019	Last response made 19.11.2019
QR codes installed	19.11.2019	
Second survey distributed	30.11.2019	Last response made 03.12.2019
Under desk heaters installed	04.12.2019	
QR code removed and changed to manual control of heaters		
Interview of 3 occupants	10.01.2019	
Third survey distributed	23.01.2019	Last response made 24.01.2019
End of study	02.02.2019	

2.2.1 Digital complaint and PCS system

The goal of the system was to be non-intrusive, affordable, and provide real-time data. For these reasons, a solution using a web page tailored for smartphones was used in combination with unique QR codes printed and attached to each work desk was chosen. No installation was needed, users could scan the

QR code with their smartphone and reach a webpage with buttons to complain about "Too cold", "Too hot", "Bad air" or "Draught". Entries into the complaint webpages were tracked and logged in a database. In Phase 3, database entries for "Too cold" were set to activate "smart plugs". The smart plugs were programmed to switch off after 30 minutes. The under desk heaters were connected to the smart plug at each desk, and glued to the underside of the desk with an adhesive backside. The heaters are 30x60 cm large and have a power range of 40-150W, but were pre-set to approximately 50W power producing surface temperatures of 40-50 C. All products were commercially available at low cost.

Figure 2. *QR code and webpage.*

Figure 3. *Under desk heater.*

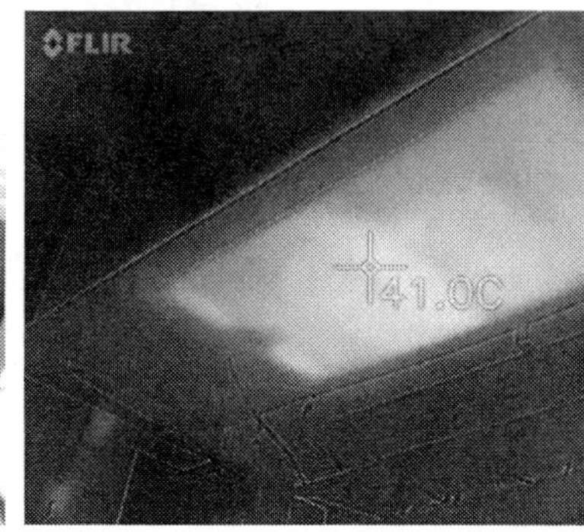

Figure 4. *IR image of infrared heater.*

2.2.2 Temperature measurements

Temperature measurements were done with 14 button sensors from Disruptive Technologies. The sensors have an absolute accuracy of 0.4 °C. The sensors were distributed throughout the space, as shown in Figure 5. There was approximately one sensor per 14 m² of floor space. Four of the sensors were placed by the floor, marked in blue in Figure 5. The others were either under desks or at approximately 1.2 m height.

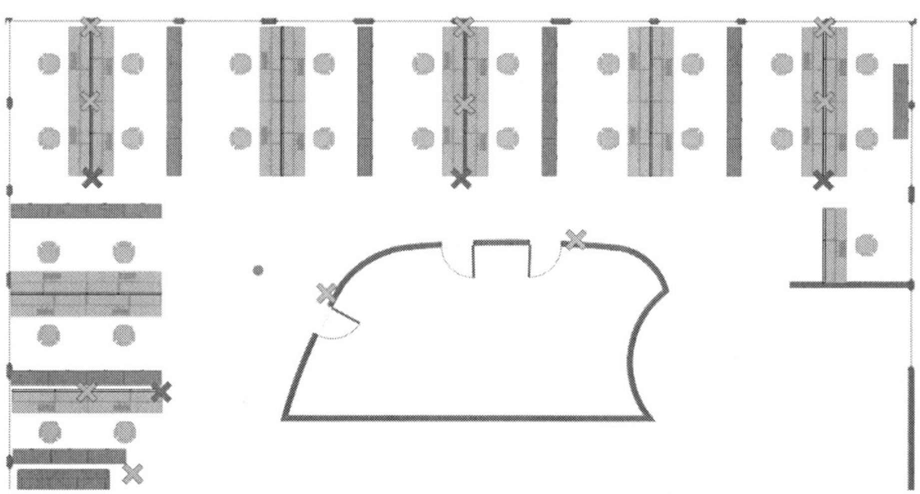

Figure 5. *Placement of temperature sensors. Four at floor level (Blue) and ten at 1.2 m height (orange).*

2.2.3 Survey

An electronic survey was distributed to the test subjects via an email link. The survey included multiple questions, but only four of the questions were relevant for this study. The respondents were asked to take the last week into consideration when answering all the questions. The relevant questions (translated from Norwegian) and answer scales are shown in Table 2.

Table 2. Survey questions

	Question	Answer scales
Whole-body thermal sensation vote (according to NS-EN 15251)	*"Where on this scale would you place your experience of the temperature?"*	*"Cold" [-3]* *"Cool" [-2]* *"Slightly cool" [-1]* *"Neutral" [0]* *"Slightly warm" [1]* *"Warm" [2]* *"Hot" [3]*
General temperature acceptability (according to NS-EN 15251)	*"How did you experience the temperature during the period?"*	*"Entirely acceptable" [1]* *"Barely acceptable" [1]* *"Not acceptable" [0]* *"Entirely unacceptable" [0]*
Perceived control	*"How much control do you think you have over your indoor climate?"*	*"No control" [0]* *"Some control" [1],* *"Much control" [2],* *"Very much control" [3]*
Portion who used QR code / heater	*"Have you used the (QR code / heater) during the period?"*	*"Yes" [1] / "No" [0]*

2.2.4 Interview
Three randomly selected respondents took part in a focus group interview. The interview questions relevant for this study were:

1. Have you used the QR codes / personal heater? If not, why?
2. What do you think of the solution with QR code / personal heater?
3. What could be done better?

2.3 Energy simulations
Potential energy savings were calculated using the software SIMIEN ver.6.0.10 and the original design phase energy model. The under desk heaters were assumed to have a corrective power of 2.8 °C, according to a study of under desk heaters with similar surface temperatures [10].

For the reference case, the year-round heating set-point was set to 22 °C. For the PCS case, the set-point was lowered to 19.2 °C. The supply air set point was also lowered from 20 °C to 19.2 °C. No other changes were made to the energy model.

3. Results

3.1 Field study

***3.1.1** Temperature measurements*
Results of the temperature measurements are given in Figure 6. White boxes show the median and 25th and 75th percentiles of the mean room temperature in each phase. The phase is set to the week preceding each of the surveys, as respondents were asked to answer for the preceding week. Blue boxes show the equivalent means for the lowest room temperature readings, and red boxes for maximum room temperature reading. The highest room temperatures were typically in the central areas of the building, (near the radiators), while the lowest room temperatures occurred near the floor and toward the corners of the building. Only daytime temperature readings between 07:00 and 18:00 were included.

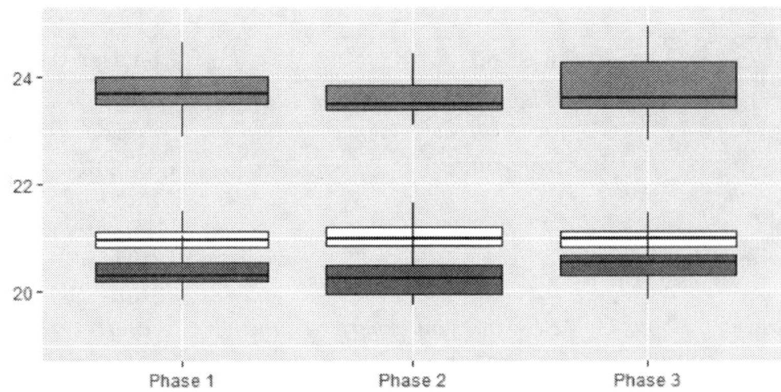

Figure 6. *Temperature measurements in the three weeks of survey. Mean room temperatures (white), max room temperatures (red), min room temperatures (blue).*

3.1.2 Survey results
Results from the survey questions are given in **Table 3**.

Table 3. Survey results

	1st survey	2nd survey	3rd survey
Number of respondents	16	10	11
Mean thermal sensation vote	-1.0	-1.2	-1.0
General temperature acceptability (% acceptable)	75 %	80 %	82 %
p-value 1st – 3rd survey			*p=0.68*
p-value 2nd – 3rd survey			*p=0.92*
Perceived control [0-3]	0.64	0.60	1.18
p-value 1st – 3rd survey			S1: p=0.07
p-value 2nd – 3rd survey			S2: p=0.10
Portion who used QR code / heater (% Yes)	-	20%	45 %
p-value 2nd – 3rd survey			*p=0.23*

3.1.3 Interview results
The focus group interviews produced the following main findings:

- Two of three have used the heater «all the time». They say several others around them do the same. They say many of them think it's generally too cold in the building.
- The third person's heater did not work, but he is content with the temperature as it is and does not want one.
- Two of them want to keep the heater. One of them claims "I love it".
- "Half an hour is way too short. That's why I have disconnected the heater from the smart plug and use the manual control. I have helped several others do the same."

3.2 Simulation results
Results of the energy simulations are provided in **Table 4**.

Table 4 Results of energy simulation

	Heating energy [kWh/m²]	Total energy [kWh/m²]	
	Net heating energy demand	Net energy demand	Gross energy use
22 °C heating set-point (Reference)	18.1	44.5	26.6

19.2 °C heating set-point	13.7	40.2	25.4

Net heating energy demand was reduced by 4.4 kWh/m², or 24%. Net total energy use was reduced by 4.3 kWh/m², or 9%. Gross energy use was reduced by 1.2 kWh/m², or 5%. Additional energy use for PCS heaters was not included.

4. Discussion

Results show that the temperature conditions across the three survey periods were similar. The survey answers for "Mean thermal sensation" confirm this as the respondents perceived all three phases as "slightly cool". Temperature acceptability is slightly higher in the 3rd survey with the PCS heater present, but this finding is not statistically significant due to the small number of respondents and relatively high temperature acceptability in the reference case. A clearer difference in acceptability may have been obtained if the test was conducted with lower ambient temperatures. Perceived control was however substantially increased with the PCS heater present, although still not significant. 45% of the respondents reported to have used the heater, compared to 20% who used the QR code while it only had the "complaint" function.

As opposed to the other studies studying satisfaction with PCS systems, this survey only asked for general thermal acceptability over the previous week. Other studies have used "right now" surveys asking for acceptability and thermal sensation "right now" while the PCS device was in use. This may be a large contributor to the fact that no significant improvement on occupant satisfaction was found in this study.

The interview results revealed that some select users were very satisfied with the PCS heater, and used it "all the time". These users also reported to be very cold under normal conditions, before the heater was introduced. It also revealed that the system using a QR code and smartphone to activate the heater for 30 minutes was not suitable, and the users preferred manual control over the device.

The energy savings by lowering the heating set-point from 22 °C to 19.2 °C was 24% net heating energy or 9% total savings. Total gross energy savings were however only 5%, due to the high efficiency of the ground source heat pump. The additional energy required for use of the PCS heater was not calculated. Assuming there are 170 occupants in the building, and 45% of these are thermally uncomfortable and use the heater, each of these occupants may use the heater for 72 whole workdays before the gross energy savings are zeroed out by heater use. If the heater is used on average 2 hours each day by each of the thermally uncomfortable occupants (77 occupants, 45% of 170) for the length of the heating season (approximately 120 workdays), this would result in an increased annual energy use of 0,5 kWh/m².

5. Conclusions

The results in this case study were not able to confirm an increase in occupant thermal acceptability due to the use of a PCS heater. Interview results however suggest that PCS heaters are a good solution for improving the satisfaction of the limited number of occupants who have special needs, preferences or are located in a place with lower temperature. The potential energy savings are dependent on heating need of the building and efficiency of the heating source. According to the calculated energy savings for Powerhouse Kjørbo of 1,2 kWh/m² and the assumed energy use for PCS heaters of 0,5 kWh/m², it is likely to assume that the heaters would provide some energy savings, even in this efficient building.

Based on the findings in the study, as well as the reviewed literature, it is likely but not proven that PCS systems can directly contribute toward realizing Plus energy buildings. However, they have been found to have the potential to provide personal control and increased thermal satisfaction to the occupants. This is especially true for those few occupants who have special needs or are placed in problematic areas of the building. In an indirect way, this may still contribute toward realizing Plus energy buildings as PCS systems can help relax the demands set to other climate installations in the building. This again can allow the use of more environmentally friendly solutions such as utilization of thermal mass, temperature stratification and natural ventilation. They may also reduce installation cost by eliminating the need for more complicated and costly measures for accommodating indoor climate demands, as well as provide valuable usage data for system learning.

The potentials of using PCS systems as a strategic contribution toward the climatization of Plus energy buildings should be investigated further. Especially in regard to its potential for dealing with problematic areas, old or existing buildings, natural climatization solutions and usage data collection.

6. Acknowledgements

Thanks to Asplan Viak for access to the field test site, to Bjørn Jenssen, Skanska for energy simulation model and to Terje Josefsen, Skanska for help with digital complaint system.

7. Bibliography

1. Fountain M, Brager G, De Dear R. Expectations of indoor climate control. Energy Build. 1996;
2. Mendell M. Indoor Thermal Factors and Symptoms in Office Workers : Findings from the Indoor Thermal Factors and Symptoms in Office Workers : Findings from the. 2009;(December):1–26.
3. Lassen N. A study of running set-points and user IEQ satisfaction perspectives in the Norwegian commercial building stock. In: 39th AIVC Conference, Smart Ventilation for buildings [Internet]. 2018. p. 180–9. Available from: www.rehva.eu
4. Karmann, Caroline, Schiavon, Stefano, Arens E. Percentage of commercial buildings showing at least 80% occupant satisfied with their thermal comfort. 2018;0–7.
5. Arens E, Humphreys MA, de Dear R, Zhang H. Are “class A” temperature requirements realistic or desirable? Build Environ. 2010;
6. Cheung T, Schiavon S, Parkinson T, Li P, Brager G. Analysis of the accuracy on PMV – PPD model using the ASHRAE Global Thermal Comfort Database II. Build Environ. 2019 Apr 15;153:205–17.
7. Zhang H, Arens E, Zhai Y. A review of the corrective power of personal comfort systems in non-neutral ambient environments. Build Environ. 2015;
8. Kim J, Bauman F, Raftery P, Arens E, Zhang H, Fierro G, et al. Occupant comfort and behavior: High-resolution data from a 6-month field study of personal comfort systems with 37 real office workers. Build Environ [Internet]. 2019;148(September 2018):348–60. Available from: https://doi.org/10.1016/j.buildenv.2018.11.012
9. Hoyt T, Arens E, Zhang H. Extending air temperature setpoints: Simulated energy savings and design considerations for new and retrofit buildings. Build Environ [Internet]. 2014;88:89–96. Available from: http://dx.doi.org/10.1016/j.buildenv.2014.09.010
10. Watanabe S, Melikov AK, Knudsen GL. Design of an individually controlled system for an optimal thermal microenvironment. Build Environ [Internet]. 2010;45(3):549–58. Available from: http://dx.doi.org/10.1016/j.buildenv.2009.07.009

IOP Conf. Series: Earth and Environmental Science **352** (2019) 012052 doi:10.1088/1755-1315/352/1/012052

POLYSOL – Thermal and electrical performance assessment of a cost-effective polygeneration system

J Soares[1], B Shahzamanian[1], S Varga[1], A Palmero-Marrero[1,2], A C Oliveira[1,2]

[1] CIENER, Institute of Science and Innovation in Mechanical and Industrial Engineering, Porto, Portugal
[2] Department of Mechanical Engineering, Faculty of Engineering of University of Porto, Porto, Portugal

joaosoares@fe.up.pt

Abstract. Buildings are responsible for a large portion of the energy consumption and harmful emissions to the environment. To change this situation, the European Performance of Buildings Directive requires new buildings to be nearly Zero Energy Buildings. To achieve this goal, energy saving measures combined with large use of renewables on-site must be applied. In this paper, a reliable and cost-effective polygeneration system for an existing test building, able to provide electricity, cooling and heating needs is presented, modelled and assessed. The system relies on solar energy as the primary source (PV and solar thermal collectors). Building heat and cooling demand is achieved with a variable geometry ejector heat pump. To reduce the mismatches between generation and consumption, it includes different types of storage (electrical and thermal). For the test building energy needs assessment, the electricity consumption of all test building consumers was measured. Thermal energy demand evaluation required the development of a dynamic numerical model using TRNSYS software. The building thermal performance was validated using experimental data. The numerical model also includes the solar field, thermal energy storage, ejector cycle subjected to know meteorological conditions and operation control. To model and simulate yearly electricity production PVSyst commercial software was applied. Regarding the thermal energy, it was found that on a yearly basis the energy supplied by the thermal collectors about six times the demand. Nevertheless, on the hourly and daily levels, there are thermal energy shortages, due to the lack of solar radiation or mismatch between production and demand. Regarding the electric energy consumption, the potential electric energy production is about 1.6 times the demand. Like for the thermal energy, on the hourly and daily levels, there are energy deficits mostly in winter, for which an electrical energy storage unit will be used. Additionally, the highest peak occurs in summer because of low solar radiation at the end of the day and high demand due to the high cooling load of the test room.

1. Introduction

Buildings are responsible for a large portion of the energy consumption and also harmful emissions to the environment. To change this situation, the European Performance of Buildings Directive [1] requires new buildings to be nearly Zero Energy Buildings. To achieve this goal energy saving measures combined with large use of renewables on-site must be applied.

In this paper, a reliable and cost-effective polygeneration system for an existing test building (SOLAR-TDF: Solar Test and Demonstration Facility [2]), able to provide electricity, cooling and heating needs is presented, modelled and assessed. The system relies on solar energy as the primary source (PV and solar thermal collectors). Building heat and cooling demand is achieved with a variable geometry ejector (VGE) heat pump. To reduce the mismatches between generation and consumption, it includes different types of storage (electrical and thermal). However, the integration of distinct renewable and storage systems with a variable daily and yearly energy demand is complex [3]. Therefore, a system prototype is under development and will be installed as a demonstration in Portugal.

The building energy assessment was carried out with a developed numerical model by the combination of two commercial software: TRNSYS [4] and PVsyst [5]. The model was validated using experimental test results. Results include daily, monthly and yearly thermal and electrical energy consumption and production of the test facility, as well as peak electrical power, peak cooling and peak heating loads.

2. Research Methodology

The SOLAR-TDF is constituted by: a solar collector field; the VGE heat pump; the thermal energy distribution (cooling and heating); air-conditioned space; and an equipment room, as shown in Figure 1. The solar field is composed by 17m^2 of evacuated tube solar collectors, providing heat to drive the VGE heat pump, which is installed in the equipment room. The VGE heat pump permits to accommodate both building heating and cooling demand. A dissipation fan coil unit is used to protect the collector field or for cooling the VGE heat pump during the cooling operation.

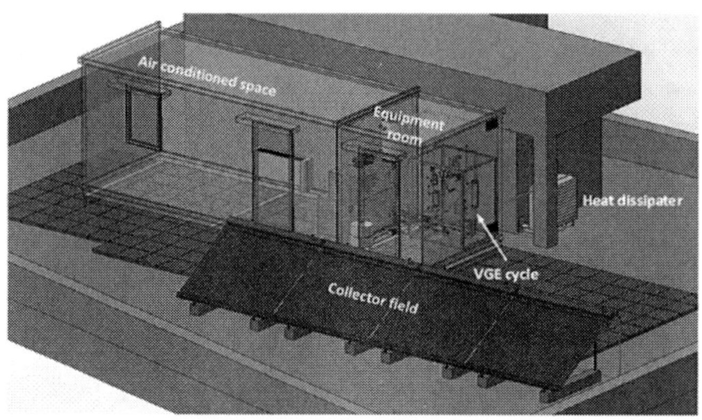

Figure 1. SOLAR-TDF test facility [6].

The assessment of the SOLAR-TDF energy needs, both electrical and thermal, was carried out through experimental and simulation approaches. Initially, the electricity consumption of each electric device was measured (section 2.1). The assessment of the thermal energy demand required the use of a dynamic mathematical model, developed using a commercial software TRNSYS (Section 2.3). To assure the numerical building model provides acceptable accuracy, results were validated (Section 3) with experimental data (Section 2.2). To model and simulate yearly electricity production by the solar panels, the PVsyst commercial software was applied (Section 2.4).

2.1. Electricity consumption assessment

The consumption of each electric device of the SOLAR-TDF determined through power, current and voltage measurements of the single-phase devices; and apparent/reactive power and power factor measurements for the three-phase consumers, using MICROVIP 3 PLUS energy analyser.

The total power consumption of the SOLAR-TDF is about 1.1 kW$_{el}$, with more than 50% due to the heat dissipation fan coil unit, used for the protection of the collector field or eventually for cooling the condenser during cooling operation. Nevertheless, during regular operation, the electrical devices do not work simultaneously. For simulation purposes, the electric devices of the test facility were grouped according to their principal function: data acquisition, solar field, dissipation fan coil unit, cooling and heating (see Figure 2).

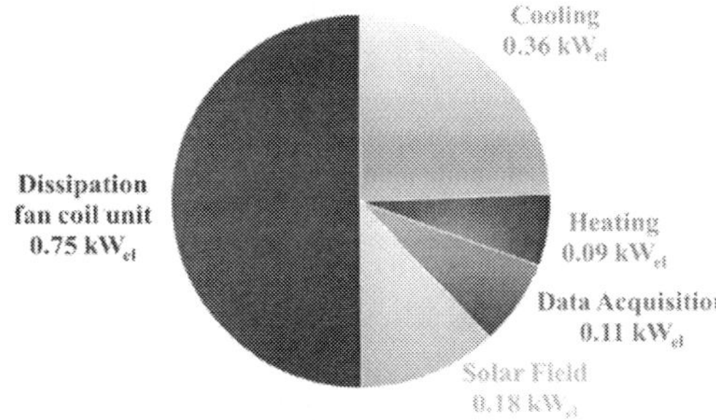

Figure 2. Electrical consumption (kW$_{el}$) by operation mode.

2.2. Experimental data collection for model validation
Experimental determination of the thermal energy demand of the SOLAR-TDF requires detailed measurement of the local climatic conditions (ambient temperature, solar radiation, wind speed and direction, etc.); indoor temperature and infiltration rate measurement for a one-year period. Therefore, it is common practice to use a dynamic mathematical model to carry out this task. In this case, the model was developed using a commercial software TRNSYS.

In order to assure that the numerical building model provides acceptable accuracy during the simulations, it is crucial to validate the results with experimental data. For this purpose, the local climatic data was measured (i.e. global horizontal solar radiation, relative humidity and ambient temperature) and the indoor temperature (5 thermocouples) was monitored without active air conditioning during an approximately one-month period from 24th of August to 19th of September. Data logging was performed in 15 min time intervals.

2.3. Development of the numerical TRNSYS model
The dynamic model was developed using TRNSYS Simulation Studio, whilst the test building was modelled using TRNBuild. In TRNBuild the building model was completely defined specifying the composition of the walls, their thermo-physical properties, their orientation and size, indoor gains, infiltration rate, etc. Some of the physical properties (e.g. thermal conductivity, density and specific heat capacity) of the building envelope were measured, some others were estimated (e.g. optical properties). Besides the building model, other components were also included in the TRNSYS model such as the Solar Field (SF), Thermal Energy Storage (TES), Ejector Cooling Cycle (ECC) subjected to know meteorological conditions and control strategy as shown in Figure 3. These other components were not used during the building model validation phase.

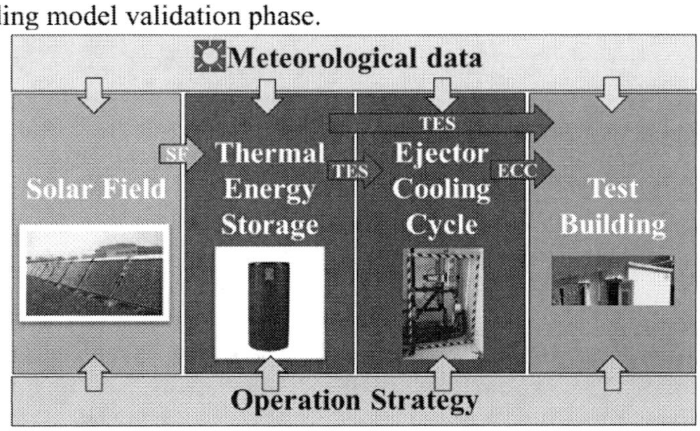

Figure 3. Model components for the thermal load simulations.

IOP Conf. Series: Earth and Environmental Science **352** (2019) 012052 doi:10.1088/1755-1315/352/1/012052

For the metrological data, a Typical Metrological Year (TMY) for Porto city from Meteonorm software was used. The annual global horizontal radiation is about 1773 kWh/m^2 with an average ambient temperature of 14.5°C. The solar field is composed by the four BAXI-AR-30 collectors in series, with a net absorber area of about 13 m^2. The efficiency curve was defined according to the manufacturer's specifications as:

$$\eta_{collector} = 0.832 - 1.14\frac{\bar{T} - T_{amb}}{I} - 0.014\, I\left(\frac{\bar{T} - T_{amb}}{I}\right)^2$$

Where I, $\bar{T}$, T_{amb} represent incident irradiation, mean collector temperature and ambient temperature, respectively. The TES tank was modelled considering water as heat transfer fluid. In order to address stratification, the tank was subdivided into 12 fully-mixed equal volume segments. Additionally, constant losses were assumed for each segment, considering a typical 10 mm thick polyurethane insulation layer (thermal conductivity of 0.03 W/m.K). The tank volume applied was highly over-sized with 5 m^3 capacity, to avoid energy dissipation during high solar radiation periods. The control strategy between the solar energy and TES was achieved using an on/off differential controller to preserve a 3°C temperature difference between the SF and TES tank outlets.

A simplified and conservative methodology was used for the ECC based on the operational experience [2]. A fixed Coefficient of Performance (COP) of 0.25 was set, i.e. the cooling cycle thermal demand is four times the thermal production. Within the test room, a heating and a cooling device were modelled to elevate or reduce the room temperature, respectively. The heat is added or removed from a flow stream at a specified rate (2.65 kW$_{th}$ for heating and 1.6 kW$_{th}$ for cooling), whenever the room temperature is lower than 22°C (heating) or higher than 26°C (cooling). A 15-minute time-step was used for the validation of the model, whilst a 30-seconds time-step was adopted for the year-around simulations due to the higher complexity of the system. As control strategy, separate heating and cooling seasons were defined (i.e. no need for heating and cooling during the same day). The heating season was set from October to May and the cooling season from June to September. Although this strategy does not permit to always achieve an indoor temperature within the predefined limits (22°C to 26°C), it was considered to be unrealistic to heat and cool the building during the same day.

Additionally, it was found that when the indoor temperature rises above 26°C during the heating season, the ambient temperature remains relatively low and thus free cooling is a valid option for assuring thermal comfort of the occupants. The electricity consumption was also simulated using the TRNSYS model. Based on the control signal for the pumps, fan coils, etc., the operation of the electric devices can be predicted for each time step. The corresponding electricity consumption was calculated according to the measured data (see Section 2.1).

2.4. Mathematical model for the electricity generation by the PV modules

To model and simulate yearly electricity production by the solar panels, the PVsyst commercial software was used. Although TRNSYS has some capabilities to simulate PV electricity generation, PVsyst is a more reliable tool and a reference application for photovoltaic system simulations with extensive component model libraries. For the simulations, the monocrystalline modules with reference 04TA_-16410989 by ONYX were considered. In order to assess the full production potential of the PV solar field, a similar approach was followed as in the case of the TES system. Namely that the generation of the panels was evaluated using a large capacity energy storage, in this case, the electric grid.

3. Model Validation

For model validation purpose, simulations were carried with the developed building model, without active heating and cooling. The measured metrological data were used as boundary conditions, and the simulated indoor temperature results were compared to the measured ones, calculated as the average of the five thermocouple readings. During these simulations, some of the estimated building properties were changed so that the deviation between simulated and measured temperatures were reduced. A

IOP Conf. Series: Earth and Environmental Science **352** (2019) 012052 doi:10.1088/1755-1315/352/1/012052

comparison between the measured and simulated indoor temperatures is shown in Figure 4 for about 26 consecutive days. It is noticeable, that after tuning of the model, a very good agreement between simulated and measured data was obtained. The maximum temperature difference found was 4.2°C, and the root-mean-square error is about 1.1°C.

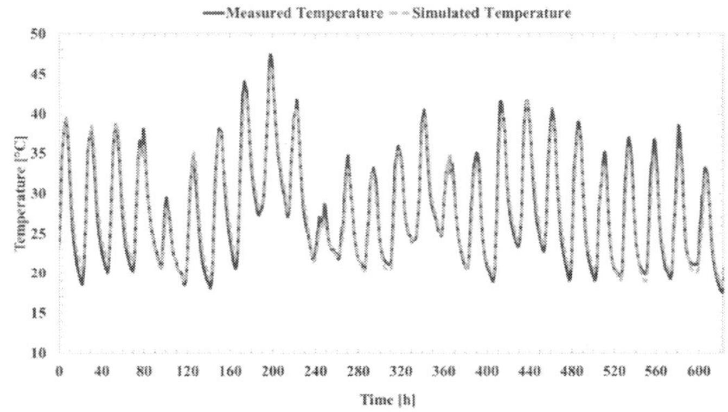

Figure 4. Comparison between measured and simulated indoor temperature for 26 days.

4. Results and discussion

4.1. Yearly thermal performance

Figure 5 shows the monthly thermal energy consumption and solar heat production of the test facility. It can be seen that the thermal demand is always satisfied by the production by the existing solar collectors. The highest heat demand occurs in the summer, as a result of the increased cooling load and in agreement to the generation. This clearly shows the benefit of using solar energy for space cooling. The maximum heat demand is in July when there is a match between the heat supplied by the solar collector and the heat consumption the generator of the cooling cycle (~450 kWh_{th}). For the rest of the year, monthly production is considerably higher than consumption.

On a yearly basis, a total heat production of about 15.7 MWh_{th} was simulated. The annual heat consumption is about 2.5 MWh_{th}, with about 1.6 MWh_{th} for cooling and 0.9 MWh_{th} for heating (see Figure 6). This significant mismatch between heat production and consumption is mostly because the SOLAR-TDF was designed to meet the space cooling need of the test room, and also there are extensive periods when nor heating nor cooling is needed to maintain indoor comfort. In a real application, overall solar energy efficiency can be improved by integrating the developed system with e.g. domestic hot water production or swimming pool heating.

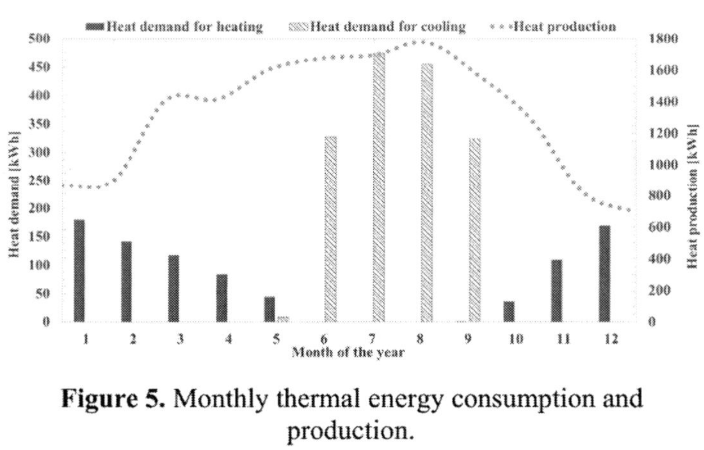

Figure 5. Monthly thermal energy consumption and production.

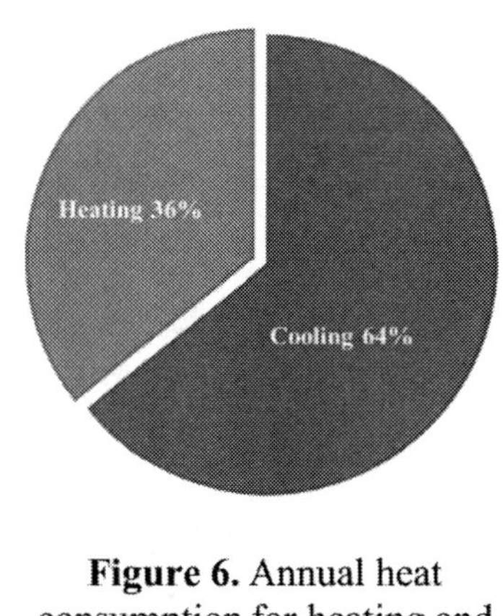

Figure 6. Annual heat consumption for heating and cooling.

It should be noted that even though thermal energy production is considerably higher in the SOLAR-TDF on the monthly and annual basis, there might be shortages on the hourly and daily level, due the lack of solar radiation or temporal mismatch between production and demand. This effect is explained in Figure 7, where the daily difference between heat production and demand is plotted for the whole year. Negative values indicated a lack of useful solar heat. The highest thermal energy deficit was found for the 15th of February, with about 10 kWh_{th}. Additionally, in the previous day (14th of February) a deficit of about 9 kWh_{th} was also obtained. In this set of two critical days, solar heat production is almost negligible (0.4 to 1.2 kWh_{th}). Thermal comfort can only be maintained with the implementation of a TES unit. TES can be charged with the excess thermal energy during the previous day (36.8 kWh_{th} on the 13th of February), and discharged when the production is reduced. With a TES unit of about 20 kWh_{th}, the thermal comfort in the test room can be assured during the entire year.

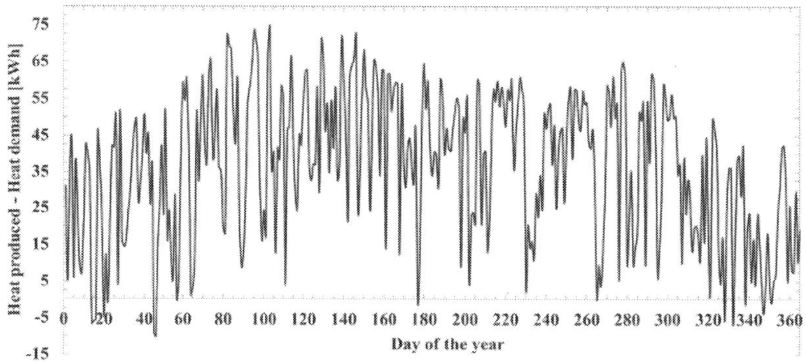

Figure 7. Daily difference between heat produced by the solar collector and consumed for indoor temperature control.

4.2. Yearly electrical performance

Two cases were assessed, one with 7 and another one with 8 modules in series, with a yearly fixed tilt angle of 35° and south orientation. The tilt angle was defined to achieve maximum annual generation. Figure 8 shows the monthly electricity consumption and production of the test facility. The electricity demand is always satisfied by the production, with the highest consumption and production occurring in the summer months. On a yearly basis, a total electricity production between 3 MWh_{el} (7 modules) and 3.3 MWh_{el} (8 modules) was simulated. The annual electricity consumption is about 1.9 MWh_{el}, with more than 50% related to the continuously operating data acquisition system (see Figure 9). The annual solar electric energy efficiency is about 60%.

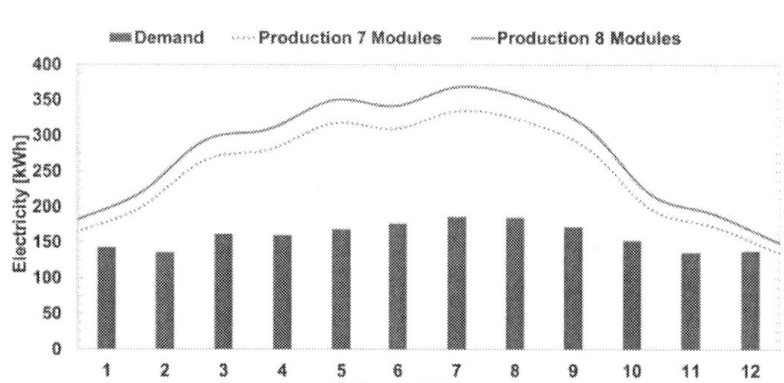

Figure 8. Monthly electrical energy consumption and production.

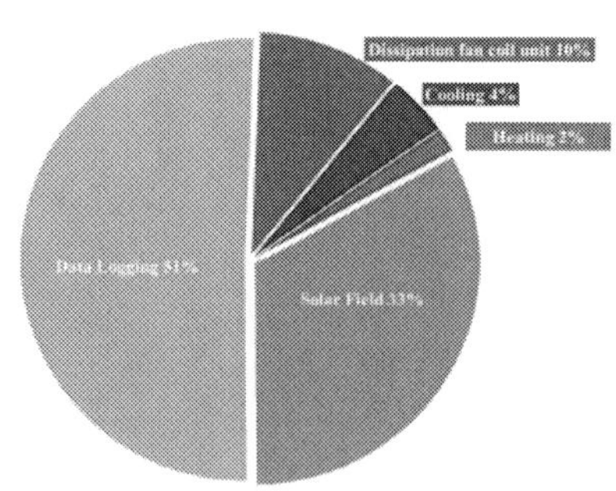

Figure 9. Annual consumption fraction of consumers groups.

The solar collector field is found to be responsible for more than 33% of annual energy consumption with the already installed solar field. Whilst the dissipation of the excess heat consumes the highest electric power (see section 2.1), it represents only about 10% of the total electrical energy consumption. It is worth noting that only about 6% of the electricity consumption is directly for heating and cooling, emphasising one of the main advantages of this thermally-driven HVAC solution.

Again, although electrical energy demand is assured by the PV modules on the yearly and monthly basis, there are periods of the year with electrical energy deficits, occurring mostly during the winter (see Figure 10) due to the time mismatch between demand and production. The maximum electrical energy deficit, about 4.4 kWh_{el}, was found for 30th of November and related to a rather low daily PV energy production (bellow 0.5 kWh_{el}). To assure the test facility's grid independence, an electrical storage unit will be implemented. Considering the cumulative data, i.e. consecutive days with a negative electrical energy balance, a maximum value of about 11 kWh_{el} was identified, corresponding to three consecutive days of energy scarcity from 1st to 3rd of January.

The power subsystem, PV and electrical battery also must handle instantaneous power peaks. Figure 11 shows the hourly electrical energy balance for the PV arrays of 7 modules. On the hourly basis, the highest peak occurs in summer (~0.8 kW_{el}), as a consequence of low solar radiation at the end of the day (see Figure 11-detail) and high demand due the high cooling load of the test room (almost all electric devices are operational at the same time).

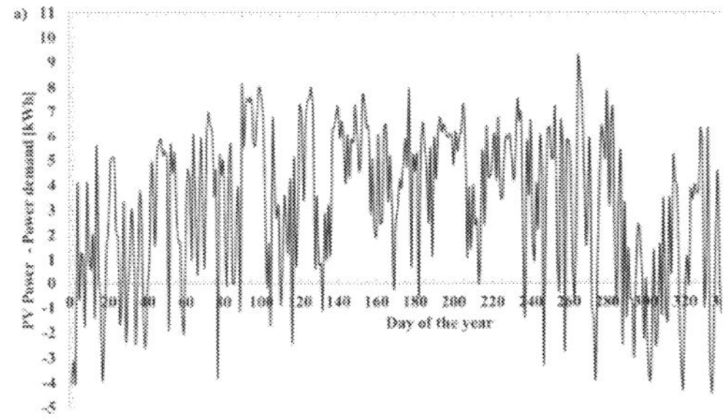

Figure 10. Daily difference between PV generation and electricity demand for 7 PV modules.

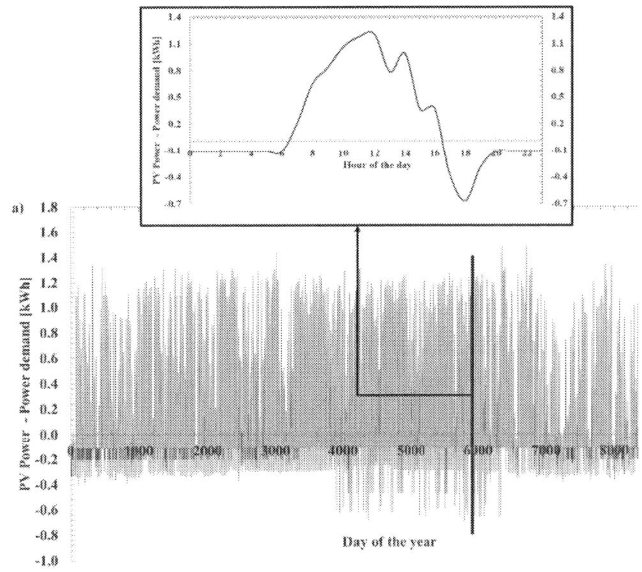

Figure 11. Hourly difference between PV generation and electricity demand for 7 PV modules.

5. Conclusions

In this paper, a reliable and cost-effective polygeneration system for an existing test building able to provide electricity, cooling and heating needs is presented, modelled and assessed. Numerical and experimental works have been carried out with the principal objective of identifying both thermal and electric energy consumption of the SOLAR-TDF throughout the year. Additionally, this energy demand was compared to the production potential of the existing solar collector field and the to be installed PV array.

Regarding the thermal energy, it was found that on a yearly basis the energy supplied by the thermal collectors (15.7 MWh_{th}) is about six times the demand, with about 64% of the total demand required for cooling. On the hourly and daily levels, there are thermal energy shortages, due to the lack of solar radiation or mismatch between production and demand. The highest cumulative thermal energy deficit (19 kWh_{th}) was found for the 14th and 15th of February. During this period, thermal comfort can be maintained with the implementation of a TES unit, using the excess thermal energy during the previous day (36.8 kWh_{th} on the 13th of February).

Regarding the electric energy consumption, the data indicated that the annual demand is about 1.9 MWh_{el}, with more than 80% of the total demand required for the data logging and the solar collector field. Regarding the potential electric energy production, it was found that the electricity produced with 7 and 8 PV modules (1.6 m^2) is about 1.6/1.7 times the demand, respectively. Whilst the maximum annual generation is attained for a 35° tilt angle, tilt angles between 25° and 55° should be assessed, i.e. maximum electrical energy production for winter and summer season respectively, in order to find the optimum layout to reduce the divergence between production and demand.

Like for the thermal energy, on the hourly and daily levels, there are energy deficits, mostly in winter, due to the lack of solar radiation or mismatch between production and demand. The maximum cumulative electrical energy deficit (~11 kWh_{el}) was found for three consecutive days of energy scarcity from 1st to 3rd of January. To assure the test facility's grid independence, an electrical storage unit will be implemented.

On the hourly basis, the highest peak occurs in summer (~0.8 kW_{el}), as a consequence of low solar radiation at the end of the day and high demand due to the high cooling load of the test room. The power subsystem, PV and electrical battery also must handle these instantaneous power peaks.

Acknowledgments

The authors gratefully acknowledge the financial support of "Fundação para a Ciência e Tecnologia" (FCT – Portugal), through the research project POCI-01-0145-FEDER-030223 ("Development of a polygeneration solar system for Zero Energy Building").

References

[1] *The Directive 2010/31/EU of the European Parliament and of the Council of 19 May 2010 on the energy performance of buildings.* Official J Eur Union 2010 53
[2] Varga S, Oliveira A C, Palmero-Marrero A and Vrba J, 2017 *Renew Energ.* **109** 83-92
[3] Lu Y, Wang S, Shan K, 2015 *Appl Energ.* **155** 463-77
[4] University of Wisconsin—Madison, Solar Energy Laboratory, 1975 *TRNSYS, a Transient Simulation Program* Madison, Wis
[5] Mermoud A, 2018 *PVSYST: Software for the study and simulation of Photovoltaic systems* Ed. 6.7.8. University of Geneva
Pereira PR, Varga S, Oliveira A C and Soares J, 2015, *Front. Mech. Eng.* **1**:7

IOP Conf. Series: Earth and Environmental Science **352** (2019) 012053 doi:10.1088/1755-1315/352/1/012053

The design process for achievement of an office living laboratory with a ZEB standard

Berit Time[a], Atle Engebø[b], Morten Christensen[b], Ola Dalby[c], Tore Kvande[b]

[a]SINTEF, Trondheim, Norway,
[b]NTNU, Trondheim, Norway
[c]Veidekke AS, Trondheim, Norway

Corresponding author; Berit.time@sintef.no

Abstract. ZEB Laboratory is a full-scale office building with a ZEB-COM ambition, searching a high degree of flexibility and where components and technical systems can be modified for research purposes. Project delivery of a living laboratory with a ZEB standard is not an easy task. The implementation of the ZEB method in a partnering contract as a project delivery model has been developed. This paper describes and elaborate the development of the project delivery and design process for ZEB Laboratory seen by the client. The ZEB Laboratory design and procurement process has given valuable insight and experience into the use of partnering and collaborative elements for planning and production of ZEB buildings.

1. Introduction

The ZEB Laboratory is an office living laboratory 4 stories high and approximately 2000 m^2 located in Trondheim, Norway. The design of the building is ongoing and construction started May the 7th 2019. ZEB Laboratory is a full-scale office building with a ZEB-COM ambition [1], searching a high degree of flexibility and where components and technical systems can be modified for research purposes. The building will form a living laboratory where the client (NTNU and SINTEF) is using it as an (ordinary) office building and for educational purposes, where the users also become an experimental parameter giving variations in loads with their use of the premises.

The funding for an office living laboratory has been achieved by an application to the Research Council of Norway. NTNU/SINTEF was awarded 63 million NOK from the Research Council, 8 million NOK from Enova, and together with in-kind contribution from NTNU and SINTEF, a total budget of 127 million NOK for the research office building with a ZEB-COM ambition is made available.

2. Vision and ambitions

The vision of the Norwegian Zero Emission Building Laboratory (ZEB Laboratory) is to be an arena where new and innovative components and solutions are developed, investigated, tested and demonstrated in mutual interaction with people.

The ZEB Laboratory shall be:

- a basis for knowledge development at an international level,
- a basis for international competitive industrial development,
- an example for new and retrofitted zero emission buildings,
- a research arena for developing zero emission buildings,
- an arena for risk reduction when implementing zero emission building technologies,
- a national resource for all research organisations in the research area,
- a tool for NTNU and SINTEF for institutionalizing the post ZEB Centre [3].

The focus is on making the initial building a relevant research case, both with regard to energy performance, climate mitigation and adaptation and mutual interaction with users-building interaction in a Nordic climate. In order to make the building a relevant research case, NTNU/SINTEF has defined and asked for a building to fulfil certain ambitions for the ZEB Laboratory [6]. These are, in prioritized order:

1. The building should be a model project and achieve ZEB-COM level (simulated in a 60 years perspective) [1]
2. Separate control and measurement systems, one for ordinary operation and one for research
3. Flexibility in design of energy and climatization systems
4. Flexibility in design of working spaces
5. Progressive selection of materials and enabling rebuilding parts of the facades
6. Adaptation of the building to climate change [5]

3. Theoretical framework

3.1. *Project delivery - Contract models*

Selecting a project's delivery method is an imperative client decision. The delivery method defines the roles and responsibilities of all project participants, how to finance the project, and it sets the framework for the delivery. In regards of different delivery methods to adopt, the owner has a wide array to choose from [19]. However, the choice is often limited by factors such as procurement laws, tradition, competence, etc. Furthermore, the project delivery method for buildings have to take into account the boundaries, framework and complexity of the particular project. For traditional building projects where the main issues are time, cost and quality *Design-Bid-Build* and *Design-Build* are the most used project delivery methods. However, as The ZEB Laboratory is a complex project with ambitions to deliver in areas beyond time, cost and quality, a new type of project delivery method is needed. A central aspect of construction projects is the allocation of risk between the actors, and one of the most prominent elements for managing risk is the use of construction contracts. In traditional project delivery methods, the contract is often extensive and carefully written, clearly defining the relationships of all parties. Furthermore, the contract determines at which phase the contractor gets (part) responsibility and has a direct impact on risk distribution and innovation in a design and construction process [4]. Due to the complexity of the project, a tradition contract for managing risk was seen as adversely towards achieving the client's ambitions. Taking this into account when crafting the right project delivery method for the ZEB Laboratory, three distinct methods were considered relevant for the project. The three methods were Design-Bid-Build, Design-Build, and lastly a collaborative form of project delivery method.

3.1.1. Design-Bid-Build contract and Design-Build.

The Design-Bid-Build is often referred to as the most prevalent project delivery method. The method is centered around the owner engaging an architectural or engineering firm to produce a set of plans, specifications, and specific project requirements. These are again used as bid documents that are being distributed to contractors. Then the contractors submit a bid to complete the work as outlined in the bid documents [9].

The Design-Build is a project delivery method in which the owner provides requirements for the project and awards a contract to the contractor who will both design and build the project [10]. Consequently, there is only one contract for the client to manage, removing some risk at the expense of influence. Both the Design-Bid-Build and Design-Build are typically used and suited if the project is clearly defined and the technology involved are well known. Under traditional contractual conditions, there are strict differences in terms of what the parties should do. The contract consequently contains detailed provisions on the placement of responsibility for something going wrong, for example. In case of delays, errors made during the execution of the work and defects discovered after delivery. Because the financial responsibility is placed solely on the "guilty one", a contradiction between the parties can easily be developed and disagreements arise with regard to who has caused the problems. The parties are naturally more concerned about document their "own excellence" and the "other's inability" than by trying to solve the problems in the best possible way.

3.1.2. Partnering/Collaborative contract

The Partnering approach may take many forms due to the lack of one consistent framework. The Partnering concept is often attributed to Oil and Gas development projects in the North Sea. The approach was later adopted by the construction industry, as Alliances in Australia and Finland, IPD in North America, and Partnering in the UK. According to the literature, there is a division between those who see partnering as an informal development and those who regard it as something that can be actively engineered. Towards the formal spectre, partnering may be archived by the use of specific contracts, and the use of incentive systems based upon risk/reward [11]

The American Institute of Architects defines Integrated Project Delivery (IPD) as a project delivery method that seeks to integrates people, systems, business structures and practices into a process that maximise the effort to optimise project results, increase value, reduce waste, and maximise efficiency through all phases [12]. The contract used in the IPD is referred to as so-called multi-party contract where the owner, architect, contractor and other key execute a single contract. Another specific approach is the Australian Alliancing method. An alliance is defined as a non-equity alliance between a client and service provider using a contract. A project alliance is formed under interorganizational cooperative arrangements for delivering a specific project or outcome and has a defined end [17].

As stated previous, selecting the project delivery method is determined by the client. Furthermore, each delivery method has its advantages and disadvantages depending on the characteristics of the projects under consideration. Some of the common features of the different collaborative project delivery methods is that they go further in seeking integration than the more traditional Design-Bid-Build and Design-Build methods. Fischer summaries key elements for archiving such integration through the following elements [18]:

- Early involvement of key participants.
- Profit disassociated from units of cost.
- Costs without profit paid to completion.
- Profit adjusted based on agreed outcomes.
- Joint setting of targets.
- Joint project management.
- Reduced liability among team members.

3.2. *ZEB design method*

The EU Commission asks for (Directive 2010/31/EU) the transformation of the building stock and the shift to more sustainable energy supply and nearly zero energy buildings. It also promotes that Member States can choose to further rise the ambitions in terms of mitigation actions and provide numerical indicators, for example for the entire building's overall energy use or greenhouse gas emissions. In the

Norwegian ZEB Centre [3] zero emission building (ZEB) concepts and technologies have been developed and five ambition levels have been defined [1], see Figure 1.

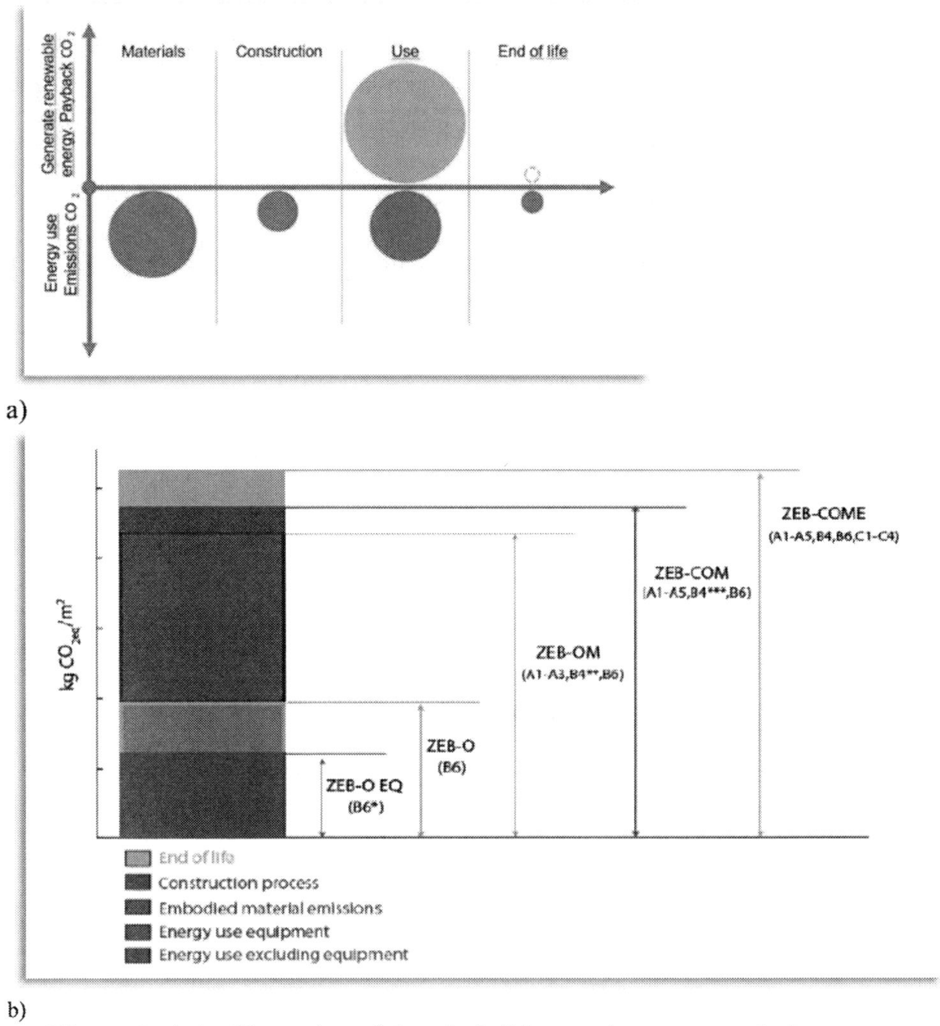

Figure 1. a) An illustration of the whole life greenhouse gas emissions are constituted by manufacturing of construction materials and building services, the construction process, building operation, and finally demolition, including reuse or recycling of materials b) The numbers in brackets refers to the included life cycle stages according to NS-EN 15978 Sustainability of construction works. Assessment of environmental performance of buildings. Calculation method. The "O" refers to emissions associated with Operational energy use. The "M" refers to embodied emissions associated with building construction Materials. The "EQ" refers to operation emissions from technical Equipment. The "C" refers to emission associated with Construction and installation, while the "E" refers to embodied emissions associated with end of life phase of the building. [1]

The ZEB design method is described in Hestnes et al. 2017 [3]. The method is recently developed and introduced in the building sector. In the former ZEB Centre [3] nine pilot building projects were performed, and an integrated energy design process (IED) was used to achieve the ZEB goal in the best possible way. In the simplest of terms the IED has these main characteristics:

1. It calls for a different approach from the very early stages of design

IOP Conf. Series: Earth and Environmental Science **352** (2019) 012053 doi:10.1088/1755-1315/352/1/012053

2. It requires a high level of general skills (energy knowledge in a broad sense) and communication within the team
3. It leads to a high level of integration and synergy of systems
4. It involves modern simulation tools where suitable

4. Results and discussion

4.1. Designing the project delivery

Project delivery of a living laboratory with a ZEB standard is not an easy task. NTNU/SINTEF went through a project delivery of ZEB Living Lab (a residential living laboratory) from 2012-2015 [2]. Some valuable lessons were learned, and experiences were made related to project delivery, i.e. project development, procurement and contracts.

The ZEB Laboratory is being built with "tomorrow's" technology. Principles and overall needs were defined in the project description for the *Norwegian Research Council*, but design and solutions for how we could best fulfil the high ambitions of the building was not determined in the application. In the initial planning phase, we draw on the research environments at NTNU and SINTEF to clarify the thoughts and needs for the use of the building/laboratory in order to obtain maximum benefit for measurement on components, as well as the development of new solutions. The users of the building also had to enter the planning early so that their need for a modern office building is satisfied. In addition, the educational aspect for students and doctoral candidates was mapped. The initial planning phases resulted in the project group at NTNU/SINTEF expressed the ambitions for the building in a note – *Level of ambitions* [6], which has followed the project and acted as a guiding document. The overall ambitions are expressed in chapter 1. Introduction.

With the high ambitions in terms of complexity and innovation the risk picture was evident. For the ZEB laboratory it would have been extremely demanding to define exactly what the delivery should be early in the project. We saw it as beneficial and an important part of the success of the project to collaborate closely with and draw on a skilled group of architects and engineers in both design and construction to achieve an excellent team within all relevant fields. Project complexity, the need for innovation, and the risk picture made the client choose a collaborative project delivery method. As opposite to traditional methods, this approach let the client focus on integration through early involvement of key participants, profit disassociated from units of cost, joint project management, and reduced liability among team members.

The first stage of the delivery of the project was initiated by a search for the right people, a team, eager to innovate and participate in a multidisciplinary group developing the laboratory. The assembled them needed to understand that changes in the project was expect and as well as understanding the client's need for innovative solutions. This also included taking care of end-user needs and influence in the upcoming planning and implementation stage. Therefore, a team of people with high level of general skills, i.e. energy knowledge in a broad sense, and evident communication skills were necessary. Not all people/personalities fit within this kind of project. Today's engineers and consultants are specialists performing very narrow and sequential tasks, which again creates fragmentation, and lowers productivity in the design process [13]. An integrated team is often proposed as a possible solution for tearing down the interfaces between the specialists. Consequently, the integrated team might aid the project in archiving its ambitious goals through allowing the skilled group to work in unison, towards some common project objective. Furthermore, the complex risk picture and interfaces in the project, made it necessary for early involvement of the contractor. This led the contractor to understands the risk picture early, leading to better planning regarding the implementation stage. A common understanding of the risks involved is also highly relevant to the cost of the project and the complex interfaces are better handled by actors with common goals.

As for the overall project delivery method, NTNU/SINTEF as client spent time to find and develop a method that could fit the current project together with an expert in project delivery method. A partnering contract structure was developed (see Figure 2) and asked for in the bid. The project delivery was divided into two phases, Phase 1; Planning and Design and Phase 2; Contract with an option to execute ZEB Laboratory (if target cost was accepted). A pre-project was developed through an integrated team approach in Phase 1 and a target cost was established. The client (NTUN/SINTEF) took the risk and they were also in charge of the decision making. The target cost principle is centered around the idea that the contractor's involvement in the pricing process improves the chances of developing an accurate estimate, as well as a useful mean for gaining commitment to project objectives [14]. In Phase 2, the contractor (Veidekke) has the responsibility for the design and the construction, the client contributes. The contractor has its expenses compensated but carries risk when exceeding the target cost. The client has decision making competence through change mechanisms.

A bid was developed focusing on the content of the contract for planning and design (Phase 1) and construction (Phase 2), competition rules, the project and overall requirements. The description with the overall ambitions for the project [6] were made as part of the bid. In the evaluation of the offers from the different contractors, the award criteria and the weighting of the different parts of the offer was:

1. tender amount 0,30
2. implementation plan 0,30
3. the project organization's expertise and experience 0.40

The selected contractor was Veidekke Entreprenør with their team; LINK Arkitektur, Aas Jakobsen, Siemens, Multiconsult, Oras/Bravida Trondheim and Vintervoll.

Figure 3 shows the partnering organization of the project. The steering group for the partnering consists of the regional leader of the contractor (Veidekke) and the property manager at NTNU. The partnering management team consists of the project leader for NTNU/SINTEF and his assistant and two responsible leaders from Veidekke. They meet weekly to sort out issues related to the partnering. The project team consists of a group of people from the contractor, its team and people from the client's core group. This team is led by a project leader from the contractor.

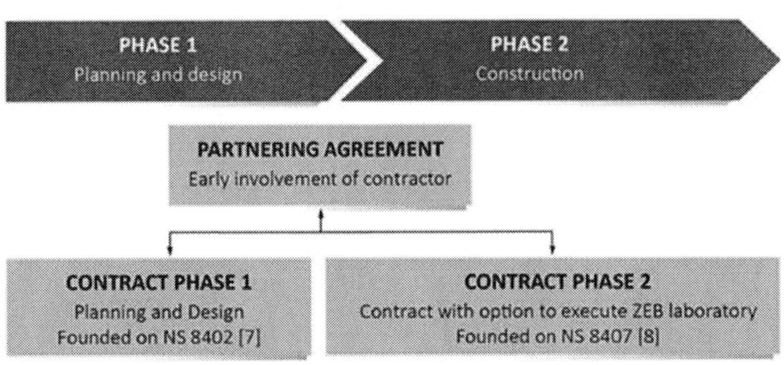

Figure 2. Contract structure. NS 8402 [7] and NS 8407 [8] refers to Norwegian contract standards

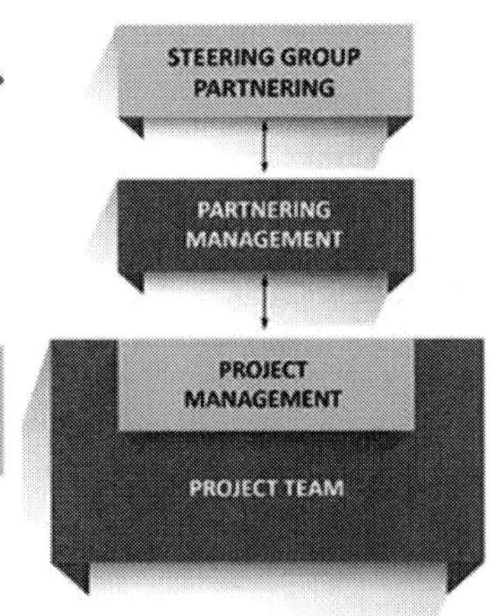

Figure 3. Partnering organization for the ZEB Laboratory project

4.2. Integration of the ZEB-method

The ZEB design method [3] is recently developed and introduced in the building sector, very few actors, i.e designers, contractors and manufacturers have learnt, experienced and are familiar with the method yet. This was also the situation for the selected ZEB Laboratory project team. The client facilitated the partnering in their BigRoom close to the building site, seeking to utilize the benefits of having a co-located project team. The BigRoom is the physical manifestation of the co-location principle, often designed as a room furnished with desks located in a horseshoe around SMART boards [15]. The contractor organized the development of the project with a two days start up seminar followed by co-location of the project team *in Integrated Concurrent Engineering (ICE)* sessions weekly, see Figure 4. While not being a radically different working method, the ICE methodology is a client-driven approach that focuses on getting the most out of the design team by relying on teamwork [16]. The main working period in Phase 1 started in the beginning of May 2017 (ICE session no. 1), ending late November 2017 (ICE session no. 19). Due to unforeseen circumstances the client had to move the building to a new site, next to the original site. Consequently all project documents needed to be revised. That meant a stop in the development of the design of the ZEB laboratory and it lasted one year between ICE session 19 and 20.

The achievement of a ZEB building calls for a different approach from the very early stages of design. We decided to organize a group of ZEB workshops integrated in the partnering. All together five workshops focusing ZEB definition, energy and emission accounts, relevant ZEB technologies for buildings and indoor environment were arranged with an extended group of people from the contractor, the consultants, the suppliers and the client as work on the design of the building prolonged in Phase 1. See also Figure 4. An important part of a ZEB process is to care for and take responsibility for the overall ZEB achievement more than only your own deliveries. A collaborative positive engagement is needed and developed collaborative skills among the people in the group are necessary. If this is not the case, changes in personnel should be done, this was also the case in this project.

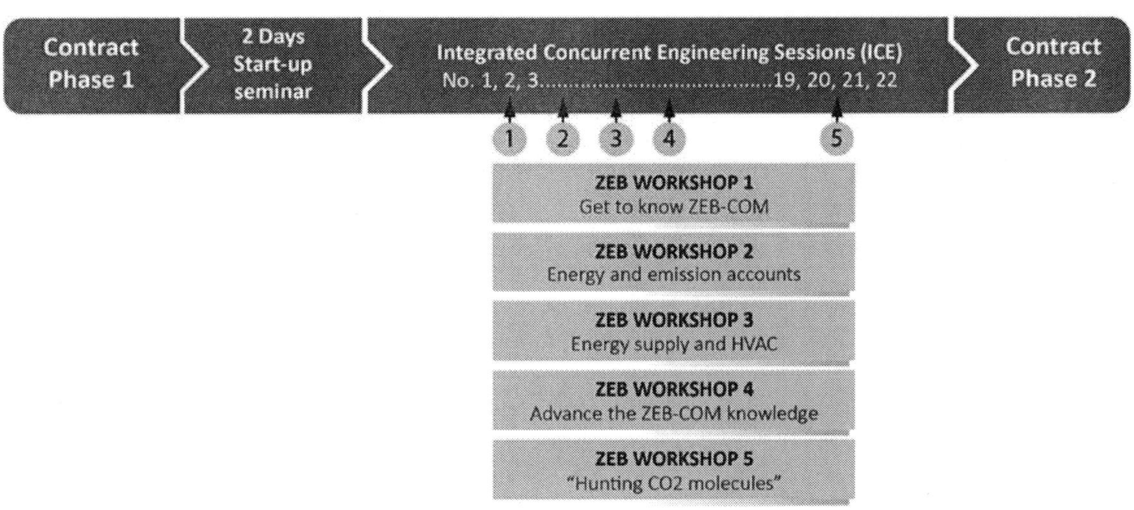

Figure 4. Sketch of implementation of measures to perform the ZEB methodology in the partnering process in the ZEB Laboratory

The project is now in Phase 2 doing the detailed design and the construction. The ICE-sessions are replaced by design meetings weekly for the project team, where relevant consultants are present in addition to the key personnel from the contractor. People from the client is also present in the design meeting, making sure the solutions from Phase 1 are implemented in the detailed design and as "consultants" in decision making. Dedicated personnel by the contractor have the responsibility for the

LCA account. Solutions and suppliers are selected based on documentation of embodied emissions in addition to cost/price. The delay of one year due to unforeseen circumstances made the contractor and some from the team of consultants change their personnel, this has been a drawback for the project and the design process.

4.3. Mutual evaluation and the use of performance measurements

Focus on improvements is essential for success working in partnering contracts with a common goal. A system for evaluation, both the process and mutual evaluation between the parties, is recommended. In order to be able to improve the process as well as keeping the productivity high in the team, the work flow was continuously measured using PPC (percent-plan-complete) [19]. PPC is calculated by dividing the number of assignments completed by the total number of assignments made for each ICE session. Such measurement helped the design manager track the process and the progression, as well as keeping the team members accountable.

4.4. Limits and potential of ZEB design processes

ZEB projects have a specific ambition aiming at a zero emission building level and a high technical quality, not only economy and progress. Achieving ZEB-level requires process iterations in architectural and technology design which you might not have in a traditional building project. The iterations are necessary to mature and optimize the building and solutions. A consequence might be a rise in initial costs and need for more time initially.

A common goal and trust between partners are a key to success for complex projects like ZEBs. The building industry do not have a reputation for a high degree of trust and confidence between partners. There is a great potential for the building industry to develop with complex projects like ZEBs.

It is the authors experience that a collaborative contract is very suitable for ZEB buildings, but also other buildings with specific targets. Compared to Design-Bid-Build contract and Design-Build we experience that the planning and design (Phase 1) of a collaborative contract takes half the time, and for Construction (Phase 2) the spent time is the same. This project has shown that there is a potential and to move forward a particular focus should be in development of project delivery. Keys to success are, bring all the actors together early to be reunited and create professional, but also social competent teams with little replacement of people through the project.

5. Conclusion

This paper reports on the design process for achievement of an office living laboratory with a ZEB standard. The implementation of the ZEB method in a partnering contract as a project delivery model has been developed. This paper describes and elaborate the development of the project delivery and design process for ZEB Laboratory through the perspective of the client. The ZEB Laboratory design and procurement process has given valuable insight and experience into the use of partnering and collaborative elements for planning and production of ZEB buildings.

6. Acknowledgement

The authors gratefully acknowledge the financial support from the Research Council of Norway to the ZEB laboratory project.

References

[1] Fufa S, Schlanbusch R, Inman M and Andresen I (2016) *A Norwegian ZEB Definition Guideline*, ZEB Project report no 29, The Research Centre on Zero Emission Buildings (ZEB).

[2] Goia F, Finocchiaro L, Gustavsen A (2015) The Living Laboratory at the Norwegian University of Science and Technology: A Zero Emission House for Engineering and Social Science Experiments. *Passivhus Norden: Sustainable Cities and Buildings*. Copenhagen, 20-21

IOP Conf. Series: Earth and Environmental Science **352** (2019) 012053 doi:10.1088/1755-1315/352/1/012053

August.
[3] Hestnes A and Eik-Nes N-L (Ed.) (2017) *Zero Emission Buildings,* Utgave 1, Fagbokforlaget.
[4] Kvande T and Christensen M (2016) *Valg av gjennomføringsmodell*, ZEB Flexible Lab Note no. 4, 102012312-1 (In Norwegian)
[5] Time B (Ed.) (2018) *Annual Report 2018*. Klima 2050 Report 8. Trondheim. ISBN: 978-82-536-1612-4
[6] Time B (Ed) (2016) *Ambisjoner*, ZEB Flexible Lab Note no. 3, 102013970 (In Norwegian)
[7] NS 8402: (2010) *General conditions of contract for consultancy commissions with remuneration on the basis of actual time taken*, Standard Norge, Utgave 1.
[8] NS 8407: (2011), *General conditions of contract for design and build contracts*, Standard Norge, Utgave 1.
[9] Sidney M L (Ed) (2010) *Chapter 1 - The genesis of a construction project, Construction Process Planning and Management,Butterworth-Heinemann*, Pages 1-11, ISBN 9781856175487, DOI: 10.1016/B978-1-85617-548-7.00001-X.
[10] Hale D R, Shrestha P P, Gibson Jr G E & Migliaccio G C (2009) Empirical comparison of design/build and design/bid/build project delivery methods. *Journal of Construction Engineering and Management*, 135(7), 579-587.
[11] Bresnen M and Marshall N (2000) Partnering in construction: a critical review of issues, problems and dilemmas, *Construction Management and Economics*, 18:2, 229-237, DOI: 10.1080/014461900370852
[12] American Institute of Architecture (2007) *Integrated Project Delivery: A Guide.The American Institute of Architects (AIA)*. Sacramento, California.
[13] Fleming Q W and Koppelman J M (1996) Integrated project development teams: another fad... or a permanent change. *International Journal of Project Management*, 14(3), 163-168, DOI: 10.1016/0263-7863(95)00068-2
[14] Bresnen M and Marshall N (2000) Motivation, commitment and the use of incentives in partnerships and alliances. *Construction Management and Economics*, 18(5), 587-598, DOI: 10.1080/014461900407392
[15] Østby-Deglum E, Svalestuen F, Drevland F et al. (2013) *Prosjekteringsledelse*. Trondheim, Norway, Tapir akademisk forlag.
[16] Love, Peter E. D. , Angappa Gunasekaran, and Heng Li (1998). Concurrent engineering: a strategy for procuring construction projects." *International Journal of Project Management* 16(6), 375-383, DOI: 10.1016/s0263-7863(97)00066-5
[17] Hauck A. J, Walker D H T, Hampson, K D and Peters, R (2004) Project alliancing at National Museum of Australia : the collaborative process. *Journal of Construction Engineering and Management*, 130(1), pp. 143-152, DOI 10.1061/(ASCE)0733-9364(2004)130:1(143) .
[18] Fischer M, Khanzode A, Ashcraft H W and Reed D (2017). *Integrating project delivery*. John Wiley & Sons.
[19] Lædre O et al. (2006) Procurement routes in public building and construction projects. *Journal of construction engineering and management* 132.7 (2006): 689-696, 10.1061/(ASCE)0733-9364(2006)132:7(689).
[20] Ballard, G., Howell, G. (1994) Implementing lean construction: stabilizing work flow. *Lean construction* pp. 101-110.

Life cycle assessment for Zero Emission Buildings – A chronology of the development of a visual, dynamic and integrated approach

A Houlihan Wiberg [1], M. Kjendseth Wiik[2], H Auklend[1], M Løkland Slåke[1], Z Tuncer[1], M Manni[1], G Ceci[1], T Hofmeister[1]

[1] Department of Architecture and Technology, Faculty of Architecture and Design, Norwegian University of Science and Technology, Trondheim, Norway

[2] SINTEF Community, Børrestuveien 3, 0373 Oslo, Norway

Corresponding author: aoife.houlihan.wiberg@ntnu.no

Abstract. Current regulations to reduce energy consumption, and GHG emissions from buildings have focused on reducing operational impacts [1] This paper addresses the specific challenge of increasing complexity and decreasing usability when dealing with the level of detail required when modelling life cycle assessment (LCA) and integrating embodied emission calculations in the design process of a ZEB. It is well known that architectural design processes inherently have high degrees of complexity, and this paper investigates how the use of information and communication technology (ICT) in the design process can have a great impact on reducing GHG emissions and increasing sustainability in ZEBs. Visualisation is an invaluable tool to communicate complex data in an interactive way that makes it easier for non-expert users to integrate LCA thinking early and throughout the design process.

The paper presents a chronology in the development of a more visual, integrated and dynamic approach involving the use of parametric LCA models for decision-support purposes. Such an approach provides the designer with a direct link between the 3D digital model and embodied emissions data contained in the ZEB Tool to perform life cycle GHG emission calculations of buildings. Integrating LCA in a more visual and easily understood way in the holistic design process can also influence more tangible material choices in terms of, for example, architectural tectonics or cultural heritage. This allows designers the possibility to choose, for example, durable or natural materials with the lowest environmental impact or innovative materials with high or low associated emissions and consider these holistically early in the design phase when the level of design freedom is greater. The extent to which existing ICT tools and User Interfaces (UI), such as dashboards, can provide dynamic visual feedback on selected parameters, including LCA, in the design process of zero emission buildings, is discussed. The paper presents two 'proof of concept' dashboards to visualise LCAs at the building (ZEB) and neighbourhood (ZEN) scale. Both approaches are currently being further developed in The ZEN research centre to visualise, analyse and model the data at different scales for different ZEN Key Performance Indicators (KPIs) using visualisation and immersive technologies, such as Extended Reality (XR) technologies including Virtual Reality (VR) and Augmented Reality (AR).

1. Introduction

The environmental impact of buildings on global energy demands and on greenhouse gas (GHG) emissions released to the atmosphere has rapidly increased during recent decades. In cooperation with the EU, Norway has committed to an ambitious climate goal – a 40 percent reduction in greenhouse gas emissions by 2030 compared to 1990 levels. However, it is increasingly recognised that this needs to be accompanied by a parallel focus on reducing embodied impact [2,3]. At the European level, the revised directive on Energy Performance of Buildings (EPBD) requires that all new buildings should be nearly zero energy by 2020 [4]. This is particularly important in Norway, since Norwegian Zero Emission

Buildings (ZEB) [5] have high-energy efficiency and the share of embodied material emissions was found to be around 60 – 75% of total emissions [6]. The most efficient way of reducing a building's environmental impact is to address it in the design stage when design freedom is the greatest, but equally uncertainty is high with an unlimited number of design options. This paper targets innovation to reduce GHG emissions in architectural design solutions and design processes for construction, renovation and transformation. The design process is where critical decisions are made.

The design process is informed by many different factors such as the architect's skills, geographical and cultural circumstances, financial aspects and technical possibilities. Architectural design processes, therefore, have a high degree of complexity, which can be difficult to sort out and communicate. The use of information and communication technology (ICT) in this phase can have a great impact on sustainability and on the performance of the building design. [7] Urban planning processes involve diverse stakeholders and there is a clear need for efficient dialogue tools to support communication in transdisciplinary environments. The Research Centre on Zero Emission Neighbourhoods in Smart Cities (ZEN) [8] develops solutions for future buildings and neighbourhoods with no greenhouse gas emissions and thereby contributes to a low carbon society. Researchers, municipalities, industry and governmental organisations work together in the ZEN research centre in order to plan, develop and run neighbourhoods with zero greenhouse gas emissions. [8, 9, 10]

The role of ICT and a more visual approach, using dynamic and integrated ICT tools, is investigated to inform early design decisions with technical scientific GHG knowledge leading to emissions reduction. Moreover, this paper investigates how the use of ICT can help communicate, involve and improve participation from those involved early and throughout the design process. The work presented here with the ZEB tool [11] and integrated parametric LCA approach [12, 13] is aligned with other ongoing international initiatives in the IEA Annex 72 [14, 15], and previously in Nordic Build Sted project [16, 17], as well as, similar initiatives such as LCA Bygg, One Click LCA, InData EPD [18, 19, 20]

The aim of this paper is to present a chronology of the early work in ZEB by Houlihan Wiberg to develop a method for the calculation of embodied carbon in buildings for use as a decision-support tool. This paper also presents a chronology of the consecutive stages by Houlihan Wiberg together with her students, to integrate the ZEB Tool to develop an integrated dynamic, parametric, dashboard and visual LCA approach, and the conclusion is presented at the end of the paper.

2. The ZEB Tool

The ZEB tool methodological development described below, is connected to the Norwegian ZEB definition and ambition levels developed by the Norwegian ZEB research centre from 2009-2016. According to the ZEB definition, a ZEB can be achieved by offsetting greenhouse gas (GHG) emissions from the entire life cycle of the building through the generation of onsite renewable energy. The system boundaries for LCA corresponding to the ZEB ambition levels are detailed in the Norwegian ZEB Definition. [21, 22]

2.1. Background to the ZEB Tool

A method for the calculation of embodied carbon for materials, later known as The ZEB tool, was created in 2010 by Houlihan Wiberg and developed by her until 2015. A key feature of this Microsoft (MS) excel based method tool was the organisation of the buildings components as separate linked sheets using 2 digit codes according to the NS 3451:2009 Table of building elements [23] These digits were ascribed a number aligned to the respective building component i.e. 21 foundation and groundworks, 22 load-bearing structure, 23 outer walls etc. (see Appendix 1) where material quantities were manually imported from REVIT building information models (BIM) or architect's drawings where a BIM mode was not available. The embodied emissions for each material quantity were calculated using emission factors from either specific EPD (Environmental Product Declarations) data [24] or generic data from Ecoinvent v.3.1. [25]. In this early version, the system boundary included A1-3 (production) and B4 (replacement). The results are summarized in a «Summary» sheet with the buildings lifetime of 60 years together with other building information. This method builds upon an excel based method for the calculation of emissions from operational energy Houlihan Wiberg created in her Ph.D. during the period 2005 to 2010. [26]. Kristjansdottir and Houlihan Wiberg harmonised the method of calculating embodied carbon for materials for use in both the ZEB residential and office concept buildings between 2013 and 2015. [27, 28, 29, 30]. In this stage of development, operational emissions (B6) were calculated separately using the software SIMIEN [31] and compensation of emissions from local, renewable energy production from photovoltaic (PV) production using the software PV SYST. [32].

2.1.1. Further development of the ZEB Tool Further development of the ZEB Tool has subsequently been led by Houlihan Wiberg in collaboration with several of her students from the M.Sc. Sustainable Architecture course during consecutive years, and through ZEB summer internships funded by the research centre on Zero Emission Buildings. These former students are listed and credited as co-authors in this paper to acknowledge their contribution to the further development of the ZEB tool and the development of a more integrated and visual approach. The first ZEB internship supervised by Houlihan Wiberg with Kjendseth Wiik (nee Inman) in the summer of 2014. During this ZEB internship, a digital library of EPDs (primarily sourced from EPD-Norge) was created and embedded in the ZEB Tool excel master document. In addition, the system boundaries were extended to include transport and construction emissions (life cycle modules A4 and A5). Here, emission data is coupled with the factory, warehouse and site locations as well as transport modes to calculate transport emissions, whereby geographical data for the calculation of transport emissions are taken from Googlemaps. [33] The ZEB tool was also adapted to automatically calculate material replacements during a 60-year building reference study period (RSP) from material and component reference service lifetimes (RSL) obtained from EPDs and Byggforskserien. During this internship, a drop-down menu feature linked with the embedded EPD library was created which improved the speed of calculating emissions. The EPD library contains a large range of building materials categorised according to building material types. Each category includes specific emission data from EPD-Norge (which can be accessed by using the EPD reference number) as well as a generic emission data from Ecoinvent v.3.1 [25]. When no EPDs were available from EPD-Norge, specific emission data was sourced from either the International EPD system Environdec [34], European EcoPlatform [35] or German EPDs registered with IBU [36].

Since 2015, the ZEB Tool has been further developed with Wiik and Schlanbusch to include other features, such as a 'construction library' sheet and an 'operation' sheet. [11]. Another sheet calculates emissions from construction activities (C), which is linked to the 'construction library' sheet where all generic and specific emission data for on-site construction activities are placed. Lastly, the 'operation' sheet (O) is structured according to energy carriers and is divided into three parts: operational energy use, delivered energy and energy produced on site, and exported energy. These emissions are calculated from energy supplier's emission factors. These enhancements were tested against GHG emission calculations for the ZEB living laboratory and Campus Evenstad buildings. The objective of each step in the development has been to analyse how a more visual, integrated and dynamic approach to life cycle assessment (LCA) can be used in a holistic way to provide dynamic feedback on construction materials, zero emission buildings (ZEB) and zero emission neighbourhoods (ZEN) early and throughout the design process. Finally, all results are summarised in the 'Summary' sheet in terms of the functional unit [11] (see Appendix 1). Information from the ZEB tool can be used to improve processes, support policy and to provide a sound basis for informed decisions.

3. A Dynamic approach to the ZEB Tool

IOP Conf. Series: Earth and Environmental Science **352** (2019) 012054 doi:10.1088/1755-1315/352/1/012054

Two dynamic approaches have been used in developing the ZEB tool, the first explores connecting the ZEB tool to Revit and Dynamo, whilst the second explores connecting the ZEB tool to Rhino and Grasshopper.

3.1. Revit ~ Dynamo ~ ZEB Tool

In 2016, as part of the Nordic Built Sted [16] funded research, Hofmeister [38] supervised by Houlihan Wiberg, created a dynamic connection between the BIM software Revit[37] and the ZEB tool [11]. The objective was to develop an integrated tool that is built upon a dynamic connection between Revit and the ZEB tool through the medium of visual programming using the plug-in Dynamo to visualise design outcomes in terms of GHGs in the early design phase (see Figures 1 and 2 below). The ZEB single family house concept study [27, 28] is used as a base case, and the 3D BIM is scheduled inside the visual programming tool Dynamo and linked to the ZEB tool. Materials are selected inside the ZEB tool and connected to Dynamo for colour coding building parts in relation to the associated emissions for each component, which enables a better understanding of emissions and ensures ease of use. In addition, every change in the design and within Revit, updates the emission results dynamically. The scope of this dynamic connection is limited to the production phase (A1-A3) and uses emission data from the EPD library. [38]

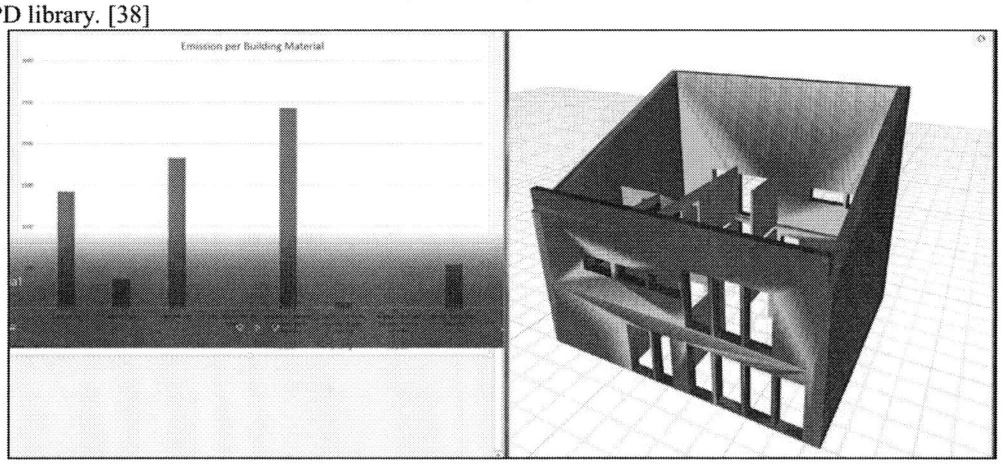

Figure 1. Screenshot from video showing emission factors in the ZEB Tool and corresponding colour coding in Dynamo. [38].

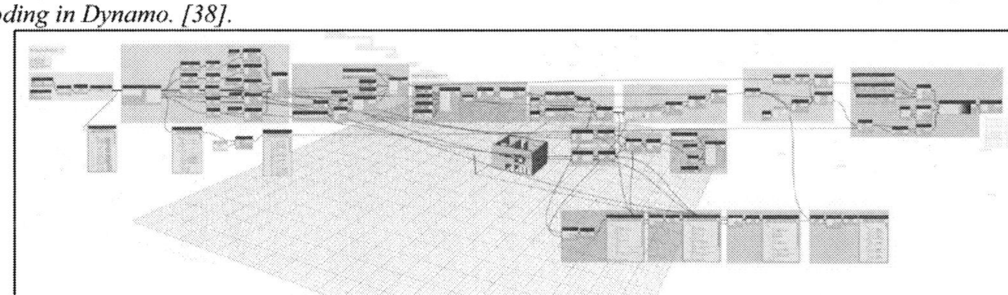

Figure 2. Screenshot from video of emission factors in the ZEB Tool and corresponding colour coding in Dynamo. [38].

3.2. Rhino ~ Grasshopper~ ZEB Tool

In 2016, in parallel to the Revit – Dynamo approach, a similar approach was created by Auklend and Slåke under the supervision of Houlihan Wiberg, to integrate the 3D modelling software Rhino, parametric plugin Grasshopper and the ZEB tool [39]. A dynamic link between Rhino and the ZEB tool was created to create a 'live' link that gives feedback on the environmental impact of material choices, which enables LCA to be considered holistically early in the design phase (see Figure 3 below). A third-party plugin Flux was used for better integration of the numerical values, and the resulting GHG emission calculations were then reported back to the Grasshopper interface. The scope of this dynamic connection is limited to the production phase (A1-A3) and uses emission data from the EPD library. Building data is transferred from Rhino to Grasshopper, and the material quantities are fed into Excel with the use of Flux, which reads the modelling information and transforms the data into a 'json' script which can be read by Excel [39].

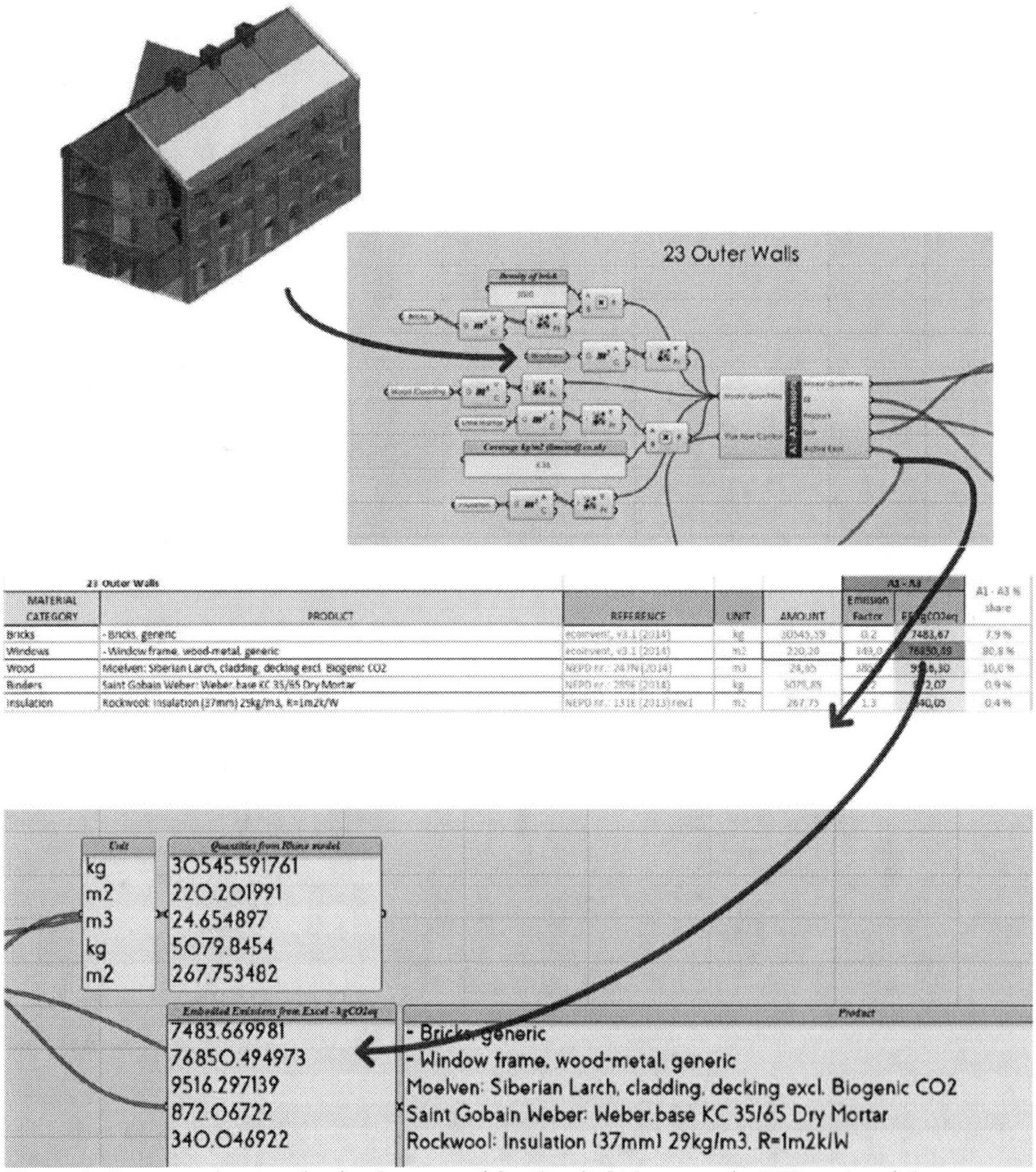

Figure 3. Diagram showing the development of the 'live link' between the ZEB tool and the Rhinoceros model using Grasshopper [39].

4. A Parametric approach to the ZEB Tool

The first stage of this work was completed in Spring 2016 as a joint master's thesis by two Erasmus students, Manni and Ceci, and focuses on combining parametric design principles with LCA to further develop the ZEB single family house concept model [12, 40]. This work presents an integrated workflow to evaluate embodied emissions of materials and the energy demand of the building in each stage of the design process. The work was developed in Rhino using plug ins such as Grasshopper, Diva for Grasshopper and Ladybug. An algorithm is created to describe the building geometry, which is then connected with the ZEB tool and sources data from the EPD library to perform LCA in Grasshopper (see Figure 5). The main advantage of this process is that it enables the integration of daylight optimisation with LCA. The optimisation of the building geometry is established by introducing an evolutionary solver which is dependent on certain parameters, such as, maximum amount of daylight factor and radiation [23, 26]. Data for radiance comes from the Lawrence Berkeley National Laboratory, and weather data is available from EnergyPlus. Dataflow is managed by Grasshopper's visual programming interface which uses a 'Python' script [12, 40].

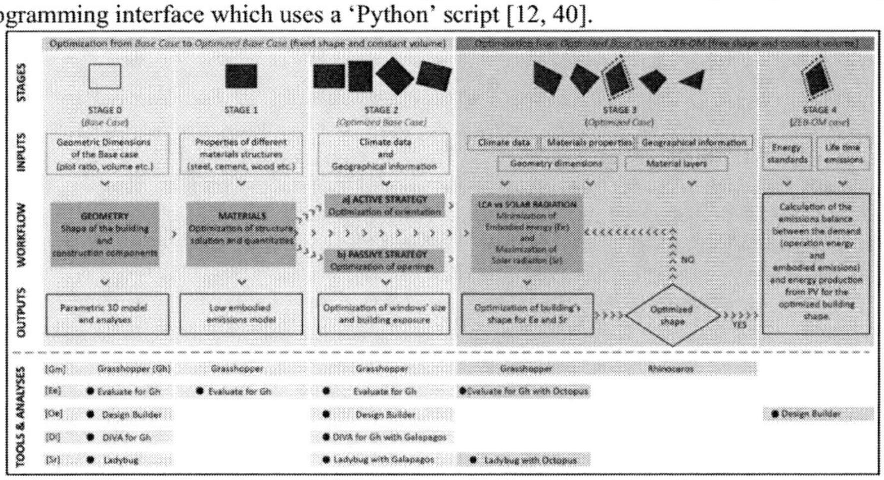

Figure 4. A flowchart of the parametric methodology used: the top part shows the plan while the bottom part shows the tools used to control the geometry (Gm) and deal with the related environmental analyses such as embodied energy (Ee) operational energy (Oe) solar radiation (Irrgl) and daylight factor (Df), conducted in each stage of the design process [12, 40]

5. A Dashboard approach to the ZEB Tool

In this work conducted by Tuncer in Spring 2017, a proof of concept ZEB Dashboard has been created by Tuncer and is used to visualise and integrate LCA early in the design process of a ZEB. This is achieved using a shared platform as a design decision support instrument that can be used collectively to ensure a broader understanding of GHG emissions for different design and material choices. A script creates a dashboard interface (shared platform) as a webpage that can be shared with multiple stakeholders [41]. The scope of this dynamic connection is limited to the production phase (A1-A3) and uses emission data from the EPD library. The Revit geometry is translated into a 'json' script inside the Flux interface and is linked to the ZEB tool in Excel. Calculated emissions from the ZEB tool are then fed back into the Flux interface (see Figure 5) The 3D Revit BIM model together with all the relevant material take-offs are connected to Flux. The Flux script uses visual programming, takes the building elements and dynamically exports the materials and quantities to the ZEB Tool. The formulas within the ZEB tool enables the use of both metric and volumetric units. [41].

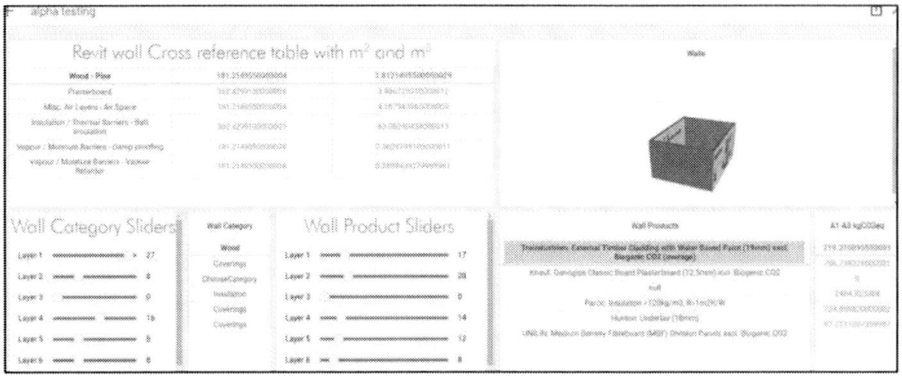

Figure 5. A screenshot of the ZEB dashboard approach. [41]

6. Visual LCA for ZEN

In this work conducted in Spring 2017, the material inventory from the ZEB Living Lab [42] is used by the students, Auklend and Slåke, to create various neighbourhood configurations within the Rhino interface using a parametric script. [43] This process is connected to the ZEB tool using Flux. GHG emission results are fed back into Flux to create a dashboard interface which is also connected to Grasshopper to control and change the neighbourhood configurations, material choices and material quantities. This work marks the further development of the student's earlier work described in section 3.2 involving the creation of a dynamic link between Rhino and the ZEB tool. This work is part of the ongoing work in ZEN to develop a User Interface (UI) in the form of a dashboard to visualise key performance indicators (KPIs), such as GHG emissions, amongst other parameters in an architectural and urban toolbox for use by diverse stakeholders to support decision making in the early design of ZENs. This work considers the scaling up from the building to neighbourhood scale in terms of material use and visualises associated embodied emissions for materials, as well, as transport emissions from the factory to a building site in Trondheim [29]. Data flow begins with Grasshopper, using Python coding, and is linked to Flux using 'json' language. Due to its interchangeable nature, the data flow can be read in excel and linked back to Flux in order to create an interactive dashboard interface [43].

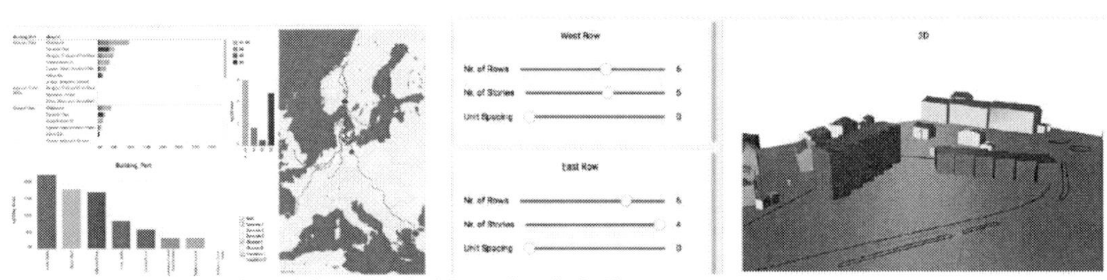

Figure 6. Screenshot of the ZEN dashboard tool. [43].

7. Discussion and Further Work

The ZEB tool marks a significant advancement in terms of integrating LCA early and throughout the design process. The tool provides a robust, reliable and transparent method of embodied emission calculations using a more visual, integrated and dynamic approach. Such an approach demonstrates its versatility and ability to provide feedback on design and material choices, largely as a result of the integrated EPD database and the interoperability of the platforms described. The ZEB tool is currently limited for use at the building scale, although a 'proof of concept' model has been created whereby it can be scaled to neighbourhood level. Different alternative configurations and associated environmental impacts can be assessed at this scale. The detailed EPD library creates a wide range of product selections

IOP Conf. Series: Earth and Environmental Science **352** (2019) 012054 doi:10.1088/1755-1315/352/1/012054

for different design options depending on the project requirements and provides a vast number of permutations in terms of GHG emission results. The organised layout of the building components according to the Table of Building Elements linked sheets provides more transparency and a faster approach to the calculation of the environmental impact of material choices. However, a limitation of the ZEB tool has been the lack of visualisation which can be problematic for most of the stakeholders and non-expert users, since many non-expert users may not understand the implications of the quantitative results in terms of their material and component design choices. Another limitation is that the tool lacks a work sharing aspect, which can limit the integration of the parties involved in the design process, making the tool available only for individual expert users. However, despite these limitations, the MS Excel based ZEB tool provides an advantage in terms of interoperability with different general design software thus enabling quick links through using third-party plugins. The ever expanding and more complete database can form the foundation of a more visual LCA platform that uses the computational and methodological qualities of the ZEB tool.

Future development should be directed towards architects for use in the early design phase, when the reduction of GHG emissions has the greatest potential for optimisation. Without the possibility for direct collaboration, this can result in the creation of 'closed box' tools only for the use of a handful of experts. Keeping the tool limited to a MS Excel format might create problems in the future, since the ever-increasing amount of available data that can be incorporate in the EPD database may be difficult to manage, thus, a more open source approach should be considered in the future, such as EPD Norge's InData project [30] In addition, it is recommended to explore other platforms to take advantage of a more visual representation of results to ensure communication and ease of use.

The Revit-Dynamo and Rhino-Grasshopper dynamic approaches provide a good foundation for future research and further development. They provide an integrated and dynamic feedback during the early design phase and a visual representation of the environmental impact of the various material and component choices using colour coding i.e. red for high emission and green for low emissions. The excel based approach provides flexibility to include emerging 'state of the art' or natural, bio-based materials not normally included in generic databases or EPD platforms. A limitation for both the Dynamic LCA approach for Rhino and Revit has been the lack of access to available data for the same LCA modules in different selected EPDs which meant the LCA analyses had to be limited to the production phase (A1-A3) only. In addition, larger models require better software specifications and only work with certain volumetric units which can create problems between different Dynamo versions. Since the majority of BIM software uses various programming languages, the data and 3D model need to be translated into a common format that can be easily read.

The parametric LCA work also represents an important step in the development of a more visual, parametric and dynamic approach to enable LCA to be considered early in the design process. This optimisation approach can be applied to a larger scale i.e. ZEN neighbourhood scale, although it would require a processor with enough capacity to handle the vast amounts of data required for different shapes and embodied emission calculations. A limitation is that the parametric approach is dependent on the MS Excel environment to conduct LCA calculations, although most of the design is managed through the Grasshopper interface. The ZEB dashboard approach demonstrates the possibilities of work sharing and interoperability between different platforms. Future work will include the further development of the ZEB tool into a web-based application within which the EPD library and embodied emission calculations are embedded as currently being conducted by Skaar et al. in ZEN. Such a web-based application would enable the possibility of customisation and collaboration, and further development of the connection between the 3D Revit integrated with more materials and components, in order to provide 'live' feedback on the embodied emissions. Furthermore, a great improvement in performance would be achieved if all data is placed inside a cloud server. The ZEN dashboard provides a proof of concept to visualise the environmental impact of scaling up from building to neighbourhood level by defining a series of base cases and then expanding them into multiple permutations. It is also worth mentioning that this process delivers a prefabricated approach on neighbourhoods. The ZEN dashboard demonstrates that visualisation can contribute to the design of ZENs by making LCA more understandable and intuitive for diverse stakeholders. It also proved that different layouts create various impacts on GHG emissions which can be easily communicated. In terms of ICT, different programmes use various coding languages, and to connect data flow in a seamless manner, an intermediary application or plugin is needed, hence the use of Flux which is structured on 'json' for its interchangeable qualities.

In the visual LCA for ZEN work, putting numbers into context is essential for all stakeholders to benefit from visualising KPIs (such as GHG emissions) early in the design process. By utilising visualisation methods which inspire and engage diverse stakeholders to explore the environment and learning by putting numbers into context, we can promote ZEN's vision of sustainable neighbourhoods more easily and to a wider audience. The next step in the development of a more integrated and visual approaches is currently being further developed in The ZEN research centre to visualise, analyse and model different ZEN KPIs at different scales using visualisation and immersive technologies, such as Extended Reality (XR) technologies including Virtual Reality (VR) and Augmented Reality (AR), as well as, the development of a user interface such as a Dashboard.

8 Conclusion

The paper provides a chronology of the development of a more visual, integrated and dynamic approach for LCA assessment early and throughout the design process. The results demonstrate how this approach can improve stakeholder participation and effectively integrate science-based knowledge on GHG emissions and other KPIs into the future development of the user-centred architectural and urban ZEN toolbox for design and planning, operation and monitoring of ZENs.

Acknowledgements

The authors gratefully acknowledge the support from the Research Council of Norway and several partners through the Research Centre on Zero Emission Neighbourhoods in Smart Cities (ZEN) and The Research Centre on Zero Emission Buildings (ZEB). The authors wish to thank the Norwegian University of Science and Technology (Trondheim, Norway) and the University of Perugia (Perugia, Italy) for having supported the collaboration between the two universities in this work, framed by the EU programme for education, training, youth and sport – 'ERASMUS+'. The authors wish to thank the Nordic Build Sted research project for having funded the internship for Tobias Hofmeister at NTNU.

References

[1] IPCC, 2014. Climate Change 2014: synthesis report. Contribution of Working Groups I, II and III to the Fifth Assessment Report of the Intergovernmental Panel on Climate Change. IPCC.

[2] EPBD, 2010. EPBD Directive 2010/31/ EU of the European Parliament and of the Council of 19 May 2010 on the energy performance of buildings. European Commission, 2010.

[3] ASHRAE, 2013. ASHRAE Energy Standard for Buildings Except Low-Rise Residential Buildings, American Society of Heating, Refrigerating, and Air-Conditioning Engineers (ASHRAE) Standards Committee 90.1,2013.

[4] Birgisdottir, H., Moncaster, A., Houlihan Wiberg, A., Chae, C., Yokoyama, K., Balouktsi, M., Seo, S., Oka, T., Lutzkendorf, T., Malmqvist, T. (2017) IEA EBC Annex 57 'Evaluation of Embodied Energy and CO2eq for Building Construction'. Energy and Buildings. vol. 154.

[5] ZEB, 2017. The Research Centre on Zero Emission Buildings (ZEB). Available at: http://zeb.no/index.php/en/

[6] Kristjansdottir, T., Houlihan Wiberg, A., Andresen, I., Georges, L., Heeren, N., Good, C., Brattebø, H. (2018) Is a net life cycle balance for energy and materials achievable for a zero emission single-family building in Norway?. Energy and Buildings. vol. 168.

[7] Houlihan Wiberg, A., Bjerregaard Jensen, L., Beim, A. LCA in early design phases. In: Informing sustainable architecture. First Edition.(pp. 124-129) ISBN 9788750211402.

Publisher: Polyteknisk Boghandel og Forlag. København, May 2018
[8] ZEN, 2019. The Research Centre on Zero Emission Neighbourhoods in Smart Cities (ZEN). Available at: https://fmezen.no/
[9] Wiik, M. K., Fufa, S. M., Baer, D., Sartori, I. & Andresen, I (2018) The ZEN definition – a guideline for the ZEN pilot areas. Version 1.0 ZEN Report 11, NTNU/SINTEF.
[10] Wiik, M. K., Fufa, S. M., Krogstie, J., Ahlers, D. Wyckmans, A., Driscoll, P., Brattebø, H., Gustavsen, A. (2018) Zero emission neighbourhoods in smart cities: Definition, key performance indicators and assessment criteria: version 1.0. Bilingual version. ZEN Report 7, NTNU/SINTEF.
[11] Wiik, M., Houlihan Wiberg, A., Schlanbusch, R., Kristjansdottir, T., 2017. ZEB Tool Manual, Technical Guide, Version 1. ZEB Project report (Internal). Norwegian University of Science & Technology, Trondheim, Norway. NTNU SINTEF.
[12] Lobaccaro, G., Houlihan Wiberg, A., Manni, M., Ceci, G., Lolli, N., Berardi, U. (2018). Parametric design to minimize the embodied GHG emissions in a ZEB. Energy and Buildings. ISSN: 0378-7788, 2018; Volume 167. p. 106-123
[13] Lolli, N., Fufa, S.M., Inman, M. (2017) A Parametric Tool for the Assessment of Operational Energy Use, Embodied Energy and Embodied Material Emissions in Building. Energy Procedia. Volume 111, (March) Available at: https://doi.org/10.1016/j.egypro.2017.03.004
[14] IEA Annex 72 (2019) - Assessing Life Cycle Related Environmental Impacts Caused by Buildings. International Energy Agency (IEA). Energy in Buildings and Communities Programme (EBC). Available at: http://annex72.iea-ebc.org/about
[15] Hollberg A., Tschetwertak J., Schneider S., Habert G. (2018) Design-Integrated LCA Using Early BIM. In: Benetto E., Gericke K., Guiton M. (eds) Designing Sustainable Technologies, Products and Policies. Springer, Cham. Available at: https://doi.org/10.1007/978-3-319-66981-6_30
[16] Nordic Build Sted (2017) Sustainable Transformation and Environmental Design. Nordic Available at: http://stednetwork.org/index.php?p=/articles
[17] Negendahl, Kristoffer. (2015). Building performance simulation in the early design stage: An introduction to integrated dynamic models. Automation in Construction. 54. 10.1016/j.autcon.2015.03.002.
[18] Birgisdottir, H., Rasmussen, F. (2019). Development of LCAbyg: A National Life Cycle Assessment Tool for Buildings in Denmark. IOP Conf. Ser.: Earth Environ. Sci. 290 012039
[19] One Click LCA (2019) Available at: https://www.oneclicklca.com/bim-based-life-cycle-assessment-guidelines/ (Accessed 05 08 18).
[20] InData project. EPD-Norge. Available: https://digi.epd-norge.no/(Accessed 05 08 18).
[21] Kristjansdottir, T., Fjeldheim, H., Selvig, E., Risholt, B., Time, B., Georges, L., Dokka, TH, Bourelle, J., Bohne, R., Cervenka, Z. A Norwegian ZEB-definition embodied emission. Sider: 41 (Illustrert) Utgivelsesår: 2014. ISBN: 978-82-536-1398-7
[22] Fufa, S.M., Dahl Schlanbusch, R., Sørnes, K., Inman, M.R., Andresen, I. A Norwegian ZEB Definition Guideline. Sider: 48 (Illustrert) Utgivelsesår: 2016 ISBN: 978-82-536-1513-4
[23] Standard Norge. (2009). NS 3451: Bygningsdelstabellen (Table of building elements).
[24] EPD Norge Environmental Product Declarations. Available : http://epd-norge.no/epder/
[25] Ecoinvent. Swiss Center for Life Cycle Inventories, EcoInvent version 3.0, available from: http://www.ecoinvent.ch/.
[26] Houlihan Wiberg, A. 2009. An Analysis Of The Performance Of Certification Schemes In The Hotel Sector In Terms Of CO2 Emissions Reduction. Ph.D. Thesis. Department of Architecture, University of Cambridge. Permanent URL http://www.dspace.cam.ac.uk/handle/1810/226841
[27] Dokka, T.H., Houlihan Wiberg, A., Georges, L., Mellegård, S., Time, Berit; Haase, M., Maltha, M., Lien, A.G., 2013. A zero emission concept analysis of a single family house. ZEB Project report 9 – 2013. ISBN 978-82-536-1324-6
[28] Houlihan Wiberg, A., Georges, L., Dokka, T.H., Haase, M., Time, B., Lien, A.G., Mellegård, S., Maltha, M. (2014) A net zero emission concept analysis of a single-family house. Energy and Buildings. vol. 74 (May).
[29] Houlihan Wiberg, A., Georges, L., Fufa, S.M., Risholt, B., Good, C., 2015. A zero emission concept analysis of a single family house: Part 2 sensitivity analysis. ZEB Project report 21–2015. ISBN 978-82-536-1449-6
[30] Georges, L., Haase, M., Houlihan Wiberg, A., Kristjansdottir, T., Risholt, B. (2015) Life cycle emissions analysis of two nZEB concepts. Building Research & Information. vol. 43 (1).
[31] ProgramByggerne. 2012. SIMIEN, SIMulation of Indoor climate and ENergy use. (Skollenborg. Retrieved

from http://www.programbyggerne.no/SIMIEN/
[32] PVsyst SA. 2013. PV-syst 5.62, Photovoltaic System Software 5.62. from http://www.pvsyst.com/en/
[33] Googlemap. Available at: https://www.google.no/maps
[34] Environdec Available at : http://www.environdec.com/
[35] Ecoplatform. Available from: http://eco-platform.org/
[36] Institutu Bau Umvelt (IBU). Available at http://ibu-epd.com/en/epdprogram/published-epds/
[37] Autodesk Revit BIM Architecture Suite.
[38] Hofmeister, T., Houlihan Wiberg, A., 2016. Emissions Visualisation ~ A dynamic, integrated approach for the ZEB Tool. (Supervisor: A. Houlihan Wiberg). Nordic Build Sted Project (unpublished). Norwegian University of Science & Technology, Trondheim, Norway.
[39] Houlihan Wiberg, A., Auklend, H. 2016. Emissions as Design Drivers. Second semester in the M.Sc. Sustainable Architecture. ISBN 9788275511308. Publisher: NTNU. Trondheim, June 2016. Available at: https://drive.google.com/file/d/0B5dx2T4jpJT5eGhPbTFxbUtaYU0/view?usp=sharing
[40] Manni, M., Ceci, G. Parametric Design Principles applied to NZEB in cold extreme climate conditions. Masters thesis submitted June 2016 (Supervisor: A. Houlihan Wiberg; G. Lobaccaro). Norwegian University of Science & Technology, Trondheim, Norway.
[41] Tuncer, Z. 2017, ZEB Tool - Transition to an Integrated and Dynamic ZEB Dashboard. Masters theory report (Supervisor: A. Houlihan Wiberg). Norwegian University of Science & Technology, Trondheim, Norway. (Online) Available: http://goo.gl/Ndgi4W(Accessed 22 02 18).
[42] Wiik, M.R, Houlihan Wiberg, A. (2017) Life Cycle Greenhouse Gas Emissions of Material Use in the Living Laboratory. Proceedings of World Sustainable Built Environment Conference 2017 Hong Kong. Transforming Our Built Environment through Innovation and Integration: Putting Ideas into Action. WSBE17.
[43] Slånke, M., Auckland, H., 2017. Visual LCA in ZENs. Interactive visualisation network for material associated GHG in preliminary neighbourhood design stages. Masters thesis submitted June 2017 (Supervisor: A. Houlihan Wiberg; Co-Supervisor: E. Reisch). Norwegian University of Science & Technology, Trondheim, Norway. (Online) Available: http://goo.gl/6sgXJB (Accessed 22 02 18).

Appendix

Scope	
Databases Used	EPDs, Ecoinvent v3.1
Lifetime of Construction (years)	60
Heated floor area - BRA (sqm)	1000
Functional Unit	1sqm over a 60 yr lifetime
Building Site	Trondheim

Informasjon
Denne fanen viser totalresultatene. Resultatene er vist for hver bygningsdel ifht. NS3451 for hver livssyklusmodul ifht. EN 15978. Arket er koplet, så det er kun nødvendig å fylle i (celle C4). Det er anbefalt å bruke en bygningslevetid på 60 år (celle C3).Oslo er valgt son byggeplass beliggenhet, men man kan også endre dette i celle C6. NB: Endring av byggep beliggenhet tar noen sekunder.

Building Element		A1 - A3	A4	A5	B4	B6	$kgCO_{2eq}$	$kgCO_{2eq}/yr$	$kgCO_{2eq}/m^2$	$kgCO_{2eq}/m^2/yr$	Contribution
2 Building	20 Building, general										
	21 Groundwork and Foundations	0	0	0	0	0	0	0	0.0	0.00	0.0 %
	22 Superstructure	0	0	0	0	0	0	0	0.0	0.00	0.0 %
	23 Outer walls	0	0	0	0	0	0	0	0.0	0.00	0.0 %
	24 Inner walls	0	0	0	0	0	0	0	0.0	0.00	0.0 %
	25 Floor Structure	0	0	0	0	0	0	0	0.0	0.00	0.0 %
	26 Outer Roof	0	0	0	0	0	0	0	0.0	0.00	0.0 %
	27 Fixed Inventory	0	0	0	0	0	0	0	0.0	0.00	0.0 %
	28 Stairs and Balconies	0	0	0	0	0	0	0	0.0	0.00	0.0 %
	29 Other	0	0	0	0	0	0	0	0.0	0.00	0.0 %
3 VVS	30 Heating, Ventilation and Sanitation, general										
	31 Sanitary	0	0	0	0	0	0	0	0.0	0.00	0.0 %
	32 Heating	0	0	0	0	0	0	0	0.0	0.00	0.0 %
	33 Fire Safety	0	0	0	0	0	0	0	0.0	0.00	0.0 %
	34 Gass and Air Pressure	0	0	0	0	0	0	0	0.0	0.00	0.0 %
	35 Process Cooling	0	0	0	0	0	0	0	0.0	0.00	0.0 %
	36 Ventilation and Air Conditioning	0	0	0	0	0	0	0	0.0	0.00	0.0 %
	37 Comfort Cooling	0	0	0	0	0	0	0	0.0	0.00	0.0 %
	38 Water Treatment	0	0	0	0	0	0	0	0.0	0.00	0.0 %
	39 Other	0	0	0	0	0	0	0	0.0	0.00	0.0 %
4 Electric Power	40 Electric Power, general										
	41 Basic Installation for Electric Power	0	0	0	0	0	0	0	0.0	0.00	0.0 %
	42 High Voltage Power	0	0	0	0	0	0	0	0.0	0.00	0.0 %
	43 Low Voltage Power	0	0	0	0	0	0	0	0.0	0.00	0.0 %
	44 Lighting	0	0	0	0	0	0	0	0.0	0.00	0.0 %
	45 Electric Heating	0	0	0	0	0	0	0	0.0	0.00	0.0 %
	46 Standby Power	0	0	0	0	0	0	0	0.0	0.00	0.0 %
	49 Other	0	0	0	0	0	0	0	0.0	0.00	0.0 %
5 Telecommunication	50 Telecommunication and Automation										
	51 Basic Installation for Tele. and Automation	0	0	0	0	0	0	0	0.0	0.00	0.0 %
	52 Integrated Communication	0	0	0	0	0	0	0	0.0	0.00	0.0 %
	53 Telephone and Paging	0	0	0	0	0	0	0	0.0	0.00	0.0 %
	54 Alarm and Signal	0	0	0	0	0	0	0	0.0	0.00	0.0 %
	55 Sound and Picture	0	0	0	0	0	0	0	0.0	0.00	0.0 %
	56 Automisation	0	0	0	0	0	0	0	0.0	0.00	0.0 %
	57 Instrumentation	0	0	0	0	0	0	0	0.0	0.00	0.0 %
	59 Other	0	0	0	0	0	0	0	0.0	0.00	0.0 %
6 Other Installations	60 Other Installation, general										
	61 Prefabricated Unit	0	0	0	0	0	0	0	0.0	0.00	0.0 %
	62 Passenger and Goods Transport	0	0	0	0	0	0	0	0.0	0.00	0.0 %
	63 Transportation Facilities for Small Goods	0	0	0	0	0	0	0	0.0	0.00	0.0 %
	64 Stage Equipment	0	0	0	0	0	0	0	0.0	0.00	0.0 %
	65 Waste and Vaccuum-cleaning	0	0	0	0	0	0	0	0.0	0.00	0.0 %
	66 Fixed Furniture	0	0	0	0	0	0	0	0.0	0.00	0.0 %
	67 Loose Furniture	0	0	0	0	0	0	0	0.0	0.00	0.0 %
	69 Other	0	0	0	0	0	0	0	0.0	0.00	0.0 %
7 Outdoor	70 Outdoor, general										
	71 Adapted Terrain	0	0	0	0	0	0	0	0.0	0.00	0.0 %
	72 Outdoor Construction	0	0	0	0	0	0	0	0.0	0.00	0.0 %
	73 Outdoor Heating, Ventilation and Sanitatio	0	0	0	0	0	0	0	0.0	0.00	0.0 %
	74 Outdoor Electric Power	0	0	0	0	0	0	0	0.0	0.00	0.0 %
	75 Outdoor Tele. and Automatisation	0	0	0	0	0	0	0	0.0	0.00	0.0 %
	76 Roads and Courtyards	0	0	0	0	0	0	0	0.0	0.00	0.0 %
	77 Parks and Gardens	0	0	0	0	0	0	0	0.0	0.00	0.0 %
	78 Outdoor Infrastructure	0	0	0	0	0	0	0	0.0	0.00	0.0 %
	79 Other	0	0	0	0	0	0	0	0.0	0.00	0.0 %
	Construction		0	0			0	0	0.0	0.00	0.0 %
	Operation					0	0	0	0.0	0.00	0.0 %
	$kgCO_{2eq}$	0	0	0	0	0	0	0	0.0	0.00	
	$kgCO_{2eq}/yr$	0	0	0	0	0					
	$kgCO_{2eq}/m^2$	0.0	0.0	0.0	0.0	0.0					
	$kgCO_{2eq}/m^2/yr$	0.00	0.00	0.00	0.00	0.00					
	Contribution	0.0 %	0.0 %	0.0 %	0.0 %	0.0 %					
				C	O	M	ZEB-COM	Production	Exported	On-site Energy	ZEB Balance
				0.00	0.00	0.00	0.00	0.00	0.00	0.00	0.00

Figure 1. An illustration of the summary sheet from the ZEB tool, where construction, operation and material emissions are summarised. Illustration includes an overview of the calculation and library sheets included in the tool [11].

Assessing cities: Applying GIS-based methods for mapping cross-scale spatial indicators

L S Rokseth[1], **B Manum**[1, 2], **T Nordström**[1, 3]

[1] Department of Architecture and Technology, Faculty of Architecture and Design, Norwegian University of Science and Technology (NTNU), Trondheim
[2] Institute of Urbanism and Landscape, The Oslo School of Architecture and Design (AHO), Oslo
[3] Spacescape AB, Stockholm

lillian.rokseth@ntnu.no

Abstract. In recent years, several systems and tools to assess energy consumption and carbon emissions at scales beyond that of merely buildings, such as LEED, CASBEE and BREEAM communities have been development. However, reviews reveal a lack of robustness in these methods both in terms of an unstructured mix of qualitative and quantitative criteria and lack of focus on urban form parameters found to influence energy consumption and carbon emissions. A promising quantitative assessment system including various urban form indicators is developed by the Urban Morphology Institute (UMI) in Paris. Within the research centre on zero emission neighbourhoods in smart cities (ZEN), a GIS-based method is applied to analyze conditions of urban form known to contribute to carbon emissions. In this paper we demonstrate how a selection of the UMI indicators describing proximity can be further specified applying GIS-based methods. The potential of applicability of urban assessment system in planning as well as design processes will increase when linked to tools that are already implemented, and map visualizations as well as data provided by these methods are highly applicable in planning and urban design. As further research, methods described in recent research within ZEN and specified measures for calculating UMI indicators, will be tested in analyses of urban development areas in Norway.

1. Introduction

> Urban stakeholders require quantitative and robust tools to implement new paths to urban sustainability. Urban form, the spatial distribution of activities and urban organization are crucial aspects of cities' sustainability. Many tools and assessment systems have been developed to improve cities' energy efficiency and environmental footprint. However, most of these tools are based on the building scale. Most urban stakeholders are now convinced that a building scale approach is not sufficient: the scale of analysis should evolve from the building to the neighbourhood, district and city scales [1, p.592].

In recent years, several systems and tools to assess energy consumption and carbon emissions at scales beyond that of merely buildings, such as LEED, CASBEE and BREEAM communities have been developed. However, reviews reveal a lack of robustness in these methods both in terms of an unstructured mix of qualitative and quantitative criteria and lack of focus on urban form parameters known to influence energy consumption and carbon emissions [2, 3]. There is still a need to develop robust tools to assess and evaluate carbon emission impacts that account for the numerous aspects of urban form that affect the energy use and carbon emission in our built-up environments. A promising assessment system including various urban form indicators is developed by the Urban Morphology Institute (UMI) in Paris [1]. In this paper we demonstrate how UMI indicators describing proximity can be further specified applying GIS-based methods.

2. Background

In recent research conducted at the research centre on zero emission neighbourhoods in smart cities (ZEN), a GIS-based method is applied to analyze conditions of urban form known to contribute to carbon emissions. Specific urban form variables related to building morphology, building densities and spatial properties of street networks are identified, mapped and analyzed in a comparative study of three design options of an urban development area in Trondheim. For each of the design options, data on population, public transport and street networks are mapped, analyzed and examined applying GIS-based methods. The analyses include various accessibility measures. Depending on the particular issue examined, proximity is measured both as "distance to" and as "amounts within distance", applying Euclidian distances as well as distances along the street network. The defined topics were selected in order to be applicable in early phase planning, and included: closeness to retail and service, closeness to public transport (Figure 1), population density, land use diversity, street connectivity and intersection density. Results were calculated as mean values within the plan area and presented in a spider chart allowing for a comparison between the design options (Figure 2). The method reveals significant differences between the design options and seem to be highly applicable in the assessment of urban planning options to optimize urban form and reduce carbon emissions, but further research is needed to evaluate more closely which topics and measures should be included [4].

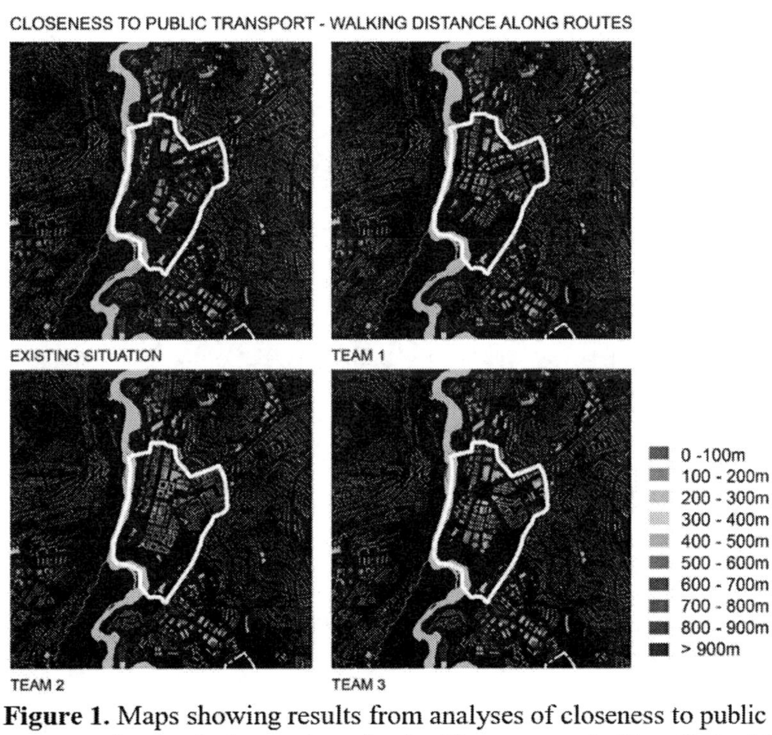

Figure 1. Maps showing results from analyses of closeness to public transport for the design options for the Sluppen area in Trondheim [4].

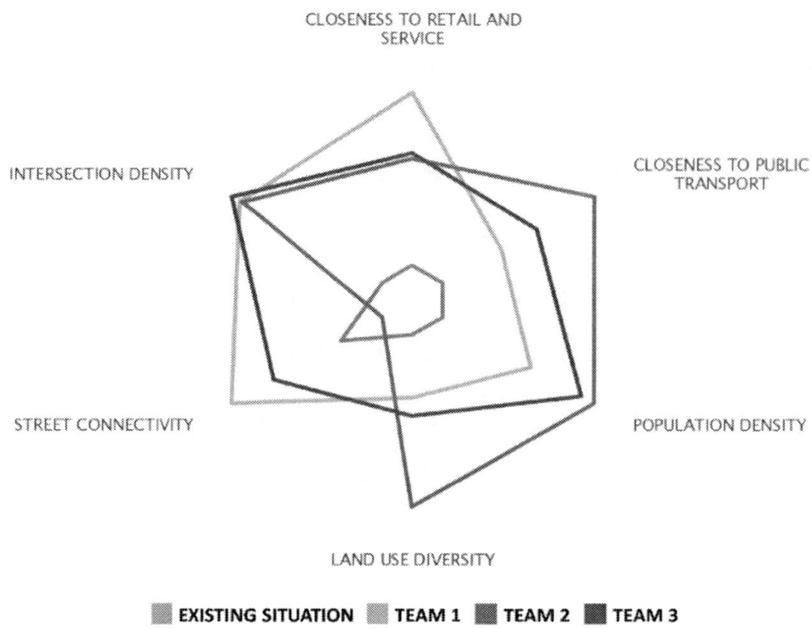

Figure 2. Spider chart of calculated mean values within the plan area for each of the topics included in the analysis of the design options [4].

An interesting approach in this respect is the work of the Urban Morphology Institute (UMI) in Paris. As a response to the lack of robustness "caused by a confusing use of qualitative and quantitative criteria" [1, p.593] in developed urban assessment systems, they propose a set of 60 quantifiable cross-scale spatial indicators for urban sustainability. They sort the indicators within the themes *land use, mobility, water, biodiversity, equity, economy, waste, culture/well-being* and, *energy and bioclimatic*. The indicators are further classified within seven indicator types: *intensity, distribution, proximity, connectivity, complexity, diversity* and *form*. Some examples of UMI indicators are:

- Human density: calculated as population per area of the selection (m^2),
- Building density, calculated as floor area (m^2) per area of the selection (m^2),
- Surface occupied by pedestrian and bicycle paths, calculated as area occupied by pedestrian and bicycle paths (m^2) per area of the selection (m^2),
- Surface occupied by road network, calculated as area of the road network (m^2) per area of the selection (m^2),
- Connectivity of the pedestrian/bike grid, calculated as the number of intersections of the pedestrian/bike grid per area of the selection (m^2),
- Cyclomatic complexity of the pedestrian/bike grid, calculated as $\mu = L - N + 1$, where L = number of links and N = number of nodes,
- Average distance between intersections in the bike/pedestrian grid,
- Proportion of the population more than 300 meters away from a public transport stop, calculated as the share of the population more than 300 meters away from public transport stop of the total population,
- Number of public transport modes accessible within 300 meters,
- Hydrological intensity, calculated as the percent of natural hydrological functions preserved or restored,
- Proportion of green fabric, calculated as green fabric area (m^2) per area of the selection (m^2),

IOP Conf. Series: Earth and Environmental Science **352** (2019) 012055 doi:10.1088/1755-1315/352/1/012055

- Proportion of jobs in relation to housing, calculated as number of jobs per number of housings, and
- Intensity of GHG emissions per resident, calculated as GHG emissions (kg eq CO_2) per GDP [1].

In Geographical Information Systems (GIS) spatially referred or geographical data can be collected, mapped, managed and analysed, and visualized in tables and thematic maps [5]. As recent research within ZEN demonstrate, GIS allow for visual and comprehensive representation of results, and data from various scales can be extracted from the GIS model. The UMI indicators can easily be mapped in GIS allowing for visualization in thematic maps and extraction of results in various scales.

Several software extensions and plugins developed for GIS such as the ESRI network analyst and Place Syntax Tool [6] allow for network-based accessibility analyses. In this paper we will demonstrate how some of the UMI indicators can be further developed by applying GIS-based methods to measure proximity, using "amount within distance" and "distance to" applying Euclidean distances as well as distances along the street network. We present a table of the UMI indicators that can be better described and analysed applying GIS-based methods and show examples of results from analyses of some of the listed UMI indicators using the methods recently developed within ZEN.

3. GIS-based proximity measures for mapping UMI indicators

Table 1 shows a selection of UMI indicators and corresponding GIS-based measures, some of which are applied in the comparative analysis of design options for the Sluppen area in Trondheim referred to in the background section of this paper, see figure 1 and 2 [4].

Table 1. GIS-based proximity measures corresponding to UMI indicators, adapted from [1].

Theme	UMI indicator	GIS-based measure
Land Use	Human density	Residents and employees within a certain walking distance (typically 1 km)
Mobility	Proportion of the population more than 300 meters away from a public transport stop	Walking distance to public transport stop. Share of the population within specific walking distances can be extracted from the GIS model
Biodiversity	Proportion of green fabric	Share of green fabric of total area within a certain distance straight-line distance (in example shown in figure 5, the distance is set to 500 meter)
Biodiversity	Distribution of green spaces	Walking distance to green space
Economy	% of residents living less than X from a convenience store	Walking distance to convenience store. Share of the population within specific walking distances can be extracted from the GIS model
Culture/Wellbeing	Proximity of leisure facilities	Walking distance to leisure facilities. Share of the population within specific walking distances can be extracted from the GIS model

For measuring proximity indicators, input data on distance between origin and destination is required. Specifying the mode of distance is here essential. For example, for the UMI indicator *proportion of the population more than 300 meters away from a public transport stop* (by UMI calculated as the share of total population), mode of distance-measure will affect the calculated results to a high degree. Figure 2 illustrate the difference between distance modes in analyses of proximity to public transport stops in the city of Trondheim. When measuring distance as straight-line, for instance the very significant barrier effect of the river will not be grasped in the analyses, while this is accounted for in a very realistic way when measuring distance along real walkable routes. The calculated result for the UMI indicator *proportion of the population more than 300 meters away from a public transport stop* for Trondheim is 80% for the straight-line distance mode, while the share is 47% if distance is measured as walking distance, illustrating the importance of carefully choosing the most relevant mode of distance-measure.

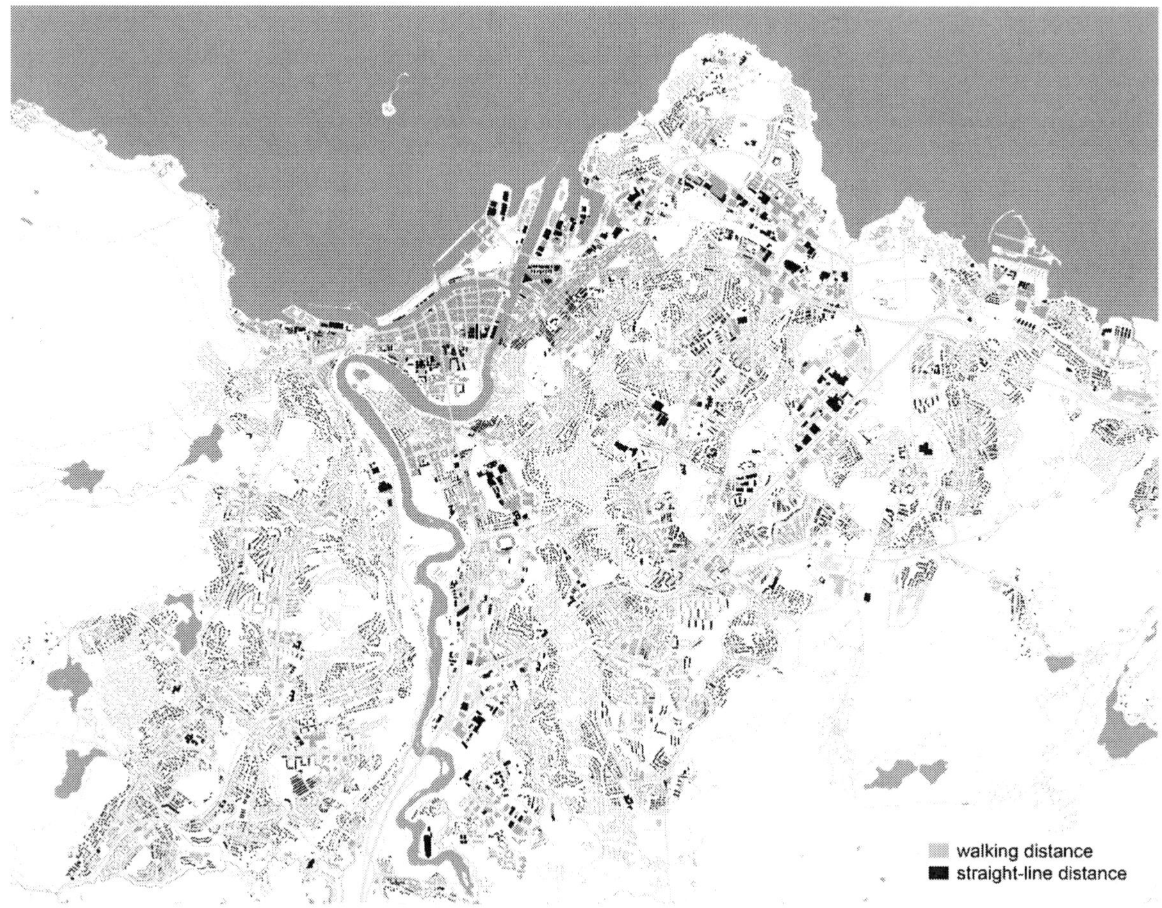

Figure 3. Map illustrating the population in Trondheim within 300 meter distance from public transport stop measuring distance as walking distance along routes (yellow). The increased share within the same distance measured as straight-line distance (black) does not afford realistic results for this specific indicator because of the very significant barrier effect of the river.

For four out of the six GIS-based measures we propose, distance is suggested measured as walking distance along route. UMI define a threshold of 300 meter distance to public transport, but for the indicators concerning proximity to convenience stores and leisure facilities thresholds are reported as X. Defining thresholds depend on what topic is addressed and what group of the population is targeted. For example, walking distance for children to playgrounds require a lower threshold compared to walking distance for active adults to leisure activities.

Figure 4 show results from analysing walking distance to convenience stores. Convenience stores are mapped manually in GIS from OpenStreetMap. The method of mapping population and street network, and of the applied analysis which is equivalent to that of the ZEN measure *walking distance*

IOP Conf. Series: Earth and Environmental Science **352** (2019) 012055 doi:10.1088/1755-1315/352/1/012055

to nearest retail cluster is previously described by Rokseth et al. [4]. By colouring the map according to walking distances, in groups of 0-100 meters, 100-200 meters, etc., geographical areas where with high and low accessibility to convenience stores are revealed in a comprehensive way. The UMI indicator formula (share of total population) can be resolved by extracting numbers from the GIS model. As an example, in the specific case of Trondheim, the share is 1 % within 100 meter walking distance, 7 % within 200 meter walking distance, 16 % within 300 meter walking distance, and 36 % within 500 meter walking distance.

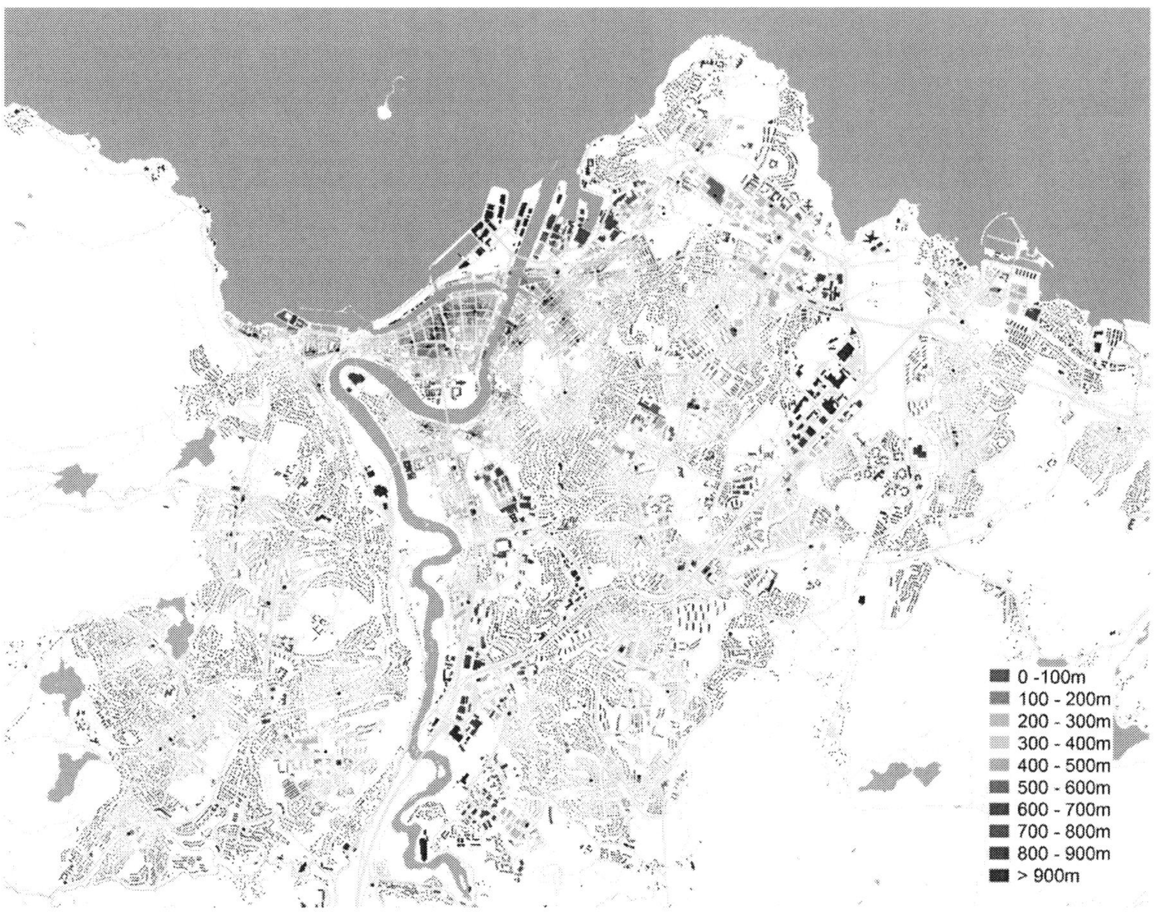

Figure 4. Walking distance from population to convenience stores in Trondheim.

Straight-line distances can be more relevant than walking distance in some cases, for example when measuring proximity from dwellings to large parks as an issue concerning availability of fresh air or the distance to motorways regarding the issue of noise and pollution. Connected to the UMI indicator *proportion of green fabric*, we propose the GIS-based measure *share of green fabric of total area within 500 meter straight-line distance.*

Input data on green fabric were retrieved form the municipality of Trondheim and included parks and recreational green spaces in several categories. Analyses were run according to procedures equivalent to the topic *land use diversity* previously described by Rokseth et al. [4], with green fabric as destination. According to UN Habitat 15-20% of the land area in "high density mixed-use urban areas" should be allocated for open public space [7, p.3]. The map in figure 5 reveal which areas that have a lower share of green than the recommendation. Out of the total population, 37% have a green fabric share of 20% or higher within 500 meter straight-line distance.

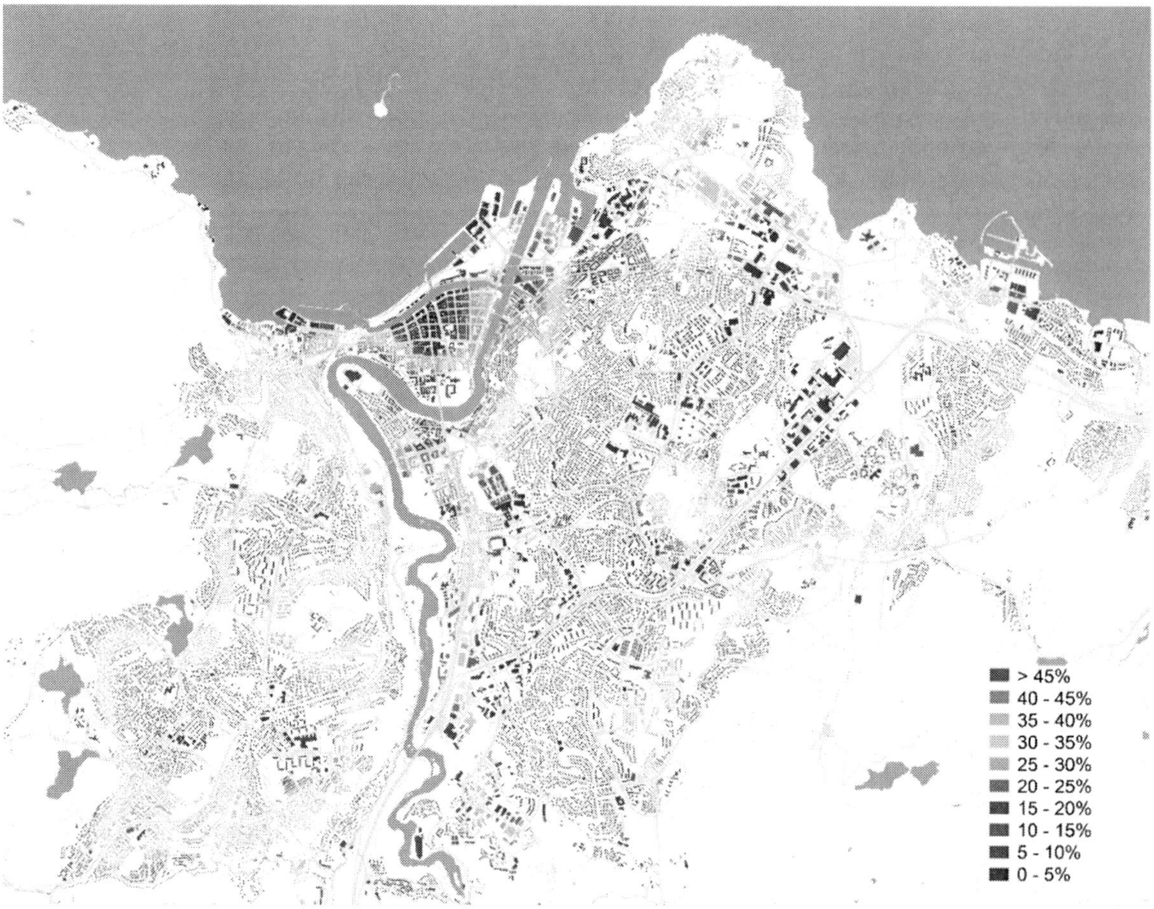

Figure 5. Share of green fabric of total area within 500 meter straight-line distance.

There are some important limitations to consider. The examples provided in this paper are analysed based on a so-called axial line map [8] representing the street network, not considering slope of the routes. Also, qualitative aspects of the street network, such as traffic safety and perceived safety and maintenance impacting route choice, are currently not grasped by the GIS model. Qualitative aspects of the green fabric will also affect whether the space will be frequented by the public. These limitations will equally apply to the UMI indicators.

4. Conclusions

GISs methods currently being developed holds a potential to support assessment systems for urban sustainability in terms of providing informative thematic maps for visualizations in different scales. In addition, GIS-based methods can be implemented to further specify which measures to apply. The GIS model will also provide data in various scales and units, allowing for several alternatives of comparing data between cities or between planning options and for a richer ground for discussions about the meaning and importance of examined topics and indicators. The potential of applicability of urban assessment system in planning as well as design processes will increase when linked to tools that are already implemented. Both map visualizations and data provided by these methods are highly applicable in planning and urban design, for instance in evaluating likely effects of alternative development alternatives as shown in the case of Sluppen.

As further research we will combine methods described in recent research within ZEN and further specify measures for calculating UMI indicators, and test these in analyses of urban development areas in Norway. Further definition of thresholds and benchmark values will be part of this work.

Acknowledgements
The authors gratefully acknowledge the support from the Research Centre on Zero Emission Neighbourhoods in Smart Cities (FME ZEN) and the Research Council of Norway.

References

1. Bourdic L, Salat S, Nowacki C. Assessing cities: A new system of cross-scale spatial indicators. Build Res Inf. 2012;40(5):592-605.
2. Bourdic L, Salat S. Building energy models and assessment systems at the district and city scales: a review. Building Research & Information. 2012;40(4):518-26.
3. Sharifi A, Murayama A. A critical review of seven selected neighborhood sustainability assessment tools. Environmental impact assessment review. 2013;38:73-87.
4. Rokseth L, Manum B, Nordström T, editors. Properties of urban form influencing carbon emissions. Implementing a GIS-based method. The 12th Space Syntax Symposium; 2019; Beijing.
5. Heywood I, Cornelius S, Carver S. An introduction to geographical information systems. 4th ed. ed. Harlow: Prentice Hall; 2011.
6. Ståhle A. Place Syntax Tool (PST). Accessibility Instruments for Planning Practice. COST Office2012. p. 173-8.
7. UN-Habitat. A New Strategy of Sustainable Neighbourhood Planning: Five principles – Urban Planning Discussion Note 3. 2014.
8. Hillier B, Hanson J. The social logic of space. Cambridge: Cambridge University Press; 1984.

Powerhouse Telemark: A plus energy building with a low exergy heating and cooling system.

C S F Tapper[1] and T H Dokka[2]
Skanska Norway P.O. Box 1175, Sentrum N-0107 Oslo,Norway
[1] *christofer.tapper@skanska.no*
[2] *Tor.helge.dokka@skanska.no*

ABSTRACT
Powerhouse Telemark (PHT) is an eleven stories high office building in Telemark in Norway, built to plus energy standard. A plus energy building that is built according to the Powerhouse definition prior to 2019 must produce more renewable, locally produced energy during the lifetime of the building, and must produce enough renewable energy to cover the total embodied energy used for the production and transportation of the building materials used in the building. The local renewable energy production must also cover the yearly net energy needs for operations, renovations and demolition of the building. The local production of renewable energy on the building is not required to cover the energy needed for plug loads in the building.

The building will have a unique diamond shape, and the roof of the building is oriented towards the south, and slopes at an angle of 24 degrees from the horizontal. The building will be a net supplier of electricity when seen over a whole year.

In addition to other "normal" measures to achieve a plus energy building, like a high performance building envelope, and an extremely efficient ventilation system, energy efficient lighting and equipment and a solar PV system, a new low exergy heating and cooling system has been implemented in the building. This low exergy system (called Lowex) is based on low temperature heating and high temperature cooling, which together with an optimized energy well design will give extremely low demand for delivered energy (electricity). To design this Lowex system, a lot of different calculations and simulations had to be done, both with commercial software packages, but also with new develop dynamic simulation models. These design procedures and models are described in this paper. This paper also describes how these results will be used for the operation of the building.

The building is currently under construction and will be commissioned in late 2019/early 2020. The main experience from the design phase and early construction phase is that the Lowex system sets strict demands on both façade design and the interior design. Transparent vs. opaque area ratios, solar shading and glazing solutions set boundaries for the architectural concept, and must be taken into account early in the design process. Flooring materials, ratio of landscape vs. office cells, and thermal zoning of the building affects the interior design. The acoustic concept and the carbon footprint of the building as well as the design of the supplementary HVAC-system are also important factors that need to be taken into account at an early stage in the design phase. To solve these interdisciplinary problems it requires a design team that can work together closely. The architect, and the engineer that is responsible for the energy performance of the building, have a central role to play at the very beginning of the design phase when the geometry and the general parameters of the building is still in flux.

INTRODUCTION

Powerhouse Telemark is a diamond shaped 11 story high office building situated close to the Skiens river. The building will have a total heated floor space of approximately 8000m^2 and will house mainly offices, but will also include a cantina, conference rooms and smaller meeting rooms. The roof of the building is inclined at a 24-degree angle facing south to maximise the incoming solar radiation to the photovoltaic (PV) modules mounted on the roof. The south facing façade is also completely covered with a building integrated PV-system.

The project has been given financial support by Enova, a government organisation owned by the Norwegian Ministry of Climate and Environment which has a goal of reducing CO2 emissions in Norway.

Powerhouse Telemark is the first Powerhouse-concept office building in southern Norway (see www.powerhouse.no). A plus energy building that is built according to the Powerhouse definition prior to 2019 must produce more renewable, locally produced energy during the lifetime of the building, and must produce enough renewable energy to cover the total embodied energy used for the production and transportation of the building materials used in the building. The local renewable energy production must also cover the yearly net energy needs for operations, renovations and demolition of the building. The local production of renewable energy on the building is not required to cover the energy needed for plug loads in the building.

This report is divided into two parts. The first part discusses the energy used for the operations of the building. The second part discusses the embodied energy used in the building materials used in the building and during renovations.

KEYWORDS
Plus energy, Power House, Low exergy heating and cooling, Thermal indoor climate.

1 METHODOLOGY

The energy performance simulation program SIMIEN v 6.009 has been used to evaluate the energy concept for Powerhouse Telemark. An embodied energy calculation has been done to decide the total amount of solar energy production need to supply the building with enough energy to cover the embodied energy stored in the buildings materials.

The effect required for heating and cooling is calculated in Simien v 6.009 for different temperature zones and the results were imported into a calculation tool created by Skanska Norway AS and that is based on the ISO 11855-2. This tool is developed to calculate supply temperatures for heating and cooling as well as the control curves that regulate flow rates in the different zones.

2 ENERGY REQUIRED FOR BUILDING OPERATIONS

This chapter discusses the energy used for the everyday operations of the building.

2.1 Heat loss through climate envelope

To realise a plus energy building, the energy losses, through the buildings climate envelope and through the ventilation system, have to be minimised. This is done by increasing the thermal insulation and minimising thermal bridges where it has the most impact on the energy performance of the building.

IOP Conf. Series: Earth and Environmental Science **352** (2019) 012056 doi:10.1088/1755-1315/352/1/012056

The building is built air tight and the air exchange rate for infiltration across the climate envelope is set to 0,4 air exchanges per hour measured at a 50 Pascals pressure difference. This is a relatively conservative estimate of what is possible to achieve, but there has to be a focus on air tightness in the design phase as well as the production phase. The building will also be tested for air leakages before the thermal envelope is completed. This ensures that points of leakage are identified and sealed before the building is completed.

The windows in the building play an important role for the whole energy concept as they effect the energy performance of the building as well as parameters like thermal indoor climate, daylight, fire safety and acoustics. Even though the building will be cooled by the LowEx system during the summers, this cooling effect is not enough to support a comfortable indoor climate on its own. Windows facing south east will be equipped with external solar blinds that limit incoming solar radiation from the sun. The facades facing west and north east have integrated wooden solar shading in the façade and thus do not require other forms of solar shading to ensure a comfortable indoor climate. The windows, behind the integrated wooden façade shading, will only be equipped with internal manually controlled blinds that protect against glare. The large glass facades will have automated internal blinds that, in combination with the glazing layers in the glass, block the majority of the incoming solar radiation. The shape of the building also provides shading for the glass façade facing south west, see architectural renditions in the figure 1 below.

Figure 1. Overhang which shades the glass façade by the building entrance

The u-values of all the windows, including the frames, is set to 0,75 W/m^{2}K and the glazing on the glass give a g-value of 37 % and a daylight transmission of 63 %. Even with these low U-values, 42 % of all the thermal energy is lost through the windows and glass facades in the building.

Table 1. U-values for climate envelope structural components

Climate envelope structural component	Average U-values (W/m^2K)
Insulated concrete slab on ground	0,11*
Wall type 1 (General wall type)	0,17
Wall type 2 (West facing façade)	0,15
Windows and doors and glass facades	0,75**
Roof	0,10

*Includes thermal resistance of the ground materials under the slab.
**Includes wooden window frames and aluminium profiles in glass facades.

2.2 *Ventilation system*
The ventilation system is a displacement ventilation system with ventilation ducts that are placed centrally in the buildings core and the main ducts branch out in the ceiling from the central core. The air is supplied to the occupancy zones and is distributed throughout the building through ventilation overflow units. The ventilation exhaust is extracted centrally in the bathrooms, closets and in the kitchen.

The variable air volume system (VAV) is controlled by CO2 and temperature sensors. The air flow rate varies between approximately 4,5 m^3/m^2h in the winter and increases to approximately 6m^3/m^2h during the summer. The maximum capacity of the ventilation system in an extreme case is limited to around 10 m^3/m^2h.

This low pressure ventilation system is highly energy efficient and has a low specific fan power (SFP) value. The average SFP value during winter months is estimated to 0,6 and is 0,75 during the summer months. The supply air is preheated during the winter months with a rotary heat exchanger that has a heat recovery efficiency of approximately 87 %. The heat exchanger is supplied with heat from the energy wells in the ground beneath the building. The heating exchanger is used for cooling in the summer. This keeps the supply temperature at approximately 18 degrees during the summer period.

2.3 *Tap water heating*
The domestic hot water (DHW) demand is low in office buildings and thus the energy savings potential is limited. The initial energy performance concept included a CO2 heat pump that would heat the DHW. The CO2 heat pump was taken out of the energy concept and was replaced with electrical boilers on each storey. The reason for this switch is that the heat losses from circulating hot water is calculated to be relatively large in comparison to the total energy used for heating the DHW in the building. The energy savings potential is too low to motivate the extra cost inferred from a CO2 heat pump specifically dedicated to the DHW system.

2.4 *Lighting*
Office buildings have traditionally had relatively high energy use for lighting. In Powerhouse Telemark the electrical use for lighting is minimised by tailoring the lighting design to the architects seating plan. This ensures sufficient light where it is necessary. The average energy use for the building during one year is calculated to be approximately 2,3 W/m^2 or 49591 kWh/year for the entire building. This is achieved by using daylight and motion activated sensors as well as using the most energy efficient commercially available LED lighting.

2.5 *Thermal energy supply*
The buildings thermal energy supply is a highly efficient heat pump connected to energy wells. The energy wells cover approximately 99 % of all heating energy needed for room and ventilation heating. The heat pump system will cover approximately 70-80 % of the peak heating load of the building. An electrical boiler covers the rest of energy needed during peak heating hours.

IOP Conf. Series: Earth and Environmental Science **352** (2019) 012056 doi:10.1088/1755-1315/352/1/012056

Heating and cooling is distributed through a system for low exergy heating and cooling which consists of pex pipes embedded into the concrete floors. The system will be referred to as the "LowEx" system in this report. This system has low supply temperatures for heating (between 23°C and 29°C) and relative high temperatures for cooling (19°C). This ensures good working conditions for the heat pump system, and a cooling system that can rely solely on free cooling from the geothermal wells in a normal year. The system is connected to 8 energy wells that are drilled approximately 300 m into the ground on the building site.

The Lowex system makes use of exposed thermal mass in the floors and the roof and the pex pipes are placed in a concrete strip in the slab that runs parallel with the façade and is approximately 4 m wide. The LowEx system will charge the wells with excess solar energy during the summer months and some of this stored heat energy will be extracted during the heating season. This results in a higher performance for the heat pump and the seasonal coefficient of power (SCOP) is improved by doing this. It is expected that it will be possible to achieve a SCOP of approximately 4,5 for heating.

In the summer season the lowex floor system is used as a self-regulating cooling system where all valves are 100 % open, and water with approximately 19 °C runs through the system. This acts as a stabilizing system which harvest surplus heat from zones with high loads (internal and/or solar) and in zones with small loads the cooling will be limited due to small or negligible temperature difference between the floor surface and the room temperature. The surplus heat form the building is heat exchanged with the brine in the geothermal wells, and will recharge the wells in the summer months, giving better working conditions for the heat pump in the winter months.

Low pressure design of the hydraulic system ensures low energy use for both brine pumps and circulation pumps. The lowex system described here has been tested in the plus energy project Lia Kindergarten (Dokka, 2019) with good results.

Due to the high thermal inertia of the lowex system, the temperature in the building is kept constant at approximately 21,5 °C even when the building is not in use (no night set back). This has a slightly negative outcome for the energy efficiency on the building, but the increase in energy use in the building is outweighed by the lower peaks for the electric power needed for heating.

2.6 Electric energy supply

The entire roof as well as the southeast facade are covered in high efficiency solar cells (22 % efficiency). The carport next to the building is also covered in solar cells. Together the system is designed to deliver a total of 243000 kWh per year of solar electric energy. During the summer months, when the production reaches its peak (210 kW), the excess energy that is produced will be exported to the electrical grid. When the solar cells do not produce enough energy to cover the operations of the building, electrical energy is purchased from the grid.

2.7 LowEx calculation model

The LowEx system is designed in Simien v 6.009. The results from the heat/cooling load simulations is then imported into the Skanska calculation tool mentioned previously.

The Lowex tool is used to calculate the necessary temperature in the floor slab that is needed to achieve optimal operative temperatures in the different rooms in the building. Due to the high thermal inertia of the system, the mass temperature in the slab is controlled using a weather forecasted weighted mean outdoor temperature for the next 48-hour period.

Figure 2 shows a typical control curve for the mass temperature as a function of the weighted mean outdoor temperature for a zone in the building. These curves will be implemented in the control system of the building. The central supply temperature (supplied by the heat pump system) will be calculated based on the zone with the highest demand for mass temperature (highest specific heat load), and the supply temperature will also be a similar control curve as a function of the weighted mean outdoor temperature. These control curves will then be tuned and calibrated to measurements that are done after the building is completed.

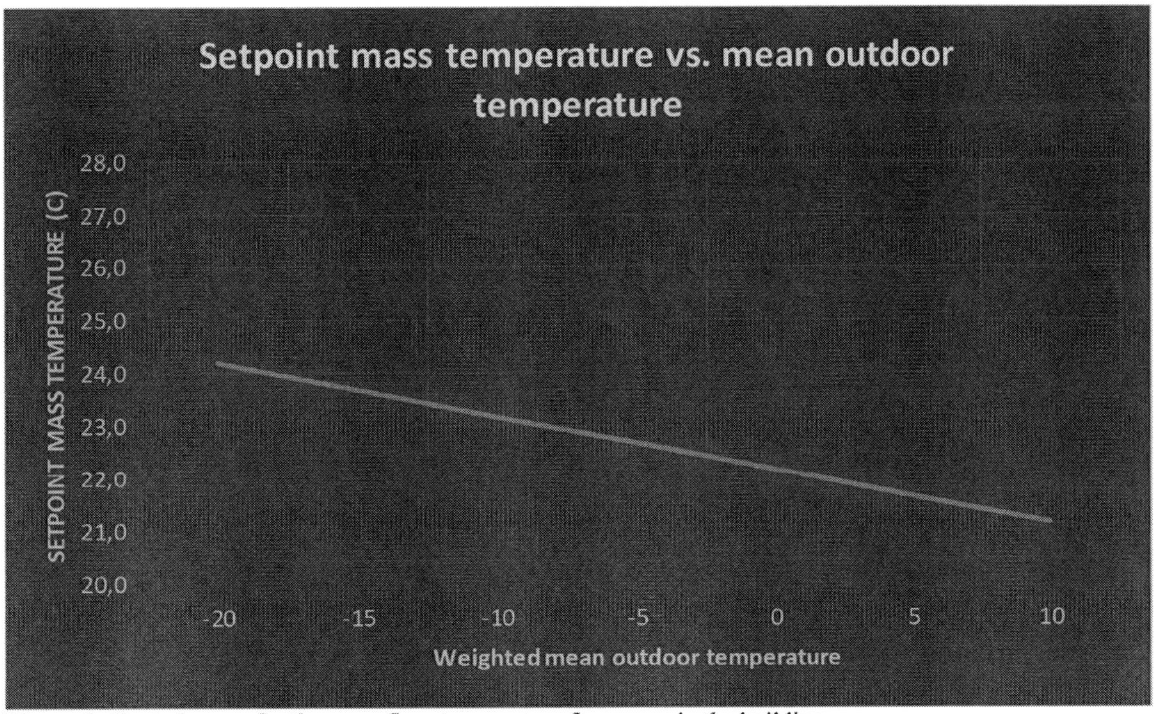

Figure 2. Control curve for the mass floor temperature for a zone in the building.

3 RESULTS AND DISCUSSION

3.1 Simulated energy use

The total net energy demand for the building and the demand for delivered energy is given in Table 2. According to the Powerhouse definition the plug load is not included in this result.

Table 2. Net energy need for Powerhouse Telemark.

	Net demand	SCOP	Delivered energy
Space and ventilation heating	14,5 kWh/m^{2}yr	4,50	3,2 kWh/m^{2}yr
DHW	1,9 kWh/m^{2}yr	0,85	2,2 kWh/m^{2}yr
Fans	3,3 kWh/m^{2}yr	1	3,3 kWh/m^{2}yr
Pumps	0,7 kWh/m^{2}yr	1	0,7 kWh/m^{2}yr
Lighting	6,3 kWh/m^{2}yr	1	6,3 kWh/m^{2}yr
Space and ventilation cooling	12,3 kWh/m^{2}yr	20	0,6 kWh/m^{2}yr
Total	39 kWh/m^{2}yr	-	16,4 kWh/m^{2}yr

3.2 Powerhouse balance calculations

The calculated embodied energy for all the building materials for including the PV-system, the energy use during construction and the embodied energy for retrofitting (after 30 years) and end of life treatment amounts to a total of 44,2 kWh/m^{2}yr (calculated as primary energy). With a primary energy factor (PEF) of 2,5, the operational energy use is calculated to 41,5 kWh/m^{2}yr. These two numbers have to be balanced with the local PV-system to achieve the Powerhouse definition. The designed PV-system has an annual production the first year of 243 000 kWh. Based on a technology scenario the PV-system is assumed replaced after 30 years by a system that is 50 % more efficient. Based on assumed primary energy factors during a 60 year time span, the specific PV-production calculated in primary energy amounts to 87,3 kWh/m^{2}yr. This balances the operational energy use and embodied energy with a surplus of 1 kWh/m^2 yr, as illustrated in figure 3.

The assumed lifetime of the powerhouse balance calculation is 60 years.

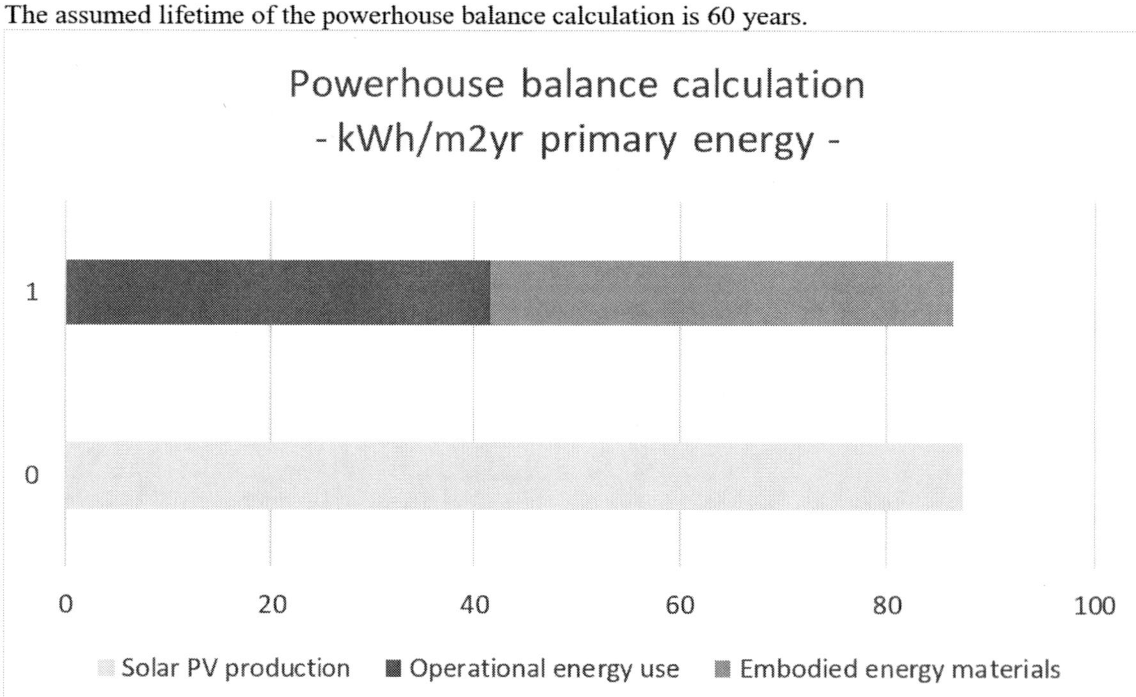

Figure 3. The powerhouse balance calculation of Powerhouse Telemark.

IOP Conf. Series: Earth and Environmental Science **352** (2019) 012056 doi:10.1088/1755-1315/352/1/012056

3.3 Challenges

The low temperature heating and high temperature cooling require large areas of exposed thermal mass in the building. This has proven difficult to achieve as this limits the area available for acoustic dampening. The Lowex system also applies restrictions for future renovations and changes in the floor plans.

Even though the building is very energy efficient, it still requires a large area for the solar PV-system. The shape of the building is designed so that the roof area is maximised for use of solar panels. Even though the roof is much larger than the base of the building the roof alone will not produce enough energy over one year to compensate for the energy used for the materials and the operation (excluding electrical energy used for plug loads) of the building as seen over the design lifespan of the building. It was thus necessary to find extra area for solar energy production. This was solved by integrating solar cells in the southeast facing façade and on the roof to the car port.

4 CONCLUSIONS

The different rooms in the building are relatively similar and can be grouped into a few different categories, for example office, landscape zones and meeting rooms. These rooms can share the same control curve, which is beneficial because it limits the amount of control curves that need to be operated. This makes the system simpler to run, and easier to tune once the building is completed.

The Lowex system supplies the room with comfort cooling but the cooling load is not sufficient to cover large thermal peak loads. It has therefore been necessary to manage the thermal peak loads for rooms with large windows. This is done by installing internal and external blinds, and by using glass with low g-values. Rooms with large internal loads also need to be complemented with ventilation cooling during the hottest summer days.

The domestic hot water (DHW) demand is very low in this building, and the heat losses from circulating hot water are relatively large. The savings made from installing a CO2 heat pump do not motivate the extra investment costs.

REFERENCES

[1] SIMIEN Version 6.009 (SIMIEN, 2018) *Simulation of Indoor Climate and Energy use, ProgramByggerne ANS*, www.programbyggerne.no

[2] ISO 11855-2:2012 *Building environment design, dimensioning, installation and control of embedded heating and cooling systems. Part 2 Determination of the design heating and cooling capacity.*

[3] Dokka TH, Myrup M, Solsem M., *"Lia Kindergarten – A plus energy building: First year experience with regard to energy use and indoor climate" Paper submitted to the first Nordic Conference on Zero Emission and Plus Energy Buildings,* Trondheim 2019.

IOP Conf. Series: Earth and Environmental Science **352** (2019) 012057 doi:10.1088/1755-1315/352/1/012057

Energy efficiency of strategies to enable temperature zoning during winter in highly-insulated residential buildings equipped with balanced mechanical ventilation

L. Georges, E. Selvnes, V. Heide and H.M. Mathisen

Energy and Process Engineering Department, Norwegian University of Science and Technology (NTNU), Trondheim, Norway
laurent.georges@ntnu.no

Abstract. To reduce the space-heating needs, balanced mechanical ventilation equipped with a heat recovery is frequently implemented in highly-insulated residential buildings. This standard ventilation strategy tends to homogenize temperature inside the building, in other words, to reduce temperature zoning. In some countries, such as Norway, many users would like colder bedrooms. It has been proved that a significant part of the occupants in Norwegian passive houses opens bedroom windows during several hours every night during winter. Dynamic simulations have shown that it strongly increases the space-heating needs and that control only is unable to create temperature zoning in an energy-efficient way. The building concept should be changed. In the present contribution, the physical processes during temperature zoning are further explained. Detailed dynamic simulations of a detached single-family house are performed using the simulation software IDA ICE for different insulation levels, construction modes (which also influence the thermal insulation in partition walls) and control strategies. Alternative mechanical ventilation strategies are compared. They manage to reduce the influence of mechanical ventilation on the increased space-heating needs due to window opening but they cannot improve the large contribution of heat conduction through partition walls between heated areas and unheated bedrooms. Among the investigated ventilation strategies, decentralized ventilation has intrinsically the best performance.

1. Introduction

The necessity to drastically reduce the space-heating (SH) needs of residential buildings in Europe has prompted the emergence of building concepts based on a super-insulated building envelope, such as the passive house (PH) standard [1]. Developed for central Europe (e.g. Germany), the PH concept has been extended to Nordic countries. In particular, Norway has elaborated a national definition with the NS 3700 standard [2]. The current Norwegian building regulation, TEK17, requires comparable energy performance for the building envelope. Besides, the proposed concepts for Norwegian Zero Emission Buildings (ZEB) are most often based on super-insulated building envelopes [3]. These buildings have a building envelope that is super- or highly-insulated leading to limited heat losses. In addition, they are equipped with a balanced mechanical ventilation with an energy-efficient heat recovery (MVHR), meaning a heat exchanger with high effectiveness, η. The ventilation is most often centralized, meaning that a single air handling unit (AHU) is installed in residential buildings, such as detached and row houses. With this standard MVHR, the supply ventilation air has the same temperature for each room in the building. This is here called a one-zone MVHR. The combination of the highly-insulated building

IOP Conf. Series: Earth and Environmental Science **352** (2019) 012057 doi:10.1088/1755-1315/352/1/012057

envelope with the one-zone MVHR limits the temperature differences between the rooms inside the building. This has been taken as an advantage for the simplification of the SH distribution. The number of heat emitters can be reduced compared to less-insulated buildings without impairing thermal comfort, see e.g. [1, 4-6]. However, recent research performed in Norway has shown that many occupants would like colder bedrooms (i.e. < 16°C) in combination with a higher temperature in the living areas, see e.g. [7, 8]. Berge et al. [9] investigated the user satisfaction and behavior using questionnaires distributed to the residents of 62 detached and terraced highly-insulated houses. Regarding bedrooms, 50% of occupants experience their temperature as too warm. During winter, approximately half of the respondents keep the bedroom window open at least a few hours per day. Results support the hypothesis that the dominant driver of bedroom window ventilation is temperature control. 65% of the respondents who keep the bedroom window open (all night or all day), have pre-set the supply air temperature to a level which requires post-heating during a significant portion of the winter season. Consequently, they do not operate the system coherently according to their desired indoor temperature. Berge et al. [10] performed another study combining field measurements with questionnaires on apartment buildings. This study confirmed that occupants would like colder bedrooms and use window opening to control the bedroom temperature (4h/day in average). It is worth mentioning that occupants often require a living room temperature of 22-24°C [7-9].

Based on these conclusions, the impact of intensive window openings on the SH needs of the building has been investigated using detailed dynamic simulations. Even though the physics to be modelled is complex, simulations suggest that opening windows to decrease bedroom temperature significantly reduce the energy efficiency [5, 6, 11]. Georges et al. [5, 6] compared different control rules for the heating and ventilation system as it has been shown that occupants do not always operate the building coherently to create the desired indoor thermal environment. None of the control strategies managed to create cold bedrooms without increasing the SH needs significantly. This suggests that this challenge cannot be addressed only using control but rather by adapting the building concept, such as the ventilation strategy. In that respect, Berge et al. [11] proposed a two-zoned ventilation concept with two distinct supply air temperatures for the living areas and bedrooms. This strategy manages to reduce the influence of window opening but the increase in SH needs is still substantial. Finally, Selvnes [12] analysed the temperature zoning in detached houses for different insulation levels and construction modes. The objective was to better understand the heat balance in the different zones and thus the physical processes explaining temperature zoning. Among other things, he showed that the phenomenon was intrinsically dynamic and cannot be fully understood using steady-state approaches. An alternative ventilation strategy based on a balanced supply and exhaust air in bedrooms was proposed as a way to reduce the influence of window opening on the SH needs. The performance of this new ventilation strategy was found to be moderate. In the remainder of the article, bedrooms will also be termed unheated zones as the local heat emission system, such as a radiator, is expected to be turned off to create lower indoor temperature. However, this is a simplified terminology as there is still active heating by the ventilation air.

As a continuation of work by Selvnes [12], the present work aims at creating at theoretical framework to understand the temperature zoning and the impact on the energy efficiency of window opening in bedrooms. It is shown that the heat conduction through the partition walls between heated and unheated areas as well as the mixing of ventilation air between zones are the two main processes of interest. Based on this framework, the influence of the construction mode can be explained and the potential of improvement using alternative ventilation strategies evaluated. In this respect, the performance of the one-zone centralized MVHR is compared with cascade ventilation and balanced airflows in bedrooms as well as with decentralized ventilation.

2. Methods

2.1. Theoretical framework

To date, most of the existing analysis about temperature zoning were based on detailed dynamic simulation (using IDA ICE) combined with parametric runs. However, a lot of the physical processes at stake occur simultaneously and are intrinsically dynamic. They are thus complex to understand so that a simplified framework of analysis would be beneficial.

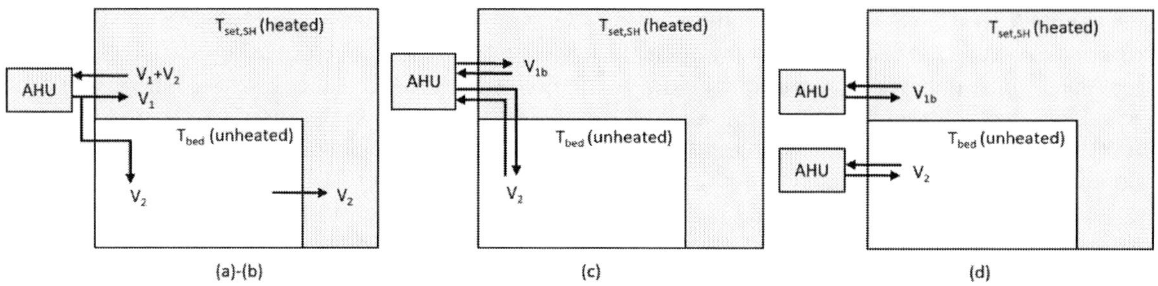

Figure 1. Simplified model for the three ventilation strategies: (a) and (b) are centralized with cascade, (c) centralized with balanced airflows in bedrooms and (d) decentralized.

In this framework, the building is modelled as two isothermal zones as shown in Figure 1: one corresponding to the heated areas and one to the unheated areas (i.e. bedrooms). The temperature in the heated area is kept at a fixed set-point ($T_{set,SH}$). The temperature of bedrooms (T_{bed}) is free-floating and depends on the thermal characteristics of the building envelope and the ventilation system. The main idea is to evaluate the difference of SH power (ΔP) in the heated zone with and without opening of windows in bedrooms. To evaluate the SH needs of buildings, the conservation of energy is usually applied by taking all the thermal zones as the control volume, thus both heated and unheated zones. In our framework, the conservation of energy is only considered in steady-state for the heated zone and the AHU (thus excluding the unheated zone).

The standard ventilation strategy is a one-zone balanced MVHR with cascade ventilation, see Figure 1(a). This means that the ventilation air supplied in bedroom is transferred to a transition zone, such as a corridor, before being extracted in a "wet" room (typically a bathroom, the kitchen or a laundry room). In Figure 1, the direction of the airflows between zones or the AHU is show by arrows. The power of the heat exchange (P) between the heated and non-heated zones has two components: firstly, the heat transferred through the partition wall with a U-value (U_p) and, secondly, the power to heat the ventilation air from the colder bedroom temperature to the living room temperature. This leads to

$$P = U_p(T_{set,SH} - T_{bed}) + \dot{V}_2 C_p (T_{set,SH} - T_{bed}) \tag{1}$$

By convention, the temperature in the heated area is unchanged when the bedroom window is opened. The thermal losses to the outdoor is thus unchanged for the heated area as well as the exhaust air temperature to the AHU. The change in SH power induced by the window opening in the bedroom is only related to the increase of the heat exchange with the unheated zone, P. Making the difference with and without window opening, the change of SH power (ΔP) is simply expressed as

$$\Delta P = U_p(T_{bed,closed} - T_{bed,open}) + \dot{V}_2 C_p (T_{bed,closed} - T_{bed,open}) \tag{2}$$

Some important conclusions can already be derived from these two simple equations:

1. For a same level of thermal insulation of the building envelope, the construction mode has a direct influence on the SH power. In lightweight buildings, the partition walls are insulated for acoustic reasons unlike heavy-weight constructions. Firstly, U_p is significantly lower in lightweight constructions. Secondly, the bedroom temperature with closed window ($T_{bed,closed}$) is lower in lightweight than in heavyweight constructions. For a same bedroom temperature with window

opening ($T_{bed,open}$), the influence of the window opening on P will be larger for heavy-weight buildings.

2. If the temperature in the living area is increased ($T_{set,SH}$), the bedroom temperature with closed windows will also be increased ($T_{bed,closed}$). Therefore, to reach a same bedroom temperature with open window ($T_{bed,open}$), a higher $T_{set,SH}$ will lead to a higher ΔP. From Equation (1), it is also clear that P increases with larger temperature differences between the living area and bedrooms.
3. A part of ΔP is due to the cascade ventilation. Therefore, it is meaningful to investigate alternative ventilation strategies to limit the impact of the window opening on the SH needs. The most obvious strategy would be to shut down the mechanical ventilation in bedrooms (i.e. removing the supply of ventilation air) when bedroom windows are open. The case is shown in Figure 1(b).

Point (3) suggests that an alternative ventilation strategy where an exhaust air terminal is placed in each bedroom (to balance exactly the amount of air supplied) would be beneficial, see Figure 1(c). The objective is to remove the ventilation contribution in P, the second term in Equation (1). With the adaptation of the ventilation strategy (without cascade), airflow rates should be changed to still comply to the building regulation. In practice, the updated airflow rates (V_{1b}) are slightly higher than the baseline value with cascade (V_1). By reducing the bedroom temperature, the opening of the bedroom window will decrease the average temperature of the exhaust air to the AHU. The mathematical expression of ΔP should then change to integrate the reduction of energy recovered by the AHU due to the window opening. For the sake of the conciseness, only the final equation is presented without the intermediate mathematical developments:

$$\Delta P = U_p (T_{bed,closed} - T_{bed,open}) + \eta \dot{V}_{1b}\, C_p\, \alpha (T_{bed,closed} - T_{bed,open}) \tag{3}$$

where the parameter α is defined as the ratio of V_{1b} to the total ventilation airflow, $V_{1b}+V_2$. Equation 3 can be reformulated in a way that can be directly compared to Equation 2:

$$\Delta P = U_p (T_{bed,closed} - T_{bed,open}) + \eta (1-\alpha) \dot{V}_2\, C_p (T_{bed,closed} - T_{bed,open}) \tag{4}$$

In the case study of Section 3, the heat recovery effectiveness is 85% and the ratio α is 0.33. For this case, the ventilation term in Equation (4) equals 56% of the ventilation term of Equation (2). Consequently, this new strategy roughly removes ~50% of the ventilation losses while heat transfer through partition walls is left unchanged. This most probably explains why Selvnes [12] found this new strategy to moderately improve the performance in his investigations using dynamic simulations. Equation (3) is based on the assumption that the AHU heating coil will not directly compensate for the reduced return air temperature created by the window opening. If the AHU heating is triggered, the energy savings will then be lower than ~50%.

To avoid these limitations, a second AHU allocated to the unheated zones can be added, see Figure 1(d). This strategy is called decentralized ventilation. In this way, a reduction of the bedroom temperature due to window opening will not impact the ventilation heat recovery of the heated zones. The only influence of the window opening on ΔP is the heat transfer through the partition walls:

$$\Delta P = U_p (T_{bed,closed} - T_{bed,open}) \tag{5}$$

In addition, the temperature set-point of the supply ventilation air ($T_{set,AH}$) can be taken different for heated and unheated zones: it can be taken lower for bedrooms so that $T_{bed,closed}$ can be closer to the desired temperature, $T_{bed,open}$. In general, this should also decrease the need for window opening to control the bedroom temperature.

2.2. Case study of detached passive house

The theoretical framework will be compared to detailed dynamic simulations in IDA ICE in order to confirm whether it provides the correct conclusions and understanding of the physics. A case study is therefore necessary and has been defined based on the work of Selvnes and Georges et al. [4, 12]. This detached house of 173 m² heated area is insulated according to the Norwegian Passive House Standard [2] and located in Oslo in an open and flat terrain without obstacles, see Figure 2. The heat transfer coefficient (U-value) is 0.14 W/m².K for external walls, 0.13 W/m².K for the external floor, 0.12 W/m².K for the roof and 0.72 W/m².K for windows. The air change rate due to infiltration 0.6/h at 50 Pa while normalized thermal bridges (ψ) are taken at 0.03 W/m².K. The heat recovery effectiveness (η) is 85 %. No frost protection of the heat exchanger has been considered. The thermal properties of the different construction modes are reported in Table 1. Detailed information about the construction can be found in [4, 12]. Internal gains are taken according to TS 3031 [13] and SH is performed using ideal heaters as the main focus is on the SH needs (i.e. not the energy use).

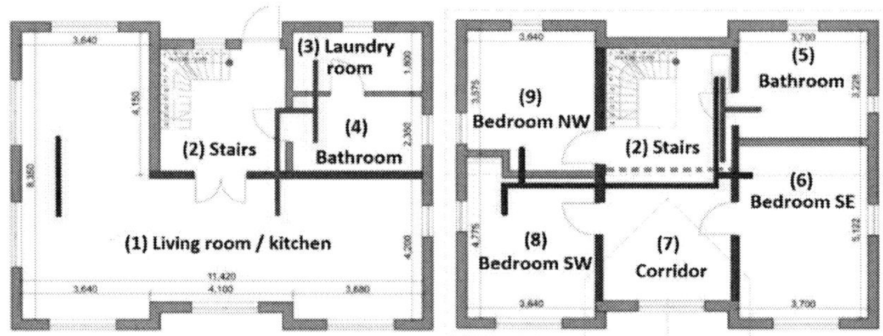

(a) First floor (b) Second floor

Figure 2. Floor plan and air distribution with cascade ventilation: ductwork for supply air are coloured in blue while the exhaust air ducts are in red [12].

Table 1. Construction modes: building thermal inertia (EN 13790), U-value of internal construction.

Construction type	Inertia level	Inertia [MJ/K]	U_{floor} [W/m².K]	U_{part} [W/m².K]	$U_{bearing}$ [W/m².K]
Masonry heavy (CM1)	*Very-heavy*	86	1.6	3.2	2.8
Wood-masonry (CM2)	*Heavy*	41	1.6	0.33	2.8
Wooden heavy (CM3)	*Medium*	35	0.23	0.33	2.8
Masonry light (CM4)	*Light*	26	0.21	0.33	1.1
Wooden light (CM5)	*Very-light*	14	0.21	0.33	0.25

The building has MVHR with constant air volumes (CAV). Airflows are described for the different ventilation strategies in Table 2 and have been defined in order to comply with the Norwegian building regulation, TEK 17. The strategy (a) is the standard ventilation with the airflow cascade principle. The strategy (b) is a small adjustment of strategy (a) where the supply air to bedrooms is stopped when the bedroom windows are open. Then, the equivalent amount of air is injected in the corridor of the second floor. The strategy (c) is the one-zone MVHR but with balanced supply and exhaust airflows in bedrooms. This lead to a new layout for the ventilation ducts, see Figure (3), that is not significantly more complex than the baseline, see Figure (2). Finally, strategy (d) is the decentralized ventilation where two AHU have a heat exchange with a same η. For ventilation strategies with balanced airflows in bedrooms, meaning (c) and (d), the gap below the bedroom doors necessary for the cascade ventilation has been removed. With cascade ventilation, the value of airflow rates is dominated by the minimum amount of supply air so that the minimum amount of extracted air should be increased to comply to TEK 17. On the contrary, with balanced airflows in bedrooms (strategies c and d), the dominant airflows are the minimum amount of extracted air so the amount of supply air should be increased compared to the minimum values. This leads to a total amount of ventilation air that is about 40% higher. In this study, the electricity use for fans is not considered.

Table 2. Ventilation airflow rates for the different ventilation strategies [12].

Zone	Room	With cascade (baseline)		Without cascade**	
		Supply [m³/h]	Return [m³/h]	Supply [m³/h]	Return [m³/h]
1	Kitchen and Living	104	40	126	36
2	Stairs	0	0	0	0
3	Technical/Laundry	0	40	0	36
4	Bathroom 1st floor	0	64	0	54
5	Bathroom 2nd floor	0	64	0	54
6	Bedroom SE	52	0	52	52
7	Corridor 2nd floor	0*	0	54	0
8	Bedroom SW	26	0	26	26
9	Bedroom NW	26	0	26	26
Total		208	208	284	284

* In strategy (b), this airflow is 104 m^3/h if the supply ventilation air in bedrooms is stopped.
** This corresponds to the strategy (c) and decentralized ventilation (d).

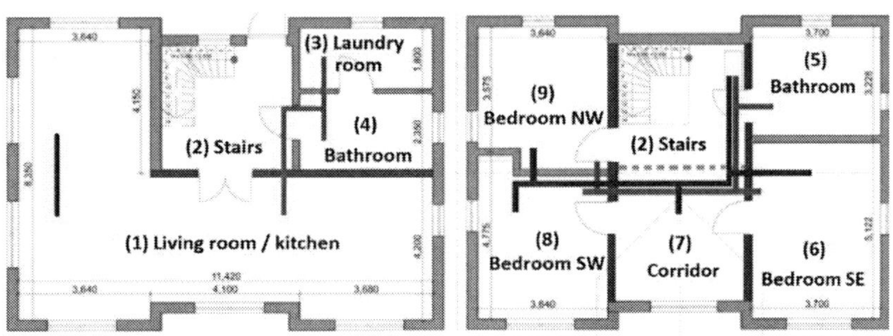

(a) First floor (b) Second floor

Figure 3. Floor plan and air distribution with balanced supply and exhaust airflows in bedrooms (strategy c) [12]: ductwork for supply air are coloured in blue while the exhaust air ducts are in red.

Table 3. Definition of the different control strategies for SH and ventilation.

Cases	Living areas	AHU	Bedrooms	Windows	Windows	Door
	$T_{set,SH}$	$T_{set,AH}$	$T_{set,bed}$	Schedule	$T_{set,win}$	Schedule
1	21 or 24°C	$T_{set,SH}$-3	$T_{set,SH}$	Closed	-	Closed
2	21 or 24°C	$T_{set,SH}$-3	None	Closed	-	Closed
3	21 or 24°C	16°C	None	Closed	-	Closed
4	21 or 24°C	14°C	None	Closed	-	Closed
5	21 or 24°C	$T_{set,SH}$-3	None	Open (Night)	16°C	Closed
6	21 or 24°C	16°C	None	Open (Night)	16°C	Closed
7	21 or 24°C	14°C	None	Open (Night)	16°C	Closed
8	21 or 24°C	$T_{set,SH}$-3	None	Open (Night)	16°C	Open (Day)

In Table 3, the different control strategies of the SH and ventilation systems are described. Different temperature set-points are applied for the SH in the heated area ($T_{set,SH}$), in bedrooms ($T_{set,bed}$) and the supply ventilation after the heating battery ($T_{set,AH}$). The area of the window opening in bedrooms is adjusted using a proportional-integral (PI) control with a cooling temperature set-point ($T_{set,win}$). This is an idealization of window control. Firstly, this removes the influence of the user behaviour as windows are here controlled automatically and, secondly, the uncertainty regarding the modelling of bi-directional airflows through windows is removed. If applied, this PI control is implemented for all the three bedrooms. Regarding time schedules, the temperature set-points are constant and windows are opened only during night-time (defined between 10 PM to 7 AM). If internal doors between bedrooms and living areas are opened, it is done during daytime (between 7 AM and 10 PM). The first strategy is the reference scenario to evaluate the SH needs. It assumes a uniform temperature for all the zones in the building. Strategies 2 to 8 try to generate colder bedroom. Strategy 2 just removes SH in bedrooms. Strategies 3 to 4 aim at decreasing bedroom temperature by reducing the temperature of the supply ventilation air but keeping windows closed. Strategies 5 to 7 resort to window opening with different temperature for the ventilation supply air. Finally, strategy 8 imposes large daily fluctuations of the bedroom temperature by opening windows during night-time and by heating bedroom during daytime using internal door opening.

3. Results

3.1. Heat balance during steady-state boundary conditions

The simplified framework of Section 2 assumes that the temperature in the living area is strictly constant with and without opening of windows. With a yearly simulation using a typical meteorological year (TMY), some periods with high solar and internal gains would lead the air temperature to exceed the set-point. Therefore, in order to be consistent with the proposed framework, the case study is first analysed with steady-state boundary conditions, meaning with a constant outdoor temperature and with constantly overcast sky. This guarantees a constant temperature in the heated zones. 30 days are used to initialize simulations corresponding to one day. It has been previously said that the physics of temperature zoning was intrinsically transient. An outdoor temperature of 5°C has been here selected to generate comparable temperature differences between the living areas and bedrooms than fully dynamic simulations.

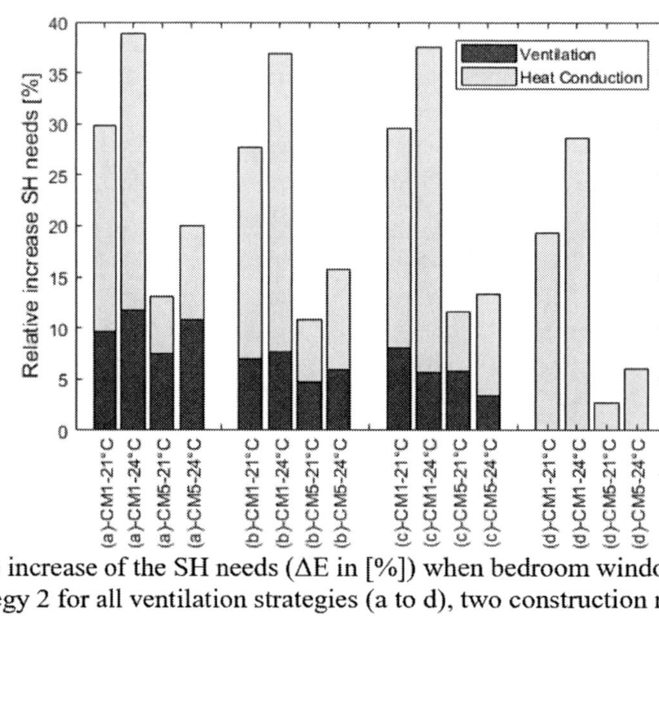

Figure 4. Relative increase of the SH needs (ΔE in [%]) when bedroom windows are open: strategy 5 compared to strategy 2 for all ventilation strategies (a to d), two construction modes (CM) and $T_{set,SH}$.

Results are shown in Figure 4 where the respective contribution of ventilation and heat conduction through partition walls are distinguished (using the heat balance of zones in IDA ICE). Starting from ventilation strategy (a), the increase of SH needs (ΔE) due to window opening strongly depends on the construction mode. The ventilation contribution is left unchanged but the heat conduction through partition walls increases significantly with higher thermal masses. The second factor of influence is the $T_{set,SH}$ as a larger temperature difference inside the building leads to a higher influence of window opening. As expected, the ventilation strategy (b) manages to reduces the ventilation losses compared to strategy (a) while the contribution from heat conduction is left almost unchanged. This reduction of the ventilation contribution is limited because the strategy (b) differs from strategy (a) only during night-time. In line with the theoretical framework, the strategy (c) only leads to a slight reduction of the ventilation component compared to the baseline strategy (a). Finally, the strategy (d) based on decentralized ventilation has only a contribution from the heat conduction. This one is lower than for the other cases because the bedroom temperature with window closed ($T_{bed,closed}$) can already be reduced compared to centralized approaches by a proper control of the AHU dedicated to bedrooms. As shown in Equation 5, this directly reduces ΔP. In conclusion, these results are in line with the simplified theoretical framework. In addition, it shows that heat conduction has a large contribution that cannot be reduced by a change of ventilation strategy. Finally, the ventilation contribution to ΔE is only slightly decreased by the alternative strategies (b) an (c). Only, the decentralized ventilation manages to remove completely the influence of mechanical ventilation on ΔE.

3.2. Yearly dynamic simulations

The analysis can now be extended to yearly dynamic simulations. For the sake of the conciseness, only the case of a lightweight building (CM5) is shown as it is most representative for Norway. In addition, a living room temperature ($T_{set,SH}$) of 24°C is assumed as it is a common temperature set-point in passive houses, see Introduction Section.

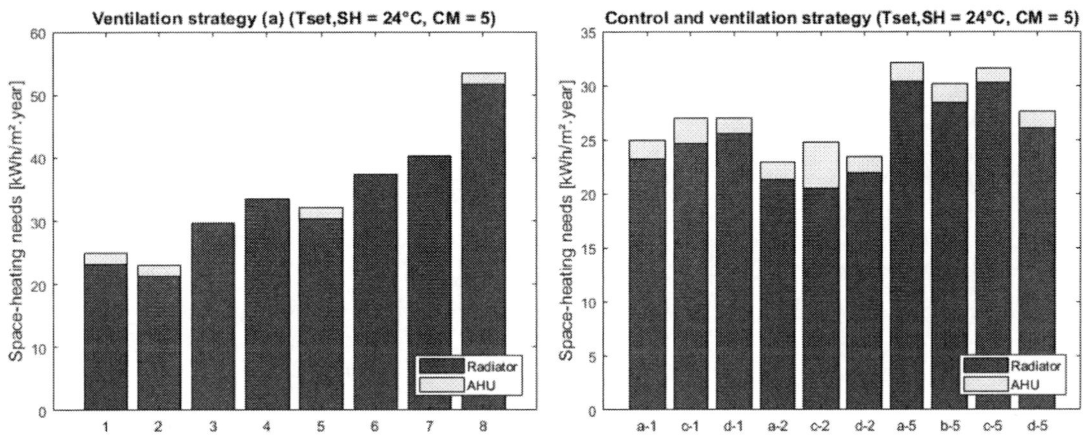

(a) (b)

Figure 5. SH needs for the baseline ventilation strategy (a) with alternative controls (left figure) and for the different ventilation strategies (right figure).

Results are shown in Figure 5. The baseline SH needs are as high as 24 kWh/m².year due to a higher indoor temperature of 24°C compared to the 21°C used by the NS3700 standard. Figure 5(a) considers the baseline ventilation strategy (a) but with different controls. Decreasing the supply air temperature leads to significantly increased SH needs while bedroom temperatures will be decreased but not up to 16°C (figure not shown here). Window opening (i.e. strategy 5) leads to an increase of 10 kWh/m².year.

This increase can be much more drastic if the bedroom is re-heated during daytime (i.e. strategy 8) leading to large temperature differences between daytime and night-time. The thermal energy stored in the building is flushed during each night and can lead to increased SH needs of 30 kWh.m².year. In conclusion, the increased of SH needs due to window opening is influenced by two user-dependant parameters: the temperature differences imposed inside the building but also the daily variations of indoor temperatures imposed to bedrooms. Figure 5(b) compares the different ventilation strategies and confirms the findings of the previous Section. Comparing control 5 between strategies (a) to (d), ventilation strategies (b) and (c) lead to a moderated improvement as they only focus on decreasing the ventilation contribution and not the heat conduction through partition walls. Unlike the theoretical framework, the energy performance of strategy (c) depends on the temperature set-point to the AHU heating coil. The decentralized ventilation, i.e. strategy (d), completely removes this ventilation contribution and gives the best performance. This work has the merit to show the relative contribution of ventilation and heat conduction through partition walls on the increase of SH needs induced by window openings. Considering alternative ventilation strategies only improves one aspect while the heat conduction remains an important factor that is left almost unchanged.

4. Conclusion

This article gave a theoretical framework to understand the increase of SH needs due to window opening in bedrooms during the SH season. Two separate effects should be considered: the effect of mechanical ventilation and the heat conduction through partition walls separating heated zones and unheated bedrooms. The influence of heat conduction is dominant for heavyweight buildings while it is comparable to mechanical ventilation for lightweight buildings (with insulated partition walls). The final increase of SH needs is strongly user-dependant. It depends on the temperature difference between heated areas and bedrooms, the daily temperature fluctuations imposed to bedrooms and the number of bedrooms where windows are opened. Alternative ventilation strategies to reduce the impact of mechanical ventilation leave the influence of heat conduction through partition walls almost unchanged. The alternative centralized ventilation strategies investigated here moderately limit the influence of ventilation. Only the decentralized ventilation completely removes the effect of ventilation on the increase of SH needs. In addition, using decentralized ventilation, the supply air temperature to bedrooms can be adjusted specifically for bedrooms and limit the need for intensive window openings in bedrooms.

Acknowledgements

The authors would like to acknowledge the Norwegian Research Council and industry partners for their support as this work was performed within the framework of the OPPTRE project and the Norwegian research center on Zero Emission Neighborhoods in Smart Cities (ZEN).

References

[1] Feist, W., et al., *Re-inventing air heating: convenient and comfortable within the frame of the Passive house concept.* Energy and Buildings, 2005. **37**: p. 1186-1203.
[2] Standard Norge, *NS 3700:2010 Criteria for passive houses and low energy houses (residential buildings)*. 2010.
[3] Houlihan Wiberg, A., et al., *A net zero emission concept analysis of a single-family house.* Energy and Buildings, 2014. **74**: p. 101-110.
[4] Georges, L., M. Berner, and H.M. Mathisen, *Air heating of passive houses in cold climate: investigation using detailed dynamic simulations.* Building and Environment, 2014. **74**: p. 1-12.
[5] Georges, L., F. Håheim, and M.J. Alonso, *Simplified Space-Heating Distribution using Radiators in Super-Insulated Terraced Houses.* Energy Procedia, 2017. **132**: p. 604-609.

[6] Georges, L., et al., *Simplified space-heating distribution using radiators in super-insulated apartment buildings.* Energy Procedia, 2016. **96**: p. 455-466.

[7] Thomsen, J., et al., *Evaluering av boliger med lavt energibehov (EBLE) – samlerapport*, in *SINTEF Fag*, S.a. forlag, Editor. 2017, SINTEF Byggforsk.

[8] Georges, L., et al., *Evaluation of Simplified Space-Heating Hydronic Distribution for Norwegian Passive Houses*, in *ZEB Project report 39*, S.a. forlag, Editor. 2017.

[9] Berge, M., J. Thomsen, and H.M. Mathisen, *The need for temperature zoning in high-performance residential buildings.* Journal of Housing and the Built Environment, 2016: p. 1-20.

[10] Berge, M. and H.M. Mathisen, *Perceived and measured indoor climate conditions in high-performance residential buildings.* Energy and Buildings, 2016. **127**: p. 1057-1073.

[11] Berge, M., L. Georges, and H.M. Mathisen, *On the oversupply of heat to bedrooms during winter in highly insulated dwellings with heat recovery ventilation.* Building and Environment, 2016. **106**: p. 389-401.

[12] Selvnes, E., *Thermal zoning during winter in super-insulated residential buildings*, in *Energy and Process Engineering Department.* 2017, Norwegian University of Science and Technology (NTNU): Trondheim.

[13] Standard Norge, *SN/TS 3031:2016 Energy performance of buildings*, in *Calculation of energy needs and energy supply.* 2016, Standard Norge.

IOP Conf. Series: Earth and Environmental Science **352** (2019) 012058 doi:10.1088/1755-1315/352/1/012058

Assessing buildings' absolute environmental sustainability performance using LCA focusing on climate change impacts

Pernille Ohms[1], Camilla Ernst Andersen[2], Freja Nygaard Rasmussen[2], Morten Ryberg[1], Michael Hauschild[1], Morten Birkved[3], Harpa Birgisdottir[2]

[1] Technical University of Denmark, Department of Management, Kongens Lyngby, Denmark
[2] Department of Energy Efficiency, Indoor Climate and Sustainability, Danish Building Research Institute, Aalborg University, Copenhagen, Denmark
[3] University of Southern Denmark, SDU Life Cycle Engineering, Department of Chemical Engineering, Biotechnology and Environmental Technology, Odense Denmark

pkroh@dtu.dk

Abstract. The building sector is of major concern when seeking to reduce the environmental impact of our society. A common tool often used in certification systems for quantification of environmental impacts is Life Cycle Assessments (LCAs). LCAs are traditionally used for relative comparisons, i.e. to assess whether one product or service performs better than another. Recently, a method for absolute evaluations based on the Planetary Boundaries, was coupled with LCA in order to define the boundary between environmental sustainability and unsustainability. In this study Planetary Boundaries-based Life Cycle Impact Assessment method has been applied to a case study of six single family stand-alone dwellings to assess whether these buildings can be considered absolute sustainable relative to the Planetary Boundaries. The results from the assessment indicate that irrespective of the design strategy used for the six houses and future increase in the use of renewables for electricity and heat production, it is unlikely that any of these houses can be regarded as sustainable in absolute terms. This underlines that more radical changes are needed in the way buildings are constructed and used in order for buildings to become environmentally sustainable.

1. State of the building industry

Every single step of a building's life cycle induces environmental impacts. Resources are extracted for the production of building components and energy is used during construction. When the building is in use, maintenance and repairs require the extraction of more resources for new building components, not to mention the energy required to keep the building heated and ventilated. Finally, when the service life of the building ends, components become waste that needs to be handled. In Europe, current construction of buildings account for 36 percent and 40 percent of the CO_2 emissions and energy consumption respectively [1]. By 2030 the global construction market is expected to increase by 85% to $15.5 trillion worldwide [2]. The Danish building industry produces 4.1 million ton waste every year, or a third of all waste produced in Denmark [3]. The reason for this is the fact that material waste from buildings is rarely used in new buildings or refurbishments. In most cases, the material waste is incinerated or landfilled and in best cases, reused or recycled [3].

However, the concept of circular economy, where fewer virgin resources are extracted and waste is considered a pool of resources, is becoming more widespread and is more often integrated into design strategies and influences the choice of materials [4]. By now some buildings are being designed for disassembly to allow for easier reuse of components and in impact-heavy building materials such as insulation and concrete the virgin raw materials are partially or completely replaced by recycled materials [5, 6]. In 2012, an adapted version of the German building sustainability certification system DGNB (Deutsche Gesellschaft für Nachhaltiges Bauen) was introduced to the Danish building industry [7]. A certification system such as DGNB offers the possibility to measure and compare the environmental performance of buildings against specified benchmarks, and this has contributed dramatically to raising awareness of sustainability in the building industry. In 2013, life cycle thinking was introduced in the building certification system LEED in the form of life cycle assessment (LCA) as an optional assessment step to improve the overall rating of a building. In DGNB a full LCA of the building is mandatory [8].

LCA is an environmental performance assessment methodology that allows for relative environmental assessment, i.e. assessment of product A's performance compared to that of product B [9]. LCA is a framework most often used to claim environmental superiority of products relative to another or to identify environmental hotspots in a product's life cycle. Building LCAs, in present certification practice, assess buildings relative to benchmarks representing (best) current building practice [10]. However, the evaluation does not relate to whether absolute sustainability is actually reached. The term sustainability is becoming increasingly broad in its meaning, and if our actions towards limiting climate change are to be successful, it is crucial that we still aim for a sustainable development as originally defined in the Brundtland report, namely "development that meets the needs of the present without compromising the ability of future generations to meet their own needs" [11]. Hence, there is a need for an absolute perspective, where we do not burden the planet with our actions more than the planet is able to cope with/compensate for.

The goal of this study is to evaluate whether or not the current approaches to sustainable building design are in fact sustainable. This is done by applying a so-called absolute environmental sustainability assessment. It should be noted that the impact scope of this paper has been limited to climate change impacts.

2. Absolute environmental sustainability assessment

In absolute environmental sustainability assessments (AESAs), environmental boundaries are used for setting the boundary between environmental sustainability and unsustainability as the environmental boundaries are used to indicate whether or not environmental systems and processes are able to cope with the environmental burden associated with a product or as in this study, a single family stand-alone dwelling. In this AESA, the Planetary Boundaries (PBs) [12] where used as absolute environmental boundaries. The Planetary Boundaries define a safe operating space that all human activities need to stay within to ensure that the carrying capacity of the Earth System processes are not compromised [13]. Impacts associated with the dwelling were quantified in the metrics of the PBs by performing an LCA and using the Planetary Boundaries-based Life Cycle Impact Assessment method [13]. Once quantified, the impact scores where related to the PBs. However, as the PBs accounts for the global Earth System processes (i.e. all people and all impacts irrespective of origin), the share of the safe operating space available for each dwelling needs to be assigned and the dwelling's impact score should be related to the assigned share of the safe operating space to determine whether the dwelling can be considered absolute sustainable [14]. There is no objective approach to share the safe operating space among activities, as it is an issue of quantifying *who* has the right to impact *how much*. In order to simplify this, only one approach for assigning the safe operating space is considered in this paper, but other approaches exist and the specific choice of sharing principle could yield different results.

In this study, the safe operating space was shared using an egalitarian and utilitarian approach. First, the global safe operating space was shared among global population (7.63×10^{-9}) using equal per capita approach (egalitarian). Next, the personal share assigned to the household was found using data on final consumption expenditure to determine the share of income that people in Denmark spend on their household relative to other products and services (giving a share of 0.571 [8]). This is an utilitarian approach as it was assumed that personal spending habits are adequate indicators of which products and services brings the largest utility to people. Next, the household share assigned to the dwelling was found (0.197 [8]), again using final consumption expenditure. Finally, the share was scaled to the number of residents (2.54 [15][16]). For further information on the sharing of the safe operating space, see [15]. As shown in Equation 1, this sharing resulted in an assigned share of 3.74×10^{-11} available to one single family stand-alone dwelling out of the entire safe operating space as defined by the Planetary Boundaries. Finally, the impact score of each house (quantified by performing an LCA) was related to the assigned share of safe operating space and if the utilized vs. available operating space ratio was less than 1, the house was considered absolute sustainable.

$$\frac{1}{7.63\cdot10^{9}}\cdot0.571\cdot0.197\cdot2.54=3.74\cdot10^{-11}\quad(Eq.1)$$

2.1. *The MiniCO2 Houses*

The methodology was applied to a case consisting of five single family stand-alone dwellings known as the MiniCO2 Houses. In 2013, the Danish foundation Realdania developed these experimental houses with the goal of assessing strategies for reducing the GHG emissions in a building's life cycle. This was achieved by introducing five different sustainability strategies focusing on different aspects of a building's life cycle and applying one strategy to each house [17]. The five sustainability strategies resulted in the five houses - Upcycle House, the Innovative Maintenance-free House, the Traditional Maintenance-free House, the Adaptable House and the Quota House.

The first MiniCO2 House is Upcycle House. The strategy with Upcycle House was to reduce the GHG emissions from the product stage by using mainly upcycled, recycled and reused materials [18]. The strategy for the Upcycle House is in line with current tendencies towards circular economy where more recycled materials are being used, thus, avoiding extraction of virgin resources.

The strategy for the Innovative Maintenance-free House is low maintenance and an extended service life, hence reducing the amounts of GHGs emitted in the product stage. The Innovative Maintenance-free House has been designed to have a service life of 150 years by using materials that are *expected* to have long service lives and low maintenance demands [19]. This strategy is in contrast to the Traditional Maintenance-free House, which provides low maintenance and a long service life by using materials that are *known* to have a long service life of 150 years and low maintenance demands. The choice of materials, architectural and technical design of the Traditional Maintenance-free House are based on experiences on family houses located in Denmark [20].

The fourth MiniCO2 House is the Adaptable House. The strategy with the Adaptable House is flexibility and variability during the building's service life. The Adaptable House is designed to accommodate the residents' changing functional requirements throughout the building's service life by, for instance, facilitating/simplifying reconstructions and extensions of the house. This will save materials and energy used for reconstruction in the buildings use stage and thereby save GHG emissions in the building's life cycle [21].

The final MiniCO2 House, the Quota House, is designed to affect the behaviour of the residents to reduce energy consumption in the building's use stage and thus reduce the associated GHG emissions. This is achieved both through the architectural design and smart technologies integrated in the house. As example, the layout of the house takes basis in the orientation of the sun and daylight and thus reduces the need for electricity. At the same time, the integrated technologies help monitor and control the consumption of water, heat and electricity. The energy consumption is measured against a quota and rendered visible so that the residents are constantly aware of their energy consumption, which influences their behaviour and should reduce their consumption. Additionally, a sixth house, the Reference House, representing a standard single family stand-alone dwelling is assessed [22].

The main difference between the MiniCO2 Houses is that they are designed to reduce impacts from different life cycle stages. The Upcycle House and Adaptable House are designed to limit impacts in the product stage, the Quota House is designed to reduce impacts in the use stage, while the two Maintenance-free Houses focus on reducing impacts in both the product stage and end of life stage.

Picture 1-5. The five MiniCO2 Houses. Picture 1, top-left: Upcycle House (UP). Picture 2, top-right: Innovative Maintenance-free House (IMF). Picture 3, mid-left: Traditional Maintenance-free House (TMF). Picture 4, mid-right: Adaptable House (AD). Picture 5, bottom-center: Quota House (Q). The Reference House (Ref) is not illustrated as it is a theoretical, standard house.

2.2. Life Cycle Assessment

The results of this study were found by performing an LCA following the standards ISO 14040:2006 and ISO 14044:2006. We defined a functional unit as to *annually house one family in a stand-alone dwelling in Denmark*. We assumed that all houses compared can fulfil this functional unit. Because the objective of the MiniCO2 Houses is to reduce GHG emissions, the impact coverage in this study is restricted to assessing climate change. Thus, conclusions about absolute sustainability are only in terms of climate change and environmental impact coverage must be expanded to provide a comprehensive AESA. All systems were modelled using the software tool SimaPro 8.9 [23] with the database ecoinvent 3.4 [24]. An outline of the system boundaries for the modelled systems are presented in Table 1 (for further information, see Supporting Information I in [15]).

Table 1. The overall system boundaries for the Life Cycle Assessment.

Life cycle stage	Included in the LCA
Product stage	All materials for constructing the buildings and waste thereof.
Use stage	District heating and electricity for ventilation. Estimations based on [25]. Replacements of building components/materials.
End of Life stage	All building materials for constructing the buildings were considered as waste.
Transportation	Average transportation processes for building materials and waste disposal in the Danish market.

The six houses are compared across two scenarios. In the first scenario, the Worst Case Scenario, all houses are assessed assuming that current practices are not improved thereby accounting for an expansion of the buildings, current consumption of heat, energy grid of 2018 (11.1 g CO2 eq per MJ heat and 668.7 g CO2 eq per kWh electricity) [26], current lifespan of materials and current End of Life routes (for further information regarding scenarios, see Supporting Information I in [15]). In the second scenario the houses are assessed according to a Best Case Scenario, i.e. no expansion of the buildings, a 25% lower consumption of heat relative to the Worst Case Scenario, an energy grid of 2030 with a larger share of renewable energy sources (9.0 g CO2 eq per MJ heat and 651.8 g CO2 eq per kWh electricity) [26], a doubled lifespan of the materials and finally a higher recycling rate for relevant materials (for specific lifespan and recycling rates of materials, see Supporting Information III in [15]). The Best Case Scenario is defined to represent a future with lower environmental impact compared to the Worst Case Scenario, representing current conditions. A full inventory for all houses can be found in [15] and [27]. The results obtained indicate the environmental performance of one house compared to the others and whether or not the houses are truly sustainable relative to an assigned share of the Planetary Boundaries For extensive results on other impact categories, see [15].

3. Results

The results of the six houses and life cycle stages obtained using the Planetary Boundaries-based Life Cycle Impact Assessment method are presented in Table 2.

Table 2. The characterised impact score obtained from the Planetary Boundaries-based Life Cycle Impact Assessment. The results are presented for each house and each life cycle stage. The life cycle stage with the highest impact score is marked in orange.
[a] 120 year service life
[b] 150 year service life

CC – Energy Imbalance [W m^{-2}]		REF[a]	UP[a]	TMF[b]	IMF[b]	AD[a]	Q[a]
Worst Case Scenario	Product stage	1.28×10^{-10}	6.06×10^{-11}	1.13×10^{-10}	1.27×10^{-10}	1.13×10^{-10}	1.42×10^{-10}
	Use stage	3.15×10^{-10}	3.05×10^{-10}	3.16×10^{-10}	3.81×10^{-10}	4.58×10^{-10}	2.92×10^{-10}
	End of Life stage	3.64×10^{-11}	4.71×10^{-11}	5.10×10^{-11}	2.72×10^{-11}	6.65×10^{-11}	3.36×10^{-11}
	Total	$\mathbf{4.80\times10^{-10}}$	$\mathbf{4.12\times10^{-10}}$	$\mathbf{4.81\times10^{-10}}$	$\mathbf{5.35\times10^{-10}}$	$\mathbf{6.38\times10^{-10}}$	$\mathbf{4.67\times10^{-10}}$
Best Case Scenario	Product stage	1.28×10^{-10}	6.06×10^{-11}	1.13×10^{-10}	1.27×10^{-10}	1.13×10^{-10}	1.42×10^{-10}
	Use stage	1.14×10^{-10}	1.06×10^{-10}	1.04×10^{-10}	1.33×10^{-10}	1.23×10^{-10}	1.01×10^{-10}
	End of Life stage	1.22×10^{-11}	1.32×10^{-11}	1.59×10^{-11}	1.83×10^{-11}	1.41×10^{-11}	1.70×10^{-11}
	Total	$\mathbf{2.54\times10^{-10}}$	$\mathbf{1.80\times10^{-10}}$	$\mathbf{2.33\times10^{-10}}$	$\mathbf{2.78\times10^{-10}}$	$\mathbf{2.50\times10^{-10}}$	$\mathbf{2.60\times10^{-10}}$

The results presented in Table 2 show that the use stage is the main contributor to the environmental impact irrespective of house and that the product stage is the second largest contributor in the Worst Case Scenario. In the Best Case Scenario, the use stage is the largest contributor for Upcycle House, Innovative Maintenance-free House and Adaptable House, while the product stage is the largest contributor for the remaining three houses. Moreover, in the Best Case Scenario it is seen that the use stage constitutes a considerably smaller part of the environmental impact compared to the Worst Case Scenario and thus the improvements introduced in the Best Case Scenario show to have an effect on the results. Furthermore, the impact potentials for the MiniCO2 Houses are compared to the share of safe operating space available to each dwelling (see Figure 1 and Table 3).

Table 3. The share of the safe operating space occupied by each of the houses in the Worst Case Scenario and the Best Case Scenario. The Planetary Boundary for the impact category Climate change (CC) – Energy Imbalance is 1 W m^{-2} and the share assigned to a single family stand-alone dwelling is 3.74×10^{-11} W m^{-2}. The darker the orange, the higher the impact.
[a] 120 year service life
[b] 150 year service life

		REF[a]	UP[a]	TMF[b]	IMF[b]	AD[a]	Q[a]
Worst Case Scenario	SoSOS occupied	32.2	27.7	32.3	36.0	42.9	31.4
Best Case Scenario	SoSOS occupied	17.1	12.1	15.6	18.7	16.8	17.4

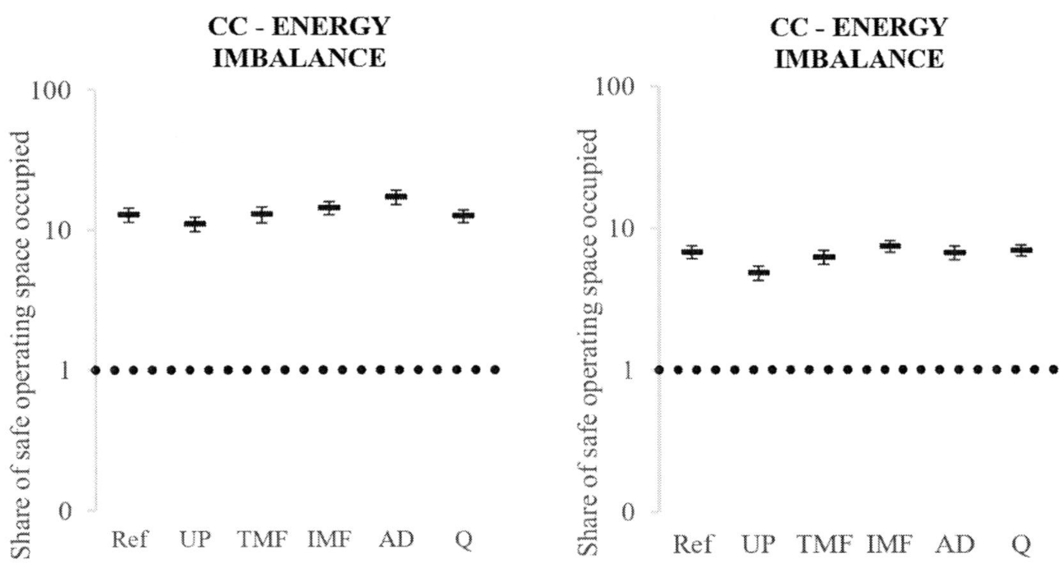

Figure 1. The impact potential for Climate Change - Energy imbalance relative to allocated share of safe operating space for each dwelling. Left and right figure reflects the Worst and Best Case Scenario, respectively. The environmental boundary is indicated by the dashed line and impact scores ≤1 can be considered sustainable. Error bars indicate the 95% confidence interval of the life-cycle inventory estimating using Monte Carlo analysis. Y-axis is on a logarithmic scale.

Figure 1 clearly shows that the impacts induced by all buildings exceed the assigned share of safe operating space irrespective of whether it is the Worst or Best Case Scenario. In the Worst Case Scenario the buildings exceed the assigned share of safe operating space by a factor 11.0 to a factor 17.0, Upcycle House being the house closest to stay within the boundary and the Adaptable House exceeding the boundary the most. In the Best Case Scenario the houses exceed the assigned share of safe operating space by a factor 4.8 to a factor 7.4, where again Upcycle House is closest to stay within the boundary and the Innovative Maintenance-free House exceeding the boundary the most. For further details on the impact contribution of processes and materials, as well as sensitivity analysis', see [15].

4. Discussion and conclusion

As described in Section 1, it is important that we keep the understanding of sustainability in mind when assessing environmental performance of products, systems or, in this case, single family stand-alone dwellings. If future generations shall be able to meet their own needs as we and previous generations have met ours, it is crucial that we do not settle for "lower" but instead strive for "low enough" in terms of induced environmental impact. While these buildings all represent an important step towards lower environmental impacts, the results presented in the previous paragraph indicate that even more effort is needed to design absolute sustainable houses. The results indicate that the focus should be on the use and product stage as these were found to be the most contributing life cycle stages.

The results presented in the previous paragraph show that the environmental impact potentials does differ between the houses depending on the approach to (environmentally) sustainable building design. However, as this difference is not extensive, it is uncertain which design strategy is the most efficient when seeking to reach a sustainable level. The results do however indicate that the changes implemented in the Best Case Scenario, i.e. energy grids, End of Life routes and consumption patterns etc. cause the houses to perform considerably better compared to the Worst Case Scenario. These observations reveals that the changes made in the Best Case Scenario does influence the results positively in terms of reaching sustainability in the built environment. However, since the houses even in the Best Case Scenario exceed the boundary considerably, the results indicate that it is unlikely that the houses can be considered absolute sustainable unless we improve dramatically on all parameters. Additionally, this conclusion is supported by uncertainty regarding the impact estimates for electricity and heat in the use stage, which are found to lower than comparable estimates [28]. This difference could cause that the impact in the use stage to be underestimated, which ultimately would result in a larger impact for each dwelling and thereby being even further from reaching a sustainable level. Hence, all stakeholders involved with the built environment need to acknowledge and take on the task of building more sustainable buildings. It is likely that this issue goes beyond the building sector and that all actors in our society, i.e. consumers, industries and politicians alike, need to mobilize to overcome this challenge.

Absolute sustainability is an approach that allows us to understand the actual distance to target in terms of environmental performance. The approach is still relatively new and some questions therefore remain unanswered in terms of application. The assignment of the safe operating space which was briefly described in Section 2 is a major challenge in assessments of this type. It has a potentially crucial impact on the results [29], but subjectivity is most likely impossible to avoid when trying to decide who has the right to impact how much. This assignment of the safe operating space should be kept in mind when interpreting the results. If a consensus could be reached on how to share the safe operating space in a fair way, it would be possible to assign the boundaries and thereby reach a per-square-meter impact allowance for buildings which would ensure absolute sustainability. Such a benchmark for environmental performance could potentially be implemented in certification systems or building regulations to facilitate that new buildings are designed to be environmentally sustainable and not just better than the existing building stock.

References

[1] European Commission, "Energy performance of buildings," [Online]. Available: https://ec.europa.eu/energy/en/topics/energy-efficiency/energy-performance-of-buildings. [Accessed: 25-Apr-2019]

[2] Global Construction Perspectives and Oxford Economics, "Global Construction 2030 – A Global forecast for the construction industry to 2030," Global Construction Perspectives and Oxford Economics, London, 2015.

[3] Green Building Council Denmark, DGNB, "Cirkulær økonomi og DGNB – Guide til cirkulære principper i DGNB bæredygtighedscertificering," 2018

[4] Eberhardt L C M, Birkved M and Birgisdóttir H, "Building life-cycle design and engineering strategies for a Circular Economy – a Systematic Literature Review," *Submitt. to J. Clean. Prod.*, 2019.

[5] Henriksen T, Juel-Hansen L, Mathiesen D, "Udredning af tekonologiske muligheder for at genbruge og genanvende beton," Miljøstyrelsen, 2015.
[6] Miljøstyrelsen, "Glas til produktion af glasuld," [Online] Available: https://www2.mst.dk/udgiv/publikationer/2003/87-7972-706-9/html/kap05.htm [Accessed: 29-Apr-2019
[7] Green Building Council Denmark, "DGNB i Danmark," [Online]. Available: https://www.dk-gbc.dk/dgnb/introduktion-til-dgnb/dgnb-i-danmark-historisk-rids/. [Accessed: 25-Apr-2019]
[8] Brejnrod K N, Kalbar P, Petersen S and Birkved M, "The absolute environmental performance of buildings," *Build. Environ.*, vol. 119, pp. 87–98, 2017.
[9] Andersen C M E and Ohms P K, "Miljøoptimering af renoveringsstrategien for Brøndby Strand," 2016.
[10] Rasmussen F N and Birgisdóttir H, "Life cycle assessment benchmarks for Danish office buildings," *IALCCE 2018 conference proceedings* (Gent), 2019.
[11] World Commission on Environment and Development, "Our common future," 1987.
[12] Steffen W, Richardson K, Rockström J, Cornell S E, Fetzer I, Bennett E M, Biggs R, Carpenter S R, de Vries W, de Wit C A et al., "Planetary boundaries: Guiding human development on a changing planet," *Science* 347, 736, 2015.
[13] Ryberg M W, Owsianiak M, Richardson K and Hauschild M Z, "Development of a life-cycle impact assessment methodology linked to the Planetary Boundaries framework," *Ecol. Indic.*, vol. 88, no. December, pp. 250–62, 2018.
[14] Ryberg M W, "Putting life-cycle indicators on an absolute scale – Integrating Planetary Boundaries into Life- Cycle Assessment," 2018.
[15] Ohms P K and Andersen C M E, "Assessment of absolute sustainability in the built environment," 2019. (Master's thesis). Retrieved from https://findit.dtu.dk/en/catalog/2444215398
[16] Frostell B M, Sinha R, Assefa G and Olsson L, "Modeling both direct and indirect environmental load of purchase decisions: A web-based tool addressing household metabolism", Environmental Modelling and Software, vol. 71, pp. 138–147, 2015
[17] Realdania Byg and B. Kleis, "The MiniCO2 houses in Nyborg – valuable lessons," 2014.
[18] Realdania Byg and B. Kleis, "Upcycle House - genbrug fra inderst til yderst," 2013.
[19] Realdania Byg and B. Kleis, "Det Vedligeholdelsesfrie Hus - fornyelse fortolker fortiden," 2014.
[20] Realdania Byg and B. Kleis, "Det Vedligeholdelsesfri Hus - Murværk holder," 2014.
[21] Realdania Byg and B. Kleis, "Det Foranderlige Hus - Bygget til at bygge om," 2013.
[22] Realdania Byg and B. Kleis, "Kvotehuset - sætter beboerne i spil," 2014.
[23] PRé, "*SimaPro, the world's leading sustainability software*", [Online]. Available at: https://www.pre-sustainability.com/sustainability-consulting/sustainable-practices/custom-sustainability-software. [Accessed: 11-Jan-2019].
[24] Ecoinvent, "Ecoinvent version 3" [Online]. Available at: https://www.ecoinvent.org/. [Accessed: 11 January 2019]
[25] Rasmussen M K and Johannsen K B, "Kvotehuset: Midtvejsevaluering," 2016
[26] Energistyrelsen, "Basisfremskrivning 2018 - Figurer og Tabeller," 2018
[27] Rasmussen F.N., Birkved M., Birgisdóttir H., "Low carbon design strategies for new residential buildings - lessons from Danish architectural practice." *Under review in* Architectural Engineering and Design Management, 2019.
[28] Cowi and Trafik- og Byggestyrelsen, "Nye emissionsfaktorer for el og fjernvarme", 2016
[29] Ryberg M W, Owsianiak M, Clavreul J, Mueller C, Sim S, King H and Hauschild M Z, "How to bring absolute sustainability into decision-making: An industry case study using a Planetary Boundary-based methodology," *Sci. Total Environ.,* 634, 1406–1416, 2018.

IOP Conf. Series: Earth and Environmental Science **352** (2019) 012059 doi:10.1088/1755-1315/352/1/012059

Overheating and daylighting evaluation for free-running classroom designs

M Kiil[1,*], R Simson[1], F de Luca[1], M Thalfeldt[1], J Kurnitski[1,2]

[1]Tallinn University of Technology, Department of Civil Engineering and Architecture, Ehitajate tee 5, 19086, Tallinn, Estonia
[2]Aalto University, School of Engineering, Otakaari 4, 02150 Espoo, Finland

[*]corresponding author: martin.kiil@taltech.ee

Abstract. Learning performance is strongly related to thermal comfort and lighting conditions of classrooms. Poor facade design can result in high indoor temperatures or insufficient access to natural light. To maintain the required temperatures and illuminance levels in such rooms may require intensive use of artificial lighting and active cooling systems, which are energy-intensive, costly to install, operate and maintain. The purpose of this study was to determine essential parameters and facade design options that ensure overheating prevention and fulfil daylight requirements in classrooms without mechanical cooling. The present study is based on simulations of a parametric room model with variable dimensions and orientations. Facade glazing solutions with optimal combination of solar factor and visible light transmittance were used to minimize overheating risk and maximize natural lighting impact. For east, south and west oriented facades, the effect of horizontal shading was also analysed. Overheating assessment through indoor temperature simulations was conducted with dynamic simulation software IDA ICE, daylighting was simulated with DIVA4 coupled with Grasshopper software. Results show that classrooms without mechanical cooling require in depth analysis to determine satisfying solutions for both overheating and daylighting criteria. The results of this paper can be used for early stage facade design guide for school buildings or similar use free-running buildings.

1. Introduction

The effects of indoor temperature and lighting conditions on schoolwork performance are relatively well researched [1-5]. Studies on lighting conditions show positive effects of natural light availability on performance and visual comfort [4]. Also, daylight utilization is an efficient way to save energy related to electric lighting [6] and heating [7], as its availability corresponds to the period during which buildings are occupied. Thus, daylighting is an important factor in classroom planning and school building design. At the same time excessive direct solar access can cause unwanted glare and solar heat gains that influence occupants' comfort and building energy use due to cooling need during warm periods [8]. Many studies have found, that higher indoor temperature has negative effect on thermal comfort and learning ability [2, 3, 9]. High indoor temperatures and overheating, specifically in temperate climate regions, are mostly recent problems, arising from paradigm shifts in architectural and energy efficiency related advances on building design [5, 10, 11]. It is essential to assess buildings in early stages of design development to properly ensure sufficient daylighting and prevent overheating.

In Estonia, daylight in building is regulated by the standard Daylight in dwelling and Offices [12]. The standard sets different minimum mean Daylight Factor (mDF) values for a series of internal spaces of buildings, of which classrooms are required to guarantee a minimum of 2% mDF. Overheating

IOP Conf. Series: Earth and Environmental Science **352** (2019) 012059 doi:10.1088/1755-1315/352/1/012059

assessment for new buildings in the design stage is required by the National Building Code by using temperature excess calculation method, based on dynamic indoor temperature simulations [13].
The present study investigates overheating and daylight performance of classroom and facade design variations for different floor dimensions, window sizes, glazing parameters and shading use. The scope was to find optimal solutions that fulfil daylight and overheating prevention requirements in Estonia.

2. Methods

We have analysed a classroom parametric model through computer simulations to assess indoor temperatures, overheating risk and daylighting. The parameters used in the simulation model creation are shown in Tables 1 and 2. The room model variations included different room widths and depths (5m, 6m, 7m, 8m and 9m) for a total of 25 room size and layout variations. The window layout was varied in accordance to the room width. For the room of 5m 2 windows of width and height 1.9x1.7m (Window-to-Wall-Ratio (WWR) 45.6%) were used, for the room width of 6m 3 windows of 1.466x1.7m (WWR 41.5%) were used, for the room width of 7m 3 windows of 1.8x1.7m (WWR 43.7%) were used, for the room of 8m width were used 4 windows of 1.45x1.7m (WWR 41.1%) and for the larger room width of 9m were used 4 windows of 1.7x1.7m (WWR 42.8%). The floor to ceiling height of the room is 3m for all the room variations. As a passive measure to reduce external heat gains from direct sunlight into the room, we used horizontal shading with a depth of 0.9m on top of the windows as an option for east, south and west orientations. Additionally, ground surface with 20% reflectance was modelled outside the room.

Table 1. Room and facade parameter combinations.

Room dimensions	Envelope	Windows	Window dimensions	Orientation	Glazing g-value	Glazing VT (%)	Shading depth (hor.)
Depth, m:	Ext. wall:	Frame fraction 0.34	Recess depth	E	0.35	0.635	-
5, 6, 7, 8, 9	Concrete 150mm	East/south/west:	0.25m		0.42	0.707	0.9m
	Exp.polystyr.	U_g 0.58W/(m²·K)		S	0.35	0.635	-
Width, m	300mm	U_{tot} 0.60W/(m²·K)	Room width,				0.9m
5, 6, 7, 8, 9	Concrete 50mm		number of	W	0.35	0.635	-
	U_{tot} 0,129W/(m²·K)	East/west with	windows-		0.42	0.707	0.9m
		shading:	width/height:	N	0.54	0.733	-
	Ext. window	U_g 0.70W/(m²·K)	5m, 2-1.9/1.7m				
	perimeter thermal	U_{tot} 0.71W/(m²·K)	6m, 3-1.466/1.7m				
	bridge: 0.1W/(m·K)		7m, 3-1.8/1.7m				
		North:	8m, 4-1.45/1.7m				
	Fixed infiltration:	U_g 0.61W/(m²·K)	9m, 4-1.7/1.7m				
	1.5m³/(h·m²)	U_{tot} 0.62W/(m²·K)					
		(north)					

Table 2. Simulation input parameters.

Schedules		Internal gains		HVAC systems			Daylighting
Internal gains	Ventilation	Occupancy	Lighting / Equipment	Temp. setpoint	Supply air temperature	CAV air exchange	Reflectance values (%)
00:00-07:00 – 0.0	00:00-08:00 – 0.036	35W/m²	5.0W/m²	+21°C	>+16°C	4.2	Walls 50
07:00-17:00 – 1.0	08:00-12:00 – 0.8	2.1m²/occ.	12.0W/m²	...	(without	l/(s·m²)	Floor 20
17:00-00:00 – 0.0	12:00-13:00 – 0.5	1.0 MET		+25°C	cooling)		Ceiling 70
	13:00-16:00 – 0.8	0.85±0.25 CLO				idle 0.15	Shading 35
	16:00-00:00 – 0.036					l/(s·m²)	Ground 20

2.1. *Overheating and indoor climate class assessment*

The overheating assessment was done according to Estonian Building Code regulations. Indoor temperature simulations were conducted with well validated building simulation software IDA ICE [14] (Figure 1). Hourly-mean indoor temperature values were used to calculate temperature excess (DH) over a set base temperature. The allowed cumulative temperature excess in case of classrooms is 100Kh and the base temperature 25°C. The simulation periods for school buildings are set from 1st of May to 15th of June and 15th of August to 30th of September. For outdoor climate input, Estonian test reference year is used, which is based on 30 year measurement data consisting of outdoor air temperature, relative humidity, wind velocity, direct and diffuse solar radiation [15].

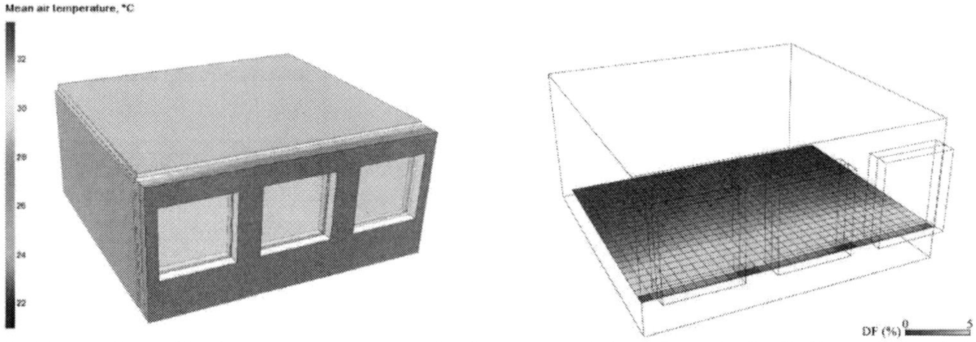

Figure 1. Examples of indoor temperature calculation model in IDA ICE (left) and daylight factor calculation model in Grasshopper using DIVA4 (right).

Aside from the overheating intensity assessment, we calculated the cumulative hours for the cooling period during which the room temperature was in bounds of specific thermal environment class according to the standard EVS-EN 15251 'Indoor environmental input parameters for design and assessment of energy performance of buildings addressing indoor air quality, thermal environment, lighting and acoustics' [16].

2.2. Daylight factor simulations

The parametric model of the classroom was built using the software Grasshopper for Rhinoceros and the analysis was carried out with daylighting design plug-in DIVA4 (Figure 1), which performs simulations through validated software Radiance [17]. Through the daylight analysis parametric model it is possible to assign reflectance (R) values to interior elements of the room (i.e. floor, wall, ceiling, external shading and ground) and visible transmittance (VT) values to the glazed surface of windows, set the simulation grid, select the simulation parameters, run the simulations and record result data. The R values used in the simulations were the same for all the classroom variations and are standard values recommended for Daylight Factor calculations, presented in Table 2. The daylight parametric model permits to associate in different ways the glazing VT values and use or not of the shading device. This procedure has been necessary to match the room variation parameters used for energy efficiency studies. Different combinations of glazing VT and use of shading has been used for the different orientations in consideration of Estonian overheating requirements. Because Daylight Factor analysis do not take into account windows orientation, daylight simulation combinations refer solely to glazing VT and use of shading. The combinations are presented in Table 1. The grid used for the simulation has a size of 0.2m, was located at 0.75m from the floor and occupies 80% of the floor area. The main Radiance parameters used in the simulations are: -aa .1 -ab 5 -ad 1024 –ar 256. As required for DF simulation the CIE overcast sky model was used. The Daylight Factor simulations were performed automatically for all the classroom size and parameters variations through an automation function of the parametric model and the values of mDF were recorded for each iteration.

3. Results and discussion

Results of overheating and daylight factor are presented for different orientations due to the different glazing g-value, VT and shading described in methods section. For each orientation a figure is composed, how rooms with different WFR and dimensions respond to requirements studied. Rooms are ordered by the decreasing value of WFR, cumulative time is shown firstly, overheating secondly and daylight factor results thirdly. The color-coded cumulative graph shows duration in percentages during which the hourly room temperature values stayed between the limits of a specific indoor climate class (IC), ranked from I (best) to IV (worst) according to the standard [16]. Overheating hours and mean daylight factor values are marked as green squares, if both criteria are met and as red if one or both of the criteria do not meet the requirements. Room result figures are divided to left and right by shading use.

For east orientation (Figure 2) rooms without shading and with WFR over 0.23 are overheated and rooms with width of 5m also do not meet the overheating requirement. Same rooms gain 1 to 2% more time out of II IC class compared to rooms, which stay below 100Kh line. All the rooms without shading are well lighted as mDF is over 2%. It is seen, that as the WFR increases and floor plan has more width and less depth, classrooms are both more naturally lighted and overheated. If shading is added and glass g-value increases from 0.35 to 0.42, room air temperature hours in III and IV IC class decrease up to 3%. Most of the rooms are underneath the overheating requirement line, only half of the rooms meeting the daylight factor criteria. Only 6 rooms, compared to 10 in the initial situation, of 25 met both criteria.

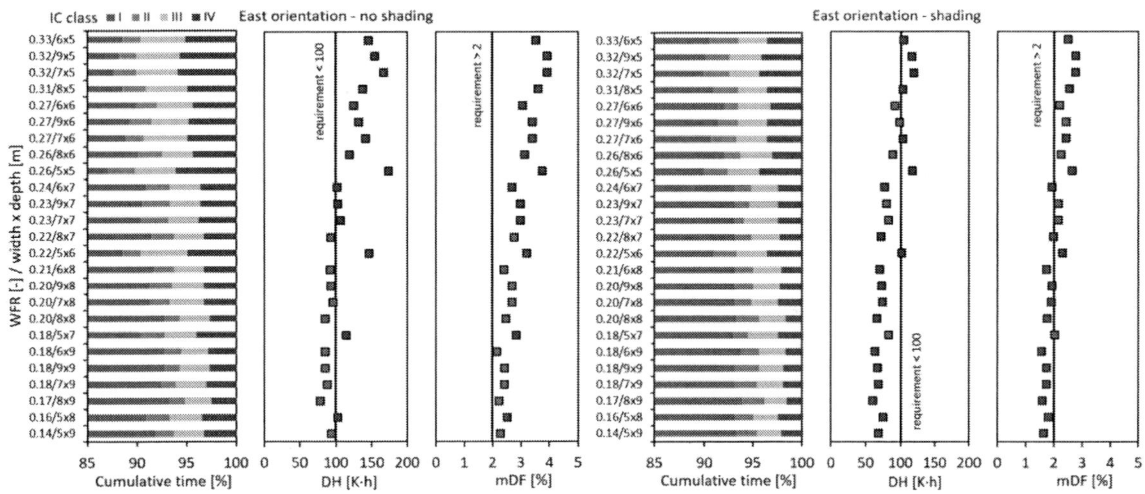

Figure 2 Simulation results for east oriented classrooms: indoor climate class (IC) cumulative time during cooling period, temperature excess (DH) and mean daylight factor (mDF) without (left) and with (right) shading.

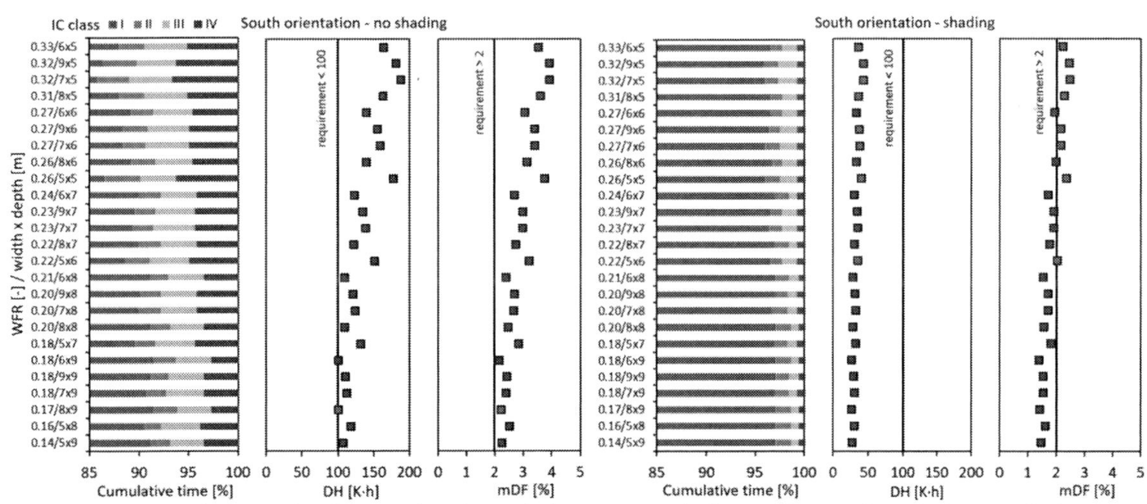

Figure 3 Simulation results for south oriented classrooms: indoor climate class (IC) cumulative time during cooling period, temperature excess (DH) and mean daylight factor (mDF) without (left) and with (right) shading.

IOP Conf. Series: Earth and Environmental Science **352** (2019) 012059 doi:10.1088/1755-1315/352/1/012059

In south orientation (Figure 3) without shading all rooms basically are overheated and properly naturally lit. Up to 10% of the time, air temperature in classrooms does not meet II IC class. After adding shading, only 2 to 3% of the time room air temperature is out of II IC class. While all the rooms now meet the overheating criteria, only 8 rooms of 25 have mDF over 2%.

For west orientation (Figure 4) 20 of 25 rooms meet both criteria without cooling, while room air temperature varies from 5 to 8% out of II IC class. Adding horizontal shading and optimized g-value of 0.42 similarly to east, all the rooms are underneath the overheating criteria, while 13 rooms with higher WFR do not meet daylight criteria. Classroom air temperatures IC classes for being out of II class also decreases between 3 to 5% of the time.

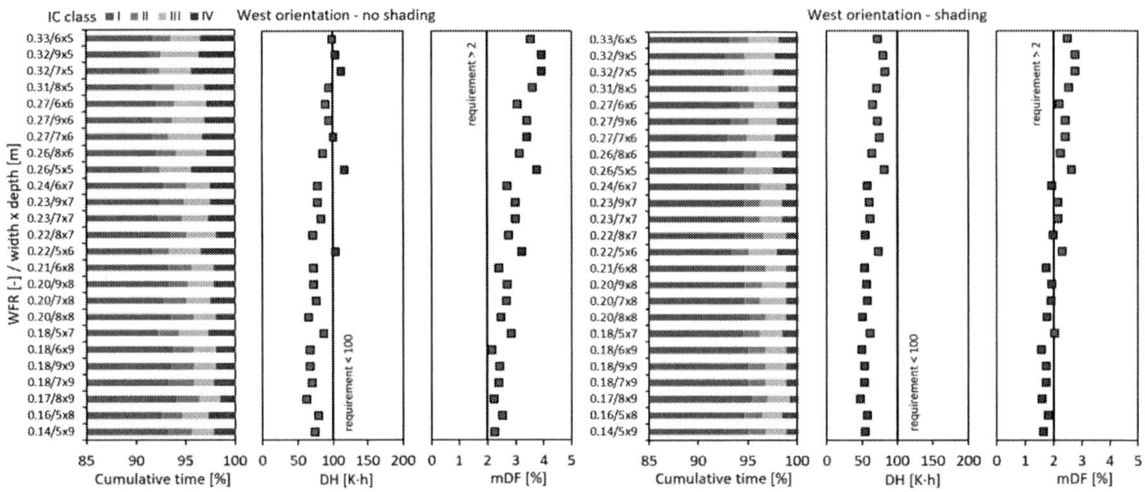

Figure 4 Simulation results for west oriented classrooms: indoor climate class (IC) cumulative time during cooling period, temperature excess (DH) and mean daylight factor (mDF) without (left) and with (right) shading.

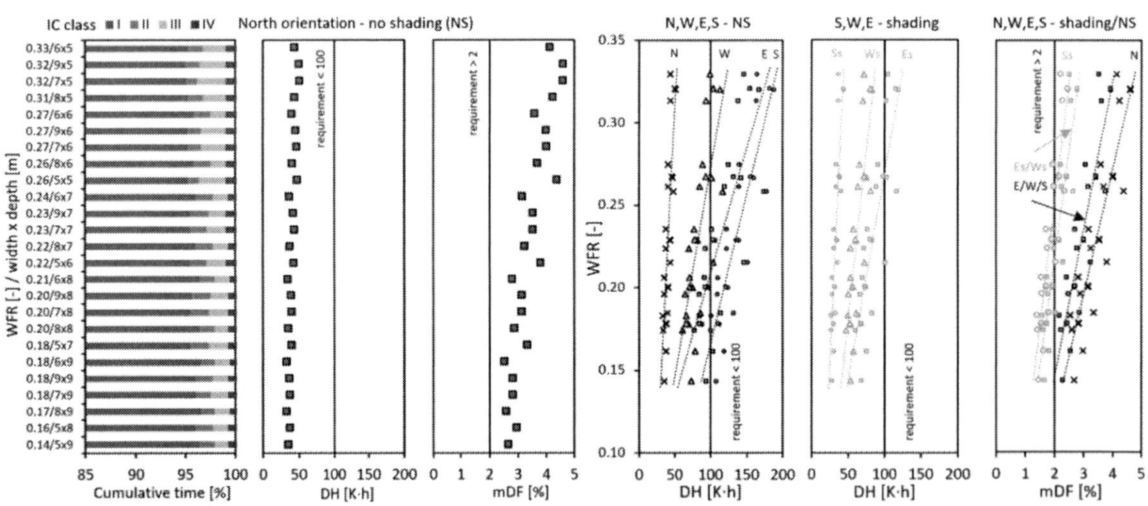

Figure 5 Simulation results for north oriented classrooms: indoor climate class (IC) cumulative time during cooling period, temperature excess (DH) and mean daylight factor (mDF) without shading (left) and for all facades, WFR correlation with DH and mDF (right).

As it is unnecessary to block direct sunlight on north façade (Figure 5), room results are presented only for the initial situation without shading. It is seen by the green squares that all the rooms meet both overheating and daylight factor criteria. Rooms with higher WFR have 1 to 2% more cumulative hours out of II IC class. Room DH values are more constant compared to higher mDF as the WFR increases.

For east, south and west orientations, rooms with wider width and shorter depth dimensions received more daylight. For north oriented facade, all the analysed cases fulfilled the overheating and daylight requirements. IC class percentages indicate that room air temperature is mainly affected by internal gains of students, electrical equipment, lighting and supply air. Results are shaped less from the direct sunlight and as the WFR increases, more diffuse lighting enters the room. Similar distribution of results is seen on south façade with shading, but the balance gap between DH and mDF is clearly smaller for maintaining both criteria requirements. On the east and west façade results are more spread out, but still parallel as WFR increases. The room air temperature is less time out of II IC class for all south, east and west orientations if shading is added to the windows of the classrooms. Right side of Figure 5 with all the facades and simulated classrooms together shows why DH and mDF should be calculated together during the building design process.

4. Conclusion

The aim of this study is to determine whether the school building classrooms could be designed without active room cooling units and cooled ventilation supply air. Passive cooling methods, like decreasing window glass g-value and external shading decrease the amount of sunlight into the rooms, may cause poorly conditions for natural lighting as a result. Therefore, both overheating and daylight parameters must be analysed jointly. Results show that as window-to-floor ratio increases, the room receives more daylight but also becomes more vulnerable to temperature rise and overheating. In the other hand, with increasing depth, overheating risk lowers and daylight level decreases.

Parametric study shows that horizontal shading is more helpful on the southern façade. Adding shading to eastern and western facades with modified window parameters, distribution of classrooms meeting both temperature excess and mean daylight factor requirement changes and for west it also decreases. The easiest balance between two criteria is on the north façade due to low amount of direct sunlight. Adding shading reduces the number of hours out of indoor climate class II, while temperature excess method illustrates more efficiently the intensity of overheating. In addition, temperature excess overheating method results correlates well with daylight result distribution.

Designing low-energy school buildings without active room cooling units or cooled mechanical supply air ventilation, facade design is critical to ensure thermal comfort and lighting conditions that directly affect students' performance. As school buildings are not used during summertime, it is possible to design classrooms to meet both overheating and daylighting requirements without the need for mechanical cooling systems. However, proper design requires skillful analysis of suitable combination of room dimensions, window sizes, glazing parameters and shading options to meet both overheating and daylight requirements.

5. Acknowledgements

This research was supported by the Estonian Centre of Excellence in Zero Energy and Resource Efficient Smart Buildings and Districts, ZEBE, grant 2014-2020.4.01.15-0016 funded by the European Regional Development Fund. The authors are grateful for the valuable help from architectural and urban studies prof. Rein Murula from Tallinn University of Technology, School of Engineering.

References

[1] Hathaway W E 1995 Effects of School Lighting on Physical Development and School Performance *J Educ Res* **88** 228-42
[2] Mendell M J and Heath G A 2005 Do indoor pollutants and thermal conditions in schools influence student performance? A critical review of the literature *Indoor Air* **15** 27-52
[3] Wargocki P and Wyon D P 2007 The effects of moderately raised classroom temperatures and classroom ventilation rate on the performance of schoolwork by children (RP-1257) *Hvac&R Res* **13** 193-220

IOP Conf. Series: Earth and Environmental Science **352** (2019) 012059 doi:10.1088/1755-1315/352/1/012059

[4] Sleegers P J C, Moolenaar N M, Galetzka M, Pruyn A, Sarroukh B E and van der Zande B 2013 Lighting affects students' concentration positively: Findings from three Dutch studies *Lighting Research & Technology* **45** 159-75
[5] Zomorodian Z S, Tahsildoost M and Hafezi M 2016 Thermal comfort in educational buildings: A review article *Renew Sust Energ Rev* **59** 895-906
[6] Delvaeye R, Ryckaert W, Stroobant L, Hanselaer P, Klein R and Breesch H 2016 Analysis of energy savings of three daylight control systems in a school building by means of monitoring *Energ Buildings* **127** 969-79
[7] Nasrollahi N and Shokri E 2016 Daylight illuminance in urban environments for visual comfort and energy performance *Renew Sust Energ Rev* **66** 861-74
[8] Gago E J, Muneer T, Knez M and Koster H 2015 Natural light controls and guides in buildings. Energy saving for electrical lighting, reduction of cooling load *Renew Sust Energ Rev* **41** 1-13
[9] Wargocki P, Wyon D P, Matysiak B and Irgens S 2005 The effects of classroom air temperature and outdoor air supply rate on the performance of school work by children *Indoor Air 2005: Proceedings of the 10th International Conference on Indoor Air Quality and Climate, Vols 1-5* 368-72
[10] McLeod R S and Swainson M 2017 Chronic overheating in low carbon urban developments in a temperate climate *Renew Sust Energ Rev* **74** 201-20
[11] Simson R, Kurnitski J and Maivel M 2017 Summer thermal comfort: compliance assessment and overheating prevention in new apartment buildings in Estonia *J Build Perform Simu* **10** 378-91
[12] EVS 2015 EVS 894:2008/A2:2015. Daylight in dwellings and offices. Estonian Centre for Standardisation)
[13] 2013 Estonian Regulation No 63: Methodology for calculating the energy performance of buildings. In: *Riigi Teataja*: Ministry of Economic Affairs and Infrastructure)
[14] Kropf S and Zweifel G 2002 Validation of the Building Simulation Program IDA-ICE According to CEN 13791 „Thermal Performance of Buildings - Calculation of Internal Temperatures of a Room in Summer Without Mechanical Cooling - General Criteria and Validation Procedures" *HLK Engineering*
[15] Kalamees T and Kurnitski J 2006 Estonian test reference year for energy calculations *Proceedings of the Estonian Academy of Sciences. Engineering* **12** 40-58
[16] 2007 EN 15251:2007. Indoor environmental input parameters for design and assessment of energy performance of buildings addressing indoor air quality, thermal environment, lighting and acoustics. Brussels: CEN (European Committee for Standardization))
[17] Reinhart C F and Walkenhorst O 2001 Validation of dynamic RADIANCE-based daylight simulations for a test office with external blinds *Energ Buildings* **33** 683-97

IOP Conf. Series: Earth and Environmental Science **352** (2019) 012060 doi:10.1088/1755-1315/352/1/012060

Co-Creation of Positive Energy Blocks

Dirk Ahlers, Patrick Driscoll, Håvard Wibe, Annemie Wyckmans

NTNU – Norwegian University of Science and Technology,
Department or Architecture and Planning,
Trondheim, Norway

E-mail: {dirk.ahlers|patrick.arthur.driscoll|havard.wibe|annemie.wyckmans}@ntnu.no

Abstract.

A main challenge in building carbon-neutral built environments is the ability to scale and replicate solutions. We examine how to develop low-carbon neighbourhoods and districts, while aiming at climate-friendly and sustainable livable urban environments. We take a view that not only scales up individual building solutions, but embraces the added complexities arising from the scale change and utilizes them for a novel approach. It includes a strong focus on co-creation and open innovation to develop sustainable solutions.

In this contribution, we present the approach of the +CityxChange project in implementing Positive Energy Blocks (PEB) through a European H2020 project from the topic of Smart Cities and Communities. A PEB comprises several connected buildings that have a averaged yearly positive energy balance between them. This definition excludes embodied emissions, but allows to focus on the infrastructure and systems between buildings as part of the built environment, and ways to implement and incorporate them within existing cities. The +CityxChange approach relies on co-creating Europe-wide deployment of Positive Energy Districts, with Integrated Planning and Design, Creation of a Common Energy Market, and CommunityxChange with all stakeholders of the city.

1. Introduction

A main challenge in carbon-neutral built environments is the ability to involve all stakeholders for sufficient buy-in and contributions to ensure well-aligned development and uptake of agreed solutions. The challenge goes hand in hand with the ability to scale up and replicate solutions out of successful small pilot deployments to become part of larger city developments. In short, successful design, development, and roll-out of smart city solutions requires cooperation of all involved stakeholders, for which we follow a quadruple helix ecosystem model. We follow this approach with a strong focus on cross-disciplinary collaboration and a co-creation approach that strongly includes cities and citizen.

In this contribution, we present the approach of the +CityxChange project in implementing Positive Energy Blocks (PEB) through a European H2020 project in the topic of Smart Cities and Communities. A PEB is a local neighbourhood-level implementation and is understood as a collection of at least 3 buildings in close proximity that have a averaged yearly positive energy balance between them achieved through a range of measures discussed in detail below. PEBs are supposed to be scaled up subsequently to net-Positive Energy Districts. The definition of PEBs so far excludes embodied emissions, but allows to focus on the infrastructure and systems between buildings as part of the built environment, and ways to implement and incorporate them within existing neighbourhoods and cities. In addition, the project aims to ensure that solutions are anchored and agreed by relevant stakeholders and citizens and thus engages them in all phases through open collaboration and co-creation processes. A PEB can thus be understood as an answer to the question of how to integrate current and future plus-energy buildings into the larger urban fabric.

We examine how to scale up from buildings to blocks, and further to low-carbon neighbourhoods and districts, while aiming at climate-friendly and sustainable livable urban environments. Our approach is part of emission reduction efforts and plus-energy building integration to reach the Paris Climate Goals. To this end, the project is piloting actionable, scaleable, and replicable solutions toward the local city-driven energy transition as part of the overall European Energy Transition by 2050.[1] Furthermore, the cities in the project have the ambition to achieve sustainable urban ecosystems that have zero emissions and to establish a 100% renewable energy city-region by 2050. The project is also contributing to the SET Plan[2] goal of 100 PEBs by 2025 in Europe[3], further supporting the ambition to replicability and scalability.

We aim for an approach that not only scales up individual building solutions, but embraces the added complexities arising from the scale change to districts or connected blocks and develops a novel approach for this challenge.

The +CityxChange vision is to enable the co-creation of the future we want to live in. To fill the slogan with life, it relies on co-creating Europe-wide deployment of Positive Energy Districts, with Integrated Planning and Design, Creation of a Common Energy Market, and CommunityxChange with all stakeholders of the city. It includes the development of frameworks, supporting tools, pilot deployments, and replication plans. This is reflected through 3 broad themes of the project:

Enabling a common energy market to translate the European Common Energy market to very local conditions, allow buildings to exchange and trade energy with their neighbours, and to integrate e-Mobility as a Service (eMaaS) to both combine the mobility and energy systems and to prototype new modes of transportation.

Creating connected communities that are enabled to be actors in the energy systems and markets, can directly benefit from the results, and both co-create the solutions and can act as market participants, democratising the energy system.

Recommendations for new policy interventions, market (de)regulations and business models to deliver positive energy communities to ensure learning can be shared and new framework conditions can be taken up. This also examines local regulations to achieve development within legal frameworks or special dispensation.

2. Positive Energy Blocks and Districts

The initial definition of a PEB is as a collection of buildings in close proximity that have a averaged yearly positive energy balance between them. A Positive Energy Block is defined by the EU as several buildings (new, retro-fitted or a combination of both) that actively manage their energy consumption and the energy flow between them and the wider energy system. They achieve an averaged annual positive energy balance between them. This happens through use, optimization, and integration of advanced materials, energy reduction, local renewable energy production and storage, smart energy grids, demand-response, energy management of electricity, heating, and cooling, user involvement, and ICT. PEBs/PEDs are designed as an integral part of

[1] 2050 Energy Strategy, European Commission
`https://ec.europa.eu/energy/en/topics/energy-strategy-and-energy-union/2050-energy-strategy`
[2] European Strategic Energy Technology Plan (SET-Plan) Action 3.2 on Smart Cities and Communities
[3] `https://setis.ec.europa.eu/system/files/setplan_smartcities_implementationplan.pdf`

IOP Conf. Series: Earth and Environmental Science **352** (2019) 012060 doi:10.1088/1755-1315/352/1/012060

the district energy system. They should be intrinsically scalable up to positive energy districts and cities and are well embedded in the spatial, economic, technical, environmental and social context, as defined by the EU call[4].

We are working to refine these definitions to be more actionable for the different cities we are working with and to take local conditions into account. Especially boundary and framework conditions and evaluation criteria are of interest.

The PEB definitions excludes embodied emissions and only focuses on energy (electricity and heating/cooling). This allows it to focus on the infrastructure and systems between buildings as part of the built environment, and ways to implement and incorporate them within existing neighbourhoods and cities. However, other definitions and buildings developed to other existing standards can be integrated, such as plus-energy buildings, low carbon buildings, zero energy buildings, etc. With the focus on the energy exchange, a PEB can be understood as a possibility to integrate the surplus energy from plus-energy buildings into the larger urban fabric.

To successfully establish a PEB, buildings with a high standards are preferred and can be an integral component. However, not all buildings in a block need to be energy-positive. An advantage of the PEB definition and development is that it can integrate existing building stock and both show an upgrade path and allow non-discriminatory integration.

As an example, within a local energy trading market or community grid, a first step is to upgrade buildings, reduce energy demand, and increase self-consumption, before balancing between buildings and energy sources and between different temporal load profiles. With increased local production, higher performing buildings can support lower performing ones, which may only be moderately upgraded. At one point, the yearly balance will turn positive and the block is on average making energy available to the larger grid. This can also be a trigger to expand the PEB. By integrating mobility aspects into the energy considerations, the energy surplus can also power local mobility needs.

3. Project Background

The +CityxChange project[5] is developing and deploying Positive Energy Blocks and Districts (PEB/PED) and scaling these out as part of the European Clean Energy Transition in cities. The project is funded by the EU H2020 Smart Cities and Communities topic SCC-1.[4] The call centers on the sustainable energy transition in cities that should realize Europe-wide deployment of Positive Energy Districts by 2050 through highly integrated and highly efficient innovative energy systems. Of interest are the direct technical solutions around PEBs, but also the interaction and integration between buildings, users, cities, the larger energy system, as well as the implications and impact on city planning, city systems, energy trading, citizen involvement, regulations, big data, digitalization, and socio-economic issues. [1]

32 partners, including 7 cities and industry and research partners will co-create these future energy systems within the 5 year project duration. It follows an integrative approach with a strong focus on city integration, open innovation and replication ability. Solutions and demonstration projects will be first piloted in the two Lighthouse Cities (Limerick, Ireland and Trondheim, Norway). They will later be replicated in five Follower Cities (Alba Iulia, Romania; Sestao, Spain; Písek, Czech Republic; Smolyan, Bulgaria and Võru, Estonia) throughout Europe to show replicability in different environments and to refine the solutions to local conditions.

Figure 1 describes the approach and framework that supports the project vision and goals through three integrated action areas:

[4] Smart Cities and Communities H2020-LC-SC3-2018-2019-2020, `https://ec.europa.eu/info/funding-tenders/opportunities/portal/screen/opportunities/topic-details/lc-sc3-scc-1-2018-2019-2020`

[5] Positive City exChange `http://cityxchange.eu/`

IOP Conf. Series: Earth and Environmental Science **352** (2019) 012060 doi:10.1088/1755-1315/352/1/012060

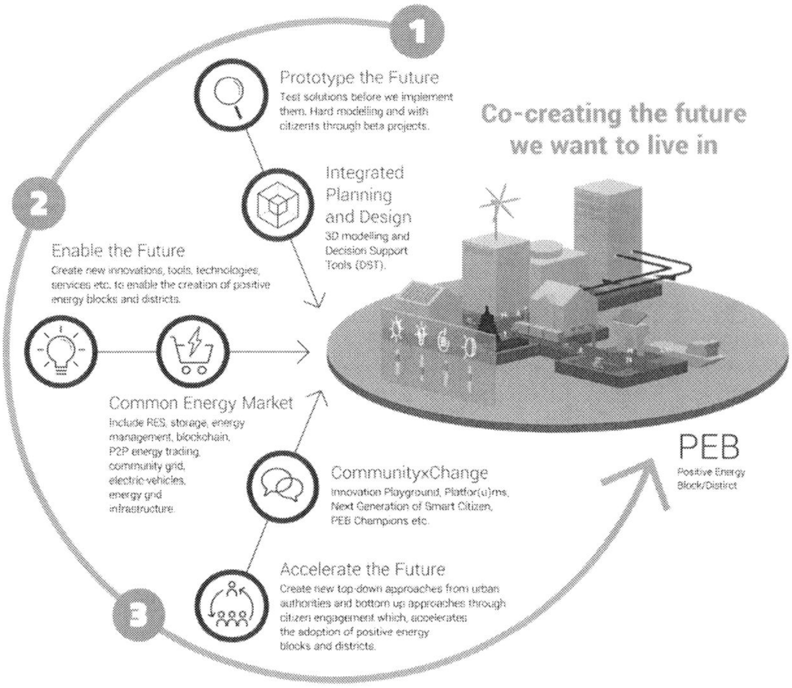

Figure 1. +CityxChange integrative approach and framework for the development of Positive Energy Blocks

- Prototype the Future – Integrated Planning and Design of Cities;
- Enable the Future – Creation and Enabling of a Common Energy Market;
- Accelerate the Future – CommunityxChange with all stakeholders of the city to create connected and engaged communities.

Within these, 11 demonstration projects in climate-friendly and sustainable urban environments are developed:

- Integrated Modeling
- Bold City Vision
- Co-Creation and Citizen Engagement
- Regulatory mechanisms and sandboxes
- Innovation Playgrounds
- Distributed Positive Energy Blocks (DPEBs)
- Micro-Grid Optimisation
- e-Mobility as a Servide
- Local Energy Trading Market
- Local Flexibility Market
- Business, Investment, and Risk Models

+CityxChange implements these 11 Demo Project in its 2 Lighthouse Cities and 5 Follower Cities. They are designed to enable the co-creation and development of Positive Energy Blocks.

The project structure mandates to take the local conditions of the different cities into account for the development, increasing broad applicability, replicability, and knowledge transfer.

We believe that citizens can and have to play a central role in the energy transition. Enabling local co-creation and collaboration for the PEBs is thus an important goal. There is a strong collaborative and strategic anchoring within the cities and participating partners through internal processes. However, reaching to stakeholders and citizens outside the project is a challenge. Experimental approaches as well as established mechanisms need to be followed to understand better local needs, and develop tailored mechanisms for engagement of citizens and stakeholders. Aspects to be trialled include citizen observatories, innovation playgrounds, hackathons, and regulatory sandboxes linked to urban living labs [2], and strategic Bold City Visions to engage civil society, local authorities, industry, and RTOs to scale up from PEBs to PEDs in an open innovation process.

4. Co-Creation and Open Innovation

Smart cities projects are understood as inherently complex and smart city evolution as an ongoing process. Based on insights around organisational and governance complexity, it is considered vital to open up processes to stakeholders in and around the city. To translate this into action, we follow an open innovation approach [3, 4, 5] and employ a quadruple helix participation model of collaboration [6]. Partners in the quadruple helix case are cities and public bodies, industry and business, research and academia, and citizens and civic society as the four main pillars. The addition of civil society is a major change from previous double or triple helix approaches. In addition, approaches of integrated innovation ecosystems move the focus from individual stakeholders to the ecosystem as a whole.

Within our framework, we aim to balance between these communities, between private and professional stakeholders, and to enable all to participate and build upon the ecosystem of smart city projects with lower entry barriers, as well as to benefit from open systems and processes. This is achieved through a range of measures, from co-creation of the project proposal, over citizen and stakeholder participation in the project, open innovation frameworks, open data architectures, and opening up large amounts of information and results, both for transfer of knowledge and experiences between cities and solutions providers within the project, and with external stakeholders. To facilitate dissemination and replication, the project assumes an "open by default" approach in line with the Open Innovation ambition. Almost all Deliverables are public and as much material, documentation, systems, and data as possible will be made available through respective channels.

On the IT side, Open data models and API specifications (Application Programming Interface, a set of functions, protocols and tools for interconnected software) have been agreed by partners where possible, not only for the internal purpose of the project, but also for replicability and portability of the services beyond the project. An API-driven approach helps ensuring interoperability and independence between existing solutions by keeping individual systems sufficiently distributed and focusing on the structured interaction between them. Privacy and GDPR issues are considered throughout the project. This is integrated through an open ICT ecosystem architecture that supports openness, access, and stakeholders' awareness [1].

During the proposal stage, the +CityxChange consortium already adopted an Open Innovation 2.0 framework [3] to secure involvement and co-creation of the partners and key stakeholders in different sectors for a joint and integrated project approach, including cities, citizens, industry partners, academia as part of the mentioned quadruple helix approach. The framework is being continued and further developed into the operational activities of the +CityxChange project and beyond [7].

To give some background, the term "open innovation" was originally defined as "the use of purposive inflows and outflows of knowledge to accelerate internal innovation, and expand the markets for external use of innovation, respectively. Open innovation is a paradigm that assumes that firms can and should use external ideas as well as internal ideas, and internal and external paths to market, as they look to advance their technology" [4, p.vii]. As a next step, Open Innovation 2.0 is then based on the principles of "integrated collaboration, co-created shared value, cultivated innovation ecosystems, unleashed exponential technologies, and rapid adoption due to network effects" [3, p.1]. It is used by the European Commission, amongst others, as a policy that contributes to the EU digital single market[6]. Its five key points are Networking; Collaboration by involving partners, competitors, universities, and users; Corporate Entrepreneurship by enhancing corporate venturing, start-ups and spin-offs; Proactive Intellectual Property Management by creating new markets for technology; Research and Development (R&D) for achieving competitive advantages in the market.

These are also integrated as key factors in our framework. In addition, +CityxChange is centred around agile, rapid and multidisciplinary experimentation by the quadruple helix ecosystem. This type of experimentation is firmly embedded within Open Innovation 2.0 and is considered a main success factor [7]. It helps to work with cases of participatory processes, where the outcome is not set yet, by being able to perform and evaluate experiments, and integrate the results of these with the broader project scope.

In addition to the overall integrated approach, a number of particular interventions and actions are built into the project to support this. A main part is a Bold City Vision, to enable each city to bridge global and local strategies with local interventions and developments. Derived from that, individual actions such as Citizen Observatories, Innovation Playgrounds, Investment Models perform tasks of citizen engagement and participation, connect local citizens and existing innovation actors, build different publicly available observatories as physical and virtual platforms and spaces for exchange, and make systems and data available through for example hackathons and open data portals to ensure a variety of technical and community-oriented interactions.

Along with agile experimentation, mechanisms need to be in place to keep open to new solutions in a fast changing research and industry environment such as smart cities. Engagement with external stakeholders as discussed above can tap into knowledge and innovation that can and will happen outside of the project and combine internal and external ideas. This links to creating a balance between exploratory and exploitative innovation [8]. While exploitative innovation typically builds on existing knowledge and activities in more structured processes, such as a project framework, explorative innovation, is close to innovation processes, such as were happening during the development of the project. Bridging this gap is an interesting challenge, which will be addressed by creating spaces and opportunity for elements of serendipity and supporting activities along the overall collaboration.

Technological innovation is a necessary condition to make a city smart. However, the challenge of smart city innovation is not only on technology, but on service transformation, integration, and improvement. A strong involvement of citizens and non-governmental associations, and the diffusion of innovative models of cooperation and social relationships are necessary [7]. The discussed methods and measures enable this involvement and will be further evaluated and refined as the project progresses.

5. Conclusion

In this contribution, we have given an overview of the +CityxChange project and of Positive Energy Blocks and Districts as a way to enable a city-driven energy transition within smart cities. Based on the PEB development needs we have discussed our framework and approach for co-creation and open innovation as a key success factor in reaching complex project goals

[6] `https://ec.europa.eu/digital-single-market/en/open-innovation-20`

through multi-actor collaboration.

Successful development and deployment of Positive Energy Blocks/Districts requires open innovation, combining knowledge and experience of a wide range of different actors in a quadruple helix ecosystem. Furthermore, managing and facilitating complex PEB/PED and other smart cities projects requires new forms of governance integrating the collaboration of various actors as a key principle. We have chosen open innovation as a relevant pathway for cities to become smarter, solve various societal challenges and improve the quality of life of its citizens by stimulating and facilitating the synergistic contribution and participation of business, university, public sector, non-for-profit partners and citizens to innovation processes.

In our ongoing work, we will refine the frameworks and processes of co-creation and collaboration as well as support the detailing out of the actual development towards the deployment of Positive Energy Blocks. All project partners are currently working on all these tasks and are showing promising initial results.

Acknowledgements

+CityxChange (Positive City ExChange) is a smart city project under the Smart Cities and Communities topic that has been granted funding from the European Union's Horizon 2020 research and innovation programme under Grant Agreement No. 824260. This paper includes work that has been performed before the project start, as part of project development and co-creation with city and partner visions and approaches. It further includes information from project deliverables and from internal project documentation.

We thank our colleagues and the whole +CityxChange project team and project preparation team of all cities and partners for valuable discussions and contributions.

References

[1] Ahlers D, Wienhofen LWM, Petersen SA, Anvaari M. A Smart City Ecosystem Enabling Open Innovation. In: 19th International Conference on Innovations for Community Services (I4CS2019). vol. 1041 of Communications in Computer and Information Science. Springer; 2019. p. 109–122.

[2] Mccormick K, Hartmann C. The Emerging Landscape of Urban Living Labs: Characteristics, Practices and Examples; 2017. Governance of Urban Sustainability Transitions (GUST) Deliverable 3.3.2.

[3] Curley M, Salmelin B. Open Innovation 2.0: The New Mode of Digital Innovation for Prosperity and Sustainability. Springer; 2018.

[4] Chesbrough HW. Open innovation: The new imperative for creating and profiting from technology. Harvard Business Press; 2003.

[5] Chesbrough HW, Vanhaverbeke W, West J. Open Innovation: Researching a New Paradigm. Oxford University Press; 2006.

[6] Carayannis EG, Campbell DF. 'Mode 3' and 'Quadruple Helix': toward a 21st century fractal innovation ecosystem. International Journal of Technology Management. 2009;46(3-4):201–234.

[7] Wyckmans A, Vandevyvere H, Gohari S, Nielsen BF, Driscoll P, Ahlers D. D9.1: Framework for intra-project collaboration. +CityxChange Project; 2019. +CityxChange Deliverable 9.1. To appear.

[8] Jansen JJ, Van Den Bosch FA, Volberda HW. Exploratory innovation, exploitative innovation, and performance: Effects of organizational antecedents and environmental moderators. Management Science. 2006;52(11):1661–1674.

IOP Conf. Series: Earth and Environmental Science **352** (2019) 012061 doi:10.1088/1755-1315/352/1/012061

Lia Kindergarten – A plus energy building: First year experience with regard to energy use

Tor Helge Dokka*[1], Maria Myrup[1], Morten Solsem[2]

[1]Skanska Norway
P.O. Box 1175, Sentrum
N-0107 Oslo, Norway
[2]ABK AS Postboks 64 Vollebekk, 0516 OSLO

*Corresponding author: tor.helge.dokka@skanska.no

Abstract. Lia kindergarten is designed to be a plus energy building. The kindergarten was opened in January 2018, and the first year has been a tune-in phase for the technical systems. The plus energy concept consist of a well insulated building envelope, a highly efficient ventilation system, a lighting system with LED-fixtures and advanced control system and a low exergy thermal energy supply based on geothermal wells and a small inverter heat pump. The energy demand is balanced with a solar PV-system on the roof to achieve the plus energy definition. A lithium-ion battery package of approximately 40 kWh has been installed for peak shaving and to utilize more of the solar energy produced on the building (reducing mismatch).

A lot of research, calculations and simulations has been done in the design phase for the low exergy system. The floor concrete slab with embedded heating (pex pipes) is designed for very low temperature heating (22-28 ^{0}C). The floor system is reversed in the summer providing high temperature cooling (18-19 ^{0}C) which also charge the geothermal wells for the heating season. The calculations and simulations has also been used to make control strategies based on model predictive control (MPC) methodology, also taking in on-line weather forecast.

1.0 Introduction

Lia kindergarten has a heated floor area of 1579 sqm and is designed as a plus energy building according to the Futurebuilt definition [1]. It is a two storey building with a flat roof, and the form of the building is basically rectangular as shown in figure 1. The kindergarten is divided into 10 departments, each having a base room approximately 30 sqm large.

Figure 1. Lia kindergarten, picture taken from the south side of the building.

IOP Conf. Series: Earth and Environmental Science **352** (2019) 012061 doi:10.1088/1755-1315/352/1/012061

1.1 Design intent

The energy- and indoor climate concept to achieve the plus energy ambition is based on experience from earlier plus energy project like Powerhouse Kjørbo [2], but with some modification. The concept have main focus on reducing the energy demand as much as possible, without sacrificing good indoor climate, before balancing this with a cost effective PV-system. Main measures in the concept is:

- A highly insulated building envelope, with high performance windows and a low air leakage number
- Ventilation is provided by a low pressure demand controlled ventilation system based on the displacement ventilation principle. The AHU is equipped with a rotary wheel heat exchanger and an internal heat pump which heats and cools the supply air when necessary.
- Lighting is provided by a modern LED-lighting system with presence- and daylight control system, giving a very low energy system.
- Room heating and room cooling is provided with a so-called "Lowex-system", meaning that low temperature heating and high temperature cooling is done with an embedded floor system. Geothermal wells together with a heat pump provide heating and DHW, while free cooling via the geothermal wells provide cooling in the summer months, which also recharge the wells before the heating season. The low temperature heating and high temperature cooling should potential give a system with very high coefficient of performance (COP) and energy efficiency ratio (EER).
- A flat roof PV-system with modules facing East and West with an angle of 10^0 is designed so the building reach the plus energy ambition.

Main design data for the plus-energy kindergarten is given in table 1. The building is close to the requirements in the Norwegian passive house standard NS3701 [3], but deviate on a couple of the minimum requirements given in the standard.

Table 1. Main design data for the plus-energy kindergarten

Component	Value
U-value external wall	0.17 W/m^{2}K
U-value roof	0.19 W/m^{2}K
U-value external wall	0.10 W/m^{2}K
U-value windows and doors	0.79 W/m^{2}K
Air leakage number	0.50 ach
Specific fan power	0.70 kW/(m^3/s)
Annual energy use lighting*	6.1 kWh/m^2
Annual energy use plug loads	5.2 kWh/m^2
SCOP heat pump DHW	3,0
SCOP heat pump heating	5,7
SEER free cooling system	60
Geothermal wells	3 wells a' 280 m
Peak power PV-system	47,2 kWp

2. Methods

This section describes briefly the simulation tools used in the design of the building, and also the different measurements and monitoring tools used to evaluate the real performance of the building.

2.1 Simulation tools

The energy building simulation tool Simien 6.0 [4] has been used to simulate design heating and cooling load, power demand, net energy demand and delivered energy use. It has also been used to simulate the PV-production and mismatch between production and demand based on a time step of 15 minutes (an electric balance with a resolution of 15 min).

The PV-simulation tool PV-sol [5] was used by the supplier of the PV-system to simulate the monthly and annual PV-production. Results is not shown in this paper, but the agreement with the Simien simulation was good.

Based on heating and cooling loads from the Simien simulations the geothermal wells was designed by using simulations in the tool EED-Earth Energy Designer [6]. A version based on a time step of 1 hour was used, giving hourly supply temperature from the wells.

The models in SIMIEN is based on the heating and cooling load is controlled after the room air set point temperature. Due to the large thermal inertia of the embedded PEX-pipes in the concrete slab, the heating load is controlled after the slab mass temperature, and the results from SIMIEN will therefore be inaccurate in many situations. A stationary model based on ISO 11588-2 [7] for embedded systems has been used to make model predictive control (MPC) algorithms implemented in the control system of Lia kindergarten. This stationary model has been extended to a transient model (hourly time step) by using electric analogy models similar to does used in SIMIEN. The transient model has been used to evaluate if the MPC-algorithm, based on the stationary model, is appropriate under different scenarios.

2.2 Measurements and monitoring tools

The energy- and indoor climate performance of the building is monitored by a few different systems. The main is the central operation system based on the Siemens Desigo CC system [8]. The Desigo system has a continuous monitoring of temperatures, mass flows, air flow rates, pressure losses and electric- and thermal energy meters. The Desigo system also have imported data from the PV-inverter system, giving the real time and aggregated PV-production. Due to a failure in the server system where monitored data are stored, we only have trend data (historic data) from 9th of January 2019. Before this date we only have aggregated energy use for the total operational period for the building (from early November, 2017). To get as much informative data as possible, we have chosen the period 1st of May 2018 to 30th of April 2019 for our analysis.

Another system used in analysis here is the building energy monitoring system Optima [9]. This system uses the data from the electricity meter which is collected in the elhub system (https://elhub.no/). The electric energy use and the surplus exported solar power can be reported on an hourly, daily, weekly, monthly or annual basis.

To analyze the heat pump system in detail the Nibe Uplink system (https://www.nibeuplink.com/) has been used. Due to problems with a firewall system only "manually" logged data directly from the heat pump system has been used for a limited period to assess the performance of the heat pump.

The battery system installed in the building has not yet been implemented in the Desigo system, and we have therefore no data on how this system performs. The battery performance is therefore not reported in this paper.

2.3 Climate data

Climate data used in the SIMIEN simulation is based on the normalized climate data set for Oslo defined in NS3031:2014 [10]. The annual average temperature in this climate file is 6,3 °C. The measured average outdoor temperature for the chosen period (May18-April19) was 8,7 °C. The summer of 2018 was unusually hot, but also the winter 2018/2019 was warmer than the normalized climate data used in the simulation. As discussed in Dokka&Grini [11] a simple degree day correction of energy use will be inaccurate for these kind of high performance buildings, and a simulation with real climate data was out of the scope for this study. Therefore no correction for the measured and simulated energy use has been done. Based on this we will expect the simulations of net space heating to be somewhat overestimated, and the simulation of the net space cooling to be somewhat underestimated.

3. Results and discussions

The different results from simulations and measured performance is shown and discussed below.

3.1 Total electricity use and export

In an all-electric-building like this kindergarten, where all the energy supply and production is electricity, the electric demand of the building is covered either by solar power used in the building or delivered electricity from the grid (bought). Figure 2 shows the measured electric balance of the building, where solar covers a large part of the electric demand from April to September, while the six months from October to March to a large degree is covered by grid electricity. Comparing the measured numbers with the simulated numbers (fig. 3) we see that the simulated numbers is less than half of the measured ones, and the difference is larger in the winter than in summer. One of the reasons for the high winter electric energy use is caused by the ventilation system "by accident" has been running 24/7 after a fire exercise from early November 2018 until it was corrected in February 2019. An analysis indicate that the electric demand in the winter months (Nov-Feb) would be reduced from 8 000-10 000 kWh down to approximately 5500-6000 kWh with correct operation of the ventilation, resulting in a reduction in electric energy use of approximately 14 400 kWh/a (or 9 kWh/m^{2}yr). Other reasons for the difference in measured and simulated results will be discussed in the later sections. The annual specific measured electricity demand is 46 kWh/m^{2}yr, while the simulated value was 22 kWh/m^{2}yr. The annual specific measured solar used in the building is 11 kWh/m^{2}yr, while the simulated value was 10 kWh/m^{2}yr. The annual specific measured electricity from grid is 35 kWh/m^{2}yr, while the simulated value was 12 kWh/m^{2}yr.

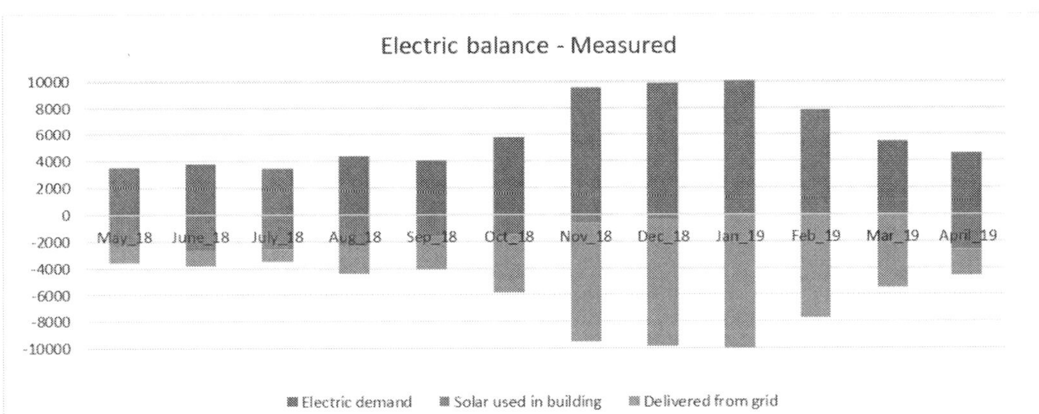

Figure 2. The measured monthly electric balance of the building.

IOP Conf. Series: Earth and Environmental Science **352** (2019) 012061 doi:10.1088/1755-1315/352/1/012061

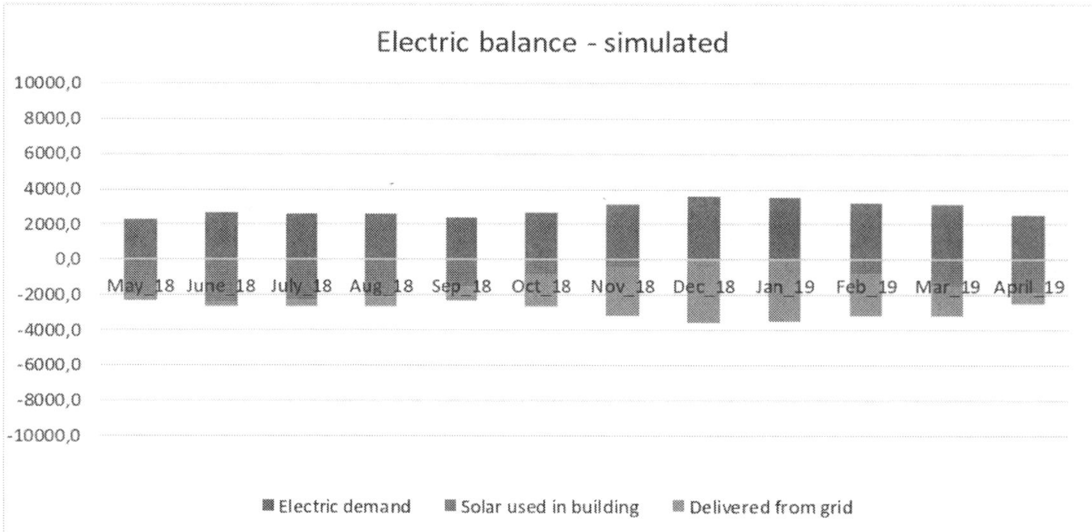

Figure 3. The simulated monthly electric balance of the building.

3.2 PV-production and mismatch demand and production

Figure 4 shows the measured monthly PV-production and the split between the use in the building and solar exported to the grid, and figure 5 shows the simulated figures. The simulated and measured numbers are overall in good agreement, however the snow cover in February and March is reducing a potential significant production in these months. How much snow cover is affecting the PV-production will of course vary from year to year, but the assumption of snow cover from 1st of December to 15th of February used in the simulations could maybe be extended one month (to 15th of March). The annual measured PV-production is 31 700 kWh, while the simulated value was 36 150 kWh. The annual measured solar used in the building is 16 780 kWh (53 %), while the simulated value was 15 890 kWh (44 %). The annual exported solar power to the grid is 14 900 kWh, while the simulated value was 20 260 kWh. That means the measured mismatch between demand and production is less than the simulated mismatch. This can probably be explained by the electric demand is higher than expected in the building. A measured performance closer to the plus energy ambition would most likely lead to more mismatch between demand and production.

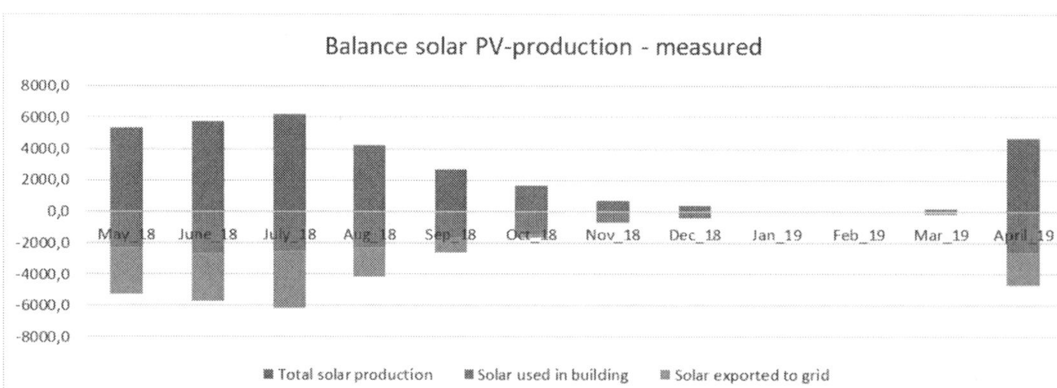

Figure 4. The measured solar PV production and split between use in building and exported to the grid.

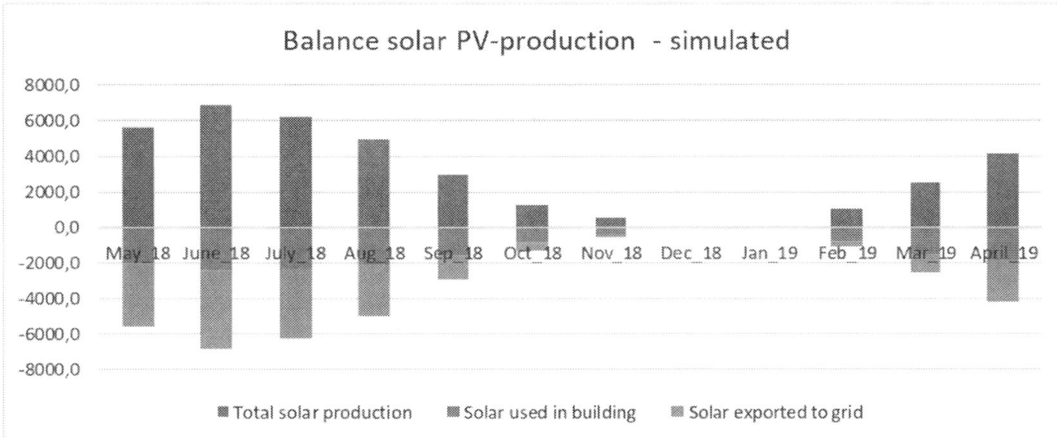

Figure 5. The simulated solar PV production and split between use in building and exported to the grid.

3.3 Net energy demand

Figure 6 shows the comparison between the measured and simulated net energy demand. The space heating from the heat pump and DHW measured by thermal meters, while the other energy items is measured by electric meters. The simulation of net energy demand does not take into account losses in distribution system.

The comparison shows that the main difference in net energy demand between simulation and measurements is due to space heating and heating coil. The primary reason for high space heating is due to the 24/7 running of the ventilation system, as described in section 3.1. Secondary reasons could be increased heat loss through the slab on ground in the first storey which the simulation model does not take into account, and also losses in the distribution system could be significant.

The net energy demand for the heating coil in ventilation system is also very much higher than the simulated value. However, we do not have reliable direct measured values of the heating demand, only electric energy use for the compressor in the internal heat pump in the air handling unit (AHU). The net demand for the heating coil is estimated by assuming a COP of 3. And in addition the compressor energy in the summer season is also included in the measured compressor energy, which due to the failure of the Desigo server system can't be deduced. Short term monitoring of the heating coil-system indicate that it "kicks in" at relatively moderate outdoor temperatures, when the rotary wheel exchanger should be more than enough to keep the desired supply set point temperature. Indicating a non-optimal control strategy of the AHU.

The net energy demand for DHW is somewhat lower than the expected (normalized) value, a trend we also have seen in other projects [2].

Energy use for fans is higher than simulated, but can largely be explained by the 24/7 running of the ventilation system from November to February. The measured specific fan power (SFP) is approximately 5 % higher than the simulated value, which corrected for the 24/7 operation is consistent with the measured energy use for the fans.

Pumps is not measured directly, but a common electric meter in the control panel measures pumps and other automation equipment. The measured energy use is therefore probably overestimated, but still quite close to the simulated value.

Energy use for lighting is measured to be appr. 9 kWh/m^{2}yr which is higher than the simulated value of 6 kWh/m^{2}yr. Short term measurements indicate that the power demand for lighting in the normal operational hours (06-18) is as designed. However, it seems like cleaning personnel trigger the light system in the evening and to some extent in the weekend, and this causes the extra energy use for lighting.

The energy use for different plug loads is not measured directly, but is the residual electric energy use when the total electricity demand (measured) is subtracted all the other measured electricity loads. The measured values are somewhat higher than the assumed plug load used in the simulation.

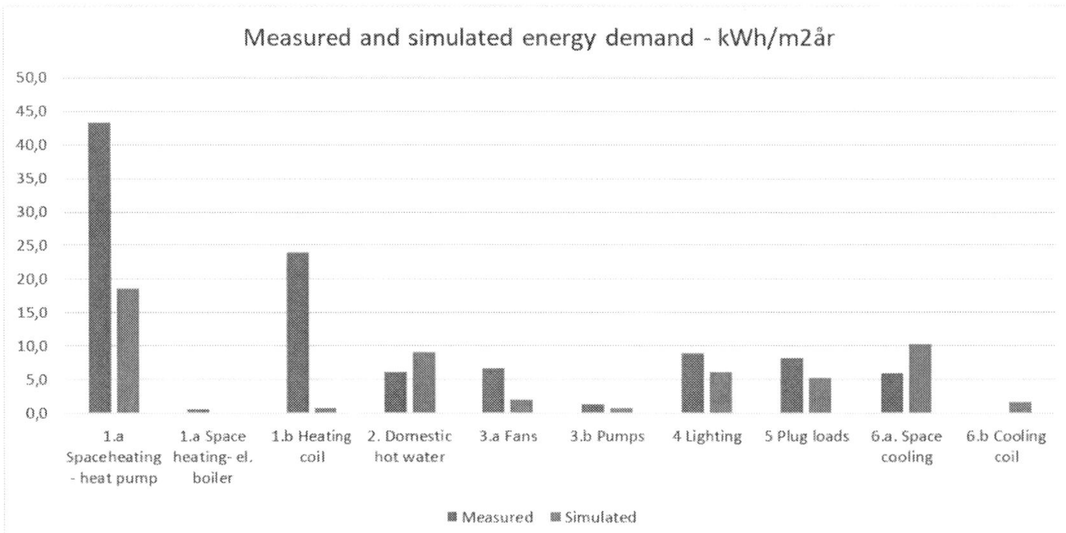

Figure 6. Measured and simulated net energy demand.

The measured space cooling load is approximately 60 % of the simulated cooling load. Reasons for this could be that solar shading and window airing is used to a larger extent than assumed in the simulations.

3.4 Performance of the lowex-system

Based on measurements of the thermal energy meters for space heating and DHW, and the electric meter for the heat pump system in the Desigo–system, an overall COP can be calculated. However this COP includes the internal circulation pumps (brine and water side) and also the internal electric element used early in the startup process of the building (late 2017). Be analyzing data from the Nibe Uplink system, this energy use for circulation pumps and the electric element could be estimated, and the COP of the heat pump (compressor) in the DHW-mode and the heating mode could be estimated.

The seasonal coefficient of performance (SCOP) for the DHW-mode of the heat pump was calculated to be 2,8. This is close to the designed SCOP of 3,0 used in the simulation.

The SCOP of the heat pump in heating mode was calculated to be 4,5. This is lower than the design calculation used in the simulation of 5,7. One reason for the lower SCOP is the supply temperature for wells in the heating season which was measured to be at average 3,0 ^{o}C instead of 5 ^{o}C calculated by the EED-simulations. The plausible explanation of the lower well temperatures is the fact that the heat balance of the geothermal wells is worse (heat dominated, lower cooling load than estimated) than used in the EED-simulations. Another factor is the supply temperature into the lowex system during the heating season, which is higher than simulated due to different flows for the external and internal circulation pumps. To balance the flows a shunt connection mixes supply and return, leading to a

IOP Conf. Series: Earth and Environmental Science **352** (2019) 012061 doi:10.1088/1755-1315/352/1/012061

demand for higher supply temperature out from the heat pump. A lower source temperature and a higher supply temperature out from the heat pump of course lead to a reduced COP of the heat pump system.

The seasonal energy efficiency ratio (SEER) of the free cooling system in summer time has not been possible to measure exactly due to unreliable pump measurements. However, short term monitoring of pressure loss and flow rates for both the brine and water side during normal operation indicate the SEER simulated for the design (SEER = 60) could be reached also in practice.

3.5 Delivered energy and net delivered energy
Figure 7 shows the need for delivered energy (electricity) for the building, where measured and simulated values is compared. The efficiency (SCOP and SEER, see section 3.4) of the thermal system is taken into account here, but the solar production of the PV-system is not included. The total measured value of 45 kWh/m^{2}yr is more than twice the simulated value of 21 kWh/m^{2}yr. The reason for the gap between the designed (simulated) and real performance is explained in the sections above.

With a measured solar production of 20 kWh/m^{2}yr, the net delivered energy to the building is 25 kWh/m^{2}yr. The simulated design intent was – 1 kWh/m^{2}yr (net exported to grid). A rough analysis indicate that it would be possible to reduce the net delivered energy to the half, around 12-13 kWh/m^{2}yr by tuning and close follow-up operation of the technical system. However, to reach the plus energy ambition different hardware updates of the system would probably be necessary.

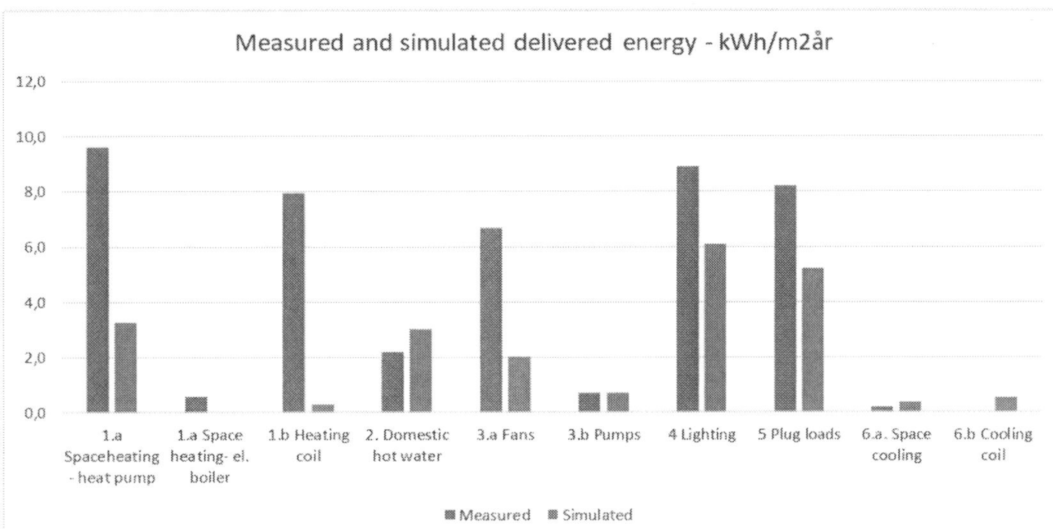

Figure 7. Measured and simulated net energy demand.

4. Conclusions
Even though the plus energy ambition is not fully met the energy performance of the kindergarten is by far the most energy efficient kindergarten the building owner (Omsorgsbygg) have. It will be possible to reduce the energy use even further than this first year of operation with tuning of the different subsystems, and close follow-up of the technical installation. To come down to the plus energy ambition in the design intent, also different hardware upgrades would be necessary.

IOP Conf. Series: Earth and Environmental Science **352** (2019) 012061 doi:10.1088/1755-1315/352/1/012061

One of the lessons learned is the internal heat pump system in the air handling unit seems not to work optimal, and a conventional coil for heating and cooling based on the geothermal system would be a better solution.

References

[1] Futurebuilt plus energy definition. www.futurebuilt.no

[2] Dokka et al. (2015) Comparison of five zero and plus energy projects in Sweden and Norway – A technical review. 7. Passivhus Norden, Copenhagen, August 2015.

[3] NS 3701:2012 - Norsk passivhusstandard for yrkesbygninger

[4] SIMIEN Version 6.009 (SIMIEN, 2018) Simulation of Indoor Climate and Energy use, ProgramByggerne ANS, *www.programbyggerne.no*

[5] PV*Sol https://www.valentin-software.com/en/products/photovoltaics/57/pvsol-premium

[6] EED- Earth Energy Designer https://buildingphysics.com/eed-2/

[7] ISO 11855-2:2012 Building environment design, dimensioning, installation and control of embedded heating and cooling systems. Part 2 Determination of the design heating and cooling capacity.

[8] Siemens Desigo CC - https://new.siemens.com/se/sv/produkter/byggnadstekniska-losningar/fastighetsautomation/desigo/building-management/desigo-cc.html

[9] Entro Optima https://www.entro.no/entro-optima/

[10] NS3031:2014 Calculation of energy performance of buildings - Method and data

[11] T.H. Dokka og C. Grini; Etterprøving av bygningers energibruk; Metodikk. SINTEF Fag 6, 2013.

Facade Integrated Photovoltaic, state of the art of Experimental Methodology

Changying Xiang and Barbara Szybinska Matusiak

NTNU, Department of Architecture and Technology, 7491 Trondheim, Norway

Changying.xiang@ntnu.no

Abstract. The concept of Building-integrated Photovoltaics (BIPV) is one of the most promising strategies to employ clean energy in the built environment. Up to now, the PVs have been applied mostly on roofs, but since the total roof area is insufficient, there is a need to integrate photovoltaics on building façades as well. This challenges not only the architectural design of a single building but also the visual image of urban environment, as photovoltaics have to harmonize with conventional building materials used on building facades as brick, concrete, wood, etc. Aiming to provide a foundation for research exploring facade-integration methods that will ensure successful architectural result, the paper presents a state of the art on façade integrated photovoltaics (FIPV) with focus on the experimental research methodology. It embraces both, theoretical research and PVs applications in building projects. As pure computer simulations are not recognized as an experimental methodology, papers conveying such generated results have not been included. In addition, the research that deals exclusively with energy aspects is omitted. The study is based on a comprehensive literature review. Advanced experimental methodologies from selected literature are described and categorized according to the scale (building or urban) and the transparency of the PVs (opaque or translucent). Then detailed features of PV experimental methods are demonstrated in structured tables for analysis and discussion. The study shows that even though solid scientific methods are used to evaluate single features of PVs, e.g. colour or reflectance, there is an obvious lack of methodology providing holistic assessment of Façade-integrated Photovoltaics, especially at the urban scale. The further research will lead toward developing of evaluation criteria framework (in interdisciplinary cooperation) and then provide a holistic methodology combining qualitative and quantitative methods for a successful FIPVs in urban context.

Keywords: BIPV, experimental methodology, urban context, architectural perspective, façade

1. Introduction

Building Integrated Photovoltaic (BIPV) is one of the most promising strategies to employ clean energy in building environment. Up to now, the PVs have been located mainly on roofs, since roof is the most efficient integration area from energy production perspective. Due to the limitation of the total available roof area, façade areas have to be utilized as well. Previous studies for façade integration mainly focus on energy aspects [1,2]. However, façade integrated photovoltaics (FIPV) require holistic architectural design other than mere consideration to maximize electricity productivity. This study demonstrates the state of art on experimental methodologies of façade integrated photovoltaic. From more than 160 literatures, 9 valuable experimental methods are selected and discussed in two scales (building scale and urban scale), aiming to provide a foundation for further research exploring systematic FIPV methods.

2. Experimental methodology

"*The experimental method is the only method of research that can truly test hypotheses concerning cause-and-effect relationships.*" — as Gay described in his book 'Educational Research: Competencies for Analysis and Application' [3], experimental methodology is a powerful empirical evaluation approach that has been widely used by researchers in various disciplines.

Experimental research usually includes: a hypothesis; one or more variables that could be manipulated by researchers in a controlled environment; experimenters and participants[4,5]. The generated result data will be measured or calculated and analysed. Researchers use experiments to establish a cause-effect relationship and prove or reject the hypothesis. There are three types of experiments [6]: *lab-experiments* (well-controlled environment), *field experiments* (in everyday outdoor environment and cannot really control extraneous variables) and *natural/quasi- experiments* (no control of independent variable in everyday environment).

In this article, computer simulation is not viewed as a type of experiments: firstly, virtual simulations are usually done in very idealizing situations which differ from real environment; Secondly, most solar design tools for Daylighting and PV simulation are using simplified methods [7], which cannot truly represent the reality. Beisbart also argues that computer simulations and experiments are different from a theoretical perspective[8]. Therefore, this study will not include computer simulation methodologies.

3. Objectives and methodology

This study aims to provide extensive list of current advanced experimental methods and approaches of BIPV in façade, the main objectives of this study is to:

- Identify and describe the most advanced experimental methods used for façade integration of PV
- Categorize the methods according to scale (building vs urban scale).
- Demonstrate the detailed features of PV experimental methods in a structured way.
- Analyze the investigation aspects/aims of presented PV experimental methods.

This paper is based on comprehensive literature review. The review methodology of this paper has been inspired by methodology from Saretta et al. [9] and consists of the following 5 parts:

1) First round literature search of recent 10 years' studies in BIPV field since elder studies rarely address aspects other than energy and economy. Data is collected from English-language literature through *ScienceDirect*, *ResearchGate* and *Google Scholar* databases. Combination of keywords that used for search are: ''*building façade integrated photovoltaic*'', ''*BIPV façade*'', ''*BIPV façade experiment*'', ''*BIPV urban context*'', ''*experimental method building integrated photovoltaic*'', ''*BIPV colour*'', ''*BIPV visual*''.

2) Filter search result according to this study's investigation scopes and investigation criteria.

3) Experimental methods from filtered literature are analysed and categorized.

4) Second round literature search of related studies in colour and daylight field is added. Data is collected from recent 10 years' English-language literature through *ScienceDirect*, *ResearchGate database* and *NTNU University Library*. Combination of keywords that used for search are: ''*façade daylight experiment*'', ''*glazing color experiment*'', ''*façade color experiment*''.

5) Comparison and synthetization of selected experimental methodologies.

Literatures without experimental methods are not included in this study scope. In addition, research that deals exclusively with energy aspects or only roof integration is also omitted. Investigation criteria are pre-set from architectural point of view, that is color and daylight performance, PV texture, pattern and dimension. More than 160 journal papers, conference proceedings, reports are searched in 2 rounds, among them only 36 papers are selected as deep review materials. 9 valuable methods from previous studies are summarized and discussed.

4. Experimental methods related with FIPV

Façade integrated photovoltaics function as normal façade materials beside the capability of producing electricity. Therefore, like conventional materials, façade integrated PVs will influence the perceived visual images from both single building and urban context levels. Previous valuable experimental studies are categorized according to **scale (building or urban)** and **transparency (translucent or opaque)** in the following paragraphs.

4.1. Methods on façade integrated translucent PV at building scale

Translucent PV products can produce electricity and at the same time be integrated into windows and façade glazing. Indoor daylight condition and visual perception is essential for integrated translucent PV products, however, most of current studies are carried out through computer simulation [10–17]. A limited number of studies have presented valuable experimental experience.

Vossen et al. studied the visual performance of a luminescent solar concentrator(LSC) windows at TU/e in Netherlands [18]. 54 participants aged 21-28 years were asked to evaluate visual experience of office environment with red LSC windows. A full-scale room with normal glazing and 3 physical models (1:6 scaled) equipped with sliding LSC covering were used for comparison study. Participants were first asked to experience the full-scale office room and filled one questionnaire. Then they observed the 2 physical models equipped with various LSC coverage percentages to get visual impression and fill another questionnaire. A third questionnaire was finished after all LSC coverage verification test. Simultaneously, interior horizontal illuminance and exterior illuminance are recorded by a lux-meter in the third scaled model and 2 Hagner SD 2 lux-meters outside the test room respectively. Five criteria are used in this test: visual comfort, naturalness, ambiance, precision and light level. The study results show that a warm color shifting is preferable and that participants prefer rooms with LSC coverage below 25%.

Aste et al. did a similar visual performance test[19]. Three types of LSCs (yellow, orange and red LSC) are tested in a 1:10 scaled physical model on a most cloudy day at the Politecnico di Milano. Experimental measurements were carried out for spectral power distribution, illuminance and correlated color temperature, which link the spatial feeling with surface colors of a real room. Results from different color LSC windows and neutral plate are compared and analysed. The study shows that yellow coloured LSC window is the most preferable due to its high luminous efficiency and the reduction of color temperature of incoming daylight.

Ghosh et al. tested the color properties and glazing factors of semi-transparent photovoltaics [20]. Firstly, the test objects are fabricated by one vacuum glazing layer and one single glazing with multi-crystalline solar cell on top. Two types of solar cells with different transparency are employed for comparison. An UV-VIS-NIR spectrophotometer was used to measure optical features of the combined PV-vacuum glazing. After that, UV transmittance, luminous transmittance, solar transmittance of PV-vacuum glazing is calculated and compared with values of vacuum and single glazing. This PV-vacuum glazing's correlated color temperature (CCT), color rendering index (CRI) were also calculated and compared with features of transparent suspended particle device (SPD). Only normal incident transmittance was measured. The study shows that semi-transparent photovoltaic-vacuum glazing can lower the daylight transmittance and provide preferable entering daylight quality and quantity (with high value of CCT and color CRI than 30% and 40% SPD).

4.2. Methods on façade integrated opaque PV at building scale

Opaque PV products are traditionally applied on building roofs. The glossy surfaces, dark-blue or black colors features and rigid shapes of typical opaque PVs make them difficult to be integrated in building façade as cladding or shading system. With the development of PV technology, new products able to tackle these barriers are emerging.

SwissINSO advertised a promising Kromatix panels for building façade integration which can overcome traditional color and glossy hinders. The Kromatrix panels co-developed by SwissINSO and EPFL (Swiss Polytechnic Institute) use a special nano-deposition technology. Various colour choices can be obtained through multi-layer interference effect (not by pigment based colouring) while nonglossy effect are achieved with outer surface etching[21]. Jolissant et al.[22] presented in their paper a series of experimental tests consist of gloss index test, electricity generation performance and color degradation as well as demonstration of real project applications. The gloss index of Kromatrix panel, c-Si PV panel and thin-film PV cells were measured by a Glossmeter at 60 °. Kromatrix panels demonstrate much lower gloss index (6 GU) than c-Si PV (90 GU) and thin-film (120 GU). Following flash test modules at SUPSI in Switzerland showed Kromatrix panels in blue and yellow have almost same electricity production efficiency as standard black panels. Kromatrix glass also passed color degradation test at EPFL-LMT laboratory and further TUV or SUPSI certificates are also obtained.

More interestingly, real façade applications at Kohlesilo in Switzerland and Copenhagen International School in Denmark show satisfying performance in energy harvesting and architecture integration at individual building scale. However, scientific experimental evaluation from indoor users and outdoor citizen's perception is lacking, and more exploration needs to be done to investigate methods and strategies of applying similar products in urban context.

Nagy et al. studied opaque PVs' application from dynamic façade shading system perspective: an advanced adaptive solar façade prototype is developed in lab and applied at the living lab at ETH[23]. The modular prototype equipped with thin-film PV is firstly tested in laboratory for its soft pneumatic actuators' functionality. Then real scale implement is carried out through ETH's house of nature source living lab. However, the lighting, shading performance is not tested in this research, potential data could be further collection from living lab study.

4.3. Methods on façade integrated PV at urban scale (mixing translucent and opaque)

Danks et al. [24] pointed out that there is a lacking of evaluation criteria and empirical study on visible and thermal impacts of reflected solar energy in urban environment. Sánchez-Pantoja et al. [25] proposed a series of aesthetic factors of solar system in urban context: color, glare, pattern-texture, fractality, visibility, integration degree etc..

Some researchers investigated BIPV/solar energy systems' visual perception in urban context with participants-involved qualitative methods. Sánchez-Pantoja et al. [25] presented a study of aesthetic perception of BIPV and BAPV prototypes in Solar Decathlon Europe 2014 Exhibition with non-verbal Self-Assessment Manikin (SAM) survey method. This SAM survey includes emotion assessment and excitation degrees. Firstly, the method of using prototype photos for assessment is validated by result comparison with data from participants' on-site visit. After that, only prototype photos are used for later SAM survey. Subjective opinion of citizens about 20 BIPV and BAPV design prototypes are collected and the result shows that BIPV systems are rated higher than BAPV systems.

Basing on the concept of 'architectural criticity' of building surfaces, Cristina Munari Probst and Roecker proposed a LESO-QSV method to support integration of solar energy system in pre-existing urban areas [26]. A building surfaces' criticity level is defined by combination of sensitivity level of the urban context and surface's visibility level from public domain. The higher the criticity level, the higher integration requirements for solar energy system integration. A LESO-QSV grid software tool[27] with qualitative assessment method is also developed to support the decision making for municipalities to install solar energy systems in urban context.

Based on the LESO-QSV method, Florio et al. present a scale-dependent methodology to assess the visibility of building surfaces with potential to harvest solar energy from territorial scale to urban neighbourhood scale. A cross-mapping application case study the city of Geneva (Switzerland) is also presented in their paper. Aims to combine solar radiation map and an estimate of the socio-cultural sensitivity for a multi-criteria decision making [28]. Multi-criteria decision-making framework for BIPV in urban context is urgently needed. The qualitative LESO-QSV method and tools could be helpful to combine quantitative experimental methods together for supporting FIPV in urban context.

4.4. Methods on façade integrated translucent PV at urban scale

IOP Conf. Series: Earth and Environmental Science **352** (2019) 012062 doi:10.1088/1755-1315/352/1/012062

Study by Lynn et al. shed light on investigation of translucent PV façade at urban scale[29]. Semi-transparent thin films (STPV)' colour performance has been investigated through experimental comparison of indoor and outdoor CRI value, providing a reference of considering façade integrated PVs' colour property in outdoor urban context. A portable measurement system was designed to realize measurement at both interior and exterior environment. Firstly, the transmitted spectral irradiance of various STPV glazing with different visible range is tested in laboratory. Then respective color rendering index (CRI) values are calculated. Secondly, outdoor on-site measurement of STPV's CRI was conducted to compare with indoor lab test result, in order to investigate the applicability of indoor lab test result in exterior built environment. The comparison shows that although correlated color temperatures are significantly different for indoor and outdoor situations, the CRI result in both situations share similarity. Lynn et al. stated that CRI is an important integration parameter for BIPV.

5. Analysis and discussion

5.1. Building scale analysis and discussion

For BIPV application research, products and building level methods can be viewed as bottom-up approaches. Table 1 summarizes the above-selected 6 experimental methods applied in building scale.

Table 1. List of state of art of methods on façade integrated PV study at building scale.

Year	Author	PV technology	Quantitative Methods	Qualitative Methods	Investigation aspects
2016	Vossen, Aarts and Debije	**Translucent** LSC	Full-scale room as reference; interior horizontal illuminance of scaled model recorded by Lux-meter; exterior vertical illuminance recorded by Lux-meter.	3 Questionnaires for 54 participants	**Color** (red color-indoor perception) CCT&CRI; Glare **Daylight** illuminance
2016	Aste et al.	**Translucent** LSC	Physical scale model with alternative LSC windows in 3 colors; Daylight condition measured by spectrophotometer	None	**Color** (multi- color) CCT **Daylight** illuminance; spectral power distribution
2018	Ghosh, Sundaram and Mallick	**Translucent** multi-crystalline PV	Glazing optical features measured by UV-VIS-NIR spectrophotometer; PV glazing from 2 transparencies are compared.	None	**Daylight** Glazing factor; **Color** CCT& CRI
2012	Lynn, Mohanty and Wittkopf	**Translucent** thin-film PV	A mobile daylight measurement system was configured to measure the transmitted spectral power distribution (SPD) through the PV; Indoor and outdoor measurements are carried out and results are compared.	None	**Color** CRI
2017	Jolissant et al.	**Opaque** Kromatix panels	gloss index measurement and compared with c-Si PV and thin-film PV; flash test of blue and yellow panels for electricity production efficiency and compared with standard black PV; color degradation test at lab; real project application	None	**Color** Color choices and outdoor stability **Texture** Glossy or nonglossy; outdoor glare risk **Energy** Electricity harvest efficiency
2016	Nagy et al.	**Opaque** thin-film PV	Lab test and living lab test	None	**Mechanical function** Dynamic adaptivity

IOP Conf. Series: Earth and Environmental Science **352** (2019) 012062 doi:10.1088/1755-1315/352/1/012062

For façade integrated PVs at building level, indoor daylight and color are current study focuses. However, no holistically investigation combining both qualitative and quantitative methods has been applied to cover all important indexes of daylight and color. Experiments under different climate conditions are also needed. For instance, study from Vossen et al. shows good combination of illuminance measurement and participants' perception survey, but multi-color aspect experiment is missing, test under more climate conditions could be desirable for further improvement. Research experience from related color science disciplines like color perception of advanced glazing [30], interaction impact between daylight and colors [31] could be good inspiration to help enrich the further BIPV experiment study. For PV integrated façade shading systems, valuable experimental approaches from architectural integration aspect is rare and similarly, knowledge and experience from related glare study and daylighting system investigations [32,33] could be valuable for reference.

5.2. Urban scale analysis and discussion

Urban scale assessment study can be viewed as top-down means, table 2 present the selected 3 façade integrated PV methods applied at urban scale.

Table 2. List of state of art of methods on façade integrated PV study at urban scale.

Year	Author	PV technology	Quantitative Methods	Qualitative Methods	Investigation aspects
2018	Sánchez-Pantoja et al.	No data	None	Self-Assessment Manikin survey through presented photos	**Emotional perception** pleasant-unpleasant; emotional intensity
2015	Florio, Roecker and Munari Probst	No data	None	3-steps quality evaluation software tool based on LESO-QSV method	**Geometry** Size/position **Materiality** Texture; Colors **Details** Module shape/size; joints
2012	Lynn, Mohanty and Wittkopf	**Translucent** thin-film PV	A mobile daylight measurement system was configured to measure the transmitted spectral power distribution (SPD) through the PV; Indoor and outdoor measurements are carried out and results are compared.	None	**Color** CRI

Compared with study at building scale, experimental methods are barely explored in urban scale, this could be due to the very lack of holistic design and assessment criteria for BIPV from architectural perspective. Current studies are mainly using qualitative survey, as IEA-SHC Task 41 "Solar Energy and Architecture" defined, the "architectural integration quality" is the result of a coherent integration from all points of view, including functional, constructive, and formal (aesthetic)[34]. Therefore, systematic experimental methodologies like mixed-methods combining both qualitative and quantitative methods are urgently needed to explore texture, dimension, outdoor color perception of BIPV products in urban context.

6. Conclusion

There is a growing demand to utilized building façade for PV integration, but there is very limited research in this field currently, especially from architectural point of view. This study shows that the current experimental methodologies on BIPV in urban context are still in its infancy. Although new PV technology create more freedom and possibility of integrating PVs on building façade, advanced research supporting BIPV for building facade is still lacking, especially in urban scale. Therefore,

IOP Conf. Series: Earth and Environmental Science **352** (2019) 012062 doi:10.1088/1755-1315/352/1/012062

future research focusing on building façade photovoltaic integration in the urban scale from architectural perspective is very essential. More advanced holistic study methodology combining quantitative and qualitative approaches, covering important architectural aspects like color, daylight, texture, dimension etc. is urgently needed.

The next research step could be establishing an evaluation criteria framework of BIPVs from the architectural point of view, supporting the holistic design and implementation of BIPVs in urban context. Interdisciplinary research activities should be carried out combining strength from PV, color and daylight science, architecture and urban design and PV industry partners to achieve the goals.

References

[1] Costanzo V, Yao R, Essah E, Shao L, Shahrestani M, Oliveira A C, Araz M, Hepbasli A and Biyik E 2018 A method of strategic evaluation of energy performance of Building Integrated Photovoltaic in the urban context *J. Clean. Prod.* **184** 82–91

[2] Brito M C, Freitas S, Guimarães S, Catita C and Redweik P 2017 The importance of facades for the solar PV potential of a Mediterranean city using LiDAR data *Renew. Energy* **111** 85–94

[3] Gay L R 1996 Educational Research: Competencies for Analysis and Application (5th ed.) *New Jersey:Prentice-Hall*

[4] Adelman L 1991 Experiments, Quasi-Experiments, and Case Studies: A Review of Empirical Methods for Evaluating Decision Support Systems *IEEE Trans. Syst. Man Cybern.*

[5] Tanner K 2002 Experimental research designs *Research Methods for Students, Academics and Professionals : Information Management and Systems* (Wagga Wagga, New South Wales :: Centre for Information Studies.) pp 125–46

[6] Diamond J and Case T J 1986 *Community Ecology* vol 61

[7] Jakica N 2018 State-of-the-art review of solar design tools and methods for assessing daylighting and solar potential for building-integrated photovoltaics *Renew. Sustain. Energy Rev.* **81** 1296–328

[8] Beisbart C 2018 Are computer simulations experiments? And if not, how are they related to each other? *Eur. J. Philos. Sci.* **8** 171–204

[9] Saretta E, Caputo P and Frontini F 2019 A review study about energy renovation of building facades with BIPV in urban environment *Sustain. Cities Soc.* **44** 343–55

[10] Knera D, Szczepańska-Rosiak E and Heim D 2015 Potential of PV Façade for Supplementary Lighting in Winter *Energy Procedia* **78** 2651–6

[11] Cannavale A, Fiorito F, Resta D and Gigli G 2013 Visual comfort assessment of smart photovoltachromic windows *Energy Build.* **65** 137–45

[12] Cannavale A, Hörantner M, Eperon G E, Snaith H J, Fiorito F, Ayr U and Martellotta F 2017 Building integration of semitransparent perovskite-based solar cells: Energy performance and visual comfort assessment *Appl. Energy* **194** 94–107

[13] Taveres-Cachat E, Bøe K, Lobaccaro G, Goia F and Grynning S 2017 Balancing competing parameters in search of optimal configurations for a fix louvre blade system with integrated PV *Energy Procedia* **122** 607–12

[14] Sun Y, Shanks K, Baig H, Zhang W, Hao X, Li Y, He B, Wilson R, Liu H, Sundaram S, Zhang J, Xie L, Mallick T and Wu Y 2018 Integrated semi-transparent cadmium telluride photovoltaic glazing into windows: Energy and daylight performance for different architecture designs *Appl. Energy* **231** 972–84

[15] Kapsis K, Dermardiros V and Athienitis A K 2015 Daylight Performance of Perimeter Office Façades utilizing Semi-transparent Photovoltaic Windows: A Simulation Study *Energy Procedia* **78** 334–9

[16] Cheng Y, Gao M, Dong J, Jia J, Zhao X and Li G 2018 Investigation on the daylight and overall energy performance of semi-transparent photovoltaic facades in cold climatic regions of China *Appl. Energy* **232** 517–26

[17] Wong P W, Shimoda Y, Nonaka M, Inoue M and Mizuno M 2008 Semi-transparent PV: Thermal performance, power generation, daylight modelling and energy saving potential in a residential application *Renew. Energy* **33** 1024–36

[18] Vossen F M, Aarts M P J and Debije M G 2016 Visual performance of red luminescent solar concentrating windows in an office environment *Energy Build.* **113** 123–32

[19] Aste N, Buzzetti M, Del Pero C, Fusco R, Testa D and Leonforte F 2017 Visual Performance of Yellow, Orange and Red LSCs Integrated in a Smart Window *Energy Procedia* **105** 967–72

[20] Ghosh A, Sundaram S and Mallick T K 2019 Colour properties and glazing factors evaluation of multicrystalline based semi-transparent Photovoltaic-vacuum glazing for BIPV application *Renew. Energy* **131** 730–6

[21] SwissINSO 2018 SwissINSO — Company - SwissINSO developer of Kromatix - Coloured Solar Panel - BIPV

[22] Jolissaint N, Hanbali R, Hadorn J-C and Schüler A 2017 Colored solar façades for buildings *Energy Procedia* **122** 175–80

[23] Nagy Z, Svetozarevic B, Jayathissa P, Begle M, Hofer J, Lydon G, Willmann A and Schlueter A 2016 The Adaptive Solar Facade: From concept to prototypes *Front. Archit. Res.* **5** 143–56

[24] Danks R, Good J and Sinclair R 2016 Assessing reflected sunlight from building facades: A literature review and proposed criteria *Build. Environ.* **103** 193–202
[25] Sánchez-Pantoja N, Vidal R and Pastor M C 2018 Aesthetic perception of photovoltaic integration within new proposals for ecological architecture *Sustain. Cities Soc.* **39** 203–14
[26] Cristina Munari Probst M and Roecker C 2015 Solar energy promotion & urban context protection : LESO-QSV (Quality-Site-Visibility)method *31th International PLEA Conference*
[27] Florio P, Roecker ; C and Munari Probst ; M C 2015 Urban acceptability of solar installations: LESO-QSV grid, a software toll to support municipalities CISBAT *CISBAT* (Lausanne: CISBAT) pp 981–6
[28] Florio P, Munari Probst M C, Schüler A, Roecker C and Scartezzini J-L 2018 Assessing visibility in multi-scale urban planning: A contribution to a method enhancing social acceptability of solar energy in cities *Sol. Energy* **173** 97–109
[29] Lynn N, Mohanty L and Wittkopf S 2012 Color rendering properties of semi-transparent thin-film PV modules *Build. Environ.* **54** 148–58
[30] Arbab S, Matusiak B S, Martinsen F A and Hauback B 2017 The impact of advanced glazing on colour perception *50-68*
[31] Matusiak B S and Anter K F 2012 *Nordic Light and Colour* ed B S Matusiak and K F Anter (Trondheim: NTNU)
[32] Moscoso C and Matusiak B 2018 Aesthetic perception of a small office with different daylighting systems *Indoor Built Environ.* **27** 1187–202
[33] Matusiak B S 2013 Glare from a translucent façade, evaluation with an experimental method *Sol. Energy* **97** 230–7
[34] Wall M, Probst M C M, Roecker C, Dubois M-C, Horvat M, Jørgensen O B and Kappel K 2012 Achieving Solar Energy in Architecture-IEA SHC Task 41 *Energy Procedia* **30** 1250–60

IOP Conf. Series: Earth and Environmental Science **352** (2019) 012063 doi:10.1088/1755-1315/352/1/012063

Design and build with straw, earth and reeds for a minus carbon and plus energy building practice

Marwa Dabaieh

Rendsburggade 14. 9000 Aalborg, DK
Mada@create-aau.dk

Abstract. The 6Zs target refers to the concept of a minus carbon and plus energy eco-cycle refugee house. A 37 m² house was designed and constructed in a participatory manner in the City of Lund, Sweden. The 6Zs include: zero emissions, zero energy, zero waste, zero cost, zero indoor air pollutants and zero impact on the environment after the shelter is demolished. The key idea of this eco-cycle house is to reach net 6Zs during all stages—material extraction, building construction, operation and maintenance—until the shelter's end of life. The main construction material is plant-based raw fibers (mainly straw and reed), which are available around the building site. This house is designed to accommodate the needs of two adults and one child. It was built with the help of 7 refugees in 11 working days through an experimental participatory urban living lab methodology. The paper discusses the 6Z design concepts and draws conclusions on the preliminary assessment of the house prototype that was built as a proof of concept. The beneficiaries of this project are not restricted to refugees but also include the majority of individuals and families seeking affordable ways of living with a low impact on the environment. The house is designed for the cold Swedish climate, but the design concept and methodological approach can be adjusted to other climates or geographical contexts.

1. Introduction

1.1. Background

Forced displacement is not a recent phenomenon. Large numbers of people flee from one place to another due to many reasons, including, but not limited to: poverty, drought, violence, persecution or armed conflict. According to UNHCR, one person is forcibly displaced every two seconds, reaching in June 2018 more than 68.5 million forcibly displaced people worldwide [1]. Finding decent, affordable, quick shelters for refugees has not only become an acute duty for hosting countries, but it is also a situation that is predicted to escalate in the near future. The idea of this refugee house project started in 2016 responding to a massive flow of migrants called at that time a 'refugee crises. The Syrian population was the highest population of forcibly displaced peoples in modern history, reaching 5.6 million who sought refuge in neighboring countries [1] and 970,000 who sought refuge in Europe[2]. Over six million Syrians are displaced within Syria and 13 million are in dire need of humanitarian aid [1]. As of 2017, over 150,000 Syrians gained asylum in Sweden in addition to 40,000 Syrian asylum seekers [3]. As Sweden has an open-door policy for refugees, hosting them in accommodation facilities rather than refugee camps, an affordable and environmentally sensitive house suits the Swedish context best.

Globally, the most commonly used refugee shelter is a lightweight tent which can be easily transported and quick to assemble [4,5]. Such tents are meant to last for short periods of time, but in reality, refugees stay in such tents for decades [5] . The majority of humanitarian tents that are currently used deteriorate quickly, leaving behind huge environmental burdens given their short life-span and the energy abusive industrial materials that are used in their manufacturing. Moreover, they don't offer an appropriate comfortable indoor environment nor do they fulfil social needs [6]. Shelters that can accommodate the urgent need to host large numbers of refugees within short notice is still a primary priority. The eco-cycle refugee house project described in this paper worked on such challenges—offering a low cost, low tech and low impact home that, according to the Swedish standards for temporary building permits, doesn't require major maintenance or renovation work for 15 years. The house design and construction utilized a transdisciplinary and participatory methodology following an occupant centered approach which involved a group of Syrian refugees in all the design and construction process.

1.2. Literature overview

Earlier experiences building emergency shelters revealed that owner-driven approached to building shelters generate better results than providing ready-built ones [7,8]. However, that is not the typical approach followed by hosting countries as refugees are seldom consulted as occupants in the design or construction process for their homes. In addition, refugees are typically not empowered to make their own choices in the emergency phase of camp construction [7]. As shelter performance, particularly energy efficiency, is still a main issue within refugee housing discourse and practice, prototypes of typical humanitarian temporary shelters were compared [9]. Murphy et al concluded that shelters built from bio-based natural materials have less impact on the environment, with a minimal carbon footprint compared to shelters constructed using industrial materials (i.e. metal-intensive structures or pre-cast concrete) [9]. However, the challenge still lies with the time required for construction as it is still faster to build with industrial materials. Another study analysed the thermal performance of refugee tents retrofitted with aerogel pads – a high-insulation material [10]. The study showed a decrease in energy demand for heating by almost half and improved indoor comfort conditions in summer – eliminating the need for mechanical cooling. But one should still consider the embodied energy and carbon in the production of high thermal insulation materials like aerogel and their environmental impacts after the building's end of life.

Applying passive solutions to humanitarian shelters, especially in a cold climate, has always been a tedious task that requires a fair amount of experience. As discussed by Crawford and his research colleagues whose research showed that it is not yet possible to depend only on passive solar heat gains to attain an adequate indoor climate [11]. Yu and his team constructed proof of concept test cell models using natural materials from the local environment for cold climates in China and monitored their thermal performance in a lab environment [12]. They tested different roof solutions and thermal insulation materials, and the monitoring results were compared to monitoring data from existing shelters which used the conventional construction techniques. Their modifications resulted in better performance outcomes using low-cost thermal insulation materials with low-tech construction methods that can still resist wind in winter and allow for ventilation in summer. A few successful modifications included the use of polypropylene sheets for walls, cardboard and air-bubble polythene sheets for roof thermal insulation, and fiberglass cement as a roof covering. Nevertheless, more research is needed on the impact of the construction of temporary humanitarian shelters, especially with regard to carbon emissions [9] or from a complete cradle-to-cradle perspective. Life cycle assessment or life cycle cost analysis haven't been extensively used either as tools to evaluate the effect of alternative building envelopes over the various life stages of buildings–from construction until final disposal–and their economic feasibility and payback time [13,14]. In the case of refugee housing, a very limited number of studies have managed to combine both integration of low impact design and the encouragement of sustainable building practices in a participatory manner with refugees. This study aims to help fill this research gap.

2. Methodology

The methodology is divided into 6 main phases:

1- The investigative participatory design part includes semi-structured interviews with 49 refugees in 5 different housing facilities in southern Sweden. Following refugee interviews, interviews were conducted with local architects, engineers, craftsmen and municipality personnel to inquire about building norms, codes and regulations. After the interviews, a preliminary sketch of the house design was prepared for discussion in a participatory focus group with the same sample of refugees. The focus group was conducted to make sure the sketch matched the preferences of the house's target inhabitants as well as to test refugees' acceptance of the eco-cycle and passive low-tech systems used. The outcome of this phase informed the building modeling and simulation phase.

2- Modeling simulation was then conducted for heating and cooling loads and daylight and natural ventilation using TRNSYS and ANSYS software. This was done to test the efficiency of the proposed passive systems with a focus on the integration of earth air heat exchanger (EAHE) coupled with three Trombe walls and a green wall. The outcome of the simulation was used to rectify the passive systems design until reaching the optimal efficiency needed.

3- Calculations were performed for the electrical power production, electricity demands, carbon footprint and payback time of the house.

4- The lab testing and test cells, which includes testing of different building materials' tension and compression strength in addition to water resistivity and fire resistance. Lab testing was also performed on several trials of different low-tech construction methods for houses that were easy to assemble and disassemble. Several test cells for the eco-cycle water and waste systems together with the different passive systems–mainly the Trombe wall and the green wall–underwent experimentation. In this phase, a soil analysis was conducted to investigate the accurate EAHE depth based on soil type. The simulation was again remodeled to test the final outcome.

5- Urban living lab construction methods were used to build the house as a proof of concept. In this phase, seven refugees were involved in a building training to learn how to build the house. They were instructed on the different building steps using participatory methods.

6- The final phase includes the monitoring and assessment of the building's performance. Temperature, humidity, CO_2 concentration, electricity production and biogas efficiency are currently being monitored. This phase is still ongoing and should be followed by a post occupancy evaluation with the tenants to test their levels of comfort and satisfaction for the eco-cycle systems used.

3. Design features of the house

The house is 37 m^2 with a bedroom for two adults, bathroom, kitchen and a living area as shown in figures (1,2&3). It accommodates the needs of two adults and one child. The main construction materials are straw, reeds, earth and wood. Straw, reeds and wood were brought from neighboring farms and surrounding forests. Clay earth is also an abundant material in the southern part of Sweden and has traditionally been used in construction as well. Passive systems like the Trombe wall, Earth Air Heat Exchanger (EAHE) and green wall are used mainly for passive heating, cooling and natural ventilation. The green wall and Trombe wall also serve to purify the air and are connected to the EAHE to eliminate overheating during summer. Window shutters painted with albedo paint for high reflectivity are used as light shelves in winter to maximize indoor daylight. The shutters are flipped to be used for shade in summer, as shown in figure (4). A hybrid solar and wind power renewable system is used to provide the house with electricity. The design is aims to provide the power supply needed for the house to be self-sufficient, with leftover energy available for sale to the main electricity grid. The house is outfitted with a biogas tank that uses organic waste from the kitchen, feces from the separating toilet in the bathroom and animal manure and agricultural waste from the neighboring farm. The biogas is used for cooking

IOP Conf. Series: Earth and Environmental Science **352** (2019) 012063 doi:10.1088/1755-1315/352/1/012063

and water heating during winter. The house is also provided with a solar water heater and a zero-energy earth fridge, the latter of which maintains food at 5 degrees year-round (similar to the root cellar concept). Detailed eco-cycle systems are shown in figures (5&6).

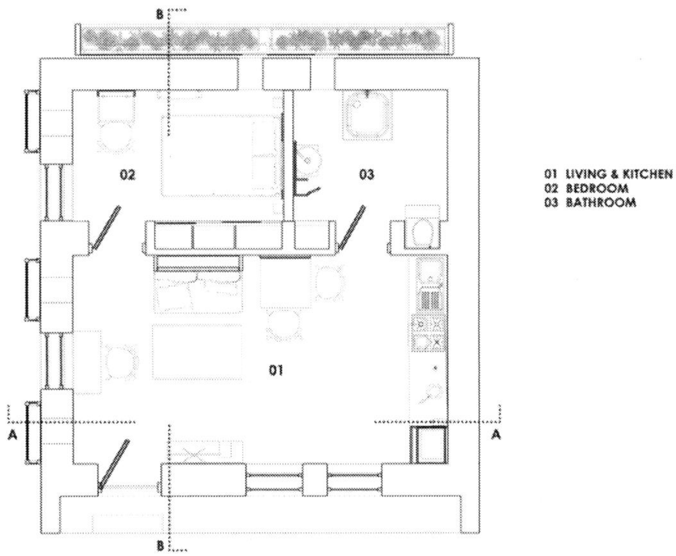

Fig. (1) Plan showing the house's main zones: bedroom, bathroom, kitchen and living areas in addition to the green wall to the west and the three Trombe walls to the south (illustration by M.Adel)

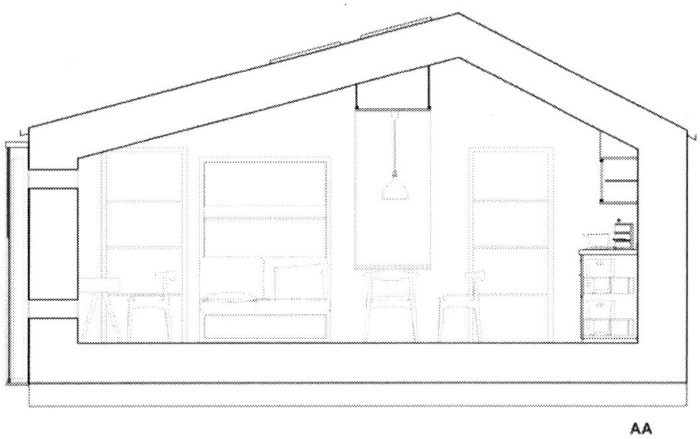

Fig. (2) Cross section showing the bedroom, living area and kitchen with the Trombe wall as a passive system for heating and air purification (illustration by M.Adel)

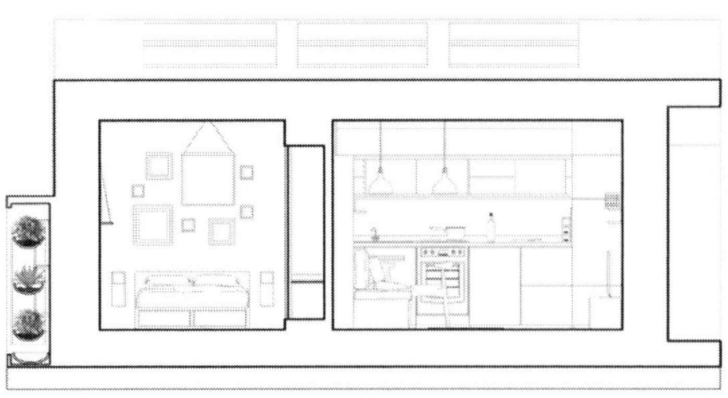

Fig. (3) Cross section showing the bedroom, living area and kitchen with the green wall as a passive system for heating, cooling and air purification (illustration by M.Adel)

Fig. (4) An interior rendering of the kitchen and living area showing the daylight efficiency of the light shelves during winter(illustration by M.Adel)

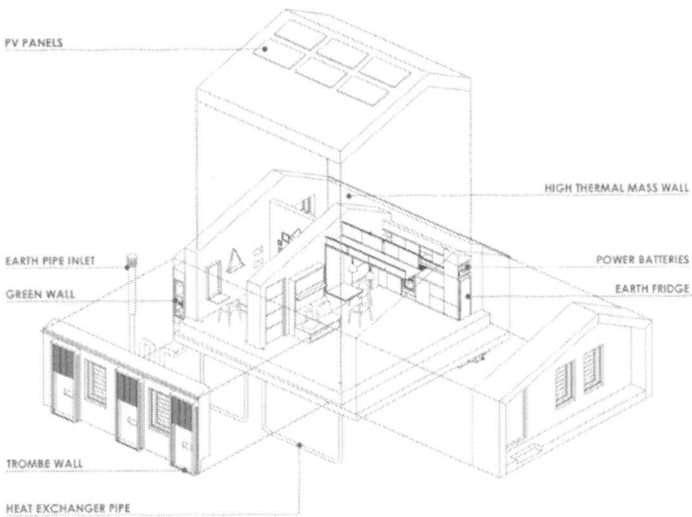

Fig. (5) Isometric drawing showing the three Trombe walls, EAHE and the earth fridge (illustration by M.Adel)

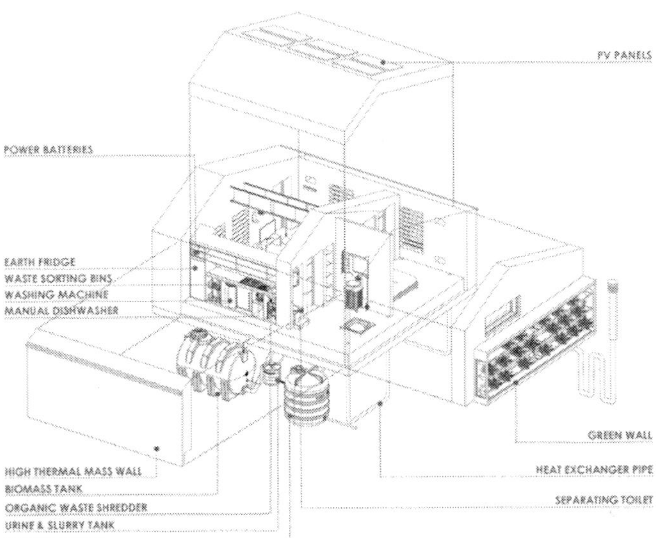

Fig. (6) Isometric drawing showing the waste water and fresh water eco-cycle systems together with the green wall connected to the open cycle EAHE (illustration by M.Adel)

4. Results and discussion

The interviews and workshop outcomes revealed that creating sustainable housing for refugees is not just a matter of taking low cost into account, or considering cultural or social needs in the planning and construction processes, it is also a matter of using materials and developing techniques that will make their shelters resilient and environmentally sustainable. The wish list assembled from the interviews with refugees and specialists informed the design of all the eco-cycle and passive systems used in the project. The skeleton of the house is designed from load-bearing compressed straw panels covered inside with reeds and clay plaster. Outside, the house is covered with wood fiberboards to serve as thermal bridges and clad with larch wood treated with beeswax and linseed oil for water proofing. The wall design reached a U value of 0.08. The roof and floors are from cross-laminated wood and plywood using air injected wood fibers as insulation. Lab tests for the building material samples confirmed that it complies with premium passive house standards and the fire safety regulations and standards in Sweden. Figure (7) shows the house during and after the participatory construction process.

Fig. (7) On the left, an interior illustration shows the house with the load-bearing straw panels and wooden roof beams. On the right, the house is depicted to show the Trombe walls to the south.

The CO_2 emissions reduction potential is calculated against conventional active means for cooling and heating- using typical residential buildings in the Swedish context as a baseline. The net savings was 86.6 kg CO_2. The reduction potential was then calculated and translated to a carbon tax reduction. Including carbon sequestering from the natural plant-based fibers (straw, reeds and wood) the building is approximately -299 kg CO_2. That means that it is not only zero carbon, it is minus carbon when calculated from a cradle-to-cradle perspective. Electrical consumption savings are calculated by comparing loads to conventional air conditioning required to achieve standard thermal conditions. The primary energy demand was 39.2 kWh/m^2. Thanks to the passive systems used for heating, cooling and natural ventilation, the house's energy consumption averaged 180.7 kWh per month. A solar water heater was used and over winter the biogas composter was used for water heating in addition to cooking. In terms of energy, the house performs beyond the typical Swedish standard for a two person occupancy residential unit which is 240 kWh per month. The grid tied photovoltaics generates the electricity needed for primary energy use and the rest is fed to the national grid. The system designed for the house includes twenty polycrystalline photovoltaic panels (320 W, dimensions 1640*990*35mm) and generates 120 kWh/m^2 to 240 kWh/m^2 per year. So, the net energy fed to the grid is 5090 kW/year. The wind turbine was not operational during the occupancy test period. Due to the low energy consumption, the house is self-sufficient and exports the extra kW produced to the grid, performing as an ultra plus energy building – even better than the targeted zero energy goal.

The simulation results for the annual thermal performance demonstrate an indoor temperature within the standard norms of 18- 26 °C for 52 % of the total occupancy time in the bedroom, 49 % in the living area and the kitchen and 50 % in the bathroom. The three passive systems used, EAHE, the Trombe wall and the green wall, managed to reduce the heating loads to a total of 175 kWh. This is considered a good contribution from the passive systems used for the cold Swedish climate where depending on passive means for heating is hard to achieve. The performance could have been improved if the EAHE was deeper; however, this was not an economically feasible solution, so the depth was restricted to 3.5 m. Cooling loads are 63 kWh, which is a bit high compared to the expected performance, but that was due to overheating from both the Trombe wall and the green wall in peak summer months. The discount payback period is 15 years and the simple payback period is 14 years and 10 months. So, the building will achieve its zero cost goal, compensating for its cost before the end of its life time.

No waste was generated during the building's construction. Leftover wood, straw and reeds were easily composted. Also, since the house became operational in January 2018, no waste has been produced. After the house's end of life, the main building skeleton and components will be degraded back into nature, leaving no waste. Zero indoor air pollutants (which are already minimal due to using natural building materials) were achieved through planting *Rhapis excels, Hedra helix* and many others in the green wall and Trombe wall vents. Such plants passively remove indoor air pollutants and can produce oxygen at night, thereby reducing CO_2 concentrations. Using clay and lime as internal plasters also enhance the indoor air quality (IAQ). More in-depth calculations and monitoring is needed as a follow up to this study to prove the zero indoor air pollutants goal.

All construction methods employed were low-tech. Basic tools, such as screwdrivers for the walls, roof and floor, were used for construction and there no deep digging, drilling or heavy machinery was necessary. The materials were sourced from the surrounding neighborhood which also reduced the total emissions. Moreover, carbon sequestering of the plant-based materials used in the construction offset what little carbon was emitted from material transportation. Initially, a green roof was planned not only to offset carbon, but also to manage storm water run-off and promote biodiversity, but in the end it was not possible due to the space required for photovoltaics.

5. Conclusion

The project paid respect to the social and cultural needs of a special group of users, in this case, refugees, and aimed to involve them in the design and construction processes while raising awareness on the feasibility of cost efficient and climate responsive temporary shelter options. The indoor air temperatures of each room were simulated and calculated together with the energy consumption and heating and cooling systems' energy savings. The energy savings are also translated to CO_2 savings and carbon credit is calculated. The outcome of both the modeling simulation and numerical calculations proved that the building in many ways is acting beyond the 6Zs target. This study proved that minus carbon and plus energy is feasible for residential units and can be a cost and time efficient alternative for refugee housing. Using straw and reeds as the main construction materials has a lot of potential for the future, especially because they are considered agricultural waste in many countries worldwide. Finally, an economic analysis helped to assess the visibility of eco-cycle and passive systems in the Swedish market for low impact temporary housing. The methodology for achieving the 6Zs can be applied in other climatic contexts as the passive systems can be adjusted based on climate parameters. The decision of which natural building materials to use should also be based on their availability in the local geographical context and will likely necessitate an adjustment to the construction techniques employed.

6. Acknowledgments

The author would like to acknowledge Crafoord, Aforsk, AQR, Stiftelsen Mellby Gård and Almi for funding together with all research assistants, the Municipality of Lund and Malmo University for hosting this project.

References

[1] UNHCR 2017 Figures At A Glance *UNHCR - Figures at a glance*

[2] Janmyr M 2018 UNHCR and the Syrian refugee response: Negotiating status and registration in Lebanon *International Journal of Human Rights*

[3] Delmi (Delegationen för migrationsstudier) 2017 *Utrikes födda personer och utländska medborgare i Sverige (2000-2017)*

[4] Manfield P, Ashmore J and Corsellis T 2004 Design of humanitarian tents for use in cold climates *Building Research and Information*

[5] Ohlson S and Melich R 2014 Designing and developing sustainable housing for refugee and disaster communities

[6] Dabaieh M and Borham A 2015 Acclimatization measures for temporary refugee shelter in hot arid climates; low-tech mobile solutions using bedouin tents. *Architecture in (R)evolution.Architecture in (R)Evolution, 31st International Conference of Passive and low Energy Architecture*. ed M Cucinella, G Pentella, A Fagnani and L D'Ambrosio (Bologna: Ass. Building Green Futures)

[7] Murphy M P and Ricks A 2013 Beyond Shelter: Architecture and Human Dignity *The Journal of Architecture*

[8] Karunasena G and Rameezdeen R 2010 Post-Disaster Housing Reconstruction: Comparative Study of Donor vs Owner-Driven Approaches *International Journal of Disaster Resilience in the Built Environment*

[9] Kuittinen M and Winter S 2015 Carbon Footprint of Transitional Shelters *International Journal of Disaster Risk Science*

[10] Cornaro C, Sapori D, Bucci F, Pierro M and Giammanco C 2015 Thermal performance analysis of an emergency shelter using dynamic building simulation *Energy and Buildings*

[11] Crawford C, Manfield P and McRobie A 2005 Assessing the thermal performance of an emergency shelter system *Energy and Buildings*

[12] Yu Y, Long E, Shen Y and Yang H 2016 Assessing the thermal performance of temporary shelters *Procedia Engineering*

[13] Atmaca A and Atmaca N 2016 Comparative life cycle energy and cost analysis of post-disaster temporary housings *Applied Energy*

[14] Islam H, Zhang G, Setunge S and Bhuiyan M A 2016 Life cycle assessment of shipping container home: A sustainable construction *Energy and Buildings*

IOP Conf. Series: Earth and Environmental Science **352** (2019) 012064 doi:10.1088/1755-1315/352/1/012064

Development of a Greenlandic sustainable index for residential buildings

Martin Erik Labansen[1] and Diane Bastien[1]

[1] University of Southern Denmark, 55 Campusvej, Odense M, 5230, Denmark

Martin.labansen@gmail.com

Abstract. Globally, there has been a rising focus on improving the sustainability of buildings. Reaching zero net greenhouse gas emissions or producing excess renewable energy throughout the entire lifetime of a building is now clearly possible and certainly worth aiming for. However, in remote Nordic regions like Greenland, current building practices are so far away from these targets that setting intermediary goals is the only option for moving forward in the construction industry. Besides, remote Nordic regions are facing different challenges that are not considered into current building certification schemes, making some practices that can be highly sustainable in some regions inapplicable or even detrimental there. It would be highly desirable to have a certification scheme more accessible for improving current building practices in remote Nordic regions. For a sustainability index to be widely adopted in such a context, it should be relatively simple and straightforward to implement. This contribution presents the Greenlandic Sustainable Index (GSI) for residential buildings, which was developed with the goal to be simple to implement while being ambitious in terms of level of performance. This index is composed of three main categories: Economical, Social and Environmental sustainability.

1. Introduction

There are currently over 50 different building certification programs all around the word. Some are regional while others have been implemented in more than 50 countries. They can vary greatly regarding their energy and sustainability goals, with some focusing on one main parameter (e.g. simulated energy consumption) and others encompassing a wide range of topics. Reaching zero net greenhouse gas emissions or producing excess renewable energy throughout the entire lifetime of a building is now clearly possible and certainly worth aiming for if one wants to alleviate environmental impacts of the construction industry. However, in remote Nordic regions like Greenland or Northern Canada, current building practices are so far away from these targets that setting intermediary goals is the only option for moving forward in the construction industry.

While some indicators of the sustainability of buildings are universal (e.g. energy-efficiency), the sustainability of some practices is strongly dependent on the local climate, degree of urbanisation and culture. Remote Nordic regions are facing specific challenges that are not considered into current building certification schemes, making some practices that can be highly sustainable in some regions inapplicable or even detrimental there (e.g. proximity to public transport or white roofs).

While it is highly valuable to have some building certification programs that have a comprehensive agenda and consider various aspects related to sustainability like the DGNB program in Denmark [1], attaining this certification is currently too time consuming to be adopted in Greenland where construction costs are already high. With the current building codes that are far from stringent in terms of energy consumption in Greenland and Northern Canada, it would be highly desirable to have a certification scheme more accessible for improving current building practices in remote Nordic regions.

Considering the currently high demand for building energy consumption allowed in building codes in Greenland and Northern Canada, there is ample room for improvement.

This contribution presents the Greenlandic Sustainable Index (GSI) for residential buildings. This index was established with the goal to be simple to implement but ambitious in terms of level of performance. This index was developed as a part of a bachelor project in collaboration with a local Greenlandic industry partner.

2. Theoretical background

The Brundtland report defined sustainable development as*" (...) development that meets the needs of the present without compromising the ability of future generations to meet their own needs. "* [2]. Herein the interpretation that materials and resources should be used in such a way that the future generations will still have access to those. Another interpretation is also that one's culture should be preserved so that future generations can have the same access to it as the current generation does.

One may then define a sustainable building as a construction that is built in a way that it has low impacts on the environment while also providing a healthy living space for its residents that supports their daily activities. A sustainable building in the Arctic region should reflect the special needs associated with this geographical location, both in terms of climate and in culture.

Luis Suter and Robert Orttung wrote in 2016 a conference paper [3] which can be considered as a stepping stone for developing a sustainable building certification scheme for the Arctic. Herein they describe and present methods for assessing sustainability that have been implemented in other countries and cover merits and flaws of different approaches.

2.1. Scope

One import decision when developing a sustainability index is related to amount of data required for proving compliance. Here is it necessary to find a balance between exhaustiveness and user-friendliness. A complex, exhaustive index require fine, detailed data, which is available only once the design is nearly finished. This largely prevents using the index as a design tool. For this index, considering the lax current building regulations (e.g. the 2006 Greenlandic Building Code [4]), high construction costs and labour shortage in Greenland, it was decided to aim for a relatively simple index which can be used at the early/mid design phase for informing more sustainable design decision.

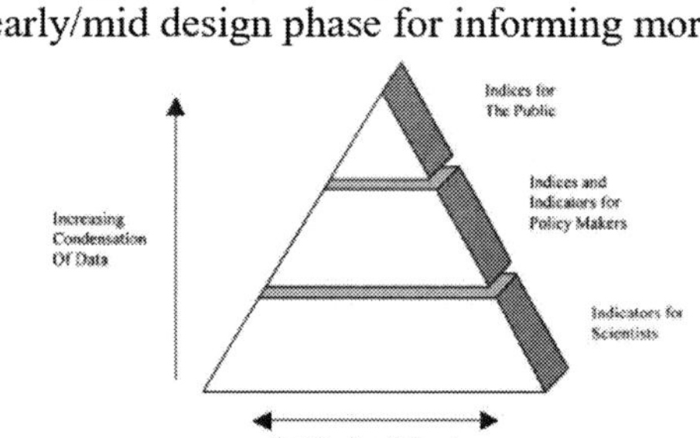

Figure 1. Pyramid illustrating the need for condensation of data for communicating with different target groups. Source: [3]

2.2. Indicators

In defining the number of indicators for an index, there is a clear trade-off between accuracy and accessibility. On one hand, the number of cities or institutions that adopt an index testify to its usefulness as a tool to compare global sustainable development policies. However, in order to make cities comparable, one also must aggregate the large number of indicators that comprise the breadth of the pillars of sustainability into a relatively few core indicators. This simplification reduces accuracy. The solution lies in selecting a set of indictors that can fulfil both the global comparative need while remaining locally relevant, which remains a pressing issue [3]. A successful index should be able to be deployed broadly across the target region, using the smallest possible number of indicators without sacrificing scientific accuracy.

IOP Conf. Series: Earth and Environmental Science **352** (2019) 012064 doi:10.1088/1755-1315/352/1/012064

3. The Greenlandic Residential Sustainable Index

Greenland is today still a part of Denmark. The culture and way of life in these two countries vary greatly even though there is a common history. Greenland today have self-government and their own building regulations.

This index is composed of three categories that encompass 13 indicators which are evaluated with 27 parameters. There are up to100 points that can be collected from the various parameters for each category. This section briefly describes the main indicators and parameters of the three categories and justify their relevance. An overview of all sustainability parameters and how they are calculated and weighted is then presented as a checklist in Table 2.

3.1. Economical sustainability

Within the economical sustainability category, the goal is to reduce expenses, both during the construction and the entire lifespan of the building, while keeping the value of a building. In Greenland the land is allocated by the municipality and therefore available at no cost. The cost of a building is therefore primarily dependent on land development, materials and workforce. A high building cost represents a scarcity of resources whether it be manpower or building materials. When this happens in the construction market it is said to be "overheating". When the construction of a building is rushed due to labour shortages, the risk of mishaps and shoddy workmanship increases which can have a significant impact on the effectiveness of the building regarding energy consumption, function and durability.

This category also addresses issues related to the quality of foundations and permafrost, which are of high importance for the building structural integrity and thus its longevity and economical sustainability.

Utilization

Area utilization

When designing and/or constructing a residential building, utilizing as much of the construction area for the residence is ideal, as you then essentially get as much real estate as possible for the area available. In DGNB the procedure of calculating the area utilization is by defining what makes a useable area (UA) and dividing this by the gross area (GA). The GA is comprised of the entire land area while in DGNB the UA is excluding stairs and hallways that are outside the residences.

Since the weather can be rather harsh in Greenland, a residential building that provides shelter for people when they access their residence is clearly beneficial and promotes safety (e.g. by avoiding injuries caused by icy stairs). This index therefore includes protected stairs and hallways in the useable area.

Room height

The height of rooms in a building can play an important factor in an eventual building retrofit that affects the building performance or room usage. Having a higher room height increases the possibilities when altering room use, as the new changes will likely involve systems like new ventilation ducting or cable trays, where a higher room height can greatly facilitate the installation of these new systems without hindering occupants' movements.

Nonbearing inner walls

During the lifetime of a residential building, room usage generally evolves over time following the needs of families. Often residents wish to expand, demolish and move room partitions in their building. If the inner walls are nonbearing, the structural integrity concerns involved when changing the placement of inner walls would be significantly less which would also reduce refurbishment costs.

Access to technical installations

The same concept is applied to the technical installations. Maintenance, repair and replacement of technical installations is easier if the installations are not cast in place or not readily accessible.

IOP Conf. Series: Earth and Environmental Science **352** (2019) 012064 doi:10.1088/1755-1315/352/1/012064

Cost

Building cost
The complete construction cost of a building project (DKK/m^2), and not a complete LCC analysis, is used in this index to assess the quality of the project, rewarding low costs over higher ones. This is both due to the intent to keep the index simple, and to the fact that little to no data is to be found on building materials life span in Greenland. The goal with this indicator is not to hinder investment in energy efficiency or renewable energy but rather to control excessive costs related to the remote location of Greenland. After consultation with local experts [5], a categorization of the cost per built area was developed and is presented in the GSI checklist.

Maintenance

Estimated lifespan of building elements
Since no arctic building element index is available, this indicator is thus based on the building element index that is found in the Danish DGNB certification program [1]. With the significant differences between the Danish and Greenlandic climates, it would be desirable to adapt estimations of building material longevity to Greenlandic conditions in the future.

Placement

Foundation anchored to bedrock
Permafrost, which is ground that has been at or below 0° C for two or more consecutive years, is present in Greenland. Thawing of the permafrost affects the structural quality of soils and sediment. Rising of the permafrost can also generate significant damage, causing water expansion and rising ground. With climate change, it is risky to build on permafrost. The safest option is to simply avoid soils with permafrost. This is best done by anchoring the foundation on bedrock.

Sheltered entrance to building
Large amounts of snowfall can potentially cover the entrances due to wind and snow drifting [6]. The annoyance, work and resources required to keep entrances passable can be significantly lessened when the predominant wind directions are considered. This index thus rewards entrances that are designed based on local assessment of wind directions.

3.2. Social sustainability

As a typical "western" house has been the template for the fabricated houses of the north during the 20th century, the focus on adapting the building template to accommodate the cultural needs in Canada have received more focus in the past decade [7].

Accessibility

Presence of entrance stairs
For promoting accessibility to all, this index rewards buildings that not have any stairs between the building and its access road. If stairs are unavoidable, the presence of a ramp or elevator is then promoted.

Indoor Air Quality

Mechanical ventilation
A study of the Indoor Air Quality (IAQ) in 79 households in Greenland showed insufficient ventilation rates and poor IAQ in many buildings [8]. The use of mechanical ventilation is slowly being introduced in Greenland. While not required in BR06, it is more common for apartment buildings than for single family houses. This index promotes the installation of mechanical ventilation and rewards systems with efficient heat recovery, flow control and low noise levels in living area.

Fire safety

Fire alarm system
As a fire alarm system is not yet mandatory for single family houses in the Greenlandic building regulations, the GSI rewards smoke detectors for promoting the safety if occupants.

Fire engineering installations
Fire engineering installations such as sprinklers, water-filled hoses and fire extinguisher can save lives and are thus promoted in this index.

Utility space

Depository available
In Greenland, hunting, fishing, foraging, trekking and camping is an important part of the culture. The interior design of a building should be made in a way to promote such activities. These generate a demand for a butchering space and large freezers, as most of the animals, game and berries are stored in freezers. Access to a depot or storage room for storing seasonal equipment, tools and chest freezers can have a tremendous importance for building occupants in Greenland.

Scullery room
A scullery room would support traditional activities such as hunting and fishing by providing an appropriate place for butchering large animals at home.

Common space

Communal area
Celebrations are an important part of Greenlandic culture. When a person celebrates a birthday, wedding or catching ones first seal, reindeer or musk ox, an open house event known as "kaffillerneq" is held where everyone remotely associated to or known by the family can come and drink coffee or tea while eating cakes, pastries and/or some local delicacy. A communal building, banquet hall or area with access to kitchen and toilet facilities would support the tradition of hosting social gatherings.

Communal residence
Having access to bedrooms or dormitories would provide accommodation for the guests of the neighboring residents. Not having to build a guestroom in every household by instead having a temporary accommodation for guests is a more effective use of space and resources.

3.3. Environmental sustainability

Energy consumption

Building energy consumption
Greenland has its own energy framework which is described in the Greenlandic building regulations BR06 [4]. Greenland is split up into two zones where zone 1 is the region south of the polar circle and zone 2 is the region north of the polar circle. For a single floor family house, the maximum permissible energy consumption is equivalent to 194 kWh/m^2 pr. year and 231 kWh/m^2 pr. year for zone 1 and 2 respectfully. For the size of an average Greenlandic home, 66m^2 [9], the comparison of the energy framework in Greenland and Denmark is presented in Table 1 below.

Table 1. Energy frameworks for a single storey 66 m^2 home in Denmark and Greenland

	Energy framework, 66 m^2 single floor house
Denmark	45 kWh/m^2/year
Greenland, zone 1	194 kWh/m^2/year
Greenland, zone 2	231 kWh/m^2/year

Greenland's energy framework for a 66 m^2 building is over four times higher the amount allowed in Denmark. Buildings with such a high energy consumption also pose concerns about thermal comfort and durability (risk of mold growth on cold surfaces). Therefore, this index will reward buildings with

IOP Conf. Series: Earth and Environmental Science **352** (2019) 012064 doi:10.1088/1755-1315/352/1/012064

a simulated energy consumption of 60 kWh/m^2 per year and less, as calculated with BE18 [10], the national program for documenting code compliance in Denmark.

Renewable electricity production
Having access to a renewable electricity or heat source not only reduces environmental impacts but also increases the resiliency of a remote settlement. This index rewards local production of electricity where points are awarded depending on the percentage of the building heating needs covered.

Renewable heat production
Similarly, the local production of renewable heat is promoted within this index.

Water and waste management

Access to water infrastructure
The state of the water and waste infrastructure in the different settlements in Greenland differ greatly. Some places have waterworks that supply clean water to the whole community all year round, while some places still utilize water tanks or water station while some places, like Qaanaaq, must even resort to harvesting ice blocks during winter with the purpose of melting them to get water [11].

Access to wastewater infrastructure
Not every home has access to sewerage. Even in the largest cities in Greenland, some residences have septic tanks or even bucket toilets. Having bucket toilets implies material consumption for the plastic bags, creating higher environmental impacts than if it was connected to a sewer system. Same could be said for septic tanks, as they require trucks to come and empty them on a regular basis. Therefore, this index promotes buildings that are equipment with municipal wastewater infrastructure.

There is currently no wastewater treatment in Greenland. This is a serious issue that should be solved at the national or local political level. The rationale for promoting wastewater infrastructure in this context it that it is arguably easier to lead wastewater to a treatment plant when an existing sewer system is already in place.

Mold index

Highest simulated mold index
Mold has a damaging effect on building materials and human health. It is an agreed consensus that Greenland is having a significant problem with mold in its national housing stock. A statement from the Greenlandic ministry from September 2015 mentions, that out of, at that time, 399 residences that stood empty because of planned renovations, 300 of those had problems with mold [12].

A methodology for calculating the mold index of a material has been established [13] and implemented in building hygrothermal simulation software. With the Arctic being a place with high risk of mold growth, such programs are a good tool to help avoid mold problems at the design stage. The WUFI Pro [14] program and the addon named “WUFI® Mold Index VTT” developed by VTT Technical Research Centre of Finland LTD can be used for estimating the mold index associated with various building elements.

Materials

Use of local or national materials
If there is any chance of using local or national resources, this would not only be preferable due to the reduced GES emissions associated with shipping, but also for developing local industries and enable Greenland to be less dependent on imported goods.

Use of Eco-labelled materials
Today more environmentally friendly products or materials can usually be identified by an ecolabel. The ecolabels that can be used to document the sustainability of the building materials used in a project must be listed in the Ecolabel index [15], which currently include 463 labels.

Table 2. Greenlandic Sustainable Index Checklist

Indicators		Criteria				Points
ECONOMICAL					(0-100)	
Utilization					(0-30)	
Area utilization		> 0,8	If yes		5	
Room height *		> 2.8 m	If yes		5	
Nonbearing inner walls		Y/N	If yes		10	
Access to technical installations		Y/N	If yes		10	
Cost					(0-25)	
Building cost *		Select one:	> DKK 26.000/m²		0	
			< DKK 26.000/m²		5	
			< DKK 25.000/m²		10	
			< DKK 24.000/m²		15	
			< DKK 23.000/m²		20	
			< DKK 22.000/m²		25	
Maintenance					(0-25)	
Estimated lifespan of building elements *	Facade					
	Load-bearing structures	Select one:	< 40 years		0	
	Foundation		40 < 80 years		2	
	Roof		> 80 years		5	
	Door and window					
Placement					(0-20)	
Foundation anchored to bedrock		Y/N	If yes		10	
Sheltered entrance to building		Y/N	If yes		10	
					Sum:	
SOCIAL					(0-100)	
Accessibility					(0-20)	
Presence of entrance stairs *		Y/N	If no		20	
IF YES	Ramp available	Y/N	If yes		10	
	Elevator available	Y/N	If yes		10	
Indoor Air Quality					(0-25)	
Mechanical ventilation	Presence of a mechanical ventilation system?	Y/N	If yes		5	
IF YES	Including minimum flow control at 3 levels?	Y/N	If yes		5	
	Including heat recovery?	Select one:	< 75 %		0	
			75 > 85 %		5	
			> 85 %		10	
	Noise below 25 dB(A) in living areas?	Y/N	If yes		5	
Fire safety					(0-10)	
Fire alarm system	Smoke detector?	Y/N	If yes		5	
	Carbon monoxide detector?	Y/N	If yes		2	
Fire engineering installations		Y/N	If yes		3	
Utility space					(0-25)	
Depository available		Y/N	If yes		15	
Scullery room		Y/N	If yes		10	
Communal					(0-20)	
Communal area		Y/N	If yes		10	
Communal residence		Y/N	If yes		10	
					Sum:	
ENVIRONMENTAL					(0-100)	
Water and waste management					(0-10)	
Access to water infrastructure		Y/N	If yes		5	
Access to wastewater infrastructure		Y/N	If yes		5	
Energy consumption					(0-50)	
Building energy consumption *		Select one:	> 60 kWh/m²		0	
			60 < 40 kWh/m²		20	
			40 < 20 kWh/m²		25	
			< 20 kWh/m²		30	
Renewable energy used for power *		Y/N	If yes		10	
Renewable energy used for heating *		Y/N	If yes		10	
Mold Index			Struct.	Int.	(0-20)	
Highest simulated mold index after 4 years		Select one:	MI = 0 < 1	MI=0<0.5	10	
			MI = 1 < 2	MI=0.5<0.8	8	
			MI = 2 < 3	MI=0.8<1	5	
			MI > 3	MI > 1	0	
Materials					(0-20)	
Use of local or national materials *		Y/N	If yes		10	
Use of Eco-labelled materials *	Walls	Y/N	If yes		2	
	Floors	Y/N	If yes		2	
	Ceilings	Y/N	If yes		2	
	Roof	Y/N	If yes		2	
	Windows/doors	Y/N	If yes		2	
					Sum:	
* Interpolation of results, if possible, is encouraged.					**Total:**	

4. Discussion and conclusion

A user manual with instructions for evaluating the different parameters is available at GSI User Manual. This index was tested on a case study that consists of an apartment building to be built in Nuuk, Greenland. This case study allowed to conclude that this index is easy to implement during the design phase, as intended. This first experience allowed to validate the usefulness of this tool for informing more sustainable design decisions in Greenland.

While more complex and exhaustive sustainable building certification programs exist, they are not implemented in Greenland because they represent a seemingly unreachable target compared with the current building practices based on the 2006 Greenlandic Building Code. With this contribution, the authors hope to start a dialogue with other actors interested in sustainability in the Arctic and look forward upon implementing this first version of the GSI in more case studies and its future refinement.

Acknowledgments
A special thanks goes to Qarsoq Tegnestue ApS whom provided the materials for the case study and the local expertise and insight regarding the Greenlandic construction industry.

References

[1] Green Building Council Denmark. *DGNB-manual for etageejendomme og rækkehuse*. 2016.
[2] World Commission on Environment and Development. *Our common future*. Oxford University Press, 1987.
[3] Suter L, Orttung R. Building an Arctic Urban Sustainability Index: A Reference to Aid in the Selection of Indicators. In: *Arctic PIRE Annual Meeting*. Washington DC, 2016.
[4] Direktorat for Boliger of Infrastruktur. Bygningsreglement. Grønlands hjemmestyre, 2006.
[5] Mario Jensen Manager/Partner at Qarsoq Tegnestue Aps. Personal communication. 2019.
[6] Semple B. House for the Canadian Arctic. *Home Energy* 2008; 24–29.
[7] Semple B. *Challenges for Remote Areas : The Design , and Construction of Sustainable Housing and Communities in the Canadian Far North*. Ottawa: [1] Semple B. Challenges for Remote Areas : The Design , and Construction of Sustainable Housing and Communities in the Canadian Far North., 2009.
[8] Kotol M, Rode C, Clausen G, et al. Indoor environment in bedrooms in 79 Greenlandic households. *Build Environ* 2014; 81: 29–36.
[9] Grønlands Statstik. Statistisk Årbog.
[10] Statens Byggeforskningsinstitut. BE18, https://sbi.dk/beregningsprogrammet/Pages/Start.aspx (2018).
[11] Qaasuitsup Kommunea. Qaasuitsup.gl, http://qaasuitsup-kp.cowi.webhouse.dk/dk/by-_og_bygdeplaner/qaanaaq/ (2018).
[12] Petersen S. sermitsiaq.ag, https://sermitsiaq.ag/node/181800 (2018).
[13] ASHRAE. *Standard 160-2016 Criteria for Design Analysis in Buildings*. Atlanta: American Society of Heating, Refrigerating, and Air-Conditioning Engineers, 2016.
[14] WUFI. *WUFI Pro 6.1 help*. topic 65: Fraunhofer Institute, 2017.
[15] Big Room inc. Ecolabelindex, http://www.ecolabelindex.com/ecolabels/ (2018).

IOP Conf. Series: Earth and Environmental Science **352** (2019) 012065 doi:10.1088/1755-1315/352/1/012065

Tracing the environmental impact origin within the existing building portfolio of prevailing building typologies

Julie Rønholt[a], Leonora Charlotte Malabi Eberhardt[b], Morten Birkved[c], Harpa Birgisdottir[d], Niki Bey[e]

[a] Technical University of Denmark, Department of Civil Engineering, Anker Engelunds Vej 1, DK-2800 Kgs. Lyngby

[b,d] The Danish Building Research Institute, Department of Energy Efficiency, Indoor Climate and Sustainability, Aalborg University, A. C. Meyers Vænge 15, Building A, DK-2450 Copenhagen SV

[c]SDU Life Cycle Engineering, Department of Chemical Engineering, Biotechnology and Environmental Technology, University of Southern Denmark, Campusvej 55, 5230 Odense-M, Denmark

[e]Technical University of Denmark, Division of Quantitative Sustainability Assessment, Department of Management Engineering, Diplomvej Building 371, 2800 Kgs. Lyngby, Denmark

Affiliation where the research was conducted: Technical University of Denmark

julie.roenholt@hotmail.com

IOP Conf. Series: Earth and Environmental Science **352** (2019) 012065 doi:10.1088/1755-1315/352/1/012065

Abstract. As circular economy (CE) is becoming a growing focus in the building industry due to the industries large resource consumption, waste production and environmental impacts a better understanding of buildings material composition, resource consumption and resulting environmental performance becomes increasingly important in order to support the transition towards CE. The research presented here is a stepping stone in order to investigating the existing building portfolio and the conducted analysis. It is used to get further information on what is included and where there is lack of information, in order to understand what is missing or what is needed to develop CE design strategies. Although, life cycle assessment (LCA) is increasingly used by the industry to assess these aspects, the lack of a systematic analytical approach as well as the high building complexity and diversity between buildings limits generalised knowledge from these studies. However, it is assumed that there are certain commonalities between the existing portfolio of prevailing building typologies which can be deducted. The study at hand is a part of a larger research project that aims at developing industry specific tools to support designers and decision makers select CE design strategies that improve the environmental performance and resource consumption of buildings. On the basis of a comprehensive systematic literature review (SLR) of whole building LCAs the paper at hand aims at tracing the environmental impact origin within the existing building portfolio of prevailing building typologies. To identify potentially important building parameters relevant to the resulting environmental impact performance and resource consumption of different building typologies. Based on 39 building LCA case studies that matched the specific inclusion and exclusion criteria of the SLR a focus on global warming and climate change was detected. It was found that even though buildings have different characteristics (size, typology, storeys, reference study period and location), for most of the buildings' environmental impacts were predominantly related to the production of structurally important concrete components e.g. the structural frame, the external envelope, floor slabs and walls. To point towards which CE design strategies should be used to improve the environmental performance of buildings; the environmental information from the studies was insufficient due to lack of detailed information.

1. Introduction

The building sector is responsible for 40% of energy consumption, 30% of greenhouse gas emission, 30% of raw materials consumption and 25% of solid waste production globally [1]. Hence the global economic growth and increasing population puts extensive pressure on the natural environment and resource reservoirs available, due to an increased demand for new construction [2]. Hence focus in the sector on resource efficiency and the circular economy (CE) has grown in recent years. Furthermore, interest in a better understanding of buildings' life cycle resource consumption and embodied environmental impacts is also growing in order to utilise the resource efficiency opportunities and seek out better design solutions [3]. A way to attain this understanding is by conducting life cycle assessment (LCA) of buildings [4]. LCA can facilitate CE decision-making by identifying the largest environmental impacts-reduction opportunities [3] and is a scientifically based method used to assess buildings' environmental impacts within a building's life cycle [5,6]. Over the last 15 years the number of published and conducted LCAs related to case studies of buildings has grown continually [7] and several standards for calculation guidelines exist. However, buildings are complex, unique, dynamic and long-lived entities, consisting of multiple components and materials, which all provide different functions, rates of replacement and degradation. Hence a building is more than the sum of its components [3] and the environmental impacts and resource consumption depend of this complex context. Systematic analysis of existing LCAs of buildings exist e.g. in Switzerland [8] and semi-systematic LCAs have been performed on buildings in Denmark for several years [9]. However, there is a need for a broader and more robust systematic analytical LCA approach to ensure that the complexity in buildings and the complex interaction between various elements are identified in appropriate manners [10]. Thus, allowing generalized learning in order to develop new design strategies and methods that help reduce environmental impacts and resource consumption and avoid focusing on optimising building components and materials of less environmental importance.

By tracing conducted LCAs and the environmental impact origins within the existing building portfolio of prevailing building typologies, a fundamental knowledge from a thorough and systematic analysis of the material composition and the typical input and output of resources within the life cycle of the buildings is achieved.

2. Methodology

A comprehensive systematic literature review (SLR) was conducted according to the method presented by [11]. The SLR focused on identifying and registering environmental hotspots and resource consumption found in existing building LCA case studies. The SLR considered both journal papers and "grey literature" (i.e. non-peer-reviewed scientific material). The outline of the SLR is shown in Figure 1. The Scopus database was used to search for relevant literature using a search string with a specific set of predefined keywords specifically aimed at titles, abstracts and keywords: TITLE-ABS-KEY ((building) AND (LCA OR "life-cycle assessment") AND ("case study")).

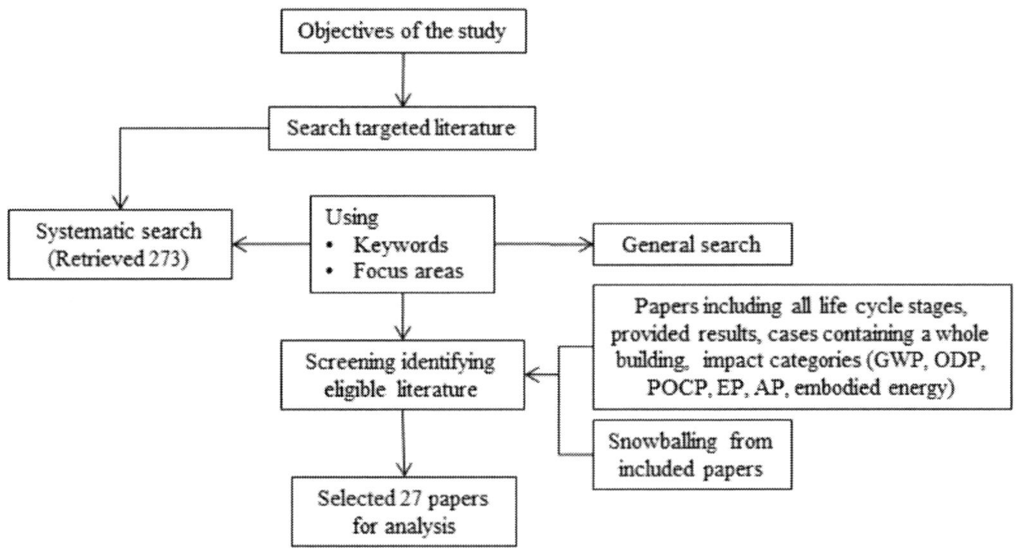

Figure 1. Outline of the SLR study methodology

2.1 Inclusion and exclusion criteria

The papers were carefully chosen using the following inclusion criteria: contain a whole building LCA of an existing building over the building's entire life cycle, which covers four stages: production, construction, use and end-of-life. Furthermore, the papers had to contain information about: 1) in which life cycle stage the highest environmental impacts were induced, 2) which building components and materials induced the highest environmental impacts and 3) at least one of the following midpoint impact categories: global warming potential (GWP), ozone depletion potential (ODP), photochemical ozone creation potential (POCP), acidification potential (AP), eutrophication potential (EP), (chosen based on their use for building LCAs in the EU/DGNB system [12]) and/or embodied energy. As focus was exclusively on the embodied environmental impacts related to the building materials, environmental impacts related to operational energy consumption were not taken into account. Therefore, if the studied LCAs showed that the "use stage" had the highest impact, investigations aimed at determining whether this was due to operational energy consumption or due to the replacement of materials throughout the life of the building.

2.2 Source selection process

All potentially relevant papers identified from the search where filtered through a screening process based on the aforementioned inclusion and exclusion criteria. From Scopus, 633 papers were identified from the search and were narrowed down to 273 by limiting the search to the three topic areas:

engineering, environmental science and material science, and in addition by excluding papers not related to buildings. The 273 papers where subsequently screened by reading the title and the abstracts in order to identify those papers that match the scope of the study presented in this paper, eventually yielding 54 papers. These 54 papers were further evaluated based on their comprehensiveness, the extent of the information provided, and whether or not this information aligned with the scope of the study presented here. Hence, papers containing whole building LCAs of existing buildings covering the entire life cycle were identified focusing on the LCA method used in the studies and the LCA results provided, yielding 15 relevant papers. By performing "snowballing" between these papers (using the citations in the papers or the reference lists to identify additional papers), eight additional papers were found. Grey literature fitting the selection criteria known *a priori* by the authors was also included, resulting in a total of 27 relevant papers (see appendix).

2.3 Information extraction
Information extracted from the papers was registered in a spreadsheet and included the following parameters potentially relevant to the resulting environmental impact performance of the case buildings studied in the papers selected: publication year, location of the conducted LCA case study, building type, material composition, number of storeys, floor area (m^2), functional unit, building lifespan, life cycle inventory (LCI) database used and life cycle impact assessment (LCIA) method applied. Furthermore, the life cycle stage as well as the building components and materials with the highest environmental impacts were registered for each of the impact categories covered. In addition, as several of the studies identified focused on embodied energy, this impact indicator was also registered in order to include more studies. If the study contained more than one case building, each of these case buildings was registered separately. During the search, it was discovered that the papers used different terms to describe the life cycle stages and system boundaries. For example, the cradle-to-gate (raw materials extraction, transportation to factory and production at factory) is described both as "raw materials extraction", "construction stage" and "A1-A3". However, closer examination of what the papers include in cradle-to-gate indicates that the papers describe more or less the same. Hence, a common name reflecting the results extracted for the SLR was chosen following the structure of [13]. The lack of LCA (terminology and modelling) alignment between the papers leading to incomparability between the cases studied has to be taken into account when analysing the results from this SLR, i.e. LCAs can only be compared when performed using the same system boundaries, functional unit as well as LCI and LCIA method [14]. Similar issues are reported by [15], who conducted an SLR also applying predefined criteria and compared the LCAs of the different building typologies on a general level.

3. Results
This study does not consider a temporal scope, although the papers found to be of relevance for this study were all published in the last 15 years. In the past five years, the number of publications has increased, potentially indicating growing environmental concern/focus within the building industry [15]. The spatial distribution of the papers is shown in Figure 2. Most of the case studies within the papers originate from Europe. 26 out of 39 of the studies are from Europe, with the majority from Italy with 7/39, while the UK and Denmark account for 5/39 and 4/39, respectively. The remaining cases are spread over the rest of the world in both developed and developing countries, although they are dominated by the U.S., Turkey, Brazil and Malaysia. Focus on sustainable solutions is essential for reviving economic growth. In line with this aim, Italy has engaged the 2030 Sustainable Development Goals in its National Sustainable Development Strategy 2017/2030 (NSDS) [16]. The increased UK focus on embodied environmental impacts in constructions during the Olympic games in 2012 yielded strategies intended to reduce the amount of waste generated and designing for reuse [17]. The Danish government published a report with political strategies for a strengthened construction industry in 2014 [18] and their CE strategies in 2018 [19].

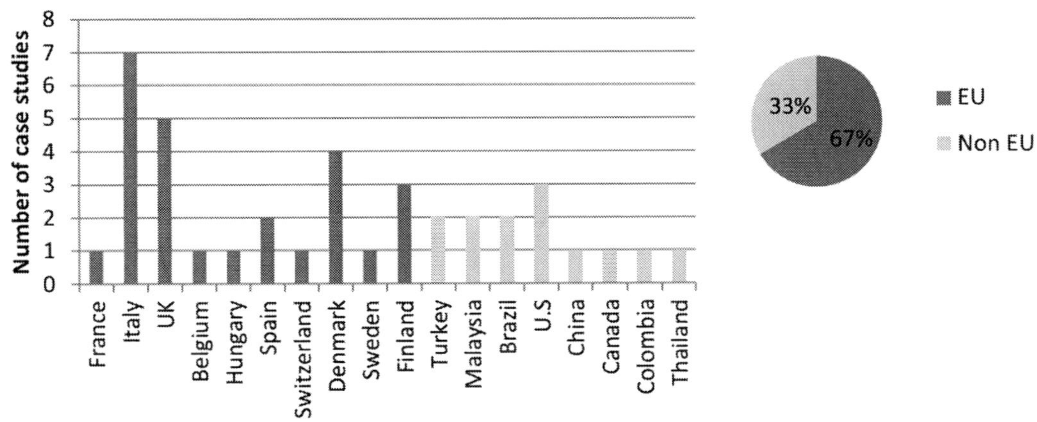

Figure 2. Selected building LCA case studies conducted in various countries. Divided into EU countries (blue, n=26) and non-EU countries (green, n=13)

Figure 3a shows the building typology in the case studies from the papers consisting of residential buildings (apartments and single family houses), offices, a few educational buildings and one opera house, 8/39, 14/39, 12/39, 4/39 and 1/39 respectively. Figure 3b shows that the reference study period of the buildings assessed ranges from 30 to 120 years. An applied reference study period of 50 years is dominant in the studied cases, with 24/39 occurrences. According to [20], the reference study period of a building plays an important role since the LCA methodology requires results for the construction, operation and end-of-life stage to be normalised on an annualised basis to allow for direct comparison. However, [20] and [14] both stress the issues of inadequately defined and poorly validated reference study periods in the LCA process. In particular, [20] highlights: no methodological documentation of any theoretical or empirical reasoning regarding the chosen selection of building reference study period, which varies between 25 and 100 years. However, 50 years is the most frequently applied building service life, typically justified by reference to other research also confirmed by [15], furthermore there is often no consistency in choice of building reference study period in relation to building type or geographical region and it is not stated whether the replacement of materials or construction elements during the building's reference study period is taken into account. These areas of concern are also found from the papers used in this study. Regarding the LCIA method, Figure 3c indicates that numerous impact assessment methods are applied in LCA studies. Figure 3c also shows that IPCC, CML and Eco-Indicator were the most common LCIA methods used, with respectively 19%, 17% and 17%. Furthermore, 19% of the studies did not specify the method used.

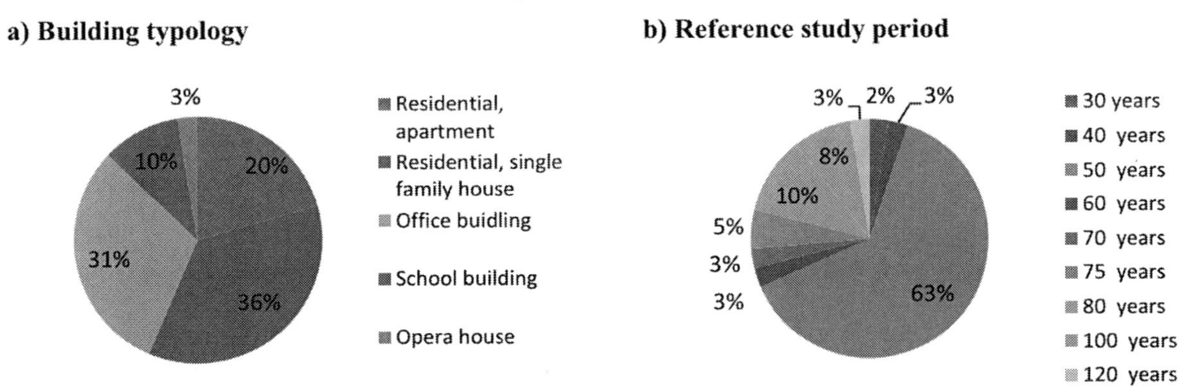

IOP Conf. Series: Earth and Environmental Science 352 (2019) 012065 doi:10.1088/1755-1315/352/1/012065

c) LCIA method

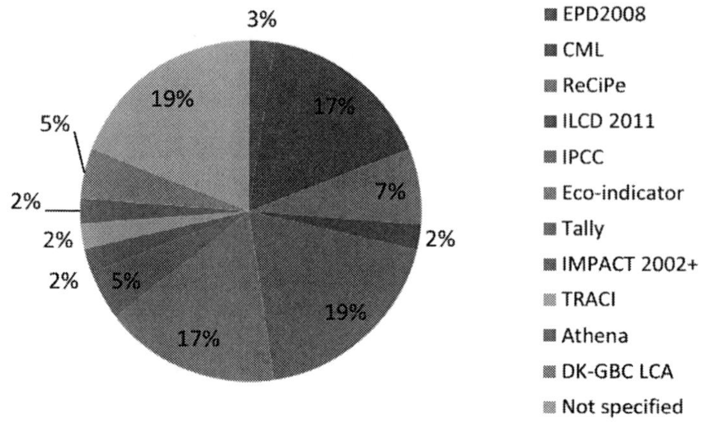

Figure 3. Mapping of the identified 39 LCA case studies from selected papers, a) Building typology, b) Reference study period, c) LCIA method applied.

Figure 4 provides an overview of the registered hotspots in the life cycle stages and hotspots for building components from all case studies. The results are only shown in overall terms and are not shown in specific units. The other life cycle stages were not registered as hotspots for any of the cases and are therefore not shown in the figure.

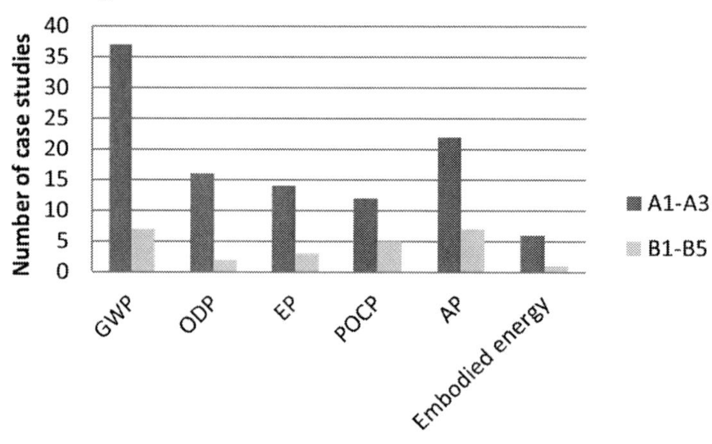

Figure 4. Number of registered life cycle stage environmental hotspots.

The results presented in Figure 5 show that the production stage (A1-A3) induces the largest impacts in all categories for by far most of the investigated cases. 7/39 cases include all of the five main impact categories for the life cycle stage hotspots. 37/39 includes climate change (GWP), indicating that many studies focus on climate change when conducting LCAs due to its status of the most well-known impact category. Maintenance during the use stage only contributes around 1-5% of the impact in several of the cases and impact categories. The vast majority of the buildings in the study, 29/38, are made of concrete structures. Perhaps indicating that concrete is the prevailing material choice in contemporary building culture. Furthermore, concrete consumption in a country can be used as an indicator of economic growth in that country [21]. As the buildings are primarily made of concrete, similar environmental impact hotspots are exhibited between the buildings, namely the production stage from the structural frame, external envelope, slabs and walls.

4. Discussion and conclusion

The SLR compiled environmental hotspots related to both life cycle stages and components for 39 case studies from 27 research papers. Several papers were found during the SLR process, but most of them were excluded due to lack of transparency, the limited inclusion of life cycle stages, or narrow focus, e.g. focus on specific building components, potentially overlooking determining factors for the overall environmental impact of the building.

Broadening of the keyword synonyms for "building" and "life cycle assessment" to include e.g. "house" or "dwelling" and "life cycle analysis" may lead to additional relevant papers. However, the research shows little or no deviation between the results of the case studies, suggesting that similar results would be found in other studies using these keywords. Hence, the conclusions found from this research are not expected to change.

The cases have different characteristics (size, typology, storeys, reference study periods, and materials) and are located in many different countries. The sample of cases is argued to be representative enough to provide a sound review of environmental hotspots in different building typologies. Although, the study at hand reveal that many LCAs on buildings exist varying reference study periods (ranging from 30 to 120 years) and choice of LCIA methods (11 different methods used) found between the studies indicate lack of a consistent LCA approach in the research field, yielding general conclusions difficult.

Consequentially, existing environmental information on buildings is insufficient to point towards which CE design strategies should be the basis for improving buildings' environmental impacts and resource consumption due to the lack of detailed information on material inputs and outputs, and the resulting environmental impacts, that change in magnitude and origin over time. Furthermore, it was discovered in the SLR that 37/39 studies include GWP, indicating that many studies mainly focus on GWP quantifies the greenhouse gas emissions when conducting LCA. This correlates with the popularity of GWP as a main metric. However, CE not only focuses on the environmental impacts associated with the consumption of resources but also the depletion of resources. [22] raise the issues of a poor correlation between GWP and other impact categories e.g. resource depletion. Consequently, a lack of focus on resource depletion was found as only 3/39 case studies included resource depletion impact categories such as abiotic depletion potential in their results.

Despite the lack of consistency between the LCAs it was, however, found from the SLR that, the largest embodied environmental impacts primarily are induced in the buildings' production stage (A1-A3) in all environmental impact categories investigated. As mentioned, the 39 case studies were dominated by concrete buildings. Due to the use of similar construction materials between the case buildings, the results between the different building types do not differ much. As a result, the environmental impacts at the production stage originate from the structurally important concrete components, e.g. structural frame, the external envelope, the ground slabs and the walls contributed with large environmental impacts in the production stage of the buildings. These studies indicate that due to the large environmental impacts in the production stage, impact reductions can be obtained by addressing more focus to the design stage. Furthermore, focusing on optimising the constructions or substituting concrete with less environmentally problematic materials would be a way to address the large environmental impacts from structural concrete elements.

IOP Conf. Series: Earth and Environmental Science **352** (2019) 012065 doi:10.1088/1755-1315/352/1/012065

Appendix: Systematic literature review (SLR) overview

		Reference	Publication Year	Year of origin	Location	Material composition	Storeys	Concept (new-build/ refurbishment)	Floor area [m2]	FU	Service life [years]	Life cycle stage hotspots: GWP	ODP	EP	POCP	AP	Embodied energy	Building component hotspots: GWP	ODP	EP	POCP	AP	Embodied energy	Database	LCIA method
1	Apartments	[7]	2016		Italy	Masonry	2	New-build	4036	1 m2.yr	100	A1-A3	A1-A3	A1-A3	A1-A3	A1-A3		Structure	Structure	Structure	Structure	Structure		Ecoinvent 2.2	EPD2008
2	Apartments	[23]	2016	2014	Turkey	In-situ reinforced concrete	5	New-build	1250	1 m2	50	A1-A3						Floors (concrete, steel)					Floors (concrete, steel)	ICE database	
3	Apartments	[24]	2015	2008	Sweden	prefab Concrete	5	new-build	11,003	1 m2	50	A1-A5						Structure (concrete)						IVL Miljödatabas Bygg)	
4	Apartments	[23]	2016	2014	Turkey	In-situ reinforced concrete	12	New-build	1200	1 m2	50	A1-A3						Floors (concrete, steel)					Floors (concrete, steel)	ICE database	
5	Apartments	[25]	2014	1975	Hungary	Prefab reinforced concrete	10	Refurbishment	6770	1 m2	80	A1-A3, B1-B5				A1-A3, B1-B5								Ecoinvent v 2.0	CML
6	Apartments	[26]	2013	2008	Italy	Reinforced concrete, bricks	4	New-build	1827	1 m2.yr	50	A1-A5, B1-B5						Envelope ext.						Ecoinvent	Eco-indicator 99, IPCC 2008
7	Apartments	[27]	2008	1965	Italy	Reinforced concrete,	10	New-build	6110	1 m2.yr	40	A1-A3	A1-A3	A1-A3	A1-A3	A1-A3		Concrete	Concrete	Concrete	Concrete	Steel	Steel	Field measured data	SEMC and Eco indicator 99
8	Apartments	[28]	2017	2012	China, in cold region	In-situ concrete	17	new-build	17558	whole building	50	A1-A3					A1-A3	Concrete, steel							IPCC 2007
9	Single family houses	[29]	2018		Malaysia	Hybrid concrete & timber	1	New-build	155	Whole building	50	A1	A1			A1								Ecoinvent + MYLCID data	ReCiPe midpoint [H]
10	Single family houses	[29]	2018		Malaysia	Hybrid steel and timber	1	New-build	155	Whole building	50	A3	A3			A3								Ecoinvent + MYLCID data	ReCiPe midpoint [H]
11	Single family houses	[30]	2018	1939	France		1	Refurbishment	59	1 m2.yr	50	A1-A3	A1-A3		A1-A3	A1-A3		Structure	Structure		Structure	Structure		Ecoinvent v. 3.3	ILCD 2011 midpoint+
12	Single family houses	[31]	2010	2007	Italy	Reinforced concrete & masonry	2 + basement	New-build	250	1 m2.yr	70	A1-A3	A1-A3	A1-A3	A1-A3, B1-B5	A1-A3								Ecoinvent 2.0 and site measurements	IPCC 2007
13	Single family houses	[32]	2005		Switzerland	Concrete	3	New-build	266	m2.yr		A1-A3, B1-B5			A1-A3, B1-B5	A1-A3, B1-B5								Ökoinventare für Energiesysteme (ESU)	CML
14	Single family houses	[33]	2015		Denmark	Concrete	1	new-build	149	1 m2.yr	120	A1-A3	A1-A3	A1-A3	A1-A3	A1-A3		Ground slab, roof	Roof	Roof. Windows	Roof	Roof, windows		ESU-CO- og Ökobaudatabaserne	
15	Single family houses	[34]	2010		Colombia	Brick, concrete and steel	2	New-build	140	1 m2	50	A1-A5, B1-B5	A1-A5, B1-B5			A1-A5, B1-B5		Concrete, Steel						Ecoinvent v 2.2	CML 2
16	Single family houses	[35]	2012		UK	Traditional build, brick and block	2	New-build	90	Whole building	50	A1-A3	A1-A3	A1-A3	A1-A3	A1-A3		Materials with bricks	Bricks	Bricks	Bricks	Bricks		Ecoinvent	CML 2001
17	Single family houses	[36]	2015		Flemish/ Belgium	Masonry brick walls	2	New-build	117	Whole building	60						A1-A3							Ecoinvent 3.1	ReCiPe Endpoint [H]
18	Single family houses	[37]	2013		UK	Brick solid wall	3	new-build	144	1 m2	50	A1-A3						Concrete precast elements						Ecoinvent	ReCiPe Midpoint and Endpoint
19	Single family houses	[35]	2012		UK	Brick and block	2	New-build	130	Whole building	50	A1-A3	A1-A3	A1-A3	A1-A3	A1-A3		Materials with bricks	Bricks	Bricks	Bricks	Bricks		Ecoinvent	CML 2001
20	Single family houses	[35]	2012		UK	Brick and block	2	New-build	60	Whole building	50	A1-A3	A1-A3	A1-A3	A1-A3	A1-A3		Materials with bricks	Bricks	Bricks	Bricks	Bricks		Ecoinvent	CML 2001
21	Single family houses	[34]	2010		Spain	Brick, concrete and steel	2	New-build	160	1 m2	50	A1-A3, B1-B5	A1-A3, B1-B5			A1-A3, B1-B5		Concrete						Ecoinvent v 2.1	CML 2
22	Single family houses	[26]	2013	2002	Italy	Reinforced concrete & bricks	3	New-build	443	1 m2.yr	50	A1-A5, B1-B5						Envelope ext., structural elements						Ecoinvent	Eco-indicator 99, IPCC 2007

IOP Conf. Series: Earth and Environmental Science **352** (2019) 012065 doi:10.1088/1755-1315/352/1/012065

23	Office	[38]	2018		Italy	Concrete structure	3	New-build	790	Whole building	100	A1-A3						Structure						Database from SimaPro	IPCC 2007
24		[39]	2004	1998-2001	Finland	Beam-column system, prefab concrete	4	New-build	17,300	1 m2	50	A1-A3		A1-A3, B1-B5	B1-B5	A1-A3, B1-B5		Structural frame		Structural frame, Ext. envelope	HVAC services, roof elements, surfaces (int)	Surfaces (int), Structural frame, external envelope		SimaPro Database	Eco-Indicator 95
25		[9]	2017	2015	Denmark	Sandwich construction, concrete and curtain walls	4 + basement	new-build	11,500	1 m2.yr	80	A1-A3					B4	Slabs, windows, ext. Envelope						Ökobau database	DK-GBC LCA
26		[9]	2017	2015	Denmark	Brick facade, reinforced concrete	6 + basement	New-build	6,200	1m2.yr	80	A1-A3					A1-A3	Slabs, inner walls						Ökobau database	DK-GBC LCA
27		[40]	2017		Brazil	Reinforced concrete floors, masonry brick walls	4	New-build (hypothetical)	2730	Whole building	50	A1-A3	A1-A3	A1-A3		A1-A3		Curtain wall mullions, walls	Floors	Walls		Walls	Walls, Curtain wall mullions	GABI database	Tally application in Revit
28		[40]	2017		Brazil	Reinforced concrete floors, masonry brick walls	4	New-build (hypothetical)	2730	Whole building	50	A1-A3	A1-A3	A1-A3		A1-A3		Walls	Floors	Walls		Floors	Walls	GABI database	Tally application in Revit
29		[33]	2015		Denmark	Reinforced concrete, wood envelope	4 + basement	new-build	3283	1 m2.yr	80	A1-A3	A1-A3	A1-A3	A1-A3	A1-A3		Slabs	Slabs	Slabs, windows	Slabs	Slabs		ESU-CO- og Ökobau-database	
30	Multi storey offices	[26]	2013	2009	Italy	Reinforced concrete structure, bricks	5	New-build	3353	1 m2.yr	50	A1-A5, B1-B5						Envelope ext.						Ecoinvent	Eco-indicator 99, IPCC 2009
31		[39]	2004	1998-2001	Finland	In-situ concrete	5	New-build	15,600	1 m2	50	A1-A3		A1-A3, B1-B5	A1-A3, B1-B5	A1-A3, B1-B5		Structural frame		Structural frame, External envelope	External envelope, structural frame, roof element, HVAC services	Structural frame, External envelope		SimaPro Database	Eco-Indicator 95
32		[39]	2004	1998-2001	Finland	Prefab reinforced concrete	9	New-build	24,000	1 m2	50	A1-A3		A1-A3, B1-B5	B1-B5	A1-A3, B1-B5		Structural frame		Structural frame	Surfaces (int), Structural frame, HVAC services	Structural frame, External envelope		SimaPro Database	Eco-Indicator 95
33		[41]	2016		UK	RC, structural frame and flat slab construction, brick facade	11	New-build	15590	Whole building	100	A1-A3					A1-A3	Floors/slabs					Floors/slabs	the GreenGuide Book (BRE)	IMPACT
34		[42]	2008		Thailand	Concrete, brick & curtain wall combination	38	New-build	60,000	Whole building	50	A1-A3			A1-A3	A1-A3		Cement, concrete			Cement, concrete	Cement, concrete		Thailand government databases	
35	Schools	[43]	2015		Canada	Brick/block building	2	New-build	2100	Whole building	30	A1-A3												Autodesk Green Building Studio (GBS).	Athena Impact Estimator (IE)
36		[44]	2013	2009	Spain	Reinforced concrete	2	New-build	1743	Whole building	50	A1-A3						Concrete, cement						Ecoinvent 2.0	IPPC 2007
37		[45]	2011		U.S.	Steel frame	3	New-build	3190	whole building	50	A1-A3					A1-A3	Steel					Steel	BEDEC database (CICT 2009)	
38		[46]	2003		U.S.	Reinforced concrete, bricks, Steel	6	new-build	7300	Whole building	75	A1-A3	A1-A3	A1-A3		A1-A3	A1-A3						Steel, cement in concrete	DEAM[TM] database	
39	Other	[47]	2017	1865	U.S.		4	Existing	2137	Whole building	75												Walls	IE4B, Athena	EPA TRACI

References

[1] United Nations Environmental Programme, Protocol for Measuring Energy Use and Reporting Greenhouse Gas Emissions from Building Operations, Sustain. Build. Clim. Initiat. (2008).

[2] United Nations, World Economic Situation and Prospects 2018, 2018. doi:10.1007/BF02929547.

[3] L.C.M. Eberhardt, H. Birgisdóttir, M. Birkved, Life cycle assessment of a Danish office building designed for disassembly, Build. Res. Inf. 0 (2018) 1–15. doi:10.1080/09613218.2018.1517458.

[4] M. Hauschild, R.K. Rosenbaum, S.I. Olsen, E. Al., Life Cycle Assessment - Theory and Practice, 2018. doi:10.1007/978-3-319-56475-3.

[5] ISO, DS/EN ISO 14044, Miljøledelse – Livscyklusvurdering – Krav og vejledning, Environmental management – Life cycle assessment – Requirements and guidelines, (2008).

[6] ISO, DS/EN ISO 14040, Miljøledelse - Livscyklusvurdering - Principper og struktur, Environmental management - Life cycle assessment - Principles and framework, (2008) 20.

[7] M. Paleari, M. Lavagna, A. Campioli, The assessment of the relevance of building components and life phases for the environmental profile of nearly zero-energy buildings: life cycle assessment of a multifamily building in Italy, Int. J. Life Cycle Assess. 21 (2016) 1667–1690. doi:10.1007/s11367-016-1133-6.

[8] V. John, Derivation of reliable simplification strategies for yje comparative LCA of individual and "typical" newly built Swiss apartment buildings, (2012).

[9] H. Birgisdóttir, S.S. Madsen, Bygningers indlejrede energi og miljøpåvirkninger, vurderet for hele bygningens livscyklus, 2017.

[10] R. Sathre, S. González-García, Life cycle assessment (LCA) of wood-based building materials, Eco-Efficient Constr. Build. Mater. Life Cycle Assess. (LCA), Eco-Labelling Case Stud. (2013) 311–337. doi:10.1533/9780857097729.2.311.

[11] C. Okoli, K. Schabram, A Guide to Conducting a Systematic Literature Review of Information Systems Research, Work. Pap. Inf. Syst. 10 (2010) 1–51. doi:10.2139/ssrn.1954824.

[12] S. Editor, K.H. Rosen, An INTRODUCTION to DGNB, Green Build. Counc. Denmark. (2007).

[13] DS Dansk Standard, DS/EN 15978 Sustainability of construction works – Assessment of environmental performance of buildings – Calculation method, (2012) 64.

[14] F.N. Rasmussen, T. Malmqvist, A. Moncaster, A.H. Wiberg, H. Birgisdóttir, Analysing methodological choices in calculations of embodied energy and GHG emissions from buildings, Energy Build. 158 (2018) 1487–1498. doi:10.1016/j.enbuild.2017.11.013.

[15] M.U. Hossain, S.T. Ng, Critical consideration of buildings' environmental impact assessment towards adoption of circular economy: An analytical review, J. Clean. Prod. 205 (2018) 763–780. doi:https://doi.org/10.1016/j.jclepro.2018.09.120.

[16] Italian Ministry for the Environment Land and Sea, Voluntary National Review of Italy: National Sustainable Development Strategy, (2017) 98.

[17] E. Macarthur Foundation, Circularity in the built environment: Case studies, (2016) 1–72. https://www.ellenmacarthurfoundation.org/assets/downloads/Built-Env-Co.Project.pdf.

[18] The Danish Government, Vejen til et styrket byggeri i Danmark - regeringens byggepolitiske strategi, Copenhagen, 2014.

[19] The Danish Government, Miljø- og Fødevareministeriet, Erhvervsministeriet, Strategi for cirkulær økonomi, (2018).

[20] R. Marsh, Building lifespan: effect on the environmental impact of building components in a Danish perspective, Archit. Eng. Des. Manag. 13 (2017) 80–100. doi:10.1080/17452007.2016.1205471.

[21] E. Davidson, Defining the trend: Cement consumption versus Gross Domestic Product, Glob. Cem. Mag. (2014) 8–14. doi:10.1146/annurev-orgpsych-041015-062429.

[22] A. Laurent, S.I. Olsen, M.Z. Hauschild, Limitations of carbon footprint as indicator of environmental sustainability, Environ. Sci. Technol. 46 (2012) 4100–4108. doi:10.1021/es204163f.

[23] A. Atmaca, Life-cycle assessment and cost analysis of residential buildings in South East of Turkey: part 2—a case study, Int. J. Life Cycle Assess. 21 (2016) 925–942. doi:10.1007/s11367-016-1051-7.

[24] C. Liljenström, T. Malmqvist, M. Erlandsson, J. Fredén, I. Adolfsson, G. Larsson, M. Brogren, Byggandets klimatpåverkan - Livscykelberäkning av klimatpåverkan och energianvändning för ett nyproducerat energieffektivt flerbostadshus i betong, 2015.

[25] S. Hrabovszky-Horváth, Z. Szalay, Environmental assessment of a precast concrete building stock in a time perspective, Environ. Eng. Manag. J. 13 (2014) 2797–2804. https://www.scopus.com/inward/record.uri?eid=2-s2.0-84930221636&partnerID=40&md5=094f8c56319bff0c38da7aa9a8de95f2.

[26] F. Asdrubali, C. Baldassarri, V. Fthenakis, Life cycle analysis in the construction sector: Guiding the optimization of conventional Italian buildings, Energy Build. 64 (2013) 73–89. doi:10.1016/j.enbuild.2013.04.018.

[27] G.A. Blengini, Life cycle of buildings, demolition and recycling potential: A case study in Turin, Italy, Build. Environ. 44 (2009) 319–330. doi:10.1016/j.buildenv.2008.03.007.

[28] X. Zhang, F. Wang, Analysis of embodied carbon in the building life cycle considering the temporal perspectives of emissions: A case study in China, Energy Build. 155 (2017) 404–413. doi:10.1016/j.enbuild.2017.09.049.

[29] A.T. Balasbaneh, A.K. Bin Marsono, S.J. Khaleghi, Sustainability choice of different hybrid timber structure for low medium cost single-story residential building: Environmental, economic and social assessment, J. Build. Eng. 20 (2018) 235–247. doi:10.1016/j.jobe.2018.07.006.

[30] C. Colli, A. Bataille, E. Antczak, F. Buyle-Bodin, Life Cycle Assessment of a French Single-Family House Refurbishment: The "bat-Eco2" Case Study, Procedia CIRP. 69 (2018) 160–165. doi:10.1016/j.procir.2017.11.043.

[31] G.A. Blengini, T. Di Carlo, Energy-saving policies and low-energy residential buildings: An LCA case study to support decision makers in piedmont (Italy), Int. J. Life Cycle Assess. 15 (2010) 652–665. doi:10.1007/s11367-010-0190-5.

[32] S. Citherlet, T. Defaux, Energy and environmental comparison of three variants of a family house during its whole life span, Build. Environ. 42 (2007) 591–598. doi:10.1016/j.buildenv.2005.09.025.

[33] F.N. Rasmussen, H. Birgisdottir, Bygningens Livscyklus: Identifikation af væsentlige bygningsdele, materialegrupper og faser i en miljømæssig vurdering., 1. udg., Statens Byggeforskningsinstitut, Aalborg Universitet, 2015.

[34] O. Ortiz-Rodríguez, F. Castells, G. Sonnemann, Life cycle assessment of two dwellings: One in Spain, a developed country, and one in Colombia, a country under development, Sci. Total Environ. 408 (2010) 2435–2443. doi:10.1016/j.scitotenv.2010.02.021.

[35] R.M. Cuéllar-Franca, A. Azapagic, Environmental impacts of the UK residential sector: Life cycle assessment of houses, Build. Environ. 54 (2012) 86–99. doi:10.1016/j.buildenv.2012.02.005.

[36] M. Buyle, A. Audenaert, J. Braet, W. Debacker, Towards a More Sustainable Building Stock: Optimizing a Flemish Dwelling Using a Life Cycle Approach, Buildings. 5 (2015) 424–448. doi:10.3390/buildings5020424.

[37] T.M. Grath, S. Nanukuttan, K. Owens, M. Basheer, P. Keig, Retrofit versus new-build house

using life-cycle assessment, Proc. Inst. Civ. Eng. - Eng. Sustain. 166 (2013) 122–137. doi:10.1680/ensu.11.00026.

[38] M. Lamperti Tornaghi, A. Loli, P. Negro, Balanced Evaluation of Structural and Environmental Performances in Building Design, Buildings. 8 (2018) 52. doi:10.3390/buildings8040052.

[39] S. Junnila, The environmental impact of an office building throughout its life cycle, 2004. http://lib.tkk.fi/Diss/2004/isbn9512272857/.

[40] M. Najjar, K. Figueiredo, M. Palumbo, A. Haddad, Integration of BIM and LCA: Evaluating the environmental impacts of building materials at an early stage of designing a typical office building, J. Build. Eng. 14 (2017) 115–126. doi:10.1016/j.jobe.2017.10.005.

[41] A. Azzouz, M. Borchers, J. Moreira, A. Mavrogianni, Life cycle assessment of energy conservation measures during early stage office building design: A case study in London, UK, Energy Build. 139 (2017) 547–568. doi:10.1016/j.enbuild.2016.12.089.

[42] O.F. Kofoworola, S.H. Gheewala, Environmental life cycle assessment of a commercial office building in Thailand, Int. J. Life Cycle Assess. 13 (2008) 498–511. doi:10.1007/s11367-008-0012-1.

[43] S.O. Ajayi, L.O. Oyedele, B. Ceranic, M. Gallanagh, K.O. Kadiri, Life cycle environmental performance of material specification: a BIM-enhanced comparative assessment, Int. J. Sustain. Build. Technol. Urban Dev. 6 (2015) 14–24. doi:10.1080/2093761X.2015.1006708.

[44] I. Zabalza, A. Valero, A. Aranda, E. Llera, Methodological aspects and design implications to achieve life cycle low emission buildings. A case study: LCA of a new university building, Strojarstvo. 55 (2013) 87–102. http://www.scopus.com/inward/record.url?eid=2-s2.0-84884891488&partnerID=tZOtx3y1.

[45] E. Wang, Z. Shen, C. Berryman, A Building LCA Case Study Using Autodesk Ecotect and BIM Model, ASC Annu. Int. Conf. Proc. 47th (2011). http://digitalcommons.unl.edu/cgi/viewcontent.cgi?article=1005&context=constructionmgmt&sei-redir=1&referer=http://scholar.google.dk/scholar?hl=da&q=Building+LCA+Case+Study+Using+Autodesk+Ecotect+and+BIM+Model&btnG=#search=%22Building LCA Case Study Using.

[46] C. Scheuer, G.A. Keoleian, P. Reppe, Life cycle energy and environmental performance of a new university building: Modeling challenges and design implications, Energy Build. 35 (2003) 1049–1064. doi:10.1016/S0378-7788(03)00066-5.

[47] M. Hu, Balance between energy conservation and environmental impact: Life-cycle energy analysis and life-cycle environmental impact analysis, Energy Build. 140 (2017) 131–139. doi:10.1016/j.enbuild.2017.01.076.

IOP Conf. Series: Earth and Environmental Science **352** (2019) 012066 doi:10.1088/1755-1315/352/1/012066

Building Bioclimatic Design in cold climate office buildings

M C Ness, I Andresen, T Kleiven

Department of Architecture and Technology, NTNU, 7491 Trondheim (Norway)
maria.coral.ness@ntnu.no

Abstract. Building Bioclimatic Design (BBCD) understands architecture as a filter between outdoor climate and indoor comfort. This way, it encourages the exploitation of freely available climatic resources, before adding any HVAC system. Therefore, BBCD represents a fundamental strategy for improving energy efficiency in buildings. The climate / comfort comparison in building design determines the passive strategies that are most suitable for a specific climatic context, as well as the level of architectural complexity. In cold climates, it would suggest the use of compact shapes and extremely airtight and insulating envelopes, in order to minimize heat losses while maximizing solar heat gains. However, when combined with high internal gains, these measures might cause overheating problems in the warm seasons. That is the case of office buildings, where cooling equipment is included as default even in cold climates, drastically increasing their energy consumption. It is therefore becoming a necessity to consider here the adoption of passive cooling strategies once identified with warmer climates. The aim of this research is to explore how the theories and tools for BBCD could be applied to cold climate office buildings. In order to study the effect of the different climatic contributors, we will use Building Performance Simulation to analyse relevant cases with EnergyPlus (in combination with DesignBuilder). This will in turn help drawing suggestions on how to adapt the Building Bioclimatic Chart (BBCC) for its application to cold climate office buildings in practice. It is well known that the earlier we apply the measures for energy efficiency, the greater their effect and with higher degree of integration. The BBCC is used in the pre-design phase to determine the most suitable passive strategies for climate adaptation and control, informing the design as early as possible along the process. This study can contribute to the development of zero emission neighbourhoods in cold climates, by improving the energy efficiency of their buildings. Additionally, it complements the existing research in BBCD by extending its application to cold climates and office buildings.

1. Introduction

Building Bioclimatic Design (BBCD) understands architecture as a filter between outdoor climate and indoor comfort [1]. In order to determine how this filter should work, it is necessary to understand the local climate, how it interacts with the building, and what our comfort needs are. BBCD encourages the exploitation of useful climatic resources before adding mechanical systems (HVAC) to further correct the indoor climate, thus representing a fundamental step for improving energy efficiency in buildings.

Traditionally, the choice of passive design strategies for climate control in buildings was experience-based (vernacular architecture). However, new building morphology, typologies (office buildings), elements and materials are challenging us to move towards research-based approaches. This is the case of the Building Bioclimatic Charts (BBCC). The most extended one is the Givoni-Milne Bioclimatic Chart [2], that studies how to reach thermal comfort within the psychrometric chart. It considers also the potential for expanding the comfort zone by means of different passive design strategies for climate regulation. In later years, Milne [3] developed Climate Consultant at UCLA for the application of the

BBCC to "envelope dominated" buildings (residential, small non-residential, schools) in mild climates.[4]. This way, the focus was on the relationship between the outdoor climate, the indoor comfort and the passive strategies for climate control. In other climates, the envelope becomes more relevant and its effect needs to be accounted for. Then other types of buildings might have more significant internal loads that should be included as well. Despite its limitations, Climate Consultant is the most complete, up-to-date and user-friendly tool for BBCD that we can find today.

In cold climates, the principles for BBCD would suggest the use of compact shapes and extremely airtight and insulating envelopes, to minimize heat losses while maximizing solar heat gains. However, when combined with high internal gains, these measures might cause overheating problems. That is the case of office buildings, where cooling equipment is included as default even in cold climates, drastically increasing energy consumption. Therefore, the adoption of passive cooling strategies once identified with warmer climates might help increasing both comfort and energy efficiency in these cases [5].

It is well known that the earlier we apply the measures for increasing environmental performance of a building, the greater their effect and with a higher degree of integration. The BBCC is used in the pre-design phase to inform the decisions as early as possible along the process, thus maximizing their effect.

An adapted BBCC for cold climates office buildings could show the passive strategies that are best suited for the indoor climate that is generated by the envelope, including also the effect of internal loads. In this way, it would complement existing research in BBCD by extending its application, earlier limited to residential buildings in warmer climates. At the same time, it could be used to reflect on the climatic consequences of choosing different envelopes, occupancy density, equipment and lighting systems.

This paper is part of a doctoral thesis on BBCD in cold climates office buildings, financed by the Faculty of Architecture and Design at NTNU and the Research Centre on Zero Emission Neighbourhoods in Smart Cities.

2. Purpose

The aim of this research is to explore how the theories and tools for Building Bioclimatic Design could be applied to cold climate office buildings to make them more sustainable.

The existing theories and tools for BBCD look at the local climate to study when the building in question will provide thermal comfort and how to extend the comfort zone to the rest of the hours, by means of passive strategies. Originally developed for warm climates, it does not include the effect of the building envelope in creating an indoor microclimate that can differ substantially from the outdoor climate, especially with highly insulating and airtight solutions. Therefore, it would suggest primarily passive heating strategies for cold climates, even though in practice those very buildings might be experiencing overheating problems.

On the other hand, there exist cases of office buildings in cold climates that make extensive use of passive strategies both for heating and cooling. Some of those, like the Baumschlager-Eberle 22/26 in Lustenau, Austria [6], has managed to substitute the conventional HVAC system for a set of passive strategies combined with sensors and automated control of the window opening and lighting system. These cases show the need to update the existing theories and tools for BBCD, to accommodate for the lessons learnt from practice. Accordingly, the BBCC should include the indoor microclimate created by the envelope (cold climates) and the internal gains (office buildings), prior to the study of thermal comfort or the application of passive strategies for climate control. Once updated, the BBCC could be used on other buildings to improve their energy efficiency and comfort through passive design, prior the implementation of active systems.

3. Method

In order to study the effect of the different climatic contributors to the indoor microclimate, we used Building Performance Simulation (BPS) to analyse two relevant cases. This helped drawing suggestions on how to adapt the Building Bioclimatic Chart for its application to cold climate office buildings in practice, thanks to a systematic parametric analysis.

IOP Conf. Series: Earth and Environmental Science **352** (2019) 012066 doi:10.1088/1755-1315/352/1/012066

The choice of BPS tools fell on EnergyPlus/DesignBuilder because of the level of accuracy, transparency, adaptability and complexity, in combination with an intuitive user interface.

The case analysed is based on Baumschlager-Eberle 22/26 (BE2226) because of its representativeness, being a rather extreme example of design optimization for cold climate, with maximal use of passive strategies. Furthermore, the availability of detailed measurements on-site made it possible to calibrate the model to resemble reality more accurately.

Then, the same model was used for two different locations with slightly different type of cold climate, in order to see how it would respond to diverse conditions. First, it was analysed in the original location, Lustenau (Austria), with a cold-temperate humid continental climate with warm summers and 2980 heating degree days (HDD). Next, it was moved to Oslo (Norway), with a cold humid continental climate with cool summers and 4428 HDD. For the original location at Lustenau, the weather file was obtained by interpolation from Meteonorm [7], while the one for Oslo was taken from the EnergyPlus site [8], corresponding to the weather station at Fornebu.

The building is conformed as a cube of around 24m long sides, with a high degree of compactness (c=1.08). In includes six floors, each with a free height of 3.4m (4.4m in the ground floor) to allow for a good air stratification, thus allowing for the used air to stay close to the ceiling and keeping the occupied volume fresher. The windows are vertical and almost from floor to ceiling, to provide with a good daylight distribution throughout the space. The envelope consists of two layers of Porotherm 38, a lighter one outside for a better insulation and a heavier one inside for structural stability and thermal mass. The rendering is with lime plaster, which helps sealing the envelope from the outside and regulating the indoor air humidity and CO2 concentration. It uses concrete slabs with glass fibre reinforced concrete rendering for the floor and lime plaster for the ceiling, ensuring the exposure of its thermal mass. Besides, it has a very open floor distribution, using secondary spaces as partitions, which keeps a clear path for air distribution, while increasing the thermal mass.

A simple model of the whole building was created for BPS analysis (figure 1), to then calculate the indoor microclimate in the open office on the second floor (figure 2).

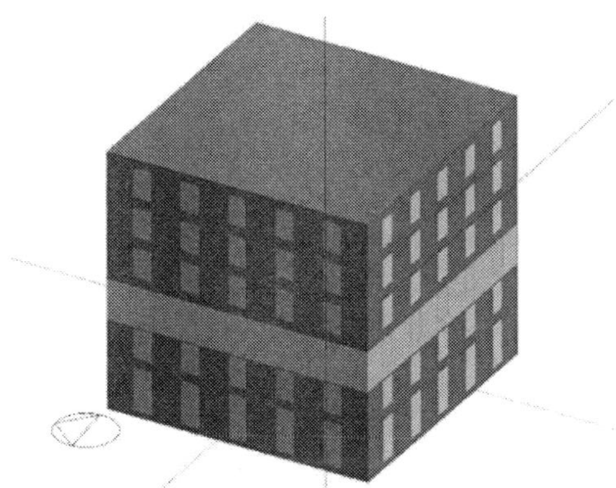

Figure 1. Axonometry of the whole office building (Design Builder)

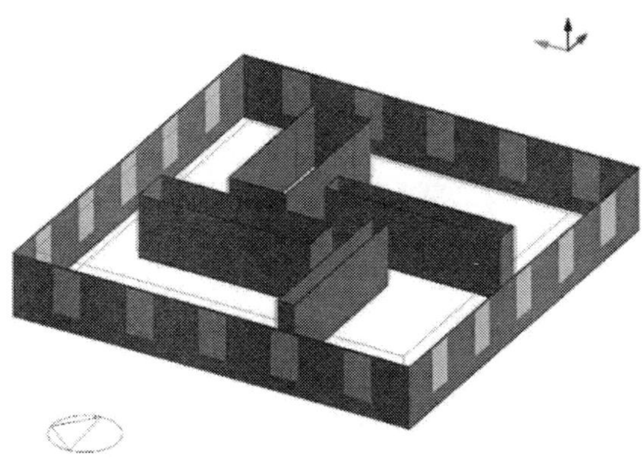

Figure 2. Model of the open office on the 2nd. floor (Design Builder)

The most relevant characteristics for the materiality of the building envelope (insulation, thermal mass and airtightness) are specified in table 1.

Table 1. Characteristics of the building envelope

wall	U-value (W/m2 K)	0.138
	Internal heat capacity (KJ/m2 K)	85.760
	Infiltration rate (ac/h)	0.040
window	U-value (W/m2 K)	0.7
	Solar transmittance factor	0.55

IOP Conf. Series: Earth and Environmental Science **352** (2019) 012066 doi:10.1088/1755-1315/352/1/012066

The internal gains for the open office on the second floor are specified in table 2. The occupancy density is rather low because of the necessity of leaving a free circulation perimeter by the façade to avoid discomfort due to draft when opening windows. The equipment consists mainly on one computer and two screens per person. As for the lighting, it corresponds to the best practice in Design Builder.

Table 2. Internal gains

Occupancy density (people/m2)	0.05
Equipment (W/m2)	11.50
Lighting (W/m2)	3.30

The simulations were carried out in different steps of complexity for the internal loads, in order to quantify the changes in the indoor microclimate for each one of them. In this way, it was done first for the empty building, to obtain the effect of the envelope alone. Next was added the occupancy, then the equipment and finally the lighting system (with response to occupancy and minimum illuminance levels 400 lux).

The indoor temperatures obtained from the simulations were then presented in two different ways. In relation to time (figure 5), to show seasonal and daily fluctuations, the amplitude of its variations and the proximity to the targeted comfort range of 22-26°C. In relation to humidity, on the Psychrometric chart, to reflect the total scattering of climatic values, thermal comfort zones for summer and winter with the percentage of hours of discomfort, and potential for extending these comfort zones using passive strategies (figure 6).

4. Case I: Office building in Lustenau (Austria)

Sited by the river Rhine, Lustenau lays on the north side of the Alps, 10 km away from the lake Constance. It is located at 47°N and 9°E, with an elevation of 404m. With an average temperature of -0.6°C for the coldest and 19.6°C for the hottest month, the climate in Lustenau is classified as Cfb, cold-temperate humid continental with warm summer, in the Köppen-Geiger scale [11]. With 2980 HDD (18°C) annually, it belongs to the ASHRAE climate zone 4A (borderline with 5A) [12].

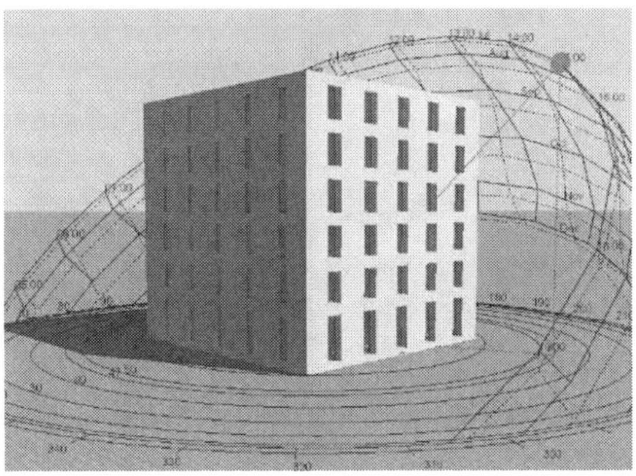

Figure 3. Sun path for Lustenau and shadows for 15th June at 15:00 (Design Builder)

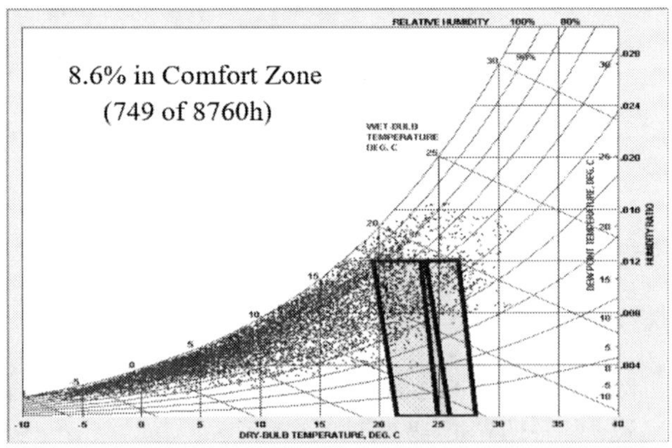

Figure 4. Hourly climatic values for Lustenau. Psychrometric Chart with thermal comfort zones for summer (right) and winter (left). (Climate Consultant)

When plotting the outdoor climatic values for Lustenau into the Psychrometric chart, only 8.6% of them fall inside the comfort zone, being most of them on the heating side of it (figure 4).

We can see in figure 5 the thermal stability created by the envelope, thanks to a combination of high insulation levels and thermal mass, though temperatures are too low throughout the year. It also creates a time lag in its thermal response to outdoor conditions. The addition of occupancy warms up the space,

IOP Conf. Series: Earth and Environmental Science **352** (2019) 012066 doi:10.1088/1755-1315/352/1/012066

so that they reach more comfortable temperatures during the summer and early autumn. After adding the equipment, temperatures experience a higher daily variation and fall within the comfort range for the mid-seasons, while there appears overheating in summer. Finally, the lighting system increases daily variations and creates an important overheating problem, reaching temperatures of more than 50°C.

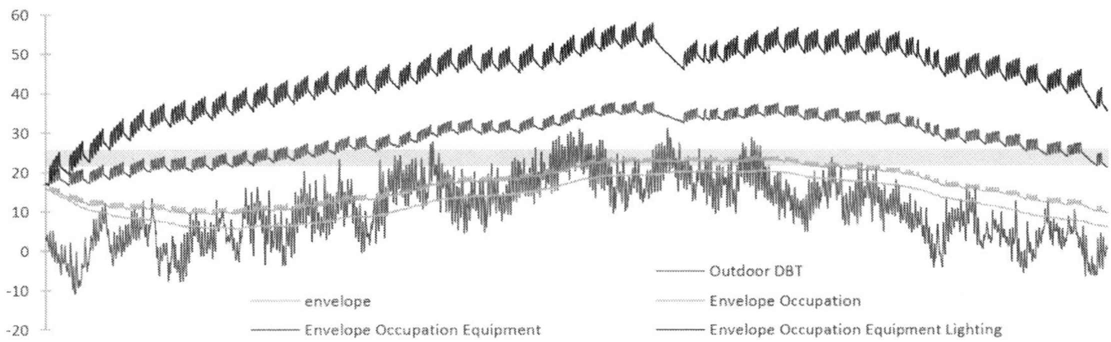

Figure 5. Annual variations in indoor temperature (°C), with cumulative internal loads. Case I, Lustenau.

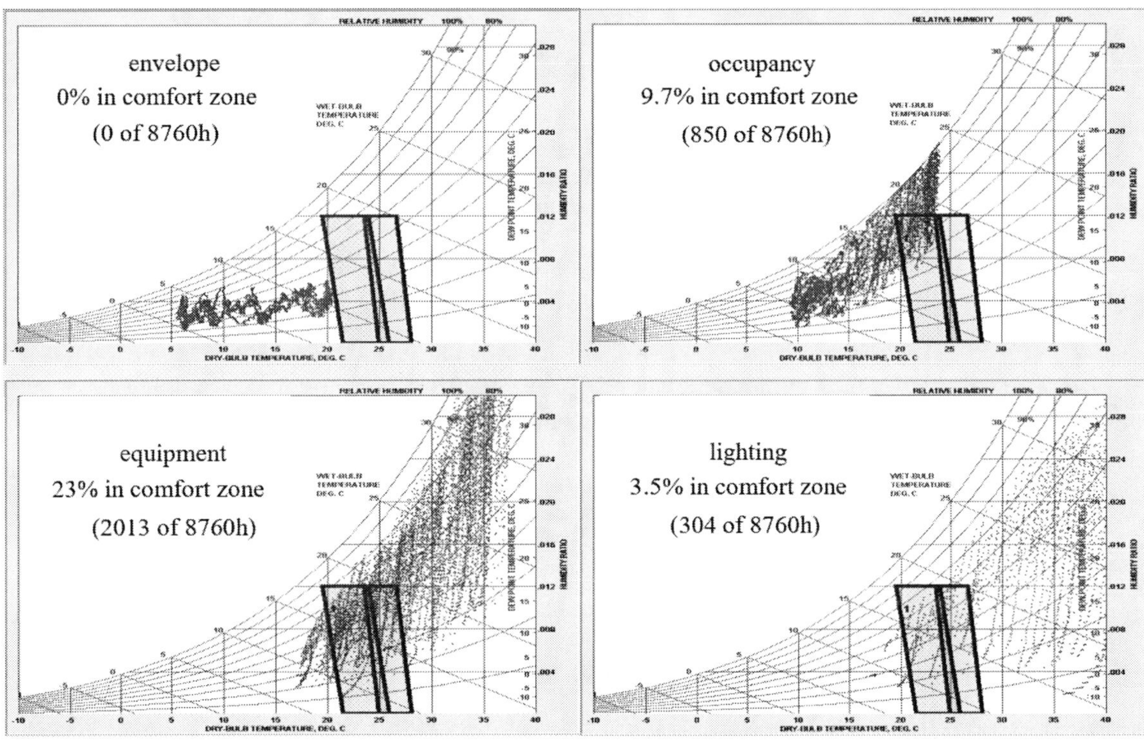

Figure 6. Psychrometric charts with thermal comfort zone for summer (right) and winter (left), for cumulative internal loads. Case I, Lustenau.

The Psychrometric charts (figure 6) clearly show the effect of the envelope in grouping the climatic points into fewer and tighter values, though all of them fall on the heating side of the comfort zone. Then there occurs again more scattering of the values, the more gains we include. At the same time, the indoor climate becomes warmer, with important overheating after adding lighting gains.

IOP Conf. Series: Earth and Environmental Science **352** (2019) 012066 doi:10.1088/1755-1315/352/1/012066

5. Case II: Office building in Oslo (Norway)

Oslo lays on the northern side of the wide Oslofjord, in the Southeast of Norway, at 59°N and 10°E, with an average elevation of 23m. The mean temperature of the coldest month is -3.8°C, 17.5°C for the hottest, which gives a cold humid continental climate with cool summer, Dfb in the Köppen-Geiger scale [11]. It is in climate zone 6A according to ASHRAE classification [12], with 4428 HDD (18°C). Having a higher latitude, the sun is lower in the horizon. This means that even though outdoor temperatures are lower throughout the year, solar gains through windows will warm up interior surfaces and increase indoor temperatures. Besides, there are fewer hours of sun in winter and more in summer, which could accentuate overheating risk in the warm season and not provide so much heat in the cold one, when it is most needed (figure 7).

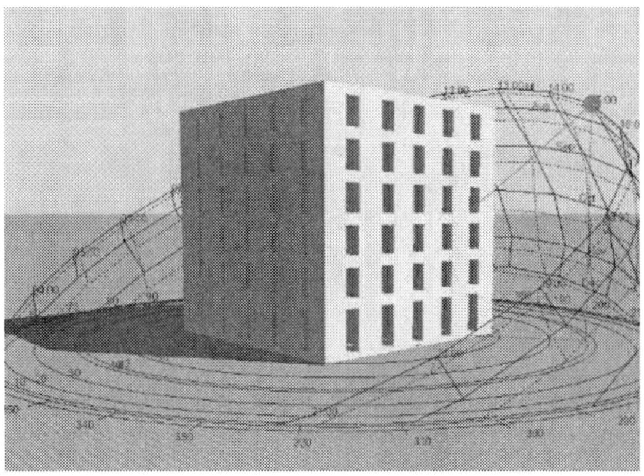

Figure 7. Sun path for Oslo and shadows for 15th June at 15:00 (Design Builder)

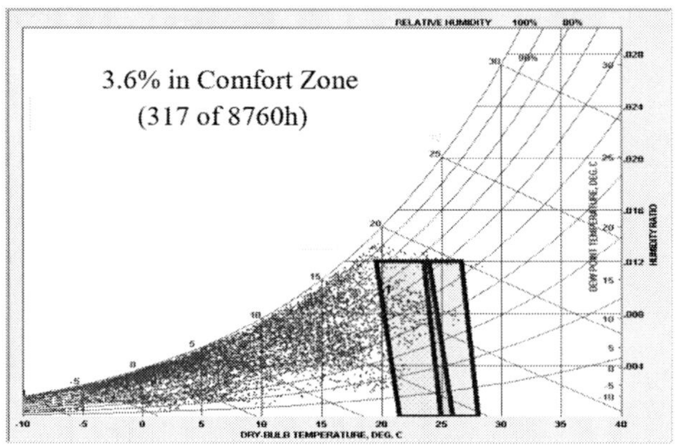

Figure 8. Hourly climatic values for Oslo. Psychrometric Chart with the thermal comfort zones for summer (right) and winter (left). (Climate Consultant)

When plotting the climatic values into the Psychrometric chart, only 3.6% of them fall inside the comfort zone, being most of them on the heating side of it (figure 8)

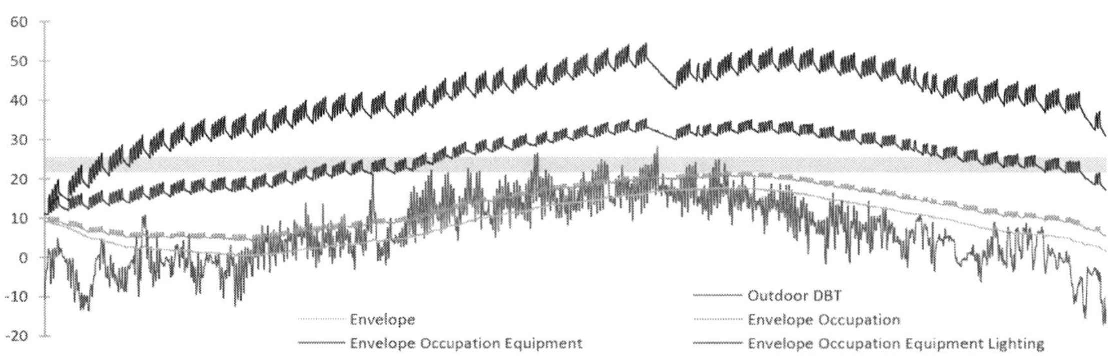

Figure 9. Cumulative variations indoor temperature (°C) through time, with internal loads. Case II, Oslo.

We can see in these graphs (figure 9) the thermal stability created by the envelope, though staying below comfort in a similar way as in Lustenau.

In the same way as for case I, in case II a moderate increase of temperature due to the internal loads is positive, bringing the curve towards more comfortable temperatures. In contrast, the total increase in indoor temperatures after adding all the internal gains can cause some important overheating problems.

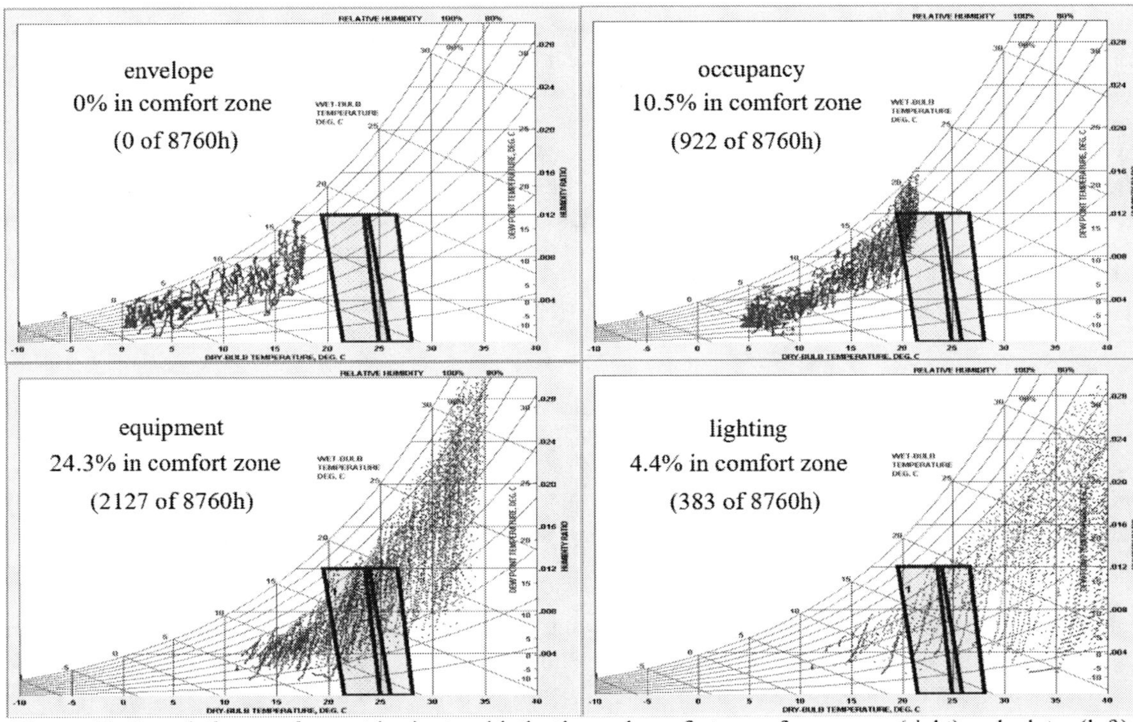

Figure 10. Cumulative psychrometric charts with the thermal comfort zone for summer (right) and winter (left), for the internal loads in case II, Oslo.

The Psychrometric charts (figure 10) show the same grouping effect of the envelope into fewer and tighter values, though all of them fall outside the comfort zone, in a similar way as they did in Lustenau. The more gains we include in the model, the more scattering of the climatic points. At the same time, the indoor microclimate becomes warmer, moving towards the right in the chart, with important overheating after adding the lighting gains. Again, it clearly shows the benefits of first bringing the climatic points closer to each other (effect of the envelope buffering extreme values) and then warming up with moderation (internal gains), in order to bring as many points as possible into the comfort zone.

6. Discussion

Comparing the psychrometric charts for the outdoor and indoor climate (figures 4 and 6 for Lustenau, 8 and 10 for Oslo), it becomes clear the relevance of considering the effect of the envelope and internal gains, for office buildings in cold climates. The climatic points change their distribution completely from the original outdoor values, with a quite homogeneous cloud on the cool side of the comfort zone (CZ) and 8.6% (case I) to 3.6% (case II) of the values inside it. After including the envelope, the indoor climatic points conform more like a path with a much lower variability and amplitude of temperatures (combined effect of insulation and thermal mass), still on the cool side of the CZ and now 0% inside it. Then, after adding the internal loads, the points get progressively more scattered and move towards the warm side of the CZ, with a growing percentage inside it but also an increased risk of overheating. Finally, once all the internal gains are included, the climatic values appear to be very spread, with a high amplitude in temperature and humidity, being most of them on the warm side of the CZ and only a very small percentage of them inside it (3.5% in case I, 4.4% in case II). The combined effect of the envelope and the internal gains changes the indoor climate from a cold one with a clear need for heating, into a hot one with and important need for cooling. This could be compensated by the use of an adequate set of passive cooling strategies, including natural ventilation and solar shading.

In figures 5 and 9, on the other hand, it is easy to appreciate the effect of the envelope in restricting the thermal variability of the indoor climate. In addition, it creates a time lag in the seasonal variations, with a delay in the warmer period. This delay allows a non-coincidence between the peak in solar gains and in outdoor temperatures, to soften their effect indoors. Thanks to the stabilizing effect of the envelope, daily fluctuations of the outdoor climate do not have much effect indoors, while changes in occupation become very relevant. It is easy to appreciate the cumulative effect of internal gains during working days, as well as the cooling effect of empty nights and weekends, and even more for the summer holidays. The operational schedule for internal gains is therefore as important as their thermal load.

The most challenging aspect in Building Bioclimatic Design is the uncertainty generated due to the occupancy (density, distribution), the complexity of the physical environment and the climate (daily and seasonal variability, climate change). Even so, despite the uncertainties and difficulties, it is important to acknowledge and utilise the potential within BBCD.

7. Conclusion

The main findings of this study show the magnitude of the microclimate that is generated in the building thanks to the envelope, with its materiality and design. Its effect is further enhanced by the addition of internal gains, transforming the original need for heating that corresponds to cold climates, into a severe need for cooling. This has lately been translated into the installation of mechanical cooling systems by default, thus increasing considerably the energy use of office buildings. On the other hand, the implementation of passive cooling strategies like natural ventilation or solar shading could help increasing their energy efficiency while improving thermal comfort.

An adapted BBCC for cold climate office buildings could significantly reduce the amount of energy and resources used nowadays in conditioning this kind of buildings, while increasing indoor comfort.

Further research is needed though, in order to quantify the effect of different types of envelopes. The indoor climate will change depending on the level of insulation, the airtightness and the thermal mass. Window size, geometry and orientation are also relevant for the incident radiation and the daylight availability, which influence the need for artificial lighting and the thermal gains. Furthermore, internal loads become more relevant for the indoor climate with a more stringent envelope. Since they depend directly on occupancy, appliances and operation, these parameters gain more importance in this type buildings and should be considered as part of their design strategy. Finally, the inclusion of a smart control system could finetune the operation of the building to respond more efficiently to changes in the climate or the occupancy, representing the active system to operate the building elements to ensure a comfortable indoor climate.

References

[1] Szokolay S. Introduction to Architectural Science. Elsevier; 2012
[2] Milne M, Givoni B. Architectural design based on climate. Watson D, McGraw-Hill; 1979
[3] Milne M. Climate Consultant 6.0. Department of Architecture and Urban Design, University of California, Los Angeles; 2015
[4] Milne M, Liggett R, Benson A, Bhattacharya Y. Additions to a design tool for visualizing the energy implications of california's climates. Berkeley, California: UC Energy Institute; 2009.
[5] Finocchiaro L, Wigenstad T, Hestnes AG. Potential of passive cooling, natural ventilation and solar control in cold climates office buildings. Tapir Academic Press; 2010. p. S. 39-48
[6] Junghans L, Concept 22/26, a high performance office building without active heating, cooling and ventilation systems. International Solar Energy Society, ISES Solar World Congress; 2015
[7] Remund J. Quality of meteonorm version 6.0. Europe. 2008;6(1.1):389.
[8] Energyplus weather data. National Renewable Energy Laboratory (NREL). U.S. Department of Energy's (DOE). Available from: https://energyplus.net/weather-location/europe_wmo_region_6/NOR//NOR_Oslo.Fornebu.014880_IWEC.
[9] Köppen W, Geiger R. Handbuch der klimatologie: Gebrüder Borntraeger Berlin, Germany; 1930
[10] Briggs RS, Lucas RG, Taylor ZT. Climate classification for building energy codes and standards: Part 1 - development process. 2003 Winter Meeting of ASHRAE; 2003; Chicago, IL

IOP Conf. Series: Earth and Environmental Science **352** (2019) 012067 doi:10.1088/1755-1315/352/1/012067

Selbukassa - A Case Study for Aiming at Low Emission Buildings through Extensive Reuse of Materials

A Plavina[1,2], M Gruner[2]

[1] Department for Architecture and Technology, NTNU, Trondheim, Norway
[2] Pir II AS, Trondheim, Norway

E-mail: alisep@stud.ntnu.no

Abstract. Recent reports from UN International Resource Panel call for double decoupling – decoupling of material use and related environmental impacts from economic growth. The construction and use of buildings constitute a significant portion of energy use, GHG emissions and extraction of materials in Europe. One of the central strategies addressing double decoupling in the construction industry is reuse and recycling of building materials and components that would limit raw material extraction and embodied emissions related to new products. A recent IPCC report states the urgency of limiting GHG emissions already by 2030 to remain below the 1.5-degree target. Given this timeframe, reducing emissions in the early stages of a buildings lifetime appears worth considering, meaning prioritizing embodied over operational emissions. The case study used for this article is an ongoing building project "Selbukassa" ("the Selbu box") situated in Svartlamon – an experimental neighbourhood for urban ecology in Trondheim, Norway. It is a bottom-up, self-build project with a strong focus on reuse of materials. The 300 m² building will house 4 families and is built reusing an old log house, as well as CLT-elements from a former pavilion used for exhibiting an art piece by the local artist Killi Olsen. Other salvaged materials and components include windows, doors, roof slate and most of the internal and external finishes. Due to the experimental status of the area and to allow for reuse of building components with lower energy performance, the fulfilment of energy requirements in the building code TEK 17 was not required by the local building authorities. The article examines the embodied and operational emissions of Selbukassa, and compares it with a reference building with no reused components and complying with energy requirements in the building code TEK17. Simien v.6.009 is used for energy simulations of the two cases, and LCA tool One Click LCA Norge for NS 3720:2018 with data sourced from Norwegian and other European EPDs is used for estimating embodied emissions. The article investigates the possible savings in emissions linked to reuse of materials, as well as the trade-offs between reused components with worse energy performance and the consequential higher energy demand. The article contributes to the discussion about the various implications of reuse of materials and more generally on how the built environment could respond more optimally to a combined emission and resource use reduction challenge.

IOP Conf. Series: Earth and Environmental Science **352** (2019) 012067 doi:10.1088/1755-1315/352/1/012067

1. Introduction

Problem statement

The latest UN IPCC report states that we have just until 2030 to limit devastating global warming and remain below the 1.5-degree target. [1] We are also facing a fast-growing global resource use with planetary boundaries pushed beyond their limits. The extraction and processing of materials, fuels and food make up about half of total global greenhouse gas emissions and more than 90 percent of biodiversity loss and water stress. [2] UNEP (United Nations Environment Programme) has therefore introduced the concept of 'double decoupling' – decoupling of material use and related environmental impacts from economic growth [3]. IRP (International Resource Panel) has recently recommended that high-income nations aim for an absolute decoupling of material use and related environmental impacts from economic growth. [4] Absolute decoupling is achieved when the related environmental pressure (e.g. resource use or emissions) either remains stable or decreases while economic activity increases. [5]. Ekins et al. state that there is no guarantee that reducing emissions will reduce material extractions and related impacts, and may even increase them. Therefore reductions in GHG emissions have to be combined with targeting reduced extractions (incl. increased resource productivity through all stages of production, and reduced resource use in consumption). [6]

In Europe, the construction and use of buildings is responsible for 42% of our final energy consumption, 35% of greenhouse gas emissions, and more than 50% of all extracted materials. European commission porposes to strengthen and complement existing policies for promoting energy efficiency and renewable energy use in buildings with policies for resource efficiency. [7]

However, most of the current sustainable building concepts (passivehouse, plushouse, nearly zero energy buildings) still focus on energy use with no or little attention paid to embodied emissions or reduction in material extraction. The Norwegian Zero Emission Building definition [8] includes emission accounting for both operational energy and embodied emissions from materials over the lifetime of a building, but it does not explicitly address resource efficiency.

The share of embodied emissions over the lifetime of residential buildings can vary between 11% and 33% in conventional buildings, with an increasing share in low energy buildings (26% – 57%) and nearly zero energy buildings (74% – 100%). [9] The largest fraction of embodied emissions occurs during construction period, with smaller fraction linked to future maintenance and replacement of elements. Some studies have shown significant emissions linked to technical systems (heating, ventilation, PV), especially if replacements were included, ranging from 18% – 46%. [10]

Already previously it has been stated that it is possible that carbon savings made at the start of a buildings lifecycle could be more valuable than predicted savings in the future [11]. Especially considering the short timeframe for emission reduction until 2030, the fraction of embodied emissions becomes more prominent for both conventional and low energy buildings. Furthermore, expected decarbonisation of the grid will lead to lower operational emissions linked to energy use and decreasing prospects for offsetting embodied emission particularly linked to higher insulation levels or technical components such as PV with energy production.

Birgisdottir et al have developed a set of strategies for reducing embodied emissions – substitution of materials (substitution with bio-based or recycled/reused materials), reduction of resource use (light-weight, more durable or recycled/reused materials), reduction of construction and end-of-life stage impacts (construction-related strategies, waste management, seasonal timing etc.)[12] All these strategies are relevant for the case study used in this article, since it addresses substitition of new materials with reused materials, while reducing resource use/ extraction and using 'waste' from other buildings. Reuse and recycling of materials has been a focus in a number of building projects – 'Återvunna huset' (recycled-house) in Sweden in 1998 [13], 'Gjenbrukshus' (reuse-house) in Norway in 2003 [14], more recently in the work of architecture studio Lendager Group in Denmark [15] to name a few, and analysed in several reports in the Nordic countries and Europe [16]. Recently with an increased interest in the 'circular economy' concept, the building industry is slowly following the trend, but procedures and necessary networks to make reuse more widespread are not yet established.

The case – Selbukassa in Svartlamon

Svartlamon is an experimental neighbourhood for urban ecology in Trondheim, Norway. The area is experimenting with participatory processes, bottom-up approach and DIY practices in building and maintenance of buildings. Experimental building is an important aspect – in 2004 the then highest CLT building in the world, several off-grid micro houses, and recently an internationally acclaimed self-build rowhousing project based on a student master work known as 'Eksperimentboliger' were all built in Svartlamon.

The latest addition to these experiments is "Selbukassa" (the "Selbu box"), a building of ca. 300 m^2 that shall provide housing for four families and wishes to showcase an alternative to the conventional housing market and building industry. As a self-build project, the future tenants are responsible for planning and construction of the house.

The project aims at low environmental impact by making use of salvaged building materials to a large extend. The main part is a log house built in the 1920s from the village of Selbustrand near Trondheim that the owner gave away for free. The house was dismantled and transported to Trondheim by the self-builders. It is the basis of the design of the new building and has been rebuilt and extended vertically by adding a third floor, and horizontally by an extension. The extension is built using reused CLT-elements from a former pavilion for the art piece "Salamander night" by the artist Killi Olsen from 2007. Other important building components such as windows, doors, roofing, internal surface materials are primarily reclaimed materials, for example through local peer-to-peer marketplaces. To reduce the use of new materials and due to the limited skills of the builders/tenants, only moderate additional insulation is planned. On the other hand, the project falls within the Norwegian building category of block of flats and has to comply with the strict fire and structural requirements, and to a lesser extend sound requirements of this building category.

Figure 1 The log building from 1920s in original state (left), log building without external cladding (center), the artists pavilion in CLT from 2007 (right).

Research questions

This work sets out to assess the environmental impact of the Selbukassa project as a case study to reuse and upgrade old log houses along with extensive reuse of other building materials and components.
The main question is if the embodied emissions savings linked to extensive reuse of materials and components with worse energy performance can compensate for higher operational energy use and related operational emissions when compared to state-of-the-art components.

In a larger context the article aims to investigate how the built environment can respond more optimally to a combined emission and resource reduction challenge both in short term (the 2030 timeframe) and over the whole lifecycle of buildings.

IOP Conf. Series: Earth and Environmental Science **352** (2019) 012067 doi:10.1088/1755-1315/352/1/012067

2. Method

Embodied and operational emissions over the entire building life are calculated for Selbukassa and a reference case of a standard new building with all-new materials and components that fulfil the current energy efficiency requirements in building code TEK17. Simien v.6.009 is used for energy simulations of the two cases. LCA tool One Click LCA Norge, customised to Norwegian requirements, is used for estimating embodied emissions. Emission calculations are based on NS 3720:2018 with data sourced primarily from Norwegian EPDs complemented with other European EPDs. Several scenarios are studied – different emission factors for energy as required by NS 3720, and different accounting periods of 60 (entire building lifetime) and 10 years (emissions accumulated until 2030).

Embodied and operational emissions are calculated over the entire building life cycle according to NS 3720:2018. Modules included in the calculation are: Product stage A1 – A3, Construction stage A4 (transport) and A5 (limited to on-site waste management), Use stage B4 and B5 (replacement and refurbishment) and B6 (energy consumption in operation), End of life stage C1- C4. Other modules were not included in the calculations due to the limited scope of the article. [17] Product service lifetimes are adopted from One Click LCA Norge, and within the ranges given in Byggforskserien 700.320 [18].

As indicator of environmental impact, Global Warming Potential (GWP) in CO_2-equivalents are used. The functional unit is 1 m^2 useable area BRA over an estimated service life time for the building of 60 years. The useable area BRA for the Selbukassa project is 311 m^2.

The cases

Case 1 – "Selbukassa". Resource saving and extensive reuse of materials and building components is given priority over energy performance. To allow for reuse of building components, the fulfilment of energy requirements in the building code TEK17 was not required by the local building authorities. Construction principles are adapted to allow for self-building. Thus timber framing and insulation levels in the external wall was reduced in dimension/ thickness. The site, however, required substantial foundation works. The gallery had to fulfill strict requirements of fire resistance and emergency escape, and will therefore be built in a steel-concrete composite structure.

Case 2 – "TEK17". The reference is a building with the same inner dimensions, calculated floor area (useable BRA), and window area. The building is made according to state-of-the-art construction with insulated timber framing for the main building. The use of non-reused logs or CLT was not considered for the external walls of the new building, since timber framing represents a more material efficient and optimal solution when reuse is not the goal. Insulation levels of external walls, roofs, and energy performance of windows is increased, balanced ventilation is installed to comply with the energy efficiency requirements in TEK17.

Embodied emissions

The embodied emissions include the main building elements of the building envelope and the major internal building elements as slabs and inner walls, as well as technical systems like ventilation and heat pump. Elements not included are internal stairs, flashings, fixed interior, appliances and sanitary objects. Due to high uncertainties, surfaces coverings (paints, varnish, but also tiles) of external and internal surfaces are not included for both cases.

In agreement with NS 3720, emissions of reused elements are assigned to the previous life of the component, i.e. not considered in the emission accounting. In case of the "Selbukassa", windows, internal and external doors, roofing tiles, as well as internal siding and flooring is omitted from the emission calculation. It is assumed that these will be salvaged similar to the Eksperimentboliger-project, and later replaced with reused elements. Replacement of reused materials and building components is assumed with similar reused materials.

Operational emissions

Normative values according to NS 3031 were used in the simulations. The energy supply system for the Selbukassa case is based on the wishes discussed with the self-builders. Balanced ventilation is unlikely, but a central extract from bathrooms and kitchens will be installed. A exhaust-air-heat pump is discussed and taken into account in the simulations and contributes to the waterborne floor heating. The TEK17 case represents state of the art balanced ventilation with heat recovery of 80 %, and waterborne floor heating. The heating is covered by an air-to-water heat pump.

In accordance with NS 3720, two emissions factors are considered for electricity – Scenario 1: Norwegian electricity only, and Scenario 2: EU28+NO, a mix of Norwegian and 28 EU countries. In both cases, emissions decrease linearly from 2020 (27 g CO2e/kWh for scenario 1, and 277 g CO2e/kWh for scenario 2, both calculated from NS 3720) to 2050 (13 g CO2e/kWh for scenario 1, and 29 g CO2e/kWh for scenario 2), and remain at that level. In this paper, calculated annual emission factors from 2020 to 2080 are used, not average emission factors over the entire lifetime.[17]

3. Results

Figure 2 presents the embodied emissions associated with phases A1-5, B4-5, and C1-3 for both cases split for the building element categories. The largest portion of emissions are linked to technical systems (41 resp. 43%), foundations (23 resp 17%), and in case of the TEK17 building to external walls (16%). While most buildings elements have largest emission in the product phase, technical installation continue to have high emissions in phase B4 replacement due to estimated service life of 25 years, i.e. double replacement during 60 years of building lifetime. Of those, most emissions are in fact linked to the electrical systems. However, the higher emissions of installations in the TEK17 case result from the more advanced ventilation system and heat pump. The higher emissions for external walls are linked to higher insulation level and emissions related to new windows. Other building elements like foundations, structure and internal floors have similar emissions since these are built with little or no reused materials in both cases.

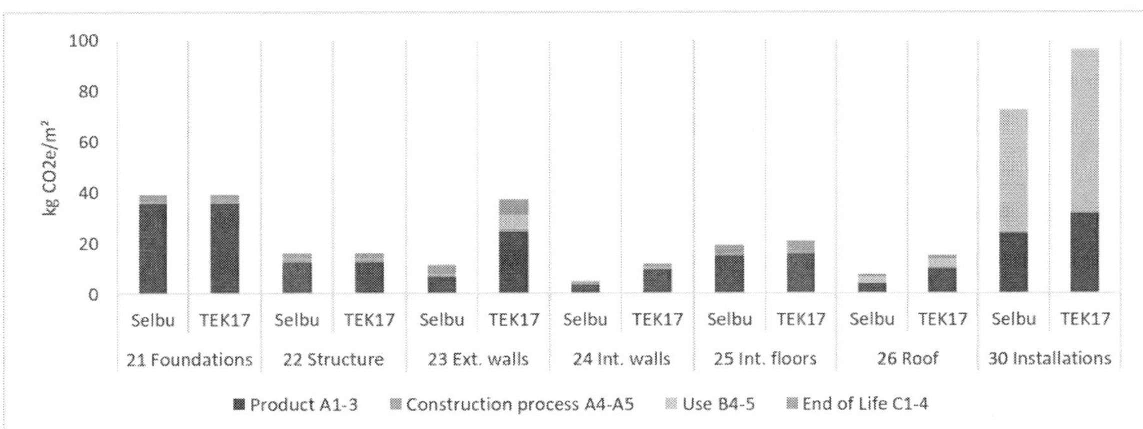

Figure 2: Embodied emissions split up for building elements

Figure 3 shows the total embodied emissions of both cases. As expected, the Selbukassa case has lower embodied emissions. Case Selbukassa has ca. 170 kg CO2e/m² (ca. 2,8 kg CO2e/m² per year for 60 years building lifetime), while case TEK17 has almost 40 % higher emissions of ca. 235 kg CO2e/m² (ca. 3,9 kg CO2e/m² per year). Although the total embodied emissions are different, the ratio of phases is similar in both cases – materials and transport (phases A1-5) account for 62 %, and replacements and end-of-life (phases B4-5, C1-4) for 38 % of total embodied emissions, which is in the range also found in other literature.

Figure 4 presents embodied emissions in relation to the two scenarios of emissions linked to energy use. In both cases and or both scenarios, embodied emissions are lower than operational emissions. In case of the scenario EU28+NO, emissions of energy use are over 5 times higher than embodied emissions of case "Selbukassa" and ca. 3 times higher than for case TEK17.

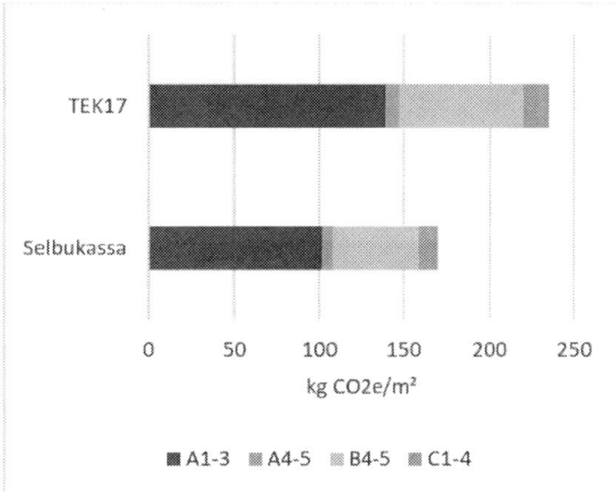

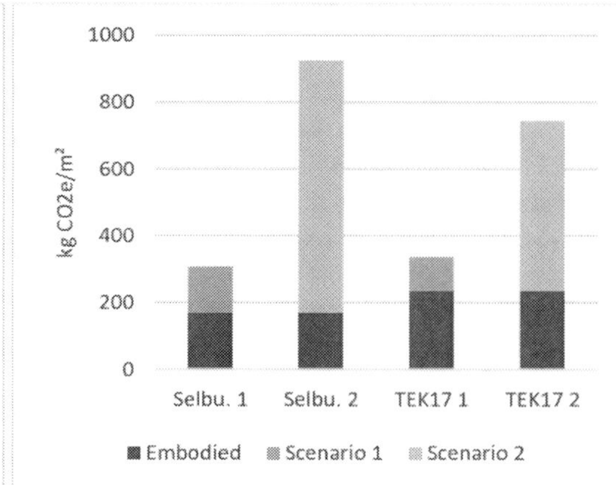

Figure 3 (left): Embodied emissions for both cases

Figure 4 (right): Total emissions for both cases for both energy scenarios

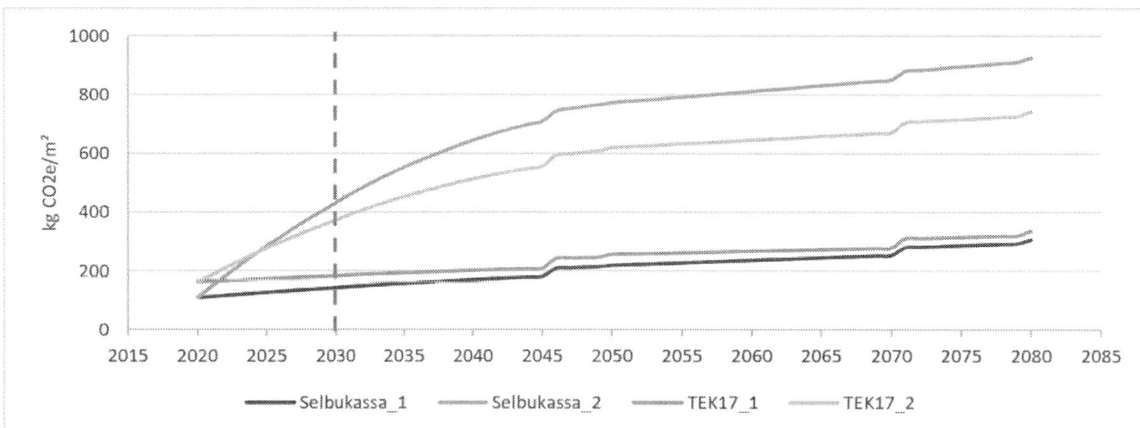

Figure 5: Cumulative emissions for both cases and both scenarios (1. Local Norwegian, 2. EU28 + Norway mix)

Figure 5 shows cumulative emissions projected over the building lifetime based on annual emission factors for electricity. Starting point is 2020 with emissions linked to phases A and B. In case of scenario 1 with the Norwegian emission factor, increase is very low, and although both cases converge, case "Selbukassa" remains below case TEK17 due to lower initial emissions. In case of scenario 2, "Selbukassa" surpasses "TEK17" after ca. 5 years, and after 60 years, total emissions are ca. 25 % higher than case TEK17. The impact of replacements is visible at 25, 30 and 50-year marks. At the critical 2030-mark, case "Selbukassa" has either ca. 22 % lower cumulative emissions than TEK17 in scenario 1, or 17 % higher emissions in scenario 2.

Embodied emissions and raw material use for both cases is shown in Figure 6. The highest use of material measured in weight is associated with the foundations, particularly the masses under the foundations and the concrete of the foundations. However, highest emissions are linked to technical installations with the lowest raw material use of all investigated building elements. No clear correlation between material use and emissions is evident.

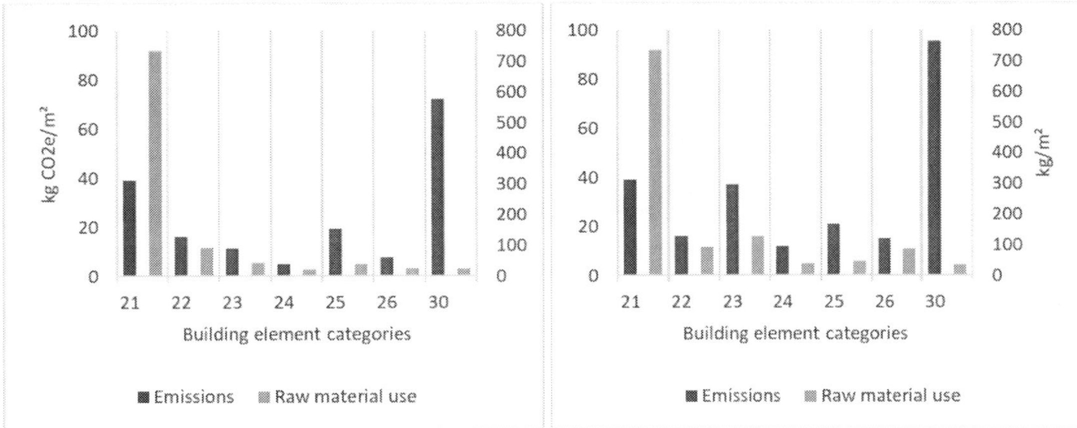

Figure 6: Embodied emissions and raw material use for case "Selbukassa" (left) and case "TEK17" (right)

4. Discussion

Over the lifetime of both Selbukassa and TEK17 cases the operational emissions are higher than embodied emissions with both scenarios for electricity emission factors (NO and EU28+NO). Since Selbukassa has lower or higher emissions than TEK17 building depending on the electricity factor used, no clear result can be obtained about the trade-offs between reused components, reduced new material use and energy efficiency.

Reuse is seen as advantageous in the context of reduced resource use, but the actual emission savings vary among different materials. The reused elements in Selbukassa like the old log building, windows and cladding might represent a large fraction of the materials used, but do not represent a large fraction of saved GHG emissions. To maximize benefits of reuse, reuse has to be carefully planned from early stages of design and it has to be project specific, prioritizing reuse of building components and materials with otherwise high environmental impact, and considering the effects of worse energy performance. A more detailed study on a single component level (e.g. windows) is necessary to optimize the reuse of components linked to energy performance of the building.

Assuming that emission reduction within 2030 is more urgent than in long-term perspective, this should be considered in the choice of indicators, e.g. using GWP20 with a shorter time horizon instead of GWP100, or introducing a relative weighting prioritizing early emission reductions. Furthermore, the principles for emission accumulation over time and sensitivity towards electricity factors and grid decarbonization should be investigated further to develop more adaptive strategies for emission reduction over the whole lifespan of the building.

Further research is necessary to develop a framework for addressing the 'absolute decoupling' targets in the building sector.

5. Acknowledgments

The authors gratefully acknowledge the support from the Norwegian Research Council and Pir II AS, as a part of the Industrial Ph.D. scheme.

IOP Conf. Series: Earth and Environmental Science **352** (2019) 012067 doi:10.1088/1755-1315/352/1/012067

References

[1] IPCC, *Summary for Policymakers. In: Global Warming of 1.5°C. An IPCC Special Report on the impacts of global warming of 1.5°C above pre-industrial levels and related global greenhouse gas emission pathways, in the context of strengthening the global response to the threat of climate change, sustainable development, and efforts to eradicate poverty.* 2018, IPCC: World Meteorological Organization, Geneva, Switzerland, 32 pp. p. 32.

[2] IRP, *Global Resources Outlook 2019: Natural Resources for the Future We Want.* , in *A Report of the International Resource Panel. United Nations Environment Programme.* 2019: Nairobi, Kenya.

[3] UNEP, *Decoupling natural resource use and environmental impacts from economic growth, A Report of the Working Group on Decoupling to the International Resource Panel.* . 2011.

[4] IRP, *Assessing global resource use: A systems approach to resource efficiency and pollution reduction*, in *A Report by the International Resource Panel. United Nations Environment Programme.* 2017, IRP: Nairobi, Kenya. p. 100.

[5] Agency, E.E., *Progress on resource efficiency and decoupling in the EU-27*, in *EEA Technical report*. 2014, European Environment Agency: Luxembourg. p. 56.

[6] Ekins, P., B. Meyer, and F. Schmidt-Bleek, *Reducing Resource Consumption - A Proposal for Global Resource and Environmental Policy, GWS Discussion Paper, No. 2009/5*. 2009, The Open Access Publication Server of the ZBW – Leibniz Information Centre for Economics.

[7] European Commission, C.f.t.C.t.t.E.P., the Council, the European Economic and Social Committee and the Committee of the Regions *Roadmap to a Resource Efficient Europe, COM/2011/0571 final.* 2011.

[8] Selamawit Mamo Fufa, R.D.S., Kari Sørnes, and I.A. Marianne Inman, *A Norwegian ZEB Definition Guideline*, in *ZEB Project report no 29* 2016 The Research Centre on Zero Emission Buildings (ZEB).

[9] Chastas, P., T. Theodosiou, and D. Bikas, *Embodied energy in residential buildings-towards the nearly zero energy building: A literature review.* Building and Environment, 2016. **105**: p. 267-282.

[10] Birgisdottir, H., A. Moncaster, A.H. Wiberg, C. Chae, K. Yokoyama, M. Balouktsi, S. Seo, T. Oka, T. Lützkendorf, and T. Malmqvist, *IEA EBC annex 57 'evaluation of embodied energy and CO2eq for building construction'.* Energy and Buildings, 2017. **154**: p. 72-80.

[11] Torhildur Kristjansdottir, H.F., Eivind Selvig, Birgit Risholt, Berit Time, Laurent Georges, Tor Helge Dokka, Julien Bourelle, Rolf Bohne and Zdena Cervenka, *A Norwegian ZEB-definition embodied emission*, in *ZEB Project report 17*. 2014, The Research Centre on Zero Emission Buildings (ZEB).

[12] Malmqvist, T., M. Nehasilova, A. Moncaster, H. Birgisdottir, F. Nygaard Rasmussen, A. Houlihan Wiberg, and J. Potting, *Design and construction strategies for reducing embodied impacts from buildings – Case study analysis.* Energy and Buildings, 2018. **166**: p. 35-47.

[13] Thormark, C., *Environmental analysis of a building with reused building materials.* International Journal of Low Energy & Sustainable Building, 2000. **Vol. 1**.

[14] Pettersen, N., *Pilotprosjekt. Gjenbrukshus i Trondheim. En bro fra destruksjon til konstruksjon.* . 2005, Trondheim kommune.

[15] Astbury, J. *Lendager Group uses recycled materials to build 20 townhouses in Copenhagen.* 2019 [cited 2019 14.05.2019]; Available from: https://www.dezeen.com/2019/04/16/upcycle-studios-townhouses-lendager-group-copenhagen-recycled-materials/.

[16] AS, A.V., *Utredning av barrierer og muligheter for ombruk av byggematerialer og tekniske installasjoner i bygg*, A.S. Nordby, Editor. 2018.

[17] Norway, S., *NS 3720:2018 Methods for greenhouse gas calculations for buildings (in English)*. 2019.

[18] SINTEF Byggforsk, *700.320 Intervaller for vedlikehold og utskifting av bygningsdeler.* Byggforskserien. 2017, Oslo: SINTEF.

IOP Conf. Series: Earth and Environmental Science **352** (2019) 012068 doi:10.1088/1755-1315/352/1/012068

Techniques for airflow measurements to determine the real efficiency of heat recovery in ventilation systems

Øystein Rønneseth[1], Peng Liu[1], Maria Justo-Alonso[1,2] and Hans Martin Mathisen[2]

[1] SINTEF Building and Infrastructure, Trondheim, Norway
[2] Norwegian University of Science and Technology, Trondheim, Norway

E-mail: Oystein.Ronneseth@sintef.no

Abstract. Heat recovery in ventilation is essential to reduce energy use and thus mitigate greenhouse gas emissions from the building sector. Heat recovery efficiency of at least 80 % in new buildings is required according to Norwegian standards. However, measurements show that the real heat recovery efficiency during operation is commonly 10-20 % lower. Measuring heat recovery efficiency in buildings is challenging, mainly due to difficulties measuring airflow rates close to the air handling unit (AHU). This study assesses the following duct airflow measurement techniques and equipment: pressure differential, velocity traversal technique, ultrasound and tracer gas. The pressure differential method can provide accurate flow rates and thus it is used as the reference measurement. However, it is not suitable for duct flow measurements due to its high pressure penalty and long straight duct requirement. Velocity traverse and tracer gas methods introduce less disturbance to the flow. Nevertheless, both methods require intensive labour work and cannot track quick changes of the airflow with time. The application of ultrasound to measure airflow is relatively novel and it can automatically measure constant and fluctuating airflows with low pressure drop and acceptable accuracy when the proper installation and minimum straight duct are provided.

1. Introduction

Heating, ventilation and air-conditioning (HVAC) systems play a vital role in achieving the desired indoor environment. Energy use for HVAC in the building sector accounts for a large proportion of the total energy use in most countries around the world. Heat recovery is employed as an efficient manner to reduce energy use in ventilation systems and thus mitigate greenhouse gas emissions from the building sector. Heat recovery effectiveness higher than 80 % is needed to meet the requirements in Norwegian building regulations for new buildings [1]. To be precise, an efficiency ≥80 % is required using the "energy measure" calculation method in Ref [1], although there is no minimum demand in the "energy frame" calculation method, an efficiency ≥80 % is commonly used and probably required in practice [2]. However, field measurements have shown that the actual heat recovery efficiency during operation is commonly 10-30 % lower than the expected efficiency. For extreme cases with improper installation, an efficiency as low as 10 % has been measured [3].

The thermal efficiency for heat exchangers can be calculated from equation (1) [4],

$$\eta_T = \frac{\dot{m}\, c_p}{(\dot{m}\, c_p)_{\min}} \frac{\theta_2 - \theta_1}{\theta_3 - \theta_1} \qquad 1)$$

Where $\dot{m}$ is the mass flow rate, c_p is the specific heat capacity of air, the subscript min is the minimum value of the supply and exhaust side, θ1 is the air temperature before the heat exchanger on the supply side, θ2 is the air temperature after the heat exchanger on the supply side and θ3 is the air temperature before the heat exchanger on the extract side.

The thermal efficiency varies as a result of the quality of the installation, system control and operating conditions associated with balanced or unbalanced airflow rates, condensation and frost formation in the heat recovery. Energy calculations with constant and overestimated thermal efficiency may lead to an underestimation of building heating demands [5]. The indoor thermal comfort will additionally be degraded by the possible cold draught problem due to under dimensioned heating coil. The discrepancy between expected and real performance is worth further investigation through field measurements. Through field airflow measurements, one can additionally identify malfunctions and improvement potential for the heat recovery, as well as leakages and recirculation.

To determine the field efficiency of heat recovery in buildings, measurements of air temperatures, relative humidity and airflows in the ducts connected to the air handling unit (AHU) are needed. In addition, several measuring points at the same cross-section are required to account for non-homogenous distribution of air temperatures, humidity and velocities in the ducts [6]. However, measuring heat recovery efficiency in buildings is challenging, mainly due to difficulties measuring accurate airflow rates close to the AHU. The equipment for measuring duct airflow rates normally requires minimum distances of straight duct up-and-downstream the measurement point to ensure fully developed velocity profiles. In most buildings, space around the AHU and ducts is insufficient to perform measurements in this manner. There is also a lack of standards for field measurements of heat recovery efficiency, as most measurement recommendations are defined for lab conditions. In practice, lab standards are difficult to follow as real systems have bends or devices disturbing the flow, making it difficult to find and access straight duct sections close to the AHU [6]. Consequently, measurement complexity increases in the field, and the accuracy is reduced due to more measuring points needed to account for the developing velocity profiles.

There are very few studies on the real efficiency of heat recovery through field measurements. Heide [7] collected examples with deviations up to 37 % compared to the rated efficiency for Norwegian residential buildings. Lassen et al. [8] refer to several studies in Norway, Switzerland and Sweden with 10-20 % lower efficiency. A study with field tests of 20 centralized and 60 decentralized ventilation systems found significantly lower heat recovery efficiencies for all units, with 65 % ±24 % for centralized systems and 70 % ±17 % for decentralized systems [9]. For this study, tracer gas was used in addition to thermal wires to determine the air temperature. Only one of the studies measured an efficiency higher than specified, with 77 % when 75 % was specified [10]. The study also pointed out challenges of performing in situ measurements in occupied buildings, such as uncontrollable boundary conditions and placing the sensors where they do not disturb the occupants. A common finding from the reviewed literature is the difficulty of measuring airflow rates with acceptable uncertainty. The provided measurement uncertainties of the temperature efficiency ranged from ±8 to ±24 %, which are higher than the typical uncertainties (±5 %) for laboratory testing [6]. Methods and protocols for rating and testing the real performance of the heat recovery for existing buildings are needed. The most significant barrier is the ability to measure duct airflow rates close enough to the AHU with an acceptable accuracy, which is assessed in this paper.

2. Methods

The focus of this study is to assess various measurement techniques and available measuring equipment to measure duct airflow rates with acceptable accuracy. During the project, researchers, equipment suppliers and contractors in this field were contacted asking for advice regarding measurement protocols. A selection of currently available methods and devices are tested through laboratory experiments at the Department of Energy and Process Engineering at NTNU in Trondheim. Field duct airflow measurements in real buildings are also performed. To compare the different equipment, they are tested in the same setup in the lab. The lab setup represents similar conditions as can be found for

AHU in buildings, but in order to have the same boundary conditions, the tests are done in the lab and not in the field. The velocity traversal method was tested in the field, but unfortunately, it was not possible to validate the results. Further details about the techniques, measuring equipment, relevant standards for measurements and the performed experiments can be found in [11].

3. Measurement techniques

Four different measurement techniques are presented and compared.

3.1. Pressure differential through an orifice plate

The pressure differential method with an orifice plate is widely used in laboratories to determine and monitor airflows. When a fluid passes through a reduction of the open area in a pipeline, it causes a static pressure drop. Based on Bernoulli's equation, the flow rate can be calculated using the measured static pressure difference before and after the orifice plate. Flow conditions should be constant or only varying slightly and slowly with time. The orifice plate must be mounted according to specified minimum straight duct lengths without disturbances or branch connections both upstream and downstream. Although highly accurate, this method introduces an additional pressure loss and has low sensitivity at very low airflow rates [12]. This method is mostly used in laboratories, due to the extra pressure loss and large intervention to ventilation systems in buildings. The airflow rate and corresponding uncertainties are calculated according to NS-EN ISO 5167-1 and NS-EN ISO 5167-2.

3.2. Velocity traversal method

By measuring the velocity at multiple designated positions over the cross-section of a duct, one can obtain the area-averaged velocity. The number of measuring points and accuracy is decided based on the flow regime, the duct diameter and velocity profile in the duct. Turbulence caused by bends, fans, branching and limited straight ductwork increases the inaccuracy and thus it may require more measuring points. Minimum straight duct length, the number of measurement points and other criteria for using velocity traversal method is specified in NS-EN 16211:2015. One significant weakness of the method is that it cannot capture the quick airflow variations with one set of measurements. The measured airflows are assumed constant within the measurement time. The sensor probe will also interfere with the velocity profile, further decreasing the measurement accuracy. The method is intuitive, but traversing the sensor needs to be conducted manually, thus requiring time-consuming labour. The measurement uncertainty is high for very low velocities and developing flows.

For this project, two different types of TSI anemometers were used, VelociCalc 8388 and VelociCalc 9565-P. These are hot-film anemometers with extendable measurement probes. VelociCalc 9565-P measures air velocity, temperature and relative humidity simultaneously. A drawback of this device is that the placement of the sensors requires that at least 75 mm of the probe is in the flow [13].

3.3. Ultrasonic measurements

Ultrasonic devices are based on the Doppler effect [14]. Through ultrasonic measurements, one can monitor both constant and time-dependent airflows at a wide range of air velocities as it responses linearly to flow velocity change. This is especially beneficial at low air velocities, and so the technique is emerging for use in VAV dampers in new buildings aiming for low energy use for fans [15, 16]. Another major benefit is that there is no additional pressure loss.

Lindab UltraLink FTMU is tested to measure airflow and temperature. Measurements are logged automatically and exported to a computer using Modbus. UltraLink also has requirements regarding minimum straight duct lengths. Its uncertainty also depends on the orientation of the device relative to disturbances as referred in the datasheet [17]. Another drawback is that the product is currently only available for relatively small duct sizes (Ø100-Ø315).

3.4. Tracer gas method

Tracer gas dilution techniques are well-known for assessing airflow rates and air distributions within buildings and ventilation systems. With the tracer gas method, the air is "marked" with one or more tracer gases, and then the concentrations are measured at various positions. By interpreting the evolution of tracer concentrations, it is possible to calculate airflow rates, ventilation efficiency, the age of air and leakage flow rates [18]. However, tracer gas measurements are difficult to perform in the field, as it requires specific knowledge for setting up the measurements and analysing the concentrations. The key factor for successful tracer gas method is the good mixing of the tracer and the air, which tends to be difficult to achieve in the ductwork where there is very limited space.

For this project, carbon dioxide (CO_2) and nitrous oxide (N_2O) are used as tracer gases, and Lumasense photoacoustic gas monitor is used for analysis of the tracer concentrations, the same device as used by [9]. Due to its background concentration, CO_2 is not ideal for analysis but was chosen for its low global warming potential and low price relative to other gases.

3.5. Comparison of measurement techniques and equipment

The selected measurement techniques and equipment used in the project are compared in table 1. The VelociCalc is the only one able to measure velocity, temperature and relative humidity, while tracer gas can determine parasitic airflows (like for example leakages in rotary heat exchangers). However, for determining the annual heat recovery efficiency, the UltraLink has a significant advantage due to its automatic logging. For the other methods, intensive manual labour is required for the measurement if the annual performance of the system is monitored. In addition, demand-controlled ventilation is becoming increasingly popular, and fluctuating airflows can be tracked by the ultrasonic measurement.

Table 1 Comparison of the selected measurement techniques and equipment used in the project.

Airflow measurement method		Velocity traverse	Ultrasound	Pressure differential	Tracer gas
Selected measuring equipment in this project		VelociCalc 9565-P [13]	Lindab UltraLink [17]	Orifice plate, TT series manometer	CO_2 and N_2O tracer gases, Lumasense photoacoustic gas monitor
Velocity	Range	0-50 m/s	0.2-15 m/s	Reynolds number ≤5000	NA
	Uncertainty	±3 % of reading or ±0.015 m/s, whichever is greater	±5 % of reading or ±X[a] l/s, whichever is greater	Depends on the flow rate, orifice plate, etc.	Uncertainty mainly depends on the mixing level
	Resolution	0.01 m/s	NA	NA	NA
Temperature	Range	-10 to 60 °C	-10 to 50 °C	NA	NA
	Uncertainty	±0.3 °C	±1.0 °C	NA	NA
	Resolution	0.1 °C	NA	NA	NA
Relative humidity	Range	5 to 95 % RH	NA	NA	NA
	Uncertainty	± 3 % RH			
	Resolution	0.1 % RH			
Automatic logging		Yes (in 1 point)	Yes	Depends on the micromanometer and the logger	For tracer concentration, yes
Duct size		Ø25 to Ø12700	Ø100-Ø315	Ø50-Ø1000 [19]	Any size
Strengths		Easy to implement and understand	Accurate, intuitive, low uncertainty for low air velocity, no flow interference	Accurate	Able to determine the main and parasitic airflows at the same time, relatively accurate
Drawbacks		High measurement uncertainty for varying flow conditions, requires manual labour	Could be difficult to install in existing ducts, not available for large duct sizes	Creates extra pressure loss, not suitable for field tests	Need good mixing, which is almost impossible in AHU, and specialized knowledge to calculate the airflows
Costs		Low equipment cost and medium/high labour work depending on measurement points	Low equipment cost and low labour work	Low equipment cost and low labour work	Very high equipment cost and high labour work

[a] X equals the distance in dm, for instance Ø100 => 1 l/s and Ø200 => 2 l/s.

IOP Conf. Series: Earth and Environmental Science **352** (2019) 012068 doi:10.1088/1755-1315/352/1/012068

4. Lab experiments

Measurements have been performed with the above techniques and the main results are briefly presented here. More detailed information can be found in the project report [11].

4.1. Comparison between velocity traversal, ultrasonic measurements and pressure differential

For this experiment, the orifice plate is mounted according to straight duct requirements and used as a reference for testing Lindab UltraLink FTMU and VelociCalc 8388. The test rig is shown in figure 1 and measurements are performed on the same straight duct section (Ø200) but at different distances from the bend. Measurements are performed with four different setups with varying distances to the bend, and with five different airflow rates for each setup. Note that the tested setups are mostly outside producers and standards recommendations for distances relative to disturbances, as we want to evaluate the UltraLink and VelociCalc in conditions similar as to what can be found in the field. This paper only presents the setup with the shortest distance from the bend to the devices: 0.05 m to VelociCalc, 0.12 m to UltraLink and 4.00 m to orifice plate (see image to the right in figure 1). For the UltraLink, the measurement uncertainty is affected by the orientation of the flow sensors relative to disturbances. Consequently, the UltraLink is tested at different orientations. For the velocity traverse method, five measuring points are used, positioned according to NS-EN 16211:2015.

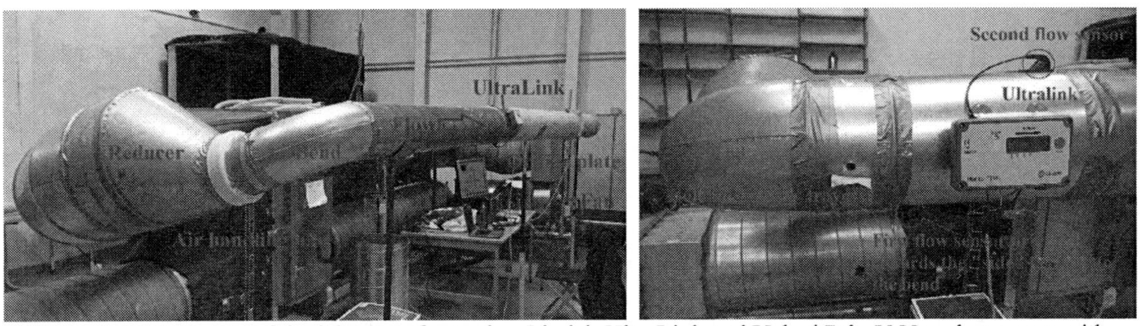

Figure 1 Left: Picture of the lab setup for testing Lindab UltraLink and VelociCalc 8388 and compare with calculated airflow rates found with the orifice plate. Right: setup used for the presented measurement results.

The results from the measurements are presented in table 2. The terms used for UltraLink orientation; *side, inner radius* and *outer radius*, refer to the placement of the first flow sensor on the device relative to the bend. Due to the very short distance from the bend, the airflow is highly turbulent and unevenly distributed over the cross-section of the duct, making it very difficult to achieve accurate measurements. An average deviation of 6.6 % when the UltraLink is oriented towards the side is observed. The deviations for the other orientations are significantly higher, with -22.8 % and -16.5 % for the inner and outer radius respectively. The results with the VelociCalc have an agreement with reference values from the orifice plate, with mean deviations between 3.8 % and 6.3 %.

Table 2 Measurement results with calculated flow rate over orifice plate, measured values for UltraLink and VelociCalc and deviations compared to the calculated value over the orifice plate.

UltraLink orientation	Measurement #	1	2	3	4	5	Avg. error compared to orifice plate	Standard deviation
	Calc. flow rate from orifice plate [m³/h]	59.9	132.1	204.8	277.9	351.4	-	
Side	UltraLink [m³/h]	62	141	222	298	376	-	
	VelociCalc 8388 [m³/h]	64.3	143.4	218.6	293.5	363.3	-	
	Deviation UltraLink	3.5%	6.7%	8.4%	7.2%	7.0%	6.6%	1.8%
	Deviation VelociCalc 8388	7.4%	8.6%	6.7%	5.6%	3.4%	6.3%	2.0%
	Calc. flow rate from orifice plate [m³/h]	59.9	132.1	204.8	277.9	351.4	-	
Inner radius	UltraLink [m³/h]	48	98	156	215	275	-	
	VelociCalc 8388 [m³/h]	64.1	144.3	215.5	285.0	361.3	-	
	Deviation UltraLink	-19.9%	-25.8%	-23.8%	-22.6%	-21.7%	-22.8%	2.2%
	Deviation VelociCalc 8388	7.0%	9.3%	5.2%	2.6%	2.8%	5.4%	2.8%
	Calc. flow rate from orifice plate [m³/h]	59.9	132.1	204.8	277.9	350.9	-	
Outer radius	UltraLink [m³/h]	50	109	171	233	295	-	
	VelociCalc 8388 [m³/h]	62.5	143.0	212.0	279.3	358.6	-	
	Deviation UltraLink	-16.5%	-17.5%	-16.5%	-16.2%	-15.9%	-16.5%	0.6%
	Deviation VelociCalc 8388	4.4%	8.3%	3.5%	0.5%	2.2%	3.8%	2.9%

Based on our measurements, including those not presented here, the UltraLink achieves the best accuracy when oriented towards the side when the distance from a bend is shorter than recommended. In general, measurements with the VelociCalc are more accurate than with the UltraLink, especially when the UltraLink is oriented towards the inner or outer radius relative to the bend. Aside from the setup in table 2, the orientation towards the inner radius was more accurate than towards the outer radius. Note that results may differ for other disturbances. The deviations for the VelociCalc are highest for low airflow rates, albeit still acceptable, and in the same range as the deviations found with the UltraLink oriented towards the side. However, for higher airflow rates, the VelociCalc is more accurate, and so the optimal choice of measurement equipment is also dependent on the range and frequency of variation of the airflow rate to be measured. However, the measurement accuracy of the ultrasonic device oriented towards the side is acceptable without additional pressure loss and interference on the velocity profile. Considering automatic logging, the ultrasonic airflow measurement technique is thus considered a superior alternative to both the velocity traversal and pressure differential techniques for field measurements, if it is properly installed.

4.2. Tracer gas measurements

The aim of this experiment is to investigate the feasibility and complexity of using the tracer gas technique to determine the main and parasitic airflow (leakage) rates in an AHU. Figure 2 shows a principle sketch of the system including tracer gas dosing and sampling spots. The test rig has the same AHU illustrated in figure 1. A LabView program is used to control the supply and extract airflows and rotary speed of the heat wheel, which are kept constant during each test scenario. The ductwork connected to the heat wheel has a dimension of Ø500, and balance between supply and exhaust airflows are verified using Testo 420 flow hood. Multiple tests are performed, including tests with two tracer gases (CO_2 and N_2O) to determine the airflows in different ducts at the same time. However, as these two gases have some overlapping wavelength adsorption intervals, the tracer concentration results obtained from the experiments are unreasonable due to the lack of cross-compensation for wave adsorption. Besides the test with two tracer gases, the airflow measurement with the single tracer gas test (CO_2) is also conducted. The single tracer gas must be applied sequentially for different ducts, with sufficient time in between to eliminate the influence from the last test. For this case, the time interval is 24 hours and the injection rate of CO_2 is constant at 4 l/min. To achieve sufficient mixing, the gas is injected through multiple holes in two tubes positioned perpendicular to the cross-section of the duct. Leakages through the ducts between the dosage point and the AHU, as well as external leakage through the AHU box, are considered negligible compared to the recirculation flow rates.

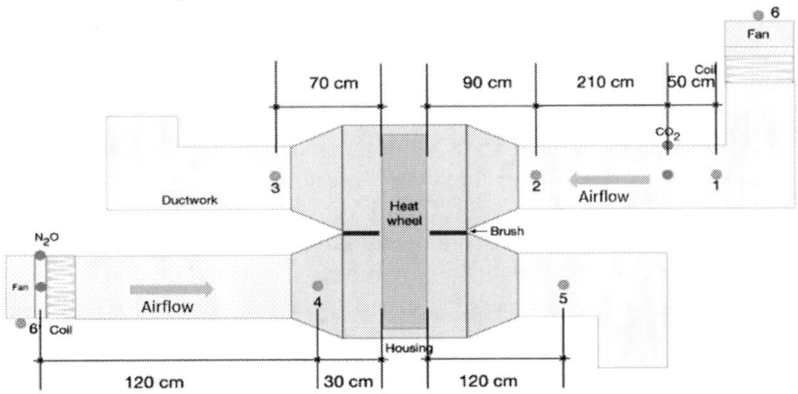

Figure 2 Principle sketch of the lab setup for tracer gas tests, with ducts, injection points in blue and measuring points in red.

Monitoring of the tracer gas concentrations is performed by infrared spectroscopy with the Lumasense photoacoustic gas monitor. The air is pumped into the tracer gas measurement chamber from the sampling points. The main and parasitic airflow rates can be determined with measured CO_2 concentrations and the injection tracer rate using tracer mass conservation. For a rotational speed of 10 rpm, the supply and extract recirculation rates for the tested AHU with rotary heat wheel are 15.2 % and 9.3 % respectively. At 20 rpm, the corresponding values are 14.3 % and 17.0 %. The extract air recirculation increases at the higher rotational speed (20 rpm), and thus more polluted air is introduced into the supply air. This may degrade the performance of the heat recovery and reduce indoor air quality. Another interesting finding is that a test with a long straight duct attached in the supply side (i.e. unbalanced static pressures between supply and extract), led to a large fraction of supply air penetrating to the exhaust side. This caused malfunction of the heat wheel and indicates that balanced duct resistance and balanced static pressure in the extract and supply sides should be pursued in practice to avoid large leakages in rotary heat recovery units.

To conclude, the single tracer gas method is able to present the main and parasitic airflows in the AHU. Good mixing between the tracer and airflow is, however, a big challenge, even when multiple injections are used. In addition to the challenges of achieving sufficient mixing of the tracer with the airflow in AHU in practical systems, the tracer gas technique is in general more difficult to perform than the other techniques. When using multiple tracer gases, the cross-compensation function of adsorption spectrums for different tracers should be activated in the gas analyser to reduce the interference between various tracer gases.

5. Conclusion

In order to reveal the field heat recovery efficiency, accurate measurements on duct airflow close to the AHU is needed. The available measuring techniques presented in this paper can be categorized as the pressure differential, velocity traversal, ultrasonic and tracer gas methods according to the measurement principle. All the equipment presented in this study requires minimum straight duct lengths to perform good measurements. On the other hand, the requirement for the available ductwork to place the sensor and provide sufficient straight length is rarely met in real buildings.

The orifice plate with pressure differential method could provide accurate airflow measurements, however, it causes high extra pressure loss and requires long straight duct. The orifice plate is not considered suitable for field duct airflow measurement and long-term monitoring due to its difficulty of connection with existing ductwork and high pressure penalty. The velocity traverse and tracer gas methods have less interference with the airflows. The velocity traverse method is intuitive and easy to be performed. The tracer gas method is of specialization to be implemented and the analysis of the tracer gas result is relatively complicated. Both methods require intensive labour. The ultrasound method, as an emerging technique for airflow measurement, is able to measure the flow rates with low cost, no disturbance to the airflow and acceptable accuracy when the proper installation is provided. The measurements with UltraLink is slightly worse than the velocity traverse method with VelociCalc 8388 in terms of measurement accuracy. However, UltraLink is more automatic and need almost no extra labour once it is installed in the ductwork. The ultrasonic method is recommended as the most promising solution for duct airflow measurement if more options on device dimension are available with its development in the market. Additionally, the clamp-on ultrasonic equipment, which is typically used for liquid flow, would be recommended to be further developed and transferred to airflow application. The clamp-on solution will avoid the duct connection and could be applied to both existing and new systems without any disturbance for the airflow. A measuring uncertainty within ±3 % for airflow rate in the duct close to the heat recovery is pursued in order to obtain accurate heat recovery efficiency.

Acknowledgements

The authors gratefully acknowledge the support from Husbanken for financing the project (Roterende varmegjenvinnere 16/5603). In addition, we would like to thank Trondheim municipality for supplying pilot buildings for measurements and Lindab for two UltraLink devices and discussion of the results.

References

[1] *Byggteknisk forskrift (TEK 17)*, 2017.
[2] A. J. Petersen, P. G. Schild, I. H. Bryn, E. N. Haugen, G. Nilsson, and K. Høydahl, "Forhold tilknyttet bruk av roterende gjenvinnere i skoler," Erichsen & Horgen AS, 2009.
[3] C.-A. Roulet, *Ventilation and Airflow in Buildings. Methods for Diagnosis and Evaluation*. Earthscan, 2008.
[4] S. Ingebrigtsen, *Ventilasjonsteknikk Del 1*. Skarland press AS, 2015.
[5] Y.-h. Choi, D. Song, D. Seo, and J. Kim, "Analysis of the variable heat exchange efficiency of heat recovery ventilators and the associated heating energy demand," *Energy and Buildings*, vol. 172, pp. 152-158, 2018/08/01/ 2018.
[6] A. B. Johnson, C. J. Simonson, and R. W. Besant, "Uncertainty analysis in testing of air-to-air heat/energy exchangers installed in buildings," presented at the 1998 ASHRAE winter meeting, San Francisco, CA (United States), 1998.
[7] V. Heide, "Miljøvennlig boligventilasjon. Sammenligning av klimagass-utslipp med naturlig- og balansert ventilasjon," Master thesis, Sustainable Architecture, NTNU, 2012.
[8] N. Lassen, A. Fylling, M. Mysen, T. H. Dokka, and L. Bordewich, "Passivbygg som forskriftskrav i 2020," Multiconsult and SINTEF Byggforsk, DiBK, 2009.
[9] A. Merzkirch, S. Maas, F. Scholzen, and D. Waldmann, "Field tests of centralized and decentralized ventilation units in residential buildings – Specific fan power, heat recovery efficiency, shortcuts and volume flow unbalances," *Energy and Buildings*, vol. 116, pp. 376-383, 2016/03/15/ 2016.
[10] E. Kamendere, G. Zogla, A. Kamenders, J. Ikaunieks, and C. Rochas, "Analysis of Mechanical Ventilation System with Heat Recovery in Renovated Apartment Buildings," *Energy Procedia*, vol. 72, pp. 27-33, 2015/06/01/ 2015.
[11] Ø. Rønneseth, P. Liu, and M. Justo Alonso, "Airflow Measurements for Air Handling Units," in "SINTEF Research 51," SINTEF Building and Infrastructure, 2018.
[12] D. Yu, H. Li, and M. Yang, "A virtual supply airflow rate meter for rooftop air-conditioning units," *Building and Environment*, vol. 46, no. 6, pp. 1292-1302, 2011/06/01/ 2011.
[13] TSI, "VelociCalc air velocity meter. Model 9565 Series, Operation and service manual," ed, 2018.
[14] K. Conrad and L. Lynnworth. (2002, 23.11). *Fundamentals of Ultrasonic Flow Meters*. Available: http://asgmt.com/paper/fundamentals-of-ultrasonic-flow-meters-2002/
[15] FläktGroup, "Optivent Ultra VAV Damper, Technical data," ed, 2017.
[16] Lindab, "Lindab UltraLink Controller FTCU - Technical information," ed, 2018.
[17] Lindab, "Lindab UltraLink Monitor, FTMU. Technical information," ed, 2018.
[18] E. Mundt, H. M. Mathisen, P. V. Nielsen, and A. Moser, "Ventilation effectiveness," in "REHVA Guidebook No. 2," 2004.
[19] *ISO 5167-2:2003 Measurement of fluid flow by means of pressure differential devices inserted in circular cross-section conduits running full — Part 2: Orifice plates*, 2003.

IOP Conf. Series: Earth and Environmental Science **352** (2019) 012069 doi:10.1088/1755-1315/352/1/012069

Simulation and control of shading systems for glazed facades

G Chaudhary[1], **F Goia**[1], **S Grynning**[2]

[1] Norwegian University of Science and Technology, Faculty of Architecture and Design, Alfred Getz vei 3, 7034 Trondheim, Norway
[2] SINTEF Building and Infrastructure, Høgskoleringen 7b, 7465 Trondheim, Norway

francesco.goia@ntnu.no

Abstract. This paper presents a simulation study on the use of shading devices in highly glazed facades in the context of zero emission buildings (ZEB) located in a Nordic climatic context. Shading devices have the task to control different uses of solar energy (solar gain for passive solar heating and daylighting) and to avoid cooling load, and to find the best strategy (i.e. to optimize the overall performance) for controlling shading devices is not trivial.
Different control strategies for the activation of the shading devices (venetian blinds) are simulated, and their performance calculated in this study through a dynamic whole-building simulation tool (EnergyPlus). The results of the simulation studies show that it is possible to achieve very high performance and balanced use of solar energy (for passive solar heating, daylighting), without occurring in too high cooling load, even in the case of fully glazed facades. The optimal strategy for shading activation is not straightforward, and different strategies should be recommended during the year to assure an overall optimal performance.

1. Introduction

Fully transparent façades based on curtain-wall systems have not been traditionally considered a sustainable solution for office buildings in the Nordic climate because heat loss through the glazed façade would have led to high energy use for heating and high thermal discomfort for the user.

Furthermore, the large transparent area, coupled to the relatively high-density of internal gain typical of an office building, would have caused cooling load not only in the summer months, but even in the mid-seasons. However, a previous studies [1] already showed that, when looking at the total energy performance of the building (including heating, cooling, and lighting), a Window-to-Wall Ratio (WWR) of up to 0.80 is not impairing the energy performance of a low energy building located in the Nordic context (for a south exposed façade), provided that it is realized with best available technologies for façade systems and it is equipped with shading devices controlled in a dynamic way.

The current development of curtain-wall façade technologies has led to market availability of outstanding façade systems, characterized by a thermal performance close to that of more conventional envelope solutions based on opaque walls and windows, coupled with shading systems capable of properly managing solar gains. Furthermore, advanced façade systems characterized by an optimal control of shading system can contribute to increase the daylight autonomy and keep the risk of glare discomfort, evaluated through the metrics $UDI_{>2000}$ [2], within a relatively safe area.

Previous investigations have thus proved the feasibility of a façade with a large WWR in a Nordic climate, as well as the need of a case-by-case investigation of the optimal control strategy for the shading systems, as increased energy demands can happen when an improper shading strategy is chosen [3]. However, there are not studies dedicated to curtain-wall solutions (so with very high

WWR) showing the implication of different control strategies for the shading, and only fully mechanical climatized buildings were investigated, while the evaluation of the performance of an office building only equipped with the heating plant has not been analyzed.
The investigation presented in this paper extends the analysis presented in the previous study [1] and looks at the operational energy performance and visual comfort of different façade configurations for a low energy office concept building. The aim has been to investigate the energy demand for heating and cooling as well as daylight conditions in the buildings as function of glazing area (conventional façade vs. curtain-wall façade) and several shading strategies. The concept building has previously been used as a case building for conceptual studies of life cycle performance in a zero-emission building context and more thorough description of the baseline building can be found in [4].

The study comprises the analysis of both a fully mechanically climatized building (i.e. with mechanical heating and cooling), and of an office building characterized by only mechanical heating, addressing the research question whether a fully glazed façade can compete or not in terms of energy and visual comfort with a more conventional façade, composed by (large) windows and opaque walls.

2. Methods

2.1. *Case study building and facade technologies*

The research activity has been carried out through the extensive use of *one* building performance simulation *tool,* EnergyPlus™ (version 8.5) [5], and the use of a case study building whose behavior is assumed to be representative of that of any conventional office building, even if the latter one is not presenting the exact same layout. This assumption is based on the adoption of a case-study building defined by other researchers [4], who developed it with the scope of making available a reference construction configuration for performance analyses a low energy office building in a Nordic context.

The case-study building used in the study is a 4-storey high office building, with a rectangular footprint of approximately 17 x 30 meters, with long facades facing south and north. Each floor has a heated floor of 495 m², giving a total area of 1980 m². The case study building is constructed according to the requirements in the passive house standard for non-residential buildings [6], with the following thermal transmittance values for the different components: external walls U = 0.12 W/m²K; external roof: U = 0.09 W/m²K; floor against basement: U = 0.11 W/m²K; windows: U = 0.75 W/m²K; doors: U = 0.75 W/m²K; normalized thermal bridge value phi = 0.03 W/m²K.

For the scope of this study, two twin buildings with a different WWR configuration have been modeled and simulated:

- a baseline WWR solution (WWR nearly 0.35) with a baseline façade, as described in [4]; and
- a fully-glazed solution representing a curtain-wall façade, with a WWR of nearly 0.95.

The baseline building has a total window and door area of 456 m². The windows used in this façade configuration are made of an aluminum insulated frame (with U-value of 1.280 W/m²K), positioned in the middle of the wall in order to reduce the thermal bridge effect. The window-wall junction thermal bridge value is 0.015 W/mK. The window is equipped with a triple-pane glazed unit, with a centre-of-glazing U-value of 0.72 W/(m²K), solar factor (g-value) of 0.716, and visible transmittance of 0.740. The U-value of the window, depending on its size, is in the range 0.65 to 0.75 W/m²K.

The curtain-wall façade building has a total window and door area of 750 m² and makes use of a curtain wall system characterized by a frame transmittance of 1.410 W/(m²K). The glazing is the same as for the reference building. Both the reference façade and the curtain-wall façade are equipped with an interpane shading system, a venetian blind that can coupled with different control strategies.

2.2. *Simulation settings*

A virtual numerical model of the case study building in the two configurations has been developed in *EnergyPlus* to assess the energy performance. The thermal zoning of the building has been realized by keeping together coherent areas but still preserving representative rooms (such as the single office room towards the south and towards the north), leading to a total of 27 zones.

The EnergyPlus model *Ideal load system* was used to calculate the room size energy use without recurring to modeling the full HVAC system (which depends on the type of building and increases the number of variables in the study). Within this system, Differential Dry Bulb Outdoor Air Economizer and a Sensible Heat recovery system was selected to simulate the adoption of best available technologies in connection to ventilation (i.e. a heat recovery system with the possibility to bypass the device). Heating was scheduled to run the all 8670 hours of the year if necessary, whereas cooling (for fully climatized building) was scheduled to run only during working hours, if necessary. The heating setpoint was 21 °C with a setback of 2 °C (for unoccupied hours in the office) and the cooling setpoint was 26 °C, but cooling was turned off during unoccupied hours so had not setback temperature.

As the plants of the building was not modelled in detail, since this would have required several additional assumptions that might have limited the generalisation features of the analysis, an ideal heating and cooling plant, characterized by constant coefficient of performance (COP) values, was assumed. The values for the COP were as follow: for heating a COPh = 3 and for cooling a COPc = 4.

To allow a simple sensitivity analysis over influence of different COPs, two additional values for COPh and COPc were considered too: a COPh = 2.5 and of 3.5; and a COPc = 3 and 6).

When the cooling is not activated because of unoccupied space, or when a configuration without mechanical cooling is tested, free cooling in the form of ventilation was simulated. This ventilation was simulated in form of night ventilation from 24:00 to 7:00 am and full 24 hours on weekends (when the office was unoccupied). V. Geros et al. [7] measured night ventilation flow rate (air change per hour, ACH) varying from 4 to 11 h^{-1} in single sided ventilation, whereas M. Kolokotroni et al. [8] mentions that low value of 5 h^{-1} for night ventilation was consistent with ventilation rates achievable through single-sided openings in offices. Considering the above studies, settings were implemented so that a (natural) ventilation flow (with outdoor air) of 2 ACH would be assured in zones if indoor temperature is above 26 °C and the outside temperature is higher than 13 °C (to avoid undercooling).

Furthermore, at the end of the study a parametric analysis was carried out to analyze the effect of varying the ACH rate for free cooling ventilation rate on overheating degree hours (ODH) during the occupied hours. The ACH was varied from 0.5 to 10 h^{-1}. This approach can be found in literature in similar studies, for example by N. Artmann et al. [9], who studied ODH above 26 °C (ODH_{26}) as a function of air change rate (0.5–32 ACH) during night-time ventilation, or by R. Ramponi et al [10] carried out a sensitivity study by imposing constant night ventilation rates varying from 0.5 to 20 h^{-1} to analyze the energy saving potential of night ventilation.

The building follows the occupancy schedule of a typical daytime office. In the occupied zones per person occupies 5 m^2 during working hours in weekdays, i.e. 08:00 to 19:00, Monday to Friday, whereas for weekends and holidays the whole building is considered unoccupied. Activity level was set to 111 W/person, following the recommendation contained in the ASHRAE Fundamental 2011 correspondent to office sitting activities, and electric equipment releasing, during occupied hours, to 4 W/m^2 [11]. The lighting in the building was on during occupied hours, with an installed lighting power density is 3 W/m^2, which is dimmed 0 to 100% depending on the daylight availability in the zone to reach the illuminance setpoint of 500 lux.

2.3. Performance metrics

The performance of glazed façade was analyzed using three different metrics referred to two domains: thermal environment and visual environment.

The direct effect of the façade configuration on energy demand of the building was analyzed by comparing annual heating, cooling and artificial lighting energy consumption expressed in kWh/m^2.

The time profile of the room indoor air temperature is used to assess the performance of the shading strategies for those scenarios where the mechanical cooling is turned off. The aim of this simple parameter is to assess whether or not the combination of natural (night-time, primarily) ventilation and high-performance façade itself is capable to maintain the required thermal comfort conditions. In the context of this study, a temperature in the range 20 °C and 26 °C is considered to be comfortable, while a temperature up to 28 °C is still considered acceptable. This assumption is supported by the body of knowledge available in the literature and standards (for example [12]) that demonstrates how temperature higher than 26 °C can still be perceived as comfortable by the user in case of a non-mechanically ventilated building, and in combination with high outdoor air temperature (i.e. during the summer season). In the scenarios without mechanical cooling, the annual hourly values of the indoor air temperature exceeding 26 °C and 28 °C were used to assess the risk of discomfort linked to overheating of the building. The number of hours, during the occupancy period (i.e. excluding time outside the set working hours and weekends), when the indoor temperature exceeded the two threshold values were counted.

Finally, the *Useful Daylight Illuminance (UDI)* [13] has been used to analyse the implication on the visual environment. The UDI measures how often the daylight on the working plane is within a specific illuminance range, and four different UDI are used, following the range limits proposed by Nabil and Mardaljevic [13]:

- $UDI_{<100}$ – insufficient to be the sole source or to contribute significantly to artificial lighting;
- $UDI_{100-500}$ – effective either as the sole source of illumination or with artificial lighting;
- $UDI_{500-2000}$ – often perceived either as desirable or at least tolerable;
- $UDI_{>2000}$ – likely to produce visual or thermal discomfort, or both.

While energy (both thermal and electric) use can easily be obtained as an output from the simulation engine, there is no direct output for UDI in EnergyPlus, and a dedicated subroutine implemented in the EnergyPlus Erl language was developed to compute the different UDI values.

2.4. *Overview of the simulations and scenarios*

The performance analysis of the two façade systems was assessed in a comparative way, using the reference façade configuration (the original one form the ZEB Concept Office Building) as a baseline to measure the impact of different configurations.

The general performance analysis is carried out by comparing four different scenarios, described in Table 1, where the following nomenclature was used: REF and FG means reference (Baseline) and fully glazed façade configuration, respectively; noSHADING and bestSHADING, refer to the lack of integration or the integration (best strategy) of shading systems, respectively.

However, the simple integration of a shading device is not enough to determine the performance of the building, as this depends on how the shading device is controlled. To deepen this aspect and to take into consideration a control strategy that leads to a (nearly) optimal performance, a parametric analysis on different control strategies was necessary. Five different shading control logics where therefore simulated and used to select the best shading control strategy.

The six different control logic scenarios (S1...S6) used in the analysis are:

1. *Shading is on when solar radiation incident on the window exceeds a SetPoint* (W/m^2) and schedule, if specified, allows shading – different setpoint values are tested.
2. *Shading is on if zone air temperature in the previous timestep exceeds a SetPoint* (°C) and schedule, if specified, allows shading – different setpoint values are tested.
3. *Shading is on if zone cooling rate in the previous timestep exceeds a Setpoint* (W) and schedule, if specified, allows shading – different setpoint values are tested.
4. *Shading is on at night and on during the day if the solar radiation incident on the window exceeds Setpoint (W/m^2) and if the zone cooling rate in the previous timestep is non-zero.*
5. *Shading is on if illuminance on a reference point is higher than a threshold value and-or zone air temperature is higher than a threshold value:* This shading scenario is not naturally implemented in EnergyPlus, and it was created in the EMS module. The threshold values used in this case are 2000 lux (illuminance-based control) and 25 °C (zone air temperature).
6. *Shading is on when scheduled.* In this shading scenario, the shading control is based on the single reading of the illuminance value on the indoor horizontal plane. In practice, such a strategy is implemented through a pre-set schedule (calculated with a separated simulation run) that identifies the hours of the year when the indoor illuminance would exceed (if shading were not lowered) the threshold value of 2000 lux.

7. Table 1. Key features of the building systems and components of the case study building

Scenario Name	WWR		Integrated shading system	
	Reference	Fully glazed	No	Yes
REF_noSHADING	✔		✔	
REF_bestSHADING	✔			✔
FG_noSHADING		✔	✔	
FG_bestSHADING		✔		✔

3. Results

3.1. Building with mechanical heating and cooling

3.1.1. Thermal environment and energy performance

In Figure 1 the overview of the results from the simulation study is presented in terms of energy use for heating, cooling, and artificial lighting. The energy use for heating (blue bar) and cooling (orange bar) is plotted for different COPs, using the error bar – the bar value is the one for the reference case, where COPh = 3 and a COPc = 5; the top value of each error bar represents the energy use for a COPh = 2.5 and COPc = 4; the bottom value of each error bar represents the energy use for a COPh = 3.5 and a COPc = 6). In Figure 1 only the results for the best shading strategy are reported.

The simulation demonstrates that the gain in reducing the energy for lighting thanks to a larger daylight income in the building with curtain wall far overcomes the very minimal worsening of the energy use for heating (almost negligible) and of the energy use for cooling (which instead almost double, though it remains below 1.5 kWh/m^2 even in the worst cases).

It is also shown that for the conventional, reference office building, the use of shading devices does not necessarily lead to a general better behavior, as the cooling energy reduction achieved through the activation of the shading devices is counterbalanced by a slight increase in the energy use for lighting, and a more noticeable increase in the energy use for heating – due to the missing solar heat gain.

Conversely, the use of shading devices is always beneficial in the case of a curtain-wall system, since it only marginally leads to an increase in the energy use for artificial lighting, but drastically reduces the energy use for cooling.

3.1.2. Shading strategy influence and visual environment performance

The efficient use of shading systems heavily relies on the identification of a correct control strategy, as these systems can provide adequate protection against cooling loads, and comfortable visual conditions only if continuously controlled, and controlled according to a suitable control logic. The identification of the most suitable control strategy is not straightforward as shading systems are dealing with competing uses of solar energy. It is very common to see that while one particular energy use is reduced by a certain shading strategy, a competing one increases.

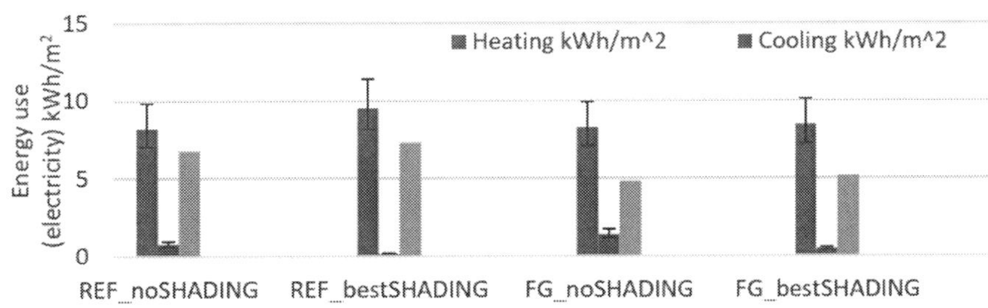

Figure 1. Comparison between Baseline building (REF), with no shading (noSHADING) and shading (bestSHADING) system, and the correspondent configuration for the fully glazed building (FG).

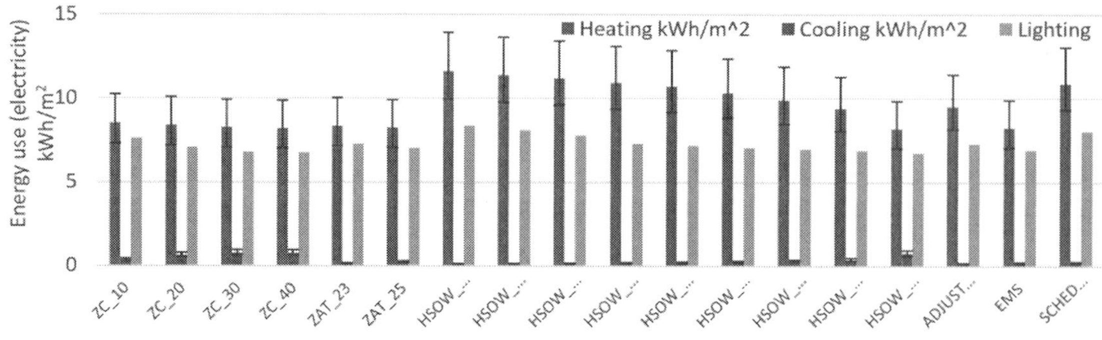

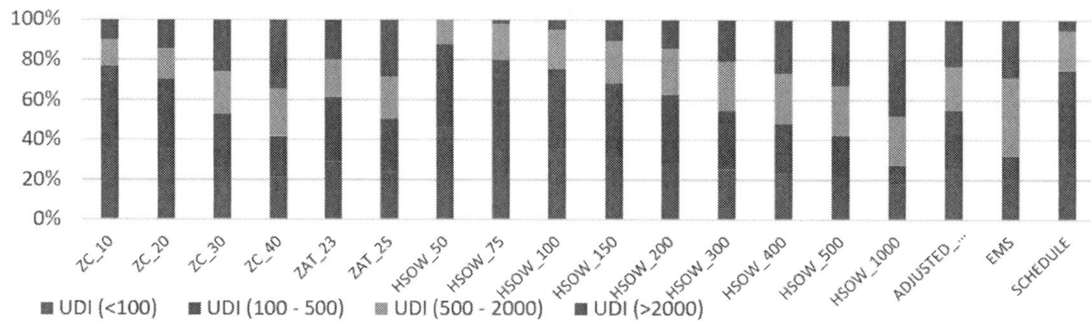

Figure 2. Comparison between different shading strategies in terms of energy use (top) and UDI (bottom), for the Baseline building with reference facade.

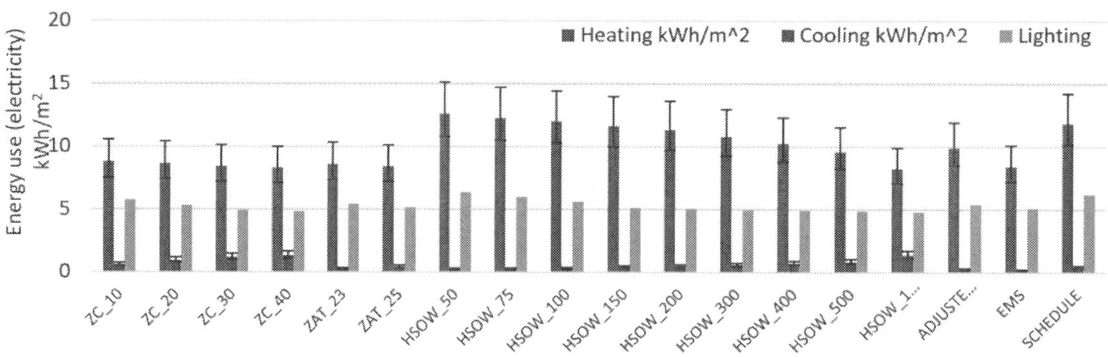

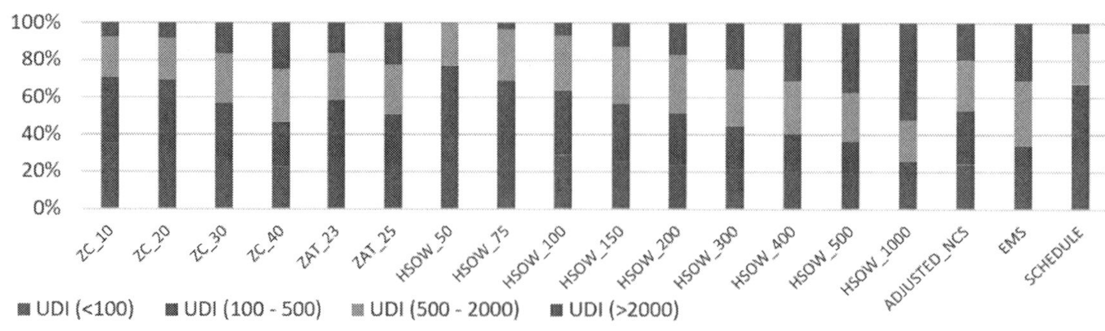

Figure 3. Comparison between different shading strategies in terms of energy use (top) and UDI (bottom), for the building with curtain wall facade.

In Figures 2 and 3, the energy use for heating, cooling, and artificial lighting for different shading strategies are compared, as well as the corresponding UDI values for the representative, south-exposed office space. By comparing the reference and the fully glazed solution in terms of shading control logic, it appears that the performance of the control logic is relatively insensitive to the WWR.

For both the cases, as expected, strategies that points towards minimisation of cooling energy use lead to higher requirements for lighting and heating, and the other way around. A suitable compromised solution, assuring both relatively low energy use for cooling, small increase in energy use for heating, and low value for $UDI_{>2000}$ (connected to the risk of glare discomfort) can be based on setting a control on the zone temperature (for example 23 °C, code ZAT_23). Strategies based on the level of irradiation on the façade (codes HSOW) are effective in controlling cooling energy use but lead to a significant increase in the energy use for heating. More advanced strategies combining controls based on thermal and lighting objectives (code EMS) just slightly outperform a simple control strategy but leads to an increased risks of glare discomfort. The strategy aiming at minimising glare discomfort risk (code Schedule) achieves the desired performance but with an increase in the energy use for heating of around 50%, compared to other optimal strategies.

3.2. Building without mechanical cooling

In Figure 4 the performance of different configurations (same as in Fig. 1) of the façade of the office building analysed are presented, for a building with only mechanical heating systems. As for the twin building equipped with the cooling plant (Fig. 1), it is shown that the fully glazed configurations are completely comparable in terms of energy performance to the reference configurations based on a large window layout, with a significant decrease in the energy for artificial lighting.

Without a cooling plant, it is however strictly necessary that the fully glazed building is equipped with shading systems, otherwise the number of hours outside the comfort zone, in a south exposed office, will be unacceptable (Fig. 4, bottom). When equipped with a shading system, it is possible to limit the discomfort hours (using the best, year-round strategy) to little less than 150 h. Almost 75% of these hours happens in July – i.e. when the office is not in full use due to summer holidays – and when a different, dedicated strategy could be implemented to completely avoid overheating. In the current simulation study the shading systems are controlled in July as during the entire year. However, if, for example, in the scenario FG_bestSHADING (with 2 ACH free night cooling), the venetian blinds are continuously closed in July, due to unoccupied spaced, the annual hours in the range 26 °C to 28 °C and exceeding 28 °C become 239 (from 439) and 91 (from 149), respectively. Hours exceeding 28 °C outside July are 41, and happens primarily in August and in the late afternoon.

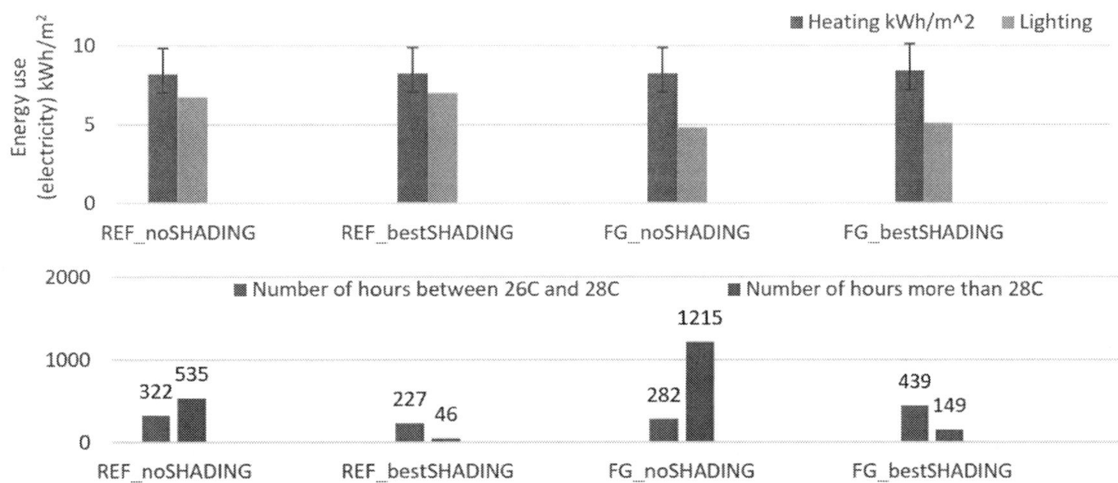

Figure 4. Energy use (top) and ODH (bottom) in the south exposed office, for the building with reference façade and curtain wall façade, in case of no mechanical cooling (2 ACH night cooling).

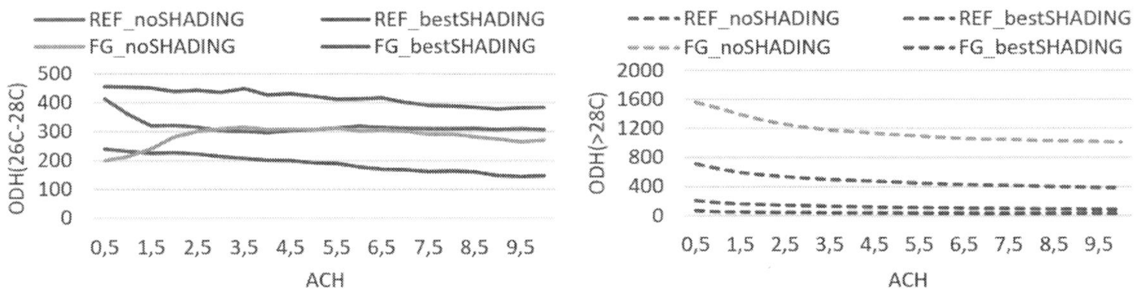

Figure 5. Overheating degree hours (ODH) for the range 26 °C to 28 °C (left) and for above 28 °C (right) as a function of air change rate in the south exposed office.

In Figure 5, the overheating degree hours between 26 °C and 28 °C, and hours above 28 °C are shown as function of ACH rate for free cooling ventilation flow, in the case of a building that only has mechanical heating systems. It can be seen that the values of ODH decrease with the increase of ACH, but not relevant changes are seen for ACH greater than 2-3 h^{-1}, especially for the cases with shading.

4. Conclusions

Fully glazed facades can be realized with current state of the art technologies in a Nordic climatic context. Integrated shading systems with suitable control logics (either based on temperature or on more advanced combinations of signals) are an important element of such constructions. If properly planned and controlled, fully glazed facades can lead to a reduced use of energy for heating, cooling, and artificial lighting, without a significant increase in the risk of discomfort due to glare.

In the case of an office building not equipped with a cooling plant, a suitable integration of passive strategies for control of indoor air temperature (free/nigh cooling) is necessary to keep the number of hours with temperature outside the comfort range within an acceptable range. An ACH rate of around 2 h^{-1} for free/night cooling seems to be a realistic target value, achievable with single side natural ventilation, or with the support of the equipment for mechanical ventilation (hybrid ventilation).

Strategies to control shading systems might change during the year to assure that the overall performance of the building is optimal. During the summer season (June to August), especially when a cooling plant is not installed, the control of solar gain should be prioritized to avoid overheating risks. With such an arrangement, highly glazed buildings without mechanical cooling can be realized in this climatic context using best available façade technologies with accurately controlled shading systems, with relatively risk of overheating leading to discomfort for the users.

Acknowledgments
This paper is part of research activities developed in the SkinTech project funded by the Research Council of Norway under grant No. 255252/E20 and the industrial partners in the project.

References

[1] Goia F 2016 *Sol. Energy* **132** 467–92.
[2] Goia F, Haase M, and Perino M 2013 *Appl Energ* **108** 515–527
[3] Grynning S, Time B, and Matusiak B 2014 *Sol Energy* **107** 182–194
[4] Georges L, Haase M, Houlihan Wiberg A, Kristjansdottir T, and Risholt B 2015 *Build Res Inf* **43** 82–93
[5] EnergyPlus – Engineering Reference, v. 8.5
[6] NS 3701:2012 - Norsk passivhusstandard for yrkesbygninger
[7] Geros V, Santamouris M, Tsangrasoulis A, and Guarracino G 1999 *Energy Build* **29** 141-154.
[8] Kolokotroni, M., & Aronis, A. 1999 *Appl energy* **63(4)** 241-253.
[9] Artmann, N., Manz, H., & Heiselberg, P. 2008 *Renewable Energy* **33(12)** 2589-2598.
[10] Ramponi, R., Angelotti, A., & Blocken, B. 2014 *Appl Energy* **123** 185-195.
[11] Handbook, Fundamentals, ASHRAE, Atlanta 2017.
[12] ASHRAE Standard 55 — Thermal environmental conditions forhuman occupancy, 2017
[13] Nabil A, Mardaljevic J 2006 *Energy Build* **38** 905–13

AUTHOR INDEX

AUTHOR INDEX

AUTHOR INDEX

Institute of Physics
Dirac House, Temple Back
Bristol BS1 6BE UK

ISSN: 1755-1307
ISBN 978-1-7138-0148-1

CURRAN ASSOCIATES INC.
proceedings.com